Peptide Hormones in Plant Development and Stress Responses

Peptide Hormones in Plant Development and Stress Responses

Edited by

Vijay Pratap Singh
Department of Botany
C.M.P. Degree College
University of Allahabad
Prayagraj
Uttar Pradesh
India

Samiksha Singh
Department of Botany
S.N. Sen B.V. Post Graduate College
Chhatrapati Shahu Ji Maharaj University
Kanpur
Uttar Pradesh
India

WILEY

This edition first published 2026

Registered Offices
John Wiley & Sons, Inc., 111 River Street, Hoboken, NJ 07030, USA
John Wiley & Sons Ltd, New Era House, 8 Oldlands Way, Bognor Regis, West Sussex, PO22 9NQ, UK

For details of our global editorial offices, customer services, and more information about Wiley products visit us at www.wiley.com.

The manufacturer's authorized representative according to the EU General Product Safety Regulation is Wiley-VCH GmbH, Boschstr. 12, 69469 Weinheim, Germany, e-mail: Product_Safety@wiley.com.

Wiley also publishes its books in a variety of electronic formats and by print-on-demand. Some content that appears in standard print versions of this book may not be available in other formats.

Library of Congress Cataloging-in-Publication Data Applied for:
Print ISBN: 9781394338153
ePDF ISBN: 9781394338221
ePUB ISBN: 9781394338184
oBook ISBN: 9781394338238

Cover Design: Wiley
Cover Image: © Larisa/stock.adobe.com

Set in 9.5/12.5pt STIXTwoText by Straive, Pondicherry, India

Contents

List of Contributors

Kamal A. M. Abo-Elyousr
Department of Agriculture
Faculty of Environmental Sciences
King Abdulaziz University
Jeddah, Saudi Arabia
and
Department of Plant Pathology
Faculty of Agriculture
University of Assiut
Assiut, Egypt

Sadia Afzal
Department of Botany
Lahore College for Women University
Lahore, Pakistan

Iqra Ali
Department of Agricultural Entomology
Centre for Plant Protection Studies
Tamil Nadu Agricultural University
Coimbatore, Tamil Nadu, India

Deva Andrews Arulsoosairaj
Division of Bio-pesticides and
Environmental Toxicology
Sri Paramakalyani Centre for Excellence
in Environmental Sciences
Manonmaniam Sundaranar University
Tirunelveli, Tamil Nadu, India

Rakesh K. Bairwa
Crop Improvement
ICAR-Indian Institute of Wheat and
Barley Research
Karnal, Haryana, India

Tanmaya Kumar Bhoi
Forest Protection Division
ICFRE-Arid Forest Research Institute
(ICFRE-AFRI)
Jodhpur, Rajasthan, India

Melike Bor
Faculty of Science
Department of Biology
Ege University
Bornova-Izmir, Türkiye

Kanagaraj Muthu-Pandian Chanthini
Department of Pharmacology
Saveetha Institute of Medical and
Technical Sciences (SIMATS)
Chennai, Tamil Nadu, India

Rutupurna Das
Department of Zoology
School of Applied Sciences
Centurion University of Technology
and Management
Bhubaneswar, Odisha, India

Weam El-Aggan
Department of Botany and Microbiology
Faculty of Science
Alexandria University
Alexandria, Egypt

Hossam S. El-Beltagi
Agricultural Biotechnology Department
College of Agriculture and Food Sciences
King Faisal University
Al-Ahsa, Saudi Arabia

Saikat Gantait
Crop Research Unit (Genetics and Plant Breeding)
Bidhan Chandra Krishi Viswavidyalaya
Mohanpur, West Bengal, India

Nihal Gören-Sağlam
Department of Biology
Faculty of Science
Istanbul University
Istanbul, Türkiye

Heeya Gupta
Department of Biotechnology
St. Xavier's College (Autonomous)
Kolkata, West Bengal, India

Om Prakash Gupta
Quality and Basic Science
ICAR-Indian Institute of Wheat and Barley Research
Karnal, Haryana, India

Mohammad Ikram
Department of Agricultural Entomology
Centre for Plant Protection Studies
Tamil Nadu Agricultural University
Coimbatore, Tamil Nadu, India

Muhammad Imran
School of Biological Engineering
Henan University of Technology
Zhengzhou, Henan, China

Aina Inam
Department of Botany
Lahore College for Women University
Lahore, Pakistan

Sumera Javad
Department of Botany
Lahore College for Women University
Lahore, Pakistan

Chaitanya Kumar Jha
Department of Microbiology and Biotechnology
Gujarat University
Ahmedabad, Gujarat, India
and
Microbiology Department
Gujarat Arts and Science College
Ahmedabad, Gujarat, India

Vijay P. Kapale
Department of Biochemistry
College of Basic Sciences and Humanities
G.B. Pant University of Agriculture and Technology
Pantnagar, Uttarakhand, India

Shehnaj Khan
Department of Biotechnology
Dr. Y.S. Parmar University of Horticulture and Forestry
Solan, Himachal Pradesh, India

Emad Hamdy Khedr
Department of Pomology
Faculty of Agriculture
Cairo University
Giza, Egypt

Nagwa Khedr
Department of Pomology
Faculty of Agriculture
Cairo University
Giza, Egypt

Konkona Lahiri
Department of Biotechnology
St. Xavier's College (Autonomous)
Kolkata, West Bengal, India

Deepak Kumar Mahanta
Forest Entomology Discipline
Forest Protection Division
Indian Council of Forestry Research and Education-Forest Research Institute (ICFRE-FRI)
Uttarakhand, India

Sanjoy Majumder
School of Biotechnology
Centurion University of Technology and Management
Bhubaneswar, Odisha, India

Farhad Nazarian-Firouzabadi
Production Engineering and Plant Genetics Department
Faculty of Agriculture
Lorestan University
Khorramabad, Iran

Parul Negi
Department of Biotechnology
Dr. Y.S. Parmar University of Horticulture and Forestry
Solan, Himachal Pradesh, India

Tülay Alp Öztürk
Faculty of Science
Department of Biology
Ege University
Bornova-Izmir, Türkiye

Gagan Kumar Panigrahi
Department of Zoology
School of Applied Sciences
Centurion University of Technology and Management
Bhubaneswar, Odisha, India

Bhavesh Parmar
Microbiology Department
Gujarat Arts and Science College
Ahmedabad, Gujarat, India

Pooja Parmar
Department of Microbiology and Biotechnology
Gujarat University
Ahmedabad, Gujarat, India
and
Microbiology Department
Gujarat Arts and Science College
Ahmedabad, Gujarat, India

Abhishek Prajapati
Department of Microbiology and Biotechnology
Gujarat University
Ahmedabad, Gujarat, India
and
Microbiology Department
Gujarat Arts and Science College
Ahmedabad, Gujarat, India

Ansh Raj
ICAR-National Research Centre on Litchi
Muzaffarpur, Bihar, India

Somayeh Rastegar
Department of Horticultural Sciences
Faculty of Agriculture and Natural Resources
University of Hormozgan
Bandar Abbas, Iran

Aryadeep Roychoudhury
Discipline of Life Sciences
School of Sciences
Indira Gandhi National Open University
New Delhi, Delhi, India

Annapurna Sahoo
Department of Zoology
Jatni College
Utkal University
Khordha, Odisha, India

Bimal Kumar Sahoo
Department of Agricultural Entomology
Centre for Plant Protection Studies
Tamil Nadu Agricultural University
Coimbatore, Tamil Nadu, India

Ipsita Samal
ICAR-National Research Centre on Litchi
Muzaffarpur, Bihar, India

Jayoti Majumder Sarkar
Department of Floriculture and Landscape Architecture
Bidhan Chandra Krishi Viswavidyalaya
Mohanpur, West Bengal, India

Kunja Bihari Satapathy
Department of Botany
School of Applied Sciences
Centurion University of Technology and Management
Bhubaneswar, Odisha, India

Sengottayan Senthil-Nathan
Division of Bio-pesticides and Environmental Toxicology
Sri Paramakalyani Centre for Excellence in Environmental Sciences
Manonmaniam Sundaranar University
Tirunelveli, Tamil Nadu, India

Ajit Singh
School of Biological and Environmental Sciences
University of Nottingham Malaysia
Semenyih, Selangor, Malaysia

Anupama Singh
Department of Biotechnology
Dr. Y.S. Parmar University of Horticulture and Forestry
Solan, Himachal Pradesh, India

Charan Singh
Crop Improvement
ICAR-Indian Institute of Wheat and Barley Research
Karnal, Haryana, India

Jyoti P. Singh
Plant Biochemistry
ICAR-National Bureau of Agriculturally Important Microorganisms
Mau, Uttar Pradesh, India

Pankaj Kumar Singh
Department of Biotechnology
University Centre for Research & Development
Chandigarh University
Mohali, Punjab, India

Zhongke Sun
School of Biological Engineering
Henan University of Technology
Zhengzhou, Henan, China

Amel Tammam
Department of Botany and Microbiology
Faculty of Science
Alexandria University
Alexandria, Egypt

Aslıhan Temel
Department of Molecular Biology and Genetics
Faculty of Science
Istanbul University
Istanbul, Türkiye

Filiz Vardar
Department of Biology
Faculty of Science
Marmara University
Istanbul, Türkiye

Preface

In the decade of 1990, the first plant peptide hormone named systemin was identified in tomato. This breakthrough discovery reported the role of peptide hormone systemin in wound responses and thus promoted research in the direction of other peptide hormones involved in regulating plant growth and development and stress responses. After this, a number of plant peptide hormone families have been identified in plants. These plant peptide hormone families include systemin, CLV3/ESR-related ("CLE") peptide family, phytosulfokine (PSK), POLARIS (PLS), rapid alkalinization factor (RALF), inflorescence deficient in abscission (IDA), ROTUNDIFOLIA4/DEVIL1 (ROT4/DVL1), SCR/SP11, ENOD40, etc. The plant peptide hormones are of two types: one is secretory and other one is nonsecretory. Recent literature showed that plant peptide hormones play a crucial role in plant growth and stress responses by integrating various plant developmental and stress-responsive signaling pathways under various environmental cues. Though plant peptide hormones chemically dissimilar with other classical plant hormones but they share more or less similar signaling pathways in plant growth, development, and stress responses.

At present, there has been an increasing evidence about the role of plant peptide hormones in regulating plant development and stress acclimation processes. The plant peptide hormones are perceived by membrane-bound receptor kinases, which convey a message to downstream targets in order to regulate plant development and stress responses locally or at distant locations in a systemic manner.

Due to impacts of different stress factors and climate change, crop productivity is getting seriously affected. This situation represents serious challenges for securing food for the global population. So, to ensure food supply for the population, measures and technologies are required to enhance crop productivity under changing scenario of climate change. Since plant peptide hormones are known to regulate various plant developmental aspects as well as stress acclimation in plants, they may be potential candidates for promoting crop productivity under stress-challenged situation.

In this book, we have compiled research done on plant peptide hormones in plants, covering both developmental and stress acclimation aspects. In this edited book, we have put plant peptide hormone research at one place, in order to make it easily available to researchers, students, academicians, etc. In this book, we have compiled total 21 book chapters. These chapters cover elaborated knowledge on plant peptide hormones–plant interaction from their biosynthesis, perception and signaling, regulation of plant development and stress responses, and their potential implications for food security. We believe that this book will be useful for the students, researchers, and academicians.

Vijay Pratap Singh
Samiksha Singh

1

Cellular Compartmentalization and the Role of Peptide Hormone

Gagan Kumar Panigrahi[1,*], *Rutupurna Das*[1], *and Annapurna Sahoo*[2]

[1] *Department of Zoology, School of Applied Sciences, Centurion University of Technology and Management, Bhubaneswar, Odisha, India*
[2] *Department of Zoology, Jatni College, Utkal University, Khordha, Odisha, India*

1.1 Introduction

1.1.1 Overview of Cellular Compartmentalization

Cellular compartmentalization is an intrinsic cellular feature of eukaryotic cells and it is responsible for separating the cellular processes into different organelles to facilitate effective biochemical reactions. Cellular homeostasis, metabolic efficiency, and unwanted interactions between biochemical pathways need to be ensured. The main organelles or major compartments of eukaryotic cells are the nucleus, mitochondria, endoplasmic reticulum (ER), Golgi apparatus, lysosomes, peroxisomes, and vesicles (Baillie et al. 2024). The nuclei are the genetic and gene controls, with the mitochondria being the power generators that generate ATP through oxidative phosphorylation. Protein synthesis, modification, and transport are all coordinated by the ER and Golgi. Degradation and detoxification reactions are carried out by the lysosomes and the peroxisomes, respectively (Arsova et al. 2018). The membrane-bound compartments keep the compartmentalization regulated since they bring up a condition where enzymic reactions can take place. Selective permeability is carried out by transport proteins, which help to give movement from one compartment to another. The photosynthetic organelles in plant cells are arranged in a form of compartmentalization as well. The removal of waste, storage, and turgor preservation are supported by vacuoles (Behera et al. 2025). The endomembrane and plasmodesmata system also assists intercellular transport and communication. At the cellular level, compartmentalization increases the effectiveness of cellular metabolism pathways, such as glycolysis, the Krebs cycle, and the Calvin cycle. Many diseases, such as cancer and neurodegenerative disorders are caused by failure in compartmentalization. It is vital to understand this phenomenon in order to advance in the science of synthetic biology, drug delivery, and metabolic engineering (Canut et al. 2016). Since plants are sessile, compartmentalization is particularly important so that they can respond to environmental changes, metabolic needs, and

*Corresponding author.

Peptide Hormones in Plant Development and Stress Responses, First Edition. Edited by Vijay Pratap Singh and Samiksha Singh.

developmental processes through efficient internal regulation. The advantage of compartmentalizing some biochemical reactions in the confines of certain organelles is not only that this can allow for the localization of steps but also that it makes it possible to accomplish such steps more efficiently. For example, chloroplasts contain systems (photosynthesis), whereas mitochondria house other systems (respiration). The cell can confine reactions to organelles, optimizing such conditions as pH, ion concentration, and enzymatic activity and thereby maximize the efficiency in metabolic pathways. In addition, plant stress response also involves compartmentalization (Chen et al. 2020). The vacuole is an example of an organelle that works to store toxins and secondary metabolites, so as to minimize damage done to essential cellular machinery. In addition, it provides the sequestration of harmful products and recycling of cellular components via autophagy, an autophagy-supporting mechanism taking place in plant cells. In addition, for hormone signaling and development, plant cells depend on compartmentalization (Chen et al. 2023). Genetic material is housed in the nucleus, yet the cytoplasm is stimulated by the nucleus to control gene expression in response to external stimuli. Like, the ER and Golgi apparatus are used to transport proteins and create proper cellular function (Corcilius et al. 2017). Finally, the benefits of compartmentalization can be summed up as meaning that compartmentalization is required to prevent cellular inefficiency, protect the cell from potential damage and to regulate fine detail in physiological processes. However, it allows plants to manage complex metabolic interactions, optimize energy production, and cope with environmental challenges in order to grow, develop, and survive (Dao et al. 2022; Zhang et al. 2024b,c).

1.1.2 Importance of Plant Peptide Hormones

Plant peptide hormones are defined as small signaling molecules with critical roles in growth regulation, development, and environmental response. Peptide hormones in many respects differ from classical phytohormones like auxins and gibberellins, whereby the latter mostly perform as active substances, substituting physiological processes (Fedoreyeva et al. 2022; Cho et al., 2014). The plant peptide hormones are synthesized as precursor proteins that are processed into bioactive peptides. They are secreted into the extracellular medium and act on receptor kinases in their plasma membrane, which trigger intracellular signal transduction cascades. We list here some of the more well-characterized plant peptide hormones, including CLAVATA (CLV), phytosulfokines (PSKs), rapid alkalinization factors (RALFs), and CLE peptides. Peptide hormones have a very important role in facilitating the maintenance of meristems and organ development. For example, CLV peptides modulate stem cell homeostasis in the shoot apical meristem by binding to CLV receptors. Indeed, PSKs promote cell growth and tissue differentiation, while CLE peptides promote the regulation of vascular patterning and root growth. Perceiving biotic and abiotic stress is also accomplished through peptide hormones (Hastwell et al. 2014). For example, plant RALF peptide modulates plant defense through balancing ions and inducing reactive oxygen species (ROS). Some peptide hormones are also involved in symbiosis with microorganisms that help to obtain nutrients and increase stress tolerance. Reproduction also requires peptide hormones (Combest et al., 2021; Coninck and De Smet, 2016). The recognition of pollen tubes directed toward the ovule depends on peptides in the Luteinizing-like Rapidly Evolved (LURE) family, which helps in the successful fertilization (Ito et al. 2006). Other peptides control seed development, dormancy, and germination. These new advances in plant biotechnology are focused on the possibility of using peptide hormones in agriculture (Goad et al., 2017; Hsu et al., 2021). This allows precision agriculture and environmentally friendly crop production to expand to include the design of synthetic peptides with specific biological functions. To discover strategies for improvement of plant health and productivity, the mechanisms of action of peptide hormones need to be clarified. More work will need to be performed to obtain a complete picture of what they can do and how they interact with other signal transduction pathways (Huang et al. 2022). The first goal of this work is to investigate cellular compartmentalization and peptide hormones in stress response and plant development. The study provides a glimpse of the spatial organization of the cellular processes and the regulatory processes of the peptide hormones and reveals aspects of plant physiology and opportunities for pragmatic application in agriculture (Kim et al., 2021; Majumder and Panigrahi et al., 2025). Cell compartmentalization guarantees the efficient metabolic pathway segregation with biochemical reaction optimization and cellular homeostasis (Zhong and Qu 2019). Here, the structure and function of basic organelles, their interaction, and their function in plant development and growth are interpreted. Further, the main classes of plant peptide hormones are discussed, as well as the biosynthetic genes involved in peptide hormone biosynthesis and their involvement in maintenance of the plant meristem, root, and tuber growth and adaptation to stress and reproductive success (Ohkubo et al. 2017). Signal transduction mechanisms, focusing on the interaction between peptide hormones and receptor kinases, and the resultant control on intracellular response and on plant growth, with special emphasis (Liu et al. 2021). The agricultural application of peptide hormones is also explored by the study, including its use in improving the productivity of a crop, enhancing its stress tolerance, and promoting higher nutrient uptake (Nguyen and Kim 2023; Igarashi et al. 2012).

1.2 Cellular Compartmentalization in Plants

1.2.1 Major Cellular Organelles and Their Functions

A number of organelles are present in plant cells perform different roles and are needed for cellular activities. These include the most prominent organelles: nucleus, chloroplasts, mitochondria, the vacuole, the ER, the Golgi apparatus, the peroxisomes, and the cytoskeleton (Hirayama and Mochida 2023). The nucleus is the control center of the cell, where genetic material, or chromatin, and gene regulatory components are stored. Inside it, the nuclear envelope is a double membrane, which contains nuclear pores so that molecules can go in and out. Ribosomal RNA (rRNA) synthesis and ribosome assembly are carried out in the nucleolus within the nucleus (Murphy et al., 2012; Panigrahi and Satapathy, 2021). Chloroplasts, which are dark green organelles specialized for photosynthesis, convert light energy into chemical energy in glucose (Guillou et al. 2022). The stroma contains the Calvin cycle, and thylakoid membranes carry out light-dependent reactions. Chloroplasts are also involved in amino acid and fatty acid synthesis as well as in plant stress responses. The powerhouse of the cell are the mitochondria, which use cellular respiration to produce ATP. The inner membrane contains the electron transport chain and ATP synthase, and the outer membrane does not permeate or cut into the inner membrane (Panigrahi et al., 2024). Along with offering mitochondrial contributions to metabolic signaling, mitochondria also participate in the regulation of programmed cell death (apoptosis) (Hagelthorn and Fletcher 2023). A vacuole is a large and membrane-bound organelle that also performs functions like the storage of water, nutrients, and waste products. It is important for the plant's rigidity and maintains turgor pressure, which is essential for cell structure. Furthermore, vacuoles are involved in detoxification by retaining harmful compounds and secondary metabolites. The ER consists of rough ER (RER) and smooth ER (SER). Ribosomes stud the RER and are involved in protein synthesis and modification as well as folding (Sahoo et al., 2024; Singh et al., 2024). The SER, which lacks ribosomes, is involved in lipid synthesis, detoxification, and calcium storage. The cell's packaging and distribution center is the Golgi apparatus. It specializes in processing and modifying proteins and lipids from the ER and moves them to their correct destinations (Kaeothip et al. 2013; Panigrahi and Satapathy 2020a). It is also required for the synthesis of cell wall components necessary for plant growth and development. Small, membrane-limited organelles are involved in the detoxification of free radical species (reactive oxygen species: ROS), the breakdown of fatty acids (e.g. QTBP3, 20268920), and the glyoxylate cycle (essential for seed germination, e.g. peroxisomes). They also associate closely with chloroplasts and mitochondria in the process of photorespiration. The cytoskeleton consists of microtubules, microfilaments, and intermediate filaments, which provide structural support and intracellular transport (Fernandez-Fernandez et al. 2023). It has a fundamental role in cell division, organelle positioning, and vesicle trafficking, thus preventing inefficient cellular communication and organization. These organelles work together and with the rest of the plant cell to provide such specialized functions as growth, energy production, metabolism, and adaptation to environmental conditions. That way, plants can jointly use and conserve resources, ensuring that life can continue (Furumizu et al. 2021; Panigrahi and Satapathy 2020b).

1.2.2 Role of Membranes in Compartmentalization

Cellular compartmentalization is achieved through membranes that act as bridges for the selective movement of molecules between one organelle and another. The most common type of these membranes is the phospholipid bilayer, embedded with proteins, which is responsible for controlling transport, signal transduction, and structural integrity (Blackburn et al. 2020). Selective permeability is one of the most crucial roles of membranes that selectively permits some molecules to enter or leave the organelles. The function of this is mediated by integral and peripheral proteins, including transporters, channels, and receptors. For instance, the metabolite exchange of chloroplast and mitochondrial membranes involves special transporters that control respective metabolite exchange for photosynthesis and respiration (Boschiero et al. 2019). The entire plant cell is surrounded by the plasma membrane, maintaining its internal environment, and acting as a mediator between the external and internal environments. These contain receptor proteins that detect signals like hormones, light, and pathogens and cause cellular responses (Chaulagain and Frugoli 2021). Furthermore, it lies connected to the plasma membrane and to the cell wall, which is responsible for giving the cell mechanical support and for regulating cell expansion (Figure 1.1).

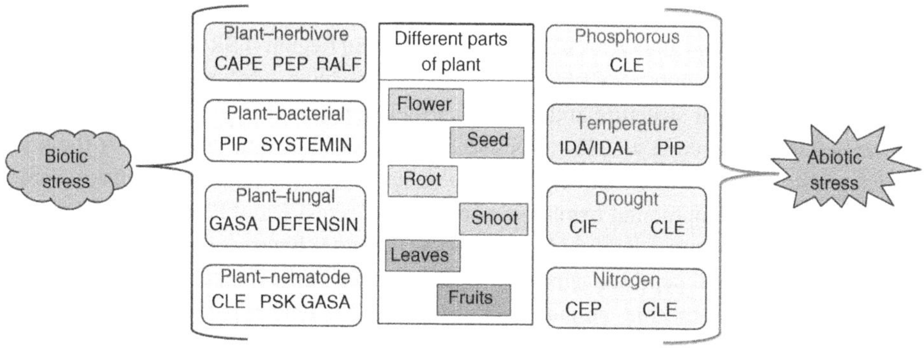

Figure 1.1 Peptide hormone signaling networks in the modulation of plant stress physiology.

1.2.3 Transport Mechanisms Between Compartments

Synthesis and trafficking of proteins and lipids are accomplished strictly within internal membranes, including the ER and the Golgi apparatus. Proteins are folded and modified in the ER membrane, and then modified and sorted in the Golgi apparatus for delivery to the appropriate part of the cell (Chapman et al. 2024; Sahoo and Satapathy 2021). Osmoregulation involves the tonoplast (vacuolar membrane), which controls ion and water movement into the vacuole. Cell turgor pressure (which affects plant rigidity and growth) is also maintained by the tonoplast (Eckardt et al. 2024). The lipid compositions of mitochondrial and chloroplast membranes support processes of energy conversion. Inner membranes have extensive folds (cristae and thylakoids, respectively), which increase the surface area for ATP production (Zhou et al. 2024). In addition, these membranes have electron transport chains which are necessary to produce energy. Additionally, membranes allow intracellular communication by hosting signaling pathways and transport via membrane organelles (Friso and van Wijk 2015). Endocytosis and exocytosis are processes by which nutrients are taken in, waste is removed, and intercellular communication is maintained, and by serving these functions, they play a very important role (Furumizu and Aalen 2023).

1.3 Peptide Hormones in Plants

Small protein-like compounds that act as signaling compounds in plants and control a large number of physiologic and developmental processes are known as peptide hormones. In contrast to other types of plant hormones, such as auxins, gibberellins, and cytokinins, which are nearly always small organic molecules, peptide hormones are made up of short amino acid sequences (Haruta et al. 2014; Panigrahi and Satapathy 2020c). These molecules are important for cell-to-cell communication and are critical to allow plants to respond to environmental cues and to regulate growth, development, and defense mechanisms. Typically, peptide hormones are classified by their structure, biosynthesis, and mode of action. Small posttranslationally modified peptides (PTMPS) and cysteine-rich peptides (CRPS) are two of the major classifications of peptides (Hirakawa and Sawa 2019). After posttranslational modifications, such as proteolysis and glycosylation, PTMPS become bioactive forms, including CLAVATA3/Embryo Surrounding Region (ESR)-related (CLE) peptides (Figure 1.1). In contrast, CRPS have conserved cysteine residues forming disulfide bonds, stabilizing their structure and increasing their functional role within the signaling pathways (Greifenhagen et al. 2024). Peptide hormones can also be classified by the hormones' physiological roles. These are some of the best-known functional categories. Some of these include developmental regulators peptides such as CLE and PSK, which influence meristem activity, organogenesis, and vascular development (Figure 1.1). Plant elicitor peptides (PEPs) and RALF peptides, which act as stress and defense mediators, can also regulate immune responses that enhance plant's ability to detect pathogen attacks and environmental stressors (Johns et al. 2021; Je et al. 2018).

Reproductive and growth modulators include LURE peptides, which activate pollen tube elongation during fertilization, and PSK, which stimulates cell proliferation and elongation. Peptide hormones have signaling pathways involving receptors. Binding to receptor-like kinases (RLKs) present on the plasma membrane leads to intracellular signaling cascades that regulate gene expression and cellular activity (Lebedeva et al. 2020; Matsubayashi 2018). These receptor–ligand interactions are important because they allow precise spatial and temporal regulation of plant growth and defense responses. Much has been learned about plant signaling networks by the discovery and classification of plant peptide hormones.

1.3.1 Key Peptide Hormones and Their Functions

A number of plants have been shown to have several peptide hormones involved in growth, development, and stress response. Among the most well-characterized are CLAVATA3 (CLV3), PSK, RALF, PEP, and LURE peptides (Matsuzaki et al. 2010; Millar et al. 2019; Ogawa-Ohnishi et al. 2022).

One critical peptide hormone in shoot and root meristem maintenance is CLV3. It regulates stem cell proliferation within the CLAVATA-WUSCHEL (CLV-WUS) signaling pathway (Okamoto et al. 2022). The signaling cascade that restricts excessive stem cell division is triggered when CLV3 binds to an RLK, CLAVATA1 (CLV1). It ensures proper meristem size so that the plant does not grow uncontrollably (Narasimhan and Simon 2022).

PSK is a small peptide that affects cell division, differentiation, and elongation. It is vital to tissue regeneration, root, and shoot growth as well as host-pathogen interactions. The precursor to PSK is synthesized and sulfated—an essential modification for bioactivity. It needs to interact with PSK receptors to stimulate growth-related pathways, therefore, it is vital for plant development (Motomitsu et al. 2015; Pastor-Fernandez et al. 2023).

RALF is a peptide hormone involved in pH and ion homeostasis. It regulates plant responses to environmental stimuli including drought, salinity, and pathogen attack. There, RALF peptides interact with RLKs to regulate gene expression and physiological responses to stress (Nakayama et al. 2017; Ohki et al. 2011; Oh et al. 2018).

PEPs are influential in plant defense. Release of these peptides occurs upon herbivore or pathogen attack and they are proalarming danger signals that elicit immune responses. PEPs then bind to specific receptors, triggering the production of defense-related compounds, such as ROS and secondary metabolites (Muller and Harrison 2019; Nakagami et al. 2024). This signaling mechanism strengthens the plant's resistance against biotic stress (Figure 1.1).

During fertilization, specialized signaling molecules, LURE peptides, guide pollen tubes toward the ovule. The synergid cells secrete these peptides in the female reproductive tissues and they serve as chemical attractants for pollen tubes (Pecenkova et al. 2024; Phour et al. 2020; Qian et al. 2018). A key feature of this system is that LURE peptides are precisely expressed in time and space to achieve successful fertilization and seed formation.

Other important peptide hormones include C-terminally encoded peptides (CEPs) serving in nitrogen homeostasis, as well as HypSys peptides that participate in wound signaling and a herbivore defense. They provide nutrition and stress adaptation, making the plant healthier in general (Figure 1.1). Peptide hormones interact with membrane-bound receptors in order to stimulate intracellular signal transduction cascades, which modulate gene expression and enzymatic activities (Okamoto et al. 2013; Pan et al. 2024; Patel et al. 2018). They act very rapidly and locally and thus are very effective tools to fine-tune plant responses to internal and external stimuli. Unlike traditional hormones that travel over long distances, peptide hormones allow precise mechanistic cell-to-cell communication to enable plants to rapidly adjust their strategies of growth and defense (Ren et al. 2024; Reichardt et al. 2020; Amano et al. 2007).

1.3.2 Biosynthesis and Transport of Peptide Hormones

The synthesis of peptide hormones in plants occurs through a multistep process involving gene transcription, translation, posttranslational modification, and secretion. In most cases, these peptides are synthesized as larger precursor proteins that have to be enzymatically processed in order to obtain the mature, bioactive form (Aggarwal et al. 2020; Bashyal et al. 2024; Chapman et al. 2024). Peptide hormones are transcribed from specific genes that encode precursor proteins which are translated into messenger RNA (mRNA). Often, these precursors have signal peptides that will lead them into the ER to continue their processing (Figure 1.2). Once in the ER, they are modified by glycosylation, sulfation, and proteolytic cleavage, which are required for activation (Bhalla et al. 2025; Avalos-Rangel et al. 2024; Bari and Jones 2009). These processing enzymes, including subtilisin-like proteases, are important in generating functional peptide hormones. Vesicular trafficking transports the peptide hormones once they are fully processed to their targets. Vesicles containing these proteins are packaged within the Golgi apparatus and transferred by the secretory pathway to the plasma membrane (Abarca et al. 2021; Chang and Xiao 2025; Ding et al. 2023). The peptides are released from the cell by exocytosis into the extracellular space and can then bind to RLKs on the surface of target cells. Plants generally transport peptide hormones between tissues via the apoplast and symplast (extracellular and cytoplasmic connections via plasmodesmata). Unlike the traditional plant hormones, auxins, for example, which use specific transport proteins to move long distances, peptide hormones tend to be localized in action. Nevertheless, some peptides, for instance PSK, can travel systemically within the phloem, affecting growth and immune systems at remote locations (Andersen et al. 2018; Carbonnel et al. 2022; de Bang et al. 2017). Mechanisms that ensure peptide hormone–triggered receptor-mediated signaling of appropriate cellular responses are described. Once these hormones bind their receptors in cells, this initiates an intracellular

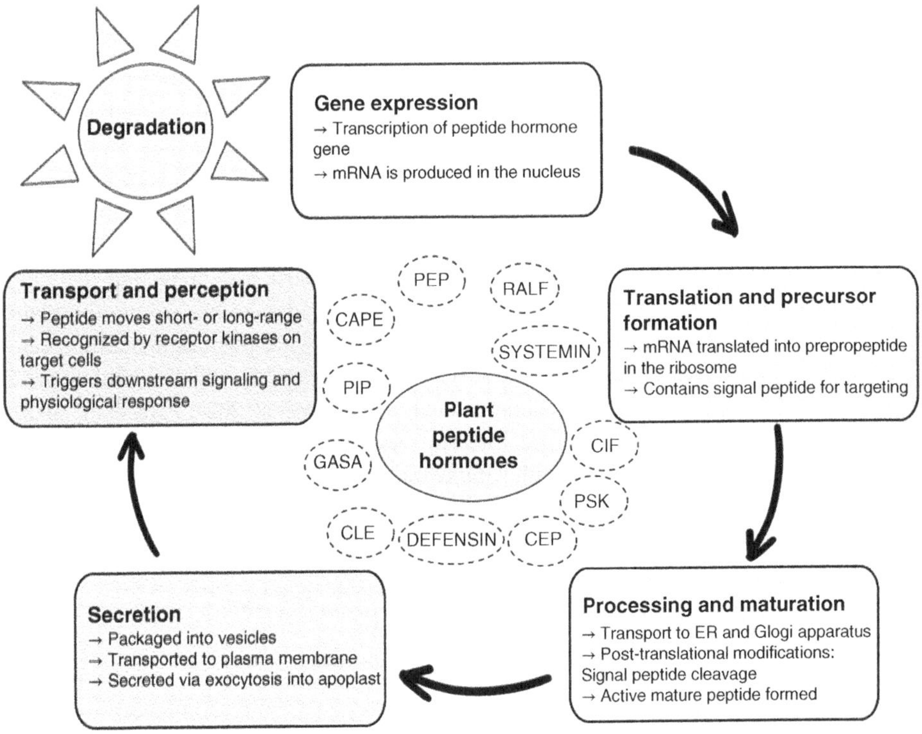

Figure 1.2 Mechanistic insights into the production, transport, and degradation of peptide hormones in plants.

signaling pathway which controls gene expression, enzyme activity, and metabolic processes. This mechanism facilitates rapid responses to developmental cues and environmental challenges (Farrokhi et al. 2008; Gancheva et al. 2019). Thus, peptide hormone biosynthesis includes precursor processing, posttranscriptional modifications, and targeted secretion (Hsiao and Yamada 2020).

1.4 Interplay Between Cellular Compartmentalization and Peptide Hormones

1.4.1 Hormone Localization Within Cellular Compartments

Peptide hormone localization and function in the plant cell are dependent on cellular compartmentalization. Unlike other classical plant hormones, whose free diffusion is common, peptide hormones are synthesized, processed, and activated in specific organelles to reach their target receptors (Kaufmann and Sauter 2019). Such compartmentalization prevents premature activation of peptide hormones by ensuring their production, storage, and transportation in a controlled, "nonreactive manner." Peptide hormones are most commonly made in the ER as precursor proteins. This organelle contains all of the molecular machinery necessary to properly fold and posttranslational modify proteins, such as glycosylation and sulfation, which are needed for hormone functionality (Kokla et al. 2024; Laffont et al. 2020; Kondo et al. 2011). These precursor peptides are conveyed from the ER to the Golgi, where they are further refined in bioactivity. Vacuoles or secretory vesicles are often involved in the storage and controlled release of peptide hormones (Figure 1.2). RALFs, like the peptide hormones, are sequestered in vacuoles before being released into the extracellular matrix. This storage prevents unregulated activation of signaling pathways and keeps hormones to remain available for release when needed (Lalun et al. 2024; Leasure and He 2012). Peptide hormone interaction with RLKs occurs mainly at the plasma membrane. These receptors are located in the membrane and discern specific peptide ligands triggering intracellular signaling cascades. Endocytosis of hormone receptor complexes is also involved in mechanisms of signal attenuation or recycling of receptors for further use, whereby the plasma membrane also participates (Kuromori et al. 2022). There are also peptide hormones involved in intracellular signaling within organelles. For instance, some peptides oversee the functioning of chloroplast and mitochondrial systems and therefore influence metabolism, including energy production and stress responses. The effect of environmental stimuli is optimized by coordination between organelles and the cytoplasm through these peptides. In summary, localization of peptide hormones within cellular compartments allows hormone synthesis, activation, and signaling to be efficiently and specifically controlled (Lyapina et al. 2024; Liu et al. 2022b).

1.4.2 Role of Vesicular Trafficking in Hormone Transport

Transport and secretion of peptide hormones require vesicular trafficking for proper localization and function. This is an intracellular transport system with membrane-bound vesicles to take hormones from the site of synthesis to their intended destinations. The pathway of a peptide hormone is revealed by a precursor protein made in the ER (Liu et al. 2022a; Mens et al. 2021). Instead, these precursors are sorted into vesicles that bud off from the ER and are delivered to the Golgi apparatus (Qiu et al. 2022). The Golgi further modifies the peptides by adding pieces of chemical tags that make them more stable and effective. Once processed, different peptide hormones are sorted into different vesicles depending on where they will be delivered. Hormone exocytosis involves the core secretory vesicle function of releasing peptide hormones into the extracellular space where they bind to their receptors. Through this regulation, the secretion of PEPs is triggered by pathogen attack, whose activation elicits immune responses in adjacent cells. Vesicular trafficking is also involved in hormone recycling as well as the degradation of hormones. The hormone receptor complex can be internalized via endocytosis, and the cell can regulate the signal duration and intensity once the hormone has bound to its receptor. The receptor-hormone complex may be directed to endosomes after endocytosis (Roman et al. 2022; Panigrahi et al. 2021a,b,c). The recycling of some of these complexes into the plasma membrane is also prevented, while others are directed toward lysosomes or vacuoles where they are degraded and no longer allow signaling to be prolonged. Long-distance hormone transport is another important aspect of vesicular trafficking. Peptide hormones mainly operate locally, but some can be transported in the phloem or xylem to distant target tissues. This movement is facilitated by vesicles, which ensure the hormone stability during transport. Vesicular trafficking is specific in that peptide hormones are able to reach their particular targets while maintaining spatial and temporal regulation (Reid et al. 2011; Ren et al. 2019).

1.4.3 Compartment-Specific Hormone Signaling Pathways

Hormone localization and transport are not the only hormone-specific signaling pathways controlled by compartmentalization. The site of peptide hormone interaction and the cellular compartment at which signaling occurs determines which signaling cascade is activated. The most common site of perception of peptide hormones is on the plasma membrane. Most peptide hormones bind to RLKs present on the plasma membrane and activate intracellular signal cascades. CLV3 peptides bind to CLV1 receptors at the plasma membrane and start a signaling pathway to control proliferation of the stem cells. Also, plasma membrane PSK receptors mediate cell expansion and growth responses. In most cases, once activated, RLKs send signals to the cytoplasm where secondary messengers, protein kinases, amplify the signal further. Additionally, these signaling molecules regulate cellular activities, including transcription factor activation, channel regulation, and changes in the cytoskeleton. More specifically, RALF peptides modulate proton pumps and their effects on ion balance and stress responses, causing cytoplasmic alkalization. Some peptide hormones also affect nuclear signaling pathways (Yan 2024; Xiao and Zhou 2023). Signal downstream of the response induced by receptor activation at the plasma membrane translocate some of the signal components to the nucleus to regulate gene expression. Peptide hormone phosphorylation cascades can activate transcription factors controlling cell differentiation, stress responses, and regulation of growth. Certain peptide hormones also act within intracellular organelles, in addition to plasma membrane signaling. Signaling peptides associated with the regulation of metabolic processes occur in chloroplasts and mitochondria. For example, some peptides control photosynthesis-related gene expression in chloroplasts and some affect mitochondrial respiration during stresses to maximize the energy production. Additionally, endosomal and vacuolar signaling affect the activity of peptide hormones. The receptors conveyed by internalized receptor–ligand complexes act as signaling hubs in endosomes, transducing signals prior to degradation or recycling (Wulf et al. 2024; Villalobos-Lopez et al. 2022). In contrast, vacuoles can be used as storage sites for inactive peptide hormones, which are then released in response to a specific stimulus. Plants restrict hormone signaling to specific compartments to achieve very highly controlled and localized responses to development and environmental cues. They achieve this compartment-specific regulation of signaling networks in order to execute fine-tuned growth, defense, and adaptation strategies.

1.5 Signal Transduction of Peptide Hormones in Plants

1.5.1 Receptors and Signal Perception

Plant peptide hormone signaling can be initiated by the recognition of specific peptides by membrane-bound receptors, the most prominent being RLKs. These receptors are embedded in the plasma membrane and act as molecular sensors of extracellular peptide hormones that activate intracellular signaling cascades. RLKs are a large family of transmembrane proteins with extracellular ligand-binding domains, a single-pass transmembrane region, and an intracellular kinase domain that is responsible for signal propagation. Ligand–receptor interactions provide the specificity in the perception of peptide hormones, and each peptide hormone binds selectively to a unique RLK or a group of closely related receptors (Verma et al. 2016; Vie et al. 2015; Xie et al. 2022). For example, CLV3 peptides can bind to the CLV1 receptor kinase to regulate shoot and root meristem maintenance; PSK peptides bind receptors to promote the growth and proliferation of cells (Figure 1.2). It similarly activates PEPs, which bind lant elicitor receptors (PEPRs) to trigger immune responses. All receptor kinases undergo autophosphorylation or transphosphorylation on ligand binding, activating downstream signaling components. Some RLKs can also co-associate with coreceptor kinases such as somatic embryogenesis receptor kinases (SERKs) for enhancing signal transduction efficiency. The function of these coreceptors is to modulate the strength and specificity of signaling pathways by enhancing receptor–ligand complex formation. Actions of peptide hormones at the plasma membrane are sensed and lead to a series of intracellular processes that include phosphorylation cascades, calcium signaling, and gene expression changes (Xu et al. 2015; Yang et al. 2023). By recognizing diverse peptide ligands, the capacity of RLKs to distinguish among them demonstrates the complexity of the regulatory mechanisms operating during plant development, stress adaptation, and recognition of biotic, including pathogen, and abiotic cues. Plants localize signal perception at the cell surface to view the exterior environment and quickly respond to detect and adapt to changes in the environment for precise growth and physiological control.

1.5.2 Intracellular Signaling Pathways

Upon binding to their receptor, peptide hormones activate intracellular signaling cascades that propagate and amplify the signal. The normal cascade conforms to the basic form of cascades in that these are usually phosphorylation events, secondary messengers, and transcriptional regulation, so that outside signals result in appropriate cellular responses. The mitogen-activated protein kinase (MAPK) cascade is one of the major peptide hormone signaling intracellular pathways. Upon activation of the receptor, the cytoplasmic kinases (e.g. brassinosteroid insensitive associated kinase1 [BAK1]) phosphorylate downstream MAPKs (Yamaguchi et al. 2016; Xu et al. 2013). The targets of these kinases share the same fate as kinases themselves, as they go on to phosphorylate receptor tyrosine kinases that regulate gene expression, further changing developmental direction, eliciting a stress response, or activating the immune system. Calmodulin and related Ca^{2+} signaling pathways are other key intracellular pathways. Peptide hormones cause the opening of the calcium channel and influx of Ca^{2+} into the cytoplasm. This stimulation of Ca^{2+} concentration causes an increase in cytoplasmic calmodulin (CAM) or calcium-dependent protein kinases (CDPKS) to activate target proteins of growth regulation and of stress adaptation (Figure 1.3). Reactive oxygen species (ROS) production also includes peptide hormone signaling in plant defense. For example, PEPs activate PEPR receptors, which generate ROS and activate an immune response to fight pathogens. Some peptide hormone pathways use ubiquitin-mediated proteolysis to regulate protein turnover. Such receptors can recruit E3

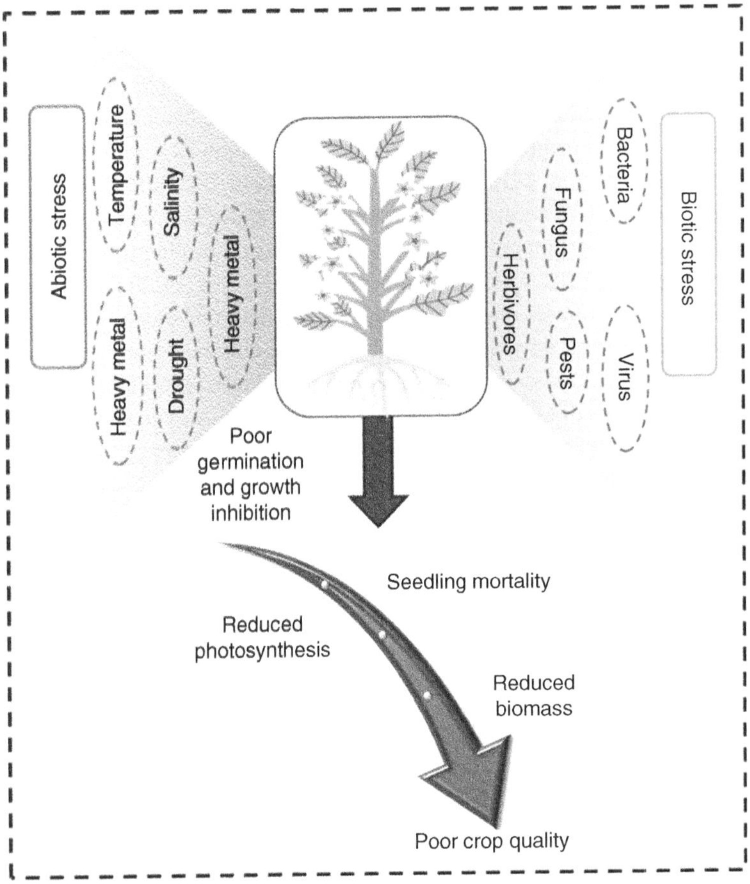

Figure 1.3 Phytohormonal control of metabolic reprogramming during plant stress responses.

ubiquitin ligases in such cases to tag specific proteins for degradation and to avoid a poorly controlled signaling response. Just as peptide hormones regulate gene expression through activation of nuclear transcription factors, they can also induce activation of nuclear transcription factors. To give an example, in the CLV signaling pathway, CLV3 perception by CLV1 represses WUSCHEL (WUS), a transcription factor that controls stem cell proliferation (Yoshida et al. 2021; Tost et al. 2021). RALF peptides function to regulate pH homeostasis and affect the activity of transcriptional regulators to change gene expression. Intracellular signaling cascades, in general, translate extracellular peptide hormone signals into a specific cellular response (Tabata and Sawa 2014; Suh et al. 2024).

1.5.3 Cross Talk Between Peptide Hormones and Other Hormones

Unlike other plant hormones, most of the peptide hormones act in concert with other plant hormones to form a highly complex regulatory network that controls growth, development, and stress responses. Such hormonal cross talk allows plants to integrate numerous signals and coordinate the physiological response in an undivided manner (Scott et al. 2024; Sang et al. 2025). Among well-known examples of cross talk is that between peptide hormones and auxins (Stintzi and Schaller 2022; Takahashi et al. 2018). Auxin signaling is used by the CLV3 peptide to modulate stem cell proliferation in response to meristem activity (Figure 1.3). As with CEPs that influence lateral root development by affecting auxin transport, peptide hormones also coordinate with auxins to regulate root architecture (Tavormina et al. 2015; Tan et al. 2024). Brassinosteroids (BRs) promote cell elongation and growth and also interact with peptide hormones. Stabilization of BR receptors by, for example, PSK signaling enhances BR responses in cells and amplifies growth-promoting signals (Royek et al. 2022; Salvi et al. 2021). The synergy between PSK and BR emphasizes cooperative actions of varying hormones in the control of plant development. PEPs work in ways to interact with jasmonic acid (JA) and salicylic acid (SA) signaling pathways during stress responses. JA and SA also contribute to systemic defense signaling, but the majority of this early systemic signaling effort is via PEPs activating immune responses to improve pathogen resistance (Schardon et al. 2016; Setty et al. 2024). This interaction ensures that the immune response is balanced in both local and systemic response mechanisms. An example of another type of cross talk is the interaction between gibberellins (GAs) and peptide hormones (Takahashi et al. 2020; Taleski et al. 2020). CRPs controlling GA signaling regulate seed germination and flowering time, and link the pathway to classical hormonal control. Plants integrate signals of peptide hormones with more traditional hormone pathways to create a dynamic, highly adaptable regulatory network. This cross communication allows them to respond efficiently to environmental conditions while balancing growth, immunity, and development simultaneously (Rzemieniewski et al. 2024; Sauter 2015).

1.6 Peptide Hormones in Plant Development and Growth

1.6.1 Regulation of Root and Shoot Development

The peptide hormones contribute significantly to the control of root and shoot architecture by influencing meristem activity, hormonal interactions, and cell differentiation. These external cues are relayed by small signaling molecules, termed key mediators of plant development, that regulate root and shoot growth. Both CEPs and root meristem growth factors (RGFs) are important peptide hormones during root development for maintaining meristem activity. GOLVEN (GLV) peptides, also called RGFs, regulate stem cell division and differentiation in the root apical meristem (RAM) (Taleski et al. 2024; Vanyushin et al. 2017). Root Growth Factor Receptor (RGFRs) and plasma membrane proton ATPases interact with and modulate auxin transport and maintain root stem cell identity. Similarly, CEPs operate in the sensing of nutrients, in particular of nitrogen under nitrogen-deficient conditions (Figure 1.3). CEPs also signal to the shoot to adjust root growth when nitrogen levels are low, making sure the roots have enough to take up. CLE peptides have a central role in the maintenance of the shoot apical meristem (SAM) in shoot development (Taleski et al. 2018; Toruno et al. 2016; Yu et al. 2024a,b).

The CLV1 RLK is known to bind the well-characterized CLE peptide CLV3, whose expression is regulated by its ligand to control WUS expression, a crucial transcription factor that governs stem cell proliferation (Yu et al. 2024a,b; Zarsky et al. 2024). The CLV3-WUS feedback loop helps maintain a balance of stem cell renewal and differentiation, ensuring that shoots outgrow and form new organs. Classical phytohormones (auxins, cytokinins, and BRs) also interact with peptide hormones for coordinating root and shoot development. For instance, auxin transport affected by RGFs regulates root hair elongation; PSK peptides act with brassinosteroids to promote shoot elongation and leaf expansion (Yuan et al. 2021; Zelman and Berkowitz 2023). Peptide hormones orchestrate meristem activity and hormone interaction, thereby maintaining a precise control over root and shoot growth to ensure plants adapt to different environmental conditions for overall development (Zelman and Berkowitz 2023; Panigrahi et al. 2021a,b,c).

1.6.2 Role in Organ Formation and Tissue Differentiation

Organogenesis depends on peptide hormones that regulate leaf, flower, vascular tissue, and other specialized structures. These molecules are used by cells to regulate cell fate determination, pattern determination, and differentiation, and proper organ development. During leaf and flower formation, peptide hormones such as epidermal patterning factor (EPF) peptides regulate stomatal development (Zhang et al. 2024a,b,c; Zhou et al. 2022). EPF1 and EPF2 bind to ERECTA family receptor kinases to control spacing between and density of stomata on the leaf surface (Figure 1.3). Gas exchange as well as water conservation require this regulation. CLE peptides such as CLV3 are involved in maintaining meristem homeostasis for the development of floral organs. However, tracheary element differentiation inhibitory factor (TDIF) peptides are critical for xylem and phloem patterning in vascular tissue differentiation. The TDIF is produced by phloem cells to interact with the TDIF receptor (TDR)/phloem intercalated with xylem (PXY) receptor-like kinase, which promotes cambial cell division but inhibits xylem differentiation. It allows a proper balance of vascular tissues synthesis as nutrients and water are transported throughout the plant. Both root hair and lateral root formation are also influenced by peptide hormones (Zhong et al. 2022; Zhang et al. 2022a,b). RALF peptides control root hair elongation by modulating proton pump activity, which alters cell wall loosening and cell expansion. Likewise, RGFs direct auxin gradients necessary for root branching, maintain stem cell activity, and regulate lateral root development. A second important class of peptides is inflorescence deficient in abscission (IDAS) peptides, which regulate organ separation processes as leaf and flower abscission plants important for developmental leaf and flower abscission processes. HAESA (HAE) and HAESA LIME (HSL) receptor kinases are IDA dependent and appear to activate cell-wall remodeling enzymes that facilitate detachment of the organ. Peptide hormones guide tissue patterning, organ differentiation, and tissue vascular development to ensure proper plant structure and function, allowing for successful plant life in a wide variety of environments (Stewart and Ronald 2022; Zhang et al. 2019, 2024a,b,c).

1.6.3 Influence on Cell Cycle and Division

Peptide hormones are important regulators of cell growth and division to facilitate controlled growth and development of the cells. These hormones influence stem cell activity, cytokinesis, and mitotic progression and thereby the rate and direction of cell expansion. PSK is one of the most well-characterized peptide hormones in cell cycle regulation. PSK acts as a growth-promoting factor that stimulates mitotic activity and cell proliferation.

PSK binds its receptor (PSKR) and activates intracellular signal pathways, which lead to the activation of cyclin-dependent kinase (CDK) necessary for cells to traverse the G1/S and G2/M phases of the cell cycle. It also provides for continual replacement of cells for organ growth and tissue expansion. CLE peptides such as CLV3 control stem cell division and differentiation only in meristematic regions (Zhang et al. 2023, 2024a,b,c). This ensures that meristem cells do not undergo uncontrolled proliferation to the extent that undifferentiated stem cells outstrip differentiated tissue cells. Similar to this, RGFs maintain root stem cell activity and division in roots. Asymmetric cell division is also coordinated by peptide hormones, which are critical for forming specific cell types. For instance, EPF peptides are known to affect stomatal precursor cells and orient their division pattern so that stomatal spacing is appropriate. TDIF signaling also controls the number of cambial cells, which produce xylem and phloem tissues at balanced rates. Some peptides control cell cycle arrest or programmed cell death under stress conditions. For example, RALF peptides can stop cell expansion in response to environmental cues as a readjustment of energy and resources (Ryslava et al. 2013; Roy et al. 2018). Like these PEP peptides, which are immunological, they influence cell cycle progression so that energy is diverted toward defense instead of growth. Peptide hormones tightly control the cell cycle so that plants have coordinated and adaptive growth whereby cell proliferation, differentiation, and environmental responses are balanced.

1.7 Peptide Hormones in Plant Stress Response

1.7.1 Role in Biotic Stress Tolerance (Pathogen Defense)

The existence of a complex immune system in plants for defense against pathogens and the importance of peptide hormones in the activation of defense responses are considered.The presence of a complex immune system in plants that defends against pathogens, and the crucial role of peptide hormones in activating defense responses, are well established. These small signaling molecules act as messengers mobilizing pathways that elicit an immune response in plants, allowing them to recognize and defend against microbial infections (Das et al. 2025). The PEP family is one of the most well-studied peptide hormones with regard to biotic stress tolerance (Panigrahi and Satapathy, 2023). On pathogen attack, PEPs are released and bind to PEPs, which activate immune responses. This interaction culminates in the production of ROS and the reinforcement of cell walls and expression of defense-related genes, which enhances resistance against bacteria and fungi (Chen et al. 2024a). Systemin is another important peptide that is principally present in Solanaceous plants, for example, in tomato and potato. Herbivore or mechanical damage results in the release of systemin, which induces the production of JA, the hormone that starts proteinase inhibitors to be produced to ward off insect feeding. This plant's wound response system is necessary for plant survival when attacked. Among other factors, RALF peptides also contribute to pathogen defense through alteration of cell wall integrity and ion homeostasis (Majumder and Panigrahi 2024). Other RALF peptides act as negative regulators during bacterial infection, allowing a certain amount of suppression that prevents excessive energy expenditure during infection (Mosher et al. 2013; Lin et al. 2023). However, some

RALF variants enhance defense by activating calcium signaling and MAPK cascades to increase pathogen resistance. In addition, plant CLE peptides are involved in plant–microbe interactions. Some CLE peptides secreted by parasitic nematodes are known to manipulate plant development, and others exist to augment immune signaling through interaction with RLKs. The peptide hormones assist plants in surviving pathogen attack by activating defense genes, ROS production, hormonal crosstalk, and immune signaling pathways.

1.7.2 Role in Abiotic Stress Adaptation (Drought, Salinity, Temperature)

A number of abiotic stress conditions, such as drought, salinity, and extreme temperatures, among others, in addition to biotic stress, will also pose stress to plants. By including their peptides in the list of candidate hormones that help plants cope with these harsh environments by regulating the water balance, osmotic stress responses, and cellular homeostasis, we contribute to precisely how those functions are established. CEPs are the key peptide hormones that play a crucial part in drought and osmotic stress adaptation. Under water-deficient conditions, CEPs signal long distances, from the root to the shoot, to inform the shoot of root water status. Stomatal conductance is adjusted by CEP perception in the shoot to enhance water uptake and root system expansion in response to prolonged drought, helping plants to survive (Ghosh and Roychoudhury 2023; Patro et al. 2024). Of similar function are contributions of PSK peptides to drought and salinity tolerance through increased antioxidant enzyme activity and enhanced cellular water retention. An interaction also occurs between PSK signaling and the key drought-response hormone abscisic acid (ABA), which is important as it regulates stomatal closure, preventing excessive water loss. CLE25 peptides are key regulators of osmotic balance and sodium ion homeostasis for salinity tolerance (Kang et al. 2022; Li et al. 2024). Under salt stress, CLE25 is transported from roots to shoots, where it triggers hormonal responses that mediate stomatal closure and reduce transpiration to minimize ion toxicity. Temperature stress adaptation also involves the participation of peptide hormones. Heat shock-induced peptides activate cellular chaperone proteins to protect structures from damage due to high temperatures. Cold-responsive peptides also increase membrane fluidity and protect cellular proteins during chilling stress (Gao et al. 2023; Das and Panigrahi 2025a,b). Since calcium-dependent signaling is critical for cell-wall remodeling under stress conditions, the regulation of RALF peptides affects calcium-dependent signaling. RALF peptides modulate ion channel activity during drought and high-salinity exposure in order to maintain cell integrity and prevent damage from oxidative stress (Lepetit and Brouquisse 2023). Peptide hormones influence water retention, ion homeostasis, stress signaling, and antioxidant defenses, and play a role in the plants' ability to survive and adapt to adverse environmental conditions.

1.7.3 Interaction with Other Hormonal Networks in Stress Conditions

Peptide hormones also work together with other phytohormones, such as ABA, SA, JA, auxins, and ethylene, to maximize plant stress responses. In this way, plant growth and defense are well balanced by this hormonal cross talk. JA signaling is one of the most well-characterized interactions under biotic stress responses, including PEP peptides. PEPR perception of PEPs amplifies JA signaling for the activation of defense genes, callose deposition, and systemic acquired resistance (SAR). This interaction strengthens the immune system of the plant and increases resistance to pathogens and insect herbivores. Like most peptide hormones, CLE25 also works with ABA in response to drought (Lee et al. 2019; Panigrahi et al. 2021b). However, CLE25 is transported from roots to shoots under water-deficit conditions, and there it increases ABA signaling to promote stomatal closure and water conservation. But this interaction prevents excessive water loss and allows a plant to survive through very long periods of drought. There is also a substantial interaction between RALF peptides and ethylene (ET). RALF peptides influence ethylene biosynthesis and regulate ethylene as a key regulator of stress responses to modulate root growth and stress cell wall integrity (Fedoreyeva 2023; Patro et al. 2023). This coordination allows plants to change root architecture and maintain cellular homeostasis under osmotic stress. Peptide hormones are also involved in growth-defense trade-offs (Gou et al. 2010; Panigrahi et al. 2021c). Growth-promoting peptides like PSK and RGFs interact with stress response hormones, e.g. SA and JA, under stress conditions in order to fine-tune allotment of energy. For example, SA-mediated immunity may impede growth-related pathways to ensure that plant resources are deployed to defense instead of elongation, as during pathogen attack. Moreover, BRs make contact with CLE peptides to promote meristem activity and tissue differentiation under stress conditions. The effect of this coordination is to start plant growth again after stress conditions are over.

1.8 Advances in Peptide Hormone Research and Biotechnology Applications

1.8.1 Genetic Engineering and Synthetic Peptides

Thus, recent developments in genetic engineering and synthetic biology have significantly expanded the boundaries of research on peptide hormones, along with opening new frontiers, including the manipulation of plant growth, stress responses, and defense mechanisms with unparalleled accuracy. Genetic modification techniques, including CRISPR-Cas9 and RNA interference (RNAi), have allowed for targeted changes in peptide hormone signaling pathways to enhance crop traits. Another research venue involves the overexpression or suppression of genes encoding peptide hormones to improve plant features. For example, it is known that overexpression of RGF peptides has been shown to improve water and nutrient uptake by increasing root biomass (Gorgues et al. 2022; Jung et al. 2020). Likewise, the architecture of the plant and the yield potential are fine-tuned through modification of CLE peptides to modulate the shoot apical meristem activity. Synthetic peptides have also found strong use in agricultural biotechnology. Successfully designed bioactive peptide analogues mimicking natural hormones are more stable and effective. For example, synthetic PEP peptides have been developed for plants in order to evoke stronger immune responses against bacterial or fungal pathogens. Furthermore, PSK peptides are modified to ensure accelerated cell division, which promotes increased plant growth and tissue regeneration. Advances in peptide delivery systems, including foliar sprays and nanoparticle-based delivery, have been made toward the external application of synthetic peptides, which decrease efforts required to produce p53 without involving genetic modification (Feng et al. 2023; Sahoo et al. 2023). These innovations offer an environmentally friendly alternative to chemical fertilizers and pesticides that are conducive to sustainable agriculture. Genetic engineering and the synthesis of new working ideal peptide libraries give researchers tools to advance plant biotechnology and continue to develop more resilient, high-yielding crops.

1.8.2 Agricultural Applications and Crop Improvement

Peptide hormones have great potential in agriculture in terms of novel ways to increase crop yield as well as their stress resilience and disease resistance. Scientists hope to introduce peptide-based technologies in farming practice to reduce dependence on conventional agrochemicals and, through synergic effects, increase plant performance (Simon and Dresselhaus 2015; Song et al. 2022). The second best application is the use of peptide hormones for stress tolerance. Specifically, CLE25 peptides that control stomatal closure can improve drought resistance by maximizing water conservation in crops (Majumder et al. 2025). Similarly, RALF peptides enable the plants to regulate the distribution of their ions under saline conditions, providing possible solutions to saline agriculture farmlands. PEP peptides have been demonstrated to be successful in inducing resistance in crops to pathogens without the use of chemical fungicides. Engineered systems in peptides have also been investigated as natural insect deterrents and therefore reduce crop damage caused by herbivorous pests. Peptide hormones also play a role in crop productivity. Researchers have been able to stimulate faster cell division and tissue regeneration by manipulating PSK signaling, which is very useful for vegetative propagation and tissue culture-based agriculture (Stewart and Ronald 2022; Zhang et al. 2022a,b). As is to be expected, by modifying RGFs and CLE peptides, root architecture and nutrient uptake have improved and biomass production has increased. The other emerging area is peptide-based biofertilizers, which improve soil microbiome interaction and nutrient assimilation to the environment without causing any harm. These biofertilizers are helping in reducing the application of synthetic fertilizers and improving sustainable crop growth. Peptide hormones are generally transforming modern agriculture and bringing innovative solutions to global challenges like food security and soil degradation.

1.9 Conclusion

Among the proteins that serve as hormones in plants, peptide hormones are important for plant growth and development as well as for responding to environmental stresses, and act as signaling molecules controlling many physiological and cellular processes. In recent years, much research on plant peptide hormones has been conducted, learning of their diverse functions and mechanisms of action. These small peptides are made in specific cellular compartments, transported across membranes into intracellular spaces, and perceived by their receptors, which activate intracellular signaling cascades. Plant growth, adaptation to environmental stressors, and coordination of developmental processes require compartmentalized signaling that can be regulated in plants. The interaction of plant peptide hormones with other phytohormonal networks has turned out to be one of the fundamental discoveries in plant peptide hormone research. Classical plant hormones such as ABA, auxins, JA, and ethylene and peptides such as PEP, CLE, PSK, and RGF work together in a balance to support growth, immunity, and stress tolerance. For example, CLE25 peptides work in concert with ABA to control intra and CLE-mediated responses to drought, while PEP peptides increase jasmonic acid-mediated defense pathways. The fact that peptide hormones and so-called traditional plant hormones do such a wonderful job of talking to each other only emphasizes the degree to which they play a crucial role in plant adaptability and survival. That peptide hormones and traditional plant hormones interact so seamlessly only reinforces how vital they are to plant adaptability and survival. Moreover, peptide hormones have had importance in agricultural biotechnology, as well as in their natural biological roles. The genetic engineering of peptide hormone pathways has advanced recently with CRISPR-based modifications in crops, which allow scientists to manipulate root growth, disease resistance, and stress tolerance. Further development of synthetic peptides has increased the potential of these molecules by enhancing their stability and effectiveness in applications for agriculture as synthetic peptides. In addition, peptide-based biofertilizers and plant-protection agents are sustainable and environmentally friendly alternatives to chemical fertilizers and pesticides. While these qualities have enabled great advances in our understanding of the complexity of peptide hormone signaling, much remains to be learned about its complexity.

References

Abarca, A., Franck, C.M., and Zipfel, C. (2021). Family-wide evaluation of rapid alkalinization factor peptides. *Plant Physiology* 187 (2): 996–1010.

Aggarwal, S., Kumar, A., Jain, M. et al. (2020). C-terminally encoded peptides (CEPs) are potential mediators of abiotic stress response in plants. *Physiology and Molecular Biology of Plants* 26: 2019–2033.

Amano, Y., Tsubouchi, H., Shinohara, H. et al. (2007). Tyrosine-sulfated glycopeptide involved in cellular proliferation and expansion in Arabidopsis. *Proceedings of the National Academy of Sciences of the United States of America* 104: 18333–18338.

Andersen, E.J., Ali, S., Byamukama, E. et al. (2018). Disease resistance mechanisms in plants. *Genes* 9 (7): 339. https://doi.org/10.3390/genes9070339.

Arsova, B., Watt, M., and Usadel, B. (2018). Monitoring of plant protein post-translational modifications using targeted proteomics. *Frontiers in Plant Science* 9: 1168.

Avalos-Rangel, A., Ruiz-Herrera, L.F., Gutierrez-Alanis, D. et al. (2024). CLE14 peptide impairs root tip regeneration and callogenesis in Arabidopsis. *Journal of Plant Growth Regulation* 43 (5): 1458–1465.

Baillie, A.L., Sloan, J., Qu, L.J., and Smith, L.M. (2024). Signalling between the sexes during pollen tube reception. *Trends in Plant Science* 29 (3): 343–354.

Bari, R. and Jones, J.D. (2009). Role of plant hormones in plant defence responses. *Plant Molecular Biology* 69: 473–488.

Bashyal, S., Gautam, C.K., and Muller, L.M. (2024). Clavata signaling in plant-environment interactions. *Plant Physiology* 194 (3): 1336–1357. https://doi.org/10.1093/plphys/kiad591.

Behera, A., Panigrahi, G.K., and Sahoo, A. (2025). Nonsense-mediated mRNA decay in human health and diseases: current understanding, regulatory mechanisms and future perspectives. *Molecular Biotechnology* 67. https://doi.org/10.1007/s12033-024-01267-7: 3374–3390.

Bhalla, H., Sudarsanam, K., Srivastava, A., and Sankaranarayanan, S. (2025). Structural insights into the recognition of RALF peptides by FERONIA receptor kinase during Brassicaceae pollination. *Plant Molecular Biology* 115 (1): 1–6.

Blackburn, M.R., Haruta, M., and Moura, D.S. (2020). Twenty years of progress in physiological and biochemical investigation of RALF peptides. *Plant Physiology* 182 (4): 1657–1666.

Boschiero, C., Lundquist, P.K., Roy, S. et al. (2019). Identification and functional investigation of genome-encoded, small, secreted peptides in plants. *Current Protocols in Plant Biology* 4 (3): e20098.

Canut, H., Albenne, C., and Jamet, E. (2016). Post-translational modifications of plant cell wall proteins and peptides: a survey from a proteomics point of view. *Biochimica Et Biophysica Acta - Proteins and Proteomics* 1864 (8): 983–990.

Carbonnel, S., Falquet, L., and Hazak, O. (2022). Deeper genomic insights into tomato CLE genes repertoire identify new active peptides. *BMC Genomics* 23 (1): 756.

Chang, S. and Xiao, F. (2025). Comprehensive review of plant small signaling peptides: from stress adaptation mechanisms to practical solutions for crop resilience. *International Journal of Biological Macromolecules* 299: 139971.

Chapman, K., Taleski, M., Frank, M., and Djordjevic, M.A. (2024). C-terminally encoded peptide (CEP) and cytokinin hormone signaling intersect to promote shallow lateral root angles. *Journal of Experimental Botany* 75 (2): 631–641.

Chaulagain, D. and Frugoli, J. (2021). The regulation of nodule number in legumes is a balance of three signal transduction pathways. *International Journal of Molecular Sciences* 22 (3): 1117.

Chen, Y.L., Fan, K.T., Hung, S.C., and Chen, Y.R. (2020). The role of peptides cleaved from protein precursors in eliciting plant stress reactions. *The New Phytologist* 225 (6): 2267–2282. https://doi.org/10.1111/nph.16241.

Chen, H., Song, Y., Li, H. et al. (2023). Enhancing the adaptability of tea plants (*Camellia sinensis* L) to high-temperature stress with small peptides and biosurfactants. *Plants (Basel)* 12 (15): 2817. https://doi.org/10.3390/plants12152817.

Chen, C., Buscaill, P., Sanguankiatticha, N. et al. (2024a). Extracellular plant subtilases dampen cold-shock peptide elicitor levels. *Nature Plants* 10 (11): 1749–1760.

Chen, X., van de Sande, J.W., Ritmejeris, J. et al. (2024b). Resolving sulfation PTMs on a plant peptide hormone using nanopore sequencing. bioRxiv

Cho, H., Ryu, H., Rho, S. et al. (2014). A secreted peptide acts on BIN2-mediated phosphorylation of ARFs to potentiate auxin response during lateral root development. *Nature Cell Biology* 16 (1): 66–76.

Combest, M.M., Moroz, N., Tanaka, K. et al. (2021). St PIP1, a PAMP-induced peptide in potato, elicits plant defenses and is associated with disease symptom severity in a compatible interaction with potato virus Y. *Journal of Experimental Botany* 72 (12): 4472–4488.

Corcilius, L., Hastwell, A.H., Zhang, M. et al. (2017). Arabinosylation modulates the growth-regulating activity of the peptide hormone CLE40a from soybean. *Cell Chemical Biology* 24 (11): 1347–1355.e7.

Dao, T.Q., Weksler, N., HMH, L. et al. (2022). Interactive CLV3, CLE16 and CLE17 signaling mediates stem cell homeostasis in the Arabidopsis shoot apical meristem. *Development* 149 (19): dev200787.

Das, R. and Panigrahi, G.K. (2025a). Complexities of salinity stress on zebrafish survivability. *Toxicology International* 32 (1): 57–63. https://doi.org/10.18311/ti/2025/v32i1/45794.

Das, R. and Panigrahi, G.K. (2025b). Messenger RNA surveillance: current understanding, regulatory mechanisms, and future implications. *Molecular Biotechnology* 67 (2): 393–409. https://doi.org/10.1007/s12033-024-01062-4.

Das, R., Singh, S.P., Sahoo, A. et al. (2025). Synergistic stress of zinc oxide nanoparticles and pH triggers gill damage and mortality in zebrafish. *Nanotechnology for Environmental Engineering* 10 (1): 1–11.

de Bang, T.C., Lundquist, P.K., Dai, X. et al. (2017). Genome-wide identification of Medicago peptides involved in macronutrient responses and nodulation. *Plant Physiology* 175 (4): 1669–1689.

De Coninck, B. and De Smet, I. (2016). Plant peptides—taking them to the next level. *Journal of Experimental Botany* 67 (16): 4791–4795. https://doi.org/10.1093/jxb/erw309.

Ding, S., Lv, J., Hu, Z. et al. (2023). Phytosulfokine peptide optimizes plant growth and defense via glutamine synthetase GS2 phosphorylation in tomato. *The EMBO Journal* 42 (6): e111858.

Eckardt, N.A., Avin-Wittenberg, T., Bassham, D.C. et al. (2024). The lowdown on breakdown: open questions in plant proteolysis. *The Plant Cell* 36 (9): 2931–2975.

Farrokhi, N., Whitelegge, J.P., and Brusslan, J.A. (2008). Plant peptides and peptidomics. *Plant Biotechnology Journal* 6 (2): 105–134.

Fedoreyeva, L.I. (2023). Molecular mechanisms of regulation of root development by plant peptides. *Plants* 12 (6): 1320. https://doi.org/10.3390/plants12061320.

Fedoreyeva, L.I., Chaban, I.A., and Kononenko, N.V. (2022). Short peptides induce development of root hair *Nicotiana tabacum*. *Plants* 11 (7): 852.

Feng, Y.Z., Zhu, Q.F., Xue, J. et al. (2023). Shining in the dark: the big world of small peptides in plants. *aBIOTECH* 4 (3): 238–256. https://doi.org/10.1007/s42994-023-00100-0.

Fernandez-Fernandez, A.D., Stael, S., and Van Breusegem, F. (2023). Mechanisms controlling plant proteases and their substrates. *Cell Death and Differentiation* 30 (4): 1047–1058. https://doi.org/10.1038/s41418-023-01120-5.

Friso, G. and van Wijk, K.J. (2015). Posttranslational protein modifications in plant metabolism. *Plant Physiology* 169 (3): 1469–1487. https://doi.org/10.1104/pp1501378.

Furumizu, C. and Aalen, R.B. (2023). Peptide signaling through leucine-rich repeat receptor kinases: insight into land plant evolution. *New Phytologist* 238 (3): 977–982.

Furumizu, C., Krabberod, A.K., Hammerstad, M. et al. (2021). The sequenced genomes of nonflowering land plants reveal the innovative evolutionary history of peptide signaling. *The Plant Cell* 33 (9): 2915–2934.

Gancheva, M.S., Malovichko, Y.V., Poliushkevich, L.O. et al. (2019). Plant peptide hormones. *Russian Journal of Plant Physiology* 66: 171–189.

Gao, Q., Wang, C., Xi, Y. et al. (2023). RALF signaling pathway activates MLO calcium channels to maintain pollen tube integrity. *Cell Research* 33 (1): 71–79.

Ghosh, P. and Roychoudhury, A. (2023). Plant peptides involved in abiotic and biotic stress responses and reactive oxygen species (ROS) signaling. *Journal of Plant Growth Regulation* 43: 1410–1427.

Goad, D.M., Zhu, C., and Kellogg, E.A. (2017). Comprehensive identification and clustering of CLV3/ESR-related (CLE) genes in plants finds groups with potentially shared function. *New Phytologist* 216 (2): 605–616.

Gorgues, L., Li, X., Maurel, C. et al. (2022). Root osmotic sensing from local perception to systemic responses. *Stress Biology* 2 (1): 36.

Gou, X., He, K., Yang, H. et al. (2010). Genome-wide cloning and sequence analysis of leucine-rich repeat receptor-like protein kinase genes in *Arabidopsis thaliana. BMC Genomics* 11: 19. https://doi.org/10.1186/1471-2164-11-19.

Greifenhagen, A., Ruwe, H., Zimmer, V. et al. (2024). The peptide hormone Pj CLE1 stimulates haustorium formation in the parasitic plant Phtheirospermum japonicum. *Proceedings of the National Academy of Sciences* 121 (42): e2414582121.

Guillou, M.C., Balliau, T., Vergne, E. et al. (2022). The PROSCOOP10 gene encodes two extracellular hydroxylated peptides and impacts flowering time in Arabidopsis. *Plants* 11 (24): 3554.

Hagelthorn, L. and Fletcher, J.C. (2023). The CLAVATA3/ESR-related peptide family in the biofuel crop pennycress. *Frontiers in Plant Science* 14: 1240342.

Haruta, M., Sabat, G., Stecker, K. et al. (2014). A peptide hormone and its receptor protein kinase regulate plant cell expansion. *Science* 343 (6169): 408–411. https://doi.org/10.1126/science.1244454.

Hastwell, A.H., Gresshoff, P.M., and Ferguson, B.J. (2014). The structure and activity of nodulation-suppressing CLE peptide hormones of legumes. *Functional Plant Biology* 42 (3): 229–238.

Hirakawa, Y. and Sawa, S. (2019). Diverse function of plant peptide hormones in local signaling and development. *Current Opinion in Plant Biology* 51: 81–87. https://doi.org/10.1016/j.pbi.2019.04.005.

Hirayama, T. and Mochida, K. (2023). Plant harmonics: a key tool for deep physiological phenotyping to improve crop productivity. *Plant and Cell Physiology* 63 (12): 1826–1839.

Hsiao, Y.C. and Yamada, M. (2020). The roles of peptide hormones and their receptors during plant root development. *Genes* 12 (1): 22. https://doi.org/10.3390/genes12010022.

Hsu, P.K., Dubeaux, G., Takahashi, Y., and Schroeder, J.I. (2021). Signaling mechanisms in abscisic acid-mediated stomatal closure. *The Plant Journal* 105 (2): 307–321. https://doi.org/10.1016/j.pmpp.2020.101568.

Huang, P., Li, Z., and Guo, H. (2022). New advances in the regulation of leaf senescence by classical and peptide hormones. *Frontiers in Plant Science* 13: 923136.

Igarashi, D., Tsuda, K., and Katagiri, F. (2012). The peptide growth factor, phytosulfokine, attenuates pattern-triggered immunity. *The Plant Journal* 71 (2): 194–204.

Ito, Y., Nakanomyo, I., Motose, H. et al. (2006). Dodeca-CLE peptides as suppressors of plant stem cell differentiation. *Science* 313 (5788): 842–845.

Je, B.I., Xu, F., Wu, Q. et al. (2018). The clavata receptor fasciated ear2 responds to distinct CLE peptides by signaling through two downstream effectors. *eLife* 7: e35673.

Johns, S., Hagihara, T., Toyota, M., and Gilroy, S. (2021). The fast and the furious: rapid long-range signaling in plants. *Plant Physiology* 185 (3): 694–706.

Jung, H.W., Panigrahi, G.K., Jung, G.Y. et al. (2020). PAMP-triggered immunity involves proteolytic degradation of core nonsense-mediated mRNA decay factors during early defense response. *The Plant Cell* 32 (4): 1081–1101. https://doi.org/10.1105/tpc.19.00631.

Kaeothip, S., Ishiwata, A., and Ito, Y. (2013). Stereoselective synthesis of Arabidopsis clavata3 (CLV3) glycopeptide, unique protein post-translational modifications of secreted peptide hormone in plant. *Organic & Biomolecular Chemistry* 11 (35): 5892–5907.

Kang, J., Wang, X., Ishida, T. et al. (2022). A group of CLE peptides regulates de novo shoot regeneration in Arabidopsis thaliana. *New Phytologist* 235 (6): 2300–2312.

Kaufmann, C. and Sauter, M. (2019). Sulfated plant peptide hormones. *Journal of Experimental Botany* 70 (16): 4267–4277.

Kim, J.S., Jeon, B.W., and Kim, J. (2021). Signaling peptides regulating abiotic stress responses in plants. *Frontiers in Plant Science* 12: 704490.

Kokla, A., Leso, M., Simura, J. et al. (2024). A long-distance inhibitory system regulates haustoria numbers in parasitic plants. bioRxiv, 2024-12

Kondo, T., Yokomine, K., Nakagawa, A., and Sakagami, Y. (2011). Analogs of the CLV3 peptide: synthesis and structure-activity relationships focused on proline residues. *Plant & Cell Physiology* 52 (1): 30–36. https://doi.org/10.1093/pcp/pcq146.

Kuromori, T., Fujita, M., Takahashi, F. et al. (2022). Inter-tissue and inter-organ signaling in drought stress response and phenotyping of drought tolerance. *The Plant Journal* 109 (2): 342–358.

Laffont, C., Ivanovici, A., Gautrat, P. et al. (2020). The NIN transcription factor coordinates CEP and CLE signaling peptides that regulate nodulation antagonistically. *Nature Communications* 11 (1): 3167.

Lalun, V.O., Breiden, M., Galindo-Trigo, S. et al. (2024). A dual function of the IDA peptide in regulating cell separation and modulating plant immunity at the molecular level. *eLife* 12: RP87912.

Leasure, C.D. and He, Z.H. (2012). CLE and RGF family peptide hormone signaling in plant development. *Molecular Plant* 5 (6): 1173–1175.

Lebedeva, M., Azarakhsh, M., Yashenkova, Y., and Lutova, L. (2020). Nitrate-induced CLE peptide systemically inhibits nodulation in *Medicago truncatula*. *Plants* 9 (11): 1456.

Lee, Z.H., Hirakawa, T., Yamaguchi, N., and Ito, T. (2019). The roles of plant hormones and their interactions with regulatory genes in determining meristem activity. *International Journal of Molecular Sciences* 20 (16): 4065.

Lepetit, M. and Brouquisse, R. (2023). Control of the rhizobium–legume symbiosis by the plant nitrogen demand is tightly integrated at the whole plant level and requires inter-organ systemic signaling. *Frontiers in Plant Science* 14: 1114840.

Li, J., Huang, Y., Yu, X. et al. (2024). Identification and application of CLE peptides for drought resistance in Solanaceae crops. *Journal of Agricultural and Food Chemistry* 72: 13869–13884.

Lin, Z.B., Zhao, Y.X., and Yang, W.C. (2023). A balanced mechanism between plant growth and stress response. *Advanced Agrochem* 2 (2): 105–106.

Liu, Y., Zuo, T., Qiu, Z. et al. (2021). Genome-wide identification reveals the function of CEP peptide in cucumber root development. *Plant Physiology and Biochemistry* 169: 119–126.

Liu, D., Shen, Z., Zhuang, K. et al. (2022a). Systematic annotation reveals CEP function in tomato root development and abiotic stress response. *Cells* 11 (19): 2935.

Liu, Q., Wu, K., Song, W. et al. (2022b). Improving crop nitrogen use efficiency toward sustainable green revolution. *Annual Review of Plant Biology* 73 (1): 523–551.

Lyapina, I., Ganaeva, D., Rogozhin, E.A. et al. (2024). Comparative analysis of small secreted peptide signaling during defense response: insights from vascular and non-vascular plants. bioRxiv, 2024-05

Majumder, S. and Panigrahi, G.K. (2024). Molecular docking and ADME evaluation of plant-based bioactive molecules targeting nonsense-mediated mRNA decay pathway factors to modulate tumorigenesis. *Plant Science Today* 11 (4): https://doi.org/10.14719/pst.3737.

Majumder, S. and Panigrahi, G.K. (2025). Advancements in contemporary pharmacological innovation: mechanistic insights and emerging trends in drug discovery and development. *Intelligent Pharmacy.* https://doi.org/10.1016/j.ipha.2024.10.001 3: 118–126.

Majumder, S., Das, R., and Panigrahi, G.K. (2025). Harnessing phytochemicals to modulate the nonsense-mediated decay pathway for anticancer drug discovery. *Research Journal of Pharmacy and Technology* 18 (2): 489–494. https://doi.org/10.52711/0974-360X.2025.00074.

Matsubayashi, Y. (2018). Exploring peptide hormones in plants: identification of four peptide hormone-receptor pairs and two post-translational modification enzymes. *Proceedings of the Japan Academy Series B, Physical and Biological Sciences* 94 (2): 59–74. https://doi.org/10.2183/pjab.94.006.

Matsuzaki, Y., Ogawa-Ohnishi, M., Mori, A., and Matsubayashi, Y. (2010). Secreted peptide signals required for maintenance of root stem cell niche in Arabidopsis. *Science* 329: 1065–1067.

Mens, C., Hastwell, A.H., Su, H. et al. (2021). Characterisation of Medicago truncatula CLE34 and CLE35 in nitrate and rhizobia regulation of nodulation. *New Phytologist* 229 (5): 2525–2534.

Millar, A.H., Heazlewood, J.L., Giglione, C. et al. (2019). The scope, functions, and dynamics of posttranslational protein modifications. *Annual Review of Plant Biology* 70 (1): 119–151.

Mosher, S., Seybold, H., Rodriguez, P. et al. (2013). The tyrosine-sulfated peptide receptors PSKR1 and PSY1R modify the immunity of Arabidopsis to biotrophic and necrotrophic pathogens in an antagonistic manner. *The Plant Journal* 73 (3): 469–482.

Motomitsu, A., Sawa, S., and Ishida, T. (2015). Plant peptide hormone signaling. *Essays in Biochemistry* 58: 115–131. https://doi.org/10.1042/bse0580115.

Muller, L.M. and Harrison, M.J. (2019). Phytohormones, miRNAs, and peptide signals integrate plant phosphorus status with arbuscular mycorrhizal symbiosis. *Current Opinion in Plant Biology* 50: 132–139.

Murphy, E., Smith, S., and De Smet, I. (2012). Small signaling peptides in Arabidopsis development: how cells communicate over a short distance. *The Plant Cell* 24 (8): 3198–3217.

Nakagami, S., Kajiwara, T., Tsuda, K., and Sawa, S. (2024). CLE peptide signaling in plant-microbe interactions. *Frontiers in Plant Science* 15: 1481650.

Nakayama, T., Shinohara, H., Tanaka, M. et al. (2017). A peptide hormone required for Casparian strip diffusion barrier formation in Arabidopsis roots. *Science* 355 (6322): 284–286.

Narasimhan, M. and Simon, R. (2022). Spatial range, temporal span, and promiscuity of CLE-RLK signaling. *Frontiers in Plant Science* 13: 906087.

Nguyen, Q.T.C. and Kim, J. (2023). PSY–PSYR peptide–receptor pairs control the trade-off between plant growth and stress response. *Plant Signaling & Behavior* 18 (1): https://doi.org/10.1080/15592324.2023.2260638.

Ogawa-Ohnishi, M., Yamashita, T., Kakita, M. et al. (2022). Peptide ligand-mediated trade-off between plant growth and stress response. *Science* 378 (6616): 175–180.

Oh, E., Seo, P.J., and Kim, J. (2018). Signaling peptides and receptors coordinating plant root development. *Trends in Plant Science* 23 (4): 337–351.

Ohki, S., Takeuchi, M., and Mori, M. (2011). The NMR structure of storage reveals the basis of stomatal density regulation by plant peptide hormones. *Nature Communications* 2: 512. https://doi.org/10.1038/ncomms1520.

Ohkubo, Y., Tanaka, M., Tabata, R. et al. (2017). Shoot-to-root mobile polypeptides involved in systemic regulation of nitrogen acquisition. *Nature Plants* 3 (4): 1–6.

Okamoto, S., Shinohara, H., Mori, T. et al. (2013). Root-derived CLE glycopeptides control nodulation by direct binding to HAR1 receptor kinase. *Nature Communications* 4 (1): 2191.

Okamoto, S., Kawasaki, A., Makino, Y. et al. (2022). Long-distance translocation of CLAVATA3/ESR-related 2 peptide and its positive effect on roots sucrose status. *Plant Physiology* 189 (4): 2357–2367. https://doi.org/10.1093/plphys/kiac227.

Pan, X., Deng, Z., Wu, R. et al. (2024). Identification of CEP peptides encoded by the tobacco (*Nicotiana tabacum*) genome and characterization of their roles in osmotic and salt stress responses. *Plant Physiology and Biochemistry* 209: 108525.

Panigrahi, G.K. and Satapathy, K.B. (2020a). *Arabidopsis* DCP5, a decapping complex protein interacts with Ubiquitin-5 in the processing bodies. *Plant Archives* 20 (1): 2243–2247.

Panigrahi, G.K. and Satapathy, K.B. (2020b). Formation of *Arabidopsis* poly(A)-specific ribonuclease associated processing bodies in response to pathogenic infection. *Plant Archives* 20 (2): 4907–4912.

Panigrahi, G.K. and Satapathy, K.B. (2020c). Sacrificed surveillance process favours plant defense: a review. *Plant Archives* 20 (1): 2551–2559.

Panigrahi, G.K. and Satapathy, K.B. (2021). *Pseudomonas syringae* pv. *syringae* infection orchestrates the fate of the Arabidopsis J domain containing cochaperone and decapping protein factor 5. *Physiological and Molecular Plant Pathology* 113: 101598. https://doi.org/10.1016/j.pmpp.2020.101598.

Panigrahi, G.K. and Satapathy, K.B. (2024). Biological synthesis of gold nanoparticles by microbes: mechanistic aspects, biomedical applications, and future prospects. *Entrepreneurship with Microorganisms* 371–388. https://doi.org/10.1016/B978-0-443-19049-0.00012-8.

Panigrahi, G.K., Sahoo, A., and Satapathy, K.B. (2021a). Differential expression of selected Arabidopsis resistant genes under abiotic stress conditions. *Plant Science Today* 8 (4): 859–864. https://doi.org/10.14719/pst.2021.8.4.1213.

Panigrahi, G.K., Sahoo, A., and Satapathy, K.B. (2021b). Insights to plant immunity: defense signaling to epigenetics. *Physiological and Molecular Plant Pathology* 113: 1–7.

Panigrahi, G.K., Sahoo, S.K., Sahoo, A. et al. (2021c). Bioactive molecules from plants: a prospective approach to combat SARS-Cov-2. *Advances in Traditional Medicine* 23. https://doi.org/10.1007/s13596-021-00599-y: 617–630.

Panigrahi, G.K., Sahoo, A., and Satapathy, K.B. (2024). The processing body component varicose plays a multiplayer role towards stress management in Arabidopsis. *Plant Physiology Reports* 29. https://doi.org/10.1007/s40502-023-00778-w: 186–192.

Pastor-Fernandez, J., Sanchez-Bel, P., Flors, V. et al. (2023). Small signals lead to big changes: the potential of peptide-induced resistance in plants. *Journal of Fungi* 9 (2): 265.

Patel, N., Mohd-Radzman, N.A., Corcilius, L. et al. (2018). Diverse peptide hormones affecting root growth identified in the *Medicago truncatula* secreted peptidome. *Molecular & Cellular Proteomics* 17 (1): 160–174.

Patro, I., Sahoo, A., Nayak, B.R. et al. (2023). Nonsense-mediated mRNA decay: mechanistic insights and physiological significance. *Molecular Biotechnology* 66. https://doi.org/10.1007/s12033-023-00927-4: 3077–3091.

Patro, A.K., Panigrahi, G.K., Majumder, S. et al. (2024). Nonsense-mediated mRNA decay: physiological significance, mechanistic insights and future implications. *Pathology, Research and Practice* 264: 155677. https://doi.org/10.1016/j.prp.2024.155677.

Pecenkova, T., Potocky, M., and Stegmann, M. (2024). More than meets the eye: knowns and unknowns of the trafficking of small secreted proteins in Arabidopsis. *Journal of Experimental Botany* 75: 3713–3730.

Phour, M., Sehrawat, A., Sindhu, S.S., and Glick, B.R. (2020). Interkingdom signaling in plant-rhizomicrobiome interactions for sustainable agriculture. *Microbiological Research* 241: 126589.

Qian, P., Song, W., Yokoo, T. et al. (2018). The CLE9/10 secretory peptide regulates stomatal and vascular development through distinct receptors. *Nature Plants* 4 (12): 1071–1081.

Qiu, Z., Zhuang, K., Liu, Y. et al. (2022). Functional characterization of C-terminally encoded peptide (CEP) family in *Brassica rapa* L. *Plant Signaling & Behavior* 17 (1): 2021365.

Reichardt, S., Piepho, H.P., Stintzi, A., and Schaller, A. (2020). Peptide signaling for drought-induced tomato flower drop. *Science* 367 (6485): 1482–1485.

Reid, D.E., Ferguson, B.J., and Gresshoff, P.M. (2011). Inoculation-and nitrate-induced CLE peptides of soybean control NARK-dependent nodule formation. *Molecular Plant-Microbe Interactions* 24 (5): 606–618.

Ren, S.C., Song, X.F., Chen, W.Q. et al. (2019). CLE25 peptide regulates phloem initiation in Arabidopsis through a CLERK-CLV2 receptor complex. *Journal of Integrative Plant Biology* 61 (10): 1043–1061.

Ren, G., Zhang, Y., Chen, Z. et al. (2024). Research progress of small plant peptides on the regulation of plant growth, development, and abiotic stress. *International Journal of Molecular Sciences* 25 (7): 4114. https://doi.org/10.3390/ijms25074114.

Roman, A.O., Jimenez-Sandoval, P., Augustin, S. et al. (2022). HSL1 and BAM1/2 impact epidermal cell development by sensing distinct signaling peptides. *Nature Communications* 13 (1): 876.

Roy, S., Lundquist, P., Udvardi, M., and Scheible, W.R. (2018). Small and mighty: peptide hormones in plant biology. *The Plant Cell* 30: tc118tt0718.

Royek, S., Bayer, M., Pfannstiel, J. et al. (2022). Processing of a plant peptide hormone precursor facilitated by posttranslational tyrosine sulfation. *Proceedings of the National Academy of Sciences of the United States of America* 119 (16): e2201195119. https://doi.org/10.1073/pnas.2201195119.

Ryslava, H., Doubnerova, V., Kavan, D., and Vanek, O. (2013). Effect of posttranslational modifications on enzyme function and assembly. *Journal of Proteomics* 92: 80–109. https://doi.org/10.1016/j.jprot.2013.03.025.

Rzemieniewski, J., Leicher, H., Lee, H.K. et al. (2024). CEP signaling coordinates plant immunity with nitrogen status. *Nature Communications* 15 (1): 1–16.

Sahoo, A. and Satapathy, K.B. (2021). Differential expression of Arabidopsis EJC core proteins under short-day and long-day growth conditions. *Plant Science Today* 8 (4): 815–819. https://doi.org/10.14719/pst.2021.8.4.1214.

Sahoo, A., Satapathy, K.B., and Panigrahi, G.K. (2023). Ectopic expression of disease resistance protein promotes resistance against pathogen infection and drought stress in Arabidopsis. *Physiological and Molecular Plant Pathology* 124: 101949. https://doi.org/10.1016/j.pmpp.2023.101949.

Sahoo, A., Satapathy, K.B., and Panigrahi, G.K. (2024). Security check: plant immunity under temperature surveillance. *Journal of Plant Biochemistry and Biotechnology* 33: 1–4. https://doi.org/10.1007/s13562-023-00846-0.

Salvi, P., Manna, M., Kaur, H. et al. (2021). Phytohormone signaling and crosstalk in regulating drought stress response in plants. *Plant Cell Reports* 40: 1305–1329.

Sang, T., Zhang, Z., Liu, G., and Wang, P. (2025). Navigating the landscape of plant proteomics. *Journal of Integrative Plant Biology* 67 (3): 740–761.

Sauter, M. (2015). Phytosulfokine peptide signalling. *Journal of Experimental Botany* 66 (17): 5161–5169.

Schardon, K., Hohl, M., Graff, L. et al. (2016). Precursor processing for plant peptide hormone maturation by subtilisin-like serine proteinases. *Science* 354 (6319): 1594–1597.

Scott, T., Correa-Lozano, A., and Foo, E. (2024). Upstairs, downstairs-conserved and divergent CLAVATA signalling in shoot meristem development and root symbioses. *Annals of Botany* mcae192.

Setty, J., Singh, P., Tantuway, G., and Vijai, P. (2024). *Recent Advancement on Peptide Research and their Application in Eco-Agriculture in Plant Elicitor Peptides: New Tool for Sustainable Agriculture*, 269–281. Singapore: Springer Nature Singapore.

Simon, R. and Dresselhaus, T. (2015). Peptides take centre stage in plant signaling preface. *Journal of Experimental Botany* 66 (17): 5135–5138. https://doi.org/10.1093/jxb/erv376.

Singh, S.P., Panigrahi, G.K., and Sahoo, A. (2024). Optimizing nanoparticles use for growth promotion in zebrafish: insights on concentration-dependent effects. *Nanotechnology for Environmental Engineering* 9. https://doi.org/10.1007/s41204-024-00393-6: 757–767.

Song, X.F., Hou, X.L., and Liu, C.M. (2022). CLE peptides: critical regulators for stem cell maintenance in plants. *Planta* 255: 1–17.

Stewart, V. and Ronald, P.C. (2022). Sulfotyrosine residues: interaction specificity determinants for extracellular protein–protein interactions. *Journal of Biological Chemistry* 298 (8): 102232.

Stintzi, A. and Schaller, A. (2022). Biogenesis of post-translationally modified peptide signals for plant reproductive development. *Current Opinion in Plant Biology* 69: 102274.

Suh, E.M., Reinhardt, J.K., and Weng, J.K. (2024). The emergence and loss of cyclic peptides in *Nicotiana* illuminate dynamics and mechanisms of plant metabolic evolution. bioRxiv, 2024-11

Tabata, R. and Sawa, S. (2014). Maturation processes and structures of small secreted peptides in plants. *Frontiers in Plant Science* 5: 311. https://doi.org/10.3389/fpls.2014.00311.

Takahashi, F., Suzuki, T., Osakabe, Y. et al. (2018). A small peptide modulates stomatal control via abscisic acid in long-distance signaling. *Nature* 556 (7700): 235–238.

Takahashi, F., Kuromori, T., Urano, K. et al. (2020). Drought stress responses and resistance in plants: from cellular responses to long-distance intercellular communication. *Frontiers in Plant Science* 11: 556972.

Taleski, M., Imin, N., and Djordjevic, M.A. (2018). CEP peptide hormones: key players in orchestrating nitrogen-demand signalling, root nodulation, and lateral root development. *Journal of Experimental Botany* 69 (8): 1829–1836.

Taleski, M., Chapman, K., Imin, N. et al. (2020). The peptide hormone receptor CEPR1 functions in the reproductive tissue to control seed size and yield. *Plant Physiology* 183 (2): 620–636.

Taleski, M., Jin, M., Chapman, K. et al. (2024). CEP hormones at the nexus of nutrient acquisition and allocation, root development, and plant–microbe interactions. *Journal of Experimental Botany* 75 (2): 538–552.

Tan, W., Nian, H., Tran, L.S.P. et al. (2024). Small peptides: novel targets for modulating plant–rhizosphere microbe interactions. *Trends in Microbiology* 32 (11): 1072–1083.

Tavormina, P., De Coninck, B., Nikonorova, N. et al. (2015). The plant peptidome: an expanding repertoire of structural features and biological functions. *The Plant Cell* 27 (8): 2095–2118. https://doi.org/10.1105/tpc.15.00440.

Toruno, T.Y., Stergiopoulos, I., and Coaker, G. (2016). Plant-pathogen effectors: cellular probes interfering with plant defenses in spatial and temporal manners. *Annual Review of Phytopathology* 54 (1): 419–441.

Tost, A.S., Kristensen, A., Olsen, L.I. et al. (2021). The PSY peptide family—expression, modification and physiological implications. *Genes* 12 (2): 218. https://doi.org/10.3390/genes12020218.

Vanyushin, B.F., Ashapkin, V.V., and Aleksandrushkina, N.I. (2017). Regulatory peptides in plants. *Biochemistry (Moscow)* 82: 89–94.

Verma, V., Ravindran, P., and Kumar, P.P. (2016). Plant hormone-mediated regulation of stress responses. *BMC Plant Biology* 16: 86. https://doi.org/10.1186/s12870-016-0771-y.

Vie, A.K., Najafi, J., Liu, B. et al. (2015). The IDA/IDA-LIKE and PIP/PIP-LIKE gene families in Arabidopsis: phylogenetic relationship, expression patterns, and transcriptional effect of the PIPL3 peptide. *Journal of Experimental Botany* 66 (17): 5351–5365.

Villalobos-Lopez, M.A., Arroyo-Becerra, A., Quintero-Jimenez, A., and Iturriaga, G. (2022). Biotechnological advances to improve abiotic stress tolerance in crops. *International Journal of Molecular Sciences* 23 (19): 12053. https://doi.org/10.3390/ijms231912053.

Wulf, K., Sun, J., Wang, C. et al. (2024). The role of CLE peptides in the suppression of mycorrhizal colonization of tomato. *Plant and Cell Physiology* 65 (1): 107–119.

Xiao, F. and Zhou, H. (2023). Plant salt response: perception, signaling, and tolerance. *Frontiers in Plant Science* 13: 1053699.

Xie, H., Zhao, W., Li, W. et al. (2022). Small signaling peptides mediate plant adaptions to abiotic environmental stress. *Planta* 255 (4): 72.

Xu, T.T., Song, X.F., Ren, S.C., and Liu, C.M. (2013). The sequence flanking the N-terminus of the CLV3 peptide is critical for its cleavage and activity in stem cell regulation in Arabidopsis. *BMC Plant Biology* 13: 1–12.

Xu, C., Liberatore, K.L., MacAlister, C.A. et al. (2015). A cascade of arabinosyltransferases controls shoot meristem size in tomato. *Nature Genetics* 47 (7): 784–792.

Yamaguchi, Y.L., Ishida, T., and Sawa, S. (2016). CLE peptides and their signaling pathways in plant development. *Journal of Experimental Botany* 67 (16): 4813–4826.

Yan, Y. (2024). Deciphering the role of the malectin-like RLK and LRR-RLKs in the Liverwort *Marchantia polymorpha*. Doctoral dissertation. Universität zu Köln)

Yang, Z., Gao, C., Zhang, Y. et al. (2023). Recent progression and future perspectives in cotton genomic breeding. *Journal of Integrative Plant Biology* 65 (2): 548–569.

Yoshida, T., Fernie, A.R., Shinozaki, K., and Takahashi, F. (2021). Long-distance stress and developmental signals associated with abscisic acid signaling in environmental responses. *The Plant Journal* 105 (2): 477–488.

Yu, B., Chao, D.Y., and Zhao, Y. (2024a). How plants sense and respond to osmotic stress. *Journal of Integrative Plant Biology* 66 (3): 394–423.

Yu, X., Huang, Z., Cheng, Y. et al. (2024b). Comparative genomics screens identify a novel small secretory peptide, SlSolP12, which activates both local and systemic immune response in tomatoes and exhibits broad-spectrum activity. *Journal of Agricultural and Food Chemistry* 72 (33): 18507–18519.

Yuan, N., Furumizu, C., Zhang, B., and Sawa, S. (2021). Database mining of plant peptide homologues. *Plant Biotechnology* 38 (1): 137–143.

Zarsky, V., Nielsen, M.E., and Blatt, M.R. (2024). Formins, cell wall integrity, ROP guanine exchange factors, secretion regulators, and small secreted peptides in plant cell exocytosis and defence. *Journal of Experimental Botany* 75 (12): 3665–3667.

Zelman, A.K. and Berkowitz, G.A. (2023). Plant elicitor peptide (Pep) signaling and pathogen defense in tomato. *Plants* 12 (15): 2856.

Zhang, L., Shi, X., Zhang, Y. et al. (2019). CLE9 peptide-induced stomatal closure is mediated by abscisic acid, hydrogen peroxide, and nitric oxide in Arabidopsis thaliana. *Plant Cell & Environment* 42 (3): 1033–1044.

Zhang, S., Li, C., Si, J. et al. (2022a). Action mechanisms of effectors in plant-pathogen interaction. *International Journal of Molecular Sciences* 23 (12): 6758.

Zhang, Y., Tan, S., Gao, Y. et al. (2022b). CLE42 delays leaf senescence by antagonizing ethylene pathway in Arabidopsis. *New Phytologist* 235 (2): 550–562.

Zhang, Y.M., Ye, D.X., Liu, Y. et al. (2023). Peptides, new tools for plant protection in eco-agriculture. *Advanced Agrochem* 2 (1): 58–78.

Zhang, C., Xie, Y., He, P., and Shan, L. (2024a). Unlocking nature's defense: plant pattern recognition receptors as guardians against pathogenic threats. *Molecular Plant-Microbe Interactions* 37 (2): 73–83.

Zhang, P., Zhao, J., Zhang, W. et al. (2024b). Sulfated peptides: key players in plant development, growth, and stress responses. *Frontiers in Plant Science* 15: 1474111. https://doi.org/10.3389/fpls.2024.1474111.

Zhang, P., Gao, W., Guo, L. et al. (2024c). Functional characterization of plant peptide-containing sulfated tyrosine (PSY) family in wheat (*Triticum aestivum* L). *International Journal of Molecular Sciences* 25 (23): 12663.

Zhong, S. and Qu, L.J. (2019). Peptide/receptor-like kinase-mediated signaling involved in male-female interactions. *Current Opinion in Plant Biology* 51: 7–14.

Zhong, S., Li, L., Wang, Z. et al. (2022). RALF peptide signaling controls the polytube block in Arabidopsis. *Science* 375 (6578): 290–296.

Zhou, H., Xiao, F., Zheng, Y. et al. (2022). Pamp-induced secreted peptide 3 modulates salt tolerance through receptor-like kinase 7 in plants. *The Plant Cell* 34 (2): 927–944.

Zhou, Y., Zheng, J., Wu, H. et al. (2024). A novel toolbox to record CLE peptide signaling. *Frontiers in Plant Science* 15: 1468763.

2

Peptide Hormone Receptors and Cellular Signaling in Diverse Responses of Plants

Aryadeep Roychoudhury[1], Konkona Lahiri[2], and Heeya Gupta[2]

[1] *Discipline of Life Sciences, School of Sciences, Indira Gandhi National Open University, New Delhi, Delhi, India*
[2] *Department of Biotechnology, St. Xavier's College (Autonomous), Kolkata, West Bengal, India*

2.1 Introduction

Over the years, plants have generated complex intercellular communication systems through immense evolutionary mechanisms. Many studies have identified several molecules, known as phytohormones, which play vital roles in the development of these processes. The discovery of mediators of several cellular pathways and receptors of said phytohormones has led to the detection of numerous peptide-based signaling molecules. Recent progress in this field suggests the involvement and contribution of these peptides in the various aspects of plant life cycle, including developmental and environmental responses, thus emphasizing similarity with the functions of canonical phytohormones (Motomitsu et al. 2015).

The first peptide discovered was systemin in tomato, and since then, a broad spectrum of such hormones and their receptors have been identified, providing evidence that their presence and function are critical to the maintenance of the plant system. Peptide hormones bind to their respective cell-surface receptors in order to trigger a cascade that involves the activity of adaptor proteins and cell surface enzymes, eventually leading to a series of proliferative and metabolic signals. Within the endosomal system, receptor functionality is controlled by decreasing vacuolar pH, augmenting ligand proteolysis, and stimulating receptor kinase dephosphorylation (Posner and Laporte 2010; Ghosh and Roychoudhury 2024; Zhang et al. 2025).

There are three types of plant peptide signaling: cysteine-rich peptides (CRPs), nonsecreted peptides, and posttranslationally modified small peptides (Hirakawa and Sawa 2019). The abundance, variety, and significance of peptides as regulators of cellular growth, differentiation, defense, and development of plants were undervalued for a long time. Peptides arose as a primary class of signals for long-range signaling and communication between cells over diminutive distances. There has been appreciable advancement in research regarding the mechanisms of perception of peptides and signal transduction

Peptide Hormones in Plant Development and Stress Responses, First Edition. Edited by Vijay Pratap Singh and Samiksha Singh.

despite the lack of knowledge about their biogenesis. Numerous posttranslational modifications (PTMs) are necessary for the maturation of peptides from inactive pre-pro-peptides, including proline hydroxylation, proteolytic processing, tyrosine sulfation, and hydroxyproline glycosylation. These PTMs may or may not be necessary for bioactivity; however, these might inactivate the peptide, its stability, or targeting, or modify its signaling specificity or its increasing binding inclination toward the receptor (Stührwohldt and Schaller 2019).

The establishment of a multitude of peptide hormones and their respective receptors has led to the formation of the Plant Peptidome—a novel and unifying system of classification of peptides to exhibit the broad diversity in the synthesis of peptides and constant expansion of this system (Tavormina et al. 2015).

This chapter delves deeper into the intricacies of the signaling pathways activated by these peptides interacting with their receptors and how these signaling pathways regulate all processes in a plant system.

2.2 Plant Peptide Hormones, Identification and Their Categorization

2.2.1 Identification of Peptide Hormones in Plants

Small, secreted polypeptides with up to 250 amino acids, collectively known as plant peptide hormones, play a variety of roles in intercellular signaling inside plants. Genomic studies have identified thousands of small secreted peptide (SSP)–encoding genes in species like *Arabidopsis*, maize, wheat, and tomato. Traditional identification methods for identifying SSPs like chemical separation and bioassays faced challenges due to their low abundance. Modern high-throughput omics (transcriptomics, proteomics, peptide omics), mass spectrometry, and machine learning have improved SSP discovery. Still, numerous unidentified SSPs, potentially functioning as peptide hormones, likely exist (Zhang et al. 2025).

The first bioactive peptide to be isolated from tomato plants in the early 1990s was systemin. This 18-amino-acid systemin peptide is proteolytically released from a 200-amino-acid precursor called prosystemin (Stührwohldt and Schaller 2019). This encourages systemic defense mechanisms against local injuries caused by insect herbivory and hence becomes a pivotal part in the wound signaling pathway in tomato and other closely related species in the Solanaceae subtribe of the nightshade family.

Numerous plant species have more than 30 families of putative peptide hormones, including rapid alkalinization factors (RALFs), phytosulfokines (PSKs), epidermal patterning factors (EPFs), C-terminally encoded peptides (CEPs), and Clavata3 (CLV3)/embryo surrounding region-related (CLE) peptides. In addition to the usual peptide hormones with an N-terminal secretion peptide signal for the secretory pathway, there are other peptides known as nonsecretory peptides that do not have an N-terminal signal peptide like systemins (SYS), *Zea mays* immune signaling peptide1 (ZIP1), plant elicitor peptides (PEPs), and grim reaper (GRI) peptide families. Both these peptides have been identified to play crucial roles in development and stress responses in plants (Matsubayashi and Sakagami 2006; Roychoudhury et al. 2024; Zhang et al. 2025).

2.3 Classification of Peptide Hormones

As classified earlier, plant peptide hormones are organized into two different types: secretory and nonsecretory peptides on the basis of secretions and localizations in plant cells (Zhang et al. 2025). In plants, secretory peptide hormones are initially synthesized as pre-proproteins within the endoplasmic reticulum (ER). These precursor proteins are then directed through the Golgi apparatus and hence ultimately transported either to the cell surface or released into the extracellular space (Corpo et al. 2024). They go through particular posttranslational processing and chemical changes during this trafficking phase. Their precursor proteins typically contain an N-terminal signal peptide, which makes it easier for them to move along the secretory pathway via exocytosis (Tavormina et al. 2015).

Further classifications suggest that secretory peptide hormones have various other subtypes on the basis of characteristics such as PTMs and the presence of multiple cysteine residues (Zhang et al. 2025). These peptides often undergo several PTMs, such as proteolytic cleavage, proline hydroxylation, tyrosine sulfation, and glycosylation of hydroxyproline residues, making them active. Peptides originate from inactive precursor proteins, known as pre-pro-peptides, typically ranging between 70 and 120 amino acids in length. These PTMs can either enhance or reduce the bioactivity of peptides, altering their signaling specificity or influencing their stability and localization, or affecting their ability to bind to specific receptors (Stührwohldt and Schaller 2019). Modifications such as hydroxylation and glycosylation of proline residues are crucial for the proper function of CLE peptides, especially within their conserved CLE domain, as these modifications influence their stability and interaction with certain receptors (Matsubayashi 2011).

Few prominent examples of PTMs, like PSKs, a group of sulfated pentapeptides, are known to promote cell division and differentiation (He et al. 2024). The sulfation of the tyrosine residue, in the core YIYTQ sequence of PSK, important for both biological activity and receptor-binding capacity, is another well-known example. Hydroxyproline-rich systemin (HypSys) peptides, well known by hydroxylated prolines, are well associated in long-distance signaling related to wound and pathogen responses (Narváez-Vásquez et al. 2007). Additional peptide hormones, including CEPs, inflorescence deficient in abscission (IDA), IDA-like proteins (IDLs), and plant peptide containing sulfated tyrosine 1 (PSY1), also depend on specific PTMs for their activation and function (Olsson et al. 2019).

Alternatively, CRP hormones are characterized by their ability to create disulfide bonds through interactions among sulfur atoms in residues of cysteine. Such peptides usually range between 2 and 16 cysteine residues, and this displays considerable diversity in length of the sequence and amino acid composition (Tavormina et al. 2015; Zhang et al. 2025). Several members of this group—such as plant defensins (PDFs), defensin-like peptides (DEFLs), RALFs, nonspecific lipid transfer proteins (nsLTPs), and knottins—are known to contribute greatly to plant immune defense mechanisms (Tang et al. 2002). Additionally, certain CRPs, including RALFs, S-locus cysteine-rich protein/S-locus protein11 (SCR/SP11), embryo SAC4 (ES4), embryo surrounding factor1 (ESF1), and LUREs, play major roles in various aspects of plant development and reproductive processes (Okuda et al. 2009; Takeuchi 2021).

The other pathway, which is of the nonsecretory peptide hormones, however, does not follow the usual secretory pathway. They are frequently found in the mitochondria, chloroplasts, nucleus, or cytoplasm. While some function within the cell in which they are produced, others are secreted by unusual pathways and translocate between cells. Proline, glycine, tyrosine, or lysine are frequently found in these peptides. Examples include SYS (18 amino acids), PEPs (23–36 amino acids), and the lysine-rich GRI (11 amino acids), all linked to plant defense (Matsubayashi 2011; Tavormina et al. 2015). Figure 2.1 represents the plant peptide hormone families according to structural and biosynthetic characteristics (Table 2.1).

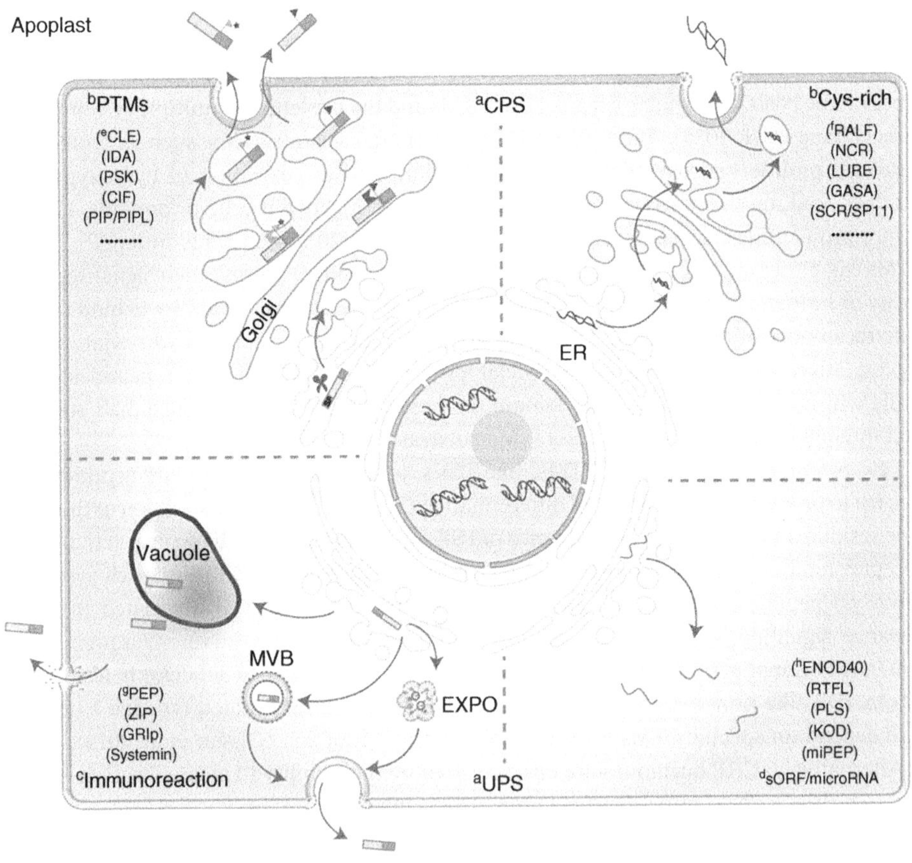

Figure 2.1 Classification of plant peptide hormone families according to structural and biosynthetic characteristics. Peptides are categorized into four groups on the basis of their structure, action, and mode of secretion. Cysteine-containing peptides that are posttranslationally modified are secreted through the conventional protein secretion (CPS) pathway, which includes the endoplasmic reticulum, Golgi body, and the apoplast. Comparatively, peptides from immunoreactive precursors (e.g. systemin, PEP), small open reading frames (sORFs), or microRNAs use the unconventional protein secretion (UPS) route, involving vacuoles, exocyst-positive organelles (EXPOs), and multivesicular bodies (MVBs). *Source:* The image has been adopted from Zhang et al. (2025) / Springer Nature / CC BY 4.0 under Creative Commons Attribution International License (http://creativecommons.org/publicdomain/zero/1.0) which permits use, sharing, adaptation, distribution, and reproduction in any medium or format, as long as you give appropriate credit to the original author(s).

Table 2.1 Classification of some secretory and nonsecretory peptides.

Types	Groups	Families	Descriptions	Biological functions
Secretory polypeptide	Pro-hydroxylation/ glycosylation	CLE	Clavata/embryo surrounding region	• *CLE* genes secrete peptides that act like intercellular signaling molecules in the extracellular space by attaching to proteins that resemble receptors on the cell surface (Betsuyaku et al. 2011) • Many developmental and physiological processes are controlled by CLE peptides, and a variety of land plants have conserved CLE signaling pathways (Betsuyaku et al. 2011)
		IDA/IDL	Inflorescence deficient in abscission	• It is a well-known peptide in *Arabidopsis thaliana* that controls cell separation. Mutants lacking IDA fail to shed floral organs, while its overexpression causes early or abnormal abscission. • Five related genes, *IDA-like 1* to *5* (*IDL1–5*), play similar roles in processes like stomatal formation, vascular development, lateral root emergence, root cap shedding, and seed release (Vie et al. 2015)
		CEP	C-terminally encoded peptide	• CEPs function by interacting with their receptors (CEPRs) and regulate nitrogen demand signaling, promote nodulation, and influence the development of lateral roots (Taleski et al. 2018)
		PIP/PIPL	PAMP-induced secreted peptide	• PAMP-induced plant immunity (PIP1) • Modifications in salt tolerance (PIP3) (Zhang et al. 2025)
		HYPSYS	Hydroxyproline-rich systemin	• Monitoring of defensive reactions (Pearce 2011)
	Tyr sulfation	GLV/RGF/ CLEL	Golven/root growth factor/CLE-Like	• This peptide family in *Arabidopsis thaliana* is encoded by 11 genes. • Members like RGF1 (GLV11), RGF2 (GLV5), and RGF3 (GLV7) play key roles in maintaining the root meristem (Fernandez et al. 2013)
		PSK	Phytosulfokine	• Phytosulfokine (PSK), a disulfated pentapeptide, controls plant growth and development and increases resistance to biotic and abiotic stress (Shen et al. 2023)

(*Continued*)

Table 2.1 (Continued)

Types	Groups	Families	Descriptions	Biological functions
		PSY	Plant peptide containing sulfated tyrosine	• They have been shown to promote cell elongation in both roots and hypocotyls. Additionally, they are involved in modulating plant immune responses (Tost et al. 2021)
		CIF	Casparian strip integrity factor	• CIFs are small peptides modified by tyrosine sulfation that are essential for the development and control of the Casparian strip. • Additionally, they have a significant contribution in how plants react to stress (Fang et al. 2025)
	Cys-rich	PDF	Plant defensin	• Plant defensins are a well-known group of positively charged peptides widely found in plants. PSY1 is linked to promoting cell growth and division, along with enhancement of plant response to stress (de Oliveira Carvalho and Gomes 2009)
		nsLTP	Nonspecific lipid transfer protein	• Higher plants have been shown to have small, basic proteins which play important roles in cellular functions, including membrane stabilization, cell wall structure, and signaling pathways (Liu et al. 2015)
		NCR	Nodule-specific cysteine-rich peptide	• NCR peptides, such as MtNCR247 and MtNCR035, promote endoreduplication, rhizobial differentiation, and show antimicrobial activity. • In *Medicago*, around 600 *NCR* genes help convert rhizobia into enlarged, polyploid, noncultivable forms, resembling antimicrobial peptides of innate immunity (Guefrachi et al. 2014)
		GASA	Gibberellic acid stimulated in Arabidopsis	• GASA proteins are commonly found across various plant species and are mainly involved in hormone signaling and responding to environmental stresses (Bouteraa et al. 2023)
		KLP	Knottin-like peptide	• KLP exhibits antifungal properties (MjAMP1/2) and is also known for its insecticidal effects (PA1a/b) (Zhang et al. 2025)
		THL	Thionin-like	• THL exhibits antifungal and antibacterial properties, as seen in CaThi (Zhang et al. 2025)

HLP	Hevein-like peptide	• Unknown (Finkina et al. 2019)
Cyclotide	Small cyclic peptide	• Cyclotides are plant-based cyclic peptides. They are often produced in high quantities (up to 1 g per kg of fresh plant tissue) and primarily serve to defend plants against pests and pathogens (Weidmann and Craik 2016)
2SA	2S Albumin	• 2S albumin mRNA is present in large amounts during the middle stage of seed development. SiAMP2 contributes to plant defense against pathogens (Chen et al. 2004)
RALF	Rapid alkalinization factor	• They are minute, cysteine-rich peptides associated with processes like the integrity of the pollen tube (AtRALF4/9), formation of lateral root (AtRALF34), and stress responses (AtRALF22/23) (Zhang et al. 2025) • Analysis of 795 RALFs among 51 plant species showed that they have evolved into four main clades (Campbell and Turner 2017)
HAIRPININS	alfa-HAIRPININ (HAIRPININ)	• SmAMP-X shows wide-range antimicrobial actions. α-hairpinins have diverse roles such as antifungal, antibacterial, trypsin inhibition, and ribosome inactivation, though their exact mechanisms are not fully clear. SmAMP-X and EcAMP1 inhibit fungal hyphal growth without damaging membranes (Slezina and Odintsova 2023)
EPF/EPFL	Epidermal patterning factor/EPF-like	• EPF2 prevents cells from becoming MMCs in a non-cell-autonomous way, helping control epidermal cell density. • EPF1, which shares a similar sequence, acts later in development to inhibit stomatal formation (Hara et al. 2009)
PCP	Pollen coat protein	• Pollen coat proteins (PCPs), essential for species-specific interactions, play a significant function in the adhesion of pollen to stigma, its recognition, hydration, and subsequent germination (Gong et al. 2015)
LURE	LURE-type cysteine-rich peptide	• LURE-type CRPs, secreted by synergid cells, guide pollen tubes in the final stage with species-specific attraction. Despite conserved cysteine patterns, they evolve rapidly, aiding species recognition during fertilization (Takeuchi 2021)

(Continued)

Table 2.1 (Continued)

Types	Groups	Families	Descriptions	Biological functions
		SCR/SP11	S-locus cysteine-rich protein/S-locus protein11	• Exhibits self-incompatibility (SI), a genetic system that resists self-pollination (Naithani et al. 2006)
		ARACIN	Aracin	• Unknown function
		ECL	Egg cell 1-like	• EC1 peptides help deliver a fusogen to sperm cells, while a link between the two sperm may prevent their unintended fusion during gamete interaction (Sprunck et al. 2012)
		ES	Embryo sac	• During megagametogenesis, the egg cell forms; sperm cells come in contact with the egg and central cell, forming a zygote in the embryo sac (Webb and Gunning 1994)
	Non-cys, non-PTM	GRP	Glycine-rich protein	• The superfamily GRPs are implicated in cellular stress responses (Czolpinska and Rurek 2018)
		PNP	Plant natriuretic peptide	• Seems to counteract the plant stress responses (Lee et al. 2020)
		SCOOP	Serine-rich endogenous peptide	• In *Arabidopsis*, P(PIN) regulated polar auxin transport to alter root growth and development (Wang et al. 2024)
	Functional precursor	CAPE	CAP-derived peptide	• Exerts both defensive and detrimental functions (Han et al. 2023)
		SUBPEP	Subtilisin-embedded plant elicitor peptide	• Protein turnover and plant development (Figueiredo et al. 2018)
		INCEPTINS	Inceptins	• Various defense mechanisms trigger in plants, such as generating reactive oxygen species (ROS), activating defense genes, emitting volatile compounds, and accumulating phytoalexins and other antimicrobial substances (Sharma and Singh 2024)

Nonsecretory polypeptide	Immunoreaction	PEP	Plant elicitor peptide	• Stimulate defense responses like hormone synthesis, upregulation of defense genes, activation of phosphorelay pathways, and production of secondary messengers in cells. (Zelman and Berkowitz 2023)
		ZIP	*Zea mays* immune signaling peptide	• Stimulate defense responses like hormone synthesis, upregulation of defense genes, activation of phosphorelay pathways, and production of secondary messengers in cells (Ajeesh Krishna et al. 2020)
		Sys	Systemin	• A wound-triggered systemic signaling pathway that activates the transcription of defense-related genes (Schaller and Ryan 1996)
		GRI	Grim reaper peptide	• Responses to oxidative stress (Zhang et al. 2025)
	sORF	ENOD40	Early nodulin 40	• Exhibits interaction between legume plants with symbiotic *Rhizobium* bacteria, supporting nodule formation (Guarnerio 2010)
		RTFL/DVL	Rotundifolia/devil	• Negatively regulates shade avoidance syndrome (SAS) (Guo et al. 2015)
		PLS	POLARIS	• Exhibits proper growth of roots along with vascular development (Chilley et al. 2006)
		KOD	KISS OF DEATH	• Activates programmed cell death pathway in *Arabidopsis thaliana* (Blanvillain et al. 2011)
		OSIP	Oxidative stress-induced peptide	• Tolerance to stress caused by metals (OSIP108) (Zhang et al. 2025)
	microRNA	miPEP	miRNA-encoded peptide	• Checks functionality of every member of miRNA families in plants (Couzigou et al. 2015)

2.4 Receptor-Mediated Perception of Peptide Hormones

2.4.1 Types of Receptors

The majority of receptors so far been discovered are receptor kinases (RKs), making up the biggest family of plant receptor proteins with over 600 encoded in the *Arabidopsis* genome. RKs exhibit a tripartite structural architecture, including a conserved intracellular kinase domain, a single-pass transmembrane region, and a diverse extracellular domain (ECD) (Shiu and Bleecker 2001). RKs are classified into 10 distinct families, with the leucine-rich repeat receptor kinases (LRR-RKs) being the largest. These and receptor-like proteins (RLPs) in plants form a major group of membrane-bound receptors that are essential for detecting external signals and initiating intracellular signal transduction pathways (Wei et al. 2022). LRR-RKs are subdivided into 13 subfamilies, designated as LRRI to LRRXIII (Zhang et al. 2016). Numerous receptor kinases of the LRRXI subfamily have been identified as key receptors for small signaling peptides, which play crucial roles in regulating plant growth and developmental processes (Matsubayashi 2014).

Signaling among peptide-RK pairings is similar. It begins with the extracellular regions of these RKs which are sufficient to recognize small signaling peptides. Secondly, this peptide bond typically triggers activation of the RK through either homodimerization or heterodimerization with another RK or, potentially, a co-receptor like RLPs (Han et al. 2014). LRR-RLPs are similar to LRR-RKs in structure but lack the cytoplasmic kinase domain, and both of them have been found to play significant roles in plant immunity and plant development (Chakraborty et al. 2019). Lastly, recognition of signaling peptides leads to the activation of specific RKs, which subsequently alter the expression profiles of the other transcription factors associated with it (Endo et al. 2014). To date, LRR-RKs such as somatic embryogenesis receptor kinases (SERKs) (Han et al. 2014) and possibly certain RLPs have been identified as coreceptors functioning alongside RKs (Zhang et al. 2016).

SERKs function as coreceptors by forming direct complexes with leucine-rich repeat receptor-like kinases (LRR-RLKs) receptors upon ligand recognition. They interact with a diverse range of primary receptors or in other words LRR-RLKs-for example, brassinosteroid insensitive1 (BRI1) in brassinosteroid (BR) signaling, peptide phytosulfokine receptor (PSKR) in PSK-induced cell expansion, root meristem growth factor receptors (RGFRs) in root meristem development, excess microsporocytes1 (EMS1) in tapetum formation, ERECTA proteins in stomatal patterning, Haesa/Haesa-like2, (HAE/HSLs) in floral organ abscission, and immune receptors such as elongation factor-Tu receptor (EFR), flagellin-sensitive2 (FLS2), and plant elicitor peptide receptors (PEPRs). Some RLPs are also known to associate constitutively with another LRR-RLK, suppressor of BAK1-interacting receptor-like kinase 1 (BIR1-1) (SOBIR1), and upon ligand detection, they recruit SERK3, resulting in the assembly of a tripartite complex consisting of SERK3, SOBIR1, and the RLP (Liu et al. 2020). A number of X-ray crystallography studies have resolved the structures of numerous plant LRR-RK receptors bound to their respective ligands, uncovering novel insights into how ligand binding triggers receptor kinase activation (Chakraborty et al. 2019). In addition to LRR-RLKs and LRR-RLPs, plant cells contain another group of leucine-rich repeat-containing proteins known as LRR-extensins (LRXs), which are localized in the cell wall (Baumberger et al. 2003). Structurally, LRXs are characterized by a short

N-terminal region with around 10 LRRs, followed by a cysteine-rich domain, and a C-terminal extensin-like domain made up of repeated pentapeptides rich in serine and hydroxyproline residues (Baumberger et al. 2001).

2.5 Activation Mechanism

Plant peptide hormones serve as key regulators in diverse processes (both physiological and developmental), by interacting with specific cell surface receptors. These hormones are generally perceived by LRR-RKs or LRR-RLPs, which constitute major classes of membrane-bound receptors. The ECDs of these receptors are specialized for ligand binding, and their structural features enable selective and high-affinity recognition of small peptide ligands (Chakraborty et al. 2019). Upon ligand binding, activation of the receptor is commonly achieved through dimerization, either via homodimerization of the receptor or heterodimerization with coreceptors. A well-characterized example is the interaction between the brassinosteroid receptor BRI1 and the coreceptor BAK1 (a member of the SERK family), where ligand binding stabilizes the receptor–coreceptor complex, triggering intracellular signaling (Chakraborty et al. 2019).

As described earlier, SERKs have emerged as versatile coreceptors, associating with numerous LRR-RKs involved in processes such as growth (e.g. BRI1 in BR signaling), development (e.g. EMS1 in tapetum formation, RGFRs in root meristem maintenance), and immunity (e.g. FLS2, EFR, and PEPRs in pathogen recognition) (Chakraborty et al. 2019; Matsubayashi and Sakagami 2006). In some cases, receptors which lack intracellular kinase domains, such as LRR-RLPs, form complexes with kinase-competent proteins like SOBIR1. Upon perception of peptide ligands, these preassembled complexes often recruit additional coreceptors like SERK3 to form a tripartite signaling unit; for instance, RLP23 constitutively interacts with SOBIR1 and recruits SERK3 upon ligand binding, enabling downstream signal transduction (Chakraborty et al. 2019). Following receptor activation, intracellular kinase domains initiate phosphorylation cascades, which propagate the signal through mitogen-activated protein kinase (MAPK) modules, ultimately modulating transcription factor activity. This leads to specific changes in gene expression tailored to the type of peptide signal perceived. An interesting case is the Clavata signaling pathway, where the peptide ligand CLV3 binds to the CLV1 receptor, modulating the expression of Wuschel and maintaining stem cell homeostasis in the shoot apical meristem (Matsubayashi and Sakagami 2006; Shiu and Bleecker 2001).

Numerous X-ray crystallography studies have revealed the resolution of various LRR-RK–peptide complexes, showing that ligand binding induces conformational changes necessary for receptor dimerization and activation, offering mechanistic insights into how these signaling pathways are initiated at the cell surface (Chakraborty et al. 2019). Altogether, the diverse combinations of receptors, coreceptors, and downstream components allow peptide hormone signaling networks in plants to be highly modular and context-specific, enabling precise control over growth, development, and environmental responses.

2.6 Signal Transduction Pathways

2.6.1 Downstream Components

Plant receptor kinases (RKs) closely resemble animal receptor tyrosine kinases (RTKs) in terms of structural layout and activation mechanism. Both possess diverse ECDs for recognition of ligand, a single transmembrane segment, and intracellular kinase domains. Activation in both cases is triggered by ligand-induced oligomerization (Han et al. 2014). They also share downstream signaling elements such as the activation of MAPK cascades and NADPH oxidase-mediated production of ROS (Liang and Zhou 2018).

Recent research studies also reveal the pivotal participation of RLCKs in signaling pathways mediated by RKs. Supporting this view, many RLCKs are found at the plasma membrane, a localization often achieved through N-myristoylation or palmitoylation (Lin et al. 2013). RLCKs work alongside RKs and RLPs to regulate various physiological and developmental processes in plants, including innate immune responses, adaptation to environmental stresses, hormone signaling, reproductive processes, stomatal development, maintenance of shoot and root meristems, vascular differentiation, petal shedding, and more (Lin et al. 2013; Liang and Zhou 2018). For example, *Botrytis*-induced kinase1 (BIK1) and its homologs directly associate with immune-related RKs to modulate defense signaling (Kong et al. 2016; Liu et al. 2013; Lu et al. 2010). Members of the RLCK-XII group, known as BR signaling kinases (BSKs), physically interact with the leucine-rich repeat receptor kinase BRI1—the key receptor for brassinosteroids—to regulate BR-related signaling in *Arabidopsis* (Liang and Zhou 2018). The RLCK-VIII protein MARIS (MRI) in *Arabidopsis* acts downstream of the peptide hormone receptors ANXUR1 (ANX1), ANX2, and Buddha's Paper Seal1/2 (BUPS1/2) to maintain pollen tube integrity (Boisson-Dernier et al. 2015). In *Brassica*, the RLCK known as M-locus protein kinase (MLPK) partners with the female-specific receptor S-locus B-lectin receptor kinase (SRK) to mediate self-incompatibility responses (Kakita et al. 2008). RLCKs also serve as crucial connectors between recognizing the pathogen and activating the plant defense mechanisms. RLK–RLCK signaling modules are involved in regulating various immune-related processes, including hormone biosynthesis and signaling, ROS generation, calcium (Ca^{2+}) signaling, MAPK cascades, and the expression of defense-related genes (Hailemariam et al. 2024). Moreover, interactions between RLKs/RLCKs and intracellular immune receptors like nucleotide-binding domain leucine-rich repeat proteins (NLRs) reveal a highly interconnected genetic framework underlying plant immunity (Liang and Zhou 2018).

Plant innate immunity is also maintained by pattern recognition receptors (PRRs) responsible for identifying pathogen-associated molecular patterns (PAMPs), initiating defense responses. RLCKs act downstream of PRRs, linking extracellular pathogen detection to intracellular signaling pathways, notably the MAPK cascades (Bi et al. 2018). Upon PAMP recognition, PRRs such as FLS2 and CERK1 (chitin elicitor receptor kinase1) associate with coreceptors like BAK1, leading to the activation of RLCKs, including BIK1 and members of the RLCK VII-4 subgroup (e.g. PBS1-LIKE 1 (PBL19), PBL20). These RLCKs phosphorylate MAPK kinase kinases (MAPKKKs), such as Mitogen-activated protein kinase kinase kinases (MAPKKK5) and MEKK1, initiating MAPK cascades that culminate in the activation of MAPKs like MPK3, MPK4, and MPK6 (Sun and Zhang 2022; Ma et al. 2022).

Activated MAPKs phosphorylate various substrates, leading to:

- **Transcriptional reprogramming**: Induction of genes linked to defense, including those that encode proteins implicated in diseases and enzymes associated with the synthesis of secondary metabolites.
- **Hormonal signaling modulation**: Regulation of hormones that coordinate defense responses, such as ethylene, salicylic acid, and jasmonic acid.
- **Reactive oxygen species (ROS) production**: Stimulation of ROS generation, contributing to pathogen restriction and signaling.
- **Calcium signaling**: Alteration of cytosolic calcium levels, further propagating defense signals.
- **Stomatal closure**: Prevention of pathogen entry by closing stomata.

These processes collectively enhance the plant's ability to resist pathogen invasion (Bi et al. 2018). RLCKs also interact with nucleotide-binding leucine-rich repeat receptors (NLRs), which detect specific pathogen effectors; for instance, in tomato, the RLCK SlMai1 interacts with MAPKKKα, facilitating MAPK activation during NLR-mediated effector-triggered immunity (ETI) (Ma et al. 2022). RLCKs serve as pivotal connectors between PRRs and MAPK cascades, orchestrating a multifaceted defense response that includes gene activation, hormonal modulation, and physiological changes to thwart pathogen attack.

2.7 Integration with Other Signaling Pathways

The integration of hormonal signaling pathways—specifically auxin and cytokinin—with ROS and calcium (Ca^{2+}) signaling orchestrates plant responses to environmental stimuli. These interconnected networks enable plants to adapt to abiotic stresses through coordinated growth, development, and defense mechanisms (Ravi et al. 2023). Auxin and cytokinin are the central regulators of plant development, often exhibiting antagonistic interactions that influence organogenesis and stress responses (Kazan 2015). The former promotes root development, while the latter favors shoot formation. Their interplay is crucial for maintaining meristem activity and overall plant architecture. Under stress conditions, the balance between auxin and cytokinin signaling is modulated to optimize growth and survival (Mohanta et al. 2018).

ROS, including hydrogen peroxide (H_2O_2), function as signaling molecules that mediate plant responses to various stresses. Upon exposure to abiotic stressors, ROS levels increase, triggering defense mechanisms, and modulating hormonal pathways (Tognetti et al. 2017). The interaction between ROS and hormonal signaling is evident in the modulation of auxin transport and signaling. Increased ROS levels can affect auxin distribution in plant tissues by influencing auxin transporter expression and location. Under stressful circumstances, adaptive growth responses depend on this modulation (García et al. 2016). Calcium ions serve as ubiquitous second messengers in plant signal transduction. Abiotic stresses often lead to transient increases in cytosolic Ca^{2+} concentrations, which are decoded by various calcium-binding proteins to elicit specific physiological responses (Li et al. 2022a,b,c). The cross talk between calcium signaling and hormonal pathways is integral to stress adaptation. For example, calcium-dependent protein kinases (CDPKs) can phosphorylate the key components of hormonal signaling cascades, thereby modulating their activity. Additionally, calcium signaling can influence the production and scavenging of ROS, further intertwining these signaling networks (García et al. 2016). The convergence of auxin, cytokinin, ROS, and calcium signaling pathways forms a complex network which enables plants to sense and respond to environmental challenges effectively. This integration ensures a coordinated regulation of growth, development, and defense mechanisms, facilitating plant resilience under stress conditions (Ravi et al. 2023). These interconnected signaling pathways provide us with insights into plant adaptive strategies and offer potential avenues for enhancing stress tolerance through targeted manipulation of key signaling components.

2.8 Some Impacts of Peptide Hormone-Receptor Interactions in Plants

2.8.1 CLE and Their Receptors

The functional relevance of peptide hormones and their respective receptors is presently being studied extensively, and recent *in silico* protein screening developments coupled with a broad spectrum of practical experiments have been fundamental in this case.

The roots of a plant are vital for providing anchorage to the soil and for the transport of minerals, nutrients, and water from the soil to the upward regions. The root apical meristem (RAM) is in charge of the production of the root tissues. The phenomenon of root development in the majority of plants spans three regions—the meristematic zone responsible for active cell proliferation, the elongation zone responsible for elongation following disruption in proliferation and the differentiation zone responsible for the ultimate fates of the cells in question following terminal differentiation (Cajero Sánchez et al. 2018). This particular temporal and spatial interaction between the maintenance of stem cells, their proliferation, and finally differentiation is controlled by the activities of small peptides such as CLE peptides and ROS in addition to phytohormones as well (Hsiao and Yamada 2020).

According to present data, there are 32 *CLE* genes coding for about 27 different CLE precursor proteins in the *Arabidopsis* genome. All of these have a conserved 14-amino-acid-long CLE motif (Yamaguchi et al., 2016). The majority of signaling pathways in plants take place through LRR-RLKs, including the ones generated by CLEs. These peptides have been established to control the development of RAM negatively, often leading to termination of the meristem in experiments. Various assays carried out with respect to root growth categorized the CLEs into two broad types: the A-type and the B-type. Of these, the A-type CLEs are more involved in the control and regulation of meristem, whereas the B-type CLEs are not involved in the region of the root (Whitford et al. 2008). A *cle40* mutant was found to form additional layers of columella stem cells and this particular phenotype could be changed by treatment with CLE40. Similar to this, *Arabidopsis crinkly4 (acr4)* and *clv1* receptor kinase mutants also have multiple columella stem cell layers. Ectopic treatment with CLE40 upregulates the transcription of *ACR4* but not *CLV1* (Stahl et al. 2013). ACR4 is critical for lateral root patterning and pericycle cell division (De Smet et al. 2008). Extensive studies involving CLE40, ACR4, and Wuschel-related homeobox protein 5 (WOX5) found that these three work in a negative feedback loop to control the fate of stem cells. This is necessary to form the final architecture of the root, which in turn influences the functioning of the apical meristems and the lateral roots. Elevation and reduction of CLE40 level have profound effects on the stem cells—increasing CLE40 level further increases the differentiation of stem cells and changes the domain of expression of WOX5, whereas reduction of the same promotes proliferation of the stem cells and disrupts differentiation. The targeted alteration in expression of WOX5 occurs due to CLE40 interacting with ACR4 kinase (Fedoreyeva 2023). The above-mentioned receptors form either heterodimers or homodimers to activate the signaling pathways whenever necessary. AtCLE25 maintains a signaling pathway to systematically modulate the physiological adaptations of the shoot when there is alteration in root water status. The transcription of AtCLE25 takes place when there is a decrease in water potential. This peptide travels to the leaves via the vasculature and makes a complex with barely any meristem (BAM) receptors. This launches a signaling pathway and increases the levels of enzyme nine cis epoxy-carotenoid dioxygenase3 (NCED3) which aids in ABA biosynthesis. The stomata close as a result of ABA accumulation and minimize the loss of water through transpiration (Takahashi et al. 2018). The levels of AtCLE19 change as a result of drought stress, and although its corresponding receptor is not known, the fact that its signaling involves mitogen-activated protein kinase (MAP3/MAP6), open stomata 1 (OST1), slow anion channel-associated1 (SLAC1), and ROS has been established (Zhang et al. 2019).

AtCLE9 and *AtCLE45* in *Arabidopsis* respond to temperature stress by their change in expression levels. Pollen tube growth at 30 °C is prolonged by *AtCLE45*. Evidence suggests that pollen tubes that have disrupted *sterility regulating kinase* (*SKM1* and *SKM2*) genes are resistant to *AtCLE45* treatment, establishing that pollen SKM1 can interact with *AtCLE45* and the AtCLE45-SKM pathway is indispensable for seed production under conditions of high temperature (Endo et al. 2013). CLE signaling is prevalent in light-dependent carbohydrate metabolism as well. When sucrose is added to the root vasculature under dark conditions, previously upregulated AtCLE3 and AtCLE2 levels are seen to be suppressed and *cle2 cle3* double mutants have lower levels of glucose and sucrose in them as compared to wild type variants. Thus, AtCLE3 and AtCLE2 are positively regulated in the context of carbon accumulation in roots and their transport from root to shoot integrates plant growth with their expression (Ma et al. 2020; Okamoto et al. 2022).

It is known that leaf senescence is associated with ROS. The signaling that commences due to AtCLE14 causes a reduction in the level of ROS by activating genes of the NAC family transcription factor JUNGBRUNNEN1 (JUB1), which are primarily ROS scavenging and this leads to a phenomenon known as the "brake signal" during leaf senescence (Zhang et al. 2022b). CLE peptides are actively involved in the autoregulation and nodulation of leguminous plants, a function that they perform because they are also sensitive to the nitrogen status of such plants. Plant-pathogen interaction experiments have also reported the presence of CLE peptides. In one such research conducted by Hanemian et al. (2016), the dependency of these peptides on a miRNA pathway was realized when *clv1* mutants exhibited resistance to pathogenesis by *Ralstonia solanacearum* and this was in association with the accumulation of Nuclear Factor Y subunit alpha (NF-YA) and microRNA mi169 transcription factors.

2.9 Tyrosine Sulfated Peptides and Their Receptors

Associated with the CLE family in maintenance of the root meristem is a family of peptides called the Root Meristem Growth Factors (RGFs) (Matsuzaki et al. 2010). The C-terminus of these peptides has a CLE-Like (CLEL) motif, and hence are also known as CLEL or Golven (GLV) (Meng et al. 2012; Whitford et al. 2012). More than half of the known peptides of the RGF family in *Arabidopsis* are expressed specifically in the innermost layer of centrally located columella cells of the root apex and diffuse via the apoplast into the meristematic system (Fernandez et al. 2013). Precursor peptides, consisting of about 100 amino acids, produce RGFs by proteolytic processing after posttranslational sulfation. Desulfation enzymes have not been found, which suggests that these modifications are stable and irreversible. RGF1 peptide retrieval was the result of an extensive search for peptides that repaired the defects of the root meristem, and in which the tyrosylprotein sulfotransferase 1 (*tpst-1*) mutant was employed. This mutant shows recovery by restoring the function of stem cells and increasing the number of meristematic cells in its system by treatment with RGF peptides (Gancheva et al. 2021; Meng et al. 2012). Moreover, TPST1 levels are regulated by auxin, due to which this particular mutant disrupts the auxin-induced as well as basal transcription of *PLETHORA* (*PLT*) genes. Transcription factors encoded by these genes interact with RGF peptides to alter the root system architecture. On the other hand, in the case of triple mutants *rgf-1*, *rgf-2*, and *rgf-3*, there is a defined decrease in the number of meristematic cells, and this results in a phenotype of short roots (Fedoreyeva 2023). There are five RGF receptors (RGFR) that have been found but RGFs bind to them with distinct affinities. This is similar to CLE signaling, aiding in the dynamic and broad signaling network to influence plant development and growth (Ou et al. 2016). The RGF signaling primarily occurs through PLT1 and PLT2, which happen to be the proximal targets of peptides. These genes help determine the stem cell patterning and amplify the cell proliferation because of their presence in the region of stem cells of the root meristem (Fedoreyeva 2023). RGFs bind to RGFRs as established earlier and become a part of a complex that also includes mitogen-activated protein kinase4 (MKK4) and MAP kinase3 (MAPK3). These downstream components subsequently modulate the levels of PLT1 and PLT2. Quintuple mutants *rgi 1, 2, 3, 4, 5* have root hair defects which can be treated with MKK4/5. The double mutant *mkk4.5* is resistant to RGF1 treatment and has a reduced root hair meristem (Lu et al. 2020; Shao et al. 2020). An important relationship between PLT and WOX5 has also been established. Expressed primarily in the quiescent center (QC), WOX5 levels, when reduced, promote the differentiation of stem cells, whereas its upregulation promotes their proliferation. There are high levels of PLT in the QC as a result of their association with WOX5. Raised PLT levels cause a reduction in the levels of WOX5 and vice versa, leading to a tight regulation of this negative feedback loop that controls stem cell proliferation and differentiation (Burkart et al. 2022; Drisch and Stahl 2015). Root geotropism is maintained by RGFs 9/6/4 by balancing auxin distribution (Meng et al. 2012). RGFs 6/9 are downregulated in leaves when infection with *Pseudomonas syringae* occurs, whereas their overexpression leads to a prominent phenotype exhibiting disease resistance. When plant leaves are pretreated with RGF6, their sensitivity to flg22 increases, leading to renewed synthesis of ethylene and *pathogenesis-related 1* (*PR1*) gene expression. Furthermore, FLS2 protein levels increase as a result of the association between RGF9 and RGFR1 (Stegmann et al. 2022).

An RGF precursor, *preRGF7*, is a part of a positive feedback loop that functions in response to pathogen attacks. The transcription factor WRKY33 binds to the promoter of this precursor and acts as a phosphorylation substrate for calcium-dependent protein kinase (CPK5/6) and MPK3/6, which in turn cause the upregulation of the transcription factor after pathogen treatment (Wang et al. 2021). PAMPs or damage-associated molecular patterns (DAMPs) can also cause alkalization of the extracellular medium, leading to inhibition of growth. This inhibition takes place due to a certain modification in a residue of the RGF1 peptide. A tyrosine residue of this peptide has a sulfite group (sY2) that undergoes protonation in acidic conditions to aid in a powerful hydrogen bond that forms via the RxGG motif between the RGF1 peptide and its receptor. As alkalization takes place, it prohibits this interaction from occurring while promoting immunity by activating other peptides (Liu et al. 2022).

The plant peptides containing sulfated tyrosine (PSY) undergo modifications such as hydroxyproline arabinosylation and tyrosine sulfation (Zhang et al. 2025). PSY1 has 18 amino acids and is capable of aiding in cell proliferation mechanisms even at nanomolar concentrations. There are eight homologues of PSY1 in *Arabidopsis*, each having a conserved domain at the C-terminal end (Zhang et al. 2024a). PSYs are critical components in the trade-off between plant growth and stress response. Research on the *tpst* mutants showed that, as a result of treatment with PSY5, the previously downregulated genes in the mutant were now abundant in stress response mechanisms such as the jasmonic acid and salicylic acid signaling pathways, as well as oxidative stress. Pathogen invasion increases cell damage and inevitably decreases the levels of PSYs by disrupting the interactions between PSYs and their receptors PSYRs (Ogawa-Ohnishi et al. 2022). The double mutants *pskr1psy1r* exhibited higher disease resistance than single mutants. In addition, the PSY1-PSY1R complex also mediates H^+ transport through proton pumps on the cell membrane and influences plant development (Fuglsang et al. 2014).

Phytosulfokine peptides are pentapeptides that are sulfated and control cell division via binding to PSK receptors (PSKR1/2). PSKs have important roles in defense response along with plant growth. They have also been known to increase the proliferation of callus cells and root growth in adventitious systems, cause the elongation of roots, promote somatic embryogenesis, maintain the identity of procambial cells, and induce the growth of pollen tubes (Zhang et al. 2025). A *pskr1* mutant shows varied levels of resistance against necrotrophic and biotrophic pathogens, hinting at a possible cross talk between jasmonic acid and salicylic acid signaling in response to pathogen attack (Mosher and Kemmerling 2013). Yang et al. (2019) found that OsPSKR1, located on the plasma membrane in cells of rice, functions as a PSKR to indeed activate the SA pathway and elevate resistance against *Xanthomonas oryzae* pv. *oryzicola* (*Xoc*). PSKs are also encoded by long noncoding RNAs such as *GblncRNA7* in order to accumulate more auxin and carry out a defined immune response (Zhang et al. 2022a). The application of PSKs and their receptors in crop improvement was also highlighted by the Clustered Regularly Interspaced Short Palindromic Repeats (CRISPR) Cas9-generated *Clpsk1* mutant that showed resistance against fungi in watermelon (Zhang et al. 2020a).

The Casparian strip integrity factor (CIF) peptides, as the name suggests, are present to carry out the developmental processes of the Casparian strip. *CIF1* and *CIF2*, found in the root stele, form an association with GASSHO1 (GSO1)/ SCHENGEN3 (SGN3) which are

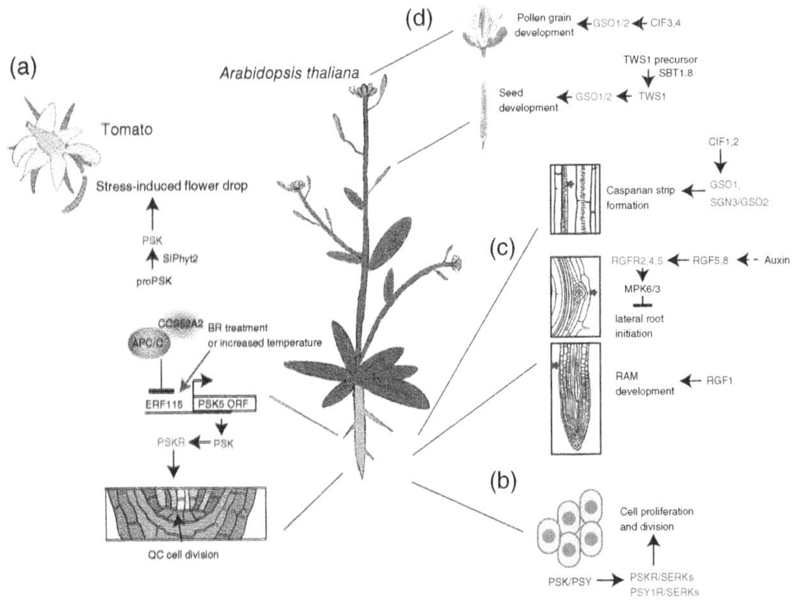

Figure 2.2 Various signaling pathways carried out by sulfated peptides in the context of plant development. (a) PSK peptides induce the abscission of tomato floral organs and the division of the Arabidopsis quiescence center. BR, brassinosteroids; QC, quiescent center. (b) PSK and PSY peptides promote cell proliferation and division. (c) RGFs mediate the initiation of lateral roots and RAM development in Arabidopsis. RAM, root apical meristem. (d) CIF peptides promote the development of pollen, seeds, and Casparian strips in Arabidopsis. *Source:* The image has been adapted from Zhang et al. (2024a) under Creative Commons Attribution license (http://creativecommons.org/publicdomain/zero/1.0), which permits the use, distribution, or reproduction in other forums provided the original author(s) are credited and that the original publication in this journal is cited.

endodermal LRR-RKs. SGN3 colocalizes with SGN1 at the outer domain of the Casparian strip. The generation of the Casparian strip comes to a halt in loss-of-function mutants of SGN1/3. Apart from this, CIFs also participate in embryo cuticle and pollen development via the aforementioned receptors and are sensitive to treatment with iron (Zhang et al. 2024a) (Figure 2.2).

2.10 CEP and Their Receptors

C-terminally encoded peptide (CEP) hormones are mobile and activate signaling pathways that involve critical developments in plants. Loci pertaining to CEPs have been detected in seed plants but not in ancient lineages of land plants, as is evident from the lack of root branching observed in species such as *Selaginella* due to an absence of these peptides (Delay et al. 2013). The *CEP* genes initially encode for a prepropeptide that is eventually processed to a smaller peptide that is functional (Taleski et al. 2018). There are several PTMs in these peptides that dictate their biological activity, for example, CEP1 in *Arabidopsis* undergoes modifications at positions P4 and P11 to hydroxy-proline (Bobay et al. 2013). There have been studies in the past which have brought forth the negative regulation of root development by CEPs via their binding to CEP receptors (CEPR), CEP DOWNSTREAMs (CEPDs), the translational repressor auxin/indole-3-acetic acid (AUX/IAA), and the cytokinin receptor *Arabidopsis* histidine kinase (AHK). Until very recently, the very minute molecular mechanisms of these interactions and their resultant downstream signaling with respect to phytohormones were not well established. A study by Zhang et al. (2024b) found that overexpressing synthetic *Arabidopsis thaliana* CEPs such as AtCEP5 negatively affected the elongation of root growth as well as the formation of lateral roots. This occurred by simultaneous auxin signaling suppression and cytokinin signaling promotion by the CEPs. CEP5 has also been found to be significantly expressed during drought and osmotic stress (Smith et al. 2020). A CEP found in *Manihot esculenta* or cassava aids in the uptake of nitrate in the roots. Cassava is an important and unique food crop because it exhibits tolerance to infertile soil. A majority of plants, including cassava, have elaborate biochemical adaptations in place to stabilize the fluctuating levels of nitrogen, which in turn influences the root system architecture (Ezui et al. 2016). The said adaptations support better uptake of nutrients while also striking a balance between plant development and stress response (Poitout et al. 2018; Roy et al. 2022). Lu et al. (2025) found that *MeCEP6* was significantly expressed in the tips of roots and was a crucial component in the response to low nitrate stress. It was already known that MeCEP6 modulates various hormones, viz., auxin, cytokinin, ethylene, and abscisic acid, that cause effective growth and development. Treatment with exogenous MeCEP6 and its overexpression promoted its binding to MeCEPR, forming MeCEPD and ultimately elevating the expression level of the nitrate transporter 2 (NRT2), along with lowering the functioning of genes that were auxin-related, promoting the uptake of nitrate and suppressing the growth of seedlings during nitrogen limitation conditions. This complex balance supposedly increases the survival of cassava by taking up and redistributing nitrate to optimal locations, sometimes at the cost of normal growth rate. It was also observed that the *MeCEP6* promoter could interact with *Manihot esculenta* WRKY transcription factors such as MeWRKY65 and MeWRKY70 to keep expression levels of MeCEP6 in check when sufficient nitrogen is available and increase the same when a deficiency is detected. This will now be further explored to ultimately achieve improved crops by targeting MeCEP6 and thereby greater yield of cassava in soil that lacks nitrogen.

SiCEP3 is a CEP derived from *Setaria italica* which aids in abscisic acid uptake and restricts root elongation, thus influencing cereal growth (Zhang et al. 2021). ZmCEP1 in maize changes the transcriptional status of genes responsible for nitrate metabolism, auxin transport, and sugar uptake and this influences kernel development (Xu et al. 2021). A high-throughput phenotypic study of the effects of small peptides such as CEPs on the utilization of various mineral nutrients revealed significant changes in sulfate and phosphate transporters in addition to the already established nitrate transporters (Roy et al. 2022). Apart from root development and nutrient uptake, CEPs hold significance in maintaining shoot development and carbon dioxide (CO_2) levels. Upregulated by an increase in the CO_2 level is the gene *AtCEP9*. The abundance of CEP14 increased in shoots under osmotic pressure and in roots as a result of high CO_2 levels. A substantial number of CEPs that are critical for the development of roots are also actively involved in the development of shoots. A *cep3* knockout mutant, when grown hydroponically, produced a different shoot phenotype. CEP1 was elevated in the shoots under nitrogen stress, whereas CEP2 and CEP13 level was diminished. Six *CEP* genes were overexpressed under the CaMV 35S promoter to identify their role in shoot phenotypes. A lesser number of rosette leaves, changed leaf morphology, and delayed flowering were seen in 5-week-old *p35S::CEP2* plants. There were defects in leaf morphology such as leaf yellowing, epinasty, and reduced rosette sizes observed in *p35S::CEP4*, but these were larger than *p35S::CEP3* lines, which had similar morphology. Two plants, *p35S::CEP6* and *p35S::CEP9,* exhibited yellowing and epinasty, but these were not as severe as others. *p35S::CEP9* plants showed reduced size of rosettes as well (Delay et al. 2013). Pleiotropic effects have been observed due to overexpression of OsCEPs-CEPs encoded by *Oryza sativa*. Of these, OsCEP6.1 stands out as a CEP, important in early as well as emerging inflorescences (Ogilvie et al. 2014). This CEP is also found during development of the panicle and induced by high levels of 6-benzylaminopurine (Sui et al. 2016). In *Medicago truncatula,* MtCEPs are crucial in nodule formation via their interaction with the receptor COMPACT ROOT ARCHITECTURE 2 (CRA2) (Kereszt et al. 2018).

2.11 RALF Family and Their Receptors

Rapid alkalinization factor (RALF) peptides were discovered as cysteine-rich polypeptides that exhibit a role in increasing pH of the medium in tobacco leaves. There are disulfide bridges between Cys-41 and Cys-47 as well as Cys-18 and Cys-28 residues, considered important for the activity of these peptides (Fedoreyeva 2023). Mutations in the Cys residues strip the peptide of its active conformation. RALFs have several motifs that are conserved and are considered ubiquitous in the plant kingdom. The active peptide has two arginine (RR) residues in the upstream region of its N-terminus. These motifs are cleaved by enzymes that are essentially involved in prohormone processing and vital in regulating immune response via *Arabidopsis thaliana* RALFs (AtRALFs). One other N-terminal YISY motif is vital for its function and aids in the peptide binding to its receptor. RALF without the YISY motif can still bind to the receptor but loses its activity (Ge et al. 2019; Liu et al. 2018; Pearce et al. 2010). The very first RALF peptide isolated from *Arabidopsis thaliana* was capable of driving up the cytosolic calcium. Bushier and shorter plants are formed when AtRALF1 and AtRALF23 are overexpressed (Fedoreyeva 2023). Expression of AtRALF1 takes place mainly in roots and *atralf1* mutants have previously shown an increase in the number of lateral roots, root, and cell length along with elongated hypocotyls (Bergonci et al. 2014). RALF1 binds to FERONIA (FER), a receptor-like kinase of the *Catharanthus roseus* family (CrRLK1L) (Zhang et al. 2020b). A phosphoproteomic approach to identify phosphorylation mediated by RALF1 identified FER as one of the candidates that responded positively to said phosphorylation. The *fer-4* mutant did not recognize RALF until high concentration was applied and thus lost the ability to enhance the cellular signaling of calcium and to prohibit the elongation of cells. The synthesis of ErbB3 binding protein 1 (EBP1) and its accumulation in the nucleus as a result of RALF1-FER signaling controls the growth of cells (Li et al. 2018). RALF1 promotes the phosphorylation of

Arabidopsis-derived H^+-ATPase2 (AHA2) at its Ser-899 position, rendering it incapable of performing its proton pump activity (Haruta et al. 2014). Li et al. (2022b,c) found three other residues whose phosphorylation could either promote proton pump activity, viz., Thr 881 and Thr 947, or inhibit it, viz., Ser 931. RALF peptides carry out a concerted regulation of cellular calcium levels and proton influx. Studies by Gjetting et al. (2020) found that the extracellular change in pH by RALF is dependent on signaling of Ca^{2+} before alkalization. Several of these peptides bind to a number of different receptors to set off Ca^{2+} signaling and thus activate H^+ pumps. While RALF inhibits acidification of the apoplast, PSYs promote it.

A majority of RALF peptides, for example, RALF 35/7/6/16/17/20/15/32/21/8/31/9/19/34/23/22/1/33, are responsible for repressing root growth and development (Liu et al. 2024). The overexpression of RALFs in several plant lines leads to the production of phenotypes that are semi-dwarf with smaller leaves, root systems, and lesser number of lateral roots (Bergonci et al. 2014; Campos et al. 2018), whereas their silencing causes longer hypocotyls and roots to form (Atkinson et al. 2013; Bergonci et al. 2014). Free RALF22 forms a complex with FER and LORELEI-LIKE-GPI-ANCHORED PROTEIN1 (LLG1) to promote pectin polymer compaction at the tip of root hairs as a means of triggering cellular responses that are adaptive in nature and form circumferential rings (Schoenaers et al. 2024). The association that takes place between RALF1 and FER is known. In this context, the presence and activity of RPM1-induced protein kinase (RIPK) have also been established by mass spectrometric (MS) analysis. *ripk* mutants exhibited similar phenotypes when compared with *fer-4* mutants, suggesting a relationship between FER and RIPK to control root growth (Du et al. 2016). Several biochemical analyses were carried out and it was observed that this aforementioned interaction was stronger when treated with RALF1, further hinting at the possibility that RALF1, after binding to FER, also recruits RIPK to this particular kinase complex, thereby beginning a cascade of phosphorylation downstream of these components, terminating in the inhibition of root growth and extracellular alkalization (Fedoreyeva 2023). This was further elaborated by experiments on transcription factors which found that the complex of RALF1-FER phosphorylates EUKARYOTIC TRANSLATION INITIATION FACTOR 4E1 (elF4E1) which subsequently influences the translation of ROOT HAIR DEFECTIVE 6-LIKE 4 (RSL4) and ROP GTPase (ROP2) to dictate polarity and root hair size. On the other hand, RSL4 also contributes to the control of RALF1 levels, establishing a negative feedback loop to further refine the root hair development (Zhu et al. 2020a,b). During the initiation of lateral roots, pericycle cell division is influenced by RALF34, which is a ligand for a member of the CrRLK1L family, THESEUS1 (THE1). The *the1* and *ralf34* mutants had distinct phenotypes and the *the1* mutant especially is observed to be insensitive to RALF34-induced surface alkalization and root growth inhibition (Gonneau et al. 2018; Murphy et al. 2016).

Besides inhibiting root development and influencing H^+ pump, the RALF1-FER pathway promotes YUCCA (YUC) synthesis and aids in the activation of TRANSPORT INHIBITOR RESPONSE1/ AUXIN SIGNALING F-BOX PROTEIN (TIR1/AFB) to ultimately enhance the levels of auxin (Li et al. 2022a). RALF peptides are significantly involved in reproductive development. In pollen, RALF communicates with mastoid cells found on stigma in order to accelerate germination. AtRALF33 and AtRALF bind to ANJEA and FERONIA (ANJ-FER) receptor kinase complex on the stigma to accumulate ROS in mastoid cells before germination takes place in *Arabidopsis*. Thereafter, when pollination occurs, there is competition between pollen coat protein class B polypeptides (PCP-Bs) and the above-mentioned RALFs to promote the transfer of water via water channels from papilla cells to the pollen grains and decrease the accumulation of ROS (Liu et al. 2021). An antagonistic relationship is observed between the stigma RALF peptides (sRALFs) and pollen RALF peptides (pRALFs). sRALFs from pistil bind to CrRLK1L family receptors and the LRX protein to form a hybridization barrier that blocks incompatible pollen tubes from penetrating. pRALFs from the affinity pollen participate in inhibition to successfully carry out intergenus/interspecies hybridization. Thus, RALFs mediate a "lock-and-key" role to regulate plant reproductive barriers (Lan et al. 2023; Liu et al. 2024). FER, ANJ, and HERCULES RECEPTOR KINASE1 (HERK1) signaling in septal epidermal cells is also activated due to pRALF, causing multitubular blockage to stop polyspermism and the formation of additional pollen tubes (Zhong et al. 2022). Two RALFs derived from *O. sativa*, OsRALF19 and OsRALF17, are crucial for the growth of pollen tubes due to binding to *O. sativa* male-gene transfer defective2 (OsMTD2) and activating ROS signaling for the maintenance of integrity (Kim et al. 2023). RALFs are actively involved in flowering. *ScRALF23* is found in the vascular bundles of fruits that are early developing and in larger fruits; upon silencing it, there is a reduction in the number of seeds. The promoter of *ScRALF23* consists of auxin regulatory elements as well, leading to the downstream regulation of early development of fruit (Chevalier et al. 2013). *RALF* genes are upregulated in response to bacterial proteins such as flg22 and are seen as potent mediators of immune responses. They regulate the levels of ROS, stabilize the lipid phase of the cell membrane and carry out necessary signaling to counteract biotic stress (Liu et al. 2024). As for abiotic stress, RALFs communicate with abscisic acid (ABA) in order to effectively curb the damaging effects of chilling stress, drought stress, or salinity (Chen et al. 2016). ABA Insensitive2 (ABI2) phosphatase is inactivated by the RALF1-FER signaling by subsequent regulation of the GEF1/4/10-ROP11 pathway and FER has been established to lower ABA response to limit ROS levels in guard cells (Chen et al. 2016; Yu et al. 2012).

2.12 Other Notable Peptides and Their Receptors

The EPF or stomagen family acts as cell-to-cell communicators and carries out various biological processes. The overexpression of EPF1/2 leads to an increase in the density of the stomata, whereas mutants of the same exhibit a phenotype with decreased stomatal density. These peptides are ligands for ERECTA Like1 (ERL1) and ERECTA (ER) receptors in the pathway that determines the development of stomata, where EPFL9 competitively binds to ER in order to further fine-tune this process. EPFL4/6 are active in the cell division of procambial cells and ensure pollen-pistil interaction by promoting stamen filament elongation. PAMP-INDUCED SECRETED PEPTIDES (PIPs) and plant elicitor peptides (PEPs) are peptides known to have roles in defense responses and resistance to pathogens. When applied exogenously, PIP1/2 rapidly increases the levels of PAMP-triggered immune responses and modulates ROS levels simultaneously. PEPs with their receptors PEPRs enhance immune signaling in alkaline environment and cause the closure of stomata. SERINE RICH

ENDOGENOUS PEPTIDES (SCOOPs) function in immune responses and root development. SCOOP 10 and SCOOP 12 work antagonistically to bind to the receptor MALE DISCOVERER1-INTERACTING RLK2 (MIK2) and control leaf senescence. EARLY NODULIN 40 (ENOD40) regulates nodulation in soybean with respect to nitrogen availability. POLARIS and ROTUNDIFOLIA via their RLKs influence leaf vascularization and morphogenesis, respectively (Zhang et al. 2025).

Table 2.2 represents the functions of some primary peptide receptors.

Table 2.2 Functions of some primary peptide-receptors.

Peptide family	Peptide receptors	Functions	References
CLE	ACR4, BAM, NCED, MAP3/MAP6, OST1, SLAC1, JUB1	Lateral root patterning, pericycle cell division, determination of the fate of stem cells, physiological adaptation of shoot, drought stress, temperature stress, carbon accumulation, modulation of ROS levels	Endo et al. (2013), Fedoreyeva (2023), Ma et al. (2020), Okamoto et al. (2022), Stahl et al. (2013), Takahashi et al. (2018), and Zhang et al. (2019, 2022b)
TSP	PLT, RGFR, PSYR, MAPK, CPK5/6, SGN1/3,	Root system architecture, stem cell proliferation and patterning, flowering, proton transport, immune response, JA/SA cross talk, Casparian strip development, oxidative stress response	Burkart et al. (2022), Drisch and Stahl (2015), Fedoreyeva (2023), Fuglsang et al. (2014), Gancheva et al. (2021), Lu et al. (2020), Meng et al. (2012), Mosher and Kemmerling (2013), Ogawa-Ohnishi et al. (2022), Shao et al. (2020), Stegmann et al. (2022), Wang et al. (2021), and Zhang et al. (2024a, 2025)
CEP	CEPR, CEPD, CRA	Root system architecture, nitrogen, sulfur, and phosphorus transport, phytohormone signaling, shoot development, CO_2-mediated response	Delay et al. (2013), Kereszt et al. (2018), Poitout et al. (2018), Roy et al. (2022), and Smith et al. (2020)
RALF	FER, EBP1, AHA2, LLG1, RIPK, THE1, HERK1, ANJEA	Calcium signaling, proton pump activity, root architecture, shoot phenotype, polarity due to auxin signaling, ROS accumulation, plant reproductive barrier control, immune response, abiotic stress response	Chen et al. (2016), Du et al. (2016), Fedoreyeva (2023), Gjetting et al. (2020), Gonneau et al. (2018), Haruta et al. (2014), Liu et al. (2021), Murphy et al. (2016), Schoenaers et al. (2024), Yu et al. (2012), Zhang et al. (2020b), and Zhong et al. (2022)

2.13 Conclusion and Future Perspectives

Our knowledge of plant signaling networks has been completely transformed in the last several decades because of the identification and study of plant peptide hormones. Since the identification of the first peptide hormone, systemin, to the detailed mapping of numerous SSPs and their receptors, research in this field has unveiled an extraordinary layer of complexity in plant physiology. Peptide hormones, previously underemphasized because of their elusive concentration and cryptic functions, are now appreciated as pivotal regulators in growth, development, stress response, immunity, and interorgan communication in plants. This chapter exhaustively covers the varied families of peptide hormones such as CLEs, RALFs, RGFs, PSKs, CEPs, and others, each acting via highly specific ligand–receptor interactions involving LRR-RLKs, LRR-RLPs, and coreceptors like SERKs. These receptor complexes, following ligand binding, trigger complex signal transduction cascades via RLCKs, MAPKs, calcium fluxes, and ROS modulation, eventually resulting in targeted transcriptional responses. The convergence of these pathways with hormonal signaling pathways including auxin, cytokinin, and abscisic acid, as well as their precise control under biotic and abiotic stress, demonstrates how peptide-mediated communication allows plants to adapt. Even after major advancements, there is still a large number of peptide-encoding genes left to be functionally characterized, and gene family redundancy makes reverse genetic analysis challenging. Detection and quantification of the peptides in planta are still technically challenging owing to their low abundance and intricate PTMs. New approaches like peptide omics, cutting-edge mass spectrometry, and single-cell resolution research are needed to bridge the gaps. It is also necessary to elucidate the structural dynamics of receptor–peptide interactions, receptor complex assemblies, and their spatiotemporal regulation under native physiological conditions. Future work will need to identify new peptide ligands and their receptors, as well as address how these molecules mediate systemic signaling, developmental plasticity, and environmental resilience. The interactions between secretory and nonsecretory peptides, their intracellular versus extracellular localization, and the possibility for cross talk between peptide signaling modules offer intriguing avenues to explore. Furthermore, using this information in agricultural biotechnology—via specific breeding, gene modification, or synthetic peptide use—is particularly promising for augmenting crop yield, stress resistance, and disease immunity in an increasingly climate-vulnerable world. In a nutshell, plant peptide hormone signaling exists at the crossroads of cellular communication, transforming developmental signals and environmental stimuli into meaningful physiological responses. As our effort for investigating these molecules continues to advance, there will be further improvement in our capacity to unravel and leverage the complete potential of peptide-mediated signaling in the overall physiology of plants.

References

Ajeesh Krishna, T.P., Maharajan, T., Victor Roch, G. et al. (2020). Structure, function, regulation and phylogenetic relationship of ZIP family transporters of plants. *Frontiers in Plant Science* 11: 662. https://doi.org/10.3389/fpls.2020.00662.

Atkinson, N.J., Lilley, C.J., and Urwin, P.E. (2013). Identification of genes involved in the response of *Arabidopsis* to simultaneous biotic and abiotic stresses. *Plant Physiology* 162 (4): 2028–2041. https://doi.org/10.1104/pp.113.222372.

Baumberger, N., Ringli, C., and Keller, B. (2001). The chimeric leucine-rich repeat/extensin cell wall protein LRX1 is required for root hair morphogenesis in *Arabidopsis thaliana*. *Genes & Development* 15 (9): 1128–1139. https://doi.org/10.1101/gad.200201.

Baumberger, N., Doesseger, B., Guyot, R. et al. (2003). Whole-genome comparison of leucine-rich repeat extensins in *Arabidopsis* and rice. A conserved family of cell wall proteins form a vegetative and a reproductive clade. *Plant Physiology* 131 (3): 1313–1326. https://doi.org/10.1104/pp.102.014928.

Bergonci, T., Ribeiro, B., Ceciliato, P.H.O. et al. (2014). *Arabidopsis thaliana* RALF1 opposes brassinosteroid effects on root cell elongation and lateral root formation. *Journal of Experimental Botany* 65 (8): 2219–2230. https://doi.org/10.1093/jxb/eru099.

Betsuyaku, S., Sawa, S., and Yamada, M. (2011). The function of the CLE peptides in plant development and plant-microbe interactions. *The Arabidopsis Book* 9: e0149. https://doi.org/10.1199/tab. 0149.

Bi, G., Zhou, Z., Wang, W. et al. (2018). Receptor-like cytoplasmic kinases directly link diverse pattern recognition receptors to the activation of mitogen-activated protein kinase cascades in *Arabidopsis*. *The Plant Cell* 30 (7): 1543–1561. https://doi.org/10.1105/tpc.17.00981.

Blanvillain, R., Young, B., Cai, Y. et al. (2011). The *Arabidopsis* peptide kiss of death is an inducer of programmed cell death. *The EMBO Journal* 30 (6): 1173–1183. https://doi.org/10.1038/emboj.2011.14.

Bobay, B.G., DiGennaro, P., Scholl, E. et al. (2013). Solution NMR studies of the plant peptide hormone CEP inform function. *FEBS Letters* 587 (24): 3979–3985. https://doi.org/10.1016/j.febslet.2013.10.033.

Boisson-Dernier, A., Franck, C.M., Lituiev, D.S., and Grossniklaus, U. (2015). Receptor-like cytoplasmic kinase MARIS functions downstream of Cr RLK1L-dependent signaling during tip growth. *Proceedings of the National Academy of Sciences* 112 (39): 12211–12216. https://doi.org/10.1073/pnas.1512375112.

Bouteraa, M.T., Ben Romdhane, W., Baazaoui, N. et al. (2023). GASA proteins: review of their functions in plant environmental stress tolerance. *Plants* 12 (10): Article 10. https://doi.org/10.3390/plants12102045.

Burkart, R.C., Strotmann, V.I., Kirschner, G.K. et al. (2022). Plethora-WOX5 interaction and subnuclear localization control *Arabidopsis* root stem cell maintenance. *EMBO Reports* 23 (6): e54105. https://doi.org/10.15252/embr.202154105.

Cajero Sánchez, W., García-Ponce, B., Sánchez, M.D.L.P. et al. (2018). Identifying the transition to the maturation zone in three ecotypes of *Arabidopsis thaliana* roots. *Communicative & Integrative Biology* 11 (1): e1395993. https://doi.org/10.1080/19420889.2017.1395993.

Campbell, L. and Turner, S.R. (2017). A comprehensive analysis of RALF proteins in green plants suggests there are two distinct functional groups. *Frontiers in Plant Science* 8: 37. https://doi.org/10.3389/fpls.2017.00037.

Campos, W.F., Dressano, K., Ceciliato, P.H.O. et al. (2018). *Arabidopsis thaliana* rapid alkalinization factor 1–mediated root growth inhibition is dependent on calmodulin-like protein 38. *Journal of Biological Chemistry* 293 (6): 2159–2171. https://doi.org/10.1074/jbc.M117.808881.

Chakraborty, S., Nguyen, B., Wasti, S.D., and Xu, G. (2019). Plant leucine-rich repeat receptor kinase (LRR-RK): structure, ligand perception, and activation mechanism. *Molecules* 24: https://doi.org/10.3390/molecules24173081.

Chen, G.Q., He, X., Liao, L.P., and McKeon, T.A. (2004). 2S albumin gene expression in castor plant (*Ricinus communis* L.). *Journal of the American Oil Chemists' Society* 81 (9): 867–872. https://doi.org/10.1007/s11746-004-0993-5.

Chen, J., Yu, F., Liu, Y. et al. (2016). FERONIA interacts with ABI2-type phosphatases to facilitate signaling cross-talk between abscisic acid and RALF peptide in *Arabidopsis*. *Proceedings of the National Academy of Sciences* 113 (37) https://doi.org/10.1073/pnas.1608449113: E5519–E5527.

Chevalier, E., Loubert-Hudon, A., and Matton, D.P. (2013). Sc RALF 3, a secreted RALF-like peptide involved in cell–cell communication between the sporophyte and the female gametophyte in a solanaceous species. *The Plant Journal* 73 (6): 1019–1033. https://doi.org/10.1111/tpj.12096.

Chilley, P.M., Casson, S.A., Tarkowski, P. et al. (2006). The POLARIS peptide of *Arabidopsis* regulates auxin transport and root growth via effects on ethylene signaling. *The Plant Cell* 18 (11): 3058–3072. https://doi.org/10.1105/tpc.106.040790.

Corpo, D.D., Coculo, D., Greco, M. et al. (2024). Pull the fuzes: processing protein precursors to generate apoplastic danger signals for triggering plant immunity. *Plant Communications* 5 (8): https://doi.org/10.1016/j.xplc.2024.100931.

Couzigou, J.-M., Lauressergues, D., Bécard, G., & and Combier, J.-P. (2015). mi RNA-encoded peptides (miPEPs): a new tool to analyze the roles of miRNAs in plant biology. *RNA Biology*, 12(11), 1178–1180. https://doi.org/10.1080/15476286.2015.1094601

Czolpinska, M. and Rurek, M. (2018). Plant glycine-rich proteins in stress response: an emerging, still prospective story. *Frontiers in Plant Science* 9: 302. https://doi.org/10.3389/fpls.2018.00302.

de Oliveira Carvalho, A. and Gomes, V.M. (2009). Plant defensins—prospects for the biological functions and biotechnological properties. *Peptides* 30 (5): 1007–1020. https://doi.org/10.1016/j.peptides.2009.01.018.

De Smet, I., Vassileva, V., De Rybel, B. et al. (2008). Receptor-like kinase ACR4 restricts formative cell divisions in the *Arabidopsis* root. *Science* 322 (5901): 594–597. https://doi.org/10.1126/science.1160158.

Delay, C., Imin, N., and Djordjevic, M.A. (2013). CEP genes regulate root and shoot development in response to environmental cues and are specific to seed plants. *Journal of Experimental Botany* 64 (17): 5383–5394. https://doi.org/10.1093/jxb/ert332.

Drisch, R.C. and Stahl, Y. (2015). Function and regulation of transcription factors involved in root apical meristem and stem cell maintenance. *Frontiers in Plant Science* 6: 505. https://doi.org/10.3389/fpls.2015.00505.

Du, C., Li, X., Chen, J. et al. (2016). Receptor kinase complex transmits RALF peptide signal to inhibit root growth in *Arabidopsis*. *Proceedings of the National Academy of Sciences* 113 (51) https://doi.org/10.1073/pnas.1609626113: E8326–E8334.

Endo, S., Shinohara, H., Matsubayashi, Y., and Fukuda, H. (2013). A novel pollen-pistil interaction conferring high-temperature tolerance during reproduction via CLE45 signaling. *Current Biology* 23 (17): 1670–1676. https://doi.org/10.1016/j.cub.2013.06.060.

Endo, S., Betsuyaku, S., and Fukuda, H. (2014). Endogenous peptide ligand–receptor systems for diverse signaling networks in plants. *Current Opinion in Plant Biology* 21: 140–146. https://doi.org/10.1016/j.pbi.2014.07.011.

Ezui, K.S., Franke, A.C., Mando, A. et al. (2016). Fertiliser requirements for balanced nutrition of cassava across eight locations in West Africa. *Field Crops Research* 185: 69–78. https://doi.org/10.1016/j.fcr.2015.10.005.

Fang, J., Liu, S., Su, Y. et al. (2025). Genome-wide identification of the CIF gene family and protein interaction with GSO1s under the p-HBA-induced continuous cropping obstacle in *Pogostemon cablin*. *International Journal of Molecular Sciences* 26 (4): Article 4. https://doi.org/10.3390/ijms26041568.

Fedoreyeva, L.I. (2023). Molecular mechanisms of regulation of root development by plant peptides. *Plants* 12 (6): 1320. https://doi.org/10.3390/plants12061320.

Fernandez, A., Drozdzecki, A., Hoogewijs, K. et al. (2013). Transcriptional and functional classification of the GOLVEN/ROOT GROWTH FACTOR/CLE-like signaling peptides reveals their role in lateral root and hair formation. *Plant Physiology* 161 (2): 954–970. https://doi.org/10.1104/pp.112.206029.

Figueiredo, J., Sousa Silva, M., and Figueiredo, A. (2018). Subtilisin-like proteases in plant defence: the past, the present and beyond. *Molecular Plant Pathology* 19 (4): 1017–1028. https://doi.org/10.1111/mpp.12567.

Finkina, E.I., Melnikova, D.N., Bogdanov, I.V., and Ovchinnikova, T.V. (2019). Peptides of the innate immune system of plants. Part II. Biosynthesis, biological functions, and possible practical applications. *Russian Journal of Bioorganic Chemistry* 45 (2): 55–65. https://doi.org/10.1134/S1068162019020043.

Fuglsang, A.T., Kristensen, A., Cuin, T.A. et al. (2014). Receptor kinase-mediated control of primary active proton pumping at the plasma membrane. *The Plant Journal* 80 (6): 951–964. https://doi.org/10.1111/tpj.12680.

Gancheva, M.S., Losev, M.R., Gurina, A.A. et al. (2021). Polymorphism of CLE gene sequences in potato. *Vavilov Journal of Genetics and Breeding* 25 (7): 746–753. https://doi.org/10.18699/VJ21.085.

García, A.C., Olaetxea, M., Santos, L.A. et al. (2016). Involvement of hormone- and ROS-signaling pathways in the beneficial action of humic substances on plants growing under normal and stressing conditions. *BioMed Research International* 2016: 1–13. https://doi.org/10.1155/2016/3747501.

Ge, Z., Cheung, A.Y., and Qu, L. (2019). Pollen tube integrity regulation in flowering plants: insights from molecular assemblies on the pollen tube surface. *New Phytologist* 222 (2): 687–693. https://doi.org/10.1111/nph.15645.

Ghosh, P. and Roychoudhury, A. (2024). Plant peptides involved in abiotic and biotic stress responses and reactive oxygen species (ROS) signaling. *Journal of Plant Growth Regulation* 43: 1410–1427.

Gjetting, S.K., Mahmood, K., Shabala, L. et al. (2020). Evidence for multiple receptors mediating RALF-triggered Ca2+ signaling and proton pump inhibition. *The Plant Journal* 104 (2): 433–446. https://doi.org/10.1111/tpj.14935.

Gong, F., Wu, X., and Wang, W. (2015). Diversity and function of maize pollen coat proteins: from biochemistry to proteomics. *Frontiers in Plant Science* 06: https://doi.org/10.3389/fpls.2015.00199.

Gonneau, M., Desprez, T., Martin, M. et al. (2018). Receptor kinase THESEUS1 is a rapid alkalinization factor 34 receptor in *Arabidopsis*. *Current Biology* 28 (15): 2452–2458.e4. https://doi.org/10.1016/j.cub.2018.05.075.

Guarnerio, C.F. (2010). Effects of enod40 overexpression in non legume plants. https://iris.univr.it/handle/11562/343215

Guefrachi, I., Nagymihaly, M., Pislariu, C.I. et al. (2014). Extreme specificity of NCR gene expression in *Medicago truncatula*. *BMC Genomics* 15 (1): 712. https://doi.org/10.1186/1471-2164-15-712.

Guo, P., Yoshimura, A., Ishikawa, N. et al. (2015). Comparative analysis of the RTFL peptide family on the control of plant organogenesis. *Journal of Plant Research* 128 (3): 497–510. https://doi.org/10.1007/s10265-015-0703-1.

Hailemariam, S., Liao, C.-J., and Mengiste, T. (2024). Receptor-like cytoplasmic kinases: orchestrating plant cellular communication. *Trends in Plant Science* 29 (10): 1113–1130. https://doi.org/10.1016/j.tplants.2024.04.006.

Han, Z., Sun, Y., and Chai, J. (2014). Structural insight into the activation of plant receptor kinases. *Current Opinion in Plant Biology* 20: 55–63. https://doi.org/10.1016/j.pbi.2014.04.008.

Han, Z., Xiong, D., Schneiter, R., and Tian, C. (2023). The function of plant PR1 and other members of the CAP protein superfamily in plant–pathogen interactions. *Molecular Plant Pathology* 24 (6): 651–668. https://doi.org/10.1111/mpp.13320.

Hanemian, M., Barlet, X., Sorin, C. et al. (2016). *Arabidopsis CLAVATA 1 and CLAVATA 2 receptors contribute to Ralstonia solanacearum pathogenicity through a miR169-dependent pathway*. *New Phytologist* 211 (2): 502–515. https://doi.org/10.1111/nph.13913.

Hara, K., Yokoo, T., Kajita, R. et al. (2009). Epidermal cell density is autoregulated via a secretory peptide, EPIDERMAL PATTERNING FACTOR 2 in Arabidopsis leaves. *Plant and Cell Physiology* 50 (6): 1019–1031. https://doi.org/10.1093/pcp/pcp068.

Haruta, M., Sabat, G., Stecker, K. et al. (2014). A peptide hormone and its receptor protein kinase regulate plant cell expansion. *Science* 343 (6169): 408–411. https://doi.org/10.1126/science.1244454.

He, L., Wu, L., and Li, J. (2024). Sulfated peptides and their receptors: key regulators of plant development and stress adaptation. *Plant Communications* 5 (6): https://doi.org/10.1016/j.xplc.2024.100918.

Hirakawa, Y. and Sawa, S. (2019). Diverse function of plant peptide hormones in local signaling and development. *Current Opinion in Plant Biology* 51: 81–87. https://doi.org/10.1016/j.pbi.2019.04.005.

Hsiao, Y.-C. and Yamada, M. (2020). The roles of peptide hormones and their receptors during plant root development. *Genes* 12 (1): 22. https://doi.org/10.3390/genes12010022.

Kakita, M., Murase, K., Iwano, M. et al. (2008). Two distinct forms of M-locus protein kinase localize to the plasma membrane and interact directly with S-locus receptor kinase to transduce self-incompatibility signaling in *Brassica rapa*. *The Plant Cell* 19 (12): 3961–3973. https://doi.org/10.1105/tpc.106.049999.

Kazan, K. (2015). Diverse roles of jasmonates and ethylene in abiotic stress tolerance. *Trends in Plant Science* 20 (4): 219–229. https://doi.org/10.1016/j.tplants.2015.02.001.

Kereszt, A., Mergaert, P., Montiel, J. et al. (2018). Impact of plant peptides on symbiotic nodule development and functioning. *Frontiers in Plant Science* 9: 1026. https://doi.org/10.3389/fpls.2018.01026.

Kim, E., Kim, J., Hong, W. et al. (2023). Rice pollen-specific OsRALF17 and OsRALF19 are essential for pollen tube growth. *Journal of Integrative Plant Biology* 65 (9): 2218–2236. https://doi.org/10.1111/jipb.13508.

Kong, Q., Sun, T., Qu, N. et al. (2016). Two redundant receptor-like cytoplasmic kinases function downstream of pattern recognition receptors to regulate activation of SA biosynthesis in *Arabidopsis*. *Plant Physiology* 171 (2): 1344–1354. https://doi.org/10.1104/pp.15.01954.

Lan, Z., Song, Z., Wang, Z. et al. (2023). Antagonistic RALF peptides control an intergeneric hybridization barrier on *Brassicaceae* stigmas. *Cell* 186 (22): 4773–4787.e12. https://doi.org/10.1016/j.cell.2023.09.003.

Lee, K.P., Liu, K., Kim, E.Y. et al. (2020). Plant natriuretic peptide A and its putative receptor PNP-R2 antagonize salicylic acid–mediated signaling and cell death. *The Plant Cell* 32 (7): 2237–2250. https://doi.org/10.1105/tpc.20.00018.

Li, C., Liu, X., Qiang, X. et al. (2018). EBP1 nuclear accumulation negatively feeds back on FERONIA-mediated RALF1 signaling. *PLoS Biology* 16 (10): e2006340. https://doi.org/10.1371/journal.pbio.2006340.

Li, L., Chen, H., Alotaibi, S.S. et al. (2022a). Ralf1 peptide triggers biphasic root growth inhibition upstream of auxin biosynthesis. *Proceedings of the National Academy of Sciences of the United States of America* 119 (31): e2121058119. https://doi.org/10.1073/pnas.2121058119.

Li, Y., Liu, Y., Jin, L., and Peng, R. (2022b). Crosstalk between Ca2+ and other regulators assists plants in responding to abiotic stress. *Plants (Basel)* 11 (10): 1351. https://doi.org/10.3390/plants11101351.

Li, Y., Zeng, H., Xu, F. et al. (2022c). H+ -ATPases in plant growth and stress responses. *Annual Review of Plant Biology* 73 (1): 495–521. https://doi.org/10.1146/annurev-arplant-102820-114551.

Liang, X. and Zhou, J.-M. (2018). Receptor-like cytoplasmic kinases: central players in plant receptor kinase–mediated signaling. *Annual Review of Plant Biology* 69: 267–299. https://doi.org/10.1146/annurev-arplant-042817-040540.

Lin, W., Ma, X., Shan, L., and He, P. (2013). Big roles of small kinases: the complex functions of receptor-like cytoplasmic kinases in plant immunity and development. *Journal of Integrative Plant Biology* 55 (12): 1188–1197. https://doi.org/10.1111/jipb.12071.

Liu, Z., Wu, Y., Yang, F. et al. (2013). BIK1 interacts with PEPRs to mediate ethylene-induced immunity. *Proceedings of the National Academy of Sciences* 110 (15): 6205–6210. https://doi.org/10.1073/pnas.1215543110.

Liu, F., Zhang, X., Lu, C. et al. (2015). Non-specific lipid transfer proteins in plants: presenting new advances and an integrated functional analysis. *Journal of Experimental Botany* 66 (19): 5663–5681. https://doi.org/10.1093/jxb/erv313.

Liu, P., Haruta, M., Minkoff, B.B., and Sussman, M.R. (2018). Probing a plant plasma membrane receptor kinase's three-dimensional structure using mass spectrometry-based protein footprinting. *Biochemistry* 57 (34): 5159–5168. https://doi.org/10.1021/acs.biochem.8b00471.

Liu, J., Li, J., and Shan, L. (2020). SERKs. *Current Biology* 30 (7): R293–R294. https://doi.org/10.1016/j.cub.2020.01.043.

Liu, C., Shen, L., Xiao, Y. et al. (2021). Pollen PCP-B peptides unlock a stigma peptide–receptor kinase gating mechanism for pollination. *Science* 372 (6538): 171–175. https://doi.org/10.1126/science.abc6107.

Liu, L., Song, W., Huang, S. et al. (2022). Extracellular pH sensing by plant cell-surface peptide-receptor complexes. *Cell* 185 (18): 3341–3355.e13. https://doi.org/10.1016/j.cell.2022.07.012.

Liu, L., Liu, X., Bai, Z. et al. (2024). Small but powerful: RALF peptides in plant adaptive and developmental responses. *Plant Science* 343: 112085. https://doi.org/10.1016/j.plantsci.2024.112085.

Lu, D., Wu, S., Gao, X. et al. (2010). A receptor-like cytoplasmic kinase, BIK1, associates with a flagellin receptor complex to initiate plant innate immunity. *Proceedings of the National Academy of Sciences* 107 (1): 496–501. https://doi.org/10.1073/pnas.0909705107.

Lu, X., Shi, H., Ou, Y. et al. (2020). RGF1-RGI1, a peptide-receptor complex, regulates *Arabidopsis* root meristem development via a MAPK signaling cascade. *Molecular Plant* 13 (11): 1594–1607. https://doi.org/10.1016/j.molp.2020.09.005.

Lu, F., Wang, X., Liu, B. et al. (2025). A c-terminally encoded peptide, MeCEP6, promotes nitrate uptake in cassava roots. *Plants* 14 (8): 1264. https://doi.org/10.3390/plants14081264.

Ma, D., Endo, S., Betsuyaku, S. et al. (2020). CLE2 regulates light-dependent carbohydrate metabolism in *Arabidopsis* shoots. *Plant Molecular Biology* 104 (6): 561–574. https://doi.org/10.1007/s11103-020-01059-y.

Ma, H., Gao, Y., Wang, Y. et al. (2022). Regulatory mechanisms of mitogen-activated protein kinase cascades in plants: more than sequential phosphorylation. *International Journal of Molecular Sciences* 23: https://doi.org/10.3390/ijms23073572.

Matsubayashi, Y. (2011). Small post-translationally modified peptide signals in *Arabidopsis*. *The Arabidopsis Book* 2011 (9): https://doi.org/10.1199/tab. 0150.

Matsubayashi, Y. (2014). Post translationally modified small-peptide signals in plants. *Annual Review of Plant Biology* 65: 385–413. https://doi.org/10.1146/annurev-arplant-050312-120122.

Matsubayashi, Y. and Sakagami, Y. (2006). Peptide hormones in plants. *Annual Review of Plant Biology* 57: 649–674. https://doi.org/10.1146/annurev.arplant.56.032604.144204.

Matsuzaki, Y., Ogawa-Ohnishi, M., Mori, A., and Matsubayashi, Y. (2010). Secreted peptide signals required for maintenance of root stem cell niche in *Arabidopsis*. *Science* 329 (5995): 1065–1067. https://doi.org/10.1126/science.1191132.

Meng, L., Buchanan, B.B., Feldman, L.J., and Luan, S. (2012). CLE-like (CLEL) peptides control the pattern of root growth and lateral root development in *Arabidopsis*. *Proceedings of the National Academy of Sciences* 109 (5): 1760–1765. https://doi.org/10.1073/pnas.1119864109.

Mohanta, T.K., Bashir, T., Hashem, A. et al. (2018). Early events in plant abiotic stress signaling: interplay between calcium, reactive oxygen species and phytohormones. *Journal of Plant Growth Regulation* 37 (4): 1033–1049. https://doi.org/10.1007/s00344-018-9833-8.

Mosher, S. and Kemmerling, B. (2013). PSKR1 and PSY1R-mediated regulation of plant defense responses. *Plant Signaling & Behavior* 8 (5): e24119. https://doi.org/10.4161/psb.24119.

Motomitsu, A., Sawa, S., and Ishida, T. (2015). Plant peptide hormone signalling. *Essays in Biochemistry* 58: 115–131. https://doi.org/10.1042/bse0580115.

Murphy, E., Vu, L.D., Van Den Broeck, L. et al. (2016). RALFL34 regulates formative cell divisions in *Arabidopsis* pericycle during lateral root initiation. *Journal of Experimental Botany* 67 (16): 4863–4875. https://doi.org/10.1093/jxb/erw281.

Naithani, S., Ripoll, D., and Nasrallah, J.B. (2006). The S-locus cysteine-rich peptide SCR/SP11. In: *Handbook of Biologically Active Peptides*, 41–47. Elsevier https://doi.org/10.1016/B978-012369442-3/50012-X.

Narváez-Vásquez, J., Orozco-Cárdenas, M.L., and Ryan, C.A. (2007). Systemic wound signaling in tomato leaves is cooperatively regulated by systemin and hydroxyproline-rich glycopeptide signals. *Plant Molecular Biology* 65 (6): 711–718. https://doi.org/10.1007/s11103-007-9242-z.

Ogawa-Ohnishi, M., Yamashita, T., Kakita, M. et al. (2022). Peptide ligand-mediated trade-off between plant growth and stress response. *Science* 378 (6616): 175–180. https://doi.org/10.1126/science.abq5735.

Ogilvie, H.A., Imin, N., and Djordjevic, M.A. (2014). Diversification of the C-TERMINALLY ENCODED PEPTIDE (CEP) gene family in angiosperms, and evolution of plant-family specific CEP genes. *BMC Genomics* 15 (1): 870. https://doi.org/10.1186/1471-2164-15-870.

Okamoto, S., Kawasaki, A., Makino, Y. et al. (2022). Long-distance translocation of CLAVATA3/ESR-related 2 peptide and its positive effect on roots sucrose status. *Plant Physiology* 189 (4): 2357–2367. https://doi.org/10.1093/plphys/kiac227.

Okuda, S., Tsutsui, H., Shiina, K. et al. (2009). Defensin-like polypeptide LUREs are pollen tube attractants secreted from synergid cells. *Nature* 458 (7236): 357–361. https://doi.org/10.1038/nature07882.

Olsson, V., Joos, L., Zhu, S. et al. (2019). Look closely, the beautiful may be small: precursor-derived peptides in plants. *Annual Review of Plant Biology* 70: 153–186. https://doi.org/10.1146/annurev-arplant-042817-040413.

Ou, Y., Lu, X., Zi, Q. et al. (2016). RGF1 INSENSITIVE 1 to 5, a group of LRR receptor-like kinases, are essential for the perception of root meristem growth factor 1 in *Arabidopsis thaliana*. *Cell Research* 26 (6): 686–698. https://doi.org/10.1038/cr.2016.63.

Pearce, G. (2011). Systemin, hydroxyproline-rich systemin and the induction of protease inhibitors. *Current Protein and Peptide Science* 12 (5): 399–408. https://doi.org/10.2174/138920311796391106.

Pearce, G., Yamaguchi, Y., Munske, G., and Ryan, C.A. (2010). Structure–activity studies of RALF, rapid alkalinization factor, reveal an essential – YISY – motif. *Peptides* 31 (11): 1973–1977. https://doi.org/10.1016/j.peptides.2010.08.012.

Poitout, A., Crabos, A., Petřík, I. et al. (2018). Responses to systemic nitrogen signaling in *Arabidopsis* roots involve trans-zeatin in shoots. *The Plant Cell* 30 (6): 1243–1257. https://doi.org/10.1105/tpc.18.00011.

Posner, B.I. and Laporte, S.A. (2010). Cellular signalling: peptide hormones and growth factors. In: *Progress in Brain Research*, vol. 181, 1–16. Elsevier https://doi.org/10.1016/S0079-6123(08)81001-1.

Ravi, B., Foyer, C.H., and Pandey, G.K. (2023). The integration of reactive oxygen species (ROS) and calcium signalling in abiotic stress responses. *Plant, Cell & Environment* 46 (7): 1985–2006. https://doi.org/10.1111/pce.14596.

Roy, S., Griffiths, M., Torres-Jerez, I. et al. (2022). Application of synthetic peptide CEP1 increases nutrient uptake rates along plant roots. *Frontiers in Plant Science* 12: 793145. https://doi.org/10.3389/fpls.2021.793145.

Roychoudhury, A., Manna, S., Banerjee, D. et al. (2024). Plant elicitor peptide mediated signalling cascades during plant–pathogen interaction. In: *Plant Elicitor Peptides* (ed. S. Singh and R. Mehrotra), 49–98. Singapore: Springer Nature.

Schaller, A. and Ryan, C.A. (1996). Systemin – a polypeptide defense signal in plants. *BioEssays* 18 (1): 27–33. https://doi.org/10.1002/bies.950180108.

Schoenaers, S., Lee, H.K., Gonneau, M. et al. (2024). Rapid alkalinization factor 22 has a structural and signalling role in root hair cell wall assembly. *Nature Plants* 10 (3): 494–511. https://doi.org/10.1038/s41477-024-01637-8.

Shao, Y., Yu, X., Xu, X. et al. (2020). The YDA-MKK4/MKK5-MPK3/MPK6 cascade functions downstream of the RGF1-RGI ligand–receptor pair in regulating mitotic activity in root apical meristem. *Molecular Plant* 13 (11): 1608–1623. https://doi.org/10.1016/j.molp.2020.09.004.

Sharma, S. and Singh, S. (2024). Inceptin: exploring its role as a peptide elicitor in plant defense mechanisms. In: *Plant Elicitor Peptides: New Tool for Sustainable Agriculture* (ed. S. Singh and R. Mehrotra), 99–111. Singapore: Springer Nature. https://doi.org/10.1007/978-981-97-6374-0_4.

Shen, X., Stührwohldt, N., and Lin, C. (2023). The research process of PSK biosynthesis, signaling transduction, and potential applications in *Brassica napus*. *Plants* 12 (17): Article 17. https://doi.org/10.3390/plants12173075.

Shiu, S.-H. and Bleecker, A.B. (2001). Receptor-like kinases from *Arabidopsis* form a monophyletic gene family related to animal receptor kinases. *Proceedings of the National Academy of Sciences* 98 (19): 10763–10768. https://doi.org/10.1073/pnas.181141598.

Slezina, M.P. and Odintsova, T.I. (2023). Plant antimicrobial peptides: insights into structure-function relationships for practical applications. *Current Issues in Molecular Biology* 45 (4): Article 4. https://doi.org/10.3390/cimb45040239.

Smith, S., Zhu, S., Joos, L. et al. (2020). The CEP5 peptide promotes abiotic stress tolerance, as revealed by quantitative proteomics, and attenuates the AUX/IAA equilibrium in *Arabidopsis*. *Molecular & Cellular Proteomics* 19 (8): 1248–1262. https://doi.org/10.1074/mcp.RA119.001826.

Sprunck, S., Rademacher, S., Vogler, F. et al. (2012). Egg cell–secreted EC1 triggers sperm cell activation during double fertilization. *Science* 338 (6110): 1093–1097. https://doi.org/10.1126/science.1223944.

Stahl, Y., Grabowski, S., Bleckmann, A. et al. (2013). Moderation of *Arabidopsis* root stemness by CLAVATA1 and ARABIDOPSIS CRINKLY4 receptor kinase complexes. *Current Biology* 23 (5): 362–371. https://doi.org/10.1016/j.cub.2013.01.045.

Stegmann, M., Zecua-Ramirez, P., Ludwig, C. et al. (2022). RGI-golven signaling promotes cell surface immune receptor abundance to regulate plant immunity. *EMBO Reports* 23 (5): e53281. https://doi.org/10.15252/embr.202153281.

Stührwohldt, N. and Schaller, A. (2019). Regulation of plant peptide hormones and growth factors by post-translational modification. *Plant Biology* 21 (S1): 49–63. https://doi.org/10.1111/plb.12881.

Sui, Z., Wang, T., Li, H. et al. (2016). Overexpression of peptide-encoding *OsCEP6.1* results in pleiotropic effects on growth in rice (*O. sativa*). *Frontiers in Plant Science* 7: 228. https://doi.org/10.3389/fpls.2016.00228.

Sun, T. and Zhang, Y. (2022). MAP kinase cascades in plant development and immune signaling. *EMBO Reports* 23 (2): e53817. https://doi.org/10.15252/embr.202153817.

Takahashi, F., Suzuki, T., Osakabe, Y. et al. (2018). A small peptide modulates stomatal control via abscisic acid in long-distance signalling. *Nature* 556 (7700): 235–238. https://doi.org/10.1038/s41586-018-0009-2.

Takeuchi, H. (2021). The role of diverse LURE-type cysteine-rich peptides as signaling molecules in plant reproduction. *Peptides* 142: 170572. https://doi.org/10.1016/j.peptides.2021.170572.

Taleski, M., Imin, N., and Djordjevic, M.A. (2018). CEP peptide hormones: key players in orchestrating nitrogen-demand signalling, root nodulation, and lateral root development. *Journal of Experimental Botany* 69 (8): 1829–1836. https://doi.org/10.1093/jxb/ery037.

Tang, W., Ezcurra, I., Muschietti, J., and McCormick, S. (2002). A cysteine-rich extracellular protein, LAT52, interacts with the extracellular domain of the pollen receptor kinase LePRK2 [W]. *The Plant Cell* 14 (9): 2277–2287. https://doi.org/10.1105/tpc.003103.

Tavormina, P., De Coninck, B., Nikonorova, N. et al. (2015). The plant peptidome: an expanding repertoire of structural features and biological functions. *The Plant Cell* 27 (8): 2095–2118. https://doi.org/10.1105/tpc.15.00440.

Tognetti, V.B., Bielach, A., and Hrtyan, M. (2017). Redox regulation at the site of primary growth: auxin, cytokinin and ROS crosstalk. *Plant, Cell & Environment* 40 (11): 2586–2605. https://doi.org/10.1111/pce.13021.

Tost, A.S., Kristensen, A., Olsen, L.I. et al. (2021). The PSY peptide family—expression, modification and physiological implications. *Genes* 12 (2): Article 2. https://doi.org/10.3390/genes12020218.

Vie, A.K., Najafi, J., Liu, B. et al. (2015). The IDA/IDA-LIKEandPIP/PIP-LIKEgene families in Arabidopsis: phylogenetic relationship, expression patterns, and transcriptional effect of the PIPL3 peptide. *Journal of Experimental Botany* 66 (17): 5351–5365. https://doi.org/10.1093/jxb/erv285.

Wang, X., Zhang, N., Zhang, L. et al. (2021). Perception of the pathogen-induced peptide RGF7 by the receptor-like kinases RGI4 and RGI5 triggers innate immunity in *Arabidopsis thaliana*. *New Phytologist* 230 (3): 1110–1125. https://doi.org/10.1111/nph.17197.

Wang, X., Chen, M., Li, J. et al. (2024). The SCOOP-MIK2 immune pathway modulates *Arabidopsis* root growth and development by regulating PIN-FORMED abundance and auxin transport. *The Plant Journal* 120 (1): 318–334. https://doi.org/10.1111/tpj.16988.

Webb, M.C. and Gunning, B.E.S. (1994). Cell biology of embryo sac development in *Arabidopsis*. In: *Genetic Control of Self-Incompatibility and Reproductive Development in Flowering Plants*, vol. 2 (ed. E.G. Williams, A.E. Clarke, and R.B. Knox), 461–485. Springer Netherlands https://doi.org/10.1007/978-94-017-1669-7_22.

Wei, X., Wang, Y., Zhang, S. et al. (2022). Structural analysis of receptor-like kinase SOBIR1 reveals mechanisms that regulate its phosphorylation-dependent activation. *Plant Communications* 3 (2): 100301. https://doi.org/10.1016/j.xplc.2022.100301.

Weidmann, J. and Craik, D.J. (2016). Discovery, structure, function, and applications of cyclotides: circular proteins from plants. *Journal of Experimental Botany* 67 (16): 4801–4812. https://doi.org/10.1093/jxb/erw210.

Whitford, R., Fernandez, A., De Groodt, R. et al. (2008). Plant CLE peptides from two distinct functional classes synergistically induce division of vascular cells. *Proceedings of the National Academy of Sciences* 105 (47): 18625–18630. https://doi.org/10.1073/pnas.0809395105.

Whitford, R., Fernandez, A., Tejos, R. et al. (2012). GOLVEN secretory peptides regulate auxin carrier turnover during plant gravitropic responses. *Developmental Cell* 22 (3): 678–685. https://doi.org/10.1016/j.devcel.2012.02.002.

Xu, R., Li, Y., Sui, Z. et al. (2021). A C-terminal encoded peptide, ZmCEP1, is essential for kernel development in maize. *Journal of Experimental Botany* 72 (15): 5390–5406. https://doi.org/10.1093/jxb/erab224.

Yang, W., Zhang, B., Qi, G. et al. (2019). Identification of the phytosulfokine receptor 1 (OsPSKR1) confers resistance to bacterial leaf streak in rice. *Planta* 250 (5): 1603–1612. https://doi.org/10.1007/s00425-019-03238-8.

Yu, F., Qian, L., Nibau, C. et al. (2012). FERONIA receptor kinase pathway suppresses abscisic acid signaling in *Arabidopsis* by activating ABI2 phosphatase. *Proceedings of the National Academy of Sciences* 109 (36): 14693–14698. https://doi.org/10.1073/pnas.1212547109.

Yamaguchi, Y.L., Ishida, T., and Sawa, S. (2016). CLE peptides and their signaling pathways in plant development. *Journal of Experimental Botany* 67 (16): 4813–4826. https://doi.org/10.1093/jxb/erw208.

Zelman, A.K. and Berkowitz, G.A. (2023). Plant elicitor peptide (pep) signaling and pathogen defense in tomato. *Plants* 12: https://doi.org/10.3390/plants12152856.

Zhang, H., Han, Z., Song, W., and Chai, J. (2016). Structural insight into recognition of plant peptide hormones by receptors. *Molecular Plant* 9 (11): 1454–1463. https://doi.org/10.1016/j.molp.2016.10.002.

Zhang, L., Shi, X., Zhang, Y. et al. (2019). CLE9 peptide-induced stomatal closure is mediated by abscisic acid, hydrogen peroxide, and nitric oxide in *ARABIDOPSIS THALIANA*. *Plant, Cell & Environment* 42 (3): 1033–1044. https://doi.org/10.1111/pce.13475.

Zhang, M., Liu, Q., Yang, X. et al. (2020a). CRISPR/Cas9-mediated mutagenesis of Clpsk1 in watermelon to confer resistance to *Fusarium oxysporum f.sp. niveum*. *Plant Cell Reports* 39 (5): 589–595. https://doi.org/10.1007/s00299-020-02516-0.

Zhang, X., Yang, Z., Wu, D., and Yu, F. (2020b). RALF–FERONIA signaling: linking plant immune response with cell growth. *Plant Communications* 1 (4): 100084. https://doi.org/10.1016/j.xplc.2020.100084.

Zhang, L., Ren, Y., Xu, Q. et al. (2021). SiCEP3, a c-terminally encoded peptide from *Setaria italica*, promotes ABA import and signaling. *Journal of Experimental Botany* 72 (18): 6260–6273. https://doi.org/10.1093/jxb/erab267.

Zhang, L., Liu, J., Cheng, J. et al. (2022a). lncRNA7 and lncRNA2 modulate cell wall defense genes to regulate cotton resistance to Verticillium wilt. *Plant Physiology* 189 (1): 264–284. https://doi.org/10.1093/plphys/kiac041.

Zhang, Z., Liu, C., Li, K. et al. (2022b). CLE14 functions as a "brake signal" to suppress age-dependent and stress-induced leaf senescence by promoting JUB1-mediated ROS scavenging in *Arabidopsis*. *Molecular Plant* 15 (1): 179–188. https://doi.org/10.1016/j.molp.2021.09.006.

Zhang, P., Zhao, J., Zhang, W. et al. (2024a). Sulfated peptides: key players in plant development, growth, and stress responses. *Frontiers in Plant Science* 15: 1474111. https://doi.org/10.3389/fpls.2024.1474111.

Zhang, Y., Duan, X., Wang, Z. et al. (2024b). CEPs suppress auxin signaling but promote cytokinin signaling to inhibit root growth in *Arabidopsis*. *Biochemical and Biophysical Research Communications* 711: 149934. https://doi.org/10.1016/j.bbrc.2024.149934.

Zhang, Z., Han, H., Zhao, J. et al. (2025). Peptide hormones in plants. *Molecular Horticulture* 5 (1): 7. https://doi.org/10.1186/s43897-024-00134-y.

Zhong, S., Li, L., Wang, Z. et al. (2022). RALF peptide signaling controls the polytubey block in *Arabidopsis*. *Science* 375 (6578): 290–296. https://doi.org/10.1126/science.abl4683.

Zhu, S., Estévez, J.M., Liao, H. et al. (2020a). The RALF1–FERONIA complex phosphorylates eIF4E1 to promote protein synthesis and polar root hair growth. *Molecular Plant* 13 (5): 698–716. https://doi.org/10.1016/j.molp.2019.12.014.

Zhu, S., Martínez Pacheco, J., Estevez, J.M., and Yu, F. (2020b). Autocrine regulation of root hair size by the RALF-FERONIA-RSL4 signaling pathway. *New Phytologist* 227 (1): 45–49. https://doi.org/10.1111/nph.16497.

3

Post-Translational Modifications of Peptide Hormones and Their Functionality

Gagan Kumar Panigrahi[1], *Sanjoy Majumder*[2], *Kunja Bihari Satapathy*[3], *and Annapurna Sahoo*[4]

[1] *Department of Zoology, School of Applied Sciences, Centurion University of Technology and Management, Bhubaneswar, Odisha, India*
[2] *School of Biotechnology, Centurion University of Technology and Management, Bhubaneswar, Odisha, India*
[3] *Department of Botany, School of Applied Sciences, Centurion University of Technology and Management, Bhubaneswar, Odisha, India*
[4] *Department of Zoology, Jatni College, Utkal University, Khordha, Odisha, India*

3.1 Introduction

Plant peptide hormones are fascinating and key signaling molecules composed of a short chain of amino acids. The production of peptide hormones from preproproteins in plants involves different significant steps including the removal of the signal peptide and folding, proteolytic processing by specific enzymes (Zhang et al. 2025). They are diverse in size, starting from a few amino acids to much larger molecules. In some cases, these peptide hormones may have lengths as long as 250 amino acids. The structural dissimilarity in length of these peptide hormones implies a wide range of biological functions and regulatory roles, due to the fact that the structure of the hormone usually reflects its receptor affinity, stability, and activity in physiological processes.

Peptide hormones usually come from larger precursor proteins that are processed in a number of intricate proteolytic steps (Figure 3.1). Within the plant, these small but powerful peptides are important at orchestrating a number of physiological and developmental processes (Zhang et al. 2025; Chen et al. 2023; Hirakawa and Sawa 2019). Peptide hormones in plants can be categorized into different types and families based on criteria including peptide sequences, secretion pathways, functions, origins, biological activities, and structural features. In general, plant peptide hormones are divided into two types: secretory or non-secretory peptides based on their secretions and localizations in plant cells.As messenger, they help plants to effectively interpret and respond to a multitude of internal cues and external environmental stimuli (Xie et al. 2022; Wang et al. 2016).The physiological processes of plants are significantly dependent on peptide hormones, which are key regulators of growth, development, stress responses, and defense mechanisms. These small peptides, which are made up of a few amino acid sequences, work using intricate signal transduction pathways to modulate various cellular functions (Chaulagain and Frugoli 2021; Corcilius et al. 2017). Certain of these peptide hormones are involved in controlling critical

Peptide Hormones in Plant Development and Stress Responses, First Edition. Edited by Vijay Pratap Singh and Samiksha Singh.

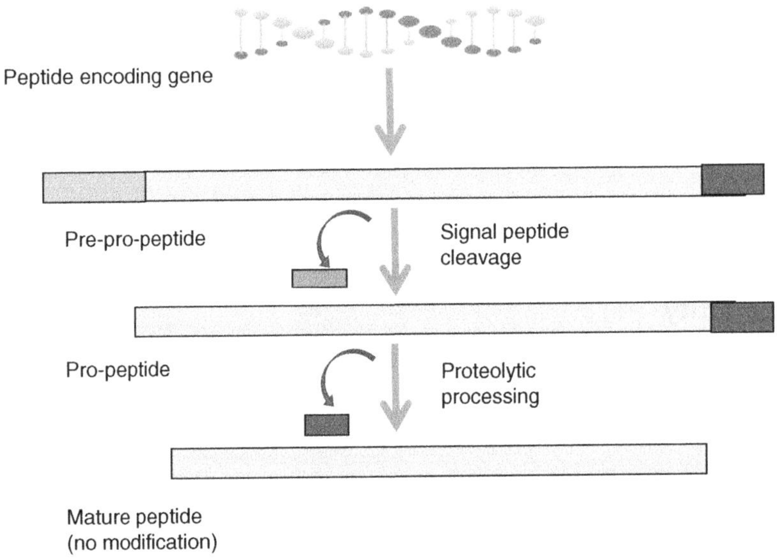

Figure 3.1 Biosynthetic pathway of peptide hormone. This diagram illustrates the sequential processing steps from a precursor protein to a mature peptide hormone through enzymatic cleavage.

developmental events including germination, flowering, and fruit ripening (Fang et al. 2024; Zhong et al. 2022; Clark et al. 1993, 1995). For example, they can regulate the differentiation of different tissue types and the formation of organs necessary for the plant's basic structure and functioning. In addition to development, peptide hormones facilitate a quick-responseto environmental stress including pest infestations, fungal infections, and abiotic stresses such as drought and salinity (Kang et al. 2022a; Panigrahi et al. 2021a). This hormone activates particular defense pathways of the plant that strengthen its ability to survive through such adversities and recover to achieve successful reproduction (Zelman and Berkowitz 2023; Kim et al. 2021; Mosher et al. 2013). Small peptides are essential signaling molecules used to do vital intercellular communication across a wide range of developmental processes in plants (Gancheva et al. 2019; Liu et al. 2022a). The first such functional peptide ligand extracted from tomato leaves was systemin, an 18-amino acid-long peptide hormone implicated in theresponse to wounding (Ryan and Pearce 2003; Zhang et al. 2016). Since then, various other peptide hormones have been discovered in plants that contribute to several developmental functions, such as CLAVATA3 (CLV3)/embryo surrounding region peptide (CLE), which plays a role in maintaining stem cells; phytosulfokine (PSK) and plant peptide containing sulfated tyrosine (PSY), which are linked to cell proliferation; RGF or CLE peptide hormones are essential for sustaining the root stem cell niche; rapid alkalinization factor (RALF), which is involved in calcium-mediated signaling and the regulation of root growth (Bashyal et al. 2024; Carbonnel et al. 2022; Stührwohldt et al. 2021; Tost et al. 2021; Fang et al. 2021).

Many interesting discoveries were made regarding peptide hormones and their receptors in multicellular organisms, which have contributed to great advances in understanding growth and development regulation. Peptides themselves are particularly act as very important signals between cells, as they are produced rapidly and possess both structural and functional diversity (De Coninck and De Smet 2016; Tavormina et al. 2015; Carbonnel et al. 2023). Post-translational modifications and proteolytic processing of these peptides often contribute to the generation of smaller oligopeptide hormones with distinct roles and activities. These small peptides, with such modifications, form the largest group of peptide hormones in the plant kingdom and represent the most important group of peptides in plant growth and for plant responses to environmental stimuli and homeostasis (Vanyushin et al. 2017; Xiao and Zhou 2023). Hormones can generally be defined as either of the following types: steroid hormones or aminoacid-based (i.e. amine, peptide, or protein). Peptide hormones differ from conventional plant hormones like auxins and cytokinins in that they are formed from precursor proteins, which undergo significant post-translational modifications (PTMs) and are enzymatically modified to become biologicallyactive. Peptide hormones are important for maintaining physiological and developmental activities in plants (Simon and Dresselhaus 2015; Xu et al. 2023; Yamaguchi et al. 2016). Despite their small sizes, peptide hormones affect significant plant processes such as growth, defense, and differentiation (Albert 2013; Panigrahi and Satapathy 2020c). Being sessile organisms, it is essential for the plants to make robust response mechanisms to their environmental and internal factors by influencing specific cellular functions through signal transduction pathways (He et al. 2024; Lv and Yang 2023).

In addition to maintaining important developmental processes, these hormones enable plants to adapt to environmental stress, ensuring their survival and reproductive success in an ever-changing world. Besides the proteolytic processing, the PTMs of amino acids also play a significant role in the maturation of peptide hormones (Kaufmann and Sauter 2019; Kaeothip et al. 2013).Tyrosine sulfation is one of the key modifications thatinvolvesadding a sulfate group to tyrosine residues so that the binding affinity of the peptides to their receptors is enhanced. Proline hydroxylation also takes place when hydroxyl groups are added to proline amino acids, to stabilize the peptide's structure and interact with other molecules (Guillou et al. 2022). Hydroxyproline arabinosylation is a key modification that follows this process and involves the attachment of arabinose sugars to hydroxyprolines (Xu et al. 2015; Shinohara and Matsubayashi 2010). These modifications work together not only to increase the level of overall stability ofthe peptides but also to optimize functional receptor binding of the peptides (Yoshida et al. 2021).

3.2 Key Functions of Peptide Hormones

3.2.1 Regulation of Growth and Development

Peptide hormones have a great influence on different physiological processes, such as growth, cell division, and differentiation in plants. They are important for plant developmental stages in such as how they grow, adjust to their environment, and form complex structures like roots, stems, leaves, and flowers (Zhang et al. 2023, 2024a; Huang et al. 2022; Figure 3.2). For instance, CLAVATA3 (CLV3) is one of the significant CLE (CLAVATA3/ESR-RELATED) peptides (Greifenhagen et al. 2024; Wang and Fiers 2010). The specific peptide hormones discussed here are important to the regulation of stem cell

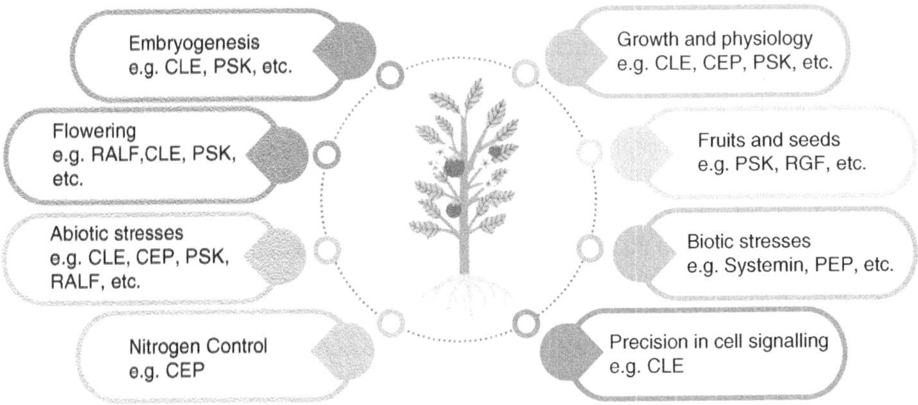

Figure 3.2 The functional diversity of plant peptide hormones. Highlighting critical involvement in growth and development, developing stress tolerance, etc.

homeostasis in the shoot apical meristem(SAM), the growth region from which stems and leaves are produced. CLV3 and CLE peptides coordinate the generation of precise signals that keep stem cells within the meristem undifferentiated (Kang et al. 2022b; Xu et al. 2013). The abilityofstem cells not to differentiate and remain in an undifferentiated state is essential for the proper development of different plant organs,asthey can provide a continuous supply of stem cells capable of differentiating into specialized tissues if necessary. Peptide hormones also play a role in the development of plants besides regulating stem cells (Yamada and Sawa 2013). They coordinate the timing and development of organs, enabling-roots to support the plant through the transport of water and nutrients, stems to providestructure,and leaves to do photosynthesis, etc.(Fedoreyeva 2023; Fedoreyeva et al. 2022; Zhang et al. 2019). Because of its necessity, the precise action of peptide hormones is crucial for the proper formation of plant organs, based on correct developmental pathways.

3.2.2 Coordination of Environmental Responses

Being stationary, plants have to deal with the challenges of living in a fluctuating environment and peptide hormones are found to play a significant role in dealing with many adverse conditions they face. These hormones help plants to adapt in response to various stress factorssuch asdrought, excessive heat, high salinity, or attack by pathogens through managing their responsesignals (Lin et al. 2023; Chen et al. 2023; Panigrahi and Satapathy 2021). For example, a peptide hormone called systemin is activated in response to physical damage caused by herbivores eating plant tissues (Andersen et al. 2018). When systemin is triggered, however, it sparks a cascade of events resulting in the expression of genes involved in plant defense. As a defensive measure, it activates the jasmonic acid (JA) pathway (Lyapina et al. 2024; Combest et al. 2021). Eventually, peptide hormones allow plants to survive and flourish despite the harshest conditions. Plants are stationary organisms, so their defensive response coordination needs to be highly precise, and thisisdone through the help of highly intricate chemical signaling pathways (Ghosh and Roychoudhury

2023; Kuromori et al. 2022). Systemin is synthesized and later released from the wounded cells when the plant is damaged mechanically, for example, by an insect feeding on its tissue (Žárský et al. 2024). The activation of this leads to the subsequent event in a cascade, which is triggered by the JA signaling pathway that is necessary for plant immune responses (Yu et al. 2024a,b; Panigrahi et al. 2021b; Sahoo et al., 2023).

Several of the defensive compounds induced as part of the JA pathway, including protease inhibitors, deter herbivores by interfering with their digestive processes (Fernández-Fernández et al. 2023; Ye et al. 2023). In addition, this also promotes the expression of various defense genes and strengthens the overall immunity of the plant (Bashyal et al. 2024; Liu et al. 2022b; Jung et al. 2020). A major part of immune signaling in plants is done by peptide hormones, which are essential key signaling molecules that help in communicating with other parts of the plant (Rzemieniewski et al. 2024; Murphy et al. 2012; Sahoo et al. 2024). As such, these hormones allow plants to respondtovarious biotic (microorganisms that attack plants, such as bacteria, fungi, and viruses) and abiotic stressors (including drought, extreme temperature, and salinity (Zhang et al. 2022; Li et al. 2024a,; Toruño et al. 2016; Ghosh and Roychoudhury 2023; Aggarwal et al. 2020). This multifaceted response comprises activation of certain specific defense genes andthe production of antimicrobial compounds that facilitate boosted resistance against threats (Pastor-Fernández et al. 2023; Igarashi et al. 2012). But, by modulating these responses, peptide hormones make a plant's ability to adapt and survive in a constantly changing environment much better (Takahashi et al. 2020; Canut et al. 2016). In general, their role is really important to guarantee plant resilience.

3.2.3 Cell to Cell Communications

The plant utilizes peptide hormones as messengers involved in mediating communication between cells. The coordination of physiological activities between different tissues and organs in the plant requires owthese hormones to maintain the functional integrity of the plant (Hirakawa and Sawa 2019; Lindsey 2001). Additionally, peptide hormone signaling between plant cells over large distances aids in an effective communication network that integrates numerous environmental and developmental cues (Johns et al. 2021). For example, coordination of complex physiology is necessary for plants to modify their patterns of growth in response to environmental stresses, against defense mechanisms, and against threat generally (Zhang et al. 2024c; Okamoto et al. 2022;Ohkubo et al. 2017). Thus, peptide hormones are important not only for the normal growth and development of the plant but also for maintaining homeostasis and resilience of plants (Dao et al. 2022). The role played by these peptides in the cell signaling system thus displays the significance of peptide hormones in plant biology.

3.2.4 Regulation of Vascular Development and Patterning

Peptide hormones are critical regulators of the development and organization of a plant's vascular system for the highly important task of transporting water, nutrients, and a variety of hormones. Peptide hormones play a crucial role in this process (Feng et al. 2023; Oldroyd and Leyser 2020; de Bang et al. 2017). A prominent group of these peptides is the CLE (CLAVATA3/EMBRYO SURROUNDING REGION) peptides, involved in vascular patterning and in the formation of vascular bundles, important for the transport network in the plant (Couto and Zipfel 2016; Reid et al. 2011). Several crucial processes rely on CLE peptides, such as setting the appropriate amount of vascular cells as well as regulating their differentiation in the tissue (Zhou et al. 2024; Mens et al. 2021; Ren et al. 2019). CLE peptides influence these processes to make sure that the vascular tissue develops correctly, which is important for effective nutrient and water transport (Wulf et al. 2024; Narasimhan and Simon 2022; Fukuda and Hardtke 2020). These peptides play a role in regulating physiological functions and keep the vascular system in a good working state, enabling plants to be resilient in theirenvironment (Yuan and Wang 2021).

3.2.5 Regulation of Root Development and Architecture

Peptide hormones affect the root growth and development (Sang et al. 2025). The root meristem (or growth region at the root tip), branching patterns, and plant response of the root to different environmental signals such as nutrient availability and soil conditions all require the action of these hormones (Lu and Xiao 2024; Patel et al. 2018; Yamada and Sawa 2013). It was found that the root tip expressed root meristem growth factor (RGF), a sulfated peptide hormone that helps in overcoming the critical tpst-1 root defects (Song et al. 2022; Lee et al. 2019; Matsuzaki et al. 2010; Kim et al. 2017). In addition, CLE (CLV3/ESR-related) peptides and systemin are important for the modulation of root growth in response to some environmental stressors (Ávalos-Rangel et al. 2024; Das et al. 2025; Lebedeva et al. 2020). For instance, in environments that lack nutrients or when drought occurs, these peptides allow the plant to change its root architecture (Ma et al. 2024). In adverse stress conditions, such as limited nutrients, these peptides help the plant to increase lateral root formation or deeper root penetration to provide additional water and nutrient sources from the soil (Taleski et al. 2024; Panigrahi et al. 2024). It is not only efficient for the uptake of nutrients and water but it also improves the survival and resilience capacity for plants in various soil complexities (Chang and Xiao 2025; Nakagami et al. 2024). These peptides enableplants to do well under diverse and often challenging conditions by fine-tuning root architecture in response to physiological and environmental cues.

3.2.6 Hormonal Cross Talk and Integration

Since peptide hormones interact with other hormones such as auxins, cytokinins, gibberellins, and abscisic acid, they are important in the complex web of plant signaling. This cross talk through multiple hormonal signals ensures that signaling pathways are harmoniously integrated together (Salvi et al. 2021; Wang et al. 2016). Because of this, plants can orchestrate the responses in a concerted and balanced fashion such that they adapt well to their environment. For example, the plant defense system is coupled by JA and ethylene to respond promptly to biotic stressors in an integrated manner with a systemin (Ogawa-Ohnishi et al. 2022; Panigrahi and satapathy 2020a; Panigrahi and Satapathy 2020c). Apart from peptides (such as CLV3) that also play a major role in essential processes like cell division and growth in the meristem, which is the place where new cells are put into the works (Kondo et al. 2011). These complex interactions enable the plant to elicit context-specific responses that are both timely, through an integration of many signaling pathways

(Oh et al. 2018; Reichardt et al. 2020). The interaction between sulfated peptides and other phytohormones helps in the complex coordination of growth and developmental processes under stress (Müller and Harrison 2019).

3.2.7 Regulation of Flowering and Reproduction

Plant peptide hormones are critical signaling molecules that are essential in regulating flowering and reproductive processes in plants. These hormones orchestrate floral organ formation including structures that are necessary for successful pollination and fertilization (Stintzi and Schaller 2022). Plant peptide hormones also modulate pollen-pistil interactions, one of which is to influence how pollen recognizes and germinates on the pistil (Taleski et al. 2020; Clark et al. 1993, 1995). Moreover, these hormones also promote the maturation of the fruits and seeds, which in turn influence the process of seed dispersal. Plant reproductive success and fertility are dependent on a proper balance of these signaling molecules and the timing of floral developmental transitions in plants (Mogg 2020). Moreover, RALF peptide hormones are involved in some physiological procedures, for example, cell expansion control, pH homeostasis, and promotion of pollen tube extension (Abarca et al. 2021; Blackburn et al. 2020; Haruta et al. 2014). All of these signaling pathways working together are necessary to coordinate the reproductive processes in flowering plants (Bhalla et al. 2025; Zhong and Qu 2019; Bedinger et al. 2010). Peptide hormones in plants fine-tune the timing of flowering and of other reproductive processes to match them with the most favorable conditions. This is important for the successful and increased seed production, which in turn contributes to the survival and continuity of plant species.

3.2.8 Control of Meristematic Activity and Stem Cell Regulation

The regulation of meristems involves numerous peptide hormones. Meristems are special areas inside the plant that continuously produce new cells (Selby and Jones 2023; Boschiero et al. 2019). These meristems are important for the plant to grow new organs and tissues, which in turn help increase the fitness of the plant in nature. Perhapsthe best-studied group of peptide hormones isthe CLV3 (CLAVATA3) peptides, and they are closely tied to stem cells in the SAM. The SAM is the region at the tip of the plant shoot, which is necessary for producing new shoots, leaves, and flowers (Scott et al. 2024; Laffont et al. 2020; Je et al. 2018). The CLV3 peptides aid in maintaining the balance between the proliferation of the stem cells and their differentiation to specialized tissues (Ogawa et al. 2008). This regulation ensures that the plant survives and can continue to grow, where it can develop structures for reproduction and survival. Besides CLV3, other peptide hormones prevent the meristematic tissues from interacting too extensively with different factors like cell division rates, differentiation pathways, and responses to environmental cues (Phour et al. 2020). Peptide hormones modulate the activity of these meristems to make sure that plants can efficiently balance cell division, i.e. replication and differentiation of cells, by which unspecialized cells become specialized cells with specific function (Ryan et al. 2002). This is crucial for the plant to continually produce new cells and tissues without losing its structural integrity.

3.2.9 Precision in Signaling

Peptide hormones are critical to the process of highly specific signaling at the molecular level within organisms. This precision allows finely tuned responses, both plants and animals to tolerate a plethora of physiological and environmental challenges (Ji et al. 2025; Panigrahi and Satapathy 2020b). For example, the peptide hormone systemin can activate the induction of defense mechanisms in a targeted way, so they can deal with a threat without wasting a lot of energy. The plant thus conserves its resources by localizing these responses, without sacrificing the protective measures (Hirakawa and Sawa 2019). The specificity of peptide hormones is so high that not only is there a fast and effective response to stimuli but itis also energetically efficient, such that the organism can thrive in changing conditions.

3.3 Post-Translational Modifications

Post-translational modification or PTM refers to a wide variety of biochemical processes whereby amino acid residues in a protein are covalently modified after the protein is translated from messenger RNA (Millar et al. 2019; Roy et al. 2018; Figure 3.3). Plant peptide hormones need PTMs that are necessary for their shape, activity, and recognition by their corresponding receptors (Arsova et al. 2018). Such modifications can enhance the efficiency and reactivity of these hormones, allowing them to play a crucial role in many physiological processes in plants (Friso and van Wijk 2015). This is an important regulatory step in terms of protein function, where there is a dynamic response depending on the physiological needs of the organism. After the transcription of gene sequences and the coding regions or exons undergo splicing, the tertiary structure of the newly formed protein is most affected by conformational fluctuations (Matsubayashi 2011). PTM is indeed interesting and one

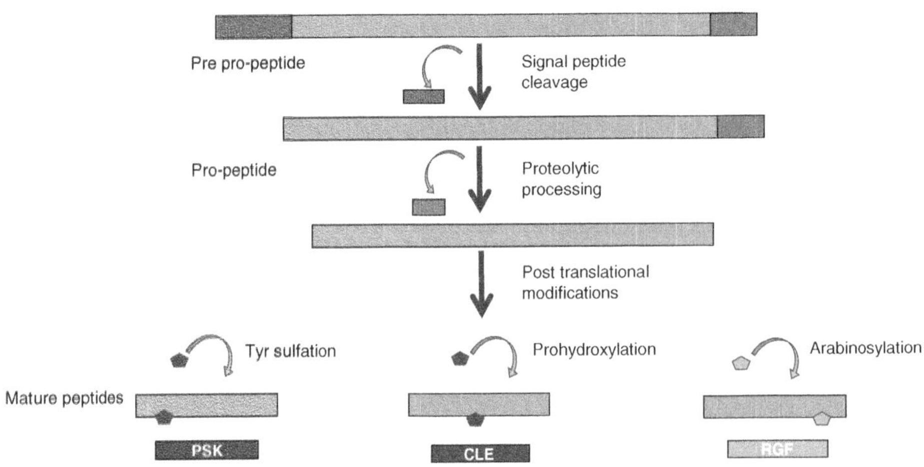

Figure 3.3 This diagram illustrates the biogenesis of plant peptide hormones through sequential enzymatic cleavages and post-translational modifications, converting inactive pre-propeptides into bioactive peptides.

subtype that stands out is the phenomenon of multisite modification, in which one protein has multiple sites treated at once or in sequence. The interplay of PTMs facilitates a more nuanced and adaptable cellular framework, critical for the evolution and adaptation of multicellular life forms (Matsubayashi et al. 2006; Matsubayashi 2012; Motomitsu et al. 2015).

Plants as stationary organisms are always under continuously changing environmental conditions. Abiotic and biotic stresses both affect growth and development in plants (Furumizu and Aalen 2023). Hormones and signaling peptides of various classes also control the postembryonic evolution of plant and their adaptation to environmental factors (Ren et al. 2024; Furumizu et al. 2021).For a long time, the role of peptides in plants, particularly in regulating such vital processes as cellular differentiation, growth, development, and defense mechanisms has been underestimated. Yet for the last two decades, peptide hormones have been recognized as a major set of signaling molecules that allow for communication between cells, both in short distances and in long-range signaling pathways. Small chains of amino acids are responsible for essential roles in regulating large cues of biological processes that provide development, cellular, and intercellular communication.

Plant peptide hormones are usually produced as larger precursor proteins that have been subsequently post-translationally modified. Proteolytic processing is a process in which these precursor proteins are cleaved (processed) to release the mature peptide forms (Olsson et al. 2019; Tabata and Sawa 2014; Srivastava et al. 2008). After proteolytic processing, these peptides generally require further post-translational processing to become biologically active (Matsubayashi 2018). The modifications can include sulfation of tyrosine residues, hydroxylation of proline, or glycosylation of hydroxyproline residues, all of which are necessary for the proper function of the peptides when they are within the plant signaling pathways (Chen et al. 2024; Amano et al. 2007).These modified peptides fine-tune the biochemical responses of plants to change, in order to allow plants to adapt to changes while maximizing their developmental potential (Stintzi and Schaller 2022). Plant peptide hormones that are post-translationally modified are key regulators in the refinement of developmental and adaptive processes in plants (Van et al., 2024; Matsubayashi 2014; Ryšlavá et al. 2013). Neighboring residues and their accessibility on the protein surface affect the types of PTMs. The 20 amino acids differ greatly with respect to general chemical reactivity, as well as types of PTMs. For instance, cysteine (Cys) and lysine (Lys) undergo various kinds of PTMs, while the N-terminal end of proteins isalso subject to diverse kinds of PTMs such as N-terminal cleavage, lipid modifications (acylation), acetylation, and so on (Setty et al. 2024; Chen et al. 2020; Cho et al. 2014). PTMs of these peptide hormones are important for the plant to overcome stress conditions. PSK and RGF peptide interactions with other key plant hormones, brassinosteroids, auxin, salicylic acid, abscisic acid, gibberellins, ethylene, and JAare described and their cooperating roles in plant physiology are explained. However, the sulfated peptides havethe potential to increase crop productivity and resilience (Li et al. 2024b; Hirayama and Mochida 2022).

3.3.1 Tyrosine Sulfation

A specific biochemical modification of tyrosine, tyrosine sulfation is the covalentattachment ofa sulfate (—SO_4) group to the hydroxyl (—OH) group of a tyrosine (Tyr) in the peptide hormone structure. Plant cell contains an enzyme tyrosylprotein sulfotransferase

(TPST), which catalyzes this important reaction in the Golgi apparatus (Royek et al. 2022). Sulfate is donated by 3′-phosphoadenosine 5′-phosphosulfate (PAPS), a high-energy donor compound for a variety of biological reactions. Hydrolysis of specific tyrosine residues, which are individually protected as tyrosine sulfopeptide precursors by the TPST enzyme, is recognized and only certain tyrosine residues are sulfated. Upon recognition, the enzyme can facilitate the transfer of the sulfate group from PAPS to the hydroxyl group of the tyrosine residue to generate a sulfated peptide. After this modification, the sulfated peptides are then typically translocated through the secretory pathway to the extracellular space, where they activate specific receptors for responses. In plants, this type of PTM in particular is important. TPST is responsible for converting PSK, PSY, RGF, and CIF peptides into their respective active forms in Arabidopsis (Zhang et al. 2024a; Nguyen and Kim 2023). Active forms influence plant growth, development, and responses to stress and they are essential forcarrying out these functions in a number of plant species. One of the features of the Arabidopsis TPST is that it functions as a single enzyme performing peptide sulfation and mutants of this enzyme also display a complement of pleiotropic phenotypes, including retarded growth, abnormal pollen tube, and grain development, reduced primary root length with increased lateral density and distorted primary and secondary vascular development (Gao et al. 2023; Nakayama et al. 2017). Closest to these phenotypic changes are four families of peptides: RGF, PSK, PSY, and CIF.

Sulfation modification of several key plant peptide hormones, including PSKs and RGFs,iscrucial for biological activity. These peptides are sulfated to strengthen their binding affinity to their correspondent receptors which is essential for a proper signal transduction (Hao et al. 2023; Wang et al. 2015; Hanai et al. 2000). For instance, enhanced binding efficiency of this receptor, as result of tyrosine sulfation, could significantly improve responsiveness of plant tissues to environmental cues and would help adaptive growth mechanisms. Sulfated peptides are important players in mediating plant responses to many abiotic stresses, for instance, drought, salinity, and extreme temperature variations (Zhou et al. 2022; Villalobos-López et al. 2022). PSK is one example of a peptide hormone that alters the expression of genes essential for stress tolerance mechanisms (Sauter 2015; Loivamaki et al. 2010). PSK regulates these genes, which areessential forbalancinggrowth andadaptation to stressful conditions in plants so they cansurvive under unfavorable environmental conditions. Consistent research has shown that sulfated peptides are critical to osmotic stress tolerance via regulation of cellular mechanisms involved with dehydration and high salinity stress (Yu et al. 2024a,; Gorgues et al. 2022). In a recent study, Arabidopsis plants that express PSK precursors have been shown to exhibit a significant increase in osmotic stress tolerance. It offers great evidence of how sulfated peptides are directly involved in the molecular pathways that affect the response to drought conditions and prove to be important in plant physiology. Sulfation of peptides allows them to enhance the interactions between hormones with their receptors, allowing for better binding and signal response. It is important to have such an improvement of hormone–receptor interactions for the regulation of a large number of physiological pathways (Hsiao and Yamada 2020; Taleski et al. 2018).Furthermore, phosphorylated tyrosyl residues in cell signaling pathways cannot be overstated. As can be seen, these phosphorylations can greatly reduce or even abolish certain cellular responsesto changes in the phosphorylation state, depending on the modifications (Tan et al. 2024; Hsu et al. 2021; Ohki et al. 2011). As a consequence, the phosphorylation status can critically influence the engagement of peptide hormones with their respective receptors and subsequent signaling cascades.

As with all peptide hormones within plants, the PSK peptide hormone regulates cell proliferation and growth to help with plant development and tissue regeneration (van de Sande et al. 2024; Ding et al. 2023). Root growth factors (RGF) are important for root meristem regulation and also for root branching and overall root architecture (Leasure and He 2012). Stomatal development is mainly controlled by the epidermal patterning factor (EPF) hormone, which acts as a signal to receptor kinases for the differentiation of epidermal cells into stomata. This process is necessary for plant gas exchange (Li et al c;Xiong et al. 2022; Takahashi et al. 2018). Plant peptide hormones are quite diverse functionally, largely due to these PTMs, particularly tyrosine sulfation, which isvital for their effect by allowing them to control multiple physiological processes in a specific and precise manner (Stührwohldt and Schaller 2019). Among the most relevant functional significances of tyrosine sulfation arethe enhancement of peptide stability, prevention of rapid degradation, enhancement of receptor-binding affinity (increasing hormone–receptor affinity), and regulation of important developmental processes such as root growth and stomatal patterning.

3.3.2 Proline Hydroxylation

Plant peptide hormones contain a large proportion of PTMs that include proline hydroxylation. The oxidation of proline residues at the fourth carbon position in this complex biochemical process is provided by the enzyme prolyl 4-hydroxylase (P4H), which specifically catalyzes this reaction(Matsubayashi 2011). This modification helps stabilizeand support the functioning of peptide hormones by the introduction of a hydroxyl group on proline and affects some physiological processes in the plant. Prolyl hydroxylases are efficient in recognizing specific proline residues located within specific peptide sequences in order to hydroxylate themat only the correct positions within a protein or peptide. In addition, hydroxylated proline can be a downstream modification target for arabinosylation that might impact protein interactions and function.Hydroxylated proline is important in enhancing structural stability and functional specificity of peptide hormones (Guillou et al. 2022). This type of PTM is crucial in secreted peptides, which include CLAVATA3 (CLV3), that play an essential role in plant development (Pečenková et al. 2024). In CLV3, proline hydroxylation blocks improper receptor binding to its related receptors on the target cell (Goad et al. 2017; Hazak and Hardtke 2016). This interaction of CLV3 produces specific signaling pathways that are necessary for the regulation of stem cell maintenance and the development of all types of organs, for example, meristems in plants. Proline hydroxylation is precise in the sense that it maintains the peptide's three-dimensional structure, as well as facilitates the peptide's interaction with its receptor for proper functioning and as such, aids in critical developmental processes and homeostasis, maintaining the plant system (Hagelthorn and Fletcher 2023; Srivastava et al. 2009). Hydroxylation is very important for some plant peptidehormones. Specifically, the hydroxylation of RALF is necessary for RALF interaction with specific receptors that regulate many plant developmental processes (Sui et al. 2023; Bedinger et al. 2010; Matos et al. 2008). As the stem cell activity of the shoot meristem is regulated by CLE peptides like CLV3, this indicates the biological importance of proline hydroxylation in plant growth

(Hu et al. 2022; Hastwell et al. 2014; Ito et al. 2006). Ultimately, this modification enables the plant to utilize complex signaling pathways to adapt and maintain homeostasis. Furthermore, proline hydroxylation has the function of a precursor for hydroxyproline arabinosylation, a PTM that can regulate the activity and distribution of diverse biologically important molecules (Schardon et al. 2016). The complexity seen in this interaction is also underscored by the fact that proline hydroxylation is not only critical formaintaining peptide hormone structure but also forfine-tuning their interaction with other key molecules in plants.

3.3.3 Hydroxyproline Arabinosylation

Arabinosylation of hydroxyproline is a particular type of *O*-linked glycosylation where arabinose sugar chains, either when single or with multiple arabinose units, are attached to hydroxyproline (Hyp) residues in peptide hormones (Shinohara and Matsubayashi 2010). The enzymes called hydroxyproline *O*-arabinosyltransferases (HPATs) catalyze this crucial biochemical reaction; they work mostly within the plant cell's Golgi apparatus. Arabinosylation is important for several plant peptide hormones. PSKs are well-studied peptides that require arabinosylation for reaching theirfull biological activity and operate different physiological processes (Yang et al. 2001, 2023). C-terminally encoded peptide (CEP) hormones are another group of arabinosylated peptide hormones that are important in nitrogen signaling to enable plant response to nutrient availability (Pan et al. 2024; Chapman et al. 2024; Lepetit and Brouquisse 2023; Qiu et al. 2022). In addition, root growth factors (RGF) control meristematic activity in roots and control root development and architecture. Hydroxyproline arabinosylation has multifunctional significance. Firstly, it increases the stability of peptide hormones by protecting them, so that they do not quickly degrade by the action of proteolysis and thus prolong their biological activity. Second, arabinosylation greatly increases their bioactivity, making them more effective in signal transduction pathways that govern these growth and development processes (Liu et al. 2021; Xu et al. 2015). In addition, this modification affects hormone transport, secretion, and processing mechanisms in plant tissues, influencing how the whole plant organism grows and reacts to its environment. Hydroxyproline arabinosylation is a key process influencing modulation of the hormone interactions with their receptors and thereby, the associated signaling pathways affecting plant growth and developmental processes. Furthermore, this PTM is essential for plants to respond to the environment in a precise way, such asdrought, salinity, or other stresses. Hydroxyproline arabinosylation stabilizes the active form of these hormones, thus helpingto create amore robust response mechanism and facilitatingplant adaptation to changing conditions (Hastwell et al. 2019).

3.4 Conclusion and Future Perspectives

Post-translational modifications of plant peptide hormones are critical regulatory processes that allow plants to modulate their growth and development as well as responsiveness to external signals. Increased PTMs, however, such as tyrosine sulfation, proline hydroxylation, and hydroxyproline arabinosylation, are critical for peptide hormones. These processes enablephytosulfokine (PSK), root growth factor (RGF), and epidermal patterning factor (EPF) to undergo structural and functional changes to carry out precise biological activities. These modifications have a significant impact on hormone stability and its interactions with receptors, as well as the subsequent signal pathways, ensuring optimal performance of the plants. It is known that peptide hormones with tyrosine sulfation act as a positive regulatory mechanism by increasing the binding strength of the peptides to their receptors, thereby providing an efficient signaling tool to promote the processes of cell proliferation by PSK, root meristem maintenance by RGF, and stomatal patterning by EPF. Functionally, proline hydroxylation and hydroxyproline arabinosylation are also utilized as means of stabilizing and activating peptides,shielding them from degradation and facilitating their passage through the plant system. Such modifications are also significant in maintaining stem cell balance, forminglateral roots, optimizing water use, enhancing stress tolerance, and enabling dynamic adjustment to the developmental and environmental challengesthat a plant encounters. PTMs of plant peptide hormones are demonstrated to have functional significance inbiological and agricultural sciences. A detailed understanding of how these modifications are regulated at the molecular level also provides researchers with a basis to develop crops with faster growth rates, better resource use, and increased stress resilience. Moreover, knowledge of peptide hormone modifications can provide useful information to develop new bioengineering techniques and plant growth regulators for enhancing the yield of crops and sustainability under changing environmental conditions. Due to the complex nature of the signaling networks, the roles of plant peptide hormones PTMs are yet to be fully understood. Recently, many advanced technologies such as next-generation sequencing (NGS) technologies have emerged thatcan be usedto identify and characterize many PTMs on a single molecule basis. Such innovations allow researchers to see in detail how particular changes affect the functioning of peptide hormones and how they control interactions between cells. For example, phosphorylation can affect the activity and stability of peptide hormone molecules,signaling pathways, and interactions with receptor proteins. Other PTMs should be investigated in the future to reveal mechanistic details to understand how peptide modifications are involved in intercellular signaling pathways and whether phytohormones interact with them. In summary, PTMs of plant peptide hormones constitute an important regulatory layer that enhances plant development as well as adaptation to environmental factors. These modifications in turn refine cellular signaling and create the potential of boosting crop resilience and agricultural sustainability, underlining how essential they are in basic plant biology and agricultural sciences. Future studiesshould focus on revealing the complex mechanisms of PTMs and their functions in different signaling pathways. Rapid progress in the fields of genome editing, synthetic biology, and molecular biotechnology has opened up opportunities to address PTM-related pathways for the advancement of plant productivity and adaptability.

References

Abarca, A., Franck, C.M., and Zipfel, C. (2021). Family-wide evaluation of RAPID ALKALINIZATION FACTOR peptides. *Plant Physiology* 187 (2): 996–1010.

Aggarwal, S., Kumar, A., Jain, M. et al. (2020). C-terminally encoded peptides (CEPs) are potential mediators of abiotic stress response in plants. *Physiology and Molecular Biology of Plants* 26: 2019–2033.

Albert, M. (2013). Peptides as triggers of plant defence. *Journal of Experimental Botany* 64: 5269–5279.

Amano, Y., Tsubouchi, H., Shinohara, H. et al. (2007). Tyrosine-sulfated glycopeptide involved in cellular proliferation and expansionin Arabidopsis. *Proceedings of the National Academy of Sciences of the United States of America* 104: 18333–18338.

Andersen, E.J., Ali, S., Byamukama, E. et al. (2018). Disease resistance mechanisms in plants. *Genes* 9 (7): 339. https://doi.org/10.3390/genes9070339.

Arsova, B., Watt, M., and Usadel, B. (2018). Monitoring of plant protein post-translational modifications using targeted proteomics. *Frontiers in Plant Science* 9: 1168.

Ávalos-Rangel, A., Ruiz-Herrera, L.F., Gutiérrez-Alanis, D. et al. (2024). CLE14 peptide impairs root tip regeneration and callogenesis in Arabidopsis. *Journal of Plant Growth Regulation* 43 (5): 1458–1465.

Bashyal, S., Gautam, C.K., and Müller, L.M. (2024). CLAVATA signaling in plant-environment interactions. *Plant Physiology* 194 (3): 1336–1357. https://doi.org/10.1093/plphys/kiad591.

Bedinger, P.A., Pearce, G., and Covey, P.A. (2010). RALFs: peptide regulators of plant growth. *Plant Signaling & Behavior* 5 (11): 1342–1346. https://doi.org/10.4161/psb.5.11.12954.

Bhalla, H., Sudarsanam, K., Srivastava, A., and Sankaranarayanan, S. (2025). Structural insights into the recognition of RALF peptides by FERONIA receptor kinase during Brassicaceae pollination. *Plant Molecular Biology* 115 (1): 1–6.

Blackburn, M.R., Haruta, M., and Moura, D.S. (2020). Twenty years of progress in physiological and biochemical investigation of RALF peptides. *Plant Physiology* 182 (4): 1657–1666.

Boschiero, C., Lundquist, P.K., Roy, S. et al. (2019). Identification and functional investigation of genome-encoded, small, secreted peptides in plants. *Current Protocols in Plant Biology* 4 (3): e20098.

Canut, H., Albenne, C., and Jamet, E. (2016). Post-translational modifications of plant cell wall proteins and peptides: a survey from a proteomics point of view. *Biochimica et Biophysica Acta (BBA)-Proteins and Proteomics* 1864 (8): 983–990.

Carbonnel, S., Falquet, L., and Hazak, O. (2022). Deeper genomic insights into tomato CLE genes repertoire identify new active peptides. *BMC Genomics* 23 (1): 756.

Carbonnel, S., Cornelis, S., and Hazak, O. (2023). The CLE33 peptide represses phloem differentiation via autocrine and paracrine signaling in Arabidopsis. *Communications Biology* 6 (1): 588.

Chang, S. and Xiao, F. (2025). Comprehensive review of plant small signaling peptides: from stress adaptation mechanisms to practical solutions for crop resilience. *International Journal of Biological Macromolecules* 299: 139971.

Chapman, K., Taleski, M., Frank, M., and Djordjevic, M.A. (2024). C-TERMINALLY ENCODED PEPTIDE (CEP) and cytokinin hormone signaling intersect to promote shallow lateral root angles. *Journal of Experimental Botany* 75 (2): 631–641.

Chaulagain, D. and Frugoli, J. (2021). The regulation of nodule number in legumes is a balance of three signal transduction pathways. *International Journal of Molecular Sciences* 22 (3): 1117.

Chen, Y.L., Fan, K.T., Hung, S.C., and Chen, Y.R. (2020). The role of peptides cleaved from protein precursors in eliciting plant stress reactions. *The New Phytologist* 225 (6): 2267–2282. https://doi.org/10.1111/nph.16241.

Chen, H., Song, Y., Li, H. et al. (2023). Enhancing the adaptability of tea plants (*Camellia sinensis* L.) to high-temperature stress with small peptides and biosurfactants. *Plants(Basel)* 12 (15): 2817. https://doi.org/10.3390/plants12152817.

Chen, X., van deSande, J.W., Ritmejeris, J.et al. (2024). Resolving sulfation PTMs on a plant peptide hormone using nanopore sequencing. bioRxiv.

Cho, H., Ryu, H., Rho, S. et al. (2014). A secreted peptide acts on BIN2-mediated phosphorylation of ARFs to potentiate auxin response during lateral root development. *Nature Cell Biology* 16 (1): 66–76.

Clark, S.E., Running, M.P., and Meyerowitz, E.M. (1993). CLAVATA1, a regulator of meristem and flower development in Arabidopsis. *Development* 119 (2): 397–418.

Clark, S.E., Running, M.P., and Meyerowitz, E.M. (1995). CLAVATA3 is a specific regulator of shoot and floral meristem development affecting the same processes as CLAVATA1. *Development* 121 (7): 2057–2067.

Combest, M.M., Moroz, N., Tanaka, K. et al. (2021). St PIP1, a PAMP-induced peptide in potato, elicits plant defenses and is associated with disease symptom severity in a compatible interaction with potato virus Y. *Journal of Experimental Botany* 72 (12): 4472–4488.

Corcilius, L., Hastwell, A.H., Zhang, M. et al. (2017). Arabinosylation modulates the growth-regulating activity of the peptide hormone CLE40a from soybean. *Cell Chemical Biology* 24 (11): 1347–1355.e7.

Couto, D. and Zipfel, C. (2016). Regulation of pattern recognition receptor signalling in plants. *Nature Reviews Immunology* 16 (9): 537–552.

Dao, T.Q., Weksler, N., Liu, H.M.H. et al. (2022). Interactive CLV3, CLE16 and CLE17 signaling mediates stem cell homeostasis in the Arabidopsis shoot apical meristem. *Development (Cambridge, England)* 149 (19): dev200787.

Das, R., Singh, S.P., Sahoo, A. et al. (2025). Synergistic stress of zinc oxide nanoparticles and pH triggers gill damage and mortality in zebrafish. *Nanotechnology for Environmental Engineering* 10 (1): 1–11.

de Bang, T.C., Lundquist, P.K., Dai, X. et al. (2017). Genome-wide identification of Medicago peptides involved in macronutrient responses and nodulation. *Plant Physiology* 175 (4): 1669–1689.

De Coninck, B. and De Smet, I. (2016). Plant peptides – taking them to the next level. *Journal of Experimental Botany* 67 (16): 4791–4795. https://doi.org/10.1093/jxb/erw309.

Ding, S., Lv, J., Hu, Z. et al. (2023). Phytosulfokine peptide optimizes plant growth and defense via glutamine synthetase GS2 phosphorylation in tomato. *The EMBO Journal* 42 (6): e111858.

Fang, Y., Chang, J., Shi, T. et al. (2021). Evolution of RGF/GLV/CLEL peptide hormones and their roles in land plant growth and regulation. *International Journal of Molecular Sciences* 22 (24): 13372. https://doi.org/10.3390/ijms222413372.

Fang, H., Zuo, J., Ma, Q. et al. (2024). Phytosulfokine promotes fruit ripening and quality via phosphorylation of transcription factor DREB2F in tomato. *Plant Physiology* 194 (4): 2739–2754.

Fedoreyeva, L.I. (2023). Molecular mechanisms of regulation of root development by plant peptides. *Plants (Basel)* 12 (6): 1320. https://doi.org/10.3390/plants12061320.

Fedoreyeva, L.I., Chaban, I.A., and Kononenko, N.V. (2022). Short peptides induce development of root hair *Nicotiana tabacum*. *Plants (Basel)* 11 (7): 852.

Feng, Y.Z., Zhu, Q.F., Xue, J. et al. (2023). Shining in the dark: the big world of small peptides in plants. *aBIOTECH* 4 (3): 238–256. https://doi.org/10.1007/s42994-023-00100-0.

Fernández-Fernández, Á.D., Stael, S., and Van Breusegem, F. (2023). Mechanisms controlling plant proteases and their substrates. *Cell Death and Differentiation* 30 (4): 1047–1058. https://doi.org/10.1038/s41418-023-01120-5.

Friso, G. and van Wijk, K.J. (2015). Posttranslational protein modifications in plant metabolism. *Plant Physiology* 169 (3): 1469–1487. https://doi.org/10.1104/pp.15.01378.

Fukuda, H. and Hardtke, C.S. (2020). Peptide signaling pathways in vascular differentiation. *Plant Physiology* 182 (4): 1636–1644.

Furumizu, C. and Aalen, R.B. (2023). Peptide signaling through leucine-rich repeat receptor kinases: insight into land plant evolution. *New Phytologist* 238 (3): 977–982.

Furumizu, C., Krabberød, A.K., Hammerstad, M. et al. (2021). The sequenced genomes of nonflowering land plants reveal the innovative evolutionary history of peptide signaling. *The Plant Cell* 33 (9): 2915–2934.

Gancheva, M.S., Malovichko, Y.V., Poliushkevich, L.O. et al. (2019). Plant peptide hormones. *Russian Journal of Plant Physiology* 66: 171–189.

Gao, Q., Wang, C., Xi, Y. et al. (2023). RALF signaling pathway activates MLO calcium channels to maintain pollen tube integrity. *Cell Research* 33 (1): 71–79.

Ghosh, P. and Roychoudhury, A. (2023). Plant peptides involved in abiotic and biotic stress responses and reactive oxygen species (ROS) signaling. *Journal of Plant Growth Regulation* 43: 1410–1427.

Goad, D.M., Zhu, C., and Kellogg, E.A. (2017). Comprehensive identification and clustering of CLV3/ESR-related (CLE) genes in plants finds groups with potentially shared function. *New Phytologist* 216 (2): 605–616.

Gorgues, L., Li, X., Maurel, C. et al. (2022). Root osmotic sensing from local perception to systemic responses. *Stress Biology* 2 (1): 36.

Greifenhagen, A., Ruwe, H., Zimmer, V. et al. (2024). The peptide hormone Pj CLE1 stimulates haustorium formation in the parasitic plant *Phtheirospermum japonicum*. *Proceedings of the National Academy of Sciences* 121 (42): e2414582121.

Guillou, M.C., Balliau, T., Vergne, E. et al. (2022). The PROSCOOP10 gene encodes two extracellular hydroxylated peptides and impacts flowering time in Arabidopsis. *Plants (Basel)* 11 (24): 3554.

Hagelthorn, L. and Fletcher, J.C. (2023). The CLAVATA3/ESR-related peptide family in the biofuel crop pennycress. *Frontiers in Plant Science* 14: 1240342.

Hanai, H., Matsuno, T., Yamamoto, M. et al. (2000). A secreted peptide growth factor, phytosulfokine, acting as a stimulatory factor of carrot somatic embryo formation. *Plant and Cell Physiology* 41 (1): 27–32.

Hao, Z., Wu, H., Zheng, R. et al. (2023). The plant peptide hormone phytosulfokine promotes somatic embryogenesis by maintaining redox homeostasis in Cunninghamia lanceolata. *The Plant Journal* 113 (4): 716–733.

Haruta, M., Sabat, G., Stecker, K. et al. (2014). A peptide hormone and its receptor protein kinase regulate plant cell expansion. *Science (New York, N.Y.)* 343 (6169): 408–411. https://doi.org/10.1126/science.1244454.

Hastwell, A.H., Gresshoff, P.M., and Ferguson, B.J. (2014). The structure and activity of nodulation-suppressing CLE peptide hormones of legumes. *Functional Plant Biology* 42 (3): 229–238.

Hastwell, A.H., Corcilius, L., Williams, J.T. et al. (2019). Triarabinosylation is required for nodulation-suppressive CLE peptides to systemically inhibit nodulation in Pisum sativum. *Plant, Cell & Environment* 42 (1): 188–197.

Hazak, O. and Hardtke, C.S. (2016). CLAVATA 1-type receptors in plant development. *Journal of Experimental Botany* 67 (16): 4827–4833.

He, L., Wu, L., and Li, J. (2024). Sulfated peptides and their receptors: key regulators of plant development and stress adaptation. *Plant Communications* 5: 100918.

Hirakawa, Y. and Sawa, S. (2019). Diverse function of plant peptide hormones in local signaling and development. *Current Opinion in Plant Biology* 51: 81–87. https://doi.org/10.1016/j.pbi.2019.04.005.

Hirayama, T. and Mochida, K. (2022). Plant hormonomics: a key tool for deep physiological phenotyping to improve crop productivity. *Plant and Cell Physiology* 63 (12): 1826–1839.

Hsiao, Y.C. and Yamada, M. (2020). The roles of peptide hormones and their receptors during plant root development. *Genes* 12 (1): 22. https://doi.org/10.3390/genes12010022.

Hsu, P.K., Dubeaux, G., Takahashi, Y., and Schroeder, J.I. (2021). Signaling mechanisms in abscisic acid-mediated stomatal closure. *The Plant Journal* 105 (2): 307–321.

Hu, C., Zhu, Y., Cui, Y. et al. (2022). A CLE–BAM–CIK signalling module controls root protophloem differentiation in Arabidopsis. *New Phytologist* 233 (1): 282–296.

Huang, P., Li, Z., and Guo, H. (2022). New advances in the regulation of leaf senescence by classical and peptide hormones. *Frontiers in Plant Science* 13: 923136.

Igarashi, D., Tsuda, K., and Katagiri, F. (2012). The peptide growth factor, phytosulfokine, attenuates pattern-triggered immunity. *The Plant Journal* 71 (2): 194–204.

Ito, Y., Nakanomyo, I., Motose, H. et al. (2006). Dodeca-CLE peptides as suppressors of plant stem cell differentiation. *Science* 313 (5788): 842–845.

Je, B.I., Xu, F., Wu, Q. et al. (2018). The CLAVATA receptor FASCIATED EAR2 responds to distinct CLE peptides by signaling through two downstream effectors. *eLife* 7: e35673.

Ji, C., Li, H., Zhang, Z. et al. (2025). The power of small signaling peptides in crop and horticultural plants. *The Crop Journal* 13: 656–667.

Johns, S., Hagihara, T., Toyota, M., and Gilroy, S. (2021). The fast and the furious: rapid long-range signaling in plants. *Plant Physiology* 185 (3): 694–706.

Jung, H.W., Panigrahi, G.K., Jung, G.-Y. et al. (2020). PAMP-triggered immunity involves proteolytic degradation of core nonsense-mediated mRNA decay factors during early defense response. *The Plant Cell* 32 (4): 1081–1101. https://doi.org/10.1105/tpc.19.00631.

Kaeothip, S., Ishiwata, A., and Ito, Y. (2013). Stereoselective synthesis of Arabidopsis CLAVATA3 (CLV3) glycopeptide, unique protein post-translational modifications of secreted peptide hormone in plant. *Organic & Biomolecular Chemistry* 11 (35): 5892–5907.

Kang, J., Peng, Y., and Xu, W. (2022a). Crop root responses to drought stress: molecular mechanisms, nutrient regulations, and interactions with microorganisms in the rhizosphere. *International Journal of Molecular Sciences* 23 (16): 9310. https://doi.org/10.3390/ijms23169310.

Kang, J., Wang, X., Ishida, T. et al. (2022b). A group of CLE peptides regulates de novo shoot regeneration in Arabidopsis thaliana. *New Phytologist* 235 (6): 2300–2312.

Kaufmann, C. and Sauter, M. (2019). Sulfated plant peptide hormones. *Journal of Experimental Botany* 70 (16): 4267–4277.

Kim, H.J., Wu, C.Y., Yu, H.M. et al. (2017). Dual CLAVATA3 peptides in Arabidopsis shoot stem cell signaling. *Journal of Plant Biology = Singmul Hakhoe chi* 60 (5): 506–512. https://doi.org/10.1007/s12374-017-0083-2.

Kim, J.S., Jeon, B.W., and Kim, J. (2021). Signaling peptides regulating abiotic stress responses in plants. *Frontiers in Plant Science* 12: 704490.

Kondo, T., Yokomine, K., Nakagawa, A., and Sakagami, Y. (2011). Analogs of the CLV3 peptide: synthesis and structure-activity relationships focused on proline residues. *Plant & Cell Physiology* 52 (1): 30–36. https://doi.org/10.1093/pcp/pcq146.

Kuromori, T., Fujita, M., Takahashi, F. et al. (2022). Inter-tissue and inter-organ signaling in drought stress response and phenotyping of drought tolerance. *The Plant Journal* 109 (2): 342–358.

Laffont, C., Ivanovici, A., Gautrat, P. et al. (2020). The NIN transcription factor coordinates CEP and CLE signaling peptides that regulate nodulation antagonistically. *Nature Communications* 11 (1): 3167.

Leasure, C.D. and He, Z.H. (2012). CLE and RGF family peptide hormone signaling in plant development. *Molecular Plant* 5 (6): 1173–1175.

Lebedeva, M., Azarakhsh, M., Yashenkova, Y., and Lutova, L. (2020). Nitrate-induced CLE peptide systemically inhibits nodulation in *Medicago truncatula*. *Plants (Basel)* 9 (11): 1456.

Lee, Z.H., Hirakawa, T., Yamaguchi, N., and Ito, T. (2019). The roles of plant hormones and their interactions with regulatory genes in determining meristem activity. *International Journal of Molecular Sciences* 20 (16): 4065.

Lepetit, M. and Brouquisse, R. (2023). Control of the rhizobium–legume symbiosis by the plant nitrogen demand is tightly integrated at the whole plant level and requires inter-organ systemic signaling. *Frontiers in Plant Science* 14: 1114840.

Li, J., Huang, Y., Yu, X. et al. (2024a). Identification and application of CLE peptides for drought resistance in Solanaceae crops. *Journal of Agricultural and Food Chemistry* 72: 13869–13884.

Li, P., Zhao, Z., Wang, W. et al. (2024b). Genome-wide analyses of member identification, expression pattern, and protein–protein interaction of EPF/EPFL gene family in Gossypium. *BMC Plant Biology* 24 (1): 554.

Li, Y., Di, Q., Luo, L., and Yu, L. (2024c). Phytosulfokine peptides, their receptors, and functions. *Frontiers in Plant Science* 14: 1326964.

Lin, Z.B., Zhao, Y.X., and Yang, W.C. (2023). A balanced mechanism between plant growth and stress response. *Advanced Agrochem* 2 (2): 105–106.

Lindsey, K. (2001). Plant peptide hormones: the long and the short of it. *Current Biology* 11 (18): R741–R743.

Liu, Y., Zuo, T., Qiu, Z. et al. (2021). Genome-wide identification reveals the function of CEP peptide in cucumber root development. *Plant Physiology and Biochemistry* 169: 119–126.

Liu, D., Shen, Z., Zhuang, K. et al. (2022a). Systematic annotation reveals CEP function in tomato root development and abiotic stress response. *Cells* 11 (19): 2935.

Liu, Q., Wu, K., Song, W. et al. (2022b). Improving crop nitrogen use efficiency toward sustainable green revolution. *Annual Review of Plant Biology* 73 (1): 523–551.

Loivamaki, M., Stührwohldt, N., Deeken, R. et al. (2010). A role for PSK signaling in wounding and microbial interactions in Arabidopsis. *Physiologia Plantarum* 139: 348–357.

Lu, S. and Xiao, F. (2024). Small peptides: orchestrators of plant growth and developmental processes. *International Journal of Molecular Sciences* 25 (14): 7627. https://doi.org/10.3390/ijms25147627.

Lv, Q.W. and Yang, Y.F. (2023). The biological functions of peptide signaling in plant and the advances on its utilization for crop improvement. *Yi chuan = Hereditas* 45 (9): 813–828. https://doi.org/10.16288/j.yczz.23-169.

Lyapina, I., Ganaeva, D., Rogozhin, E.A.et al. (2024). Comparative analysis of small secreted peptide signaling during defense response: insights from vascular and non-vascular plants. bioRxiv, 2024-05.

Ma, Y., Zhang, Y., Xu, J. et al. (2024). Research on the mechanisms of phytohormone signaling in regulating root development. *Plants (Basel)* 13 (21): 3051.

Matos, J.L., Fiori, C.S., Silva-Filho, M.C., and Moura, D.S. (2008). A conserved dibasic site is essential for correct processing of the peptide hormone AtRALF1 in Arabidopsis thaliana. *FEBS Letters* 582 (23–24): 3343–3347.

Matsubayashi, Y. (2011). Post-translational modifications in secreted peptide hormones in plants. *Plant & Cell Physiology* 52 (1): 5–13.

Matsubayashi, Y. (2012). Recent progress in research on small post-translationally modified peptide signals in plants. *Genes to Cells* 17: 1–10.

Matsubayashi, Y. (2014). Posttranslationally modified small-peptide signals in plants. *Annual Review of Plant Biology* 65: 385–413. https://doi.org/10.1146/annurev-arplant-050312-120122.

Matsubayashi, Y. (2018). Exploring peptide hormones in plants: identification of four peptide hormone-receptor pairs and two post-translational modification enzymes. *Proceedings of the Japan Academy. Series B, Physical and Biological Sciences* 94 (2): 59–74. https://doi.org/10.2183/pjab.94.006.

Matsubayashi, Y., Ogawa, M., Kihara, H. et al. (2006). Disruption and overexpression of Arabidopsis phytosulfokine receptor gene affects cellular longevity and potential for growth. *Plant Physiology* 142 (1): 45–53.

Matsuzaki, Y., Ogawa-Ohnishi, M., Mori, A., and Matsubayashi, Y. (2010). Secreted peptide signals required for maintenance of root stem cell niche in Arabidopsis. *Science* 329: 1065–1067.

Mens, C., Hastwell, A.H., Su, H. et al. (2021). Characterisation of Medicago truncatula CLE34 and CLE35 in nitrate and rhizobia regulation of nodulation. *New Phytologist* 229 (5): 2525–2534.

Millar, A.H., Heazlewood, J.L., Giglione, C. et al. (2019). The scope, functions, and dynamics of posttranslational protein modifications. *Annual Review of Plant Biology* 70 (1): 119–151.

Mogg, S. (2020). *Investigating the Diversity of CLE Proteins and Their Function in Arabidopsis thaliana*. The University of Manchester (United Kingdom).

Mosher, S., Seybold, H., Rodriguez, P. et al. (2013). The tyrosine-sulfated peptide receptors PSKR1 and PSY1R modify the immunity of Arabidopsis to biotrophic and necrotrophic pathogens in an antagonistic manner. *The Plant Journal* 73 (3): 469–482.

Motomitsu, A., Sawa, S., and Ishida, T. (2015). Plant peptide hormone signalling. *Essays in Biochemistry* 58: 115–131. https://doi.org/10.1042/bse0580115.

Müller, L.M. and Harrison, M.J. (2019). Phytohormones, miRNAs, and peptide signals integrate plant phosphorus status with arbuscular mycorrhizal symbiosis. *Current Opinion in Plant Biology* 50: 132–139.

Murphy, E., Smith, S., and De Smet, I. (2012). Small signaling peptides in Arabidopsis development: how cells communicate over a short distance. *The Plant Cell* 24 (8): 3198–3217.

Nakagami, S., Kajiwara, T., Tsuda, K., and Sawa, S. (2024). CLE peptide signaling in plant-microbe interactions. *Frontiers in Plant Science* 15: 1481650.

Nakayama, T., Shinohara, H., Tanaka, M. et al. (2017). A peptide hormone required for Casparian strip diffusion barrier formation in Arabidopsis roots. *Science* 355 (6322): 284–286.

Narasimhan, M. and Simon, R. (2022). Spatial range, temporal span, and promiscuity of CLE-RLK signaling. *Frontiers in Plant Science* 13: 906087.

Nguyen, Q.T.C. and Kim, J. (2023). PSY–PSYR peptide–receptor pairs control the trade-off between plant growth and stress response. *Plant Signaling & Behavior* 18 (1): https://doi.org/10.1080/15592324.2023.2260638.

Ogawa, M., Shinohara, H., Sakagami, Y., and Matsubayashi, Y. (2008). Arabidopsis CLV3 peptide directly binds CLV1 ectodomain. *Science* 319 (5861): 294–294.

Ogawa-Ohnishi, M., Yamashita, T., Kakita, M. et al. (2022). Peptide ligand-mediated trade-off between plant growth and stress response. *Science* 378 (6616): 175–180.

Oh, E., Seo, P.J., and Kim, J. (2018). Signaling peptides and receptors coordinating plant root development. *Trends in Plant Science* 23 (4): 337–351.

Ohki, S., Takeuchi, M., and Mori, M. (2011). The NMR structure of stomagen reveals the basis of stomatal density regulation by plant peptide hormones. *Nature Communications* 2: 512. https://doi.org/10.1038/ncomms1520.

Ohkubo, Y., Tanaka, M., Tabata, R. et al. (2017). Shoot-to-root mobile polypeptides involved in systemic regulation of nitrogen acquisition. *Nature Plants* 3 (4): 1–6.

Okamoto, S., Shinohara, H., Mori, T. et al. (2013). Root-derived CLE glycopeptides control nodulation by direct binding to HAR1 receptor kinase. *Nature Communications* 4 (1): 2191.

Oldroyd, G.E. and Leyser, O. (2020). A plant's diet, surviving in a variable nutrient environment. *Science* 368 (6486): eaba0196.

Olsson, V., Joos, L., Zhu, S. et al. (2019). Look closely, the beautiful may be small: precursor-derived peptides in plants. *Annual Review of Plant Biology* 70 (1): 153–186.

Pan, X., Deng, Z., Wu, R. et al. (2024). Identification of CEP peptides encoded by the tobacco (Nicotiana tabacum) genome and characterization of their roles in osmotic and salt stress responses. *Plant Physiology and Biochemistry* 209: 108525.

Panigrahi, G.K. and Satapathy, K.B. (2020a). *Arabidopsis* DCP5, a decapping complex protein interacts with ubiquitin-5 in the processing bodies. *Plant Archives* 20 (1): 2243–2247.

Panigrahi, G.K. and Satapathy, K.B. (2020b). Formation of *Arabidopsis* poly(A)-specific ribonuclease associated processing bodies in response to pathogenic infection. *Plant Archives* 20 (2): 4907–4912.

Panigrahi, G.K. and Satapathy, K.B. (2020c). Sacrificed surveillance process favours plant defense: a review. *Plant Archives* 20 (1): 2551–2559.

Panigrahi, G.K., and Satapathy, K.B. (2021). *Pseudomonas syringae*pv. syringae infection orchestrates the fate of the Arabidopsis J domain containing cochaperone and decapping protein factor 5. *Physiological and Molecular Plant Pathology*113101598https://doi.org/10.1016/j.pmpp.2020.101598.

Panigrahi, G.K., Sahoo, A., and Satapathy, K.B. (2021a). Differential expression of selected Arabidopsis resistant genes under abiotic stress conditions. *Plant Science Today* 8 (4): 859–864. https://doi.org/10.14719/pst.2021.8.4.1213.

Panigrahi, G.K., Sahoo, A., and Satapathy, K.B. (2021b). Insights to plant immunity: defense signaling to epigenetics. *Physiological and Molecular Plant Pathology* 113: 1–7. https://doi.org/10.1016/j.pmpp.2020.101568.

Panigrahi, G.K., Sahoo, A., and Satapathy, K.B. (2024). The processing body component varicose plays a multiplayer role towards stress management in Arabidopsis. *Plant Physiology Reports* 29. https://doi.org/10.1007/s40502-023-00778-w: 186–192.

Pastor-Fernández, J., Sánchez-Bel, P., Flors, V. et al. (2023). Small signals lead to big changes: the potential of peptide-induced resistance in plants. *Journal of Fungi* 9 (2): 265.

Patel, N., Mohd-Radzman, N.A., Corcilius, L. et al. (2018). Diverse peptide hormones affecting root growth identified in the Medicago truncatula secreted peptidome. *Molecular & Cellular Proteomics* 17 (1): 160–174.

Pečenková, T., Potocký, M., and Stegmann, M. (2024). More than meets the eye: knowns and unknowns of the trafficking of small secreted proteins in Arabidopsis. *Journal of Experimental Botany* 75: 3713–3730.

Phour, M., Sehrawat, A., Sindhu, S.S., and Glick, B.R. (2020). Interkingdom signaling in plant-rhizomicrobiome interactions for sustainable agriculture. *Microbiological Research* 241: 126589.

Qiu, Z., Zhuang, K., Liu, Y. et al. (2022). Functional characterization of C-terminally encoded peptide (CEP) family in *Brassica rapa* L. *Plant Signaling & Behavior* 17 (1): 2021365.

Reichardt, S., Piepho, H.P., Stintzi, A., and Schaller, A. (2020). Peptide signaling for drought-induced tomato flower drop. *Science* 367 (6485): 1482–1485.

Reid, D.E., Ferguson, B.J., and Gresshoff, P.M. (2011). Inoculation-and nitrate-induced CLE peptides of soybean control NARK-dependent nodule formation. *Molecular Plant-Microbe Interactions* 24 (5): 606–618.

Ren, S.C., Song, X.F., Chen, W.Q. et al. (2019). CLE25 peptide regulates phloem initiation in Arabidopsis through a CLERK-CLV2 receptor complex. *Journal of Integrative Plant Biology* 61 (10): 1043–1061.

Ren, G., Zhang, Y., Chen, Z. et al. (2024). Research progress of small plant peptides on the regulation of plant growth, development, and abiotic stress. *International Journal of Molecular Sciences* 25 (7): 4114. https://doi.org/10.3390/ijms25074114.

Roy, S., Lundquist, P., Udvardi, M., and Scheible, W.R. (2018). Small and mighty: peptide hormones in plant biology. *Plant Cell* 30: tc118tt0718.

Royek, S., Bayer, M., Pfannstiel, J. et al. (2022). Processing of a plant peptide hormone precursor facilitated by posttranslational tyrosine sulfation. *Proceedings of the National Academy of Sciences of the United States of America* 119 (16): e2201195119. https://doi.org/10.1073/pnas.2201195119.

Ryan, C.A. and Pearce, G. (2003). Systemins: a functionally defined family of peptide signals that regulate defensive genes in Solanaceae species. *Proceedings of the National Academy of Sciences of the United States of America* 100 (Suppl. 2): 14577–14580. https://doi.org/10.1073/pnas.1934788100.

Ryan, C.A., Pearce, G., Scheer, J., and Moura, D.S. (2002). Polypeptide hormones. *The Plant Cell* 14 (suppl_1): S251–S264.

Ryšlavá, H., Doubnerová, V., Kavan, D., and Vaněk, O. (2013). Effect of posttranslational modifications on enzyme function and assembly. *Journal of Proteomics* 92: 80–109. https://doi.org/10.1016/j.jprot.2013.03.025.

Rzemieniewski, J., Leicher, H., Lee, H.K. et al. (2024). CEP signaling coordinates plant immunity with nitrogen status. *Nature Communications* 15 (1): 1–16.

Sahoo, A., Satapathy, K.B., and Panigrahi, G.K. (2023). Ectopic expression of disease resistance protein promotes resistance against pathogen infection and drought stress in Arabidopsis. *Physiological and Molecular Plant Pathology* 124 (101949): 1–7. https://doi.org/10.1016/j.pmpp.2023.101949.

Sahoo, A., Satapathy, K.B., and Panigrahi, G.K. (2024). Security check: plant immunity under temperature surveillance. *Journal of Plant Biochemistry and Biotechnology* 33: 1–4. https://doi.org/10.1007/s13562-023-00846-0.

Salvi, P., Manna, M., Kaur, H. et al. (2021). Phytohormone signaling and crosstalk in regulating drought stress response in plants. *Plant Cell Reports* 40: 1305–1329.

Sang, T., Zhang, Z., Liu, G., and Wang, P. (2025). Navigating the landscape of plant proteomics. *Journal of Integrative Plant Biology* 67: 740–761.

Sauter, M. (2015). Phytosulfokine peptide signalling. *Journal of Experimental Botany* 66 (17): 5161–5169.

Schardon, K., Hohl, M., Graff, L. et al. (2016). Precursor processing for plant peptide hormone maturation by subtilisin-like serine proteinases. *Science* 354 (6319): 1594–1597.

Scott, T., Correa-Lozano, A., and Foo, E. (2024). Upstairs, downstairs-conserved and divergent CLAVATA signalling in shoot meristem development and root symbioses. *Annals of Botany* mcae192.

Selby, R. and Jones, D.S. (2023). Complex peptide hormone signaling in plant stem cells. *Current Opinion in Plant Biology* 75: 102442.

Setty, J., Singh, P., Tantuway, G., and Vijai, P. (2024). Recent advancement on peptide research and their application in eco-agriculture. In: *Plant Elicitor Peptides: New Tool for Sustainable Agriculture*, 269–281. Singapore: Springer Nature Singapore.

Shinohara, H. and Matsubayashi, Y. (2010). Arabinosylated glycopeptide hormones: new insights into CLAVATA3 structure. *Current Opinion in Plant Biology* 13: 515–519.

Simon, R. and Dresselhaus, T. (2015). Peptides take centre stage in plant signaling. Preface. *Journal of Experimental Botany* 66 (17): 5135–5138. https://doi.org/10.1093/jxb/erv376.

Song, X.F., Hou, X.L., and Liu, C.M. (2022). CLE peptides: critical regulators for stem cell maintenance in plants. *Planta* 255: 1–17.

Srivastava, R., Liu, J.X., and Howell, S.H. (2008). Proteolytic processing of a precursor protein for a growth-promoting peptide by a subtilisin serine protease in Arabidopsis. *The Plant Journal* 56 (2): 219–227.

Srivastava, R., Liu, J.X., Guo, H. et al. (2009). Regulation and processing of a plant peptide hormone, AtRALF23, in Arabidopsis. *The Plant Journal* 59 (6): 930–939.

Stintzi, A. and Schaller, A. (2022). Biogenesis of post-translationally modified peptide signals for plant reproductive development. *Current Opinion in Plant Biology* 69: 102274.

Stührwohldt, N. and Schaller, A. (2019). Regulation of plant peptide hormones and growth factors by post-translational modification. *Plant Biology (Stuttgart, Germany)* 21 (Suppl. 1): 49–63. https://doi.org/10.1111/plb.12881.

Stührwohldt, N., Bühler, E., Sauter, M., and Schaller, A. (2021). Phytosulfokine (PSK) precursor processing by subtilase SBT3. 8 and PSK signaling improve drought stress tolerance in Arabidopsis. *Journal of Experimental Botany* 72 (9): 3427–3440.

Sui, J., Xiao, X., Yang, J. et al. (2023). The rubber tree RALF peptide hormone and its receptor protein kinase FER implicates in rubber production. *Plant Science* 326: 111510.

Tabata, R. and Sawa, S. (2014). Maturation processes and structures of small secreted peptides in plants. *Frontiers in Plant Science* 5: 311. https://doi.org/10.3389/fpls.2014.00311.

Takahashi, F., Suzuki, T., Osakabe, Y. et al. (2018). A small peptide modulates stomatal control via abscisic acid in long-distance signalling. *Nature* 556 (7700): 235–238.

Takahashi, F., Kuromori, T., Urano, K. et al. (2020). Drought stress responses and resistance in plants: from cellular responses to long-distance intercellular communication. *Frontiers in Plant Science* 11: 556972.

Taleski, M., Imin, N., and Djordjevic, M.A. (2018). CEP peptide hormones: key players in orchestrating nitrogen-demand signalling, root nodulation, and lateral root development. *Journal of Experimental Botany* 69 (8): 1829–1836.

Taleski, M., Chapman, K., Imin, N. et al. (2020). The peptide hormone receptor CEPR1 functions in the reproductive tissue to control seed size and yield. *Plant Physiology* 183 (2): 620–636.

Taleski, M., Jin, M., Chapman, K. et al. (2024). CEP hormones at the nexus of nutrient acquisition and allocation, root development, and plant–microbe interactions. *Journal of Experimental Botany* 75 (2): 538–552.

Tan, W., Nian, H., Tran, L.S.P. et al. (2024). Small peptides: novel targets for modulating plant–rhizosphere microbe interactions. *Trends in Microbiology* 32: 1072–1083.

Tavormina, P., De Coninck, B., Nikonorova, N. et al. (2015). The plant peptidome: an expanding repertoire of structural features and biological functions. *The Plant Cell* 27 (8): 2095–2118. https://doi.org/10.1105/tpc.15.00440.

Toruño, T.Y., Stergiopoulos, I., and Coaker, G. (2016). Plant-pathogen effectors: cellular probes interfering with plant defenses in spatial and temporal manners. *Annual Review of Phytopathology* 54 (1): 419–441.

Tost, A.S., Kristensen, A., Olsen, L.I. et al. (2021). The PSY peptide family—expression, modification and physiological implications. *Genes* 12 (2): 218. https://doi.org/10.3390/genes12020218.

van de Sande, J.W., Streefkerk, D.E., Immink, R.G. et al. (2024). Phytosulfokine peptide library: chemical synthesis and biological evaluation on protoplast regeneration. *New Journal of Chemistry* 48 (18): 8055–8063.

van Wijk, K.J., Leppert, T., Sun, Z. et al. (2024). Detection of the Arabidopsis proteome and its post-translational modifications and the nature of the unobserved (dark) proteome in PeptideAtlas. *Journal of Proteome ResearchJournal of proteome research* 23 (1): 185–214.

Vanyushin, B.F., Ashapkin, V.V., and Aleksandrushkina, N.I. (2017). Regulatory peptides in plants. 82 (2): 89–94.

Villalobos-López, M.A., Arroyo-Becerra, A., Quintero-Jiménez, A., and Iturriaga, G. (2022). Biotechnological advances to improve abiotic stress tolerance in crops. *International Journal of Molecular Sciences* 23 (19): 12053. https://doi.org/10.3390/ijms231912053.

Wang, G. and Fiers, M. (2010). CLE peptide signaling during plant development. *Protoplasma* 240 (1–4): 33–43. https://doi.org/10.1007/s00709-009-0095-y.

Wang, J., Li, H., Han, Z. et al. (2015). Allosteric receptor activation by the plant peptide hormone phytosulfokine. *Nature* 525: 265–268.

Wang, G., Zhang, G., and Wu, M. (2016). CLE peptide signaling and crosstalk with phytohormones and environmental stimuli. *Frontiers in Plant Science* 6: 1211.

Wulf, K., Sun, J., Wang, C. et al. (2024). The role of CLE peptides in the suppression of mycorrhizal colonization of tomato. *Plant and Cell Physiology* 65 (1): 107–119.

Xiao, F. and Zhou, H. (2023). Plant salt response: perception, signaling, and tolerance. *Frontiers in Plant Science* 13: 1053699.

Xie, H., Zhao, W., Li, W. et al. (2022). Small signaling peptides mediate plant adaptions to abiotic environmental stress. *Planta* 255 (4): 72.

Xiong, L., Huang, Y., Liu, Z. et al. (2022). Small EPIDERMAL PATTERNING FACTOR-LIKE2 peptides regulate awn development in rice. *Plant Physiology* 190 (1): 516–531. https://doi.org/10.1093/plphys/kiac278.

Xu, T.T., Song, X.F., Ren, S.C., and Liu, C.M. (2013). The sequence flanking the N-terminus of the CLV3 peptide is critical for its cleavage and activity in stem cell regulation in Arabidopsis. *BMC Plant Biology* 13: 1–12.

Xu, C., Liberatore, K.L., MacAlister, C.A. et al. (2015). A cascade of arabinosyltransferases controls shoot meristem size in tomato. *Nature Genetics* 47 (7): 784–792.

Xu, K., Tian, D., Wang, T. et al. (2023). Small secreted peptides (SSPs) in tomato and their potential roles in drought stress response. *Molecular Horticulture* 3 (1): 17.

Yamada, M. and Sawa, S. (2013). The roles of peptide hormones during plant root development. *Current Opinion in Plant Biology* 16 (1): 56–61.

Yamaguchi, Y.L., Ishida, T., and Sawa, S. (2016). CLE peptides and their signaling pathways in plant development. *Journal of Experimental Botany* 67 (16): 4813–4826.

Yang, H., Matsubayashi, Y., Nakamura, K., and Sakagami, Y. (2001). Diversity of Arabidopsis genes encoding precursors for phytosulfokine, a peptide growth factor. *Plant Physiology* 127 (3): 842–851.

Yang, Z., Gao, C., Zhang, Y. et al. (2023). Recent progression and future perspectives in cotton genomic breeding. *Journal of Integrative Plant Biology* 65 (2): 548–569.

Ye, M., Liu, C., Li, N. et al. (2023). A constitutive serine protease inhibitor suppresses herbivore performance in tea (*Camellia sinensis*). *Horticulture Research* 10 (10): uhad178. https://doi.org/10.1093/hr/uhad178.

Yoshida, T., Fernie, A.R., Shinozaki, K., and Takahashi, F. (2021). Long-distance stress and developmental signals associated with abscisic acid signaling in environmental responses. *The Plant Journal* 105 (2): 477–488.

Yu, B., Chao, D.Y., and Zhao, Y. (2024a). How plants sense and respond to osmotic stress. *Journal of Integrative Plant Biology* 66 (3): 394–423.

Yu, X., Huang, Z., Cheng, Y. et al. (2024b). Comparative genomics screens identify a novel small secretory peptide, SlSolP12, which activates both local and systemic immune response in tomatoes and exhibits broad-spectrum activity. *Journal of Agricultural and Food Chemistry* 72 (33): 18507–18519.

Yuan, B. and Wang, H. (2021). Peptide signaling pathways regulate plant vascular development. *Frontiers in Plant Science* 12: 719606. https://doi.org/10.3389/fpls.2021.719606.

Žárský, V., Nielsen, M.E., and Blatt, M.R. (2024). Formins, cell wall integrity, ROP guanine exchange factors, secretion regulators, and small secreted peptides in plant cell exocytosis and defence. *Journal of Experimental Botany* 75 (12): 3665–3667.

Zelman, A.K. and Berkowitz, G.A. (2023). Plant elicitor peptide (Pep) signaling and pathogen defense in tomato. *Plants* 12 (15): 2856.

Zhang, H., Han, Z., Song, W., and Chai, J. (2016). Structural insight into recognition of plant peptide hormones by receptors. *Molecular Plant* 9 (11): 1454–1463.

Zhang, L., Shi, X., Zhang, Y. et al. (2019). CLE9 peptide-induced stomatal closure is mediated by abscisic acid, hydrogen peroxide, and nitric oxide in Arabidopsis thaliana. *Plant, Cell & Environment* 42 (3): 1033–1044.

Zhang, S., Li, C., Si, J. et al. (2022). Action mechanisms of effectors in plant-pathogen interaction. *International Journal of Molecular Sciences* 23 (12): 6758.

Zhang, Y.M., Ye, D.X., Liu, Y. et al. (2023). Peptides, new tools for plant protection in eco-agriculture. *Advanced Agrochem* 2 (1): 58–78.

Zhang, P., Gao, W., Guo, L. et al. (2024a). Functional characterization of plant peptide-containing sulfated tyrosine (PSY) family in wheat (Triticum aestivum L.). *International Journal of Molecular Sciences* 25 (23): 12663.

Zhang, P., Zhao, J., Zhang, W., Guo, Y., &Zhang, K. (2024b). Sulfated peptides: key players in plant development, growth, and stress responses. *Frontiers in Plant Science*, 15, 1474111. https://doi.org/10.3389/fpls.2024.1474111.

Zhang, C., Xie, Y., He, P., and Shan, L. (2024c). Unlocking nature's defense: plant pattern recognition receptors as guardians against pathogenic threats. *Molecular Plant-Microbe Interactions* 37 (2): 73–83.

Zhang, Z., Han, H., Zhao, J. et al. (2025). Peptide hormones in plants. *Molecular Horticulture* 5 (1): 7. https://doi.org/10.1186/s43897-024-00134-y.

Zhong, S. and Qu, L.J. (2019). Peptide/receptor-like kinase-mediated signaling involved in male–female interactions. *Current Opinion in Plant Biology* 51: 7–14.

Zhong, S., Li, L., Wang, Z. et al. (2022). RALF peptide signaling controls the polytubey block in Arabidopsis. *Science* 375 (6578): 290–296.

Zhou, H., Xiao, F., Zheng, Y. et al. (2022). Pamp-induced secreted peptide 3 modulates salt tolerance through receptor-like kinase 7 in plants. *The Plant Cell* 34 (2): 927–944.

Zhou, Y., Zheng, J., Wu, H. et al. (2024). A novel toolbox to record CLE peptide signaling. *Frontiers in Plant Science* 15: 1468763.

4

Plant Peptide Hormones and Plant Development

Emad Hamdy Khedr[1], *Nagwa Khedr*[1], *and Hossam S. El-Beltagi*[2]

[1] *Department of Pomology, Faculty of Agriculture, Cairo University, Giza, Egypt*
[2] *Agricultural Biotechnology Department, College of Agriculture and Food Sciences, King Faisal University, Al-Ahsa, Saudi Arabia*

4.1 Introduction

Plant morphology, the physical structure and organization of a plant, is governed by a sophisticated interplay of genetic programs and environmental signals. Classical plant hormones such as auxins, gibberellins, cytokinins, ethylene, and abscisic acid (ABA) have historically dominated our understanding of how growth and development are orchestrated. However, over the past few decades, small peptide hormones have emerged as critical players in plant development, offering a new lens through which we can understand the dynamic regulation of plant architecture (Ali et al. 2024). Unlike classical hormones, peptide hormones are short chains of amino acids that function as mobile signaling molecules. They are typically secreted into the apoplast and perceived by plasma membrane-localized receptors, particularly leucine-rich repeat receptor-like kinases (LRR-RLKs), to activate downstream signaling cascades (Escocard de Azevedo Manhães et al. 2021). These molecules play essential roles in intercellular communication, coordinating diverse processes including meristem maintenance, organ patterning, vascular differentiation, and responses to environmental stimuli (Cui et al. 2022).

The first functional peptide ligand discovered in plants was Systemin, an 18-amino-acid polypeptide from tomato involved in wound signaling (Pearce et al. 1991). Since then, many other peptide families have been identified, including CLAVATA3/embryo surrounding region-related (CLE), C-terminally encoded peptides (CEP), inflorescence deficient in abscission (IDA), rapid alkalinization factor (RALF), phytosulfokine (PSK), and root meristem growth factor (RGF), each with specialized and often tissue-specific roles (Ren et al. 2024). A seminal discovery in this field was the identification of CLV3, a CLE peptide that restricts stem cell proliferation in the shoot apical meristem (SAM) by repressing the expression of WUSCHEL (WUS), a transcription factor that promotes stem cell identity. This CLV3–WUS negative feedback loop exemplifies how peptide hormones mediate precise spatial control of developmental programs (Yamaguchi et al. 2016). In roots, peptide

Peptide Hormones in Plant Development and Stress Responses, First Edition. Edited by Vijay Pratap Singh and Samiksha Singh.

hormones such as RGF and Casparian strip integrity factor (CIF) are essential for maintaining root apical meristem (RAM) organization and barrier formation, respectively (Hsiao and Yamada 2020).

Plants, as sessile organisms, are constantly exposed to changing environmental conditions and are restricted to the environment where they germinate. Abiotic and biotic stresses influence the growth and development of plants (Khedr et al. 2023; Khedr et al., 2025; EL-Beltagi and Khedr, 2025; Rabiae et al., 2025). Postembryonic development and adaptation of the plant body plan to the prevailing environmental conditions are orchestrated by several classes of hormones and hundreds of signaling peptides. For a long time, the number, diversity, and significance of peptides as regulators of cellular differentiation, growth, and development were highly underestimated. During the last two decades, it has become more and more clear how abundant and important peptides are as signaling molecules during all stages of growth and development. Peptides emerged as an important class of signaling molecules for cell-to-cell communication over short distances, and also for long-range signaling. We refer to them as peptide growth factors and hormones, respectively. In *Arabidopsis thaliana*, there are more than 1000 genes encoding potential peptide hormones (Takahashi et al. 2019). However, only a few of them have had their molecular or physiological functions elucidated.

Plant development depends heavily on the maintenance of stem cell populations within meristems, specialized zones of continuous cell division. In both the SAM and RAM, the balance between stem cell maintenance, proliferation, and differentiation is tightly regulated. Asymmetric cell divisions and positional cues orchestrate this balance, ensuring robust growth and adaptation to environmental conditions (Pillitteri et al. 2016). Positional signals, rather than lineage, often determine cell fate, and many plant cells can switch identity depending on new positional information (Pierre-Jerome et al. 2018). This chapter explores how small peptide hormones function as key regulators of plant morphology, particularly in the context of meristem dynamics. We discuss their biosynthesis, signaling pathways, receptor interactions, and the regulatory networks that control their activity. By highlighting recent advances, we aim to illustrate the emerging potential of peptide hormones not only in developmental biology but also in agricultural biotechnology for crop improvement and stress resilience.

4.2 Classification of Peptide Hormones and Growth Factors

Peptide hormones and growth factors in plants can be broadly classified based on their structural features and biosynthetic origins. According to Stührwohldt et al. (2015), these signaling molecules fall into two primary categories: (i) peptides that arise from the proteolytic processing of larger precursor proteins, and (ii) those that are directly encoded by small peptide-coding genes, or translated from short open reading frames (sORFs) located within the 5′ untranslated regions (UTRs) of mRNA transcripts, or primary microRNA sequences. These diverse origins reflect the complexity of peptide-mediated signaling in plants.

Among the precursor-derived peptides, Khan and Luqman (2025) further categorize them into two subclasses: cysteine-rich peptides, which rely on the formation of disulfide bonds to achieve their active conformations, and a larger group of peptide hormones that undergo various posttranslational modifications (PTMs). The latter group,posttranslationally modified peptides — has garnered significant interest due to their diverse roles in developmental regulation and stress responses. Typically, these modified peptides are synthesized as prepropeptides consisting of approximately 70 to 120 amino acids. They are characterized by an N-terminal signal peptide that targets them for secretion. Once the precursor enters the endoplasmic reticulum (ER), the signal sequence is cleaved, and the remaining propeptide undergoes additional modifications as it traverses the secretory pathway. Key PTMs include tyrosine sulfation, proline hydroxylation, and hydroxyproline O-glycosylation, which are essential for proper peptide maturation and functionality (Stührwohldt and Schaller 2019). While many of the enzymes responsible for these modifications have been identified, the precise biological implications of PTMs on peptide activity remain incompletely understood. These modifications can influence a peptide's bioactivity, receptor-binding specificity, stability, and even its subcellular targeting. In some cases, PTMs are critical for activating peptide function, while in others, they may alter or even suppress signaling potential.

4.2.1 Systemin

Systemin, discovered in the early 1990s, marked a breakthrough as the first identified bioactive peptide in plants (Pearce et al. 1991). It plays a crucial role in mediating wound-induced systemic defense responses in tomato and related Solanaceae species. Initially believed to be the elusive long-distance wound hormone, subsequent research revealed that systemin functions locally at the site of injury to amplify jasmonic acid (JA) biosynthesis, which then acts as the mobile signal for systemic defense activation (Ryan and Moura 2002). This localized action draws a parallel between systemin and animal cytokines rather than classical hormones, highlighting its role as a key signal amplifier rather than a systemic messenger.

Systemin is derived from a larger precursor protein known as prosystemin, consisting of approximately 200 amino acids. Unlike most peptide hormone precursors, prosystemin lacks an N-terminal signal peptide, which raises questions about how the mature systemin peptide is transported into the apoplast for perception by membrane-bound receptors (Pearce and Ryan 2003). The receptor initially proposed for systemin was the LRR-RLK SR160, later identified as the brassinosteroid receptor BRI1, thus disqualifying it as the true systemin receptor (Scheer and Ryan 2002). More recent studies have identified two LRR-RLKs, SYR1 and SYR2, as bona fide systemin receptors in tomato. These receptors exhibit different binding affinities and, while dispensable for general wound signaling, are essential for mounting effective defense responses against insect herbivory (Scheer and Ryan 2002).

4.2.2 The CLV3 and CLE (CLV3/Endosperm Surrounding Region [ESR]-Related) Peptides

CLAVATA3 (CLV3) and its related CLE peptides represent a distinct class of small signaling peptides primarily involved in the regulation of meristem activity and cellular differentiation. CLV3 is a 13-amino-acid peptide processed from a 96-amino-acid precursor and requires specific PTMs, including proline hydroxylation and O-arabinosylation, to achieve full biological activity (Ohyama et al. 2009). In *A. thaliana*, CLV3 is a central component in maintaining the balance between stem cell proliferation and differentiation within the SAM. It operates through two main signaling branches: one involving the receptor kinase CLV1 and CLV3-INSENSITIVE KINASE (CIK) coreceptors, and another involving the receptor-like protein CLV2 in complex with the pseudokinase CRN (Hu et al. 2018).

CLV3 was the first member discovered in the broader CLE peptide family, which encompasses 32 genes in Arabidopsis encoding at least 27 distinct peptides sharing a conserved 14-amino-acid CLE-box motif (Mortier et al. 2010). Among these, CLE40 is a notable paralog expressed in the root stele and columella, where it regulates root stem cell differentiation via interactions with CLV1, ARABIDOPSIS CRINKLY4 (ACR4), and the transcription factor WOX5 (Fedoreyeva 2023). The ability of CLE40 to restrict shoot meristem size when ectopically expressed suggests functional redundancy with CLV3 in meristem regulation (Hobe et al. 2003). Similarly, CLE19 has been shown to inhibit RAM activity when overexpressed, pointing to the presence of a CLE-mediated signaling network controlling both shoot and root development (Wang et al. 2025).

Figure 4.1 provides an overview of how various peptide hormones coordinate root development by interacting with specific receptors and intracellular signaling pathways. CLE peptides, for instance, require PTMs for full activity and receptor specificity. RGF peptides promote root meristem maintenance through ROS-mediated stabilization of key transcription factors, while RALFs regulate cell expansion through complex pathways involving Ca^{2+} signaling and pH changes. This integrated network underpins precise spatial and

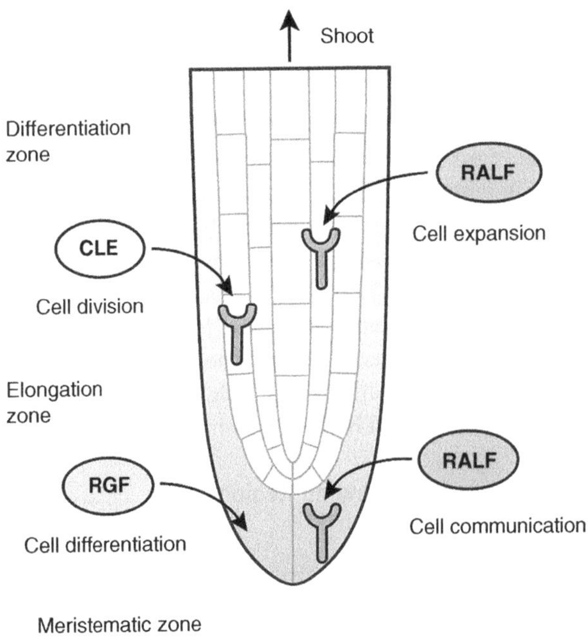

Figure 4.1 Peptide hormones in root development and their signaling mechanisms.

temporal control of root growth, underscoring the complexity and importance of peptide-mediated signaling in plant developmental biology.

Beyond meristem maintenance, CLE peptides also contribute significantly to vascular development. CLE41/44, also known as tracheary element differentiation inhibitory factor (TDIF), promotes procambial cell proliferation while inhibiting xylem differentiation. This function is mediated through interaction with the LRR-RLK and TDIF Receptor (TDR), also called PXY (phloem intercalated with xylem) (Wang et al. 2013). Additionally, CLE45 regulates protophloem development through the Barely any meristem 3 (BAM3) receptor-like kinase (Hu et al. 2022). These findings underscore the multifunctionality of CLE peptides in orchestrating plant development through finely tuned spatial and temporal signaling.

4.2.3 Phytosulfokine (PSK)

Among the diverse groups of plant peptide hormones, tyrosine-sulfated peptides represent a specialized class that requires PTM for full biological activity. This group includes PSKs, plant peptides containing sulfated tyrosine (PSYs), root meristem growth factors (RGFs), and CIFs (Stührwohldt and Schaller 2019). Among them, PSKs were the first to be identified and characterized. Initially discovered in *Asparagus officinalis* as a mitogenic factor capable of inducing cell division in mesophyll cell cultures, PSKs were later found to have multifaceted roles in plant growth and development (Stührwohldt and Schaller 2019). These roles include promoting root and hypocotyl elongation primarily through cell expansion (Kutschmar et al. 2009) and directing pollen tube growth (Stührwohldt et al. 2015).

Phytosulfokine receptor 1 (PSKR1) functions at the plasma membrane as part of a multi-component signaling module that includes the cyclic nucleotide-gated channel 17 (CNGC17), the plasma membrane Arabidopsis H^+-ATPase 1 (AHA1) and Arabidopsis H+-ATPases ATPase 2 (AHA2), and the coreceptor kinase BRI1-associated receptor kinase 1 (BAK1) (Pei et al. 2022). This module orchestrates changes in ion flux and cellular expansion in response to PSK perception. Despite significant progress in understanding the PSK signaling mechanism, the biosynthetic pathway of PSK, including the specific enzymes involved in its cleavage and sulfation, remains largely uncharacterized. This knowledge gap highlights the need for further research into the maturation and secretion processes of tyrosine-sulfated peptides in plants.

4.2.4 Tyrosine-Sulfated Peptides: PSYs

PSY peptides represent another group of tyrosine-sulfated plant peptide hormones, sharing structural and functional similarities with PSKs, yet exhibiting distinct features in their biogenesis and signaling mechanisms. The prototype member, PSY1, is an 18-amino-acid glycopeptide modified not only by tyrosine sulfation but also by the addition of a tri-arabinosyl chain to one of its hydroxyproline residues, an uncommon and highly specific PTM (Pandita et al. 2023). PSY1 is synthesized as part of a 75-amino-acid precursor, with two closely related homologues identified in *A. thaliana*, reflecting a level of gene redundancy within the PSY family (Stührwohldt and Schaller 2019).

Functionally, PSY1 is known to stimulate both cell proliferation and expansion, contributing to overall plant growth and development. Similar to PSKs, PSY1 also plays a role in modulating plant immune responses, suggesting a shared involvement in balancing growth and defense mechanisms (Mosher and Kemmerling 2013). Despite its similarities to PSKs, PSY1 operates through a distinct signaling pathway. It is perceived by its cognate receptors PSY1R, and LRR-RLK structurally related to PSKR1 and PSKR2. However, PSY1R exhibits unique functionality by directly phosphorylating and activating the plasma membrane H^+-ATPases AHA1 and AHA3, key regulators of apoplastic pH (Li et al. 2022).

Activation of AHA1 and AHA3 by PSY1R leads to extracellular acidification, which promotes cell wall loosening and cell expansion, thereby facilitating growth responses. This signaling mechanism underlines a distinctive mode of action for PSYs, emphasizing their direct control over proton pump activity and cell expansion dynamics. While the receptor–ligand interaction and downstream targets of PSY1 are becoming increasingly well defined, further investigation is needed to fully uncover the roles of PSY homologues and the broader implications of PSY signaling in various developmental and stress-related contexts (Li et al. 2022).

4.2.5 Tyrosine-Sulfated Peptides: CIFs

Casparian strip integrity factor (CIF) peptides represent a specialized class of tyrosine-sulfated peptides that play a crucial role in root endodermis development, particularly in the formation of the Casparian strip, a lignin-based apoplastic diffusion barrier. The two well-characterized members, CIF1 and CIF2, are 21-amino-acid peptides derived from approximately 80-amino-acid precursors. While the exact proteolytic processing mechanisms responsible for generating the mature, bioactive peptides from their precursors remain elusive, the necessity of posttranslational tyrosine sulfation for their biological function is well established (Nakayama et al. 2017).

CIF peptides are specifically expressed in the stele of the root, but their functional site of action is the endodermis, where they are perceived by the LRR-RLKs SCHENGEN3 (SGN3), also known as GASSHO1 (GSO1), and its close homologue GSO2. These receptors are expressed in the endodermal cell layer and act as critical components in detecting CIF signaling to initiate and maintain Casparian strip integrity (Doblas et al. 2017a,b). The high-affinity interaction between CIFs and their receptors requiring tyrosine sulfation for nanomolar binding strength is essential for the spatially regulated deposition of lignin, ensuring that the Casparian strip forms correctly and functions as a selective barrier.

This peptide-receptor communication exemplifies a non-cell-autonomous signaling pathway, in which CIFs, produced in one tissue layer (stele), signal to another (endodermis) to coordinate root barrier formation. Such spatial signaling precision underscores the importance of sulfated peptides like CIFs in developmental patterning and organogenesis. While considerable progress has been made in identifying the receptors and biological roles of CIFs, further studies are required to elucidate their full processing pathway and possible interactions with other regulatory modules involved in endodermal differentiation (Bloom et al. 2016).

4.2.6 Tyrosine-Sulfated Peptides: RGFs

Root meristem growth factors (RGFs) are a family of tyrosine-sulfated peptides that play essential roles in postembryonic root development. There are 11 known RGF precursors in Arabidopsis, with a range of 86–163 amino acids. These precursors are subject to various

PTMs, which are crucial for the activity and receptor interactions of the mature peptides. RGFs are often referred to by different names, including CLE-Like (CLEL) and Golven (GLV) peptides, due to findings from multiple independent studies that have generated overlapping and somewhat conflicting nomenclature (Hastwell et al. 2024). RGFs are involved in important developmental processes in plants, particularly in maintaining the root stem cell niche and controlling gravitropic responses (Fang et al. 2021). The bioactive, secreted RGFs are typically 13–18 amino acids long, and they exhibit two key PTMs: the crucial tyrosine sulfation, which is necessary for receptor binding and biological activity, and the hydroxylation of the proline residue at the carboxy-terminus of the peptide (Tan et al. 2024).

The identification of RGF receptors has been a subject of extensive research, with five RGF receptors identified by three independent research groups using various approaches. These receptors are LRR-RLKs, a characteristic feature of many plant peptide receptors (Furumizu and Aalen 2023). These receptors are integral in mediating the RGF signal, facilitating the regulation of root growth, and maintaining the integrity of the root stem cell niche. Overall, the RGF signaling pathway exemplifies a sophisticated mechanism in plants where peptide hormones, with specific modifications like tyrosine sulfation and proline hydroxylation, regulate critical aspects of root development and response to environmental cues such as gravity (Hsiao and Yamada 2020). Understanding the full scope of RGF receptor interactions and their downstream effects will offer valuable insights into root developmental biology and plant growth regulation.

4.2.7 Inflorescence Deficient in Abscission (IDA)

Inflorescence deficient in abscission (IDA) is a critical gene in Arabidopsis responsible for regulating the process of abscission, specifically the detachment of sepals, petals, and stamens after fertilization (Kim 2014). The IDA precursor peptide undergoes processing to produce a 14-mer peptide, which plays a crucial role in this process. One significant PTM of the IDA peptide is the hydroxylation of proline at position 9, a modification essential for the peptide's bioactivity and its ability to bind to its receptor (Virág et al. 2020).

The IDA peptide is perceived by LRR-RLKs, specifically HAESA (HAE) and its homolog HAE-like 2 (HSL). These receptors, in collaboration with somatic embryogenesis receptor kinase (SERK) coreceptors, initiate a signaling cascade involving MAP kinases. This cascade activates the expression of various hydrolytic and cell wall–modifying enzymes, which are crucial for the breakdown of the pectin-rich middle lamella and the separation of cells in the abscission zone (Pautot et al. 2025).

The IDA-mediated abscission process is integral for various stages of plant reproduction and growth, and understanding the molecular mechanisms underlying this pathway can provide valuable insights into the regulation of flower organ separation and overall plant development. Additionally, the role of IDA in promoting abscission highlights the complex interaction between peptide hormones and their receptor systems in regulating important developmental transitions.

4.2.8 C-Terminally Encoded Peptides (CEPs)

C-Terminally encoded peptides (CEPs) were first identified through an *in silico* screen aimed at discovering peptide precursors sharing a conserved C-terminal sequence

(Aggarwal et al. 2020). Phylogenetic analyses suggest that CEPs are widely distributed across flowering plants and gymnosperms, indicating their fundamental role in plant biology (Ogilvie et al. 2014). In Arabidopsis, 15 CEP genes have been identified, each displaying distinct developmental and organ- or tissue-specific expression patterns, suggesting that these peptides have diverse physiological functions in different parts of the plant (Aggarwal et al. 2020).

In legumes, CEPs play a significant role in regulating nodulation and lateral root development through the CEP receptor compact root architecture 2 (CRA2), which shares similarities with CEPR1 from Arabidopsis (Mohd-Radzman et al. 2016). In *Medicago truncatula*, CEP1 was shown to promote nodulation via CRA2 in an ethylene-dependent manner, while the inhibition of lateral root development through CRA2 occurs independently of ethylene (Mohd-Radzman et al. 2016). Interestingly, tri-arabinosylation at Pro11 impairs this nodulation activity, which contrasts with the requirement of tri-arabinosylation for the maximum activity of CLE peptides (Stührwohldt and Schaller 2019).

The regulatory roles of CEPs in nitrogen uptake and root development, as well as their involvement in nodulation and lateral root growth in legumes, underscore their importance in adapting plant growth to environmental conditions. These findings also highlight the complexity of peptide signaling in plants and the influence of PTMs on peptide bioactivity.

4.2.9 Rapid Alkalinization Factors (RALFs)

Rapid alkalinization factors (RALFs) were first identified in tobacco as a 49-amino acid precursor-derived cysteine-rich polypeptide that can induce extracellular alkalinization in a cell culture bioassay (Stührwohldt and Schaller 2019). The first Arabidopsis RALF, AtRALF1, was isolated for its ability to induce a rapid increase in cytosolic calcium levels (Haruta et al. 2014). Since then, more than 30 RALF and RALF-related genes have been identified in Arabidopsis, and their functions are gradually being elucidated. In addition to its role in root growth regulation, FER has been shown to act as a scaffold for immune receptors such as FLAGELLIN-SENSITIVE 2 (FLS2) and EF-TU RECEPTOR (EFR), which are involved in pattern-triggered immunity (Rzemieniewski and Stegmann 2022). The complex formation of FLS2 and EFR with their coreceptor BAK1 is enhanced by FER and is antagonized by RALF23, a ligand of FER.

Figure 4.2 illustrates the mechanism by which RALF peptides regulate root development. Upon binding to CrRLK1L-type receptors such as FERONIA (FER), often in association with LRX proteins, RALF peptides trigger downstream MAPK signaling. This cascade influences cytosolic calcium (Ca^{2+}) levels, pH modulation, and cellular expansion, collectively regulating root growth. Understanding these interactions helps clarify how RALF peptides contribute to root architecture and development.

RALF peptides play crucial roles in plant root development by mediating complex signaling pathways that involve receptor kinases, calcium signaling, and pH regulation. Upon perception by CrRLK1L receptors like FER, RALF peptides activate MAP kinase cascades, modulate cytoplasmic Ca^{2+}, and adjust apoplastic pH processes essential for cell expansion. These mechanisms underscore the significance of RALF-receptor interactions in dictating root patterning and growth, yet much remains to be uncovered about the temporal and spatial specificity of these peptide-mediated events.

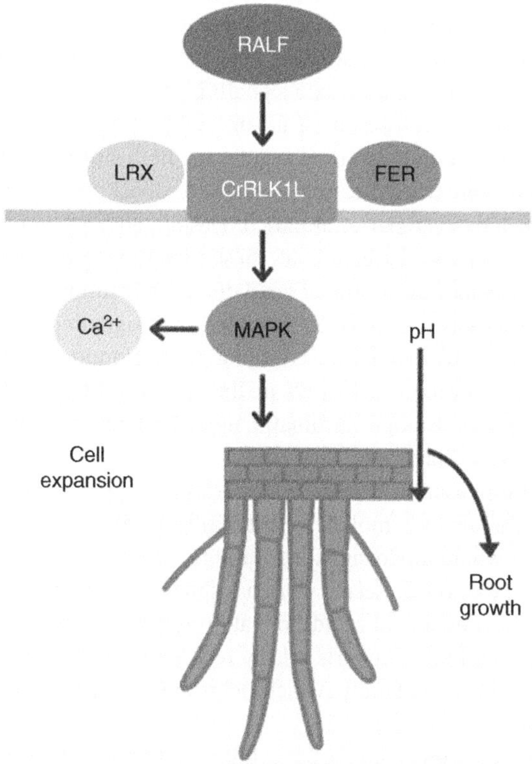

Figure 4.2 Role of RALF peptides in root development.

Pollen-specific RALFs, such as RALF4 and RALF19, bind a receptor complex consisting of ANXUR1/2 and BUPS1/2, and this interaction is crucial for pollen tube integrity. Ovule-derived RALF34 can displace RALF4/19 from the receptor complex, resulting in pollen tube burst and the release of sperm cells for fertilization. Thus, RALFs are involved in a variety of key physiological processes, including root growth regulation, immune signaling, and reproductive processes like pollen tube integrity and fertilization. The diverse roles of RALFs highlight their importance in plant development and stress responses.

4.3 Hormones and Patterning in Plants

Hormones are essential for defining cellular patterns in both plants and animals. In plants, the biosynthesis, movement, and signaling of hormones create gradients that regulate the fate of cells. Among the numerous hormones involved, auxin and cytokinin are crucial for maintaining and positioning meristem activity. The balance between these two hormones is vital for the equilibrium between cell division and differentiation, with their interaction leading to a stable state that governs cellular identity (Khedr and Khedr 2024a,b). This balance is fundamental to a cell's ability to respond to hormonal signals (Weyers and Paterson 2001).

In the root, auxin controls the division of stem cells, while cytokinin promotes differentiation. In contrast, in the shoot, auxin accumulation triggers organ formation, while cytokinins facilitate cell proliferation. These hormones activate specific transcription factors that lead to significant changes in gene expression. Auxin signaling, for example, destabilizes its repressors to activate downstream transcription factors, while cytokinin uses a phosphorelay system to activate transcriptional response regulators. Both hormones also regulate the expression of their own repressors, creating negative feedback that controls the intensity and duration of the response (Azizi et al. 2015). Despite advancements in understanding these pathways, the genetic dissection of these signaling mechanisms remains complex due to the redundancy of transcriptional regulators. While transcriptional reporters are used to track hormone activity, deciphering the network of signaling components remains challenging. Additionally, the role of transcriptional repression in these processes is still not fully understood. Many hormone pathways are linked to TOPLESS (TPL), a corepressor that influences transcription through chromatin modifications, particularly involving histone deacetylases (HDACs) (Plant et al. 2021).

In embryogenesis, auxin and cytokinin establish spatial boundaries through complementary signaling patterns, although the precise molecular mechanisms remain unclear. Genetic circuits have been identified where auxin signaling influences cytokinin's regulators of biosynthesis and signaling, and vice versa. Recent work has shown that auxin forms a minimum at the boundary between the meristematic and elongation zones of the root, guiding the transition to differentiation. Cytokinin supports this by repressing auxin signaling and inactivating auxin through the *GH3* gene family, which helps define the area where differentiation occurs (Vielba 2019).

The ability of plants to regenerate also relies on auxin and cytokinin. By manipulating the balance of these hormones, organ formation can be induced in explants. This has been a key finding since the 1950s, when cytokinins were first identified (Letham 2019). For example, high auxin to cytokinin ratios promote callus formation, while high cytokinin to auxin ratios induce shoot formation (Kurepa and Smalle 2022). The flexibility of organ identity is demonstrated by experiments where cytokinin and auxin treatments can convert lateral root primordia into shoot meristems without forming callus tissue. This transformation involves a characteristic shift in hormone domains. For instance, auxin accumulates at the root tip, while cytokinin localizes in the central vasculature. As the shoot meristem develops, new auxin maxima form at leaf primordia, and a broad cytokinin domain is established in the meristem's inner layers. Studies on root regeneration show similar patterns of auxin and cytokinin interaction. After the root tip is excised, cells from various tissues acquire stem cell identity, demonstrating the regenerative potential of auxin and cytokinin signaling (Ćosić and Raspor 2022). By manipulating these hormones, the patterning of the root tip can be altered, highlighting the importance of hormone gradients in establishing the stem cell niche and patterning during both embryogenesis and regeneration.

4.4 Spatial Organization of Plant Stem Cells

The regulation of stem cell activity in both the SAM and the RAM follows a similar principle, with opposing gradients of differentiation along the primary axis of the plant. In both cases, a delicate balance between stem cell maintenance and differentiation dictates the formation of organs and tissues (Uchida and Torii 2019).

The SAM is responsible for producing all aerial organs of the plant, including the stem, leaves, and flowers. At its very tip, a small group of cells serves as the organizing center (OC), which maintains the stem cell population in the overlying central zone (CZ). In this region, stem cells divide, and their daughter cells are displaced either to the peripheral zone (PZ) at the edges of the meristem or to the rib meristem (RM) beneath the CZ. The displacement of daughter cells enables them to acquire the competence to differentiate into various organ primordia, such as leaves or flowers in the PZ, or into different cell types of the elongating stem in the RM. This dynamic organization allows for the continuous production of aerial organs in a coordinated fashion (Gaillochet and Lohmann 2015).

4.5 Regulatory Feedback in the Shoot and Root Stem Cell Niches

The maintenance of stem cell homeostasis in both the SAM and RAM involves intricate regulatory feedback mechanisms. Members of the WOX gene family, notably WUS in the shoot and WOX5 in the root, are crucial for maintaining these niches (Jha et al. 2020). The expression of WUS and WOX5 marks the early stages of stem cell niche establishment and is restricted to few cells at the center of each niche, which function as organizers. The loss of either WUS or WOX5 leads to the collapse of the respective stem cell niche (Dolzblasz et al. 2016). This tight regulation of WUS and WOX5 expression within the organizing cells is controlled by complex feedback mechanisms that ensure proper patterning and development.

In the SAM, WUS expression is regulated by a gradient of cytokinin. Specifically, an enzyme that activates cytokinin is expressed in the uppermost stem cell layer, which facilitates the diffusion of cytokinin to the lower layers of the SAM (Agarwal et al. 2022). The organizing cells of the shoot meristem, which harbor the WUS domain, have high levels of cytokinin receptors, sensitizing them to cytokinin and resulting in high cytokinin signaling (Wang et al. 2017). This signaling activates WUS expression, while WUS, in turn, promotes further cytokinin signaling by repressing negative regulators of cytokinin. This creates a positive feedback loop where WUS expression triggers the production of more WUS and elevates cytokinin signaling levels, ensuring the continuous maintenance of stem cells in the shoot (Yadav et al. 2010).

WUS expression in the OC is essential for determining stem cell fate in the overlying cells of the CZ (Lin et al. 2016). Although WUS is only expressed in the OC, its protein is mobile and can move into neighboring CZ stem cells via plasmodesmata (Kitagawa and Jackson 2017). Within the CZ, the WUS protein activates the expression of its own negative regulator, the small, secreted peptide CLAVATA3 (CLV3) (Hirakawa 2021). Interestingly, WUS binding on the regulatory region of CLV3 suggests that WUS regulates both activation and repression, depending on its concentration. At lower concentrations, WUS binds to the cis elements of CLV3 as monomers, whereas at higher concentrations, it binds as dimers. This differential binding accounts for the activation of CLV3 transcription only in the CZ, where WUS levels are lower, limiting the expression domain of WUS. In contrast, in the OC, where WUS levels are high, CLV3 transcription is repressed, and WUS expression becomes self-sustaining. This mechanism ensures that WUS is only expressed in the OC, and its expression in the CZ is tightly controlled by the activation of CLV3. The interplay between WUS and CLV3 thus serves as a key regulatory feedback loop that ensures the proper patterning of the stem cell niche and the maintenance of stem cell populations in both the shoot and root meristems (Hirakawa 2021).

4.6 Intercellular Trafficking Through Plasmodesmata

Plasmodesmata are essential channels that facilitate communication between plant cells, playing a crucial role in development and stem cell maintenance by allowing the passage of various signaling molecules. These channels are plasma membrane-lined, containing a strand of endoplasmic reticulum, and are much larger than animal gap junctions, with a size exclusion limit of around 10 kDa and a diameter of 50–60 nm (Li et al. 2021). Plasmodesmata are selective, allowing some molecules to pass through while blocking others, and their permeability can be regulated during cell maturation. In young, recently divided cells, the gap between the plasma membrane and endoplasmic reticulum in plasmodesmata is minimal. However, as cells mature and elongate, this gap increases, possibly to accommodate larger molecules (Bayer and Benitez-Alfonso 2024). An example of intercellular signaling via plasmodesmata is the movement of WUS and WOX5 transcription factors, which play key roles in the maintenance of stem cell niches in the shoot and root, respectively. In the shoot, WUS is expressed in the OC and moves to neighboring stem cells through plasmodesmata, promoting stem cell identity in the overlying CZ. Similarly, in the root, WOX5 is confined to the quiescent center (QC) but moves to the surrounding stem cells, particularly in the columella, to repress differentiation (Forzani et al. 2014). This intercellular movement of WOX5, driven by the WUS promoter, highlights the ability of WOX5 to migrate in the SAM and its endogenous mobility in the RAM (Fambrini et al. 2022).

In both cases, the movement of these transcription factors is essential for maintaining stem cell populations and their proper differentiation. The relationship between WOX5 and the surrounding stem cells is regulated by a gradient of auxin in the root. WOX5 expression coincides with a peak in auxin levels in the QC, and this feedback loop ensures the sustained expression of WOX5 in the QC, where it represses differentiation in the neighboring columella stem cells (Wang et al. 2020). Outside of the QC, auxin signaling repressors regulate WOX5 expression to restrict its activity to the QC. Additionally, chromatin regulation plays a significant role in limiting the expression domains of WUS and WOX5. The Sucrose Non-Fermenting 2 (SNF2) chromatin remodeling complex, including SPLAYED (SYD), regulates WUS expression in the SAM (Wu et al. 2012), while the REPRESSOR OF WUSCHEL1 (ROW1) confines both WUS and WOX5 expression to specific regions. ROW1 represses WOX5 expression in the proximal meristem by binding to H3K4me3 chromatin marks in the WOX5 promoter (Fambrini et al. 2022). This repression is critical for restricting the movement of WOX5 and WUS to their appropriate expression domains, ensuring proper stem cell maintenance and development.

Furthermore, WUS and WOX5 are highly conserved within the WOX gene family, with overlapping roles in stem cell regulation in both the shoot and root. While WUS and WOX5 differ in certain aspects of their function, they share a significant degree of functional overlap. The HAIRYMERISTEM (HAM) family, which acts as a cofactor for WUS, further exemplifies the complex interactions between WOX family members and their cofactors in regulating stem cell niches. These interactions between WOX and HAM family members are vital for maintaining stem cell identity and promoting proper development across plant tissues (Dolzblasz et al. 2016). The trafficking of key transcription factors like WUS and WOX5 through plasmodesmata is an essential mechanism for maintaining stem cell niches in plants, ensuring coordinated growth and development (Tvorogova et al. 2021). This intercellular communication is regulated by a combination of signaling molecules, chromatin modifications, and feedback loops, which together coordinate the complex processes of stem cell maintenance and differentiation in plant meristems.

4.7 Peptide Hormones in Meristem Development

Peptide hormones, such as those from the CLE family, are critical signaling molecules that regulate development and differentiation in plants, much like their roles in animal development (Fletcher 2020). A prominent example of such signaling in plants is the well-established relationship between WUS and CLV3 in the SAM, though the downstream mechanisms of CLV3 signaling remain less understood (Kuznetsova et al. 2023). In the wild-type SAM, CLV3 acts to repress WUS, thereby promoting differentiation in the PZ of the meristem (Han et al. 2020). This repression of WUS by CLV3 ensures that stem cells remain in the OC, while the surrounding cells differentiate to form various plant tissues.

The signaling mechanism of CLV3 is known to involve its movement to the peripheral meristem, where it activates a receptor signaling cascade. CLV3 binds to CLAVATA1 (CLV1), a leucine-rich repeat receptor kinase, which is a key component of the signaling complex (Hirakawa 2021). This binding initiates a cascade of signaling events that ultimately results in the repression of WUS expression. Despite the identification of CLV1 and other receptor complexes involved in this pathway, the precise molecular mechanisms that lead to WUS repression are still unclear (Fletcher 2018). Furthermore, different receptor complexes likely have distinct roles in stem cell regulation, with each complex contributing uniquely to meristem maintenance. In the root meristem, the function of CLE family peptides is similarly crucial for regulating stem cell populations. CLE40, a peptide closely related to CLV3, plays an important role in repressing WOX5 expression in the stem cells surrounding the QC (Song et al. 2022). While CLV3 is expressed in the shoot stem cells, CLE40 is expressed in differentiated columella cells and acts on receptors expressed in the columella stem cells, thereby regulating the stem cell pool in the root.

Similar to the function of CLV3 in the shoot, CLE40 in the root also signals through a receptor complex involving CLV1 and ACR4, another receptor-like kinase (Fletcher 2020). This complex senses CLE40 and activates downstream signaling pathways that regulate the maintenance of columella stem cells. Interestingly, other receptors that are involved in CLV3 signaling in the SAM also seem to play a role in detecting CLE peptides in the RAM (Yamaguchi et al. 2016). Beyond the root and shoot meristems, CLE family peptides also influence other aspects of root development. They regulate the differentiation of root vascular tissue and play crucial roles in lateral root development, affecting various stages of lateral root initiation and differentiation (Oh et al. 2018). This indicates that the CLE family of peptides has a broader role in root development, beyond just stem cell maintenance. Peptide hormones like CLE3 and CLE40 are essential for regulating stem cell maintenance and differentiation in both shoot and root meristems. These peptides function through receptor signaling complexes that activate downstream pathways, with significant effects on meristem size, cell differentiation, and overall development. Despite some mechanistic gaps, especially in the context of CLV3 signaling, CLE peptides continue to be key players in coordinating plant growth and tissue patterning (Ren et al. 2024).

Peptide hormones play a significant role in various aspects of plant differentiation, particularly in regulating specialized cell lineages and organ development. The epidermal patterning factor (EPF) family of peptides, for instance, is critical in regulating stomatal patterning and differentiation in the epidermis (Jia et al. 2024). Stomatal development begins with asymmetric divisions of the meristemoid mother cell, followed by a series of divisions that ultimately lead to guard cell formation. Peptides like EPF1 and EPF2 function by inhibiting transitions to specific stages of this development, such as EPF1 limiting the formation of guard mother cells and EPF2 controlling the initial asymmetric divisions of meristemoid mother cells (Jangra et al. 2021). This regulation helps ensure proper spacing and patterning of guard cells and pavement cells. The signaling pathway downstream of these peptides involves the ERECTA (ER) family of receptor kinases and the too many mouths (TMM) coreceptor, and activation is mediated through the mitogen-activated protein kinase (MAPK) pathway (Jordá et al. 2016). Beyond stomatal development, peptide hormones also regulate differentiation in the root, such as in the formation of the endodermis. The endodermis forms a waterproof barrier, the Casparian strip, to control solute uptake. Peptides like CIF1 and CIF2 are involved in proper Casparian strip formation, and their activity is dependent on the SCHENGEN3 (SGN3) receptor-like kinase in the endodermis (Doblas et al. 2017a,b). Peptide hormones are critical players in regulating plant differentiation, patterning, and organ formation by interacting with receptor kinases and signaling pathways. These peptides are not only involved in the regulation of meristem activity but also in more specialized differentiation processes, showing the intricate and diverse roles of peptide signaling in plant development.

Mobile transcription factors (TFs) are essential regulators of plant development, guiding cell fate decisions and patterning in various tissues. These factors, often responsive to hormonal cues, move between cells to coordinate developmental processes across long distances within the plant (Cohen and Efroni 2024). A prominent example of this is the PLETHORA (PLT) family of transcription factors, which are crucial for regulating the RAM. In response to sustained auxin accumulation, PLTs are expressed in the stem cells of the RAM, with their levels gradually decreasing in the shootward direction (Bitonti and Chiappetta 2011). Root growth factors (RGFs), a family of peptide hormones, regulate the PLT gradient by modulating PLT expression and facilitating meristem maintenance (Fang et al. 2021).

One of the best-characterized mobile transcription factors is shortroot (SHR), a key player in root patterning. SHR, a member of the GRAS family of transcription factors, is expressed in the vascular tissue and moves into adjacent cell layers, including the endodermis, stem cell progenitors, and the QC (Pierre-Jerome et al. 2018). SHR's movement is critical for the specification and differentiation of the endodermis, and it ensures proper root development by maintaining stem cell populations. The shr mutants exhibit severe defects in root growth and patterning, mainly due to impaired QC and stem cell function in the absence of SHR (Della Rovere et al. 2015). Additionally, SHR movement is vital for asymmetric division in the endodermal and cortex stem cells. Studies have shown that the proper division of these cells is tightly regulated by SHR and its interaction with the RETINOBLASTOMA (RB) protein, which controls cell cycle progression (Roufayel et al. 2021). Auxin signaling is thought to regulate the timing of these divisions, acting in concert with SHR and RB to initiate cell division at the correct spatial and temporal locations (Roufayel et al. 2021). The movement of transcription factors like SHR is not only a prerequisite for maintaining proper root patterning but also critical for nuclear-cytoplasmic compartmentalization. In SHR's case, while it is expressed both in the nucleus and cytoplasm of the vascular tissue, it requires nuclear localization to move into the neighboring endodermis, where it becomes sequestered in the nucleus to initiate transcriptional regulation (Kim et al. 2017).

The mobile transcription factor MUTE is another example of how TF mobility impacts plant development. MUTE regulates stomatal differentiation and cell division in leaf patterning, and its mobility has been a subject of recent interest. While MUTE is not mobile in Arabidopsis, its homolog in the grass species Brachypodium exhibits mobility, suggesting a role in coordinating additional supporting cells in stomatal development in monocots (Serna 2020). This difference in MUTE mobility could help explain the observed variations in stomatal patterning between dicots and monocots, further emphasizing how evolutionary changes in the mobility of transcription factors can contribute to species-specific developmental outcomes (Serna 2020). These examples illustrate the essential role of mobile transcription factors in plant development, where their movement and regulation across tissues help orchestrate complex processes such as root and shoot patterning, stem cell maintenance, and differentiation. The ability of transcription factors like PLTs, SHR, STM, and MUTE to move between cells allows for the precise coordination of cellular activities, ensuring proper development and adaptation to environmental cues. The study of these factors and their movement mechanisms offers valuable insights into the fundamental processes that govern plant growth and development.

4.8 The Role of CLE Peptides and Their Receptors During Root Development

Currently, 32 CLE genes, encoding 27 distinct CLE precursor proteins with a conserved 12- to 14-amino-acid CLE motif, have been identified in the Arabidopsis genome (Fletcher 2020). These secreted CLE peptides induce various intracellular signaling through LRR-RLKs during plant development (Furumizu and Aalen 2023). Due to the lack of visible phenotypes in cle null and knockdown mutants in the roots, overexpression and exogenous supplement experiments are commonly used to study CLE peptide functions (Liu et al. 2016).

The CLE peptide signaling pathway is crucial for maintaining the balance between stem cell maintenance and differentiation in the root meristem. Upon maturation through PTMs, CLE peptides interact with specific receptor-like kinases at the plasma membrane,

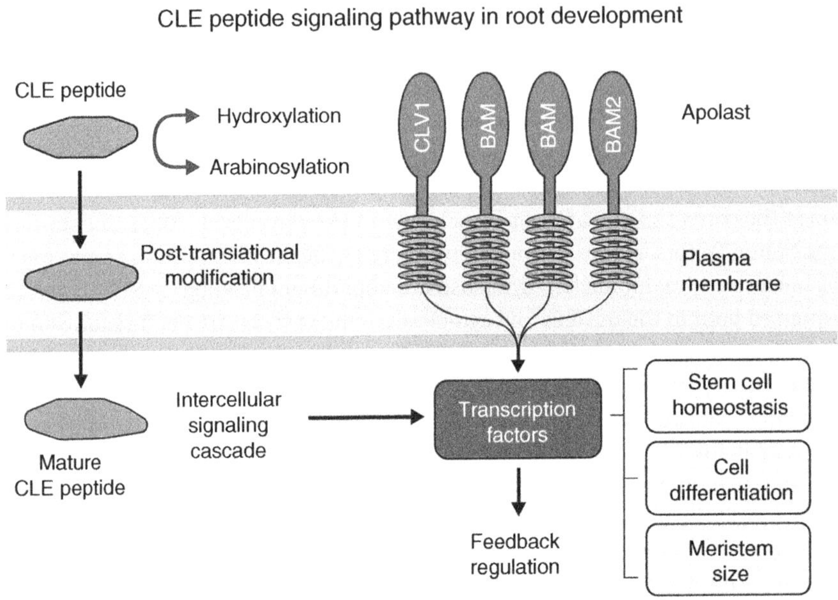

Figure 4.3 CLE peptide signaling pathway in root development.

triggering a transcriptional network that governs meristem activity. Disruption in this pathway, often due to genetic redundancy or impaired receptor interaction, can significantly impact root architecture (Figure 4.3).

4.9 The Roles of Tyrosine-Sulfated Peptides and Their Receptors in Root Development

Tyrosine-sulfated peptides play a crucial role in promoting root meristem development, contrasting with the negative regulation exerted by CLE peptides on meristematic activity. These peptides, including PSKs, PSYs, and RGFs, require tyrosine sulfation for their maturation, a PTM catalyzed by the enzyme TYROSYL-PROTEIN SULFOTRANSFERASE (TPST) (Yadav 2024). Mutants deficient in TPST, such as tpst-1, exhibit pleiotropic phenotypes, including severely shortened roots due to a reduction in the size of both the meristematic zone and the RAM (Yadav 2024). The root defects in tpst-1 mutants can be partially rescued by the exogenous application of PSY1 or PSK peptides, supporting the role of these peptides in promoting cell proliferation within the meristematic zone (Shinohara 2021). Furthermore, PSK is involved in the regulation of QC cell division and the differentiation of distal stem cells. Similarly, RGF1 peptide application restores root meristem size defects in the tpst-1 mutant, and the rgf1,2,3 triple mutant displays a short-root phenotype similar to that of tpst-1 mutants (Shinohara 2021). These findings suggest that PSK, PSY1, and RGF1 are all modified by TPST and are necessary for proper root meristem development (Shinohara 2021).

In addition to PSKs and PSYs, RGF peptides are vital for root meristem maintenance. Eleven RGF genes have been identified in the *Arabidopsis* genome (Zhang et al. 2024). RGF1 is specifically expressed in the QC cells and columella stem cells within the RAM, while RGF2 and RGF3 are primarily expressed in the innermost layer of columella cells and QC cells (Hsiao and Yamada 2020). These peptides help maintain the root stem cell niche and regulate meristematic activity (Hsiao and Yamada 2020).

The interaction between RGFR1 or RGFR2 and SOMATIC EMBRYOGENESIS RECEPTOR-LIKE KINASE (SERK1/2)/BAK1 has been demonstrated biochemically, with evidence showing that these interactions are crucial for RGF1-mediated signaling and meristem development. Mutants of serk and bak, which lack these coreceptors, display smaller meristems and shorter roots than wild-type plants, further underscoring the importance of these coreceptors in RGF signaling (Hsiao and Yamada 2020).

Recent studies have elucidated the downstream signaling cascade of RGF1-RGI1, which regulates root meristem maintenance. After RGF1 treatment, MITOGEN-ACTIVATED PROTEIN KINASE KINASE 4 (MKK4) and MAP KINASE 3 (MPK3) were identified as key components in the RGI1 complex (Lu et al. 2020). Constitutive expression of active MKK4 or MKK5 partially rescues the root defects in rgi1,2,3,4,5 quintuple mutants, indicating their involvement in the same pathway as RGF1 signaling. MPK3 and MPK6 play redundant roles in regulating mitotic activity in the root meristem, and their phosphorylation in response to sulfated RGF1 peptide is dependent on RGI and MKK4/5 (Lu et al. 2020). These findings highlight the importance of the MKK4/5-MPK3/6-PLT1/2 pathway in root meristem development.

Ubiquitination-mediated degradation also plays a critical role in modulating protein stability during root meristem development. UBP12 and UBP13, two closely related ubiquitin-specific proteases, act as positive regulators of root meristem development by stabilizing RGFR1 (Wang et al. 2024). The ubp12,13 double mutant shows reduced meristematic cell numbers and arrested root growth, and the decreased RGFR1 protein levels in this mutant result in insensitivity to RGF1 treatment. These findings indicate that UBP12/13 regulates RGF1 signaling by preventing RGFR1 degradation, thus ensuring proper root meristem function (Wang et al. 2024).

The RGF1-RGFR signaling pathway has also been implicated in regulating reactive oxygen species (ROS) distribution along the developmental zones of the root. RGF1 treatment increases $O2^{\bullet-}$ levels in the meristematic zone while decreasing H_2O_2 levels in the elongation and differentiation zones, thereby modulating root meristem size (Liu et al. 2024). Genetic analyses have shown that the RGF1 receptor pathway is essential for maintaining proper gradients of PLT proteins in the root meristem (Liu et al. 2024). Furthermore, the RGF1-induced expression of RGF1 INDUCIBLE TRANSCRIPTION FACTOR1 (RITF1) in the meristematic zone further supports its role in regulating root meristem development. Overexpression of RITF1 mimics the effects of RGF1 treatment, whereas ritf1 mutants show shorter roots and reduced meristem size, highlighting the critical role of RITF1 in mediating RGF1 signaling (Liu et al. 2024).

4.10 The Roles of RALF Peptides and Their Receptors in Root Development

RALF peptides and their receptors play crucial roles in regulating root development, with a particular focus on root growth, lateral root formation, and root hair development. RALF was first discovered in tobacco leaf as a cysteine-rich polypeptide that induces an increase in the pH of the surrounding medium. The disulfide bridges between cysteine residues are essential for the formation of active RALF, which in turn inhibits root growth. The first Arabidopsis RALF peptide was identified for its ability to elevate cytosolic calcium. Over 30 RALF genes have since been identified in Arabidopsis, with highly diverse expression patterns and roles in plant development. For instance, overexpression of AtRALF1 and AtRALF23 results in shorter and bushier plants, whereas knocking down AtRALF1 expression leads to increased root length, lateral root number, and hypocotyl elongation (Blackburn et al. 2020).

AtRALF1 is perceived by FER, a member of the *Catharanthus roseus* RLK1-like (CrRLK1L) family. FER was originally identified as a regulator of male and female gametophyte communication during fertilization (Zhang et al. 2020). Upon binding RALF1, FER initiates a signaling cascade involving increased cytosolic calcium, inhibition of cell elongation, and phosphorylation of the proton pump AHA2, which further reduces cell elongation. This cascade highlights the critical role of FER in root growth regulation. In addition to FER, the RPM1-induced protein kinase (RIPK) also associates with FER when RALF1 is present. RIPK mutants exhibit a short root hair phenotype and are insensitive to RALF1 treatment, similar to fer-4 mutants. Overexpressing RIPK can partially rescue the root hair defects in the fer-4 mutant, suggesting that FER and RIPK work together to regulate root growth (Zhang et al. 2020).

Further analysis of RALF peptides revealed that 20 out of 32 synthetic AtRALF peptides significantly inhibited root growth. Interestingly, fer-4 mutants were insensitive to 16 of these RALF peptides, indicating that FER is involved in the signaling pathway of most RALFs that inhibit root growth (Ji et al. 2020). Moreover, FER also interacts with RopGEFs, such as RopGEF4 and RopGEF10, which regulate ROS-mediated root hair development. The RALF1–FER complex also promotes the phosphorylation of eIF4E1, which regulates the translation of ROOT HAIR DEFECTIVE 6-LIKE 4 (RSL4) and ROP GTPase (ROP2; Rho-related GTPase from Plants 2), important players in determining root hair size and polarity. Notably, RSL4 negatively regulates RALF1 expression, creating a feedback loop that fine-tunes root hair development (Ji et al. 2020).

RALF34, another peptide, is perceived by THESEUS1 (THE1), a CrRLK1L member that helps regulate pericycle cell division during lateral root initiation. Mutants of RALF34 and THE1 display similar lateral root defects, and THE1 mutants are insensitive to RALF34-induced root growth inhibition (Hsiao and Yamada 2020). Additionally, FER plays a key role in maintaining the integrity of the actin cytoskeleton, which is necessary for the polar localization of PIN-FORMED 2 (PIN2) proteins. This localization contributes to lateral root development and gravitropic responses. Overall, these findings highlight the complex role of RALF peptides and their receptors in controlling root development, with FER serving as a central hub in various signaling pathways that regulate root growth and morphology.

4.11 Peptide Hormones That Regulate Leaf Senescence

Peptide hormones play a significant role in regulating various aspects of plant growth and development, including leaf senescence. In addition to classical phytohormones like auxin, cytokinins, and ethylene, small polypeptide signaling molecules are critical for intercellular communication in plants (El-Beltagi et al. 2025a,b). These peptides regulate numerous physiological processes such as root growth, organ abscission, and responses to environmental stresses like drought and salinity. For example, the peptide systemin induces the production of proteinase inhibitors in response to wounding (Chao et al. 1999), while RALF peptides are involved in processes such as root development and stress responses (Blackburn et al. 2020). Despite the known roles of these peptides in various developmental processes, their specific functions in leaf senescence remain largely unexplored.

Recent studies have shed light on how peptide hormones regulate leaf senescence. CLAVATA3/ESR-related 14 (CLE14), a small, secreted peptide, has been shown to delay leaf senescence by activating JUB1-dependent ROS scavenging genes, thereby reducing the levels of ROS in leaves (Zhang et al. 2022). CLE14 expression is induced by factors such as age, high salinity, drought, and hormones like ABA, salicylic acid (SA), and JA. This suggests that CLE14 acts as a "brake signal" to modulate age-dependent and stress-induced leaf senescence El-Beltagi et al. 2025c (Zhang et al. 2022). Similarly, CLE42 also delays leaf senescence by antagonizing the ethylene signaling pathway. CLE42 suppresses ethylene biosynthesis, which leads to the accumulation of EBF proteins and the decreased function of the EIN3 transcription factor, which is involved in senescence promotion (Zhang et al. 2022). In addition to CLE peptides, other peptide hormones such as PSK and PSY1 also play a role in leaf senescence regulation. PSK is perceived by the PSKR receptor, and loss-of-function mutants of PSKR show premature leaf senescence (Sauter 2015). Interestingly, PSKR homologs also contribute to PSY1 perception and exhibit overlapping functions with PSKR. Mutants lacking all three receptors show an enhanced senescence phenotype, suggesting a complex network of peptide hormone signaling in the regulation of senescence (Huang et al. 2022). Additionally, PTMs, such as tyrosine sulfation catalyzed by tyrosylprotein sulfotransferase (TPST), are crucial for the biological activity of PSK and PSY1.

The biological activities of PSK and PSY1 suggest that they may delay senescence by influencing chlorophyll content under heat stress (Jaiswal et al. 2022) or by elevating the expression of ROS scavenging genes to reduce oxidative damage. Peptide hormones interact with classical hormones to regulate leaf senescence, and their precise roles in this process are a subject of ongoing research. As more peptide hormones and their receptors are identified, a deeper understanding of their functions in leaf senescence is expected, potentially offering new insights for agricultural practices aimed at prolonging plant productivity and reducing postharvest losses.

4.12 Conclusions and Future Perspectives

Peptide hormones play essential roles in regulating various aspects of plant root development, including maintaining the root meristem, promoting cell expansion, and initiating lateral root and root hair formation. Despite increasing research on the roles of signal peptides in root development, many functions of these peptides remain poorly understood. A key challenge lies in the limited understanding of the receptor recognition mechanisms, intracellular signaling pathways, and the downstream ligand-receptor interactions that are central to peptide hormone functions. Therefore, further investigation into the molecular functions of these peptide hormones is necessary to fully understand their roles in root development.

One of the major challenges in studying the functions of CLE (CLAVATA3/ENDOSPERM) peptides is genetic redundancy. Many CLE mutants, which lack functional CLE peptides, do not exhibit distinct phenotypic changes, making it difficult to characterize their individual roles. Several unresolved questions remain, including identifying the specific functions of each CLE peptide, the receptors they interact with under different environmental and developmental conditions, and the effects of PTMs on their activity. PTMs, such as arabinosylation of hydroxyproline in the CLV3 peptide, are critical for the maturation and activation of peptide hormones. This modification enhances the peptide's ability to bind to receptors and inhibits SAM activity. However, only a few of the 32 CLE peptides have been studied in terms of modification. Previous studies using chemically synthesized CLE peptides without PTMs have demonstrated that these peptides can bind to multiple receptors. Mutants of the corresponding receptors show resistance to the inhibition activity of several synthetic peptides, but they may remain sensitive to the mature, modified forms of CLE peptides. Understanding how PTMs influence CLE peptide activity and their interaction with receptors is essential to elucidate their role in root meristem development.

Another important class of signaling peptides in root development is the RGF peptides. These peptides positively regulate root meristem development by enhancing the stability of the PLT1/2 proteins via ROS. However, the exact regulatory components involved in the RGF-RGFR-RITF1 signaling cascade and how ROS influence the PLT1/2 proteins at the translational level remain unclear. ROS have emerged as critical regulators of root cell division and differentiation, yet the precise mechanisms by which ROS modulate PLT1/2 function, potentially through PTMs like phosphorylation or ubiquitinylation, require further investigation. Identifying the receptors involved in transducing ROS signals will be key to advancing our understanding of how ROS regulates root development. Future research should focus on clarifying the role of ROS and their interplay with peptide hormones in regulating root meristem function. The RALF peptides are involved in regulating cell expansion and root development by modulating Ca^{2+} signaling, MAP kinase signaling, and pH levels. However, the exact sequence of events and interactions that lead to pH regulation by RALFs remains poorly understood. It is also important to consider how pH may affect the binding affinity between RALFs and their receptors. While some RALFs have been shown to interact with their CrRLK1Ls receptors, many potential interactions remain unexplored. Investigating the specificity of RALF-receptor associations and how individual RALFs coordinate complex signaling responses will be crucial for understanding their role in root development. Peptide hormones are critical in regulating plant root development, but many of their mechanisms remain unclear. Research efforts should focus on unraveling the molecular details of receptor recognition, signaling pathways, and the impact of PTMs. Additionally, understanding the complex interactions between peptide hormones, ROS, and other signaling molecules will provide deeper insights into the regulation of root development and overall plant growth.

Acknowledgments

We thank to Deanship of Scientific Research, Vice Presidency for Graduate Studies and Scientific Research, King Faisal University, Saudi Arabia (KFU251672), for supporting this research work.

References

Agarwal, Y., Shukla, B., Manivannan, A., and Soundararajan, P. (2022). Paradigm and framework of WUS-CLV feedback loop in stem cell niche for SAM maintenance and cell identity transition. *Agronomy* 12 (12): 3132.

Aggarwal, S., Kumar, A., Jain, M. et al. (2020). C-terminally encoded peptides (CEPs) are potential mediators of abiotic stress response in plants. *Physiology and Molecular Biology of Plants* 26: 2019–2033.

Ali, J., Mukarram, M., Ojo, J. et al. (2024). Harnessing phytohormones: advancing plant growth and defence strategies for sustainable agriculture. *Physiologia Plantarum* 176 (3): e14307.

Azizi, P., Rafii, M.Y., Maziah, M. et al. (2015). Understanding the shoot apical meristem regulation: a study of the phytohormones, auxin and cytokinin, in rice. *Mechanisms of Development* 135: 1–15.

Bayer, E.M. and Benitez-Alfonso, Y. (2024). Plasmodesmata: channels under pressure. *Annual Review of Plant Biology* 75: 291–317.

Bitonti, M.B. and Chiappetta, A. (2011). Root apical meristem pattern: hormone circuitry and transcriptional networks. *Progress in Botany* 72: 37–71.

Blackburn, M.R., Haruta, M., and Moura, D.S. (2020). Twenty years of progress in physiological and biochemical investigation of RALF peptides. *Plant Physiology* 182 (4): 1657–1666.

Bloom, J., Sun, S., and Al-Abed, Y. (2016). MIF, a controversial cytokine: a review of structural features, challenges, and opportunities for drug development. *Expert Opinion on Therapeutic Targets* 20 (12): 1463–1475.

Chao, W.S., Gu, Y.Q., Pautot, V. et al. (1999). Leucine aminopeptidase RNAs, proteins, and activities increase in response to water deficit, salinity, and the wound signals systemin, methyl jasmonate, and abscisic acid. *Plant Physiology* 120 (4): 979–992.

Cohen, I. and Efroni, I. (2024). Mobile signals, patterning, and positional information in root development. *Plant Physiology* 196 (4): 2175–2183.

Ćosić, T. and Raspor, M. (2022). The role of auxin and cytokinin signaling components in de novo shoot organogenesis. In: *Auxins, Cytokinins and Gibberellins Signaling in Plants*, 47–75. Cham: Springer International Publishing.

Cui, Y., Lu, X., and Gou, X. (2022). Receptor-like protein kinases in plant reproduction: current understanding and future perspectives. *Plant Communications* 3 (1): 100273.

Della Rovere, F., Fattorini, L., D'Angeli, S. et al. (2015). Arabidopsis SHR and SCR transcription factors and AUX1 auxin influx carrier control the switch between adventitious rooting and xylogenesis in planta and in in vitro cultured thin cell layers. *Annals of Botany* 115 (4): 617–628.

Doblas, V.G., Geldner, N., and Barberon, M. (2017a). The endodermis, a tightly controlled barrier for nutrients. *Current Opinion in Plant Biology* 39: 136–143.

Doblas, V.G., Smakowska-Luzan, E., Fujita, S. et al. (2017b). Root diffusion barrier control by a vasculature-derived peptide binding to the SGN3 receptor. *Science* 355 (6322): 280–284.

Dolzblasz, A., Nardmann, J., Clerici, E. et al. (2016). Stem cell regulation by Arabidopsis WOX genes. *Molecular Plant* 9 (7): 1028–1039.

El-Beltagi, H.S., Abdel-Haleem, M., EL-Mogy, M.M., and Khedr, E.H. (2025a). Resilience of forests in a changing climate: responses, mechanisms, and adaptations. *Notulae Botanicae Horti Agrobotanici Cluj-Napoca* 53 (1): 14360. https://doi.org/10.15835/nbha53114360.

El-Beltagi, H.S., Gad, M., Abdel-Haleem, M. et al. (2025b). Triacontanol: a multifunctional growth regulator for enhancing stress tolerance. *Journal of Crop Health* 77: 78. https://doi.org/10.1007/s10343-025-01144-6.

El-Beltagi, H.S., Abdel-Haleem, M., Rezk, A.A., and Khedr, E.H. (2025c). Progesterone as a Plant Physiochemical Regulator: Mechanisms and Efficacy in Alleviating Abiotic Stress. *Journal of Crop Health* 77 (2). https://doi.org/10.1007/s10343-025-01125-9.

EL-Beltagi, H.S. & Khedr, E.H. (2025). Biosensors and nanosensors: Innovative tools for precision fruit quality assessment. *Journal of Food Composition and Analysis*, 146, 2025,107957. https://doi.org/10.1016/j.jfca.2025.107957.

Escocard de Azevedo Manhães, A.M., Ortiz-Morea, F.A., He, P., and Shan, L. (2021). Plant plasma membrane-resident receptors: surveillance for infections and coordination for growth and development. *Journal of Integrative Plant Biology* 63 (1): 79–101.

Fambrini, M., Usai, G., and Pugliesi, C. (2022). Induction of somatic embryogenesis in plants: different players and focus on WUSCHEL and WUS-RELATED HOMEOBOX (WOX) transcription factors. *International Journal of Molecular Sciences* 23 (24): 15950.

Fang, Y., Chang, J., Shi, T. et al. (2021). Evolution of RGF/GLV/CLEL peptide hormones and their roles in land plant growth and regulation. *International Journal of Molecular Sciences* 22 (24): 13372.

Fedoreyeva, L.I. (2023). Molecular mechanisms of regulation of root development by plant peptides. *Plants (Basel)* 12 (6): 1320.

Fletcher, J.C. (2018). The CLV-WUS stem cell signaling pathway: a roadmap to crop yield optimization. *Plants (Basel)* 7 (4): 87.

Fletcher, J.C. (2020). Recent advances in Arabidopsis CLE peptide signaling. *Trends in Plant Science* 25 (10): 1005–1016.

Forzani, C., Aichinger, E., Sornay, E. et al. (2014). WOX5 suppresses CYCLIN D activity to establish quiescence at the center of the root stem cell niche. *Current Biology* 24 (16): 1939–1944.

Furumizu, C. and Aalen, R.B. (2023). Peptide signaling through leucine-rich repeat receptor kinases: insight into land plant evolution. *New Phytologist* 238 (3): 977–982.

Gaillochet, C. and Lohmann, J.U. (2015). The never-ending story: from pluripotency to plant developmental plasticity. *Development* 142 (13): 2237–2249.

Han, H., Liu, X., and Zhou, Y. (2020). Transcriptional circuits in control of shoot stem cell homeostasis. *Current Opinion in Plant Biology* 53: 50–56.

Haruta, M., Sabat, G., Stecker, K. et al. (2014). A peptide hormone and its receptor protein kinase regulate plant cell expansion. *Science* 343 (6169): 408–411.

Hastwell, A.H., Chu, X., Liu, Y., and Ferguson, B.J. (2024). The parallel narrative of RGF/GLV/CLEL peptide signalling. *Trends in Plant Science* 29: 1342–1355.

Hirakawa, Y. (2021). CLAVATA3, a plant peptide controlling stem cell fate in the meristem. *Peptides* 142: 170579.

Hobe, M., Müller, R., Grünewald, M. et al. (2003). Loss of CLE40, a protein functionally equivalent to the stem cell restricting signal CLV3, enhances root waving in Arabidopsis. *Development Genes and Evolution* 213: 371–381.

Hsiao, Y.C. and Yamada, M. (2020). The roles of peptide hormones and their receptors during plant root development. *Genes* 12 (1): 22.

Hu, C., Zhu, Y., Cui, Y. et al. (2018). A group of receptor kinases are essential for CLAVATA signalling to maintain stem cell homeostasis. *Nature Plants* 4 (4): 205–211.

Hu, C., Zhu, Y., Cui, Y. et al. (2022). A CLE–BAM–CIK signalling module controls root protophloem differentiation in Arabidopsis. *New Phytologist* 233 (1): 282–296.

Huang, P., Li, Z., and Guo, H. (2022). New advances in the regulation of leaf senescence by classical and peptide hormones. *Frontiers in Plant Science* 13: 923136.

Jaiswal, P., Sahi, A.N., and Barthakur, S. (2022). Cytokinin seed priming mediated induction of terminal heat stress tolerance and expression profiling of SKP1 transcripts, a component of ubiquitin proteasome system in bread wheat (*Triticum aestivum* L.): a transgenerational analysis. *Plant Growth Regulation* 98 (2): 259–280.

Jangra, R., Brunetti, S.C., Wang, X. et al. (2021). Duplicated antagonistic EPF peptides optimize grass stomatal initiation. *Development* 148 (16): dev199780.

Jha, P., Ochatt, S.J., and Kumar, V. (2020). WUSCHEL: a master regulator in plant growth signaling. *Plant Cell Reports* 39: 431–444.

Ji, D., Chen, T., Zhang, Z. et al. (2020). Versatile roles of the receptor-like kinase feronia in plant growth, development and host-pathogen interaction. *International Journal of Molecular Sciences* 21 (21): 7881.

Jia, M., Wang, Y., Jin, H. et al. (2024). Comparative genomics analysis of the Populus epidermal pattern factor (EPF) family revealed their regulatory effects in *Populus euphratica* stomatal development. *International Journal of Molecular Sciences* 25 (18): 10052.

Jordá, L., Sopeña-Torres, S., Escudero, V. et al. (2016). ERECTA and BAK1 receptor like kinases interact to regulate immune responses in Arabidopsis. *Frontiers in Plant Science* 7: 897.

Khan, S.S. and Luqman, S. (2025). Antimicrobial plant peptides: structure, classification, mechanism and therapeutic potential. *Current Topics in Medicinal Chemistry* 25 (27): 3103–3156.

Khedr, E.H. and Khedr, N. (2024a). Enhancing productivity, modifying biochemical parameters, and regulating the phenylpropanoid pathway in 'Le-Conte' pears through optimal protocatechuic acid treatments. *BMC Plant Biology* 24 (1): 50.

Khedr, E.H. and Khedr, N. (2024b). Unraveling the mechanisms behind salt stress alleviation and enhanced fruit storability in pomegranate through triacontanol treatment. *Scientia Horticulturae* 325 (325): 112634.

Khedr, E.H., Khedr, N., and Abdel-Haleem, M. (2023). Harnessing the metabolic modulatory and antioxidant power of 1-(3-phenyl-propyl) cyclopropane and melatonin in maintaining mango fruit quality and prolongation storage life. *BMC Plant Biology* 23 (1): 464. https://doi.org/10.1186/s12870-023-04485-4.

Khedr, E.H., El-Mogy, M.M., Abdel-Haleem, M., and El-Beltagi, H.S. (2025). Encapsulated GABA coatings as a postharvest strategy to enhance Mango chilling tolerance. *Plant Physiology and Biochemistry* 229: 110644. https://doi.org/10.1016/j.plaphy.2025.110644.

Kim, J. (2014). Four shades of detachment: regulation of floral organ abscission. *Plant Signaling & Behavior* 9 (11): e976154.

Kim, G., Dhar, S., and Lim, J. (2017). The SHORT-ROOT regulatory network in the endodermis development of Arabidopsis roots and shoots. *Journal of Plant Biology* 60: 306–313.

Kitagawa, M. and Jackson, D. (2017). Plasmodesmata-mediated cell-to-cell communication in the shoot apical meristem: how stem cells talk. *Plants (Basel)* 6 (1): 12.

Kurepa, J. and Smalle, J.A. (2022). Auxin/cytokinin antagonistic control of the shoot/root growth ratio and its relevance for adaptation to drought and nutrient deficiency stresses. *International Journal of Molecular Sciences* 23 (4): 1933.

Kutschmar, A., Rzewuski, G., Stührwohldt, N. et al. (2009). PSK-α promotes root growth in Arabidopsis. *New Phytologist* 181 (4): 820–831.

Kuznetsova, K., Efremova, E., Dodueva, I. et al. (2023). Functional modules in the meristems: "Tinkering" in action. *Plants (Basel)* 12 (20): 3661.

Letham, D.S. (2019). Cytokinins as phytohormones—sites of biosynthesis, translocation, and function of translocated cytokinin. In: *Cytokinins*, 57–80. CRC Press.

Li, Z.P., Paterlini, A., Glavier, M., and Bayer, E.M. (2021). Intercellular trafficking via plasmodesmata: molecular layers of complexity. *Cellular and Molecular Life Sciences* 78 (3): 799–816.

Li, J., Guo, Y., and Yang, Y. (2022). The molecular mechanism of plasma membrane H+-ATPases in plant responses to abiotic stress. *Journal of Genetics and Genomics* 49 (8): 715–725.

Lin, T.F., Saiga, S., Abe, M., and Laux, T. (2016). OBE3 and WUS interaction in shoot meristem stem cell regulation. *PLoS One* 11 (5): e0155657.

Liu, Y., Yang, S., Song, Y. et al. (2016). Gain-of-function analysis of poplar CLE genes in Arabidopsis by exogenous application and over-expression assays. *Journal of Experimental Botany* 67 (8): 2309–2324.

Liu, H., Mu, Y., Xuan, Y. et al. (2024). Hydrogen peroxide signaling in the maintenance of plant root apical meristem activity. *Antioxidants* 13 (5): 554.

Lu, X., Shi, H., Ou, Y. et al. (2020). RGF1-RGI1, a peptide-receptor complex, regulates Arabidopsis root meristem development via a MAPK signaling cascade. *Molecular Plant* 13 (11): 1594–1607.

Mohd-Radzman, N.A., Laffont, C., Ivanovici, A. et al. (2016). Different pathways act downstream of the CEP peptide receptor CRA2 to regulate lateral root and nodule development. *Plant Physiology* 171 (4): 2536–2548.

Mortier, V., Den Herder, G., Whitford, R. et al. (2010). CLE peptides control *Medicago truncatula* nodulation locally and systemically. *Plant Physiology* 153 (1): 222–237.

Mosher, S. and Kemmerling, B. (2013). PSKR1 and PSY1R-mediated regulation of plant defense responses. *Plant Signaling & Behavior* 8 (5): e24119.

Nakayama, T., Shinohara, H., Tanaka, M. et al. (2017). A peptide hormone required for Casparian strip diffusion barrier formation in Arabidopsis roots. *Science* 355 (6322): 284–286.

Ogilvie, H.A., Imin, N., and Djordjevic, M.A. (2014). Diversification of the C-TERMINALLY ENCODED PEPTIDE (CEP) gene family in angiosperms, and evolution of plant-family specific CEP genes. *BMC Genomics* 15: 1–15.

Oh, E., Seo, P.J., and Kim, J. (2018). Signaling peptides and receptors coordinating plant root development. *Trends in Plant Science* 23 (4): 337–351.

Ohyama, K., Shinohara, H., Ogawa-Ohnishi, M., and Matsubayashi, Y. (2009). A glycopeptide regulating stem cell fate in *Arabidopsis thaliana*. *Nature Chemical Biology* 5 (8): 578–580.

Pandita, D., Bhat, J.A., Wani, S.H. et al. (2023). Mobile signaling peptides: secret molecular messengers with a mighty role in plant life. *Journal of Plant Growth Regulation* 42 (10): 6801–6834.

Pautot, V., Crick, J., and Hepworth, S.R. (2025). Abscission zones: cellular interfaces for the programmed separation of organs. *Annals of Botany* 136: 29–48.

Pearce, G. and Ryan, C.A. (2003). Systemic signaling in tomato plants for defense against herbivores: isolation and characterization of three novel defense-signaling glycopeptide hormones coded in a single precursor gene. *Journal of Biological Chemistry* 278 (32): 30044–30050.

Pearce, G., Strydom, D., Johnson, S., and Ryan, C.A. (1991). A polypeptide from tomato leaves induces wound-inducible proteinase inhibitor proteins. *Science* 253 (5022): 895–897.

Pei, D., Hua, D., Deng, J. et al. (2022). Phosphorylation of the plasma membrane H+-ATPase AHA2 by BAK1 is required for ABA-induced stomatal closure in Arabidopsis. *The Plant Cell* 34 (7): 2708–2729.

Pierre-Jerome, E., Drapek, C., and Benfey, P.N. (2018). Regulation of division and differentiation of plant stem cells. *Annual Review of Cell and Developmental Biology* 34 (1): 289–310.

Pillitteri, L.J., Guo, X., and Dong, J. (2016). Asymmetric cell division in plants: mechanisms of symmetry breaking and cell fate determination. *Cellular and Molecular Life Sciences* 73: 4213–4229.

Plant, A.R., Larrieu, A., and Causier, B. (2021). Repressor for hire! The vital roles of TOPLESS-mediated transcriptional repression in plants. *New Phytologist* 231 (3): 963–973.

Rabiae, M., Abdelglil, M.I., Foda, N., Sayedahmed, M.M., Khedr, E.H., Abdel-Haleem, M. (2025). Leveraging endophytic fungi for enhancing plant resilience to abiotic stresses. *Journal of Crop Health* 77, 141. /https://doi.org/10.1007/s10343-025-01204-x

Ren, G., Zhang, Y., Chen, Z. et al. (2024). Research progress of small plant peptides on the regulation of plant growth, development, and abiotic stress. *International Journal of Molecular Sciences* 25 (7): 4114.

Roufayel, R., Mezher, R., and Storey, K.B. (2021). The role of retinoblastoma protein in cell cycle regulation: an updated review. *Current Molecular Medicine* 21 (8): 620–629.

Ryan, C.A. and Moura, D.S. (2002). Systemic wound signaling in plants: a new perception. *Proceedings of the National Academy of Sciences* 99 (10): 6519–6520.

Rzemieniewski, J. and Stegmann, M. (2022). Regulation of pattern-triggered immunity and growth by phytocytokines. *Current Opinion in Plant Biology* 68: 102230.

Sauter, M. (2015). Phytosulfokine peptide signalling. *Journal of Experimental Botany* 66 (17): 5161–5169.

Scheer, J.M. and Ryan, C.A. Jr. (2002). The systemin receptor SR160 from *Lycopersicon peruvianum* is a member of the LRR receptor kinase family. *Proceedings of the National Academy of Sciences* 99 (14): 9585–9590.

Serna, L. (2020). The role of grass MUTE orthologues during stomatal development. *Frontiers in Plant Science* 11: 55.

Shinohara, H. (2021). Root meristem growth factor RGF, a sulfated peptide hormone in plants. *Peptides* 142: 170556.

Song, X.F., Hou, X.L., and Liu, C.M. (2022). CLE peptides: critical regulators for stem cell maintenance in plants. *Planta* 255 (1): 5.

Stührwohldt, N. and Schaller, A. (2019). Regulation of plant peptide hormones and growth factors by post-translational modification. *Plant Biology* 21: 49–63.

Stührwohldt, N., Dahlke, R.I., Kutschmar, A. et al. (2015). Phytosulfokine peptide signaling controls pollen tube growth and funicular pollen tube guidance in *Arabidopsis thaliana*. *Physiologia Plantarum* 153 (4): 643–653.

Takahashi, F., Hanada, K., Kondo, T., and Shinozaki, K. (2019). Hormone-like peptides and small coding genes in plant stress signaling and development. *Current Opinion in Plant Biology* 51: 88–95.

Tan, W., Nian, H., Tran, L.S.P. et al. (2024). Peptides hormones: novel targets for modulating plant–rhizosphere microbe interactions. *Trends in Microbiology*.

Tvorogova, V.E., Krasnoperova, E.Y., Potsenkovskaia, E.A. et al. (2021). What does the WOX say? Review of regulators, targets, partners. *Molecular Biology* 55: 311–337.

Uchida, N. and Torii, K.U. (2019). Stem cells within the shoot apical meristem: identity, arrangement and communication. *Cellular and Molecular Life Sciences* 76 (6): 1067–1080.

Vielba, J.M. (2019). Identification and initial characterization of a new subgroup in the GH3 gene family in woody plants. *Journal of Plant Biochemistry and Biotechnology* 28 (3): 280–290.

Virág, D., Dalmadi-Kiss, B., Vékey, K. et al. (2020). Current trends in the analysis of post-translational modifications. *Chromatographia* 83: 1–10.

Wang, J., Kucukoglu, M., Zhang, L. et al. (2013). The Arabidopsis LRR-RLK, PXC1, is a regulator of secondary wall formation correlated with the TDIF-PXY/TDR-WOX4 signaling pathway. *BMC Plant Biology* 13: 1–11.

Wang, J., Tian, C., Zhang, C. et al. (2017). Cytokinin signaling activates WUSCHEL expression during axillary meristem initiation. *The Plant Cell* 29 (6): 1373–1387.

Wang, J., Su, Y., Kong, X. et al. (2020). Initiation and maintenance of plant stem cells in root and shoot apical meristems. *aBIOTECH* 1: 194–204.

Wang, X., Liu, X., Song, K., and Du, L. (2024). An insight into the roles of ubiquitin-specific proteases in plants: development and growth, morphogenesis, and stress response. *Frontiers in Plant Science* 15: 1396634.

Wang, S., Liang, Y., and Jiang, Y. (2025). Plant CLE peptides: functions, challenges, and future prospects. *Plant Hormones* 1: e006.

Weyers, J.D. and Paterson, N.W. (2001). Plant hormones and the control of physiological processes. *New Phytologist* 152 (3): 375–407.

Wu, M.F., Sang, Y., Bezhani, S. et al. (2012). SWI2/SNF2 chromatin remodeling ATPases overcome polycomb repression and control floral organ identity with the LEAFY and SEPALLATA3 transcription factors. *Proceedings of the National Academy of Sciences of the United States of America* 109 (9): 3576–3581.

Yadav, V. (2024). Endogenous peptides involved in plant growth and development. In: *Plant Elicitor Peptides: New Tool for Sustainable Agriculture*, 113–130. Singapore: Springer Nature Singapore.

Yadav, R.K., Tavakkoli, M., and Reddy, G.V. (2010). WUSCHEL mediates stem cell homeostasis by regulating stem cell number and patterns of cell division and differentiation of stem cell progenitors. *Development (Cambridge, England)* 137 (21): 3581–3589.

Yamaguchi, Y.L., Ishida, T., and Sawa, S. (2016). CLE peptides and their signaling pathways in plant development. *Journal of Experimental Botany* 67 (16): 4813–4826.

Zhang, X., Yang, Z., Wu, D., and Yu, F. (2020). RALF–FERONIA signaling: linking plant immune response with cell growth. *Plant Communications* 1 (4): 100084.

Zhang, Z., Liu, C., Li, K. et al. (2022). CLE14 functions as a "brake signal" to suppress age-dependent and stress-induced leaf senescence by promoting JUB1-mediated ROS scavenging in Arabidopsis. *Molecular Plant* 15 (1): 179–188.

Zhang, P., Zhao, J., Zhang, W. et al. (2024). Sulfated peptides: key players in plant development, growth, and stress responses. *Frontiers in Plant Science* 15: 1474111.

5

Meristem Activity Regulation by Peptide Hormones

Vijay P. Kapale[1], *Rakesh K. Bairwa*[2], *Jyoti P. Singh*[3], *Om Prakash Gupta*[4], *and Charan Singh*[2]

[1] *Department of Biochemistry, College of Basic Sciences and Humanities, G.B. Pant University of Agriculture and Technology, Pantnagar, Uttarakhand, India*
[2] *Crop Improvement, ICAR-Indian Institute of Wheat and Barley Research, Karnal, Haryana, India*
[3] *Plant Biochemistry, ICAR-National Bureau of Agriculturally Important Microorganisms, Mau, Uttar Pradesh, India*
[4] *Quality and Basic Science, ICAR-Indian Institute of Wheat and Barley Research, Karnal, Haryana, India*

5.1 Introduction

Meristems are the fundamental tissues in plants that sustain postembryonic growth and give rise to all plant organs throughout the plant's life cycle. These highly dynamic regions consist of undifferentiated stem cells that divide continuously, producing new cells that either retain their meristematic identity or differentiate into specialized tissues and organs. As such, meristems are the central hubs of development, driving the formation of the plant body and determining overall plant architecture. Their critical role in plant growth and form makes them key targets for both basic research and agricultural innovation (Lindsay et al. 2024).

Meristems are beneficially located in the plant body. The vertical growth of a plant is accomplished via the root apical meristem (RAM) and shoot apical meristem (SAM), which are located at the tips of the root and shoot, respectively. The plant's conducting system originates from the procambium meristematic tissues, which are located in vascular tissues between xylem and phloem. The epidermis, ground tissues, xylem, and phloem are among the many tissue types that are produced by the closely controlled division and differentiation of stem cells within these areas.

Cereal species, in contrast to the eudicot model *Arabidopsis thaliana*, exhibit a more elaborate meristem structure. Additional meristems, including seminal root meristems, are initiated by their embryos and are believed to contribute to adaptive features that were chosen during domestication (Yu et al. 2016; McKim 2019; Viana et al. 2022). In grasses, the identity of the meristem is defined by the organs it produces. For example, inflorescence meristems (IMs) are produced during the reproductive transition, whereas vegetative SAMs and RAMs produce leaves and roots, respectively. Floral organs are eventually produced by

Peptide Hormones in Plant Development and Stress Responses, First Edition. Edited by Vijay Pratap Singh and Samiksha Singh.

the IMs' subsequent differentiation into branch meristems (BMs), spikelet meristems (SMs), and floret meristems (FMs). A characteristic early sign of this developmental transition is a noticeable increase in SAM size (Bommert and Whipple 2018; Koppolu et al. 2022). Beyond canonical meristems, additional meristematic-like cells contribute to plant development. Archesporial cells (ARs), which act as precursors to germ cells, share regulatory mechanisms with SAM stem cells (Zhao et al. 2017). Similarly, developing leaves contain transient meristematic cells called meristemoids, which transform into specialized cell types like guard cells (Nunes et al. 2020).

The spatial arrangement of the stem cell niche is a crucial aspect of meristematic activity. Self-renewing stem cells in the central zone and developing cells in the peripheral or transition zones are balanced in both SAM and RAM. The peripheral or transition zones give rise to new tissues and organs. This delicate equilibrium is essential to sustain continuous growth and development, and it is governed by a highly coordinated network of gene regulatory interactions. These networks are influenced by a combination of several intrinsic genetic programs and extrinsic signaling pathways, including hormonal and environmental signals (Umeda et al. 2021).

Historically, research has focused on classical phytohormones such as auxin and cytokinin as primary regulators of meristem function. Auxin is known to promote cell division, organ initiation, and patterning, while cytokinin is essential for regulating meristem homeostasis and stem cell identity. However, new research has brought attention to the crucial functions of small signaling peptides that work in concert with classical hormones to regulate meristem activity.

5.2 Identification of Peptide Hormones in Plants

Peptide hormones are small polypeptide molecules that are released by plants and are crucial for intercellular communication (Lease and Walker 2006). Peptide hormones have historically been identified using a variety of analytical approaches, including chromatographic detection, chemical separation methods, phenotype-based genetic investigations, and bioassays (Fletcher et al. 1999). However, peptide hormones pose major difficulties for purification, separation, and biochemical analysis due to their incredibly low quantities in plant tissues (Hou et al. 2014). Researchers are gradually using high-throughput "omics" techniques, such as transcriptomics, proteomics, and more recently, peptideomics, to tackle these issues. The ability to identify and characterize peptide hormones in different plant species has significantly increased due to different methods and sophisticated detection instruments, including high-throughput mass spectrometry (MS) and *in vitro* activity assays (Yan et al. 2022). Furthermore, the identification of potential peptide hormones across several genomes has been expedited by recent developments in bioinformatics and machine learning (Hou et al. 2014).

A large number of peptide hormones are synthesized from small open reading frames (sORFs), which give rise to small secreted peptides (SSPs). These peptides generally exhibit common traits: they are usually less than 250 amino acids in length, have an N-terminal signal for secretion, include a conserved motif responsible for peptide function, and do not contain transmembrane regions or signals for retention in the endoplasmic reticulum (ER). In *Arabidopsis thaliana*, researchers identified 979 potential genes encoding SSPs, each ranging from 50 to 150 amino acids in length (Matsubayashi 2011). A similar analysis in maize (*Zea mays*) revealed 1491 candidate SSP genes with protein lengths under 200 amino acids (Li et al. 2014). With the increasing availability of plant genome sequences, the discovery of SSP-encoding genes has significantly grown. For instance, 4439 such genes have been found in *Medicago truncatula* (Boschiero et al. 2020), 4981 in *Triticum aestivum* (wheat) (Tian et al. 2022), and 1050 in *Solanum lycopersicum* (tomato) (Xu et al. 2023).

In addition to annotated SSPs, recent studies have focused on discovering peptides encoded by previously unannotated small ORFs, which may have been overlooked in earlier genome annotations. These hidden ORFs could potentially encode a vast number of undiscovered peptide hormones. In *Arabidopsis thaliana*, re-annotation and comprehensive genome analysis predicted 33,809 small ORFs that could encode SSPs (Lease and Walker 2006). In *Populus deltoides* (poplar), researchers have identified 12,852 sORFs that encode peptides between 10 and 200 amino acids in length (Yang et al. 2011). In rice (*Oryza sativa*), an impressive 101,048 small ORFs were found to potentially encode SSPs (Pan et al. 2012). These findings suggest that earlier studies focusing only on annotated protein-coding genes may have significantly underestimated the number of functional SSPs in plant genomes. However, there are some caveats. While large numbers of predicted ORFs have the potential to encode SSPs, many may be false positives due to incomplete annotations or a lack of functional validation. For instance, a recent study on tomato found that out of 61,306 predicted small ORFs, only 59 had the essential features of SSPs (Xu et al. 2023). This highlights the need for careful validation and more accurate predictive tools. Nonetheless, it remains likely that a substantial number of biologically relevant SSPs—and possibly peptide hormones—await discovery in plant genomes. So far, over 30 distinct families of peptide hormones have been discovered in various plant species. Some of the well-characterized families include:

- CLAVATA3/embryo surrounding region-related peptides (CLEs)
- C-terminally encoded peptides (CEPs)
- Phytosulfokines (PSKs)
- Epidermal patterning factors (EPFs)
- Rapid alkalinization factors (RALFs)

Most peptide hormones contain an N-terminal signal peptide, which enables their transport via the classical secretory pathway. However, this is not a universal feature. Certain peptide hormones lack this signal sequence and are therefore categorized as nonsecretory peptides. Examples of these include systemins (SYS), plant elicitor peptides (PEPs), *Zea mays* immune signaling peptide1 (ZIP1), and GRIM REAPER (GRI) peptides. Despite lacking classical secretion signals, nonsecretory peptides are also involved in key biological processes and can function through alternative signaling mechanisms.

Secretory and nonsecretory peptides are both essential for the regulation of various aspects of plant growth, development, and adaptation to environmental conditions. They participate in numerous biological processes, such as maintaining stem cell balance, shaping organ development, facilitating pollination and fertilization, supporting embryo and endosperm formation, guiding cell differentiation, initiating senescence, and mediating interactions with pathogens, insects, and environmental stressors. With ongoing advancements in genome sequencing, functional genomics, and peptide detection technologies, it is expected that many more peptide hormones will be discovered in the coming years. These findings will continue to deepen our understanding of plant signaling networks and may contribute to innovations in agriculture and crop improvement.

5.3 Classification of Peptide Hormones

Plant peptide hormones represent a diverse class of small signaling molecules that can be grouped according to several factors, such as their amino acid composition, biosynthesis pathways, secretion mechanisms, biological roles, origin, structural characteristics, and posttranslational modifications (PTMs). In this review, we broadly divide them into two primary categories—secretory peptides and nonsecretory peptides—based on their cellular localization and method of transport.

5.3.1 Secretory Peptide Hormones

Secretory peptide hormones are initially produced as precursor proteins called preproproteins, which are synthesized in the ER. These precursors are then processed and transported through the Golgi apparatus before being secreted into the extracellular space or targeted to the plasma membrane. Throughout this process, the peptides commonly undergo proteolytic cleavage along with various PTMs that are crucial for their stability and biological function (Hou et al. 2014). These peptides are among the most extensively studied and frequently identified in plants. They typically possess an N-terminal signal peptide that guides them through the classical secretory pathway via exocytosis.

Secretory peptides can be further categorized based on their biochemical properties, primarily into those that undergo PTMs and those enriched in cysteine residues (Tavormina et al. 2015).

5.3.2 Posttranslationally Modified Peptides

Posttranslational modifications are essential for the proper maturation, stability, receptor-binding capability, and overall biological function of secretory peptides. These modifications influence a variety of processes, including plant growth, defense mechanisms, and hormone signaling pathways (Taleski et al. 2018; Fletcher 2020). For instance, CLE peptides experience hydroxylation and glycosylation on proline residues, particularly within their conserved CLE motif. Such modifications are vital for peptide processing and interaction with their receptors (Matsubayashi 2011). PSKs, which are sulfated pentapeptides, are well known for their role in stimulating cell proliferation and differentiation. A key PTM in PSKs is the sulfation of a tyrosine residue in the conserved YIYTQ sequence, a modification crucial for their biological activity and receptor binding (He et al. 2024).

Hydroxyproline-rich systemins (HypSys) represent another class distinguished by extensive proline hydroxylation. These peptides play important roles in coordinating systemic responses to injury and pathogen attack in plants (Bhattacharya et al. 2013). Other examples of peptides reliant on PTMs for their function include C-terminally encoded peptides (CEPs), inflorescence deficient in abscission (IDA) and its related IDL proteins (IDA-like), as well as plant peptides containing sulfated tyrosine 1 (PSY1), all of which require specific modifications for their maturation and activity (Olsson et al. 2019).

5.3.3 Cysteine-Rich Peptides

Cysteine-rich peptides represent a significant group of secretory hormones characterized by the presence of 2 to 16 cysteine residues. These cysteines form disulfide bonds that are essential for proper peptide folding, maintaining structural stability, ensuring biological function, and protecting against proteolytic breakdown (Hou et al. 2014). Prominent families of cysteine-rich peptides include plant defensins (PDFs), defensin-like peptides (DEFLs), rapid alkalinization factors (RALFs), nonspecific lipid transfer proteins (nsLTPs), and knottin-like peptides. Many of these peptides play critical roles in regulating immune responses and mediating interactions between plants and pathogens (Missaoui et al. 2022; Tang et al. 2022). Beyond their immune functions, certain cysteine-rich peptides also contribute to plant growth and reproduction. For example, peptides such as RALFs, S-locus cysteine-rich protein/S-locus protein 11 (SCR/SP11), embryo sac4 (ES4), embryo surrounding factor1 (ESF1), and LURE peptides are involved in cell signaling pathways that are crucial for fertilization, embryogenesis, and tissue differentiation (Takeuchi 2021).

5.3.4 Nonsecretory Peptide Hormones

Unlike secretory peptide hormones, nonsecretory peptides do not utilize the classical ER–Golgi secretory pathway. These peptides generally lack an N-terminal signal peptide and are found within intracellular locations such as the cytoplasm, nucleus, chloroplasts, or mitochondria. Although some nonsecretory peptides may be secreted through nonconventional pathways (as described in Section 5.4), many function exclusively within the cells where they are produced. Nonsecretory peptides often have amino acid sequences enriched in proline (Pro), glycine (Gly), tyrosine (Tyr), or lysine (Lys). Despite their intracellular localization, they are biologically active and play important roles in physiological processes, particularly those related to stress responses and immunity. Prominent families of nonsecretory peptide hormones include:

- **Systemins (SYS)**: small peptides approximately 18 amino acids long,
- **Plant elicitor peptides (PEPs)**: peptides ranging from 23 to 36 amino acids,
- **GRIM REAPER (GRI) peptides**: lysine-rich peptides consisting of 11 amino acids.

These peptides are known to be key players in triggering plant immune defenses, boosting resistance to herbivores and pathogens, and coordinating defense signaling pathways (Tavormina et al. 2015).

5.4 Maturation of Peptide Hormones

The activation of plant peptide hormones from their inactive precursor forms involves a series of carefully regulated steps, including the removal of signal peptides, correct folding, proteolytic cleavage, and various PTMs. Mature peptides are usually shorter than 20 amino acids, contain few or no cysteine residues, and often feature modified amino acids like hydroxyproline (Hyp) or sulfated tyrosine (Tyr). Once processed, these peptides leave the cell either through the conventional secretory pathway, which includes the ER, Golgi apparatus, and other endomembrane compartments, or via alternative, nonclassical routes such as secretory vesicles, multivesicular bodies, vacuoles, or exocyst-positive organelles (EXPO).

5.4.1 Proteolytic Processing and Enzyme Roles

For many peptides, accurate proteolytic cleavage is critical for their biological function. For instance, the CLE40 preproprotein is processed by subtilisin-like serine proteases—specifically SBT1.4, SBT1.7, and SBT4.13—to produce the mature peptide responsible for regulating root development. Likewise, peptides from the IDA/IDL family, which are involved in organ abscission, are activated through cleavage by SBT4.12, SBT4.13, and SBT5.2, resulting in functional 14-amino-acid peptides. Important amino acid residues, such as proline at the second position and tyrosine at the fourth, are crucial for recognition by SBT4.13, with changes to these sites impairing the ability to rescue abscission defects in IDA mutants. In the case of PEP1, its maturation depends on cleavage by calcium-dependent type II metacaspases (MCs). An increase in cytosolic Ca^{2+} following wounding activates MC4, which then cleaves the PEP1 proprotein. Mutations affecting the catalytic cysteine residue of MC4 or alterations to the conserved arginine in the precursor prevent proper maturation of PEP1.

5.4.2 Posttranslational Modifications (PTMs)

The PTMs critically shape peptide stability, receptor binding, and biological function:

- Tyrosine sulfation is essential for the function of PSKs, which originate from precursors of 80 to 110 amino acids and are processed into sulfated pentapeptides. This sulfation is carried out by tyrosylprotein sulfotransferase (TPST) in the cis-Golgi, followed by proteolytic cleavage by SBT1.1 or SBT3.8 enzymes in the apoplast. TPST modifies PSKs, PSYs, RGFs, and CIF peptides and is essential for their interaction with cognate receptors.
- Hydroxyproline conversion and arabinosylation occur in families like CLEs, CEPs, RGFs, PSYs, systemins, and Serine-rich endogenous peptides (SCOOPs). Most hydroxylation leads to arabinosylation, which generally enhances receptor binding and bioactivity (Okamoto et al. 2013). However, exceptions exist: in *Medicago truncatula* CEP1-D1, tri-arabinosylation reduces biological activity by affecting nodule formation and lateral root formation. Also, in the Arabidopsis CLE40 peptide, although hydroxylation and arabinosylation at Pro4 do not affect root growth inhibition, modifications alter cleavage specificity by SBT proteases, thereby influencing the generated peptide species.

Hydroxyproline formation is catalyzed by prolyl-4-hydroxylases (P4Hs), and arabinosylation is mediated by arabinosyltransferases (ArafTs). While their involvement is known, the specific enzymes responsible for several peptide families remain elusive.

5.4.3 Structural Features of Peptide Hormones

Peptide hormones in plants display diverse structural characteristics that are vital for their biological activity, stability, and specificity. These features include variations in amino acid composition, peptide length, and the three-dimensional (3D) conformation of mature peptides. Structural elements influence not only the physical stability of the peptides but also their ability to bind to receptor kinases and other cellular targets. This modification thereby influences key physiological processes, including growth, development, and responses to stress.

I) **Small Motifs and Hydroxyproline Modifications**
A large number of peptide hormones originate from conserved C-terminal regions of their precursor proteins. For example, CLAVATA3/ESR-related (CLE) peptides are typically composed of 12–13 amino acids containing 1–3 conserved proline (Pro) residues. These Pro residues are frequently hydroxylated to hydroxyproline (Hyp), a PTM essential for full bioactivity. For CLV3, a thoroughly studied CLE peptide, the attachment of a tri-arabinoside sugar chain to a hydroxyproline residue induces a conformational change in the peptide backbone, greatly improving its biological activity (Shinohara and Matsubayashi 2013). Although the importance of these modifications for physiological function is well recognized, the exact molecular details underlying how they influence peptide structure are still not fully understood.

II) **CEP Family and β-Turn Formation**
Members of the CEP family display significant structural characteristics linked to PTMs. For instance, the mature Arabidopsis CEP1 (AtCEP1) peptide, which is 15 amino acids long, contains two proline residues that are hydroxylated. These modifications facilitate the formation of a β-turn-like secondary structure, which is crucial for effective receptor interaction (Bobay et al. 2013). Comparable structural changes have been suggested for MtCEP1 from *Medicago truncatula* and MhCEP1 from the root-knot nematode *Meloidogyne hapla*, indicating a conserved relationship between structure and function across different species.

III) **Cysteine-Rich Peptides and Complex Folding**
Cysteine-rich peptides (CRPs), such as SCR/SP11, which participate in self-incompatibility mechanisms in *Brassica*, have complex 3D structures. The mature SCR/SP11 peptide contains distinct structural features, including an L1 loop, an α-helix, and a β-sheet. Both the L1 loop and α-helix are key for receptor recognition and binding (Mishima et al. 2003).

Another group of CRPs, the nsLTPs, displays a compact architecture composed of four α-helices connected by three loops. These structural elements are reinforced by four disulfide bonds formed from eight conserved cysteine residues, which provide the protein with significant stability and enable its role in lipid transport (Hou et al. 2014).

IV) **Unique Structures in EPF Family Peptides**

Peptides belonging to the epidermal patterning factor (EPF) family possess unique structural characteristics. For instance, stomagen, an EPF-related peptide, features a loop and two antiparallel β-strands stabilized by three disulfide bonds, which together preserve its structure and ensure its biological activity (Ohki et al. 2011).

5.4.4 Challenges in Structural Characterization

Despite major advances in bioinformatics, peptide modeling, and MS, the detailed molecular structures of most plant peptide hormones remain poorly characterized. This is primarily due to their low natural abundance, rapid turnover, and the transient nature of many mature peptide forms in plant tissues. As a result, relatively few peptide hormones have been fully resolved at the molecular level, and much remains to be uncovered regarding the structure–function relationships that underlie their diverse physiological roles.

5.5 The Roles of Peptide Hormones in Regulation of Meristem Activity in Plants

5.5.1 Role of CLE Peptides in Meristem Development

CLE peptides are crucial regulators of stem cell balance in several meristematic tissues in plants, including the SAM, RAM, and vascular procambium (Song et al. 2022; Selby and Jones 2023).

5.5.1.1 CLE Peptides in the Shoot Apical Meristem (SAM)

In the SAM, CLE peptides—particularly CLAVATA3 (CLV3)—serve as crucial regulators that balance stem cell proliferation and differentiation. Loss-of-function mutations in the *clv3* gene cause an overaccumulation of stem cells, resulting in an enlarged SAM (Fletcher et al. 1999). CLV3 is recognized by a receptor complex that includes CLAVATA1 (CLV1) and its homologs BARELY ANY MERISTEMs (BAMs), the CLV2-CORYNE (CRN) complex, RECEPTOR-LIKE KINASE 2 (RPK2), along with coreceptors such as CLV3 insensitive kinases (CIKs) and CLERK (DeYoung and Clark 2008; Kinoshita et al. 2010). These receptors function to suppress WUSCHEL (WUS), a homeodomain transcription factor critical for sustaining stem cell identity. Conversely, WUS activates CLV3 expression, creating a tightly controlled negative feedback loop that maintains stem cell homeostasis in the SAM (Schoof et al. 2000; Ren et al. 2019). Research has shown that the 14-amino acid CLE motif of CLV3 alone is sufficient to regulate stem cells (Fiers et al. 2006). Specific amino acids within this motif, such as glycine, are essential for function, as replacing them with residues like threonine produces dominant-negative effects in Arabidopsis (Song et al. 2013). In addition to CLV3, other CLE peptides, including CLE16, CLE17, and CLE27, also contribute to regulating stem cells in both vegetative and reproductive SAMs (Jun et al. 2010). Although single or double mutants of these peptides may not display strong phenotypes, they act redundantly to support stem cell balance (Gregory et al. 2018).

Interestingly, CLE40, which is expressed in differentiated SAM regions in a pattern opposite to CLV3, promotes stem cell expansion by enhancing WUS expression through BAM1 signaling (Schlegel et al. 2021). This suggests two antagonistic CLE pathways: the CLV3-CLV1 pathway limits stem cell proliferation, whereas the CLE40-BAM1 pathway encourages stem cell maintenance and growth. In tomato, functional redundancy among CLE peptides is evident. When SlCLV3 is knocked out, its paralog SlCLE9 becomes upregulated, compensating to preserve stem cell homeostasis and maintain meristem function (Rodriguez-Leal et al. 2019).

5.5.1.2 CLE Peptides in the Root Apical Meristem (RAM)

CLE peptides play vital roles in root development, particularly in maintaining the columella stem cell (CSC) population and shaping overall root architecture. Application of synthetic CLE peptides or overexpression of their conserved CLE motifs leads to disruption of RAM activity, resulting in shortened primary roots and impaired lateral root formation (Kinoshita et al. 2007; Wang and Fiers 2010). These peptides, often referred to as "root-active CLEs," are recognized by specific receptors such as CLV2, CRN, and RPK2. Mutants lacking these receptors show resistance to the typical root phenotypes induced by CLE peptide treatments (Miwa et al. 2008; Muller et al. 2008; Hazak et al. 2017). Within the RAM, CLE40 plays a central role in modulating CSC numbers. Loss-of-function mutants in *cle40* accumulate excessive CSCs, whereas external application of CLE40 peptide suppresses their proliferation (Stahl et al. 2009). CLE40 signaling involves receptors like ARABIDOPSIS CRINKLY4 (ACR4), CLV1, RPK2, and coreceptors including CLERK and CIKs, and operates through regulation of the WOX5 transcription factor (Stahl et al. 2013; Ren et al. 2019). Additional CLEs, such as CLE16 and CLE17, also participate in CSC differentiation and lateral root initiation, acting via CLV1 and ACR4 (Zhang et al. 2022b). Other CLE members—specifically CLE1 through CLE7—are implicated in suppressing lateral root initiation and elongation (Araya et al. 2014; Nakagami et al. 2023). Notably, plants with mutations in all seven of these genes (the *cle1–cle7* septuple mutant) show significantly increased lateral root length and density (Nakagami et al. 2023). In terms of root hair formation, CLE14 promotes excessive root hair development by downregulating GLABRA2 (GL2) expression under normal growth conditions (Hayashi et al. 2018).

In vascular tissues, CLE peptides are critical regulators of procambial and cambial stem cell behavior (Fukuda and Hardtke 2020). For example, phloem-derived peptides like TDIF (also known as CLE41/CLE44) stimulate cambial proliferation while simultaneously inhibiting xylem differentiation through interaction with the TDIF RECEPTOR (TDR)/PHLOEM INTERCALATED WITH XYLEM (PXY) pathway, affecting WOX4 and WOX14 expression (Hirakawa et al. 2010; Etchells and Turner 2010). Conversely, other phloem-expressed peptides, such as CLE25, CLE26, CLE33, and CLE45, function to repress phloem formation and maintain proximal root meristem identity. Their activity is mediated through receptors like CLV2/CRN, BAM1/BAM3, and CLERK/CIK (Depuydt et al. 2013; Carbonnel et al. 2023; Ren et al. 2019; Hu et al. 2022).

CLE peptides are also involved in xylem development. For instance, overexpression of CLE19 in *Brassica napus* disrupts normal xylem formation, producing fragmented xylem structures and vascular islands in floral tissues (Fiers et al. 2004). In *Arabidopsis thaliana*,

CLE9 and CLE10 suppress protoxylem formation by inhibiting the expression of ARR5 and ARR6 (Kondo et al. 2011).

Due to their central role in meristem maintenance, CLE peptides are strongly implicated in plant regeneration processes. These involve dedifferentiation and reestablishment of new meristems. Several CLE genes are actively expressed in callus tissues, which possess regenerative potential. CLE1–CLE7 peptides inhibit de novo shoot regeneration, whereas cle1-cle7 mutants demonstrate enhanced shoot-forming ability. This regulation occurs through suppression of WUSCHEL (WUS) expression via CLV1 and BAM1 receptors (Kang et al. 2022).

In reproductive development, CLE peptides also have important functions. CLE8 is expressed in early-stage embryos and endosperms. Mutations in this gene result in reduced seed size and developmental defects, likely linked to disruptions in WOX8 expression (Fiume and Fletcher 2012). CLE19, localized in the embryonic epidermis—especially in cotyledons and hypocotyls—plays a role in seed and embryo development. Plants expressing dominant-negative CLE19 variants exhibit seed abortion and delayed endosperm proliferation (Xu et al. 2015). Furthermore, CLE19 contributes to normal pollen development through its interaction with the PXY-LIKE1 (PXL1) receptor, which regulates tapetal gene expression (Yu et al. 2023b). CLE peptides also influence the timing of leaf senescence. For instance, cle14 mutants show early leaf senescence, whereas overexpression of CLE14 or peptide application delays senescence. This effect is mediated by upregulation of JUB1, a transcriptional repressor of senescence, through enhanced reactive oxygen species (ROS) scavenging (Han et al. 2022). Similarly, CLE42 extends leaf longevity by interfering with the ethylene biosynthesis pathway (Zhang et al. 2022a), highlighting that CLE14 and CLE42 regulate senescence through distinct mechanisms.

In response to environmental cues, CLE peptides are dynamically involved in stress adaptation and nutrient sensing (Xie et al. 2022; Bashyal et al. 2024). Their expression is regulated by hormones and external factors. For example, CLV3 may be recognized by the FLS2 receptor, initiating immune responses in the SAM (Lee et al. 2011), and may also function through CLV1 and CLV2 to mediate immunity against *Ralstonia solanacearum* via the miR169/NF-YA module (Hanemian et al. 2016).

Under drought conditions, CLE9 and CLE25 promote stomatal closure, improving water-use efficiency. Loss-of-function mutants for these peptides show increased drought sensitivity (Takahashi et al. 2018). CLE45 plays a role in reproductive thermotolerance by facilitating pollen tube elongation under high-temperature conditions, without affecting germination, through SKM1/SKM2 receptors (Endo et al. 2013). Temperature changes also influence floral development, with CLV3 and CLE25 contributing to flower outgrowth under shifting thermal regimes (John et al. 2023).

Nutrient deficiencies also impact CLE gene expression. Under nitrogen scarcity, CLE1, CLE3, CLE4, and CLE7 are upregulated in the roots, with CLE3 signaling through phloem-localized CLV1 to inhibit lateral root formation (Araya et al. 2014). In phosphate-limited conditions, CLE14 is induced and signals through CLV2 and PEPR2 to reduce meristem activity by repressing SHR, SCR, and WOX5 (Gutiérrez-Alanís et al. 2017). Sulfur deficiency downregulates CLE2 and CLE3, resulting in reduced lateral root density. However, this phenotype is absent in clv1 mutants, implicating the CLE–CLV1 pathway in sulfur-dependent root development (Dong et al. 2019).

CLE genes are highly conserved across land plants (Oelkers et al. 2008; Goad et al. 2017) and even found in plant-parasitic nematodes (Wang et al. 2005). CLV3 and its orthologs govern stem cell regulation via conserved feedback loops (Somssich et al. 2015; Hirakawa and Sawa 2019). In *Marchantia polymorpha*, MpCLE1–MpTDR negatively regulates apical stem cell activity, while MpCLE2–MpCLV1 has an opposing, activating role (Hirakawa et al. 2020). In rice, homologs such as FON2/CLV3 and FCP1 maintain shoot meristems through distinct mechanisms (Suzaki et al. 2008). In maize, loss of ZmCLE7 or ZmFCP1 results in meristem enlargement and abnormal organogenesis (Je et al. 2018). Similarly, in tomato, a mutation in a CLV3-like gene leads to the fasciated phenotype characterized by extra floral locules (Rodriguez-Leal et al. 2019).

In legumes, CLE peptides are key regulators of autoregulation of nodulation (AON). In *Medicago truncatula* and *Lotus japonicus*, AON triggers CLE peptide expression, which is detected by the SUNN and HAR1 receptors, respectively, to suppress further nodule development (Nishida and Suzaki 2018; Mens et al. 2021; Moreau et al. 2021).

5.5.2 Role of CEP Peptides in Regulation of Meristem Activity

C-terminally encoded peptides (CEPs) are conserved across many plant species (Delay et al. 2013; Ogilvie et al. 2014) and are key regulators of root architecture, nitrogen signaling, nodulation, and stress responses (Roberts et al. 2013; Taleski et al. 2018). Application or overexpression of CEPs typically inhibits primary root elongation and reduces lateral root formation in a receptor-dependent manner, primarily via CEP receptors (CEPRs) (Ohyama et al. 2008; Roberts et al. 2016). In contrast, cep mutants show enhanced primary root growth and increased lateral root density (Huang et al. 2023). In *Medicago truncatula*, MtCEP1 specifically restricts lateral root development (Imin et al. 2013; Mohd-Radzman et al. 2016), and sucrose induces CEP expression, further suppressing lateral roots (Chapman et al. 2019).

CEP peptides also shape overall root system architecture. In both Arabidopsis and *M. truncatula*, they promote the formation of shallow root systems (Chapman et al. 2020). Interestingly, in species like *Brassica rapa* and *Cucumis sativus*, certain CEPs have opposite effects—enhancing primary root elongation instead (Liu et al. 2021).

Beyond local root regulation, CEPs serve as systemic signals that optimize nitrogen uptake and transport. Under nitrogen deficiency, root-expressed CEPs are transported to shoots, where they are recognized by CEPRs in leaves. This triggers expression of nitrate transporters and enhances nitrogen acquisition (Tabata et al. 2014; Ohkubo et al. 2017; Ota et al. 2020). This CEP–CEPR system is conserved in various species, including rice, apple, and *M. truncatula* (Bourion et al. 2014; Sui et al. 2016).

In legumes, CEPs also regulate nodulation. Nitrogen-starved *M. truncatula* plants upregulate CEP expression, which signals through the COMPACT ROOT ARCHITECTURE 2 (CRA2) receptor to enhance nodule formation by modulating ethylene signaling (Huault et al. 2014; Luo et al. 2022). CEP-deficient mutants generated via CRISPR show impaired nodulation and reduced lateral root formation, while MtCEP peptide application increases nodule numbers (Laffont et al. 2020; Zhu et al. 2021; Ivanovici et al. 2023).

CEPs are also responsive to abiotic stress. For instance, CEP3 is highly induced by salinity; cep3 mutants perform better under salt, nitrogen limitation, acidity, and osmotic stress

(Delay et al. 2013). Exogenous CEP treatment helps alleviate salt stress in cucumber and tomato (Liu et al. 2022a; Shen et al. 2023). Similarly, CEP4 is upregulated under nitrogen stress and contributes to immune responses (Rzemieniewski et al. 2022), while CEP5, induced by osmotic stress, enhances drought tolerance likely through modulation of auxin signaling via both CEPR-dependent and -independent mechanisms (Smith et al. 2020). In tomato, synthetic peptides SlCEP10 and SlCEP11b improve drought resilience (Xu et al. 2023).

5.5.3 Role of RALF Peptides in Regulation of Meristem Activity

Rapid alkalinization factor peptides are important regulators of root meristem function, influencing processes such as cell expansion, hormone pathways, and changes in extracellular pH (see Table 5.1). These peptides were first recognized for their capacity to cause a swift increase in the pH of the extracellular environment. Among them, RALF1 has been particularly noted for its ability to suppress root growth and reduce meristematic activity in *Arabidopsis thaliana* (Pearce et al. 2001; Wang et al. 2022).

Overexpression of RALF1, RALF8, and RALF23 results in compact, semi-dwarf plant phenotypes characterized by smaller leaves and markedly shorter roots. These developmental changes rely on the presence of the receptor-like kinase FERONIA (FER) (Srivastava et al. 2009; Atkinson et al. 2013; Bergonci et al. 2014). In contrast, ralf1 loss-of-function mutants show enhanced root meristem function, including longer primary roots, more lateral roots, elongated hypocotyls, and expanded root cells (Bergonci et al. 2014). These contrasting phenotypes suggest that RALF1 acts as a negative regulator of root growth and cell expansion within the meristem.

Table 5.1 Key structural and functional features of PIP/PRIME and RALF peptides.

Feature	PIP/PRIME peptides	RALF peptides	References
Typical size (mature)	15–25 amino acids	80–120 AA	Yu et al. (2023a) and Campbell and Turner (2017)
Key conserved motifs	A highly conserved SGP-rich C-terminus motif	The mature peptide features a single RR motif, a YISY motif, and four highly conserved cysteine residues	Yu et al. (2023a) and Liu et al. (2024)
Processing	Cleaved from PROPEPs and signal peptide (SP) removal	Cleaved from preproprotein, signal peptide removal, and protease cleavage	Yu et al. (2023a) and Liu et al. (2024)
Primary receptors	PEPRs	FERONIA (FER)	Yamaguchi et al. (2010) and Du et al. (2016)
Key functions	Intercellular communication, Plant elicitor, induces/amplifies defense	Rapid alkalinization; cell expansion, root/pollen tube growth	Ryan et al. (2007) and Campbell and Turner (2017)

RALF1 also promotes root hair elongation, a process that depends on its interaction with the FER receptor (Zhu et al. 2020). The RALF1–FER signaling axis modulates meristem activity by adjusting levels of ROS and interacting with abscisic acid (ABA) signaling pathways (Duan et al. 2010; Chen et al. 2016). Upon binding to FER, RALF1 triggers phosphorylation of the H^+-ATPase AHA2 at the plasma membrane, which inhibits proton pumping, elevates apoplastic pH, and ultimately restricts root cell expansion and meristem function (Haruta et al. 2014).

In addition, the inhibitory effects of RALF peptides on root development are closely linked to intact auxin synthesis and signaling, positioning them as key players in the hormonal control of meristem dynamics (Li et al. 2022). RALF1 also raises cytoplasmic Ca^{2+} levels (Haruta et al. 2008), and calcium influx is essential for the alkalinization response. Blocking Ca^{2+} channels suppresses this effect, supporting a central role for Ca^{2+} signaling in RALF-induced inhibition of root growth (Gjetting et al. 2012).

Together, these studies point to RALF1 as a critical inhibitor of root meristem activity, acting through a multifaceted mechanism involving apoplastic pH shifts, ROS and calcium signaling, and hormone cross talk—all largely dependent on FER-mediated perception.

Beyond internal signaling networks, exogenous RALF-like peptides also influence root development and meristem regulation. For instance, the RALF23–FER module contributes to rhizosphere equilibrium by regulating ROS production, thereby impacting both immune responses and redox-mediated maintenance of the meristem (Song et al. 2021; Tang et al. 2022).

Remarkably, several pathogens—including fungi and nematodes—produce RALF-mimicking peptides that hijack host signaling pathways to their advantage. These mimic peptides engage FER receptors to modulate plant immunity and may disrupt meristem integrity as a secondary effect. For example, root-knot nematodes (*Meloidogyne incognita*) secrete peptides such as MiRALF1 and MiRALF3, which interact with host FER to dampen immune signaling and potentially impact root development via altered growth-defense dynamics (Yimer et al. 2023). Similarly, the fungal pathogen *Fusarium* releases F-RALF to boost its virulence in tomato, likely affecting root architecture and function (Masachis et al. 2016).

Interestingly, beneficial microbes also utilize RALF-like signaling to mediate host interactions. The root endophyte Colletotrichum tofieldiae produces CtRALF, which directly engages FER to promote symbiosis (Liao et al. 2023). Although primarily involved in beneficial interactions, such FER-dependent signaling also influences root development through changes in hormone pathways and cell wall properties.

Altogether, the available evidence highlights a central role for both plant-derived and exogenous RALF peptides in controlling root meristem activity. These peptides exert their effects primarily through FER-mediated pathways, orchestrating changes in apoplastic pH, ROS, and calcium signaling, and the balance between growth and defense.

5.5.4 Role of RGF/GLV/CLEL Peptides in Regulation of Meristem Activity

The RGF/GLV/CLEL family consists of 11 tyrosine-sulfated peptides encoded in the *Arabidopsis thaliana* genome (Meng et al. 2012; Whitford et al. 2012). These peptides are primarily recognized by the leucine-rich repeat receptor-like kinases RGF1 INSENSITIVE

(RGI) or RGF RECEPTOR (RGFR), and they are mainly expressed in the RAM and leaves (Fernandez et al. 2013).

RGF peptides are key regulators of root meristem development and maintenance across various plant species (Fang et al. 2021). While individual knockouts of rgf1, rgf2, or rgf3 do not lead to major root abnormalities, plants lacking all three genes exhibit significantly shorter roots. This phenotype can be reversed by applying the synthetic RGF1 peptide, highlighting its functional redundancy and importance (Matsuzaki et al. 2010). On the other hand, constitutive expression of RGF genes results in an enlarged RAM, reduced lateral root formation, and irregular anticlinal cell divisions in the pericycle, all of which are dependent on RGFR-mediated signaling (Fernandez et al. 2015).

At the mechanistic level, RGF peptides help preserve the root stem cell niche by regulating both the expression and protein stability of PLETHORA1 (PLT1) and PLETHORA2 (PLT2), two transcription factors essential for RAM identity (Matsuzaki et al. 2010). RGF1 also induces the transcription factor RGF1-INDUCIBLE TRANSCRIPTION FACTOR 1 (RITF1), which enhances superoxide anion production. This oxidative signal contributes to PLT2 stabilization, further supporting meristem function (Yamada et al. 2020).

More recent findings suggest that acidic extracellular conditions enhance RGF signaling. Low pH strengthens the interaction between RGF1 and its receptor RGI1, thereby boosting meristem activity. Interestingly, this improved signaling under acidic conditions also dampens immune responses, suggesting a trade-off mechanism that reallocates resources toward growth (Liu et al. 2022b).

In conclusion, RGF/GLV/CLEL peptides act as vital developmental signals in root architecture by orchestrating cell division patterns, stem cell maintenance, and spatial organization of the RAM. Their finely tuned signaling pathways ensure a resilient and adaptable root system responsive to environmental changes.

5.5.5 Role of Epidermal Patterning Factor (EPF)/Stomagen Peptides in Regulation of Meristem Activity

The EPF/EPF-LIKE (EPFL)/stomagen peptide family functions as key non-cell-autonomous signaling molecules that regulate various aspects of plant development, including meristem activity, organ patterning, and cellular differentiation (Zeng et al. 2020). Several members of this peptide family are particularly involved in processes that influence SAM behavior, stem growth, and reproductive meristem function.

Among them, EPF1 and EPF2 act as negative regulators of stomatal development by limiting cell division within the stomatal lineage. Loss-of-function mutants for these peptides display elevated stomatal density, whereas overexpression reduces stomatal formation (Hara et al. 2007, 2009). In contrast, EPFL9 (also known as stomagen) serves an opposing role by promoting stomatal differentiation (Hunt and Gray 2009). All three peptides exert their effects via the receptor kinases ERECTA (ER) and ERECTA-LIKE 1 (ERL1), which are also expressed in meristematic regions, indicating a close connection between stomatal patterning and broader developmental programs originating from meristem tissues. Notably, EPFL9 competes with EPF2 for binding to ER-family receptors, providing a regulatory mechanism that balances proliferation and differentiation within the epidermis (Lee et al. 2015).

Beyond their role in stomatal control, EPFL4, EPFL5, and EPFL6 contribute to procambial cell division and stem elongation, highlighting their function in vascular meristem activity and vertical growth of the shoot (Ikematsu et al. 2017; Fischer and Teichmann 2017). Modulating the expression of these peptides can impact the timing and extent of shoot elongation, directly influencing plant architecture.

EPFL peptides also play essential roles in reproductive development. EPFL2 and EPFL9 jointly regulate the initiation of ovules, which determines final seed count and fruit size (Kawamoto et al. 2020). Additionally, EPFL1 through EPFL6 exhibit redundant activity in promoting integument elongation—a developmental process originating from the ovule meristem (Li et al. 2023). Furthermore, EPFL4, EPFL5, and EPFL6 contribute to the elongation of stamen filaments, a process critical for effective pollination, particularly under fluctuating temperature conditions (Negoro et al. 2023; He et al. 2023).

In rice, homologs of the EPF/EPFL peptide family are deeply integrated into SAM function and organogenesis. For instance, overexpressing certain EPF genes results in reduced plant stature and enlarged SAMs, suggesting a regulatory role in meristem size and organ initiation potential (Mohammed et al. 2019). Conversely, suppression of OsEPFL1 and OsEPFL2, which encode the GRAIN LENGTH AND AWN DEVELOPMENT1 (GAD1) peptide, impairs spikelet formation, reduces grain size, and compromises seed germination (Jin et al. 2023). Mutations in other rice EPFL genes—such as OsEPFL6, OsEPFL7, OsEPFL8, and OsEPFL9—lead to severe defects in panicle structure, spikelet fertility, and grain yield, underscoring their critical role in reproductive meristem regulation (Guo et al. 2023).

In summary, the EPF/EPFL/stomagen peptide family orchestrates a wide range of meristem-related processes, including cell division control, organogenesis, and reproductive development. By integrating developmental signals with environmental inputs, these peptides help shape meristem function and overall plant architecture.

5.5.6 Role of Phytosulfokine (PSK) Peptides in Regulation of Meristem Activity

Phytosulfokines are sulfated pentapeptides recognized for their crucial role in stimulating cell division and supporting meristem activity in plants. Their signaling functions are mediated through two key receptor kinases, PSK RECEPTOR 1 (PSKR1) and PSKR2, which together form the core of the PSK signaling pathway (Matsubayashi et al. 2002).

Within the RAM, PSKs are vital for sustaining active cell proliferation. They promote root elongation by stimulating divisions in both the quiescent center (QC) and adjacent stem cell populations. This activity involves interaction with the transcription factor ETHYLENE RESPONSE FACTOR 115 (ERF115), a central regulator of meristem cell division under both normal growth and regenerative conditions (Heyman et al. 2013).

Beyond the RAM, PSKs contribute to maintaining procambial cell identity, which is essential for the organization and functionality of vascular meristems. This indicates a broader role for PSKs in preserving the identity and structure of meristematic tissues (Holzwart et al. 2018). PSKs also enhance callus formation and support somatic embryogenesis, reflecting their capacity to initiate new meristematic activity and reprogram differentiated cells toward a meristem-like state (Eun et al. 2003).

On a mechanistic level, PSK signaling leads to a rapid increase in cyclic guanosine monophosphate (cGMP), a secondary messenger involved in regulating cell division pathways within meristem zones (Aghdam and Luo 2022).

In summary, PSKs act as key molecular signals that regulate meristem function by coordinating cell proliferation, maintaining cellular identity, and enabling developmental plasticity through the initiation of new meristematic growth.

5.5.7 Role of PSY Peptides in Regulation of Meristem Activity

Plant peptides containing sulfated tyrosine (PSY) are posttranslationally modified signaling molecules that play a key role in regulating cell proliferation and meristem activity, particularly in root tissues. In *Arabidopsis thaliana*, nine PSY genes have been identified, with PSY1 being the most extensively characterized for its function in promoting meristematic development (Ogawa-Ohnishi et al. 2022). PSY1 acts as a positive regulator of root meristem activity and operates in opposition to RALF peptides. While RALFs suppress root growth by increasing apoplastic pH, PSY1 facilitates root elongation by promoting apoplastic acidification—an environment that favors cell expansion and division in meristematic regions (Fuglsang et al. 2014).

Plants overexpressing PSY1 (35S:PSY1) exhibit expanded root meristems, elongated primary roots, and enlarged cotyledons, highlighting its role in enhancing cell proliferation and organ growth (Amano et al. 2007). Similarly, exogenous application of synthetic PSY peptides results in increased root length, further supporting their function in maintaining active meristem zones (Ogawa-Ohnishi et al. 2022).

Although certain loss-of-function mutants, such as the psy5 psy6 psy8 triple knockout, display no pronounced developmental defects, gain-of-function studies strongly indicate that PSY peptides contribute to meristem expansion and support overall plant growth under favorable conditions.

5.5.8 Role of PIP Peptides in Regulation of Meristem Activity

PAMP-induced secreted peptides (PIPs) have emerged as important regulators at the intersection of plant immunity and development, with demonstrated roles in modulating meristem activity and root growth. In *Arabidopsis thaliana*, 11 PIP and PIP-LIKE (PIPL) genes have been identified, several of which are implicated in controlling root meristem function (Hou et al. 2014; Vie et al. 2015).

Experimental application of synthetic PIP1 and PIP2 peptides, or their overexpression in transgenic lines, results in markedly reduced primary root length, suggesting that these peptides act as negative regulators of root meristem activity (Yu et al. 2023a). Supporting this, plants overexpressing PIP2 or PIP3 exhibit shortened primary roots, while pip3 single mutants and pip2 pip3 double mutants also develop short roots. These findings point to a complex role for PIPs in regulating meristem-dependent root elongation (Hussain et al. 2021).

Interestingly, not all PIP family members function identically. Overexpression of PIPL3 reduces lateral root number without altering primary root growth, indicating a more specific role in lateral meristem regulation (Toyokura et al. 2019). Additionally, PIP2 overexpression enhances hypocotyl elongation, suggesting a possible connection to shoot meristem-controlled cell elongation. However, this effect appears to be gene-specific, as similar phenotypes are not consistently observed in other PIP mutants (Hussain et al. 2021).

Overall, PIP peptides act as modulators of root meristem activity by regulating both primary and lateral root development. In some contexts, they function as growth suppressors, potentially linking immune signaling pathways with developmental programs to fine-tune plant growth and stress responses.

5.6 Future Perspectives on Peptide Regulation of Meristem Activity

The last decade has revealed a complex landscape where small signaling peptides—CLE, CEP, RALF, RGF, CIF, EPF, PSK, PSY, and PIP—intersect to finely regulate meristem formation, maintenance, and regeneration. These insights open several promising avenues for research and application:

1) **Unraveling Receptor–Peptide Specificity and Cross Talk**
   - **Dissecting receptor complexes**
     Many peptides—such as CLE, RALF, and EPFL—share receptors (e.g. FER, ER/ERL, CLV1/BAM), yet produce distinct developmental outcomes. Advanced structural biology (e.g. cryo-EM, crystallography) coupled with mutagenesis will be key to understanding how receptor specificity and ligand competition drive diverse meristem outcomes.
   - **Signaling pathway integration**
     These peptides influence overlapping downstream effectors—ROS, Ca^{2+}, pH, and hormones like auxin, cytokinin, and ABA. Using systems biology and mathematical modeling, we can build integrated pathways that predict how meristem behavior shifts under combined peptide cues or environmental stimuli.
2) **Peptide Application for Targeted Meristem Manipulation**
   - **Agronomic trait enhancement**
     Exploring controlled application of PSK, PSY, RGF, or PSK analogs in crops could enhance root architecture, shoot regeneration, or tissue culture efficiency, offering routes to optimize growth or propagation.
   - **Selective peptide modulation**
     Small molecules or peptide mimetics that inhibit or enhance specific receptors (e.g. agonists for RGI, antagonists for CLE–CLV1) may enable precision modulation of meristem size or activity, potentially increasing yields or facilitating graft compatibility.
3) **Engineering Peptide Signaling Pathways**
   - **Synthetic peptide engineering**
     Create designed peptides with enhanced stability (e.g. noncanonical amino acids, altered sulfation/glycosylation) that resist degradation for sustained meristem modulation—or peptides tailored to specific receptor binding affinities.

- **Transgenic receptor modulation**
  Overexpressing or editing (via CRISPR) peptide receptors in crop species could amplify beneficial meristem responses, such as enhanced shoot regeneration in tissue culture or improved root system resilience.

4) **Decoding Environmental Integration**
   - **Peptide-stress cross talk**
     Many peptides serve dual roles: regulating meristems while modulating stress (e.g. RALF–ABA, PSK–salinity). Dissecting the molecular switches that pivot peptide signaling from growth to defense is vital for breeding stress-resilient crops without compromising development.
   - **Nutrient signaling roles**
     CEP and RGF peptides respond to nitrogen and phosphate cues. Detailed mapping of how these peptides integrate nutrient availability with meristem architecture promises precision-rooting strategies, optimizing nutrient uptake efficiency.
5) **Meristem Regeneration and De Novo Organogenesis**
   - **Reprogramming differentiated cells**
     CLE and PSK peptides regulate callus formation and shoot regeneration. Systematic combinations with auxins, cytokinins, or RGF/CLE mimetics might greatly improve plant regeneration protocols, expanding the toolkit for recalcitrant species.
   - **Cross-species peptide utility**
     Testing whether Arabidopsis-derived peptides (e.g. RGF1, PSK) can induce meristem formation in nonmodel or woody species may revolutionize clonal propagation and genetic transformation pipelines.
6) **Biosensors and Visualization Tools**
   - ***In vivo* peptide tracking**
     Developing fluorescent peptide biosensors or peptide–receptor interaction reporters will enable live imaging of peptide dynamics in the SAM and RAM, enhancing understanding of spatial–temporal patterns during growth or regeneration.
   - **Phosphoproteomics in the meristem**
     Profiling phosphorylation changes upon peptide treatment in meristem tissues will elucidate downstream targets of receptors (e.g. AHA2 by RALF, PLT2 by RGF, PSKR via cGMP), revealing key regulatory nodes for enhanced developmental control.
7) **Translational Opportunities for Agriculture**
   - **Root system engineering**
     Modifying CEP or RGF signaling may yield crops with deeper, more branched root systems—improving drought resilience and nutrient uptake.
   - **Enhanced regeneration and propagation**
     Leveraging CLE and PSK pathways to boost tissue culture success could accelerate propagation of high-value or endangered plants via more efficient regeneration.
   - **Peptide-based agrochemicals**
     Formulating stable peptide sprays—or small molecules that mimic peptide signaling—offers a novel class of plant biostimulants, potentially reducing fertilizer or hormone reliance.

References

Aghdam, M.S. and Luo, Z. (2022). Harnessing cGMP signaling pathways for improving fruits and vegetables marketability. *Scientia Horticulturae* 291: 110587.

Amano, Y., Tsubouchi, H., Shinohara, H. et al. (2007). Tyrosine-sulfated glycopeptide involved in cellular proliferation and expansion in Arabidopsis. *Proceedings of the National Academy of Sciences* 104 (46): 18333–18338.

Araya, T., Miyamoto, M., Wibowo, J. et al. (2014). CLE-CLAVATA1 peptide-receptor signaling module regulates the expansion of plant root systems in a nitrogen-dependent manner. *Proceedings of the National Academy of Sciences* 111 (5): 2029–2034.

Atkinson, N.J., Lilley, C.J., and Urwin, P.E. (2013). Identification of genes involved in the response of Arabidopsis to simultaneous biotic and abiotic stresses. *Plant Physiology* 162 (4): 2028–2041.

Bashyal, S., Gautam, C.K., and Müller, L.M. (2024). CLAVATA signaling in plant–environment interactions. *Plant Physiology* 194 (3): 1336–1357.

Bergonci, T., Ribeiro, B., Ceciliato, P.H. et al. (2014). *Arabidopsis thaliana* RALF1 opposes brassinosteroid effects on root cell elongation and lateral root formation. *Journal of Experimental Botany* 65 (8): 2219–2230.

Bhattacharya, R., Krishna Koramutla, M., Negi, M. et al. (2013). Hydroxyproline-rich glycopeptide signals in potato elicit signalling associated with defense against insects and pathogens. *Plant Science* 207: 88–97.

Bobay, B.G., DiGennaro, P., Scholl, E. et al. (2013). Solution NMR studies of the plant peptide hormone CEP inform function. *FEBS Letters* 587 (24): 3979–3985.

Bommert, P. and Whipple, C. (2018). Grass inflorescence architecture and meristem determinacy. *Seminars in Cell & Developmental Biology* 79: 37–47. https://doi.org/10.1016/j.semcdb.2017.10.004.

Boschiero, C., Dai, X., Lundquist, P.K. et al. (2020). MtSSPdb: the Medicago truncatula small secreted peptide database. *Plant Physiology* 183 (1): 399–413.

Bourion, V., Martin, C., De Larambergue, H. et al. (2014). Unexpectedly low nitrogen acquisition and absence of root architecture adaptation to nitrate supply in a Medicago truncatula highly branched root mutant. *Journal of Experimental Botany* 65 (9): 2365–2380.

Campbell, L. and Turner, S.R. (2017). A comprehensive analysis of RALF proteins in green plants suggests there are two distinct functional groups. *Frontiers in Plant Science* 8: 37.

Carbonnel, S., Cornelis, S., and Hazak, O. (2023). The CLE33 peptide represses phloem differentiation via autocrine and paracrine signaling in Arabidopsis. *Communications Biology* 6 (1): 588.

Chapman, K., Taleski, M., Ogilvie, H.A. et al. (2019). CEP–CEPR1 signalling inhibits the sucrose-dependent enhancement of lateral root growth. *Journal of Experimental Botany* 70 (15): 3955–3967.

Chapman, K., Ivanovici, A., Taleski, M. et al. (2020). CEP receptor signalling controls root system architecture in Arabidopsis and Medicago. *New Phytologist* 226 (6): 1809–1821.

Chen, J., Yu, F., Liu, Y. et al. (2016). FERONIA interacts with ABI2-type phosphatases to facilitate signaling cross-talk between abscisic acid and RALF peptide in Arabidopsis. *Proceedings of the National Academy of Sciences* 113 (37): E5519–E5527.

Delay, C., Imin, N., and Djordjevic, M.A. (2013). CEP genes regulate root and shoot development in response to environmental cues and are specific to seed plants. *Journal of Experimental Botany* 64 (17): 5383–5394.

Depuydt, S., Rodriguez-Villalon, A., Santuari, L. et al. (2013). Suppression of Arabidopsis protophloem differentiation and root meristem growth by CLE45 requires the receptor-like kinase BAM3. *Proceedings of the National Academy of Sciences* 110 (17): 7074–7079.

DeYoung, B.J. and Clark, S.E. (2008). BAM receptors regulate stem cell specification and organ development through complex interactions with CLAVATA signaling. *Genetics* 180 (2): 895–904.

Dong, W., Wang, Y., and Takahashi, H. (2019). Cle-CLAVATA1 signaling pathway modulates lateral root development under sulfur deficiency. *Plants (Basel)* 8 (4): 103.

Du, C., Li, X., Chen, J. et al. (2016). Receptor kinase complex transmits RALF peptide signal to inhibit root growth in Arabidopsis. *Proceedings of the National Academy of Sciences* 113 (51): E8326–E8334.

Duan, Q., Kita, D., Li, C. et al. (2010). FERONIA receptor-like kinase regulates RHO GTPase signaling of root hair development. *Proceedings of the National Academy of Sciences* 107 (41): 17821–17826.

Endo, S., Shinohara, H., Matsubayashi, Y., and Fukuda, H. (2013). A novel pollen-pistil interaction conferring high-temperature tolerance during reproduction via CLE45 signaling. *Current Biology* 23 (17): 1670–1676.

Etchells, J.P. and Turner, S.R. (2010). The PXY-CLE41 receptor ligand pair defines a multifunctional pathway that controls the rate and orientation of vascular cell division. *Development* 137 (5): 767–774.

Eun, C.H., Ko, S.M., Matsubayashi, Y. et al. (2003). Phytosulfokine-α requires auxin to stimulate carrot non-embryogenic cell proliferation. *Plant Physiology and Biochemistry* 41 (5): 447–452.

Fang, Y., Chang, J., Shi, T. et al. (2021). Evolution of RGF/GLV/CLEL peptide hormones and their roles in land plant growth and regulation. *International Journal of Molecular Sciences* 22 (24): 13372.

Fernandez, A., Drozdzecki, A., Hoogewijs, K. et al. (2013). Transcriptional and functional classification of the GOLVEN/ROOT GROWTH FACTOR/CLE-like signaling peptides reveals their role in lateral root and hair formation. *Plant Physiology* 161 (2): 954–970.

Fernandez, A., Drozdzecki, A., Hoogewijs, K. et al. (2015). The GLV6/RGF8/CLEL2 peptide regulates early pericycle divisions during lateral root initiation. *Journal of Experimental Botany* 66 (17): 5245–5256.

Fiers, M., Hause, G., Boutilier, K. et al. (2004). Mis-expression of the CLV3/ESR-like gene CLE19 in Arabidopsis leads to a consumption of root meristem. *Gene* 327 (1): 37–49.

Fiers, M., Golemiec, E., Van Der Schors, R. et al. (2006). The CLAVATA3/ESR motif of CLAVATA3 is functionally independent from the nonconserved flanking sequences. *Plant Physiology* 141 (4): 1284–1292.

Fischer, U. and Teichmann, T. (2017). The ERECTA and ERECTA-like genes control a developmental shift during xylem formation in Arabidopsis. *New Phytologist* 213 (4): 1562–1563.

Fiume, E. and Fletcher, J.C. (2012). Regulation of Arabidopsis embryo and endosperm development by the polypeptide signaling molecule CLE8. *The Plant Cell* 24 (3): 1000–1012.

Fletcher, J.C. (2020). Recent advances in Arabidopsis CLE peptide signaling. *Trends in Plant Science* 25 (10): 1005–1016.

Fletcher, J.C., Brand, U., Running, M.P. et al. (1999). Signaling of cell fate decisions by CLAVATA3 in Arabidopsis shoot meristems. *Science* 283 (5409): 1911–1914.

Fuglsang, A.T., Kristensen, A., Cuin, T.A. et al. (2014). Receptor kinase-mediated control of primary active proton pumping at the plasma membrane. *The Plant Journal* 80 (6): 951–964.

Fukuda, H. and Hardtke, C.S. (2020). Peptide signaling pathways in vascular differentiation. *Plant Physiology* 182 (4): 1636–1644.

Gjetting, S.K., Ytting, C.K., Schulz, A., and Fuglsang, A.T. (2012). Live imaging of intra-and extracellular pH in plants using pHusion, a novel genetically encoded biosensor. *Journal of Experimental Botany* 63 (8): 3207–3218.

Goad, D.M., Zhu, C., and Kellogg, E.A. (2017). Comprehensive identification and clustering of CLV3/ESR-related (CLE) genes in plants finds groups with potentially shared function. *New Phytologist* 216 (2): 605–616.

Gregory, E.F., Dao, T.Q., Alexander, M.A. et al. (2018). The signaling peptide-encoding genes CLE16, CLE17 and CLE27 are dispensable for Arabidopsis shoot apical meristem activity. *PLoS One* 13 (8): e0202595.

Guo, T., Lu, Z.Q., Xiong, Y. et al. (2023). Optimization of rice panicle architecture by specifically suppressing ligand–receptor pairs. *Nature Communications* 14 (1): 1640.

Gutiérrez-Alanís, D., Yong-Villalobos, L., Jimenez-Sandoval, P. et al. (2017). Phosphate starvation-dependent iron mobilization induces CLE14 expression to trigger root meristem differentiation through CLV2/PEPR2 signaling. *Developmental Cell* 41 (5): 555–570.e3.

Han, H., Zhuang, K., and Qiu, Z. (2022). CLE peptides join the plant longevity club. *Trends in Plant Science* 27 (10): 961–963.

Hanemian, M., Barlet, X., Sorin, C. et al. (2016). Arabidopsis CLAVATA 1 and CLAVATA 2 receptors contribute to *Ralstonia solanacearum* pathogenicity through a miR169-dependent pathway. *New Phytologist* 211 (2): 502–515.

Hara, K., Kajita, R., Torii, K.U. et al. (2007). The secretory peptide gene EPF1 enforces the stomatal one-cell-spacing rule. *Genes & Development* 21 (14): 1720–1725.

Hara, K., Yokoo, T., Kajita, R. et al. (2009). Epidermal cell density is autoregulated via a secretory peptide, EPIDERMAL PATTERNING FACTOR 2 in Arabidopsis leaves. *Plant and Cell Physiology* 50 (6): 1019–1031.

Haruta, M., Monshausen, G., Gilroy, S., and Sussman, M.R. (2008). A cytoplasmic Ca2+ functional assay for identifying and purifying endogenous cell signaling peptides in Arabidopsis seedlings: identification of AtRALF1 peptide. *Biochemistry* 47 (24): 6311–6321.

Haruta, M., Sabat, G., Stecker, K. et al. (2014). A peptide hormone and its receptor protein kinase regulate plant cell expansion. *Science* 343 (6169): 408–411.

Hayashi, N., Tetsumura, T., Sawa, S. et al. (2018). CLE14 peptide signaling in Arabidopsis root hair cell fate determination. *Plant Biotechnology* 35 (1): 17–22.

Hazak, O., Brandt, B., Cattaneo, P. et al. (2017). Perception of root-active CLE peptides requires CORYNE function in the phloem vasculature. *EMBO Reports* 18 (8): 1367–1381.

He, Y., He, X., Wang, X. et al. (2023). An EPFL peptide signaling pathway promotes stamen elongation via enhancing filament cell proliferation to ensure successful self-pollination in *Arabidopsis thaliana*. *New Phytologist* 238 (3): 1045–1058.

He, L., Wu, L., and Li, J. (2024). Sulfated peptides and their receptors: key regulators of plant development and stress adaptation. *Plant Communications* 5 (6): 100918.

Heyman, J., Cools, T., Vandenbussche, F. et al. (2013). ERF115 controls root quiescent center cell division and stem cell replenishment. *Science* 342 (6160): 860–863.

Hirakawa, Y. and Sawa, S. (2019). Diverse function of plant peptide hormones in local signaling and development. *Current Opinion in Plant Biology* 51: 81–87.

Hirakawa, Y., Kondo, Y., and Fukuda, H. (2010). TDIF peptide signaling regulates vascular stem cell proliferation via the WOX4 homeobox gene in Arabidopsis. *The Plant Cell* 22 (8): 2618–2629.

Hirakawa, Y., Fujimoto, T., Ishida, S. et al. (2020). Induction of multichotomous branching by CLAVATA peptide in *Marchantia polymorpha*. *Current Biology* 30 (19): 3833–3840.e4.

Holzwart, E., Huerta, A.I., Glöckner, N. et al. (2018). BRI1 controls vascular cell fate in the Arabidopsis root through RLP44 and phytosulfokine signaling. *Proceedings of the National Academy of Sciences* 115 (46): 11838–11843.

Hou, S., Wang, X., Chen, D. et al. (2014). The secreted peptide PIP1 amplifies immunity through receptor-like kinase 7. *PLoS Pathogens* 10 (9): e1004331.

Hu, C., Zhu, Y., Cui, Y. et al. (2022). A CLE–BAM–CIK signalling module controls root protophloem differentiation in Arabidopsis. *New Phytologist* 233 (1): 282–296.

Huang, A., Cui, T., Zhang, Y. et al. (2023). CRISPR/Cas9-engineered large fragment deletion mutations in Arabidopsis CEP peptide–encoding genes reveal their role in primary and lateral root formation. *Plant and Cell Physiology* 64 (1): 19–26.

Huault, E., Laffont, C., Wen, J. et al. (2014). Local and systemic regulation of plant root system architecture and symbiotic nodulation by a receptor-like kinase. *PLoS Genetics* 10 (12): e1004891.

Hunt, L. and Gray, J.E. (2009). The signaling peptide EPF2 controls asymmetric cell divisions during stomatal development. *Current Biology* 19 (10): 864–869.

Hussain, S., Wang, W., Ahmed, S. et al. (2021). PIP2, an auxin induced plant peptide hormone regulates root and hypocotyl elongation in Arabidopsis. *Frontiers in Plant Science* 12: 646736.

Ikematsu, S., Tasaka, M., Torii, K.U., and Uchida, N. (2017). ERECTA-family receptor kinase genes redundantly prevent premature progression of secondary growth in the Arabidopsis hypocotyl. *New Phytologist* 213 (4): 1697–1709.

Imin, N., Mohd-Radzman, N.A., Ogilvie, H.A., and Djordjevic, M.A. (2013). The peptide-encoding CEP1 gene modulates lateral root and nodule numbers in *Medicago truncatula*. *Journal of Experimental Botany* 64 (17): 5395–5409.

Ivanovici, A., Laffont, C., Larrainzar, E. et al. (2023). The Medicago SymCEP7 hormone increases nodule number via shoots without compromising lateral root number. *Plant Physiology* 191 (3): 2012–2026.

Je, B.I., Xu, F., Wu, Q. et al. (2018). The CLAVATA receptor FASCIATED EAR2 responds to distinct CLE peptides by signaling through two downstream effectors. *eLife* 7: e35673.

Jin, J., Xiong, L., Gray, J.E. et al. (2023). Two awn-development-related peptides, GAD1 and OsEPFL2, promote seed dispersal and germination in rice. *Molecular Plant* 16 (3): 485–488.

John, A., Smith, E.S., Jones, D.S. et al. (2023). A network of CLAVATA receptors buffers auxin-dependent meristem maintenance. *Nature Plants* 9 (8): 1306–1317.

Jun, J., Fiume, E., Roeder, A.H. et al. (2010). Comprehensive analysis of CLE polypeptide signaling gene expression and overexpression activity in Arabidopsis. *Plant Physiology* 154 (4): 1721–1736.

Kang, J., Wang, X., Ishida, T. et al. (2022). A group of CLE peptides regulates de novo shoot regeneration in *Arabidopsis thaliana*. *New Phytologist* 235 (6): 2300–2312.

Kawamoto, N., Del Carpio, D.P., Hofmann, A. et al. (2020). A peptide pair coordinates regular ovule initiation patterns with seed number and fruit size. *Current Biology* 30 (22): 4352–4361.e4.

Kinoshita, A., Nakamura, Y., Sasaki, E. et al. (2007). Gain-of-function phenotypes of chemically synthetic CLAVATA3/ESR-related (CLE) peptides in *Arabidopsis thaliana* and *Oryza sativa*. *Plant and Cell Physiology* 48 (12): 1821–1825.

Kinoshita, A., Betsuyaku, S., Osakabe, Y. et al. (2010). RPK2 is an essential receptor-like kinase that transmits the CLV3 signal in Arabidopsis. *Development* 137 (22): 3911–3920.

Kondo, Y., Hirakawa, Y., Kieber, J.J., and Fukuda, H. (2011). CLE peptides can negatively regulate protoxylem vessel formation via cytokinin signaling. *Plant and Cell Physiology* 52 (1): 37–48.

Koppolu, R., Chen, S., and Schnurbusch, T. (2022). Evolution of inflorescence branch modifications in cereal crops. *Current Opinion in Plant Biology* 65: 102168. https://doi.org/10.1016/j.pbi.2021.102168.

Laffont, C., Ivanovici, A., Gautrat, P. et al. (2020). The NIN transcription factor coordinates CEP and CLE signaling peptides that regulate nodulation antagonistically. *Nature Communications* 11 (1): 3167.

Lease, K.A. and Walker, J.C. (2006). The Arabidopsis unannotated secreted peptide database, a resource for plant peptidomics. *Plant Physiology* 142 (3): 831–838.

Lee, H., Chah, O.K., and Sheen, J. (2011). Stem-cell-triggered immunity through CLV3p–FLS2 signalling. *Nature* 473 (7347): 376–379.

Lee, J.S., Hnilova, M., Maes, M. et al. (2015). Competitive binding of antagonistic peptides fine-tunes stomatal patterning. *Nature* 522 (7557): 439–443.

Li, Y.L., Dai, X.R., Yue, X. et al. (2014). Identification of small secreted peptides (SSPs) in maize and expression analysis of partial SSP genes in reproductive tissues. *Planta* 240: 713–728.

Li, L., Chen, H., Alotaibi, S.S. et al. (2022). RALF1 peptide triggers biphasic root growth inhibition upstream of auxin biosynthesis. *Proceedings of the National Academy of Sciences of the United States of America* 119 (31): e2121058119.

Li, M., Lv, M., Wang, X. et al. (2023). The EPFL–ERf–SERK signaling controls integument development in Arabidopsis. *New Phytologist* 238 (1): 186–201.

Liao, Y., Wen, X., Zheng, J. et al. (2023). RALF-like peptide improves the colonization of endophytic *Colletotrichum tofieldiae* through interacting with plant receptor-like kinase. *Plant Pathology* 72 (9): 1649–1661.

Lindsay, P., Swentowsky, K.W., and Jackson, D. (2024). Cultivating potential: harnessing plant stem cells for agricultural crop improvement. *Molecular Plant* 17 (1): 50–74. https://doi.org/10.1016/j.molp.2023.12.014.

Liu, Y., Zuo, T., Qiu, Z. et al. (2021). Genome-wide identification reveals the function of CEP peptide in cucumber root development. *Plant Physiology and Biochemistry* 169: 119–126.

Liu, D., Shen, Z., Zhuang, K. et al. (2022a). Systematic annotation reveals CEP function in tomato root development and abiotic stress response. *Cells* 11 (19): 2935.

Liu, L., Song, W., Huang, S. et al. (2022b). Extracellular pH sensing by plant cell-surface peptide-receptor complexes. *Cell* 185 (18): 3341–3355.e13.

Liu, L., Liu, X., Bai, Z. et al. (2024). Small but powerful: RALF peptides in plant adaptive and developmental responses. *Plant Science* 112085.

Luo, Z., Moreau, C., Wang, J. et al. (2022). NLP1 binds the CEP1 signalling peptide promoter to repress its expression in response to nitrate. *New Phytologist* 234 (5): 1547–1552.

Masachis, S., Segorbe, D., Turrà, D. et al. (2016). A fungal pathogen secretes plant alkalinizing peptides to increase infection. *Nature Microbiology* 1 (6): 1–9.

Matsubayashi, Y. (2011). Post-translational modifications in secreted peptide hormones in plants. *Plant and Cell Physiology* 52 (1): 5–13.

Matsubayashi, Y., Ogawa, M., Morita, A., and Sakagami, Y. (2002). An LRR receptor kinase involved in perception of a peptide plant hormone, phytosulfokine. *Science* 296 (5572): 1470–1472.

Matsuzaki, Y., Ogawa-Ohnishi, M., Mori, A., and Matsubayashi, Y. (2010). Secreted peptide signals required for maintenance of root stem cell niche in Arabidopsis. *Science* 329 (5995): 1065–1067.

McKim, S.M. (2019). How plants grow up. *Journal of Integrative Plant Biology* 61 (3): 257–277. https://doi.org/10.1111/jipb.12786.

Meng, L., Buchanan, B.B., Feldman, L.J., and Luan, S. (2012). CLE-like (CLEL) peptides control the pattern of root growth and lateral root development in Arabidopsis. *Proceedings of the National Academy of Sciences* 109 (5): 1760–1765.

Mens, C., Hastwell, A.H., Su, H. et al. (2021). Characterisation of *Medicago truncatula* CLE34 and CLE35 in nitrate and rhizobia regulation of nodulation. *New Phytologist* 229 (5): 2525–2534.

Mishima, K., Tanaka, T., Pu, F. et al. (2003). Vitamin E isoforms α-tocotrienol and γ-tocopherol prevent cerebral infarction in mice. *Neuroscience Letters* 337 (1): 56–60.

Missaoui, K., Gonzalez-Klein, Z., Pazos-Castro, D. et al. (2022). Plant non-specific lipid transfer proteins: an overview. *Plant Physiology and Biochemistry* 171: 115–127.

Miwa, H., Betsuyaku, S., Iwamoto, K. et al. (2008). The receptor-like kinase SOL2 mediates CLE signaling in Arabidopsis. *Plant and Cell Physiology* 49 (11): 1752–1757.

Mohammed, U., Caine, R.S., Atkinson, J.A. et al. (2019). Rice plants overexpressing OsEPF1 show reduced stomatal density and increased root cortical aerenchyma formation. *Scientific Reports* 9 (1): 5584.

Mohd-Radzman, N.A., Laffont, C., Ivanovici, A. et al. (2016). Different pathways act downstream of the CEP peptide receptor CRA2 to regulate lateral root and nodule development. *Plant Physiology* 171 (4): 2536–2548.

Moreau, C., Gautrat, P., and Frugier, F. (2021). Nitrate-induced CLE35 signaling peptides inhibit nodulation through the SUNN receptor and miR2111 repression. *Plant Physiology* 185 (3): 1216–1228.

Muller, R., Bleckmann, A., and Simon, R. (2008). The receptor kinase CORYNE of Arabidopsis transmits the stem cell–limiting signal CLAVATA3 independently of CLAVATA1. *The Plant Cell* 20 (4): 934–946.

Nakagami, S., Notaguchi, M., Kondo, T. et al. (2023). Root-knot nematode modulates plant CLE3-CLV1 signaling as a long-distance signal for successful infection. *Science Advances* 9 (22): eadf4803.

Negoro, S., Hirabayashi, T., Iwasaki, R. et al. (2023). EPFL peptide signalling ensures robust self-pollination success under cool temperature stress by aligning the length of the stamen and pistil. *Plant, Cell & Environment* 46 (2): 451–463.

Nishida, H. and Suzaki, T. (2018). Nitrate-mediated control of root nodule symbiosis. *Current Opinion in Plant Biology* 44: 129–136.

Nunes, T.D., Zhang, D., and Raissig, M.T. (2020). Form, development and function of grass stomata. *The Plant Journal* 101 (4): 780–799. https://doi.org/10.1111/tpj.14552.

Oelkers, K., Goffard, N., Weiller, G.F. et al. (2008). Bioinformatic analysis of the CLE signaling peptide family. *BMC Plant Biology* 8: 1–15.

Ogawa-Ohnishi, M., Yamashita, T., Kakita, M. et al. (2022). Peptide ligand-mediated trade-off between plant growth and stress response. *Science* 378 (6616): 175–180.

Ogilvie, H.A., Imin, N., and Djordjevic, M.A. (2014). Diversification of the C-TERMINALLY ENCODED PEPTIDE (CEP) gene family in angiosperms, and evolution of plant-family specific CEP genes. *BMC Genomics* 15: 1–15.

Ohki, S., Takeuchi, M., and Mori, M. (2011). The NMR structure of stomagen reveals the basis of stomatal density regulation by plant peptide hormones. *Nature Communications* 2 (1): 512.

Ohkubo, Y., Tanaka, M., Tabata, R. et al. (2017). Shoot-to-root mobile polypeptides involved in systemic regulation of nitrogen acquisition. *Nature Plants* 3 (4): 1–6.

Ohyama, K., Ogawa, M., and Matsubayashi, Y. (2008). Identification of a biologically active, small, secreted peptide in Arabidopsis by in silico gene screening, followed by LC-MS-based structure analysis. *The Plant Journal* 55 (1): 152–160.

Okamoto, S., Shinohara, H., Mori, T. et al. (2013). Root-derived CLE glycopeptides control nodulation by direct binding to HAR1 receptor kinase. *Nature Communications* 4 (1): 2191.

Olsson, V., Joos, L., Zhu, S. et al. (2019). Look closely, the beautiful may be small: precursor-derived peptides in plants. *Annual Review of Plant Biology* 70 (1): 153–186.

Ota, R., Ohkubo, Y., Yamashita, Y. et al. (2020). Shoot-to-root mobile CEPD-like 2 integrates shoot nitrogen status to systemically regulate nitrate uptake in Arabidopsis. *Nature Communications* 11 (1): 641.

Pan, B., Sheng, J., Sun, W. et al. (2012). OrysPSSP: a comparative platform for small secreted proteins from rice and other plants. *Nucleic Acids Research* 41 (D1): D1192–D1198.

Pearce, G., Moura, D.S., Stratmann, J., and Ryan, C.A. Jr. (2001). RALF, a 5-kDa ubiquitous polypeptide in plants, arrests root growth and development. *Proceedings of the National Academy of Sciences* 98 (22): 12843–12847.

Ren, S.C., Song, X.F., Chen, W.Q. et al. (2019). CLE25 peptide regulates phloem initiation in Arabidopsis through a CLERK-CLV2 receptor complex. *Journal of Integrative Plant Biology* 61 (10): 1043–1061.

Roberts, I., Smith, S., De Rybel, B. et al. (2013). The CEP family in land plants: evolutionary analyses, expression studies, and role in Arabidopsis shoot development. *Journal of Experimental Botany* 64 (17): 5371–5381.

Roberts, I., Smith, S., Stes, E. et al. (2016). CEP5 and XIP1/CEPR1 regulate lateral root initiation in Arabidopsis. *Journal of Experimental Botany* 67 (16): 4889–4899.

Rodriguez-Leal, D., Xu, C., Kwon, C.T. et al. (2019). Evolution of buffering in a genetic circuit controlling plant stem cell proliferation. *Nature Genetics* 51 (5): 786–792.

Ryan, C.A., Huffaker, A., and Yamaguchi, Y. (2007). New insights into innate immunity in Arabidopsis. *Cellular Microbiology* 9 (8): 1902–1908.

Rzemieniewski, J., Leicher, H., Lee, H. K., Broyart, C., Nayem, S., Wiese, C., ... & Stegmann, M. (2022). CEP signaling coordinates plant immunity with nitrogen status. bioRxiv; 2022. https://doi.org/10.1101/2022.12.20.521212.

Schlegel, J., Denay, G., Wink, R. et al. (2021). Control of Arabidopsis shoot stem cell homeostasis by two antagonistic CLE peptide signalling pathways. *eLife* 10: e70934.

Schoof, H., Lenhard, M., Haecker, A. et al. (2000). The stem cell population of Arabidopsis shoot meristems is maintained by a regulatory loop between the CLAVATA and WUSCHEL genes. *Cell* 100 (6): 635–644.

Selby, R. and Jones, D.S. (2023). Complex peptide hormone signaling in plant stem cells. *Current Opinion in Plant Biology* 75: 102442.

Shen, Z., Zuo, T., Xia, H. et al. (2023). Synthetic CsCEP3 peptide attenuates salinity stress via ROS and ABA signaling in cucumber primary root. *Horticulturae* 9 (8): 921.

Shinohara, H. and Matsubayashi, Y. (2013). Chemical synthesis of Arabidopsis CLV3 glycopeptide reveals the impact of hydroxyproline arabinosylation on peptide conformation and activity. *Plant and Cell Physiology* 54 (3): 369–374.

Smith, S., Zhu, S., Joos, L. et al. (2020). The CEP5 peptide promotes abiotic stress tolerance, as revealed by quantitative proteomics, and attenuates the AUX/IAA equilibrium in Arabidopsis. *Molecular & Cellular Proteomics* 19 (8): 1248–1262.

Somssich, M., Ma, Q., Weidtkamp-Peters, S. et al. (2015). Real-time dynamics of peptide ligand–dependent receptor complex formation in planta. *Science Signaling* 8 (388): ra76.

Song, X.F., Guo, P., Ren, S.C. et al. (2013). Antagonistic peptide technology for functional dissection of CLV3/ESR genes in Arabidopsis. *Plant Physiology* 161 (3): 1076–1085.

Song, Y., Wilson, A.J., Zhang, X.C. et al. (2021). FERONIA restricts Pseudomonas in the rhizosphere microbiome via regulation of reactive oxygen species. *Nature Plants* 7 (5): 644–654.

Song, X.F., Hou, X.L., and Liu, C.M. (2022). CLE peptides: critical regulators for stem cell maintenance in plants. *Planta* 255 (1): 5.

Srivastava, R., Liu, J.X., Guo, H. et al. (2009). Regulation and processing of a plant peptide hormone, AtRALF23, in Arabidopsis. *The Plant Journal* 59 (6): 930–939.

Stahl, Y., Wink, R.H., Ingram, G.C., and Simon, R. (2009). A signaling module controlling the stem cell niche in Arabidopsis root meristems. *Current Biology* 19 (11): 909–914.

Stahl, Y., Grabowski, S., Bleckmann, A. et al. (2013). Moderation of Arabidopsis root stemness by CLAVATA1 and ARABIDOPSIS CRINKLY4 receptor kinase complexes. *Current Biology* 23 (5): 362–371.

Sui, Z., Wang, T., Li, H. et al. (2016). Overexpression of peptide-encoding OsCEP6. 1 results in pleiotropic effects on growth in rice (*O. sativa*). *Frontiers in Plant Science* 7: 228.

Suzaki, T., Yoshida, A., and Hirano, H.Y. (2008). Functional diversification of CLAVATA3-related CLE proteins in meristem maintenance in rice. *The Plant Cell* 20 (8): 2049–2058.

Tabata, R., Sumida, K., Yoshii, T. et al. (2014). Perception of root-derived peptides by shoot LRR-RKs mediates systemic N-demand signaling. *Science* 346 (6207): 343–346.

Takahashi, F., Suzuki, T., Osakabe, Y. et al. (2018). A small peptide modulates stomatal control via abscisic acid in long-distance signalling. *Nature* 556 (7700): 235–238.

Takeuchi, H. (2021). The role of diverse LURE-type cysteine-rich peptides as signaling molecules in plant reproduction. *Peptides* 142: 170572.

Taleski, M., Imin, N., and Djordjevic, M.A. (2018). CEP peptide hormones: key players in orchestrating nitrogen-demand signalling, root nodulation, and lateral root development. *Journal of Experimental Botany* 69 (8): 1829–1836.

Tang, J., Wu, D., Li, X. et al. (2022). Plant immunity suppression via PHR1-RALF-FERONIA shapes the root microbiome to alleviate phosphate starvation. *The EMBO Journal* 41 (6): e109102.

Tavormina, P., De Coninck, B., Nikonorova, N. et al. (2015). The plant peptidome: an expanding repertoire of structural features and biological functions. *The Plant Cell* 27 (8): 2095–2118.

Tian, D., Xie, Q., Deng, Z. et al. (2022). Small secreted peptides encoded on the wheat (*Triticum aestivum* L.) genome and their potential roles in stress responses. *Frontiers. Plant Science* 13: 1000297.

Toyokura, K., Goh, T., Shinohara, H. et al. (2019). Lateral inhibition by a peptide hormone-receptor cascade during Arabidopsis lateral root founder cell formation. *Developmental Cell* 48 (1): 64–75.e5.

Umeda, M., Ikeuchi, M., Ishikawa, M. et al. (2021). Plant stem cell research is uncovering the secrets of longevity and persistent growth. *The Plant Journal* 106 (2): 326–335. https://doi.org/10.1111/tpj.15184.

Viana, W.G., Scharwies, J.D., and Dinneny, J.R. (2022). Deconstructing the root system of grasses through an exploration of development, anatomy and function. *Plant, Cell & Environment* 45 (3): 602–619. https://doi.org/10.1111/pce.14270.

Vie, A.K., Najafi, J., Liu, B. et al. (2015). The IDA/IDA-LIKE and PIP/PIP-LIKE gene families in Arabidopsis: phylogenetic relationship, expression patterns, and transcriptional effect of the PIPL3 peptide. *Journal of Experimental Botany* 66 (17): 5351–5365.

Wang, G. and Fiers, M. (2010). CLE peptide signaling during plant development. *Protoplasma* 240: 33–43.

Wang, X., Mitchum, M.G., Gao, B. et al. (2005). A parasitism gene from a plant-parasitic nematode with function similar to CLAVATA3/ESR (CLE) of *Arabidopsis thaliana*. *Molecular Plant Pathology* 6 (2): 187–191.

Wang, P., Clark, N.M., Nolan, T.M. et al. (2022). Integrated omics reveal novel functions and underlying mechanisms of the receptor kinase FERONIA in *Arabidopsis thaliana*. *The Plant Cell* 34 (7): 2594–2614.

Whitford, R., Fernandez, A., Tejos, R. et al. (2012). GOLVEN secretory peptides regulate auxin carrier turnover during plant gravitropic responses. *Developmental Cell* 22 (3): 678–685.

Xie, H., Zhao, W., Li, W. et al. (2022). Small signaling peptides mediate plant adaptions to abiotic environmental stress. *Planta* 255 (4): 72.

Xu, T.T., Ren, S.C., Song, X.F., and Liu, C.M. (2015). CLE19 expressed in the embryo regulates both cotyledon establishment and endosperm development in Arabidopsis. *Journal of Experimental Botany* 66 (17): 5217–5227.

Xu, K., Tian, D., Wang, T. et al. (2023). Small secreted peptides (SSPs) in tomato and their potential roles in drought stress response. *Molecular Horticulture* 3 (1): 17.

Yamada, M., Han, X., and Benfey, P.N. (2020). RGF1 controls root meristem size through ROS signalling. *Nature* 577 (7788): 85–88.

Yamaguchi, Y., Huffaker, A., Bryan, A.C. et al. (2010). PEPR2 is a second receptor for the Pep1 and Pep2 peptides and contributes to defense responses in Arabidopsis. *The Plant Cell* 22 (2): 508–522.

Yan, S., Bhawal, R., Yin, Z. et al. (2022). Recent advances in proteomics and metabolomics in plants. *Molecular Horticulture* 2 (1): 17.

Yang, X., Tschaplinski, T.J., Hurst, G.B. et al. (2011). Discovery and annotation of small proteins using genomics, proteomics, and computational approaches. *Genome Research* 21 (4): 634–641.

Yimer, H.Z., Luu, D.D., Coomer Blundell, A. et al. (2023). Root-knot nematodes produce functional mimics of tyrosine-sulfated plant peptides. *Proceedings of the National Academy of Sciences* 120 (29): e2304612120.

Yu, P., Gutjahr, C., Li, C., and Hochholdinger, F. (2016). Genetic control of lateral root formation in cereals. *Trends in Plant Science* 21 (11): 951–961. https://doi.org/10.1016/j.tplants.2016.07.011.

Yu, X.S., Wang, H.R., Lei, F.F. et al. (2023a). Structure and functional divergence of PIP peptide family revealed by functional studies on PIP1 and PIP2 in *Arabidopsis thaliana*. *Frontiers in Plant Science* 14: 1208549.

Yu, Y., Song, W., Zhai, N. et al. (2023b). PXL1 and SERKs act as receptor–coreceptor complexes for the CLE19 peptide to regulate pollen development. *Nature Communications* 14 (1): 3307.

Zeng, S.M., Lo, E.K., Hazelton, B.J. et al. (2020). Effective range of non-cell autonomous activator and inhibitor peptides specifying plant stomatal patterning. *Development* 147 (17): dev192237.

Zhang, Y., Tan, S., Gao, Y. et al. (2022a). CLE42 delays leaf senescence by antagonizing ethylene pathway in Arabidopsis. *New Phytologist* 235 (2): 550–562.

Zhang, L., Yang, Y., Mu, C. et al. (2022b). Control of root stem cell differentiation and lateral root emergence by CLE16/17 peptides in Arabidopsis. *Frontiers in Plant Science* 13: 869888.

Zhao, X.A., Bramsiepe, J., Van Durm, M. et al. (2017). RETINOBLASTOMA RELATED1 mediates germline entry in Arabidopsis. *Science* 356 (6336): eaaf6532. https://doi.org/10.1126/science.aaf6532.

Zhu, S., Estevez, J.M., Liao, H. et al. (2020). The RALF1–FERONIA complex phosphorylates eIF4E1 to promote protein synthesis and polar root hair growth. *Molecular Plant* 13 (5): 698–716.

Zhu, Y., Hu, C., Cui, Y. et al. (2021). Conserved and differentiated functions of CIK receptor kinases in modulating stem cell signaling in Arabidopsis. *Molecular Plant* 14 (7): 1119–1134.

6

Regulation of Plant Morphology by Peptide Hormones

Parul Negi, Shehnaj Khan, and Anupama Singh

Department of Biotechnology, Dr. Y.S. Parmar University of Horticulture and Forestry, Solan, Himachal Pradesh, India

6.1 Introduction

Phytohormones are naturally occurring compounds that regulate the growth, development, and various cellular processes in plants at very low concentrations. They include auxins, cytokinins (CK), gibberellins (GA), abscisic acid (ABA), ethylene (ETH), brassinosteroids (BR), salicylates (SA), jasmonates (JA), and strigolactones (SL). Besides this several small peptides exhibit a similar regulatory role as classical plant hormones. The proteinaceous hormones are present in both the animal and plant kingdoms. The first peptide hormone was insulin, detected in 1921, in animals. Peptide hormones in plants were long overlooked because non-peptidic hormones, auxins and cytokinins, were discovered earlier. In 1991, the first signal protein implicated in defense response in tomato, named "systemin," was found in plants (Gancheva et al. 2016). Peptide hormones are tiny secreted molecules that are based on polypeptides and serve as intercellular communication signals. These peptide hormones are encoded by nuclear genes and frequently undergo posttranslational modifications (PTMs) and proteolytic processing of preproproteins. They play a part in pollination, self-incompatibility, fertilization, development of embryos and endosperm, stem cell homeostasis, cell differentiation, organogenesis, dehiscence, senescence, plant–pathogen and plant–insect interactions, and stress reactions (Zhang et al. 2025). Peptide hormones can be as short as a few amino acids or as long as 250 amino acids. In the past, the identification of peptide hormones was based on a number of analytical techniques, including chromatographic detection, phenotype-based genetic studies, chemical separation, and bioassays (Fletcher et al. 1999; Butenko et al. 2003; Skripnikov 2023). However, peptide hormones are present in extremely low concentrations in plants, posing significant challenges for separations, purifications, and biochemical analyses. To overcome these obstacles, plant peptide hormones have been identified using high-throughput omics techniques, including peptidomics, transcriptomics, and proteomics, in conjunction with detection techniques like high-throughput mass

Peptide Hormones in Plant Development and Stress Responses, First Edition. Edited by Vijay Pratap Singh and Samiksha Singh.

spectrometry (MS) and *in vitro* tests (Yang et al. 2022). Furthermore, the discovery of potential peptide hormones in various plant species has been made possible by developments in bioinformatics tools and machine-learning techniques (Zhang et al. 2025). Since peptide hormones are assumed to play a role in plant intercellular signaling, numerous studies have focused on finding small open reading frames encoding short secreted peptides (SSPs). Several plant species have so far been found to contain over 30 families of putative peptide hormones, including RAPID ALKALINIZATION FACTORS (RALFs), PHYTOSULFOKINES (PSKs), C-TERMINALLY ENCODED PEPTIDES (CEPs), EPIDERMAL PATTERNING FACTORS (EPFs), and CLAVATA3 (CLV3)/EMBRYO SURROUNDING REGION-RELATED peptides (CLEs). Although most peptide hormones have an N-terminal secretion peptide signal for the secretory pathway, certain peptide hormones are devoid of this signal peptide. They are classified as nonsecreted peptides and include the GRIM REAPER (GRI) PEPTIDE family, PLANT ELICITOR PEPTIDES (PEPs), ZEA MAYS IMMUNE SIGNALING PEPTIDE 1 (ZIP1), and SYSTEMINS (SYS) families. Plant growth, development, and stress responses have been demonstrated to be significantly influenced by both secreted and nonsecreted peptides. Typically, these peptide hormones attach to membrane-localized receptors, changing their structure and triggering downstream signaling to alter a variety of biological processes. Among the primary peptide signal receptors are LEUCINE-RICH-REPEAT RECEPTOR-LIKE KINASES (LRR-RLKs); *Arabidopsis* contains more than 200 members of the LRR-RLK family (Hsiao and Yamada 2021). Peptide hormones are classified into small peptides, secreted peptides, and signaling peptides. These hormones are integral to root development, coordinating signals that influence root architecture and growth. They interact with reactive oxygen species (ROS), linking environmental stress responses to developmental cues, and thereby affecting root morphology (Hsiao and Yamada 2021). Certain peptide hormones also regulate plant immunity, influencing gene expression related to immune responses, which can indirectly affect plant morphology by altering growth patterns under stress conditions (Luo et al. 2012). In contrast, peptide hormones are recognized for their roles in development, traditional phytohormones like auxins and cytokinins continue to dominate discussions on plant morphology, suggesting a need for integrated studies that encompass both classes of hormones to fully understand their combined effects on plant form and function (Gancheva et al. 2019).

6.2 Role of Peptide Hormones in Developmental Processes

The shoot apical meristems (SAMs) serve as a location of ongoing organogenesis, meaning that a little pool of pluripotent stem cells regenerates into lateral organs. Maintaining the size and structure of SAM depends on the coordination of intracellular and intercellular networks, which also promotes lateral organ development and patterning. Starting at the SAM's flanks, leaves grow into flattened structures of varying sizes and shapes. The transcriptional regulators and mechanical characteristics that control leaf development are mostly responsible for this process.

6.3 Classification of Peptide Hormones

6.3.1 Post-transcriptionally Modified Peptides

6.3.1.1 CLAVATA3/EMBRYO SURROUNDING REGION-RELATED (CLE)

Most research has been done on the CLE peptide hormones. Monocots, dicots, mosses, and algae are among the plants that have been reported to contain genes for CLE peptides and their receptors (Gancheva et al. 2016). The functional portion of the CLE peptide is the CLE domain of full-length precursor peptides situated close to the C-terminus. The CLE peptides undergo proteolytic processing right after translation, leading to the formation of the CLE domain from full-length precursor peptides. Because of this, the mature CLE peptide has 12–13 amino acids and contains 1–3 highly conserved proline residues. As a key regulator of the activity of the SAM, CLV3, the first member of the CLE family, suppresses the expression of the WUSCHEL (WUS) gene, which encodes a homeodomain-containing TF, a regulator of the meristem organizing center.

A number of the WUSCHEL-RELATED HOMEOBOX (WOX) transcription factors have been identified as downstream targets in CLE signaling pathways and receptor kinases are known receptors for CLE peptides. The stem cells are found in the central zone (CZ) of the SAM, where CLV3 is generated and expelled. CLV3 suppresses the expression of WUSCHEL (WUS) in the organizing center (OC) of the SAM by signaling through its receptors. WUS stimulates the CZ stem cell accumulation by non–cell-autonomous action. SAM stem cell number is controlled by the CLV3/WUS signaling pathway via a negative feedback loop (Schoof et al. 2000).

6.3.1.2 Root Growth Factor/GOLVEN/CLE-like (RGF/GLV/CLEL)

The length of RGF/GLV/CLEL peptides is 13–18 AA and they include sulfated tyrosine and hydroxylated proline. After initially being discovered in *Arabidopsis thaliana*, these peptides were later discovered in other plants (Matsuzaki et al. 2010). There are 11 members of the RGF/GLV/CLEL peptide family in *A. thaliana*. Proteolytic processing transforms an 86–163 AA precursor peptide into mature RGF peptides, which match the sequence of a conserved GLV domain. The TPST enzyme catalyzes the sulfation of the precursor peptide at conservative tyrosine residues (Amano et al. 2007). The peptide RGF1 is known to play a role in preserving the quiescent center cells' undifferentiated state in the apical root meristem and the synchronization of the columella stem cells' divisions. The root meristem and the expression levels of the *PLT1* and *PLT2* genes expressing TF PLETHORA, whose gradient regulates the destiny of root stem cells, both rise with exogenous treatment with RGF1. The regulation of polar auxin and root gravitropism is another role of RGF peptides; overexpression of the *GLV3*, *GLV6*, and *GLV9* genes causes the growth of wavy roots, hence the family's name, "GOLVEN," which translates to "waves" in Dutch (Shinohara et al. 2012).

6.3.1.3 Plant Peptides Containing Sulfated TYROSINE1 (*PSY1*)

PSY1 is an 18 AA glycopeptide containing one sulfated tyrosine and two hydroxylated prolines, functionally similar to PSK. As in the case of PSK, treatment with PSY1 stimulates cell division in suspension culture as well as cell growth by enlargement in the root and stem.

There are two genes encoding the *PSY1* precursor peptides in the genome of *A. thaliana*. They are expressed in the whole plant with maximum expression in the SAM and root elongation zone (Amano et al. 2007). PSY1 precursor peptides are subjected to tyrosine sulfation by the tyrosylprotein sulfotransferase enzyme and proteolytic processing, which yields the 18 amino acid sequence corresponding to the PSY1 domain from the precursor. Apart from development and cell division, PSK and PSY1 peptides that operate similartly and also important in regulating plant–pathogen interactions. Therefore, it was demonstrated that infection with *Pseudomonas syringae* pv. tomato induces the expression of the *PSKR1*, *PSK1*, *2*, *4*, and *PSY1* genes.

By increasing jasmonic acid signaling and suppressing salicylic acid signaling (response to biotrophic pathogens), PSKR1 and PSY1R receptor activation improves defense against necrotrophic pathogens and damage responses (Oehlenschlæger et al. 2017).

6.3.1.4 SYSTEMIN

The first plant peptide hormone to be identified was systemin, a protective peptide that was extracted from tomato leaves in 1991 (Pearce et al. 1991). Systemin is made up of 18 amino acid residues and plays a role in enhancing and activating the plant's defensive response to phytophage consumption. Two hundred amino acids make up prosystemin, the precursor of systemin, which lacks an N-terminal signal sequence. Prosystemin is produced and compartmentalized in the cytoplasm of phloem parenchymal cells, according to immunocytochemical studies. The membranes of *Solanum peruvianum* cells contained a 160-kD protein known as SR-160 that binds specifically to systemin (SURFACE RECEPTOR OF SYSTEMIN CELL). Mass spectrometry was used to further identify this protein as LRR-RLK, a homolog of the BRI1 brassinosteroid receptor. Following systemin's binding to the receptor, phospholipase is triggered, releasing linolenic acid from the cell membrane's phospholipids. This is a precursor to the manufacture of jasmonic acid, one of the primary "protective" plant hormones. Protective genes, such as the prosystemin gene, are activated by jasmonic acid–isoleucine conjugates; it is thought that injury to a single leaf causes systemin mRNA to accumulate across the entire plant (Li et al. 2017).

6.3.1.5 EMBRYO SURROUNDING FACTOR 1 (ESF1)

ESF1 is an additional CRP that accumulates in the female gametophyte's cells. At nanomolar quantities, the mature ESF1 peptide, which has 68 amino acids, can cause the suspensor cells to elongate. Only the central cell and, following fertilization in its descendants, the endosperm cells surrounding the embryo express the three genes that encode the ESF1 peptide of *A. thaliana* (*ESF1.1*, *1.2*, and *1.3*). While RNA interference of all three *ESF1* genes in *A. thaliana* results in significant defects in embryo morphology, mutations in one of the ESF1 genes have no discernible effect. Additionally, the expression pattern of the WOX8 gene, which is involved in the establishment of the embryo's apical-basal axis, changes, and the PIN1 gene, which codes for the auxin transporter, has a decreased expression level. Based on these findings, it appears that ESF1 regulates the suspensor and auxin distribution during the early phases of embryonic development (Costa et al. 2014).

6.3.1.6 POLARIS

A. thaliana contains POLARIS (PLS), an NSP with a length of 36 AA that is created as a precursor with a length of roughly 100 amino acids and subsequently undergoes proteolytic processing. Shorter roots, short, broad leaves with few vessels, and a decreased response to exogenous auxin and cytokinin are the hallmarks of the pls mutant. Therefore, it is thought that the PLS peptide is required for both the development of the leaf's vascular system and for coordinating plant growth along the vertical axis. The *PIN1* and *PIN2* genes, which encode auxin polar proteins, had lower expression levels in the pls mutant (Liu et al. 2013). As one of the most crucial mechanisms of cytokinin and auxin antagonistic interactions in regulating ontogenetic processes is the regulation of PIN gene expression, it is observed that the PLS peptide is required to preserve auxin and cytokinin balance, particularly given that the pls mutant exhibits an increased level of cytokinin and a decreased level of auxin biosynthesis.

The ctr1 mutant (CONSTITUTIVE TRIPLE RESPONSE 1) exhibits a constitutive response to ethylene and a normal phenotype when the *PLS* gene is overexpressed, whereas the pls mutant also exhibits supersensitivity to ethylene and the constitutive triple response of seedlings even in the absence of ethylene. The PLS peptide is therefore found to be a negative response regulator of ethylene, and furthermore, the *PLS* gene is negatively regulated by ethylene (Chilley et al. 2006).

6.3.1.7 PAMP-Induced Secreted Peptides (PIPs)

The PAMP-induced secreted peptides (PIPs) family comprises 11 peptides that exhibit a significant degree of homology to both IDA and CEP. Nonetheless, while IDA and CEP play critical roles in facilitating plant developmental processes, the function of PIP is strictly confined to the modulation of responses to an extensive array of pathogens (Vie et al. 2015). In the context of plant biology, the initiation of an immune response occurs at the plasma membrane's surface and is predominantly mediated by receptors—principally the LRR-RLK receptor protein kinases—that interact with elicitors, which are molecules synthesized by pathogens, such as pathogen-associated molecular patterns (PAMPs), including bacterial flagellin and fungal chitin, or arise in response to cellular damage—termed damage-associated molecular patterns (DAMPs), exemplified by oligogalacturonides and certain peptides, inclusive of signal peptides. The genes responsible for encoding the secreted peptides of PIP have been identified among the genetic sequences whose expression is promptly induced upon the recognition of PAMPs. Orthologs of the PIP gene have been discerned in various dicotyledonous and monocotyledonous species, including, but not limited to, soybean, grapes, maize, and rice. In *A. thaliana*, the expression of the *PIP1* gene is significantly augmented upon infection by *P. syringae*, *Fusarium oxysporum* or f. sp. *conglutinans*, as well as during exposure to the phytohormone that plays a pivotal role in the plant's immune responses—salicylic acid and its methyl ester. The precursor peptides, denoted as pre-PIP, comprise 50 to 80 amino acids and contain multiple conserved C-terminal Sulfated GlycoPeptide Signaling (SGPS) domains of 13 to 14 amino acids (designated based on the most conserved tetrapeptide motif), akin to the EPIP domain found in IDA peptides (Gancheva et al. 2019). The PIP receptor is identified as receptor-like kinase 7 (RLK7). PIP peptides function as integral constituents of the plant's systemic immune response: the transcription of genes within the PIP group is activated by the binding of PAMP flagellin to the FLS2 receptor kinase (Flagellin-Sensitive 2) and by the PEP peptide (discussed below), which serves as a DAMP, alongside PEPR receptor kinases (PEP RECEPTOR). The interaction of PIP with the receptor kinase RLK7 instigates the activation of the mitogen-activated protein (MAP) kinase cascade, leading to the production of ROS, the deposition of callose, and the upregulation of several gene regulators associated with the immune response. Expression of the PIP1 gene is notably detected in conductive tissues, implying that PIP1 acts as a mobile signaling entity participating in the systemic immune response (Hou et al. 2017).

6.3.1.8 LURE

In dicotyledonous plants, the LURE (LURE1 and LURE2) peptides, initially identified in the ornamental species *Torenia fournieri* (Okuda et al. 2009), exhibit a functional role analogous to that of egg cell-secreted attractants (EA). Subsequently, five homologous variants of LURE1, designated as AtLURE1.1-1.5, were discovered in *A. thaliana*. The mature LURE peptides comprise 62–70 amino acids in length and contain six cysteine residues. These peptides are produced by synergid cells and play a crucial role in guiding the growth of the pollen tube upon its arrival at the micropyle of the ovule. Recent studies have elucidated that AtPRK6, a leucine-rich repeat (LRR) kinase known as Pollen Receptor Kinase 6, serves as the receptor for AtLURE. AtPRK6 is capable of interacting with the nucleotide exchange factor ROP GEF12 (a ROP GTP exchange factor) as well as receptor-like cytoplasmic kinases LIP1/2 (LOST IN POLLEN TUBE GUIDANCE) that are anchored to the inner membrane of the pollen tube. It is postulated that LIP1/2 function as coreceptors for AtPRK6; however, the precise signaling cascade activated within the pollen tube by LURE peptides remains to be elucidated (Takeuchi and Higashiyama 2016) (Figure 6.1).

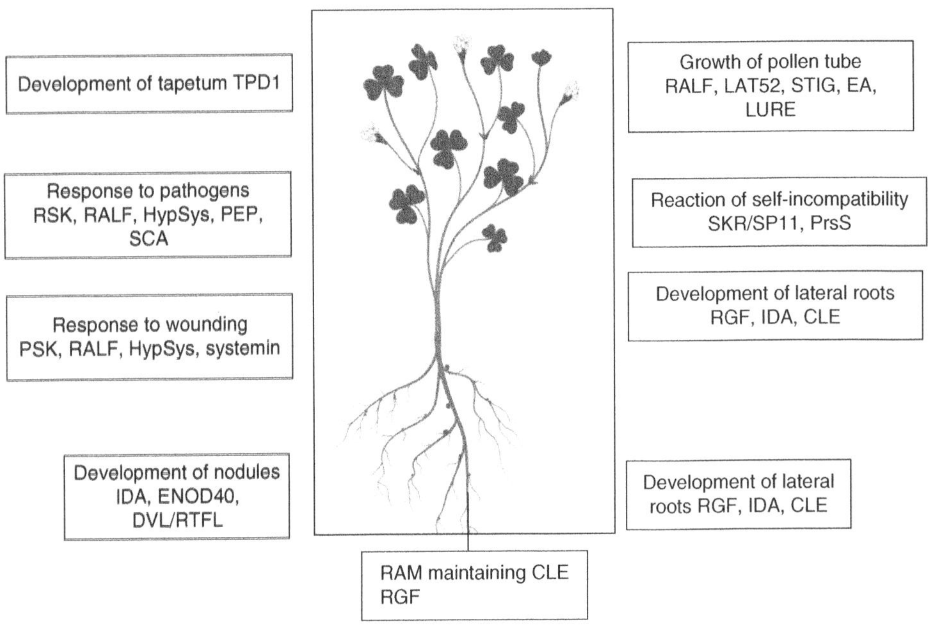

Figure 6.1 Functions of peptide hormones in plant development.

6.4 Maturation of Peptide Hormones

The biosynthesis of peptide hormones from preproproteins in botanical entities encompasses several pivotal stages, including the excision of the signal peptide and subsequent folding, proteolytic cleavage, and PTMs facilitated by specific enzymes. Ultimately, these peptides are expelled from the cellular environment via either the classical protein secretion (CPS) pathway, which necessitates the involvement of endoplasmic reticulum (ER), Golgi apparatus, and endomembrane system, or the nonclassical protein secretion (UPS) pathways, which include secretory vesicles or granules, secretory multivesicular bodies, vacuoles, and exocyst-positive organelles (EXPO) (Corpo et al. 2024). Peptide hormones synthesized subsequent to PTMs typically comprise fewer than 20 amino acid residues, and exhibit either an absence or a limited presence of cysteine (Cys) residues, while containing PTMs on amino acids such as proline (Pro) and tyrosine (Tyr). The maturation of these peptides necessitates reactions that are catalyzed by specific enzymes. For example, the PTMs facilitated by proline hydroxylation and C-terminal processing are imperative for the functional activation of the CLE40 peptide. In this maturation process, the CLE40 preproprotein undergoes proteolytic cleavage by subtilisin-like serine proteases (SBTs), specifically SBT1.4, SBT1.7, and SBT4.13, ultimately resulting in the liberation of the active peptide, as reported by Stührwohldt et al. (2020). Furthermore, the IDA/IDL family of peptide hormones, which are crucial for various aspects of plant development, particularly in the process of organ abscission, are enzymatically processed by subtilisin-like proteinases (SBT4.12, SBT4.13, and SBT5.2) to yield their mature forms consisting of 14 amino acid residues, as indicated by Schardon et al. (2016). Critical residues for the maturation of IDA peptides include proline and tyrosine at positions 2 (P2) and 4 (P4), respectively, as substitutions of these residues hinder the recognition of the IDA precursor peptide (proIDA) by the enzyme SBT4.13, thereby significantly reducing its capacity to ameliorate the abscission defects manifested in the *ida* mutants, as elucidated by Stenvik et al. (2008). Peptide hormone proteolytic processing may be a crucial signal transduction regulation point. PEP1 peptide maturation requires the proteolytic cleavage of the PEP1 preproprotein by Ca^{2+} -dependent type-II metacaspases (MCs). High Ca^{2+} levels are both required and sufficient for PEP1 maturation and activate MC4 upon injury (Hander et al. 2019). According to another study, the catalytic Cys residue in MC4 or the conserved Arg residue that comes before the PEP1 domain in the precursor protein might be moved to inhibit processing of the PEP1 peptide (Shen et al. 2019).

6.5 Conclusion

Peptide hormones serve as fundamental developmental regulators across all terrestrial plant taxa. Recent investigations into nonvascular plant species indicate that several peptide families exhibit highly conserved functions in stem cell signaling processes. The extent of conservation concerning developmental mechanisms that operate downstream of peptide signaling throughout various plant lineages remains to be elucidated; nonetheless, it is plausible that at least CLE signaling within meristematic tissues is intertwined with critical phytohormone signaling pathways, such as auxin, across all plant lineages examined thus far.

An evolutionary pattern has emerged, suggesting extensive gene expansions within both CLE and RGF gene families, although such expansions are not exclusively confined to these two peptide hormone families. The duplication of genes that encode peptide hormones may lead to a proportional increase in the quantity of each peptide synthesized within a species, thereby creating a remarkably intricate genetic framework upon which evolutionary processes can act. These expansions within gene families have significantly contributed to the intricate forms of genetic compensation and redundancy observed during angiosperm development. Ongoing research within diverse land plant lineages is anticipated to yield further insights into the ubiquity of compensatory genetic mechanisms in plant development and to illuminate the processes by which novel gene regulation emerges.

Abbreviations

AA	amino acids
AtPRK6	*Arabidopsis thaliana* pollen receptor kinase 6
CEP	C-TERMINALLY ENCODED PEPTIDE
CLE	CLAVATA3/EMBRYO SURROUNDING REGION-RELATED
DEV	DEVIL (family name for small peptides involved in developmental regulation)
ENOD 40	EARLY NODULIN 40
HypSys	hydroxyproline-rich systemin
IDA	INFLORESCENCE DEFICIENT IN ABSCISSION
IDA-L	INFLORESCENCE DEFICIENT IN ABSCISSION-LIKE
LURE	class of small cysteine-rich peptides
NSP	nonsecreted peptides
PEP1	plant elicitor peptide 1
PrsS	Pistil S-locus protein S
PSKR1	phytosulfokine receptor 1
RALF	rapid alkalinization factor
RGF	root meristem growth factor
RSK	rapid stress response kinase
RTFL	ROT-FOUR-LIKE
SCA	serine-rich cysteine-rich acidic peptide
SP 11	S-locus protein 11
TPD 1	tapetum determinant 1
TPST	tyrosylprotein sulfotransferase

References

Amano, Y., Tsubouchi, H., Shinohara, H. et al. (2007). Tyrosine-sulfated glycopeptide involved in cellular proliferation and expansion in *Arabidopsis*. *Proceedings of the National Academy of Sciences* 104: 18333–18338.

Butenko, M.A., Patterson, S.E., Grini, P.E. et al. (2003). Inflorescence deficient in abscission controls floral organ abscission in *Arabidopsis* and identifies a novel family of putative ligands in plants. *Plant Cell* 15: 2296–2307.

Chilley, P.M., Casson, S.A., Tarkowski, P. et al. (2006). The POLARIS peptide of *Arabidopsis* regulates auxin transport and root growth via effects on ethylene signaling. *The Plant Cell* 18: 3058–3072.

Costa, L.M., Marshall, E., Tesfaye, M. et al. (2014). Central cell-derived peptides regulate early embryo patterning in flowering plants. *Science* 344: 168–172.

Corpo DD, Coculo D, Greco M, Lorenzo GD. And Lionetti V. 2024. Pull the fuzes: Processing protein precursors to generate apoplastic danger signals for triggering plant immunity. *Plant Communications.* 5: 100931.

Fletcher, J.C., Brand, U., Running, M.P. et al. (1999). Signaling of cell fate decisions by CLAVATA3 in *Arabidopsis* shoot meristems. *Science* 283: 1911–1914.

Gancheva, M.S., Dodueva, I.E., and Lebedeva, M.A. (2016). Identification, expression, and functional analysis of CLE genes in radish (*Raphanus sativus* L.) storage root. *BMC Plant Biology* 16: 7.

Gancheva, M.S., Dodueva, I.E., Lebedeva, M.A. et al. (2019). Plant peptide hormones. *Russian Journal of Plant Physiology* 1: 7.

Hander, T., Fernández-Fernández, Á.D., and Kumpf, R.P. (2019). Damage on plants activates Ca2+-dependent metacaspases for release of immunomodulatory peptides. *Science* 80: 363.

Hou S, Wang X, Chen D, Yang X, Wang M, Turrà D, P.A. Zhang W (2017) The secreted peptide PIP1 amplifies immunity through receptor-like kinase 7. *PLoS Pathogens* 10: e1004331.

Hsiao, Y.C. and Yamada, M. (2021). The roles of peptide hormones and their receptors during plant root development. *Genes* 12 (1): 22.

Li, Z., Wang, Y., Huang, J. et al. (2017). Two SERK receptor-like kinases interact with EMS1 to control anther cell fate determination. *Plant Physiology* 173: 326–337.

Liu, J., Mehdi, S., Topping, J. et al. (2013). Interaction of PLS and PIN and hormonal crosstalk in *Arabidopsis* root development. *Frontiers in Plant Science.* 4: 75.

Luo, J., Fishburn, J., Han, S., and Ranish, J. (2012). An integrated chemical cross-linking and mass spectrometry approach to study protein complex architecture and function. *Molecular & Cellular Proteomics* 11: M111–M008318.

Matsuzaki, Y., Ogawa-Ohnishi, M., Mori, A., and Matsubayashi, Y. (2010). Secreted peptide signals required for maintenance of root stem cell niche in *Arabidopsis. Science* 329: 1065–1067.

Oehlenschlæger, C.B., Gersby, L.B.A., Ahsan, N. et al. (2017). Activation of the LRR receptor-like kinase PSY1R requires transphosphorylation of residues in the activation loop. *Frontiers in Plant Science* 8: https://doi.org/10.3389/fpls.2017.02005.

Okuda, S., Tsutsui, H., Shiina, K. et al. (2009). Defensin-like polypeptide LUREs are pollen tube attractants secreted from synergid cell. *Nature* 458: 357–361.

Pearce, G., Strydom, D., Johnson, S., and Ryan, C.A.A. (1991). Polypeptide from tomato leaves induces wound inducible proteinase inhibitor proteins. *Science* 253: 895–897.

Schardon, K., Hohl, M., and Graff, L. (2016). Precursor processing for plant peptide hormone maturation by subtilisin-like serine proteinases. *Science* 354: 159.

Schoof, H., Lenhard, M., Haecker, A. et al. (2000). The stem cell population of Arabidopsis shoot meristems in maintained by a regulatory loop between the CLAVATA and WUSCHEL genes. *Cell* 100: 635–644.

Shen, W., Liu, J., and Li, J.F. (2019). Type-II metacaspases mediate the processing of plant elicitor peptides in *Arabidopsis. Molecular Plant* 12: 1524–1533.

Shinohara, H., Moriyama, Y., Ohyama, K., and Matsubayashi, Y. (2012). Biochemical mapping of a ligand-binding domain within *Arabidopsis* BAM1 reveals diversified ligand recognition mechanisms of plant LRR-RKS. *Plant Journal* 70: 845–854.

Skripnikov, A. (2023). Bioassays for identifying and characterizing plant regulatory peptides. *Biomolecules* 13 (12): 1795.

Stührwohldt, N., Scholl, S., and Lang, L. (2020). The biogenesis of Cle l peptides involves several processing events in consecutive compartments of the secretory pathway. *eLife* 9: e55580.

Stenvik, G.E., Tandstad, N.M., Guo, Y. et al. (2008). The EPIP peptide of INFLORESCENCE DEFICIENT IN ABSCISSION is sufficient to induce abscission in Arabidopsis through the receptor-like kinases HAESA and HAESA-LIKE2. *Plant Cell* 20: 1805–1817.

Takeuchi, H. and Higashiyama, T. (2016). Tip-localized receptors control pollen tube growth and LURE sensing in *Arabidopsis*. *Nature* 531: 245–248.

Vie, A.K., Najafi, J., Liu, B. et al. (2015). The IDA/IDA-LIKE and PIP/PIP-LIKE gene families in Arabidopsis: phylogenetic relationship, expression patterns, and transcriptional effect of the PIPL3 peptide. *Journal of Experimental Botany* 66: 5351–5365.

Yang, Y., Li, P., and Liu, C. (2022). Systematic analysis of the non-specific lipid transfer protein gene family in *Nicotiana tabacum* reveal its potential roles in stress responses. *Plant Physiology and Biochemistry* 172: 33–47.

Zhang, Z., Han, H., Zhao, J. et al. (2025). Plant peptide hormones. *Molecular Horticulture* 5: 7. https://doi.org/10.1186/s43897-024-00134-y.

7

Peptide Hormones in Flowering Regulation

Jayoti Majumder Sarkar[1] *and Saikat Gantait*[2]

[1] *Department of Floriculture and Landscape Architecture, Bidhan Chandra Krishi Viswavidyalaya, Mohanpur, West Bengal, India*
[2] *Crop Research Unit (Genetics and Plant Breeding), Bidhan Chandra Krishi Viswavidyalaya, Mohanpur, West Bengal, India*

7.1 Introduction

7.1.1 What Is a Peptide Hormone (PH)?

Proteins and PHs consist of amino acid (AA) chains of varying lengths. PHs are small, secreted polypeptide-based molecules that function as intercellular communication signals (Lease and Walker 2006). The apoplast serves as a free diffusional outer space between cells and plays a crucial role in intercellular signaling, and is essential for plant growth. Among the key signaling components in this space are secreted peptides, including PHs, which regulate various biological processes by binding to specific receptors (Galiullina et al. 2015). PHs have been gradually discovered across various plant species, highlighting their crucial roles in both short- and long-range signal transduction processes. Their identification by their receptors on the plasma membrane of the target cell triggers them to operate extracellularly. These peptide signals or plant PHs are divided into three categories based on mature peptide structures and trafficking mechanisms into the extracellular space: secreted cysteine-rich peptides, nonsecreted short peptides, and secreted small peptides (Matsubayashi 2014).

7.1.2 History of PH Studies in Plants and Flowers

Communication between cells is crucial for multicellular organisms' reactions to environmental factors and developmental signals. In multicellular organisms, effective intercellular communication is vital for responding to developmental cues and environmental stimuli. Traditionally, classical phytohormones—auxin, abscisic acid (ABA), cytokinin, ethylene, and gibberellin (GA)—are considered the primary regulators of intercellular signaling in plants. Over time, additional molecules, such as salicylic acid, brassinosteroids, and jasmonate, have been recognized as important plant growth regulators (Lindsey et al. 2002; Vanstraelen and Benková 2012). Over the past decade, researchers have revealed that, in addition to plant hormones, other molecules play vital roles in these cell–cell interaction

Peptide Hormones in Plant Development and Stress Responses, First Edition. Edited by Vijay Pratap Singh and Samiksha Singh.

networks. These molecules include small RNAs, signaling peptides, and transcription factors, which contribute to the regulation of various physiological and developmental stages (Lindsey et al. 2002; Van Norman et al. 2011; Murphy et al. 2012). However, the first signaling peptide identified in plants was tomato systemin, which was discovered in 1991 for its role in defense responses (Gancheva et al. 2019).

The discovery of tomato systemin in 1991 inspired plant biologists to search for additional peptide signals. In 1996, phytosulfokines (PSKs) were identified as regulators of cell division. Notably, PSKs were the first plant signaling peptides found to undergo posttranslational modification (PTM) (Matsubayashi and Sakagami 1996). For flowering, in 1999, it was suggested that the CLAVATA3 protein is secreted and, by interacting with its receptor, regulates the size of the pool of stem cells in the shoot apical meristem (SAM) (Fletcher et al. 1999). The HypSys families of peptides, which are plant protection signal peptides, were discovered in tobacco in 2001 as a result of the quest for systemin-like peptides in other species of plant (Gancheva et al. 2019). Putative PHs, which were discovered and identified later in various plant species, play key roles in short- and long-range signal transduction and are involved in a number of floral developmental processes, such as embryo development, pollination, fertilization, dehiscence, and senescence (Zhang et al. 2025).

7.1.3 Place of Origin and Signaling in Plants and Flowers

The SAM is vital for postembryonic growth, generating leaves, stems, and flowers (Miwa et al. 2009). It maintains equilibrium between stem cell renewal and organ differentiation and is regulated by genetic and hormonal signaling networks. The WUSCHEL (WUS) and CLAVATA (CLV) pathways are key regulators, with WUS promoting stem cell identity and CLV restricting stem cell proliferation to prevent overgrowth (Song et al. 2022; Selby and Jones 2023). Phytohormones such as auxin and cytokinin guide cell division and differentiation. As the plant transitions to reproductive development, FLOWERING LOCUS T (FT) and SUPPRESSOR OF OVEREXPRESSION OF CONSTANS1 (SOC1) regulate this shift (Song et al. 2022; Selby and Jones 2023). CLAVATA3 (CLV3), a CLE peptide, binds to the CLAVATA1 (CLV1) receptor to regulate stem cell activity in the SAM, maintaining stem cell population stability (Jun et al. 2010). CLE40, another CLE peptide, promotes stem cell fate through the BARELY ANY MERISTEM 1 (BAM1) receptor (Gregory et al. 2018; Dao et al. 2022), thereby counteracting CLV3 signaling (Song et al. 2013). The functional redundancy among CLE peptides (CLE16, CLE17, and CLE27) contributes to SAM regulation, ensuring balanced growth and organogenesis.

7.1.4 Overall Activities of Peptides

In light of their role in important developmental processes like meristem maintenance, cell elongation, cell proliferation, organ abscission, differentiation, gravitropism, and defense against biotic stress factors, secreted peptides are identified as molecular messengers in plants (Matsuzaki et al. 2010; Whitford et al. 2012). Self-incompatibility, pollination, fertilization, formation of embryos and endosperm, organogenesis, cell differentiation, senescence, communication between plants and pathogens, between plants and insects, and responses to stress are among the events in which these genes are implicated. A survey of the significant findings, roles, chemical makeup, transcriptional controls, and PTMs of plant PHs is intended to be given in this chapter. Additionally, downstream signal transduction pathways and peptide-sensing receptor kinases have received interest (Zhang et al. 2025).

Once flowering is initiated, specific peptide signals regulate floral meristem identity and organ differentiation, such as ROOT GROWTH FACTOR (RGF) peptides (Keerthana et al. 2025). Although RGF peptides were originally studied in the context of root development, they also play a role in floral organ formation by modulating auxin distribution in the meristem. EPIDERMAL PATTERNING FACTOR-LIKE (EPFL) peptides such as EPFL2 and EPFL9 (STOMAGEN) are involved in floral organ patterning and inflorescence architecture, as they interact with ER family receptors to regulate floral organ growth and spacing (Tameshige et al. 2017). INFLORESCENCE DEFICIENT IN ABSCISSION PEPTIDES (IDAP) influence floral organ peptide signals and do not act in isolation; they interact with plant hormones such as GAs, auxins, cytokinins, and ethylene to fine-tune flowering, as CLE peptides and cytokinin coordinate stem cell maintenance in the floral meristem. FTIP1 with GAs regulates FT protein movement, linking GA levels to flowering time. EPFL peptides with auxins modulate floral organ positioning and growth.

7.2 Types of PHs for Flowering

This fundamental distinction between the central zone (CZ) and peripheral zone (PZ) is essential for balancing stem cell maintenance and organ initiation. The CZ harbors a self-renewing population of stem cells, whereas the PZ serves as the site where cells are recruited for differentiation, leading to the formation of lateral organs such as leaves, flowers, and branches (Pazourek et al. 1992). The CZ/PZ distinction allows plants to adapt their growth to environmental conditions, ensuring a continuous supply of organs throughout their lifespan. The dynamic interplay between genetic, hormonal, and peptide signaling networks enables the shoot meristem to remain functional and responsive. In different plant species, over 30 families of PHs have been found. These peptides are classified as secretory peptides (which contain an N-terminal secretion signal), which undergo PTMs before being released into the apoplast, where they interact with specific receptor-like kinases (RLKs) to regulate signaling pathways. CLAVATA3 (CLV3)/embryo surrounding region-related peptides (CLEs) that regulate stem cell homeostasis in the SAM, e.g. CLV3 maintains the balance between meristem proliferation and differentiation. In addition, C-terminally encoded peptides (CEPs), PSKs, epidermal patterning factors (EPFs), and rapid alkalinization factors (RALFs) are involved in nutrient signaling and root architecture modulation, such as regulate nitrate-dependent root development. Nonsecretory peptides (which lack this signal function within the cytoplasm or after cell lysis), such as systemins (SYS), trigger jasmonic acid (JA)—mediated defense responses against herbivores. PHs associated with floral organ development and the flowering process are described below.

7.2.1 CLE Peptides (CLAVATA3/Embryo Surrounding Region-Related)

The main source of cells that produce the above-ground anatomy of higher plants is stem cell reservoirs found in floral and shoot meristems (Somssich et al. 2016). Signals from the underlying organizing center (OC) sustain the identity and activity of stem cells, which are found at the apex of meristems in a specialized area known as the CZ (Mayer et al. 1998). During the meristematic activity for vegetative and reproductive phases, the SAM remains indeterminate and continuously produces leaves and lateral organs on its flanks. The CLAVATA-WUSCHEL (CLV-WUS) signaling pathway is essential for regulating stem cell homeostasis in the SAM. CLAVATA3 (CLV3), a CLE peptide, is discharged by stem cells and binds to CLV1, CLV2-CRN, and BAM receptors, restricting WUS expression to control meristem size (Somssich et al. 2016). Upon transitioning to reproductive development, the SAM converts into an inflorescence meristem (IM) and then differentiates into floral meristems (FMs), which produce floral organs. LEAFY (*LFY*) and APETALA1 (*AP1*) promote floral meristem identity and suppress vegetative fate (Weigel and Nilsson 1995). AGAMOUS (AG), a MADS-box transcription factor, terminates stem cell activity in floral meristems by repressing WUS expression, ensuring proper flower development (Lohmann et al. 2001). The dynamic activity of stem cells in floral and shoot meristems involves genetic feedback loops, peptide signaling, and hormonal interactions, ensuring continuous organogenesis during plant growth and reproduction.

7.2.2 Phytosulfokines

Phytosulfokines (PSKs) are a class of PHs that are distinguished by tyrosine sulfation alterations. In low-density cultivated *Asparagus officinalis* mesophyll cells, the first PSK was shown to stimulate proliferation (Matsubayashi and Sakagami 1996). For almost thirty years, studies have shown that PSKs are essential for biotic and abiotic stress responses, plant reproduction, root nodule formation in legumes, and cell division and organ growth. *In vitro* germination of pollen and growth of pollen tubes are usually reliant on population density, a phenomenon termed the "pollen population effect" (Pasonen and Kapyla 1998). Exogenous application of PSK-α peptides to low-density cultures promotes germination of pollen and growth of pollen tubes (Chen et al. 2000; Kou et al. 2020), indicating that PSK signaling is essential for pollen function. Several PSK precursor and receptor genes exhibit tissue- and organ-specific expression patterns during reproductive development. In *Arabidopsis*, *AtPSK2* is highly expressed in pollen (Stuhrwohldt et al. 2015). Similarly, in maize, *ZmPSK1* is identified in central cells, pollen, and egg cells (Lorbiecke et al. 2005), and in pear, *PbrPSK2* is highly expressed in pollen (Kou et al. 2020). Genetic studies further confirmed the importance of PSK signaling in plant fertility. *Arabidopsis* PSK receptor double mutants (pskr1 pskr2) and TPST mutants (that impede PSK signaling and maturation) exhibit declined fertility due to defective funicular pollen tube guidance (Stuhrwohldt et al. 2015). The overexpression of *AtPSK4* leads to sterility, impairing fibrous cell wall thickening in anther endothecium (Yu et al. 2016). PSK peptides also regulate seed growth, as in soybean; *GmPSKγ1* is explicitly expressed in developing seeds, and its overexpression enhances embryo cell expansion and seed growth (Yu et al. 2019). In wheat, the overexpression of *TaPSK5*, a wheat PSK precursor gene, considerably amplifies the number and size of rice grains (Geng et al. 2020). Collectively, these findings suggest that PSK peptides are essential for pollen function, fertilization, and seed development and play a vital role in crop yield improvement. PSKs have been identified as potent inducers of somatic embryogenesis. In carrot (*Daucus carota*) and *Cryptomeria japonica*, PSKs stimulate somatic embryo (SE) formation (Hanai et al. 2000; Igasaki et al. 2003). *C. japonica* plantlets regenerated via SE develop typically, confirming the role of PSK in embryo development (Igasaki et al. 2003). PSK treatment dramatically enhances SE efficiency, overcoming regeneration recalcitrance in legumes (Ochatt et al. 2018). These findings indicate that PSK peptides enhance regeneration competence, suggesting promising applications in plant biotechnology, genetic transformation, and molecular breeding.

7.2.3 Rapid Alkalinization Factor (RALF)

Flowering plants have developed a special double-fertilization mechanism to accomplish sexual reproduction. The male gametophyte of angiosperms, pollen, is made up of two stationary sperm cells and facilitates the interaction between males and females. Pollen instantly germinates when it lands on a flower's stigma, creating a highly polarized and quickly expanding pollen tube that transports the sperm cells to the female gametes located deep within the ovules (Dresselhaus et al. 2016). The pollen tubes and the ovule's communication and signaling are strictly controlled at several stages, both temporally and spatially, to ensure successful fertilization. Over the past 20 years, studies have revealed that interactions between polymorphic peptides discharged from female gametes and receptors on the pollen tube surface play vital parts in directing pollen tubes (Okuda et al. 2009; Wang et al. 2016). While numerous factors regulating pollen tube integrity have been identified, the mechanisms controlling pollen tube rupture remain elusive (Muschietti and Wengier 2018). Recent research has shed light on signaling molecules that trigger pollen tube burst and sperm discharge, significantly advancing our understanding of how pollen tube integrity is regulated to confirm fruitful double fertilization (Mecchia et al. 2017). These findings contribute to the growing knowledge of the complex signaling interactions that orchestrate communication between male and female gametes in plants.

7.2.4 INFLORESCENCE DEFICIENT IN ABSCISSION (IDA) Peptides

IDA is a small secreted peptide that plays a vital role in the abscission of floral organs in *Arabidopsis*. It acts as a signaling molecule that regulates the detachment of floral organs when they are no longer needed, typically after fertilization or when the plant needs to conserve resources. Loss-of-function mutations in the *IDA* gene result in flowers and floral organs remaining attached, suggesting that IDA is necessary for initiating abscission. On the other hand, overexpression of IDA leads to premature shedding of floral organs, indicating that IDA can promote abscission when it is overproduced. The full-length IDA peptide undergoes proteolytic processing, and a conserved 20-AA stretch, termed extended PCNA interacting protein (EPIP), is identified as the active region that is required for IDA function. This peptide fragment is sufficient to rescue the loss-of-function phenotype of IDA mutants (Stenvik et al. 2008). The EPIP sequence contains a highly conserved dodecamer peptide (12 AA), which can activate HAESA and HAESA-Like 2 (HSL2) receptors in transient assays, demonstrating its functional importance (Butenko et al. 2014).

HAESA and HSL2 are leucine-rich repeat receptor-like kinases (LRR-RLKs) that redundantly control floral abscission. They are essential for sensing the IDA signal and activating the downstream signaling pathway that leads to the detachment of floral organs. HAESA and HSL2 are thought to work in tandem with IDA to regulate the timing of abscission, although their precise interaction with IDA is still not fully understood. The HAESA receptor might directly recognize and bind the EPIP peptide, triggering receptor activation and signaling that induces the cell wall modifications necessary for organ separation. The EPIP peptide within IDA contains a central proline (Pro) residue, which is presumed to be post-translationally modified to hydroxyproline (Hyp), a modification that might be crucial for the peptide's ability to interact with its receptor. This modification may impact the peptide's structure and flexibility, influencing its interaction with the receptor (Butenko et al. 2014). While genetic and biochemical evidence suggests that IDA and its receptors HAESA and HSL2 work together to control floral abscission, a detailed understanding of how IDA is directly sensed by the HAESA receptor and how the binding of IDA to the receptor activates the signaling cascade is lacking. This area remains an important focus for further research. Butenko et al. (2003) reported IDA and its role in floral abscission and Stenvik et al. (2008) investigated the loss-of-function and overexpression phenotypes of IDA in *Arabidopsis*.

7.3 Signaling of PHs

7.3.1 PH Signaling in Flowering Plants

CLE peptides function as ligands that bind to LRR-RLKs to regulate downstream transcription factors and hormonal pathways. These interactions govern stem cell maintenance, vascular differentiation, and stress responses (Yamaguchi et al. 2016). CLAVATA3 (CLV3) is essential for balancing stem cell homeostasis in the SAM. CLV3 binds to CLV1, CLV2-CRN, and BAM receptors, forming a negative feedback loop with WUS to regulate meristem size (Nimchuk et al. 2015; Somssich et al. 2016). CLE40 works antagonistically to CLV3 by interacting with BAM1 in the SAM to increase stem cell populations (Schlegel et al. 2021). CLE45 regulates root growth by modulating ABA signaling and root meristem activity (Takahashi et al. 2018). CLE peptides play crucial roles in floral meristem maintenance by acting as signaling molecules that bind to LRR-RLKs, regulating stem cell homeostasis, vascular differentiation, and stress responses (Yamaguchi et al. 2016).

PSK, another important peptide signaling molecule, plays a significant role in floral organ growth by binding to PSK receptors (PSKRs), which belong to the X subclass of LRR-RLKs (Matsubayashi et al. 2006). The first PSKR (DcPSKR1) was identified in carrot and was found to exhibit high-affinity binding to PSK, with structural analyses revealing an extracellular domain rich in leucine-rich repeats (Matsubayashi et al. 2006). The *Arabidopsis* genome encodes two PSK receptors, AtPSKR1 and AtPSKR2, where AtPSKR1 is ubiquitously expressed, whereas AtPSKR2 is predominantly found in flowers (Matsubayashi et al. 2006). Functional studies indicate that PSKRs possess both kinase and guanylate cyclase (GC) activity, generating cyclic guanosine monophosphate (cGMP) upon PSK binding, which in turn activates the cyclic nucleotide–gated channel (CNGC17), promoting cell expansion and growth (Kwezi et al. 2011). In tomato, PSK signaling is linked to plant immunity through interactions with calcium-dependent protein kinase (CPK28) and E3 ubiquitin ligases (PUB12/13), which regulate receptor stability via PTMs (Zhang et al. 2008; Hu et al. 2022).

In pollen tube growth and fertilization, RALF peptides, particularly RALF4 and RALF19, have been identified as essential regulators that interact with RLKs such as Buddha's Paper Seals 1 and 2 (BUPS1/2) and ANXUR1/2 (ANX1/2) (Ge et al. 2017). These receptors, members of the CrRLK1L family, maintain pollen tube integrity and prevent early rupture during pollen tube elongation. Loss-of-function mutants of BUPS1 and BUPS2 exhibit pollen tube defects, leading to sterility (Ge et al. 2017). Additionally, ovule-derived RALF34 binds to BUPS1/2 and ANX1/2 at nanomolar concentrations, inducing pollen tube burst and sperm discharge at the site of fertilization. Interestingly, a set of leucine-rich repeat extensions (LRXs) also interacts with RALF4/19, influencing cell wall remodeling by regulating pectin composition, which is essential for controlled pollen tube expansion (Mecchia et al. 2017). These findings collectively highlight the critical role of peptide signaling in coordinating meristem maintenance, floral organ development, and reproductive success in plants.

7.3.2 Peptide Receptors Involved in Different Fertilization Processes Post-flowering

7.3.2.1 Self-Incompatibility

Self-incompatibility serves as a reproductive block in many plant species, preventing self-pollination and interspecific hybridization. When pollen grains land on the stigma, they rapidly interact with papilla cells ahead of germination, leading to a recognition process that enforces both interspecific and self-incompatibility, ultimately reducing self-fertilization (Takayama and Isogai 2005). Self-incompatibility mechanisms are classified into two types: gametophytic self-incompatibility (GSI) and sporophytic self-incompatibility (SSI). The SSI response is regulated by male and female components determined by S-locus genes, where interactions between proteins of the same haplotype inhibit pollen tube growth. S-locus Cys-rich protein/S-locus protein 11 (SCR/SP11), a small cysteine-rich protein (CRP), is secreted into the pollen sac after translation, where it adheres to the pollen surface. Upon pollination, it interacts with S locus receptor kinase (SRK) expressed in the stigma's papillary cells. In *Brassica napus*, this process is mediated by the small peptide ligand SCR/SP11 expressed in pollen and the corresponding receptor kinase (SRK) present on the stigma (Stein et al. 1991). The binding of SCR/SP11 to SRK triggers a phosphorylation cascade, leading to the recruitment of ubiquitin ligases, which degrade the protein Exo70A1. Since Exo70A1 plays a crucial role in water uptake and pollen hydration, its degradation ultimately inhibits pollen tube germination, reinforcing the self-incompatibility response (Samuel et al. 2009).

7.3.2.2 Pollen Germination and Tube Polar Growth

In flowering plants, sperm cells depend on pollen tube growth to arrive at the embryo sac, a process that requires coordinated signaling between the pollen and pistil tissues. Pollen germination and tube elongation depend on molecular signals from the carpel, which interact with receptors on the pollen tube, triggering intracellular changes, including cytoskeletal dynamics that establish polarity (Guan et al. 2013). The pollen coat proteins PCP-Bα/β/γ/δ are crucial for pollen germination, as their absence leads to defects in pollen adhesion, hydration, and tube growth *in vivo* (Wang et al. 2017). The CRP late anther tomato 52

(LAT52) in tomato, which is secreted from pollen, binds specifically to the receptor kinase *Lycopersicon esculentum* pollen receptor kinase 2 (LePRK2) on the pollen tube surface. This interaction is strongest during germination and diminishes as the tube elongates (Tang et al. 2002). Once germinated, the pollen tube requires signals from the pistil for sustained growth. The stigma-expressed CRP *Lycopersicon esculentum* stigma-specific protein 1(LeSTIG1) facilitates pollen tube elongation by binding to LePRK2 and accumulating in the tube. The LePRK2 intracellular domain networks with plant-specific Rop GTPase guanylate exchange factor (GEF) KPP, potentially regulating tube growth via downstream ROP signaling (Zhang et al. 2008). Additionally, the C-terminal cysteine-rich domain of LeSTIG1 binds to phospholipids such as PI(3)P, contributing to intracellular redox regulation (Huang et al. 2014). Homologs of LeSTIG1 in petunia and tobacco may play additional roles, including stigma secretion regulation (Verhoeven et al. 2005). Once inside the style, the pollen tube must adhere to the transmitting tract for proper guidance toward the ovule. Several CRPs, for instance, stigma-style cysteine-rich adhesin (SCA), play crucial roles in adhesion by binding to pectic polysaccharides. This interaction is necessary for pollen tube attachment to other tubes and the style matrix *in vitro* (Lord 2000). Within the tube, the SCA undergoes endocytosis at the tip and moves through an intracellular transport pathway, potentially triggered by ligand–receptor interactions, although the specific receptor and downstream signaling pathways involved remain unclear (Kim et al. 2006).

Gradients of pistil-secreted proteins, such as plantacyanin, further regulate tube elongation. Plantacyanin, a secreted protein (~10 kDa), exhibits a gradient from the stigma to the ovule and is likely controlled by the miRNA pathway (Maunoury and Vaucheret 2011). Overexpression disrupts this gradient, leading to random pollen tube growth in papilla cells and loss of polarity. The copper-binding properties and high redox potential of this plantacyanin suggest that it has a role in reactive oxygen species (ROS) metabolism. In lilies, the homologous protein chemocyanin acts as a chemotactic factor, guiding tube growth (Kim et al. 2006).

Environmental factors, particularly temperature, also influence pollen tube growth. The small peptide CLV3/ESR-related 45 (CLE45) helps to maintain seed yield at high temperatures. At 22 °C, CLE45 is expressed mainly in the stigma, but at 30 °C, its expression extends into the transmitting tract. While synthetic CLE45 promotes pollen tube elongation *in vitro* at 30 °C, its *in vivo* role is to prolong rather than accelerate tube growth. CLE45 interacts with the receptor kinases STERILITY-REGULATING KINASE MEMBER1 (SKM1) and SKM2, forming the CLE45-SKM1/SKM2 signaling pathway, which preserves pollen viability and ensures successful fertilization under heat stress (Endo et al. 2013).

7.3.2.3 Pollen Tube Guidance and Reception

Recent studies have established that pollen tubes are guided by signals from the embryo sac. Using laser ablation, Higashiyama et al. (2001) reported that in *Torenia fournieri*, a single pollen tube penetrates a synergid cell, releasing gametes upon cell rupture. The effective attraction distance (100–200 μm) suggests the involvement of short-range diffusible small peptides. LURE1 and LURE2, produced by synergid cells, act as key attractants (Okuda et al. 2009). *Arabidopsis* has more than 300 defensin-like (DEFL) genes involved in male–female interactions. AtLURE1 peptides, which are secreted from synergid cells, guide pollen tubes via interactions with the pollen tube receptors PRK6 and male discoverer

1-interacting receptor-like kinase (MDIS1-MIK) (Wang et al. 2016). In monocots, *Zea mays*, *EGG APPARATUS1* (*ZmEA1*) is essential for pollen tube attraction. Downregulation of *ZmEA1* disrupts pollen tube guidance, leading to ovule sterility (Marton et al. 2005). After secretion, *ZmEA1* is identified by the pollen tube and subsequently degraded (Marton et al. 2012). In addition to guidance, synergid cells regulate pollen tube reception. The receptor-like kinase FERONIA (FER) in synergid cells ensures pollen tube rupture for sperm release (Rotman et al. 2003). Its pollen-expressed homolog, ANXUR1/2 (ANX1/2), redundantly controls pollen tube integrity, with *anx1 anx2* mutants exhibiting premature rupture (Boisson-Dernier et al. 2009). RALF peptides interact with FER, inducing Ca^{2+} signaling crucial for pollen tube reception (Haruta et al. 2014). In maize, the defensin-like peptide ZmES1-4 triggers pollen tube rupture and sperm release, likely through KZM1, a pollen tube-expressed K^+ channel (Amien et al. 2010).

7.3.2.4 Fertilization

Following the disruption of pollen tubes, fusion occurs between one sperm cell and the egg, and the other sperm cell and the central cell, ensuring species isolation. *EGG CELL 1* (EC1), a small CRP stored in egg cell vesicles, triggers gamete activation upon sperm arrival by relocating HAPLESS 2/GENERATIVE CELL SPECIFIC 1 (HAP2/GCS1) to the surface of sperm, enabling fusion (Sprunck et al. 2012). In *Arabidopsis*, *hap2/gcs1* mutants release sperm but fail to fuse, causing sterility (von Besser et al. 2006). A similar mechanism may occur in the central cell. In a *glauce (glc)* mutant, fusion between the egg and one sperm cell occurs, although the fusion between the second sperm cell and the central cell remains unsuccessful, leading to single fertilization. The rescue of this defect by a Benzylalcohol O-acetyltransferase, Anthocyanin O-hydroxycinnamoyl transferase, Hydroxycinnamoyl anthranilate N-benzoyl transferase, and Deacetylvindoline 4-O-acetyltransferase (BAHD) transferase suggests central cell-specific signals, akin to those of EC1 (Leshem et al. 2012).

7.4 Conclusion

Recent advances in peptide signaling have improved our understanding of male–female gametophyte interactions in plant reproduction. However, many aspects, such as the downstream phosphorylation cascades and gene regulation involved, remain unclear. Mitogen-activated protein kinase (MAPK) components, such as mitogen-activated protein kinase 3 (MPK3) and MPK6, are crucial for the direction of pollen tube movement in *Arabidopsis*. Peptide–receptor interactions, such as LAT52/LeSTIG1 and LePRK1/2 in tomato, influence the germination of pollen and the subsequent growth of pollen tubes. RALF peptide exchange with the BUPS/ANX receptor complex regulates pollen tube structural conformation in *Arabidopsis*. Environmental stresses, including heat and drought, affect male reproductive development, with the CLE45–SKM1/2 pathway playing a role in heat tolerance. Further research is needed to explore how peptide signaling responds to environmental stress and regulates reproduction. Clustered regularly interspaced short palindromic repeats/CRISPR-associated protein 9 (CRISPR/Cas9) multiplexing could help uncover the functions of redundant peptides, while secretome analysis of reproductive cells may identify new regulatory peptides.

Conflict of Interest

The authors declare that they have no conflict of interest.

Abbreviations

AA	amino acid
ABA	abscisic acid
CLV	*CLAVATA*
CRP	cysteine-rich protein
FER	FERONIA
GSI	gametophytic self-incompatibility
Hyp	hydroxyproline
IDA	INFLORESCENCE DEFICIENT IN ABSCISSION
LRR-RLKs	leucine-rich repeat receptor-like kinases
LRX	leucine-rich repeat extensions
PH	peptide hormone
PSK	phytosulfokines
RALF	rapid alkalinization factor
SAM	shoot apical meristem
SCA	stigma-style cysteine-rich adhesin
SE	somatic embryo
SSI	sporophytic self-incompatibility
WUS	*WUSCHEL*

References

Amien, S., Kliwer, I., Marton, M.L. et al. (2010). Defensin-like ZmES4 mediates pollen tube burst in maize via opening of the potassium channel KZM1. *PLoS Biology* 8: e1000388. https://doi.org/10.1371/journal.pbio.1000388.

Boisson-Dernier, A., Roy, S., Kritsas, K. et al. (2009). Disruption of the pollen-expressed FERONIA homologs ANXUR1 and ANXUR2 triggers pollen tube discharge. *Development* 136: 3279–3288. https://doi.org/10.1242/dev.040071.

Butenko, M.A., Patterson, S.E., Grini, P.E. et al. (2003). Inflorescence deficient in abscission controls floral organ abscission in *Arabidopsis* and identifies a novel family of putative ligands in plants. *Plant Cell* 15: 2296–2307. https://doi.org/10.1105/tpc.014365.

Butenko, M.A., Wildhagen, M., Albert, M. et al. (2014). Tools and strategies to match peptide-ligand receptor pairs. *Plant Cell* 26: 1838–1847. https://doi.org/10.1105/tpc.113.120071.

Chen, Y.F., Matsubayashi, Y., and Sakagami, Y. (2000). Peptide growth factor phytosulfokine-α contributes to the pollen population effect. *Planta* 211: 752–755. https://doi.org/10.1007/s004250000370.

Dao, T.Q., Weksler, N., Liu, H.M.H. et al. (2022). Interactive CLV3, CLE16, and CLE17 signaling mediates stem cell homeostasis in the *Arabidopsis* shoot apical meristem. *Development* 149: dev200787. https://doi.org/10.1242/dev.200787.

Dresselhaus, T., Sprunck, S., and Wessel, G.M. (2016). Fertilization mechanisms in flowering plants. *Current Biology* 26: R125–R139. https://doi.org/10.1016/j.cub.2015.12.032.

Endo, S., Shinohara, H., Matsubayashi, Y., and Fukuda, H. (2013). A novel pollen-pistil interaction conferring high-temperature tolerance during reproduction via CLE45 signaling. *Current Biology* 23: 1670–1676. https://doi.org/10.1016/j.cub.2013.06.060.

Fletcher, J.C., Brand, U., Running, M.P. et al. (1999). Signaling of cell fate decisions by *CLAVATA3* in *Arabidopsis* shoot meristems. *Science* 283: 1911–1914. https://doi.org/10.1126/science.283.5409.1911.

Galiullina, R.A., Kasperkiewicz, P., Chichkova, N.V. et al. (2015). Substrate specificity and possible heterologous targets of phytaspase, a plant cell death protease. *The Journal of Biological Chemistry* 290: 24806–24815. https://doi.org/10.1074/jbc.M115.675819.

Gancheva, M.S., Malovichko, Y.V., Poliushkevich, L.O. et al. (2019). Plant peptide hormones. *Russian Journal of Plant Physiology* 66: 171–189. https://doi.org/10.1134/S1021443719010072.

Ge, Z., Bergonci, T., Zhao, Y. et al. (2017). *Arabidopsis* pollen tube integrity and sperm release are regulated by RALF-mediated signaling. *Science* 358: 1596–1600. https://doi.org/10.1126/science.aao3642.

Geng, Y., Jian, C., Xu, W. et al. (2020). miR164-targeted *TaPSK5* encodes a phytosulfokine precursor that regulates root growth and yield traits in common wheat (*Triticum aestivum* L.). *Plant Molecular Biology* 104: 615–628. https://doi.org/10.1007/s11103-020-01064-1.

Gregory, E.F., Dao, T.Q., Alexander, M.A. et al. (2018). The signaling peptide encoding genes *CLE16*, *CLE17*, and *CLE27* are dispensable for *Arabidopsis* shoot apical meristem activity. *PLoS One* 13: e0202595. https://doi.org/10.1371/journal.pone.0202595.

Guan, Y., Guo, J., Li, H., and Yang, Z. (2013). Signaling in pollen tube growth: crosstalk, feedback, and missing links. *Molecular Plant* 6: 1053–1064. https://doi.org/10.1093/mp/sst070.

Hanai, H., Matsuno, T., Yamamoto, M. et al. (2000). A secreted peptide growth factor, phytosulfokine, acting as a stimulatory factor of carrot somatic embryo formation. *Plant & Cell Physiology* 41: 27–32. https://doi.org/10.1093/pcp/41.1.27.

Haruta, M., Sabat, G., Stecker, K. et al. (2014). A peptide hormone and its receptor protein kinase regulate plant cell expansion. *Science* 343: 408–411. https://doi.org/10.1126/science.1244454.

Higashiyama, T., Yabe, S., Sasaki, N. et al. (2001). Pollen tube attraction by the synergid cell. *Science* 293: 1480–1483. https://doi.org/10.1126/science.1062429.

Hu, C., Zhu, Y., Cui, Y. et al. (2022). A CLE–BAM–CIK signalling module controls root protophloem differentiation in *Arabidopsis*. *The New Phytologist* 233: 282–296. https://doi.org/10.1111/nph.17791.

Huang, W.J., Liu, H.K., McCormick, S., and Tang, W.H. (2014). Tomato pistil factor STIG1 promotes in vivo pollen tube growth by binding to phosphatidylinositol 3-phosphate and the extracellular domain of the pollen receptor kinase LePRK2. *Plant Cell* 26: 2505–2523. https://doi.org/10.1105/tpc.114.123281.

Igasaki, T., Akashi, N., Ujino-Ihara, T. et al. (2003). Phytosulfokine stimulates somatic embryogenesis in *Cryptomeria japonica*. *Plant & Cell Physiology* 44: 1412–1416. https://doi.org/10.1093/pcp/pcg161.

Jun, J., Fiume, E., Roeder, A.H.K. et al. (2010). Comprehensive analysis of CLE polypeptide signaling gene expression and overexpression activity in *Arabidopsis*. *Plant Physiology* 154: 1721–1736. https://doi.org/10.1104/pp.110.163683.

Keerthana, K., Ramakrishnan, M., Ahmad, Z. et al. (2025). Root-derived small peptides: key regulators of plant development, stress resilience, and nutrient acquisition. *Plant Science* 354: 112433. https://doi.org/10.1016/j.plantsci.2025.112433.

Kim, S.T., Zhang, K., Dong, J., and Lord, E.M. (2006). Exogenous free ubiquitin enhances lily pollen tube adhesion to an *in vitro* stylar matrix and may facilitate endocytosis of SCA. *Plant Physiology* 142: 1397–1411. https://doi.org/10.1104/pp.106.086801.

Kou, X., Liu, Q., Sun, Y. et al. (2020). The peptide *pbrPSK2* from phytosulfokine family induces reactive oxygen species (ROS) production to regulate pear pollen tube growth. *Frontiers in Plant Science* 11: 601993. https://doi.org/10.3389/fpls.2020.601993.

Kwezi, L., Ruzvidzo, O., Wheeler, J.I. et al. (2011). The phytosulfokine (PSK) receptor is capable of guanylate cyclase activity and enabling cyclic GMP-dependent signaling in plants. *The Journal of Biological Chemistry* 286: 22580–22588. https://doi.org/10.1074/jbc.M110.168823.

Lease, K.A. and Walker, J.C. (2006). The *Arabidopsis* unannotated secreted peptide database, a resource for plant peptidomics. *Plant Physiology* 142: 831–838. https://doi.org/10.1104/pp.106.086041.

Leshem, Y., Johnson, C., Wuest, S.E. et al. (2012). Molecular characterization of the *glauce* mutant: a central cell-specific function is required for double fertilization in *Arabidopsis*. *Plant Cell* 24: 3264–3277. https://doi.org/10.1105/tpc.112.096420.

Lindsey, K., Casson, S., and Chilley, P. (2002). Peptides: new signalling molecules in plants. *Trends in Plant Science* 7: 78–83. https://doi.org/10.1016/S1360-1385(01)02194-X.

Lohmann, J.U., Hong, R.L., Hobe, M. et al. (2001). A molecular link between stem cell regulation and floral patterning in *Arabidopsis*. *Cell* 105: 793–803. https://doi.org/10.1016/s0092-8674(01)00384-1.

Lorbiecke, R., Steffens, M., Tomm, J.M. et al. (2005). Phytosulphokine gene regulation during maize (*Zea mays* L.) reproduction. *Journal of Experimental Botany* 56: 1805–1819. https://doi.org/10.1093/jxb/eri169.

Lord, E. (2000). Adhesion and cell movement during pollination: cherchez la femme. *Trends in Plant Science* 5: 368–373. https://doi.org/10.1016/S1360-1385(00)01744-1.

Marton, M.L., Cordts, S., Broadhvest, J., and Dresselhaus, T. (2005). Micropylar pollen tube guidance by egg apparatus 1 of maize. *Science* 307: 573–576. https://doi.org/10.1126/science.1104954.

Marton, M.L., Fastner, A., Uebler, S., and Dresselhaus, T. (2012). Overcoming hybridization barriers by the secretion of the maize pollen tube attractant ZmEA1 from *Arabidopsis* ovules. *Current Biology* 22: 1194–1198. https://doi.org/10.1016/j.cub.2012.04.061.

Matsubayashi, Y. (2014). Posttranslationally modified small-peptide signals in plants. *Annual Review of Plant Biology* 65: 385–413. https://doi.org/10.1146/annurev-arplant-050312-120122.

Matsubayashi, Y. and Sakagami, Y. (1996). Phytosulfokine, sulfated peptides that induce the proliferation of single mesophyll cells of *Asparagus officinalis* L. *Proceedings of the National Academy of Sciences of the United States of America* 93: 7623–7627. https://doi.org/10.1073/pnas.93.15.7623.

Matsubayashi, Y., Ogawa, M., Kihara, H. et al. (2006). Disruption and overexpression of *Arabidopsis* phytosulfokine receptor gene affects cellular longevity and potential for growth. *Plant Physiology* 142: 45–53. https://doi.org/10.1104/pp.106.081109.

Matsuzaki, Y., Ogawa-Ohnishi, M., Mori, A., and Matsubayashi, Y. (2010). Secreted peptide signals required for maintenance of root stem cell niche in *Arabidopsis*. *Science* 329: 1065–1067. https://doi.org/10.1126/science.1191132.

Maunoury, N. and Vaucheret, H. (2011). AGO1 and AGO2 act redundantly in miR408-mediated Plantacyanin regulation. *PLoS One* 6: e28729. https://doi.org/10.1371/journal.pone.0028729.

Mayer, K.F., Schoof, H., Haecker, A. et al. (1998). Role of *WUSCHEL* in regulating stem cell fate in the *Arabidopsis* shoot meristem. *Cell* 95: 805–815. https://doi.org/10.1016/s0092-8674(00)81703-1.

Mecchia, M.A., Santos-Fernandez, G., Duss, N.N. et al. (2017). RALF4/19 peptides interact with LRX proteins to control pollen tube growth in *Arabidopsis*. *Science* 358: 1600–1603. https://doi.org/10.1126/science.aao5467.

Miwa, H., Kinoshita, A., Fukuda, H., and Sawa, S. (2009). Plant meristems: *CLAVATA3/ESR*-related signaling in the shoot apical meristem and the root apical meristem. *Journal of Plant Research* 122: 31–39. https://doi.org/10.1007/s10265-008-0207-3.

Murphy, E., Smith, S., and De Smet, I. (2012). Small signaling peptides in *Arabidopsis* development: how cells communicate over a short distance. *Plant Cell* 24: 3198–3217. https://doi.org/10.1105/tpc.112.099010.

Muschietti, J.P. and Wengier, D.L. (2018). How many receptor-like kinases are required to operate a pollen tube. *Current Opinion in Plant Biology* 41: 73–82. https://doi.org/10.1016/j.pbi.2017.09.008.

Nimchuk, Z.L., Zhou, Y., Tarr, P.T. et al. (2015). Plant stem cell maintenance by transcriptional cross-regulation of related receptor kinases. *Development* 142: 1043–1049. https://doi.org/10.1242/dev.119677.

Ochatt, S., Conreux, C., Mcolo, R.M. et al. (2018). Phytosulfokine-alpha, an enhancer of *in vitro* regeneration competence in recalcitrant legumes. *Plant Cell, Tissue and Organ Culture* 135: 189–201. https://doi.org/10.1007/s11240-018-1455-0.

Okuda, S., Tsutsui, H., Shiina, K. et al. (2009). Defensin-like polypeptide LUREs are pollen tube attractants secreted from synergid cells. *Nature* 458: 357–361. https://doi.org/10.1038/nature07882.

Pasonen, H.L. and Kapyla, M. (1998). Pollen-pollen interactions in *Betula pendula in vitro*. *The New Phytologist* 138: 481–487. https://doi.org/10.1046/j.1469-8137.1998.00135.x.

Pazourek, J.T.A., Steeves, I.M., and Sussex, I.M. (1992). Patterns in plant development. *Folia Geobotanica et Phytotaxonomica* 27: 136. https://doi.org/10.1007/BF02856244.

Rotman, N., Rozier, F., Boavida, L. et al. (2003). Female control of male gamete delivery during fertilization in *Arabidopsis thaliana*. *Current Biology* 13: 432–436. https://doi.org/10.1016/S0960-9822(03)00093-9.

Samuel, M.A., Chong, Y.T., Haasen, K.E. et al. (2009). Cellular pathways regulating responses to compatible and self-incompatible pollen in *Brassica* and *Arabidopsis* stigmas intersect at Exo70A1, a putative component of the exocyst complex. *Plant Cell* 21: 2655–2671. https://doi.org/10.1105/tpc.109.069740.

Schlegel, J., Denay, G., Wink, R. et al. (2021). Control of *Arabidopsis* shoot stem cell homeostasis by two antagonistic CLE peptide signalling pathways. *eLife* 10: e70934. https://doi.org/10.7554/eLife.70934.

Selby, R. and Jones, D.S. (2023). Complex peptide hormone signaling in plant stem cells. *Current Opinion in Plant Biology* 75: 102442. https://doi.org/10.1016/j.pbi.2023.102442.

Somssich, M., Je, B.I., Simon, R., and Jackson, D. (2016). CLAVATA-WUSCHEL signaling in the shoot meristem. *Development* 143: 3238–3248. https://doi.org/10.1242/dev.133645.

Song, X.-F., Guo, P., Ren, S.C. et al. (2013). Antagonistic peptide technology for functional dissection of *CLV3/ESR* genes in *Arabidopsis*. *Plant Physiology* 161: 1076–1085. https://doi.org/10.1104/pp.112.211029.

Song, X.F., Hou, X.L., and Liu, C.M. (2022). CLE peptides: critical regulators for stem cell maintenance in plants. *Planta* 255: 5. https://doi.org/10.1007/s00425-021-03791-1.

Sprunck, S., Rademacher, S., Vogler, F. et al. (2012). Egg cell-secreted EC1 triggers sperm cell activation during double fertilization. *Science* 338: 1093–1097. https://doi.org/10.1126/science.1223944.

Stein, J.C., Howlett, B., Boyes, D.C. et al. (1991). Molecular cloning of a putative receptor protein kinase gene encoded at the self-incompatibility locus of *Brassica oleracea*. *Proceedings of the National Academy of Sciences of the United States of America* 88: 8816–8820. https://doi.org/10.1073/pnas.88.19.8816.

Stenvik, G.E., Tandstad, N.M., Guo, Y. et al. (2008). The EPIP peptide of INFLORESCENCE DEFICIENT IN ABSCISSION is sufficient to induce abscission in *Arabidopsis* through the receptor-like kinases HAESA and HAESA-LIKE2. *Plant Cell* 20: 1805–1817. https://doi.org/10.1105/tpc.108.059139.

Stuhrwohldt, N., Dahlke, R.I., Kutschmar, A. et al. (2015). Phytosulfokine peptide signaling controls pollen tube growth and funicular pollen tube guidance in *Arabidopsis thaliana*. *Physiologia Plantarum* 153: 643–653. https://doi.org/10.1111/ppl.12270.

Takahashi, F., Suzuki, T., Osakabe, Y. et al. (2018). A small peptide modulates stomatal control via abscisic acid in long-distance signaling. *Nature* 556: 235–238. https://doi.org/10.1038/s41586-018-0009-2.

Takayama, S. and Isogai, A. (2005). Self-incompatibility in plants. *Annual Review of Plant Biology* 56: 467–489. https://doi.org/10.1146/annurev.arplant.56.032604.144249.

Tameshige, T., Ikematsu, S., Torii, K.U., and Uchida, N. (2017). Stem development through vascular tissues: EPFL–ERECTA family signaling that bounces in and out of phloem. *Journal of Experimental Botany* 68: 45–53. https://doi.org/10.1093/jxb/erw447.

Tang, W., Ezcurra, I., Muschietti, J., and McCormick, S. (2002). A cysteine-rich extracellular protein, LAT52, interacts with the extracellular domain of the pollen receptor kinase LePRK2. *Plant Cell* 14: 2277–2287. https://doi.org/10.1105/tpc.003103.

Van Norman, J.M., Breakfield, N.W., and Benfey, P.N. (2011). Intercellular communication during plant development. *Plant Cell* 23: 855–864. https://doi.org/10.1105/tpc.111.082982.

Vanstraelen, M. and Benková, E. (2012). Hormonal interactions in the regulation of plant development. *Annual Review of Cell and Developmental Biology* 28: 463–487. https://doi.org/10.1146/annurev-cellbio-101011-155741.

Verhoeven, T., Feron, R., Wolters-Arts, M. et al. (2005). STIG1 controls exudate secretion in the pistil of petunia and tobacco. *Plant Physiology* 138: 153–160. https://doi.org/10.1104/pp.104.054809.

von Besser, K., Frank, A.C., Johnson, M.A., and Preuss, D. (2006). *Arabidopsis HAP2* (GCS1) is a sperm-specific gene required for pollen tube guidance and fertilization. *Development* 133: 4761–4769. https://doi.org/10.1242/dev.02683.

Wang, T., Liang, L., Xue, Y. et al. (2016). A receptor heteromer mediates the male perception of female attractants in plants. *Nature* 531: 241–244. https://doi.org/10.1038/nature16975.

Wang, L., Clarke, L.A., Eason, R.J. et al. (2017). PCP-B class pollen coat proteins are key regulators of the hydration checkpoint in *Arabidopsis thaliana* pollen-stigma interactions. *The New Phytologist* 213: 764–777. https://doi.org/10.1111/nph.14162.

Weigel, D. and Nilsson, O. (1995). A developmental switch sufficient for flower initiation in diverse plants. *Nature* 377: 495–500. https://doi.org/10.1038/377495a0.

Whitford, R., Fernandez, A., Tejos, R. et al. (2012). GOLVEN secretory peptides regulate auxin carrier turnover during plant gravitropic responses. *Developmental Cell* 22: 678–685. https://doi.org/10.1016/j.devcel.2012.02.002.

Yamaguchi, Y.L., Ishida, T., and Sawa, S. (2016). CLE peptides and their signaling pathways in plant development. *Journal of Experimental Botany* 67: 4813–4826. https://doi.org/10.1093/jxb/erw208.

Yu, L., Liu, Y., Li, Q. et al. (2016). Overexpression of phytosulfokinea induces male sterility and cell growth by regulating cell wall development in *Arabidopsis*. *Plant Cell Reports* 35: 2503–2512. https://doi.org/10.1007/s00299-016-2050-7.

Yu, L., Liu, Y., Zeng, S. et al. (2019). Expression of a novel PSK-encoding gene from soybean improves seed growth and yield in transgenic plants. *Planta* 249: 1239–1250. https://doi.org/10.1007/s00425-019-03101-w.

Zhang, D., Wengier, D., Shuai, B. et al. (2008). The pollen receptor kinase LePRK2 mediates growth-promoting signals and positively regulates pollen germination and tube growth. *Plant Physiology* 148: 1368–1379. https://doi.org/10.1104/pp.108.124420.

Zhang, Z., Han, H., Zhao, J. et al. (2025). Peptide hormones in plants. *Molecular Horticulture* 5: 7. https://doi.org/10.1186/s43897-024-00134-y.

8

Regulation of Plant Cell Death by Peptide Hormones

Pankaj Kumar Singh

Department of Biotechnology, University Centre for Research & Development, Chandigarh University, Mohali, Punjab, India

8.1 Introduction

Programmed cell death (PCD) is an important physiological process in the development and stress responses of plants. It guarantees the elimination of unwanted or damaged cells and is essential for tissue differentiation, defense responses, and responses to environmental cues. Peptide hormones are known to play important roles in regulating plant PCD. These peptides regulate the induction of cell death at receptor recognition and/or downstream signaling modules by interacting with receptor-like kinases (RLKs) and regulating the response signaling pathways. This chapter focuses on the functions and actions of peptide hormones controlling PCD in plants, contributing to our understanding of these hormones in plant agriculture and biotechnology. Peptide hormones in plants have come under intense scrutiny, not only with respect to their involvement in developmental processes, such as maintenance of the root meristem, preservation of the stem cell niche, and floral organ development, but also for their enigmatic involvement in activation and repression of cell death pathways in relation to both endogenous and exogenous stimuli (Matsubayashi 2014). This dual function demonstrates their plastic and essential role in plant biology. The context dependency of PCD control by peptide hormones is diverse, ranging from inducing local cell death during xylem specification to actively suppressing necrosis during pathogen attack. Plant PCD can be classified into two main types, developmental PCD (dPCD) and environmental PCD (ePCD). During plant morphogenesis, dPCD is essential for the development of tracheary elements, leaf senescence, and breakdown of the aleurone layer during seed germination. On the contrary, ePCD is frequently induced by biotic and abiotic stresses such as pathogen infection (mechanical stress), drought, salinity, and oxidative stress (van Doorn et al. 2011). dPCD is a genetically controlled and organized process, but ePCD has a more heterogeneous and dynamic nature that is often regulated by signaling molecules, including reactive oxygen species (ROS), phytohormones, and peptides. In contrast to animals, plants do not have homologs to caspases and Bcl-2 family proteins other than cell death sustainable 1 (cds1). Instead, they encode functional analogs, such as

Peptide Hormones in Plant Development and Stress Responses, First Edition. Edited by Vijay Pratap Singh and Samiksha Singh.

metacaspases and vacuolar processing enzymes (VPEs), and depend on extracellular signals to control PCD. Peptide hormones are of increasing interest as powerful regulators of cell death, acting locally and at a distance to change the cellular setting and determine cell fate.

Peptide hormones are expressed as precursor proteins and must be posttranslationally modified, i.e. proteolytically processed and tyrosine sulfated, in order to be biologically functional. They are produced into the apoplast and recognized by the membrane-localized leucine-rich repeat receptor-like kinases (LRR-RLKs) or receptor-like proteins (RLPs). This ligand-receptor binding activates a series of intracellular signaling, which causes transcriptional reprogramming (Tavormina et al. 2015). Several families of peptide hormones have been identified, including CLAVATA3/EMBRYO SURROUNDING REGION (CLE), PHYTOSULFOKINE (PSK), RAPID ALKALINIZATION FACTOR (RALF), INFLORESCENCE DEFICIENT IN ABSCISSION (IDA), and PLANT PEPTIDE CONTAINING SULFATED TYROSINE (PSY). These families are classified based on their sequences and biological functions. Members of each family have been linked to regulating cell death, either directly or indirectly. For example, CLE peptides like CLE41 and CLE44 help keep vascular stem cells from changing into other types of cells and dying (Ito et al. 2006). PSK has been shown to protect cells from oxidative stress by encouraging them to divide and stopping them from dying (Kutschmar et al. 2009). RALF peptides, on the other hand, can quickly lower the pH of the cytoplasm and have been linked to the death of cells when they are attacked by pathogens (Pearce et al. 2001). These different roles show how complicated and context-dependent peptide hormones are when it comes to controlling plant PCD.

There is a lot of evidence that peptide hormones play a role in dPCD, especially in processes like xylem differentiation and the growth of lateral roots, where selective cell death helps shape and function of organs. For instance, the TDIF (TRACHEARY ELEMENT DIFFERENTIATION INHIBITORY FACTOR) peptide, which is part of the CLE family, binds to the TDR/PXY receptor to stop procambial cells from differentiating too soon and going through PCD. This keeps the vascular meristem active (Hirakawa et al. 2008). On the other hand, peptide hormones can either stop or help stress-induced PCD. Some peptides help the hypersensitive response (HR), which is a localized form of PCD that stops pathogens from spreading when they attack. Researchers have found that IDA-like peptides make HR-like symptoms worse in infected tissues, which suggests that they may play a role in controlling immune-related cell death (Butenko et al. 2003). Sulfated peptides like PSY1 have also been linked to the suppression of ROS-induced PCD, which adds to the already complicated hormonal control that happens when the body adapts to stress (Amano et al. 2007).

The integration of peptide hormone signaling with classic plant defense hormones such as salicylic acid (SA), jasmonic acid (JA), and ethylene further broadens their regulatory impact. This cross talk fine-tunes the PCD outcome, balancing development and defense in a resource-efficient manner. In summary, peptide hormones serve as essential regulators of plant PCD, linking developmental signals with environmental responses. Their flexibility to either promote or hinder cell death depending on context makes them valuable instruments for protecting plant integrity. With the rising availability of high-throughput omics techniques and functional genomics approaches, our understanding of peptide hormone-mediated PCD regulation is expanding, giving interesting targets for crop improvement and stress resilience measures.

8.2 Overview of Plant Programmed Cell Death

In plants, PCD is a genetically controlled, ordered process in which cells die on purpose as part of development or in response to environmental signals. PCD is a controlled and necessary type of cell death that occurs when a plant is badly hurt. Necrosis, on the other hand, is an uncontrolled and frequently harmful type of cell death. It is very important for processes including embryogenesis, organ abscission, xylem differentiation, and fighting off infections (van Doorn et al. 2011). There are different proteins and processes that control PCD in plants at the molecular level. Plants do not have the same apoptotic regulators that animals do, like caspases. Instead, they have functional analogs like metacaspases, VPEs, and papain-like cysteine proteases (Zhou et al. 2014). The vacuole is very important for plant PCD because it often releases hydrolytic enzymes that break down parts of cells during cell death. ROS and calcium signaling are also important secondary messengers that bring together developmental cues and environmental signals to start PCD pathways (Petrov et al. 2015). Hormonal signals have been getting more attention in the last several years for their role in changing plant PCD. Depending on the situation, classic plant hormones, including SA, ethylene (ET), JA, and abscisic acid (ABA), can either help or stop PCD. Emerging data also points to a key role for peptide hormones as regulators of PCD, especially in developmental transitions and defense responses. Understanding the regulation of PCD in plants is vital for crop improvement, particularly in boosting stress tolerance and disease resistance. As research advances, merging genetic, biochemical, and cell biology techniques continues to show the complexity of plant cell death systems and their critical role in plant adaptation and survival.

8.2.1 Types of Cell Death in Plants

There are two main types of plant PCD: dPCD and ePCD. dPCD happens during normal growth, like when the tapetum breaks down in anthers, tracheary elements form, and leaves and petals die (Huysmans et al. 2017). On the other hand, ePCD is often caused by biotic or abiotic stresses, like pathogens attacking the plant, oxidative damage, or very high or low temperatures. The HR is a type of localized cell death at the site of infection that stops the spread of pathogens. It is one of the most important signs of ePCD (Coll et al. 2011).

The dPCD is a necessary part of plant growth and morphogenesis that is controlled by genes. As part of normal development, it shapes tissues, gets rid of old structures, and helps recycle nutrients. The breakdown of the aleurone layer during seed germination, the degeneration of the tapetum during pollen development, and the autolysis of cells in the root cap and tracheary elements to help roots grow and move water are all well-known examples (van Doorn et al. 2011; Huysmans et al. 2017). dPCD is very tightly controlled in space and time, often by developmental cues and hormonal signals. This is different from stress-induced cell death. For example, the differentiation of xylem vessels involves a carefully planned sequence in which cell death follows secondary wall thickening to make hollow conduits (Fukuda 2000). Proteases like VPEs and transcription factors like the NAM, ATAF, and CUC (NAC) and Myeloblastosis (MYB) families are important for dPCD. Importantly, dPCD is not just a passive result of differentiation; it is an active process carried out by the dying cell itself, making sure that the development goes exactly as planned without triggering defense responses (Minina et al. 2017). Recent research also suggests that peptide hormones, such as CLE and IDA family peptides, may act as signals that change or start dPCD during organ abscission and tissue remodeling. This adds another layer of control to this important process.

Plants use ePCD as a defense and adaptation mechanism in response to external stressors such as UV radiation, pathogen attack, drought, salinity, and extreme temperatures. ePCD is frequently quick and dynamic, with the goal of protecting the organism in harsh environments, in contrast to dPCD, which is genetically predetermined and a normal aspect of growth. The HR, a localized type of cell death that takes place at infection sites to stop the spread of biotrophic pathogens, is one of the most researched examples (Coll et al. 2011). ROS, calcium signaling, and defense-related hormones such as SA mediate this response, which frequently activates VPEs and other proteases that break down the cell from the inside out (Hatsugai et al. 2004). Furthermore, ePCD can be brought on by abiotic stressors like drought and high salinity by activating stress-responsive transcription factors, accumulating ROS, and causing mitochondrial dysfunction (Petrov et al. 2015). Excessive or improperly managed ePCD can be harmful, resulting in tissue damage and yield loss, even though it is often protective. Remarkably, new research indicates that peptide hormones, which have historically been associated with developmental functions, might also play a part in regulating ePCD by combining environmental cues with intracellular death pathways (Tavormina et al. 2015). Gaining an understanding of ePCD mechanisms is essential for enhancing plant resistance to biotic stresses and climate change.

8.2.2 Regulatory Mechanisms

Programmed cell death (PCD) in plants is controlled by a sophisticated web of molecular signals that combine environmental stimuli and developmental cues to ensure timely and localized cell death (Figure 8.1). Unlike animals, plants do not have canonical apoptotic regulators like caspases; however, they do have functionally similar proteases like papain-like cysteine proteases, VPEs, and metacaspases that coordinate the breakdown of cellular contents during PCD (Minina et al. 2017). As signaling mediators in both developmental and stress-induced PCD pathways, ROS and calcium ions (Ca^{2+}) are essential for the regulation of PCD (Petrov et al. 2015). Another important factor is transcriptional regulation, whereby particular transcription factors belonging to the WRKYGQK motif (WRKY), Basic leucine zipper (bZIP), and NAC families alter the expression of genes linked to PCD (Coll et al. 2011). Furthermore, phytohormones like ET, SA, JA, and ABA are important modulators that, depending on the cellular context and type of stress, either promote or suppress PCD (Danon et al. 2000). According to a research, peptide hormones like IDA and CLE (CLAVATA3/ESR-RELATED) peptides play a crucial role in regulating PCD processes during defense responses and tissue remodeling (Tavormina et al. 2015). Cross talk among these pathways ensures a balanced cellular response, preventing unnecessary cell death and allowing the plant to adapt and endure in a changing environment. In addition to advancing our knowledge of plant biology, analyzing the regulatory mechanisms of PCD provides fresh approaches to boosting crop productivity and resilience.

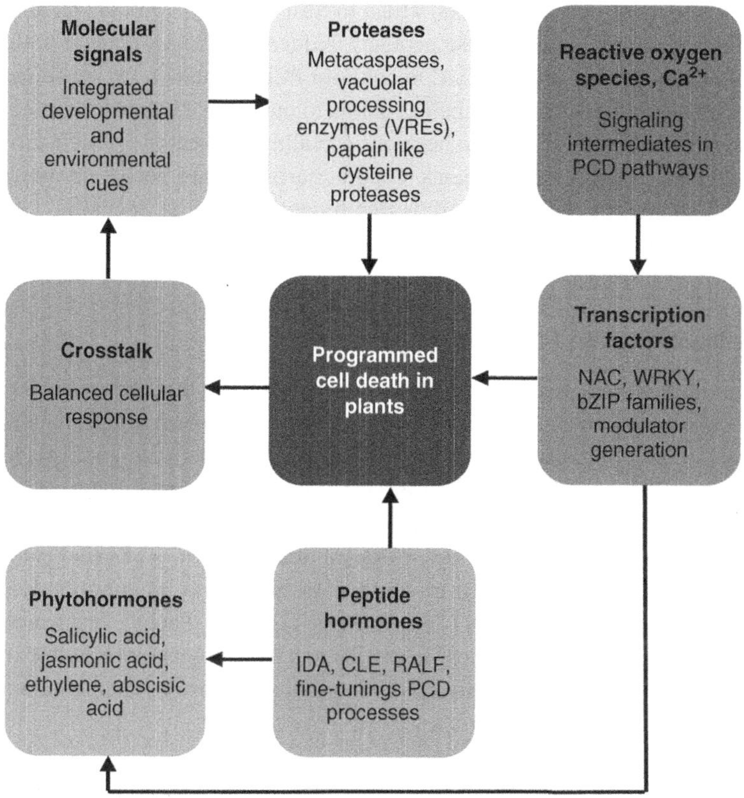

Figure 8.1 Regulatory mechanisms of programmed cell death in plants. This diagram displays the multilayered regulation of plant cell death, showing how hormonal signaling, both phytohormones and peptide hormones, along with proteolytic enzymes, redox signals, and transcriptional control, combine to coordinate precise cell fate decisions. The cross talk among these pathways ensures that programmed cell death (PCD) contributes effectively to plant development, immunity, and adaptation without causing unnecessary damage.

8.3 Peptide Hormones in Plant Cell Death

Peptide hormones, which function as signaling molecules that coordinate cellular communication during development and stress responses, are becoming recognized as important regulators of PCD in plants. Peptide hormones, in contrast to traditional phytohormones, are tiny, secreted proteins that frequently work by activating particular RLKs on the plasma membrane to initiate intracellular signaling cascades. The IDA peptide and its family members are a well-characterized example; they promote localized PCD in boundary cells by interacting with the HAE/HSL2 receptors, thereby regulating cell separation and organ abscission (Butenko et al. 2003). The regulation of tissue patterning and meristem maintenance, which entails selective PCD for appropriate developmental outcomes, has also been linked to CLE peptides (Yamaguchi et al. 2016). It has been demonstrated that specific RALF peptides in defense responses modify immunity and cell death in response to pathogen attack by modifying pH and ion fluxes, which in turn affect ROS production and cellular stability (Stegmann et al. 2017). These peptides frequently work in tandem with other signaling molecules, like ethylene and SA, to enhance PCD regulation by providing specificity and spatial resolution. Because it enables plants to maintain a balance between growth and defense, precise control of peptide hormone signaling is essential for their survival in both normal and stressful environments. Peptide hormones are becoming more widely acknowledged as growth regulators and as fine-tuners of cell fate, including the onset and containment of PCD, as research progresses.

8.3.1 Characteristics of Peptide Hormones

In plants, peptide hormones are a varied class of tiny, secreted proteins that are crucial for controlling stress reactions, growth, and development. These peptides, which usually have a length of 5–50 amino acids, are created as larger precursor proteins and then go through posttranslational changes like glycosylation, hydroxylation, sulfation, or proteolytic cleavage to become physiologically active (Matsubayashi 2014). Peptide hormones work through ligand-receptor interactions, frequently attaching to membrane-localized RLKs that start downstream signaling cascades (Figure 8.2), in contrast to traditional phytohormones, which are low-molecular-weight substances (Tavormina et al. 2015). Peptide hormones stand out for their high specificity and local action, which enable precise temporal and spatial regulation of developmental processes like cell differentiation, meristem maintenance,

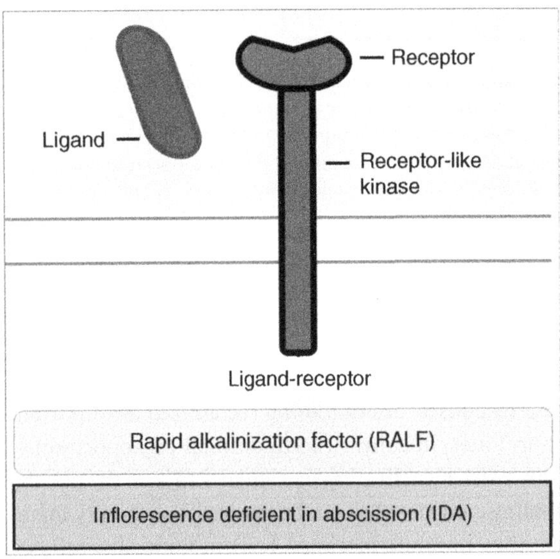

Figure 8.2 A simplified visual representation of ligand–receptor signaling involving peptide hormones in plants. It specifically shows two important peptide ligands, rapid alkalinization factor (RALF) and inflorescence deficient in abscission (IDA). These ligands bind to a specific receptor, a membrane-bound receptor-like kinase (RLK), on the cell surface of plants. Once the ligand binds, the receptor initiates intracellular signaling cascades. This interaction is crucial for activating downstream responses in plant cells, such as those related to development, defense, or PCD.

and PCD. Plants contain a number of families of peptide hormones, such as RALF peptides, IDA peptides, and CLE-related peptides. Each of these families has a unique receptor partner and biological function (Li et al. 2024). Furthermore, peptide hormone expression is strictly controlled, frequently in reaction to environmental cues or developmental signals, allowing plants to precisely adjust their defense and growth tactics. Peptide hormones are important molecular messengers in plant signaling, and their structural diversity, receptor specificity, and integration into larger hormonal networks demonstrate their complexity and adaptability.

8.3.2 Classes of Peptide Hormones Involved in PCD

In both plants and animals, PCD is a strictly controlled process that is essential for growth and stress reactions. Particularly in plants, where tiny signaling peptides coordinate cell death during development and defense, peptide hormones are important in regulating PCD. RALFs, PSKs, CLE peptides, and IDA-like peptides are the main classes of peptide hormones implicated in PCD. For example, it has been demonstrated that RALFs affect the integrity of the cell wall and pH variations linked to defense-induced cell death (Blackburn et al. 2020). PSKs are known to play a role in immunity and growth, and depending on the signaling context, they may either promote cell death or survival (Sauter 2015). Stem cell regulation involves CLE peptides, especially CLE19 and CLE41, which may have an indirect effect on PCD during organogenesis (Czyzewicz et al. 2013). Organ abscission and related cell separation events involving localized PCD are mediated by IDA and IDA-like peptides (Estornell et al. 2013). These peptide hormones play a critical role in the temporal and spatial regulation of PCD by interacting with particular RLKs to start signaling cascades that control cell fate.

8.3.2.1 Rapid Alkalinization Factor (RALF) Peptides

The class of small cysteine-rich plant signaling molecules known as RALF peptides gets its name from the fact that they cause an abrupt rise in extracellular pH. To become physiologically active, these peptides, which are normally 49–53 amino acids long, go through posttranslational changes, such as proteolytic processing. Membrane-localized RLKs that are involved in RALF perception include FERONIA (FER), a member of the CrRLK1L family that plays a crucial role in regulating RALF-mediated signaling pathways. When RALFs bind to FER and its co-receptors, they start signaling cascades that control the production of ROS, cytoplasmic calcium levels, and cell wall integrity (Stegmann et al. 2017; Haruta et al. 2014). Physiological reactions like quick apoplast alkalinization, ion channel modulation, and changes in growth and defense-related gene expression are ultimately caused by this signaling. RALF peptides have a variety of functional roles in immunity, growth, and development in plants. They frequently have context-dependent effects and are involved in pollen tube integrity, root elongation inhibition, and cell expansion. RALF peptides alter the cell wall's mechanical characteristics during developmental processes, which affect organ formation and patterning (Blackburn et al. 2020). Functional divergence within the RALF family is evident in the way that some members suppress pattern-triggered immunity (PTI), which negatively regulates defense responses in plants, while others may enhance defense under particular circumstances (Stegmann et al. 2017). Because too much or too little RALF signaling activity can upset cellular homeostasis and result in abnormal growth or weakened immunity, it is crucial to maintain a balance. RALF peptides are an essential connection between extracellular stimuli and intracellular reactions in plants because of their pivotal function in regulating signaling networks at the cell surface.

8.3.2.2 Systemin

One of the first plant peptide hormones to be identified, systemin is found in tomatoes (*Solanum lycopersicum*) and is essential for the systemic wound response. It is made from an 18-amino acid peptide that is broken down by proteolysis from prosystemin, a 200-amino acid precursor protein. Systemin is released in response to wounding or herbivore attack, and it attaches itself to the systemin receptor 1 (SYR1), an LRR-RLK on the plasma membrane (Wang et al. 2018). ROS, calcium influx, and mitogen-activated protein kinases (MAPKs) are examples of secondary messengers involved in the signaling cascade that is triggered by receptor activation. This cascade ultimately results in the activation of JA biosynthesis and downstream gene expression (Pearce et al. 1991). The plant can quickly react to localized damage thanks to this mechanism, which also allows it to start defense signaling in distant, undamaged tissues. In terms of function, systemin is a key modulator of defense responses triggered by wounds in solanaceous plants. It promotes the transcriptional activation of genes linked to defense, especially those that encode proteinase inhibitors (PIs), which prevent herbivory by interfering with insect digestive enzymes (Ryan 2000). In addition to providing direct defense, systemin serves as a systemic warning signal by making the plant more vulnerable to subsequent attacks, thereby preparing it for future stressors. Systemin-like peptides and systemin-mediated pathways were first discovered in tomatoes, but they have since been found in other Solanaceae family members, suggesting a more widespread evolutionary function. Recent research highlights systemin's integrative function in coordinating intricate defense networks by pointing to possible cross talk between it and other hormonal pathways like ethylene and SA (Beloshistov et al. 2018).

8.3.2.3 Phytosulfokine (PSK) Peptides

Phytosulfokine (PSK), a sulfated pentapeptide (Tyr(SO_3H)-Ile-Tyr(SO_3H)-Thr-Gln), acts as a plant growth factor and stress modulator by binding to specific membrane-localized leucine-rich repeat receptor kinases, mainly PSKR1 and PSKR2. When PSK binds to PSKR1, it autophosphorylates and forms a signaling complex with co-receptors like BAK1 and CNGC17, which starts downstream signaling cascades (Ladwig et al. 2015; Kaufmann et al. 2017). These pathways change the amount of calcium inside cells, turn on H^+-ATPases in the plasma membrane, and control transcription factors that control how cells grow and respond to stress. This interaction between a ligand and a receptor is very important for keeping cells healthy and stable, especially when they are under stress from things like heat or cold. One of the most important protective functions of PSK signaling is to stop the buildup of ROS, which is a common response of cells to stress that can cause oxidative damage and PCD. Research has shown that PSK signaling increases the expression of antioxidant enzymes like superoxide dismutase (SOD) and catalase (CAT), which lowers the amount of ROS in plant cells (Igarashi et al. 2012). Furthermore, PSK enhances membrane stability and cellular integrity by bolstering the antioxidative defense system, thereby promoting cell survival during pathogen assaults or environmental stressors like drought or salinity (Sauter 2015). So, the

PSK–PSKR module is becoming increasingly well-known as a key regulator of both developmental and defense-related pathways in higher plants. PSK peptides are very important for controlling PCD, which is a tightly controlled process that happens during development and in response to different biotic and abiotic stresses. PSK signaling helps cells stay alive when the environment is bad, like when there is oxidative stress, pathogens invade, or the salinity is too high. It does this by stopping or delaying PCD mechanisms. This effect is caused by PSK binding to its receptors (PSKR1/2), which starts signaling pathways inside cells that turn on antioxidant systems, calcium influx, and the expression of genes that help cells stay alive (Ladwig et al. 2015; Sauter 2015). According to studies, PSK application dramatically lowers stress-induced cell death, especially by reducing the buildup of ROS, a major cause of PCD. Exogenous PSK, for instance, suppresses lesion formation and cell death in Arabidopsis after exposure to pathogens and oxidative stress, indicating that it serves as a pro-survival signal (Igarashi et al. 2012). Furthermore, PSK signaling contributes to plant resilience by interacting with calcium-dependent and hormonal pathways that control the ratio of survival to death. With potential uses in enhancing crop stress tolerance, PSK's anti-PCD function emphasizes its importance in plant adaptive responses.

8.3.2.4 C-TERMINALLY ENCODED PEPTIDE (CEP) Family

A class of tiny signaling peptides called C-TERMINALLY ENCODED PEPTIDES (CEPs) is derived from roots and is essential for systemic nutrient signaling and nitrogen sensing. Certain CEP genes are transcriptionally upregulated in roots when nitrogen is scarce, and the peptides they encode are transported to the shoot through the xylem and attach to CEP receptors (CEPR1 and CEPR2). This sets off a feedback loop that effectively controls root architecture and nitrogen uptake efficiency by increasing the expression of nitrogen transporter genes, particularly in root zones with higher nitrate availability (Tabata et al. 2014; Ohkubo et al. 2017). CEPs are linked to the regulation of PCD under stress in addition to nutrient signaling, most likely via interactions with redox and hormonal signaling pathways. Roberts et al. (2013) reported that CEP signaling reduces stress-induced PCD, potentially by maintaining nitrogen homeostasis and preventing excessive ROS accumulation, which is a major cause of cell death in nutrient-stressed environments. By incorporating environmental cues into developmental and defense programs, CEPs optimize nutrient acquisition under limiting circumstances while also promoting cell survival and adaptive plasticity. CEPs are essential systemic signaling molecules that control plant nutrient homeostasis, especially when nitrogen levels are low. CEPs are created in nitrogen-starved roots and move through the xylem to the shoot, where receptor kinases like CEPR1 and CEPR2 detect them. In order to fine-tune the architecture of the root system and nutrient uptake, this perception sets off a long-distance feedback mechanism that modifies the expression of the nitrate transporter gene in distal root tissues with higher nitrate availability (Tabata et al. 2014; Ohkubo et al. 2017). CEPs indirectly support preserving metabolic balance under stress by maximizing nitrogen distribution within the plant. Remarkably, CEPs' control over nutrient signaling indirectly affects PCD. ROS buildup from nutrient stress, especially nitrogen deficiency, can cause PCD. CEP signaling reduces oxidative stress and cellular nutrient deprivation by coordinating nitrogen acquisition and allocation, which lowers the risk of PCD initiation (Roberts et al. 2013). CEPs do not directly stop PCD, but they do help keep the internal nutrient environment stable and make plants more resistant to abiotic stress. This is especially helpful during important stages of plant growth and development.

8.3.2.5 Small Peptides in Pathogen Response (SPR)

Alongside traditional defense pathways, small peptides involved in pathogen response (SPR) are becoming powerful modulators of plant innate immunity that help regulate immune signaling during pathogen invasion. When a pathogen is recognized, these peptides, which are usually shorter than 100 amino acids, are either constitutively expressed or quickly induced. When pathogen-associated molecular patterns (PAMPs) are recognized, precursor proteins release plant elicitor peptides (PEPs), a well-characterized subgroup of SPRs, which attach to PEPR receptors and trigger mitogen-activated protein kinase (MAPK) cascades, transcriptional reprogramming, and ROS generation (Huffaker et al. 2006; Yamaguchi et al. 2010). Immune responses, such as the strengthening of cell walls and the expression of genes linked to defense, are enhanced by this process. Additionally, SPR peptides have the ability to alter immune priming, which improves the plant's capacity to respond to subsequent attacks more quickly and powerfully. Certain SPRs, for instance, function as damage-associated molecular patterns (DAMPs), bridging the gap between biotic stress responses and abiotic damage. A strong immune network is produced by their activity, which also interacts with hormonal pathways like SA and JA signaling and amplifies defense signaling (Albert 2013). SPRs are therefore essential molecular messengers that guarantee the efficient mobilization of both systemic and local immune responses in the event of pathogen stress. In addition to improving immune signaling, SPR directly contributes to the induction of HR-associated PCD, a localized defense mechanism that limits the spread of pathogens. Systemin-like peptides and Peps are two important peptide groups that have been demonstrated to enhance defense signaling cascades that lead to HR. Certain plasma membrane-localized receptors, such as PEPR1/PEPR2, recognize these peptides, which are frequently released upon pathogen detection or tissue damage. These receptors then trigger downstream MAPK cascades, ROS generation, and defense gene expression, ultimately resulting in cell death at infection sites (Huffaker et al. 2006; Yamaguchi et al. 2010). In order to stop biotrophic pathogens from colonizing host tissues, infected cells are quickly destroyed by the HR-PCD that is set off by SPR peptides. This tightly controlled cell death frequently involves transcriptional activation of HR-associated genes, cross talk with SA signaling, and calcium influx (Albert 2013). Additionally, according to recent research, SPRs may interact with R-gene-mediated signaling pathways to strengthen effector-triggered immunity (ETI) and reinforce PCD via feedback loops. SPR peptides are therefore essential mediators that strengthen plant resistance to pathogen attack by bridging early immune perception with irreversible defense outcomes like HR-PCD.

8.4 Mechanisms of Peptide Hormone Action in PCD

In plants, peptide hormones have become key players in controlling PCD, combining stress reactions and developmental cues to promote the best possible growth, survival, and defense. Peptide hormones are gene-encoded signaling molecules that undergo posttranslational processing and are secreted to produce paracrine or systemic effects, in contrast to conventional plant hormones, which are frequently small molecules. They function as precisely calibrated messengers in PCD, influencing immune responses, redox balance, hormonal cross talk, and transcriptional regulation to decide cell fate (Matsubayashi 2014; Olsson et al. 2019). The PSK family is a well-known class that is known to attach to particular plasma membrane receptors, such as PSKR1/2. The PSK signaling preserves cell viability and postpones death under stress by reducing the excessive buildup of ROS, a common cause of PCD. By controlling Ca^{2+} influx and activating plasma membrane H^{+}-ATPases, PSK maintains cellular ion homeostasis and inhibits apoptotic signaling (Ladwig et al. 2015; Sauter 2015). As a result, PSKs protect against pathogen- or stress-induced PCD, promoting tissue survival in harsh environments. On the other hand, as a component of immunological or developmental programs, other peptide hormones actively support PCD. For instance, when a pathogen attacks, Peps are released and attach to PEPR receptors, activating transcription factors and MAPK cascades linked to HR-type PCD. Especially against biotrophic pathogens, this localized cell death creates a physical barrier that stops the spread of the pathogen (Yamaguchi et al. 2010; Albert 2013). The role of peptide hormones in dPCD is also demonstrated by the IDA peptide, which, via its receptor HAE/HSL2, coordinates cell separation and death during abscission and lateral root emergence (Stenvik et al. 2006). By controlling nutrient signaling, especially nitrogen availability, other families, like CEP, have an indirect impact on PCD. CEPs lessen stress-induced ROS accumulation and, consequently, the risk of premature PCD by optimizing resource allocation in nutrient-limited environments (Tabata et al. 2014). This is a prime example of the intricate relationship between environmental sensing, metabolic regulation, and hormonal peptide signaling, which shapes plant resilience by dynamically modifying cell survival and death programs. Peptide hormones play a critical role in the spatiotemporal regulation of plant development and defense, as evidenced by their mechanistic versatility in modulating PCD. Their dual roles as PCD inducers and inhibitors highlight a sophisticated regulatory system that strikes a balance between immunity and growth, ensuring plant survival in changing conditions.

8.4.1 Receptor-Like Kinases (RLKs)

In order to control PCD, RLKs are essential plasma membrane receptors that detect extracellular peptide hormones and start intracellular signaling cascades. Certain members of the large RLK family, including FER, phytosulfokine receptor 1 (PSKR1), and S-receptor kinase (SRK), have been thoroughly studied for their functions in regulating plant cell fate via peptide hormone interaction. A multipurpose RLK, FER plays a crucial mediating role in the integration of stress reactions and developmental cues. It identifies small cysteine-rich peptides called rapid alkalinization factors (RALFs), which affect the regulation of cell growth, immunity, and cell death (Stegmann et al. 2017). By controlling calcium signaling and ROS accumulation, FER partially modulates PCD by striking a balance between cell survival and defense activation (Duan et al. 2010). It also aids in the formation of immune receptor complexes and improves defense signaling, which can result in PCD linked to HR to infections. The sulfated peptide hormone PSK, which suppresses excessive ROS accumulation, a common inducer of PCD, and promotes cell survival, is specifically bound by the PSKR1 receptor (Matsubayashi 2014). By stabilizing cellular homeostasis under stress, PSKR1 activation initiates phosphorylation cascades and ion flux adjustments, which delay or inhibit PCD and encourage tissue regeneration (Ladwig et al. 2015; Sauter 2015). This receptor serves as an example of how RLK-mediated peptide hormone signaling can provide protection in the face of both biotic and abiotic stress. RLKs' ability to control PCD through reproductive signaling peptides is demonstrated by SRK, which is mainly investigated in Brassicaceae self-incompatibility responses (Nasrallah and Nasrallah 2014). In order to ensure genetic diversity, SRK detects S-locus cysteine-rich peptides (SCR/SP11), which trigger downstream signaling that results in the selective PCD of incompatible pollen tubes. This reveals a specific but crucial setting where PCD is tightly regulated by the peptide-RLK interaction for successful reproduction. Together, FER, PSKR1, and SRK demonstrate the flexible ways in which RLKs decipher peptide hormone signals to control PCD in a variety of physiological settings, from development and reproduction to immune defense and stress adaptation.

8.4.2 ROS and Calcium Signaling

Peptide hormones control calcium influx and ROS generation, two important mediators of PCD. For instance, PSK peptides alter calcium-dependent survival pathways, whereas RALF peptides affect ROS bursts. Peptide hormones are essential for modifying the dynamics of ROS and calcium ions (Ca^{2+}), which are essential second messengers in the control of plant PCD. Initiating and carrying out PCD, especially during stress and immune responses, depends on the production and spatial-temporal regulation of ROS bursts (Zurbriggen et al. 2010). By interacting with their RLKs, particularly FER, peptide hormones like RALFs have been demonstrated to control the production of ROS. In order to avoid excessive cell death, FER activation by RALF peptides balances defense signaling and cell survival by fine-tuning ROS levels (Stegmann et al. 2017). In order to minimize collateral damage to host tissues and limit pathogen spread through the HR, this modulation is essential. In order to coordinate PCD, calcium signaling and ROS dynamics are closely linked. One important upstream signal that starts downstream cascades, such as ROS production by NADPH oxidases, is calcium influx through plasma membrane channels (Choi et al. 2014). To increase cell survival, peptide hormones like PSK affect calcium-dependent pathways. The activation of calcium-permeable channels by PSK binding to its receptor PSKR1 promotes cellular homeostasis and prevents stress-induced PCD (Ladwig et al. 2015). PSK signaling delays the onset of PCD during environmental challenges by controlling calcium levels, which also helps to suppress ROS accumulation and mitigate oxidative damage. A complex regulatory network in which exact control of these second messengers dictates cell fate is exemplified by the interaction between ROS and calcium signaling mediated by peptide hormones. By calibrating PCD, this cross talk allows plants to respond dynamically to biotic and abiotic stress, guaranteeing an ideal balance between defense and survival (Suzuki et al. 2013). Through targeted manipulation of PCD pathways, an understanding of the molecular mechanisms by which peptides modulate ROS and calcium fluxes holds promise for enhancing crop stress resilience.

8.4.3 Hormonal Cross Talk

In plants, peptide hormones interact intricately with traditional phytohormones like ethylene, ABA, and JA to modify PCD in response to stressors and developmental cues. Plants are able to balance growth, defense, and survival by fine-tuning PCD pathways thanks to this integration.

One important regulator that encourages PCD during processes like pathogen-induced HR and leaf senescence is ethylene. PCD responses are amplified or modulated by the interaction of peptide hormone signaling with ethylene biosynthesis and signaling pathways. For example, peptide hormones such as Peps can increase the production of ethylene, which speeds up PCD to stop the spread of pathogens (Meng and Zhang 2013). ROS signaling and ethylene play a closely related role in PCD, frequently occurring after peptide hormone receptor activation (Van de Poel et al. 2015). A gaseous plant hormone, ethylene plays a key role in controlling PCD, especially during fruit ripening, leaf senescence, and defense mechanisms like the HR against infections. To optimize plant fitness, peptide hormones integrate developmental and environmental cues by actively modulating ethylene biosynthesis and signaling, which fine-tunes these PCD processes. The Peps family, which are tiny peptides released in response to pathogen attack or cellular damage, is a well-known example. According to Bartels and Boller (2015), pep peptides attach to their cognate receptors, PEPR1 and PEPR2, which sets off defense signaling cascades that increase the expression of genes involved in ethylene biosynthesis, including ACC synthase (ACS) and ACC oxidase (ACO). By encouraging localized PCD as a component of the HR, this ethylene production amplification successfully limits pathogen invasion. Additionally, ROS bursts and Pep-induced ethylene signaling work together to strengthen immune gene activation and cell death execution (Yamaguchi et al. 2010). The RALF peptides are another family of peptides that interact with ethylene signaling. The receptor-like kinase FER, which RALFs bind to, affects downstream signaling components and modulates ethylene responses. It has been demonstrated that FER controls the expression of ethylene-responsive transcription factors that govern genes linked to PCD (Stegmann et al. 2017). By balancing ethylene-mediated PCD through this pathway, RALF-FER signaling inhibits excessive cell death during stress responses and, when necessary, promotes cell survival. Additionally, after herbivore attack or mechanical damage, systemin, a wound-induced peptide hormone present in solanaceous plants, increases the production of ethylene (Ryan 2000). Ethylene biosynthesis genes are activated by systemin signaling, resulting in PCD in injured tissues that act as obstacles to the spread of pests. By coordinating with JA and other hormonal pathways, this peptide-mediated ethylene enhancement guarantees the prompt execution of PCD, aiding in effective defense. In conclusion, peptide hormones precisely regulate PCD outcomes during development and stress by modulating ethylene pathways at several levels, ranging from receptor-mediated signaling to biosynthesis gene regulation. The complex hormonal cross talk required for plants to precisely control cell death in response to various physiological demands is exemplified by this layered regulation.

PCD brought on by oxidative stress, salinity, and drought is largely regulated by ABA, which also mediates abiotic stress responses. The timing and severity of PCD can be influenced by peptide hormones' synergistic or antagonistic modulation of ABA signaling pathways. For instance, root meristem activity and cell death under stress are modulated by the interaction of CLE peptides with ABA signaling components (Hazak and Hardtke 2016). Furthermore, peptide-triggered cascades and ABA-induced calcium signaling frequently combine to regulate the survival or death of cells (Takahashi et al. 2019). A crucial phytohormone, ABA, controls how plants react to abiotic stressors like cold, salinity, and drought.

It also frequently controls PCD to maximize survival in harsh environments. In order to fine-tune stress-induced PCD and cellular adaptation, peptide hormones actively modulate ABA biosynthesis and signaling pathways. Modulation of ABA signaling to affect PCD during stress responses has been linked to a number of peptide families, most notably CLE-related peptides. CLE peptides have an impact on downstream signaling cascades and ABA sensitivity through their interactions with RLKs like CLV1 and BAMs (Hazak and Hardtke 2016). For instance, under drought or osmotic stress, certain CLE peptides can change ABA-mediated stomatal closure and root meristem activity, two processes that are inherently connected to ABA-regulated PCD. Plants can balance cell death and survival in tissues that are essential for growth maintenance and water conservation, thanks to this modulation. Furthermore, calcium signaling pathways that converge with ABA responses to regulate cell fate decisions are influenced by peptide hormones like PSK. Calcium influx is triggered by PSK binding to its receptor PSKR1, and this interaction is known to interact with ABA-dependent signaling components, such as calcium-dependent protein kinases (CDPKs) (Ladwig et al. 2015). This integration promotes cellular homeostasis by reducing excessive ROS accumulation and preventing premature PCD under stress. Furthermore, transcription factors like ABI5 and NAC domain proteins are frequently activated in ABA-induced PCD, and these factors can be indirectly influenced by peptide hormone signaling networks (Takahashi et al. 2019). Through cross-talk mechanisms, peptide hormones may control the expression or activity of these transcription factors, thereby adjusting the degree and timing of cell death to maximize stress tolerance. In conclusion, peptide hormones are important modulators of stress-induced PCD because they complexly alter signal transduction pathways and ABA biosynthesis. Through this hormonal interaction, plants are able to balance growth and survival in the face of harsh environmental conditions by adaptively controlling cell death.

A key component of defense-related PCD, especially during herbivore attack and pathogen infection, is JA. Peptide hormone networks and JA signaling pathways interact to control the expression of defense genes and localized cell death. As part of the immune response, peptides like systemin can cause JA biosynthesis, which in turn encourages PCD (Ryan 2000). A strong and prompt PCD response that limits pathogen progression and minimizes tissue damage is ensured by the synergistic interaction between JA and peptide signaling. One important plant hormone involved in defense responses, especially those related to wounding, herbivory, and pathogen attack, is JA. As part of the plant's defense mechanism, JA signaling is essential for mediating PCD. Peptide hormones modulate defense-related PCD through complex interactions with JA pathways, guaranteeing accurate temporal and spatial control of cell death during biotic stress. Systemin and other peptide hormones play a key role in JA-mediated defense signaling. By activating JA biosynthesis through a receptor-mediated cascade, systemin, a wound-induced peptide primarily present in solanaceous plants, causes JA and its bioactive derivatives to accumulate (Ryan 2000). This increase in JA causes PCD in injured tissues, which promotes localized cell death and stops the spread of pathogens or herbivores. One example of how peptides function as upstream activators of hormone biosynthesis is the systemin-JA signaling axis, which coordinates quick defense reactions involving PCD execution. Furthermore, it has been demonstrated that Peps improve JA signaling. When Peps bind to their receptors (PEPRs), downstream transcription factors and MAP kinase cascades are activated, which increases the production of JA and signaling genes (Yamaguchi et al. 2010). This amplification loop effectively inhibits pathogen colonization by promoting defense gene expression and HR-associated PCD. The cross talk between JA and peptide hormone signaling also involves modulation of ROS and calcium dynamics, which are vital for PCD initiation (Wasternack and Hause 2013). Peptide hormones can fine-tune JA-dependent ROS production, thereby adjusting the intensity of cell death and defense responses. This regulatory network ensures that PCD is tightly controlled to maximize defense while minimizing detrimental effects on healthy tissues. Therefore, peptide hormones play a key role in regulating JA biosynthesis and signaling, which helps to precisely coordinate defense-related PCD. The complex hormonal interactions that support plants' resistance to biotic stress are best illustrated by this interaction.

In conclusion, a complex regulatory network is established by the hormonal cross talk between peptide hormones and traditional plant hormones such as ethylene, ABA, and JA. By dynamically adjusting PCD in response to environmental cues and developmental programs, this network improves plant resilience and adaptability.

8.5 Applications of Peptide Hormones

Peptide hormones are useful tools for agricultural biotechnology and crop improvement since they have been shown to be essential regulators of plant growth, development, and stress responses. Their distinct signaling pathways and modes of action present special chances to modify plant physiology for increased yield, disease resistance, and stress tolerance. Enhancing crop resistance to abiotic stresses like drought, salinity, and temperature extremes is one of the main uses of peptide hormones. It has been demonstrated that peptides such as PSK and CLE peptides can alter stress-induced PCD, postponing cell death and enhancing survival in adverse circumstances (Matsubayashi 2014). Root architecture and water-use efficiency can be improved by exogenous application of, or genetic modification to, peptide hormone pathways, offering long-term solutions to lessen the effects of climate change on agriculture (Tabata 2023). Peptides like Systemin and Peps stimulate immune signaling pathways in disease resistance, including the HR, which limits pathogen spread through localized PCD (Bartels and Boller 2015). By using breeding or bioengineering to harness these peptides, innate immunity can be increased, lowering the need for chemical pesticides and promoting environmentally friendly plant protection. Additionally, peptide hormones are essential for controlling development. For instance, RALF peptides affect cell division and proliferation, whereas CLE peptides control meristem activity and vascular development (Murphy and De Smet 2014). There are opportunities to regulate plant architecture and maximize crop biomass production by modifying these peptides. Despite their potential, peptide hormone delivery in field conditions is still difficult because peptides degrade easily and need to be applied with specific techniques. These obstacles are about to be removed by developments in peptide stability, formulation technologies, and gene editing tools like CRISPR/Cas9, which will allow for precise modification of peptide hormone pathways in crops (Ladwig et al. 2015). Peptide hormones, in conclusion, offer flexible and promising instruments for promoting sustainable agriculture by providing focused control over growth, development, and stress reactions to enhance crop performance in the face of global obstacles.

8.5.1 Agricultural Implications

By controlling pathogen-induced PCD, a mechanism essential to limiting the spread of pathogens, peptide hormones have emerged as important modulators in boosting plant immune responses. The HR, a type of localized PCD that confines biotrophic pathogens, is one of the defense signaling cascades that are triggered by small peptides like Systemin and Peps, which function as endogenous danger signals (Bartels and Boller 2015; Ryan 2000). These peptides enhance immune signaling pathways by activating RLKs like PEPRs. This results in the buildup of defense-related hormones, ROS, and nitric oxide, which together coordinate effective pathogen resistance (Yamaguchi et al. 2010). By using these peptide signaling pathways biotechnologically, either by overexpressing them or by applying them exogenously, it is possible to create crop varieties with increased resistance, which will lessen the need for chemical pesticides and encourage sustainable farming methods (Bartels and Boller 2015). Extreme temperatures, salinity, and drought are examples of abiotic stresses that cause oxidative damage and PCD in plants, which frequently lead to large yield losses. Peptide hormones, such as CLE family peptides and PSK, are becoming recognized as significant regulators that inhibit PCD brought on by excessive stress, improving cellular survival and stress tolerance (Matsubayashi 2014; Hazak and Hardtke 2016). These peptides affect vital signaling elements that are essential for stress perception and subsequent reactions, including calcium influx and ROS homeostasis (Tabata et al. 2014). For example, when PSK binds to its receptor PSKR1, calcium-dependent pathways are activated, thereby reducing the accumulation of ROS and preventing cells from dying too soon under stress (Ladwig et al. 2015). Plants can sustain growth and productivity in challenging conditions by improving root architecture and water-use efficiency through genetic engineering or exogenous peptide treatments that manipulate peptide hormone pathways (Tabata et al. 2014). A promising strategy for creating climate-resilient crops is the modification of peptide hormone signaling.

These uses collectively demonstrate the dual potential of peptide hormones in agriculture: boosting immunity by promoting defense-related PCD and boosting resilience by inhibiting stress-induced PCD. Considering the problems facing global food security, these complex regulatory functions highlight the importance of peptide hormones as targets for creative crop improvement techniques.

8.5.2 Biotechnological Applications

By modifying peptide hormone pathways, synthetic biology provides revolutionary methods for reprogramming plant development and stress responses. Researchers can adjust PCD for increased plant resilience and productivity by creating synthetic peptide sequences or altering preexisting receptors. To increase immune responses without causing growth penalties, for instance, modified forms of Peps with improved stability or changed receptor affinity have been developed (Segonzac and Monaghan 2019). Additionally, synthetic biology makes it possible to create synthetic receptors or chimeric RLKs that react to specific peptides, enabling precise temporal and spatial control of PCD in response to environmental stimuli (Santiago et al. 2016). These developments improve agricultural efficiency by making it easier to create crops that respond differently to abiotic stresses or pathogens.

Peptide hormone signaling pathways can now be dissected and rewired with the help of gene editing tools, particularly CRISPR/Cas9. Researchers can alter PCD-related processes to provide stress tolerance or disease resistance by focusing on important genes that encode peptide ligands, receptors, or downstream transcription factors. For example, it has been demonstrated that CRISPR-mediated knockout of negative regulators in the PSK signaling pathway improves survival in the face of oxidative stress (Ghorbani et al. 2016). According to Rodríguez-Leal et al. (2017), altering the CLE-CLAVATA signaling module can also affect the dynamics of cell death and meristem activity, which can have an impact on biomass accumulation and developmental plasticity. Peptide hormone engineering is a feasible approach for next-generation crop improvement because of the accuracy and effectiveness of CRISPR/Cas9, which enables targeted interventions in plant genomes.

A viable and expandable method of controlling PCD in crops is the use of synthetic peptides as agrochemicals. Depending on the situation, exogenous administration of peptides like PSK, RALF, and Peps has been shown to either suppress stress-induced cell death or initiate defense responses (Matsubayashi 2014; Bartels and Boller 2015). To produce desired physiological effects, such as increased immunity or improved drought tolerance, these peptides can be prepared for foliar sprays, seed coatings, or root drenching. Peptide-based agrochemicals, which are biodegradable and biologically derived, provide a sustainable substitute for traditional pesticides and stress protectants. Their commercialization potential in integrated crop management practices is further supported by developments in peptide stabilization and delivery systems (Balmer et al. 2018).

8.6 Challenges and Research Opportunities

Determining the specificity and selectivity of peptide-receptor interactions is one of the main obstacles in the study of plant peptide hormones. The underlying molecular basis for the binding specificity and signal discrimination of numerous peptide ligands and their corresponding RLKs is still unknown (Okuda et al. 2020). Although structurally conserved peptide motifs are recognized by certain receptors, like PSKR1 and PEPRs, minor alterations in peptide sequence can significantly change downstream signaling outputs (Tang et al. 2015). Comprehending this selectivity is essential for basic biology as well as the logical development of synthetic peptides or modified receptors intended to modify context-dependent PCD responses. These complex ligand-receptor architectures can be resolved through the use of cutting-edge structural biology techniques like molecular dynamics simulations and cryo-electron microscopy.

Although peptide hormone signaling frequently demonstrates exceptional tissue specificity, little is known about how this spatial regulation relates to PCD. Some peptides, such as RALFs or CLEs, have opposing functions in various tissues; in some situations, they promote survival, while in others, they cause cell death (Ghorbani et al. 2016; Czyzewicz et al. 2013). This dichotomy makes it difficult to predict the results of peptide manipulation in transgenic crops and to interpret experimental data. For mapping tissue-specific expression and signaling landscapes, methods like promoter-reporter fusions, spatial proteomics, and single-cell transcriptomics are becoming increasingly useful (Denyer et al. 2019). In order to enable more accurate uses of peptide hormones in agriculture, future studies should attempt to link these tissue-specific patterns with PCD phenotypes during development or under stress.

The entire range of peptide diversity in the plant kingdom is still mostly unexplored, even though the number of characterized peptides in model plants like *Arabidopsis thaliana* is increasing. According to new research, peptides that are species- or lineage-specific may have special functions in environmental adaptation and developmental plasticity, including controlling PCD (Delay et al. 2013). To find new peptides and their receptors in a variety of taxa, especially in crop relatives that are resilient to stress in the wild, comparative genomics and peptidomics techniques are crucial. Crop improvement techniques like introducing advantageous peptide signaling traits into elite cultivars could benefit from this natural variation. Furthermore, synthetic biology techniques targeted at creating reliable, orthogonal signaling modules for precision agriculture may benefit from an understanding of evolutionary advancements in peptide-receptor coevolution.

8.7 Conclusion and Future Prospects

Peptide hormones are increasingly recognized as important players in how plants control cell death, a process essential for development and stress responses. Although some peptide families like CLE, RALF, and PSK are already known to influence growth and defense, their role in directly managing when and where plant cells die is still not fully understood. Peptide hormones play a crucial role in controlling plant PCD by coordinating a complex network of signals and interacting with other phytohormones. They provide promising tools for enhancing plant resilience and productivity by regulating stress-induced and developmental cell death. Novel uses in plant biotechnology and sustainable agriculture will be made possible by further investigation into the molecular mechanisms underlying peptide hormone signaling. In the coming years, researchers are likely to discover new peptide signals and their receptors that help fine-tune cell death during events like pathogen attacks, drought, or even normal tissue development. With the help of advanced tools like single-cell analysis, gene editing (e.g. CRISPR), and real-time imaging, researchers can better map how these peptides interact with other known cell death regulators such as ROS and specific proteins that trigger the death process. This growing knowledge could lead to practical applications, such as designing crops that are better at surviving stress without compromising their growth.

References

Albert, M. (2013). Peptides as triggers of plant defence. *Journal of Experimental Botany* 64: 5269–5279.

Amano, Y., Tsubouchi, H., Shinohara, H. et al. (2007). Tyrosine-sulfated glycopeptide involved in cellular proliferation and expansion in Arabidopsis. *Proceedings of the National Academy of Sciences of the United States of America* 104: 18333–18338.

Balmer, A., Pastor, V., Glauser, G., and Mauch-Mani, B. (2018). Tricarboxylates induce defense priming against bacteria in *Arabidopsis thaliana*. *Frontiers in Plant Science* 9: 1221.

Bartels, S. and Boller, T. (2015). Quo vadis, Pep? Plant elicitor peptides at the crossroads of immunity, stress, and development. *Journal of Experimental Botany* 66: 5183–5193.

Beloshistov, R.E., Dreizler, K., Galiullina, R.A. et al. (2018). Phytaspase-mediated precursor processing and maturation of the wound hormone systemin. *The New Phytologist* 218: 1167–1178.

Blackburn, M.R., Haruta, M., and Moura, D.S. (2020). Twenty years of progress in physiological and biochemical investigation of RALF peptides. *Plant Physiology* 182: 1657–1666.

Butenko, M.A., Patterson, S.E., Grini, P.E. et al. (2003). Inflorescence deficient in abscission controls floral organ abscission in Arabidopsis and identifies a novel family of putative ligands in plants. *Plant Cell* 15: 2296–2307.

Choi, W.G., Toyota, M., Kim, S.H. et al. (2014). Salt stress-induced Ca2+ waves are associated with rapid, long-distance root-to-shoot signaling in plants. *Proceedings of the National Academy of Sciences of the United States of America* 111: 6497–6502.

Coll, N.S., Epple, P., and Dangl, J.L. (2011). Programmed cell death in the plant immune system. *Cell Death and Differentiation* 18: 1247–1256.

Czyzewicz, N., Yue, K., Beeckman, T., and De Smet, I. (2013). Message in a bottle: small signalling peptide outputs during growth and development. *Journal of Experimental Botany* 64: 5281–5296.

Danon, A., Delorme, V., Mailhac, N., and Gallois, P. (2000). Plant programmed cell death: a common way to die. *Plant Physiology and Biochemistry* 38: 647–655.

Delay, C., Imin, N., and Djordjevic, M.A. (2013). CEP genes regulate root and shoot development in response to environmental cues and are specific to seed plants. *Journal of Experimental Botany* 64: 5383–5394.

Denyer, T., Ma, X., Klesen, S. et al. (2019). Spatiotemporal developmental trajectories in the Arabidopsis root revealed using high-throughput single-cell RNA sequencing. *Developmental Cell* 48: 840–852.e5.

Duan, Q., Kita, D., Li, C. et al. (2010). FERONIA receptor-like kinase regulates RHO GTPase signaling of root hair development. *Proceedings of the National Academy of Sciences of the United States of America* 107: 17821–17826.

Estornell, L.H., Agustí, J., Merelo, P. et al. (2013). Elucidating mechanisms underlying organ abscission. *Plant Science* 199–200: 48–60.

Fukuda, H. (2000). Programmed cell death of tracheary elements as a paradigm in plants. *Plant Molecular Biology* 44: 245–253.

Ghorbani, S., Hoogewijs, K., Pečenková, T. et al. (2016). The SBT6.1 subtilase processes the GOLVEN1 peptide controlling cell elongation. *Journal of Experimental Botany* 67: 4877–4887.

Haruta, M., Sabat, G., Stecker, K. et al. (2014). A peptide hormone and its receptor protein kinase regulate plant cell expansion. *Science* 343: 408–411.

Hatsugai, N., Kuroyanagi, M., Yamada, K. et al. (2004). A plant vacuolar protease, VPE, mediates virus-induced hypersensitive cell death. *Science* 305: 855–858.

Hazak, O. and Hardtke, C.S. (2016). CLAVATA 1-type receptors in plant development. *Journal of Experimental Botany* 67: 4827–4833.

Hirakawa, Y., Shinohara, H., Kondo, Y. et al. (2008). Non-cell-autonomous control of vascular stem cell fate by a CLE peptide/receptor system. *Proceedings of the National Academy of Sciences of the United States of America* 105: 15208–15213.

Huffaker, A., Pearce, G., and Ryan, C.A. (2006). An endogenous peptide signal in Arabidopsis activates components of the innate immune response. *Proceedings of the National Academy of Sciences of the United States of America* 103: 10098–10103.

Huysmans, M., Lema, A.S., Coll, N.S., and Nowack, M.K. (2017). Dying two deaths-programmed cell death regulation in development and disease. *Current Opinion in Plant Biology* 35: 37–44.

Igarashi, D., Tsuda, K., and Katagiri, F. (2012). The peptide growth factor, phytosulfokine, attenuates pattern-triggered immunity. *The Plant Journal* 71: 194–204.

Ito, Y., Nakanomyo, I., Motose, H. et al. (2006). Dodeca-CLE peptides as suppressors of plant stem cell differentiation. *Science* 313: 842–845.

Kaufmann, C., Motzkus, M., and Sauter, M. (2017). Phosphorylation of the phytosulfokine peptide receptor PSKR1 controls receptor activity. *Journal of Experimental Botany* 68: 1411–1423.

Kutschmar, A., Rzewuski, G., Stührwohldt, N. et al. (2009). PSK-α promotes root growth in Arabidopsis. *The New Phytologist* 181: 820–831.

Ladwig, F., Dahlke, R.I., Stührwohldt, N. et al. (2015). Phytosulfokine regulates growth in arabidopsis through a response module at the plasma membrane that includes CYCLIC NUCLEOTIDE-GATED CHANNEL17, H+-ATPase, and BAK1. *Plant Cell* 27: 1718–1729.

Li, S., Fu, H., Sharif, Y. et al. (2024). Genome-wide identification, functional characterization, and stress-responsive expression profiling of subtilase (sbt) gene family in peanut (*Arachis hypogaea* L.). *International Journal of Molecular Sciences* 25: 13361.

Matsubayashi, Y. (2014). Posttranslationally modified small-peptide signals in plants. *Annual Review of Plant Biology* 65: 385–413.

Meng, X. and Zhang, S. (2013). MAPK cascades in plant disease resistance signaling. *Annual Review of Phytopathology* 51: 245–266.

Minina, E.A., Coll, N.S., Tuominen, H., and Bozhkov, P.V. (2017). Metacaspases versus caspases in development and cell fate regulation. *Cell Death and Differentiation* 24: 1314–1325.

Murphy, E. and De Smet, I. (2014). Understanding the RALF family: a tale of many species. *Trends in Plant Science* 19: 664–671.

Nasrallah, J.B. and Nasrallah, M.E. (2014). S-locus receptor kinase signalling. *Biochemical Society Transactions* 42: 313–319.

Ohkubo, Y., Tanaka, M., Tabata, R. et al. (2017). Shoot-to-root mobile polypeptides involved in systemic regulation of nitrogen acquisition. *Nature Plants* 3: 17029.

Okuda, S., Fujita, S., Moretti, A. et al. (2020). Molecular mechanism for the recognition of sequence-divergent CIF peptides by the plant receptor kinases GSO1/SGN3 and GSO2. *Proceedings of the National Academy of Sciences of the United States of America* 117: 2693–2703.

Olsson, V., Joos, L., Zhu, S. et al. (2019). Look closely, the beautiful may be small: precursor-derived peptides in plants. *Annual Review of Plant Biology* 70: 153–186.

Pearce, G., Strydom, D., Johnson, S., and Ryan, C.A. (1991). A polypeptide from tomato leaves induces wound-inducible proteinase inhibitor proteins. *Science* 253: 895–897.

Pearce, G., Moura, D.S., Stratmann, J., and Ryan, C.A. Jr. (2001). RALF, a 5-kDa ubiquitous polypeptide in plants, arrests root growth and development. *Proceedings of the National Academy of Sciences of the United States of America* 98: 12843–12847.

Petrov, V., Hille, J., Mueller-Roeber, B., and Gechev, T.S. (2015). ROS-mediated abiotic stress-induced programmed cell death in plants. *Frontiers in Plant Science* 6: 69.

Roberts, I., Smith, S., De Rybel, B. et al. (2013). The CEP family in land plants: evolutionary analyses, expression studies, and role in Arabidopsis shoot development. *Journal of Experimental Botany* 64: 5371–5381.

Rodríguez-Leal, D., Lemmon, Z.H., Man, J. et al. (2017). Engineering quantitative trait variation for crop improvement by genome editing. *Cell* 171: 470–480.e8.

Ryan, C.A. (2000). The systemin signaling pathway: differential activation of plant defensive genes. *Biochimica et Biophysica Acta* 1477: 112–121.

Santiago, J., Brandt, B., Wildhagen, M. et al. (2016). Mechanistic insight into a peptide hormone signaling complex mediating floral organ abscission. *eLife* 5: e15075.

Sauter, M. (2015). Phytosulfokine peptide signalling. *Journal of Experimental Botany* 66: 5161–5169.

Segonzac, C. and Monaghan, J. (2019). Modulation of plant innate immune signaling by small peptides. *Current Opinion in Plant Biology* 51: 22–28.

Stegmann, M., Monaghan, J., Smakowska-Luzan, E. et al. (2017). The receptor kinase FER is a RALF-regulated scaffold controlling plant immune signaling. *Science* 355: 287–289.

Stenvik, G.E., Butenko, M.A., Urbanowicz, B.R. et al. (2006). Overexpression of INFLORESCENCE DEFICIENT IN ABSCISSION activates cell separation in vestigial abscission zones in Arabidopsis. *Plant Cell* 18: 1467–1476.

Suzuki, N., Miller, G., Salazar, C. et al. (2013). Temporal-spatial interaction between reactive oxygen species and abscisic acid regulates rapid systemic acclimation in plants. *Plant Cell* 25: 3553–3569.

Tabata, R. (2023). Regulation of the iron-deficiency response by IMA/FEP peptide. *Frontiers in Plant Science* 14: 1107405.

Tabata, R., Sumida, K., Yoshii, T. et al. (2014). Perception of root-derived peptides by shoot LRR-RKs mediates systemic N-demand signaling. *Science* 346: 343–346.

Takahashi, F., Hanada, K., Kondo, T., and Shinozaki, K. (2019). Hormone-like peptides and small coding genes in plant stress signaling and development. *Current Opinion in Plant Biology* 51: 88–95.

Tang, J., Han, Z., Sun, Y. et al. (2015). Structural basis for recognition of an endogenous peptide by the plant receptor kinase PEPR1. *Cell Research* 25: 110–120.

Tavormina, P., De Coninck, B., Nikonorova, N. et al. (2015). The plant peptidome: an expanding repertoire of structural features and biological functions. *Plant Cell* 27: 2095–2118.

Van de Poel, B., Smet, D., and Van Der Straeten, D. (2015). Ethylene and hormonal cross talk in vegetative growth and development. *Plant Physiology* 169: 61–72.

van Doorn, W.G., Beers, E.P., Dangl, J.L. et al. (2011). Morphological classification of plant cell deaths. *Cell Death and Differentiation* 18: 1241–1246.

Wang, L., Einig, E., Almeida-Trapp, M. et al. (2018). The systemin receptor SYR1 enhances resistance of tomato against herbivorous insects. *Nature Plants* 4: 152–156.

Wasternack, C. and Hause, B. (2013). Jasmonates: biosynthesis, perception, signal transduction and action in plant stress response, growth and development. An update to the 2007 review in Annals of Botany. *Annals of Botany* 111: 1021–1058.

Yamaguchi, Y., Huffaker, A., Bryan, A.C. et al. (2010). PEPR2 is a second receptor for the Pep1 and Pep2 peptides and contributes to defense responses in Arabidopsis. *Plant Cell* 22: 508–522.

Yamaguchi, Y.L., Ishida, T., and Sawa, S. (2016). CLE peptides and their signaling pathways in plant development. *Journal of Experimental Botany* 67: 4813–4826.

Zhou, J., Yu, J.Q., and Chen, Z. (2014). The perplexing role of autophagy in plant innate immune responses. *Molecular Plant Pathology* 15: 637–645.

Zurbriggen, M.D., Carrillo, N., and Hajirezaei, M.R. (2010). ROS signaling in the hypersensitive response: when, where and what for? *Plant Signaling & Behavior* 5: 393–396.

9

Regulation of Plant Senescence by Peptide Hormones

Aslıhan Temel[1], Filiz Vardar[2], and Nihal Gören-Sağlam[3]

[1] *Department of Molecular Biology and Genetics, Faculty of Science, Istanbul University, Istanbul, Türkiye*
[2] *Department of Biology, Faculty of Science, Marmara University, Istanbul, Türkiye*
[3] *Department of Biology, Faculty of Science, Istanbul University, Istanbul, Türkiye*

9.1 Introduction

Hormones in higher plants are small, lipophilic compounds rather than their peptide analogues in animals (Matsubayashi and Sakagami 2006). The term hormone indeed arose from the work of Starling on mammalian tissues (Bayliss and Starling 1902; Starling 1905). Fitting (1909) and Paál (1919) were among the first to describe the hormone concept in plants. Particular characteristics, e.g. the presence of the cell wall, of plants were suggested to explain the differences in synthesis, transport, regulation, and action mechanisms of plant hormones, which have several synonyms, such as "plant regulators," "plant growth regulators," and "plant growth substances" (Weyers and Paterson 2001). Auxins (AUXs), cytokinins (CKs), gibberellins (GAs), ethylene, abscisic acid (ABA), along with brassinosteroids (BRs), polyamines (PAs), strigolactones (SLs), salicylic acid (SA), and jasmonic acid (JA) comprise plant hormones (Matsubayashi and Sakagami 2006; Bajguz and Piotrowska-Niczyporuk 2023). These hormones are present in low amounts and either act locally or are transported over long distances (Santner et al. 2009). AUXs, ethylene, PAs, SA, and melatonin are amino acid-derived plant hormones, while ABA, BRs, CKs, GAs, and SLs are isoprenoid derived and JAs are lipid derived (Bajguz and Piotrowska-Niczyporuk 2023).

AUXs are the first-discovered and the most-studied plant hormone class and are involved in all processes in plants (Santner et al. 2009; Bajguz and Piotrowska-Niczyporuk 2023). CKs are adenine derivatives that stimulate cell division together with auxin. They are necessary for the growth of the shoot apical meristem and repress the root apical meristem through cell differentiation (Schaller et al. 2014). GAs were first identified in rice plants infected with the fungus *Gibberella fujikuroi*, whose secretions caused sterility and abnormal growth in the infected plants (Bajguz and Piotrowska-Niczyporuk 2023). They are important regulators of plant architecture as semi-dwarf plants developed by the Green Revolution have mutations in GA signaling and exhibit improved crop yield (Wu et al. 2021a,b).

Peptide Hormones in Plant Development and Stress Responses, First Edition. Edited by Vijay Pratap Singh and Samiksha Singh.

In addition to their functions in vegetative and reproductive development, GAs are essential for the breakdown of seed dormancy (Castro-Camba et al. 2022). ABA, which is responsible for stomatal closure, is commonly associated with stress response, although its functions in growth, leaf senescence, and germination (Chen et al. 2020; Bajguz and Piotrowska-Niczyporuk 2023). Its effects on plant yield under stress depend on the duration of the stress (Kavi Kishor et al. 2022). Ethylene is a volatile phytohormone and is synthesized from *S*-adenosyl-L-methionine. It was first identified due to illuminating gas leaks that caused defoliation of nearby plants. Its most prominent functions are the induction of senescence and fruit ripening. Its biosynthesis is stimulated under stress conditions (Bleecker and Kende 2000).

BRs were discovered as growth-promoting substances in *Brassica napus* pollen. They are found in very low amounts and in immature seeds and pollen (Clouse and Sasse 1998). They are involved in cell elongation, cell cycle progression, increased mitotic activity, and chromosomal aberrations (Kartal et al. 2009; Fridman and Savaldi-Goldstein 2013); due to their protective roles against abiotic stress (Hafeez et al. 2021), BRs can be potential drivers of the next Green Revolution (Yang et al. 2024). SAs were first purified from the bark of a willow tree, Salix, in the form of salicin and were later released by the Bayer company as acetylsalicylic acid under the trade name Aspirin, which treats and prevents heart attack, thrombosis, and reduces pain and fever (Raskin 1992). PAs are small, organic, polycationic alkylamines that are synthesized from ornithine, arginine, or decarboxylated *S*-adenosyl-L-methionine and the first product of PA synthesis is putrescine. They interact with nucleic acids and phospholipids and decrease the activities of enzymes related to senescence (Chen et al. 2019; Blázquez 2024). Because they are found at high levels in cells, they can be classified as hormone-like substances rather than hormones (Sobieszczuk-Nowicka et al. 2019; Blázquez 2024). Strigolactones (SLs) are a new type of plant hormone; they are present in root exudates. SLs are involved in symbiotic relationships between plants and soil microorganisms and are essential for the germination of a parasitic plant, *Striga* (Janeeshma et al. 2024). SA is a phenolic compound synthesized through the phenylpropanoid pathway in the cytoplasm or the isochorismate pathway in the chloroplast and found in modified forms, e.g. methylated. It is rather known for its role in plant immunity, yet also involved in several plant processes such as seed germination and particularly senescence, of which markers such as decreased photosynthetic activity and increased reactive oxygen species (ROS) levels are caused by the accumulation of SA (Rivas-San Vicente and Plasencia 2011). JA is synthesized from α-linolenic acid, present in the membranes of chloroplasts, where JA biosynthesis is initiated. An intermediate is transferred to the peroxisome to finally yield JA (Ghorbel et al. 2021). Although it is mainly known for its role in defense against biotic stress, JA also alleviates the toxic effects of abiotic stress and regulates plant development, e.g. senescence (Ghorbel et al. 2021; Rehman et al. 2023). The functions of SA and JA during senescence are summarized in detail in the Section 9.4.

In higher plants, intercellular communication is primarily mediated by small lipophilic signaling molecules known as plant hormones. However, recent studies have highlighted the significant contribution of both secretory and nonsecretory peptide signals in regulating various stages of plant growth and development. Most of these peptides have been identified through biochemical and molecular approaches, while in silico genome analyses have

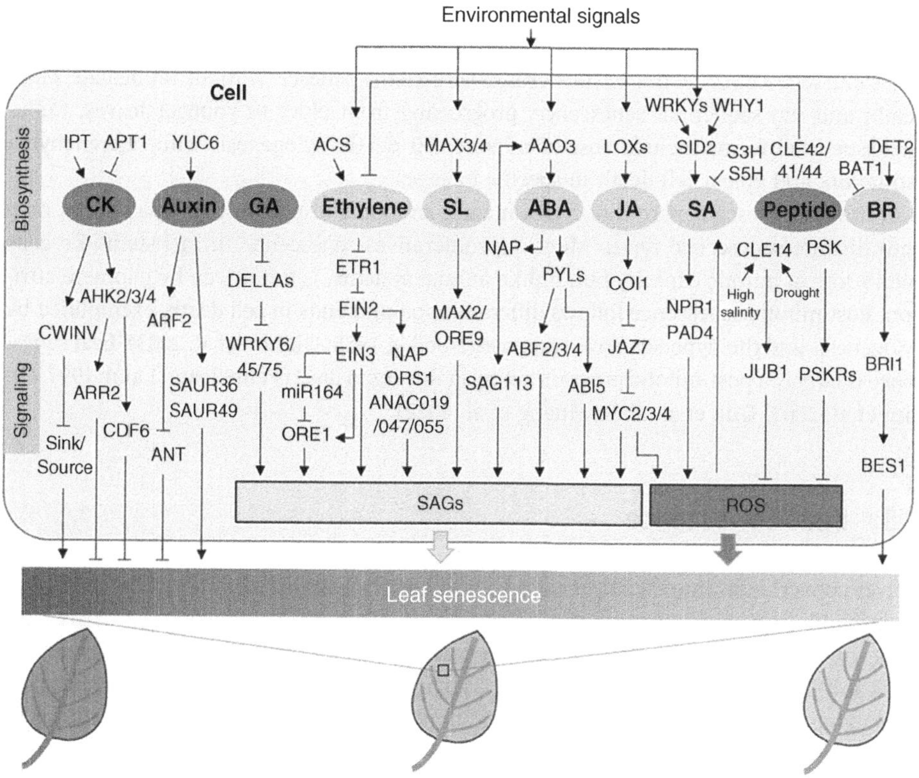

Figure 9.1 Leaf senescence and relations of classical and peptide hormones. *Source:* Huang et al. (2022) / Frontiers Media S.A / CC BY 4.0.

further revealed the presence of numerous genes encoding short open reading frames homologous to these peptides. In recent years, increasing attention has focused on the role of peptide hormones, signal molecules, in senescence, a key developmental stage in plants. This chapter overviews the roles of classical plant hormones and biologically active peptides, now recognized as a novel class of plant hormones, in regulating plant senescence, with particular emphasis on leaf senescence (Figure 9.1).

9.2 Types of Plant Senescence

Senescence represents the terminal developmental phase at cellular, tissue, organ, or whole-plant scales, culminating in death but ensuring nutrient recovery and allocation to sinks (e.g. nitrogen, sulfur, phosphorus, potassium), thereby sustaining growth and reproduction. Although genetically programmed under optimal conditions, senescence can be prematurely activated by drought, heat, nutrient scarcity, low light, disease, or pathogens. Thus, senescence is a controlled, signal-integrating program rather than simple degeneration.

Leaves exhibit diverse senescence modes: (i) seasonal senescence in deciduous species; (ii) monocarpic senescence in once-flowering plants culminating in whole-plant death; (iii) stem senescence, common in herbs where stems senesce without immediate plant death; and (iv) sequential senescence, progressing from older to younger leaves. Cellular/tissue-level forms include abscission-zone cell death, xylogenesis with sclerenchyma formation, and apical cell death in specific lineages.

At the cellular scale, plant life cycles include mitotic and post-mitotic phases that map onto distinct senescence types. Mitotic (proliferative) senescence in meristem-like cells entails loss of mitotic capacity but, unlike animal systems, is not driven by telomere attrition. Post-mitotic senescence follows differentiation and ends in cell death, exemplified by xylogenesis and the hypersensitive response (Nooden 1988; Ok Lim et al. 2003). Leaf senescence is largely a post-mitotic program acting at the organ level (Pennell and Lamb 1997; Ok Lim et al. 2003; Guo et al. 2021; Huang et al. 2022).

9.3 Leaf Senescence

Leaves convert light into chemical energy via chloroplast-localized chlorophylls. With age, photosynthetic capacity wanes, chloroplasts dismantle, and macromolecules (lipids, proteins, nucleic acids) are catabolized, with remobilized nutrients supporting developing sinks and enhancing fitness (Guo et al. 2021; Huang et al. 2022). In annuals, nutrient export from aging leaves fuels reproductive growth; in perennials (e.g. deciduous trees), nitrogen-rich proteins from senescent leaves are stored as bark proteins in phloem during winter and later reused for shoot/flower growth. Because ~70% of leaf protein resides in chloroplasts, their degradation is central to N and C recycling. Both premature and delayed senescence can depress yield and quality (Su et al. 2017; Piao et al. 2019; Guo et al. 2021; Huang et al. 2022).

Senescence progression is actively controlled by senescence-associated genes (SAGs), whose expression increases with age and is regulated at transcriptional, posttranscriptional, translational, and posttranslational levels (Woo et al. 2019; Guo et al. 2021). Multiomics approaches now reveal dynamic, network-like control with spatial and temporal structure (Kim et al. 2020; Guo et al. 2021). While developmentally programmed, senescence is modulated by internal and external signals integrated into age-associated networks (Guo et al. 2021; Figure 9.2).

Abiotic drivers—DNA damage, drought, heat, salinity, darkness, and nutrient deprivation—and biotic challenges—pathogens and phloem feeders—contribute to senescence (Sade et al. 2018; Guo et al. 2021). DNA damage from endogenous and genotoxic agents is an emerging determinant (Li et al. 2020; Guo et al. 2021). Calcium (Ca^{2+}) operates as a ubiquitous second messenger coordinating enzyme and protein activities; exogenous Ca^{2+} can delay senescence in detached leaves. Elevated CO_2 often increases sugars but lowers leaf nitrogen, perturbing the C/N ratio and predisposing to premature senescence (Agüera and De la Haba 2018; Guo et al. 2021). Importantly, these factors cross talk extensively, often via hormones, embedding senescence in a multilayered regulatory web rather than a linear cascade (Guo et al. 2021).

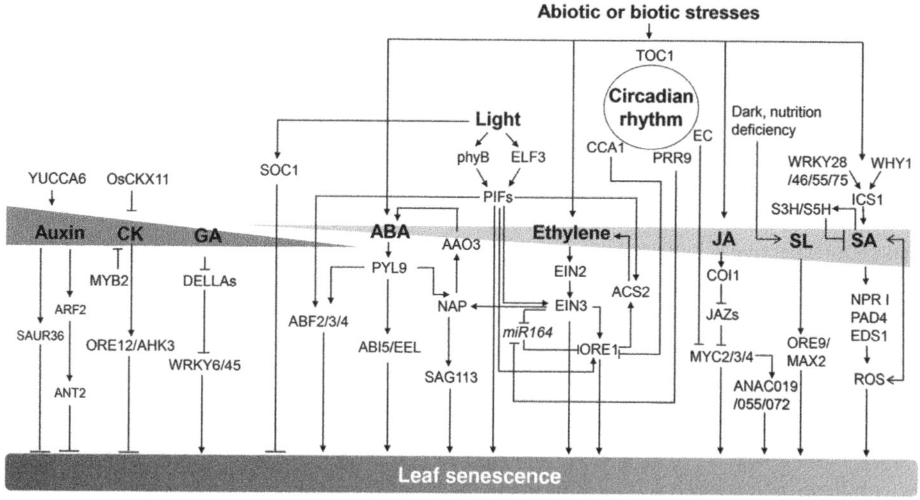

Figure 9.2 Hormonal and environmental regulation of leaf senescence. *Source:* Guo et al. (2021) / Springer Nature / CC BY 4.0.

9.4 Effects of Classical Plant Hormones in Plant Senescence

Phytohormones integrate developmental and environmental information to initiate, pace, and complete leaf senescence (Guo et al. 2021; Huang et al. 2022). Auxin, CKs, and GAs generally delay senescence, whereas ethylene, ABA, SA, JA, BRs, and SLs predominantly promote it (Zhu et al. 2015; Mao et al. 2017; Guo et al. 2021; Huang et al. 2022). Hormones shape senescence by (i) influencing leaf growth state; (ii) modulating signaling and gene expression during the senescence program; and (iii) translating environmental inputs into developmental outputs. Peptide hormones add a further regulatory tier (Aghdam et al. 2021; Zhang et al. 2022; Huang et al. 2022).

9.4.1 Plant Hormones That Delay Leaf Senescence

9.4.1.1 Auxin

Auxin coordinates leaf initiation and morphogenesis and is broadly considered a negative regulator of senescence, though context matters (Vanneste and Friml 2009). IAA levels can double in senescent leaves compared with young tissue, while exogenous IAA suppresses SAGs (van der Graaff et al. 2006), implying a restraining role. Time-resolved analyses indicate that IAA can also advance late-stage transcriptional changes depending on developmental timing and cross talk with ethylene and JA (Goren-Saglam et al. 2020; Guo et al. 2021; Huang et al. 2022). Enhanced IAA biosynthesis via YUCCA6 overexpression lowers ROS and SAG expression and postpones senescence (Kim et al. 2007, 2020; Cha et al. 2016). Early auxin-responsive *SAUR* genes modulate senescence: saur36 loss-of-function delays, whereas overexpression accelerates senescence; SAUR49 promotes SARK-mediated signaling while repressing SSPP (Xu et al. 2011; Wen et al. 2020). Collectively, auxin likely delays senescence by counteracting senescence-promoting pathways and by integrating with environmental inputs.

9.4.1.2 Gibberellins

GAs regulate elongation, expansion, dormancy release, and stress responses (Gao and Chu 2020). GA levels decline with leaf age; exogenous GA_3 slows chlorophyll loss, and GA inhibitors hasten senescence (Li et al. 2010). Upregulation of GA2OX2 during senescence suggests reduced active GA contributes to the onset (van der Graaff et al. 2006). The TCP factor BrTCP21 activates GA biosynthesis and delays senescence (Xiao et al. 2019). GA signaling through GID1 relieves DELLA repression; intriguingly, loss of certain DELLAs causes earlier senescence, underscoring genotype- and context-dependent roles (Hedden and Sponsel 2015; Chen et al. 2014).

9.4.1.3 Cytokinins

CKs antagonize senescence: external CKs delay, while endogenous CKs decline with age (Gan and Amasino 1996; Buchanan-Wollaston et al. 2005). A landmark proSAG12:IPT system dramatically postpones senescence across crops (Gan and Amasino 1995). APT1 converts CK bases to nucleotides and positively regulates senescence; reduced APT1 elevates CKs and delays senescence (Zhang et al. 2013). CKs also influence source–sink relations via cell-wall invertase (CWINV); SAG12-CWINV expression delays senescence (Balibrea Lara et al. 2004). CK–ABA antagonism is evident: CKs suppress ABA biosynthesis genes and promote ABA catabolism, contributing to delayed senescence (Zhang et al. 2021a).

9.4.2 Plant Hormones That Accelerate Leaf Senescence

9.4.2.1 Ethylene

Transcriptome studies show extensive induction of ethylene pathway genes during senescence; chemical inhibition of ethylene formation retards senescence (van der Graaff et al. 2006; Guo et al. 2021; Huang et al. 2022). Disruption of ACS genes (ethylene biosynthesis) markedly delays senescence (Tsuchisaka et al. 2009). Classic signaling mutants (e.g. etr1-1, ein2, ein3, eil1) show prolonged leaf longevity, whereas etr1 ers1 accelerates senescence (Oh et al. 1997; Qu et al. 2007). EIN2 upregulates ORE1/NAC092 and represses miR164, forming an EIN2–EIN3–miR164–ORE1 module that promotes chlorophyll catabolism and senescence, while additional EIN2-dependent routes also contribute (Kim et al. 2009; Li et al. 2013; Qiu et al. 2015; Yang et al. 2020). WRKY71 further enhances EIN2/ORE1 expression (Yu et al. 2021a).

9.4.2.2 Abscisic Acid

ABA integrates development with stress signaling and often accelerates leaf yellowing. Elevated ABA in rice psf and es3 mutants correlates with enhanced SAG expression and PSII damage (Wang et al. 2016; Su et al. 2017). TFs such as CsHB5, OsNAC2, and CDF4 raise ABA biosynthesis, promoting senescence (Xu et al. 2020; Zhang et al. 2021b); OsMYB102/OsNAC2 also impact ABA catabolism (Piao et al. 2019). The ABA–NAP–SAG113 module links ABA to stomatal behavior during senescence (Yang et al. 2014; Zhang and Gan 2012). Curiously, an aba2 mutant with reduced ABA can senesce faster (Song et al.

2016), reflecting ABA's dual protective and promotive roles depending on context. Overexpressing PYL8/9 receptors accelerates ABA- or dark-induced senescence, whereas a pyl duodecuple mutant is ABA-insensitive (Lee et al. 2015; Zhao et al. 2016, 2018). Downstream ABRE TFs (ABF2/3/4) directly activate chlorophyll catabolic genes (Gao et al. 2016). ABA cross talks antagonistically with CKs, e.g. via OsCKX11 upregulation to degrade CKs (Zhang et al. 2021b).

9.4.2.3 Salicylic Acid

SA accumulates during senescence and promotes aging; reducing SA (e.g. NahG or sid2 mutants) delays senescence (Abreu and Munne-Bosch 2009; Zhang et al. 2017). NPR1 and PAD4 are central to SA signaling; PAD4 drives SA accumulation and is required for senescence in saul1 and in aphid-infested leaves (Morris et al. 2000; Makandar et al. 2015; Vogelmann et al. 2012). Nuclear/chloroplast forms of WHY1 balance SA biosynthesis (Lin et al. 2020). SA catabolic enzymes S3H/S5H prevent SA over-accumulation; double loss leads to precocious senescence (Zhang et al. 2017). Numerous WRKY TFs bind the SID2 promoter to enhance SA and accelerate senescence; WRKY46 interacts with NPR1 to induce WRKY6, advancing SA-dependent aging (Tian et al. 2020; Wang et al. 2020; Zhang et al. 2021c). SA and ethylene converge through NPR1–EIN3 cooperation to drive SAG expression (Wang et al. 2021; Yu et al. 2021b).

9.4.2.4 Jasmonic Acid

JAs rise during senescence; LOX-pathway genes display differential regulation (He et al. 2002; Seltmann et al. 2010). KAT2 is strongly induced in natural and dark senescence; lowering KAT2 delays aging (Castillo and Leon 2008). Wheat TaWRKY13-A/42-B enhances LOX expression and JA, promoting senescence; miR139 overexpression delays aging, likely via JA modulation (Schommer et al. 2008; Zhao et al. 2020; Qiao et al. 2021). JA perception via COI1–JAZ enables 26S-proteasome removal of JAZ repressors; coi1 abolishes JA-induced precocious senescence (Castillo and Leon 2008; Li et al. 2021). JAZ7 limits dark-induced senescence by constraining MYC2/3/4; jaz7 mutants yellow rapidly (Yu et al. 2016). MYC factors promote ROS accumulation (e.g. by repressing CAT2) and activate chlorophyll catabolic genes (PAO, NYC1, NYE1); myc2/3/4 show stay-green phenotypes (Zhu et al. 2015; Zhang et al. 2020). Additional regulators (ANAC019/055/072; bHLH IIId) modulate JA outcomes (Qi et al. 2015; Zhu et al. 2015). A MYC2–Dof2.1–MYC2 feed-forward loop operates in dark- and age-dependent senescence; OsERF101 enhances OsMYC2/OsCOI1a, whereas apple MdBT2 antagonizes JA signaling by destabilizing MdMYC2 and stabilizing MdJAZ2 (Zhuo et al. 2020; Lim et al. 2020; An et al. 2021). JA interacts with auxin via WRKY57 and JAZ4/8–IAA29 complexes, and with SA through ESR/ESP–WRKY53 antagonism (Jiang et al. 2014; Miao and Zentgraf 2007).

9.4.2.5 Brassinosteroids

Genetic or biochemical attenuation of BR levels/signaling—det2, UGT73C6 overexpression, P450su1 transgenics, drl1-D, or AtBAT1 overexpression—delays leaf senescence across species (Li et al. 1996; Husar et al. 2011; Dasgupta et al. 2011; Zhu et al. 2013; Han et al. 2017). The bHLH factor AIF2 restrains dark- and BR-induced senescence; BRs reduce AIF2 to promote aging (Kim et al. 2020). BR actions interface with auxin and CK pathways, linking developmental phenotypes to age- and stress-triggered senescence.

9.4.2.6 Strigolactones

SL biosynthesis genes CCD7/CCD8 (MAX3/MAX4 orthologs) and signaling via the F-box MAX2/ORE9 promote senescence; mutants show extended greenness and reduced ion leakage (Liu et al. 2013; Stirnberg et al. 2007; Yan et al. 2007; Ueda and Kusaba 2015). SL actions integrate with ethylene (GR24 amplifies ethylene-driven senescence; ethylene induces MAX3/MAX4) and with nutrient status (phosphate deficiency raises SLs) and carbon starvation (Xu et al. 2021; Yamada and Umehara 2015; Tian et al. 2018). Exogenous sugars can mitigate SL-induced senescence in darkness.

9.5 Plant Peptide Hormones

In addition to the plant hormones, small and lipophilic compounds briefly outlined in the Introduction, both secreted and nonsecreted peptides also mediate cell-to-cell communication in plants (Matsubayashi and Sakagami 2006). They are termed as "small peptides," "small molecule signaling peptides," "peptide signals," "peptide ligands," or "plant peptide hormones, PPH" (Hirakawa et al. 2017; Lu and Xiao 2024; Lalun and Butenko 2025). They may be either secreted or nonsecreted.

9.5.1 Secreted Peptide Hormones

Most PPHs are secreted due to the presence of an N-terminal signal sequence that guides their polypeptide precursor through the endoplasmic reticulum and Golgi to undergo cleavage and modifications, and then through the endomembrane system to be released from the cell (Zhang et al. 2025a). Their receptors are leucine-rich repeat receptor kinases (LRR-RKs), the largest family of transmembrane receptor-like kinases with extracellular LRRs in plants (Hirakawa et al. 2017; Ogawa-Ohnishi et al. 2022). The binding of a peptide hormone allows the receptor to recruit its co-repressor. LRR-RKs may interact with transmembrane LRR proteins lacking a kinase domain or with the kinases without LRR (Hirakawa et al. 2017). Secreted peptide hormones, which make up the majority of PPHs, can be further divided into two major groups as cysteine-rich secreted peptides (CRPs) and peptide hormones with posttranslational modifications (PTMs) (Zhang et al. 2025a). In addition to plants, even phytopathogens can produce plant peptide mimics, as in the case of soybean cyst nematode *Heterodera glycines*, whose CLE homologs aid its parasitic effects (Hirakawa et al. 2017).

CRPs consist of more amino acids (50–100) than other secreted peptides, which usually have 5–20 amino acids (Hirakawa et al. 2017; Zhang et al. 2025a). The cysteine residues of CRPs form disulfide bonds, which are often necessary for peptide stability. Most CRPs, such as plant defensins, are involved in immune responses, however, there are several CRPs, e.g. epidermal patterning factor/EPF-Like, that mediate development (Zhang et al. 2025a). Plant defensins represent an intriguing group of CRPs. Defensins are present in different species, including insects, plants, and mammals, which have three types of defensins.

Their cationic and amphiphilic nature allows them to penetrate the bacterial membranes, contributing to immune defense in mammals (Qi et al. 2025). They were originally called γ-thionins in plants but were later named defensins due to their structural similarities to mammalian defensins. Their 3D structure is constructed by the disulfide bridges (Thomma et al. 2002). In plants, they are expressed continuously and found in a range of tissues. Their main activity is antifungal, yet they provide protection against bacteria and viruses. Their modes of action are disruption of cellular membranes and inhibition of enzymes, ion channels, and protein synthesis (Kovaleva et al. 2020). Most of the plant defensins are upregulated under abiotic stresses (Guo et al. 2025). Defensins from different plant species such as pea (*Pisum sativum*) and alfalfa (*Medicago sativa*) offer treatment strategies against fungal infection (Bogdanov et al. 2023) and multidrug resistance (Pandurangi et al. 2023) and exert selective cytotoxicity against cancer cells (do Amaral et al. 2020). In all these cases, glucosylceramide was a common target. The therapeutic effects of plant defensins in mammalian systems occur locally and without toxicity (do Amaral et al. 2020).

PTM peptides are better characterized than CRPs and their most prominent members are CLAVATA3 (CLV3)/embryo surrounding region-related peptide (CLE), phytosulfokine (PSK), and plant peptide containing sulfated tyrosine (PSY) (Lu and Xiao 2024). They obtain their stability through various PTMs, e.g. tyrosine sulfation, proline hydroxylation, and hydroxyproline arabinosylation (Matsubayashi 2011; Hirakawa et al. 2017; Zhang et al. 2025a). PSKs are pentapeptides and were first characterized as cell growth inducers. Their precursors are 80–120 amino acids long; they undergo tyrosine sulfation catalyzed by tyrosylprotein sulfotransferases (TPST) in the Golgi and are cleaved by subtilisin serine proteases in the apoplast to generate the mature peptides. PSKs promote not only cell division and elongation but also enhance resistance to biotic and abiotic stresses (Li et al. 2024b). Regulation of subtilisin might be more important for the phenotype than the expression of PSKs (Lalun and Butenko 2025).

9.5.2 Nonsecreted Peptide Hormones

Knowledge about nonsecreted peptides is more limited than that of secreted peptides. Certain peptides, e.g. plant elicitor peptide (PEP) and systemin, lack the sequences necessary for protein secretion and function within the cell, although they can be released from the cell through different mechanisms (Lu and Xiao 2024; Zhang et al. 2025a,b). Due to their roles in the immune system, these two peptides, along with others, are called damage-associated molecular patterns or immunomodulatory peptides (Hander et al. 2019; Zhang et al. 2025a,b). They can also harbor PTMs (Li et al. 2024a). The best characterized group among the nonsecreted peptides is the PEP family. They exhibit sequence diversity and were first identified in *A. thaliana* as AtPEP1 (Zelman and Berkowitz 2023). Under normal conditions, PEP1 is anchored to the vacuolar membrane. An increase in cytoplasmic Ca^{2+} concentration following a cellular injury activates METACASPASE4 to process PROPEP, the precursor of PEP1, to its mature form. Then, the mature PEP1 is released from the cell, whose membrane integrity is compromised, and binds to its receptor on neighboring cells to activate a defense response (Hander et al. 2019; Deng et al. 2025). The expression of PROPEP1 is induced upon perception of bacterial infection (Bartels et al. 2013). PEPs confer resistance to biotic, as well as abiotic stresses (Zhang et al. 2025a,b) and regulate the root vascular system (Dhar et al. 2021).

9.6 Effect of PPHs in Plant Senescence

Interactions among plant cells are fundamental for coordinating growth and development in multicellular organisms. In plants, such intercellular communication is largely mediated by phytohormones, which regulate a wide range of physiological and developmental processes. Recent findings, however, have revealed that in addition to traditional hormones, numerous small polypeptide-based signaling molecules play indispensable roles in mediating cellular communication (Matsubayashi and Sakagami 2006; Grienenberg and Fletcher 2015). These peptide signals are involved not only in developmental regulation but also in the modulation of plant responses to environmental challenges (Matsubayashi and Sakagami 2006; Marmiroli and Maestri 2014; Grienenberg and Fletcher 2015). For instance, systemin has been reported to enhance the synthesis of proteinase inhibitors I and II, thereby contributing to wound signaling in leaf tissues (Pearce et al. 1991; Lee and Howe 2003). Additionally, the rapid alkalinization factor (RALF) peptides have been implicated in diverse physiological processes, including root morphogenesis, pollen tube elongation, and adaptive responses to salinity stress (Pearce et al. 2001; Mecchia et al. 2017; Zhu et al. 2020; Zhao et al. 2021; Zhong et al. 2022). Collectively, these observations demonstrate that peptide molecules are critical regulators of plant metabolic pathways and justify their designation as peptide hormones due to their hormone-like signaling properties. Despite the established roles of these molecules in growth and developmental control, their contribution to plant aging processes—particularly leaf senescence—remains insufficiently characterized.

Although studies addressing the connection between peptide hormones and leaf senescence are still limited, research in this field has recently intensified. Elevated levels of PSK have been shown to mitigate heat-induced senescence in seedlings (Yamakawa et al. 1999). Biochemical analyses revealed that PSK signaling is perceived through the membrane-localized leucine-rich repeat receptor kinase PSKR1 (Matsubayashi et al. 2002). The simultaneous loss of PSKR1 and its homologs in *Arabidopsis thaliana* results in a pleiotropic phenotype characterized by reduced root elongation, smaller leaves, and premature onset of senescence (Matsubayashi et al. 2006; Amano et al. 2007).

Among the various proteolytic enzymes associated with the processing of peptide hormones (PPHs), subtilases (SBTs) represent a major class (Schaller and Ryan 1994; Schaller 2004; Djordjevic et al., 2011). Unlike S1P/SBT6.1 and SBT6.2, which function as preprocessing enzymes, most of the 56 identified *Arabidopsis* SBTs are secreted into the extracellular matrix, where they colocalize with peptide precursors either within the secretory pathway or in the cell wall (Rautengarten et al. 2005; Schaller et al. 2018). These enzymes serve as essential regulators across the plant life cycle, influencing embryogenesis, seed maturation and germination, epidermal differentiation, vascular development, organ abscission, programmed cell death, senescence, and the plant's adaptive responses to both biotic and abiotic stimuli (Schaller et al. 2018).

Loss of function in TPST results in pronounced developmental abnormalities, including dwarfism, shortened roots and hypocotyls, chlorotic leaves, impaired venation, accelerated senescence, reduced reproductive output, defective pollen tube guidance, and weakened pathogen resistance (Komori et al. 2009; Stührwohldt et al. 2011, 2015; Zhang et al. 2018). Similarly, hydroxyproline arabinosylation—catalyzed by hydroxyproline *O*-arabinosyltransferase (HPAT)—constitutes a common PTM in cell wall-associated peptides and growth regulators (Kieliszewski et al. 2011). This enzyme catalyzes the initial step of sequential arabinosylation, generating a characteristic triarabinoside chain linked β-1,2 to hydroxyproline residues. Simultaneous disruption of multiple *hpat* genes leads to abnormalities in hypocotyl elongation, altered cell wall thickness, delayed flowering, premature senescence, and impaired pollen tube elongation (Ogawa-Ohnishi et al. 2013).

Recent discoveries have expanded our understanding of how CLE peptides regulate the senescence of leaves. Mutant plants lacking CLE14 or CLE42 exhibit early senescence, whereas transgenic lines overexpressing these peptides display delayed aging. Mechanistic studies have shown that CLE14 mitigates senescence progression by reducing the accumulation of ROS. In contrast, CLE42 interferes with ethylene-mediated signaling by suppressing ethylene biosynthesis in young leaves, increasing the abundance of EIN3-BINDING F-BOX PROTEIN (EBF1), and promoting the degradation of the ETHYLENE-INSENSITIVE 3 (EIN3) transcription factor. Although these findings clarify parts of the molecular framework, further investigation is required to unravel the receptor-level signaling interactions governing CLE peptide function (Wang et al. 2025). Moreover, under saline stress conditions, AtCLE14 has been shown to regulate leaf senescence through ROS homeostasis. Overexpression of AtCLE14 or its synthetic peptide analog (AtCLE14p) delays senescence under salinity stress, whereas cle14 mutants exhibit accelerated leaf aging under both stress and nonstress conditions (Zhang et al. 2025b).

Similarly, research involving the *Salvia miltiorrhiza* peptide SmCLE25 demonstrated that its overexpression in *Arabidopsis thaliana* not only retards leaf senescence but also enhances drought tolerance (Zhao et al. 2025). Collectively, plant-derived polypeptide hormones represent a distinctive class of signaling molecules crucial for regulating developmental and stress-adaptive responses. Evidence from recent investigations suggests that polypeptide hormones such as CLE14 and CLE42 act as negative modulators of leaf senescence in *Arabidopsis*, emphasizing their involvement in the intricate regulatory networks that determine plant longevity. Furthermore, microRNA-encoded peptides (miPEPs) have recently emerged as novel modulators of plant development by amplifying the transcription of their corresponding primary miRNAs, thereby downregulating specific target genes (Zhao et al. 2024). Despite these advancements, the precise biochemical functions and the broader physiological implications of miPEPs remain to be fully elucidated.

9.7 Conclusions and Future Perspectives

PPHs are products of protein translation and, in addition, exhibit greater diversity than plant hormones, which are produced via metabolic pathways. PPHs are usually present in low concentrations yet are able to affect several processes. For these reasons, manipulation of the expression of PPHs and discovery of novel plant peptides can improve yield. PPHs can also be manipulated to gain the ability to reach longer distances and achieve higher stability, as the majority of them are modified posttranscriptionally and tend to remain in close proximity to the cell of origin. In addition to agronomic uses, they can be utilized as drugs. Leaf senescence, as the last step of plant development, provides nutrients from dying cells to actively growing cells. Suppression or delay of leaf senescence increases stress tolerance in plants. Identification of PPHs involved in leaf senescence and the monitoring and characterization of their structure and molecular functions hold a promising approach for crop improvement.

References

Abreu, M.E. and Munne-Bosch, S. (2009). Salicylic acid deficiency in NahG transgenic lines and sid2 mutants increases seed yield in the annual plant *Arabidopsis thaliana*. *Journal of Experimental Botany* 60: 1261–1271. https://doi.org/10.1093/jxb/ern363.

Aghdam, M.S., Ebrahimi, A., and Sheikh-Assadi, M. (2021). Phytosulfokine alpha (PSKalpha) delays senescence and reinforces SUMO1/SUMO E3 ligase SIZ1 signaling pathway in cut rose flowers (*Rosa hybrida* cv Angelina). *Scientific Reports* 11: 23227. https://doi.org/10.1038/s41598-021-02712-2.

Agüera, E. and De la Haba, P. (2018). Leaf senescence in response to elevated atmospheric CO2 concentration and low nitrogen supply. *Biologia Plantarum* 62: 401–408.

Amano, Y., Tsubouchi, H., Shinohara, H. et al. (2007). Tyrosine-sulfated glycopeptide involved in cellular proliferation and expansion in Arabidopsis. *Proceedings of the National Academy of Sciences of the USA* 104: 18333–18338. https://doi.org/10.1073/pnas.0706403104.

Amaral, V.S.G.D., Santos, S.A.C.S., and de Andrade, P.C. (2020). *Pisum sativum* Defensin 1 eradicates mouse metastatic lung nodules from B16F10 melanoma cells. *International Journal of Molecular Sciences* 21: 2662. https://doi.org/10.3390/ijms21082662.

An, J.P., Wang, X.F., Zhang, X.W. et al. (2021). Apple BT2 protein negatively regulates jasmonic acid-triggered leaf senescence by modulating the stability of MYC2 and JAZ2. *Plant, Cell & Environment* 44: 216–233. https://doi.org/10.1111/pce.13913.

Bajguz, A. and Piotrowska-Niczyporuk, A. (2023). Biosynthetic pathways of hormones in plants. *Metabolites* 13: 884.

Balibrea Lara, M.E., Gonzalez Garcia, M.C., Fatima, T. et al. (2004). Extracellular invertase is an essential component of cytokinin-mediated delay of senescence. *The Plant Cell* 16: 1276–1287. https://doi.org/10.1105/tpc.018929.

Bartels, S., Lori, M., Mbengue, M. et al. (2013). The family of Peps and their precursors in *Arabidopsis*: differential expression and localization but similar induction of pattern-triggered immune responses. *Journal of Experimental Botany* 64: 5309–5321. https://doi.org/10.1093/jxb/ert330.

Bayliss, W.M. and Starling, E.H. (1902). On the causation of the so-called periferal reflex secretion of the pancreas. Preliminary communication. *Proceedings of the Royal Society London* 69: 352–353.

Blázquez, M.A. (2024). Polyamines: their role in plant development and stress. *Annual Review of Plant Biology* 75 (1): 95–117. https://doi.org/10.1146/annurev-arplant-070623-110056.

Bleecker, A.B. and Kende, H. (2000). Ethylene: A gaseous signal molecule in plants. *Annual Review of Cell and Developmental Biology* 16: 1–18. https://doi.org/10.1146/annurev.cellbio.16.1.1.

Bogdanov, I.V., Fateeva, S.I., Voropaev, A.D. et al. (2023). Immunomodulatory effects of the pea defensin Psd1 in the Caco-2/immune cells co-culture upon *Candida albicans* infection. *International Journal of Molecular Sciences* 24 (9): 7712. https://doi.org/10.3390/ijms24097712.

Buchanan-Wollaston, V., Page, T., Harrison, E. et al. (2005). Comparative transcriptome analysis reveals significant differences in gene expression and signalling pathways between developmental and dark/starvation-induced senescence in *Arabidopsis*. *The Plant Journal* 42: 567–585. https://doi.org/10.1111/j.1365-313X.2005.02399.x.

Castillo, M.C. and Leon, J. (2008). Expression of the beta-oxidation gene 3-ketoacyl-CoA thiolase 2 (KAT2) is required for the timely onset of natural and dark-induced leaf senescence in *Arabidopsis*. *Journal of Experimental Botany* 59: 2171–2179. https://doi.org/10.1093/jxb/ern079.

Castro-Camba, R., Sánchez, C., Vidal, N., and Vielba, J.M. (2022). Plant development and crop yield: the role of gibberellins. *Plants* 11 (19): 2650. https://doi.org/10.3390/plants11192650.

Cha, J.Y., Kim, M.R., Jung, I.J. et al. (2016). The thiol reductase activity of YUCCA6 mediates delayed leaf senescence by regulating genes involved in auxin redistribution. *Frontiers in Plant Science* 7: 626. https://doi.org/10.3389/fpls.2016.00626.

Chen, M., Maodzeka, A., Zhou, L. et al. (2014). Removal of DELLA repression promotes leaf senescence in *Arabidopsis*. *Plant Science* 219–220: 26–34. https://doi.org/10.1016/j.plantsci.2013.11.016.

Chen, D., Shao, Q., Yin, L. et al. (2019). Polyamine function in plants: metabolism, regulation on development, and roles in abiotic stress responses. *Frontiers in Plant Science* 9: 1945. https://doi.org/10.3389/fpls.2018.01945.

Chen, K., Li, G.J., Bressan, R.A. et al. (2020). Abscisic acid dynamics, signaling, and functions in plants. *Journal of Integrative Plant Biology* 62: 25–54. https://doi.org/10.1111/jipb.12899.

Clouse, S.D. and Sasse, J.M. (1998). BRASSINOSTEROIDS: essential regulators of plant growth and development. *Annual Review of Plant Physiology and Plant Molecular Biology* 49: 427–451. https://doi.org/10.1146/annurev.arplant.49.1.427.

Dasgupta, K., Ganesan, S., Manivasagam, S., and Ayre, B.G. (2011). A cytochrome P450 monooxygenase commonly used for negative selection in transgenic plants causes growth anomalies by disrupting brassinosteroid signaling. *BMC Plant Biology* 11: 67. https://doi.org/10.1186/1471-2229-11-67.

Deng, Z., Yang, Y., Marowa, P. et al. (2025). Processing of plant peptide signals: critical steps mediated by subtilisins. *Plant Hormones* 1: e003. https://doi.org/10.48130/ph-0025-0004.

Dhar, S., Kim, H., Segonzac, C., and Lee, J.Y. (2021). The danger-associated peptide pep1 directs cellular reprogramming in the *Arabidopsis* root vascular system. *Molecules and Cells* 44: 830–842. https://doi.org/10.14348/molcells.2021.0203.

Djordjevic, M.A., Oakes, M., Wong, C.E. et al. (2011). Border sequences of *Medicago truncatula* CLE36 are specifically cleaved by endoproteases common to the extracellular fluids of *Medicago* and soybean. *Journal of Experimental Botany* 62: 4649–4659. https://doi.org/10.1093/jxb/err185;

Fitting, H. (1909). Die Beeinflussing der Orchideenblüten Durch die Bestäubung und Durch Andere Umstände. *Zeitschrift für Botanik* 1: 1–86.

Fridman, Y. and Savaldi-Goldstein, S. (2013). Brassinosteroids in growth control: how, when and where. *Plant Science* 209: 24–31. https://doi.org/10.1016/j.plantsci.2013.04.002.

Gan, S. and Amasino, R.M. (1995). Inhibition of leaf senescence by autoregulated production of cytokinin. *Science* 270: 1986–1988. https://doi.org/10.1126/science.270.5244.1986.

Gan, S. and Amasino, R.M. (1996). Cytokinins in plant senescence: from spray and pray to clone and play. *BioEssays* 18: 557–565. https://doi.org/10.1002/bies.950180707.

Gao, S. and Chu, C. (2020). Gibberellin metabolism and signaling: targets for improving agronomic performance of crops. *Plant & Cell Physiology* 61: 1902–1911. https://doi.org/10.1093/pcp/pcaa104.

Gao, S., Gao, J., Zhu, X. et al. (2016). ABF2, ABF3, and ABF4 promote ABA-mediated chlorophyll degradation and leaf senescence by transcriptional activation of chlorophyll catabolic genes and senescence-associated genes in *Arabidopsis*. *Molecular Plant* 9: 1272–1285. https://doi.org/10.1016/j.molp.2016.06.006.

Ghorbel, M., Brini, F., Sharma, A. et al. (2021). Role of jasmonic acid in plants: the molecular point of view. *Plant Cell Reports* 40: 1471–1494. https://doi.org/10.1007/s00299-021-02687-4.

Goren-Saglam, N., Harrison, E., Breeze, E. et al. (2020). Analysis of the impact of indole-3-acetic acid (IAA) on gene expression during leaf senescence in *Arabidopsis thaliana*. *Physiology and Molecular Biology of Plants* 26: 733–745. https://doi.org/10.1007/s12298-019-00752-7.

Grienenberger, E. and Fletcher, J.C. (2015). Polypeptide signaling molecules in plant development. *Current Opinion in Plant Biology* 23: 8–14. https://doi.org/10.1016/j.pbi.2014.09.013.

Guo, Y., Ren, G., Zhang, K. et al. (2021). Leaf senescence: progression, regulation, and application. *Molecular Horticulture* 1: 5.

Guo, M., Chen, X., Li, S. et al. (2025). Identification of the plant defensin (MsPDF) gene family in medicago sativa and analysis of expression patterns under abiotic stress. *Plants* 14 (9): 1312. https://doi.org/10.3390/plants14091312.

Hafeez, M.B., Zahra, N., Zahra, K. et al. (2021). Brassinosteroids: molecular and physiological responses in plant growth and abiotic stresses. *Plant Stress* 2: 100029. https://doi.org/10.1016/j.stress.2021.100029.

Han, Y.J., Kim, Y.S., Hwang, O.J. et al. (2017). Overexpression of *Arabidopsis thaliana* brassinosteroid-related acyltransferase 1 gene induces brassinosteroid-deficient phenotypes in creeping bentgrass. *PLoS One* 12: e0187378. https://doi.org/10.1371/journal.pone.0187378.

Hander, T., Fernández-Fernández, Á.D., Kumpf, R.P. et al. (2019). Damage on plants activates Ca2+-dependent metacaspases for release of immunomodulatory peptides. *Science* 363: eaar7486. https://doi.org/10.1126/science.aar7486.

He, Y., Fukushige, H., Hildebrand, D.F., and Gan, S. (2002). Evidence supporting a role of jasmonic acid in *Arabidopsis* leaf senescence. *Plant Physiology* 128: 876–884.

Hedden, P. and Sponsel, V. (2015). A century of gibberellin research. *Journal of Plant Growth Regulation* 34: 740–760. https://doi.org/10.1007/s00344-015-9546-1.

Hirakawa, Y., Torii, K.U., and Uchida, N. (2017). Mechanisms and strategies shaping plant peptide hormones. *Plant & Cell Physiology* 58 (8): 1313–1318. https://doi.org/10.1093/pcp/pcx069.

Huang, P., Li, Z., and Guo, H. (2022). New advances in the regulation of leaf senescence by classical and peptide hormones. *Frontiers in Plant Science* 13: 923136. https://doi.org/10.3389/fpls.2022.923136.

Husar, S., Berthiller, F., Fujioka, S. et al. (2011). Overexpression of the UGT73C6 alters brassinosteroid glucoside formation in *Arabidopsis thaliana*. *BMC Plant Biology* 11: 51. https://doi.org/10.1186/1471-2229-11-51.

Janeeshma, E., Habeeb, H., Shackira, A.M. et al. (2024). Strigolactone and analogues: a new generation of plant hormones with multifactorial benefits in environmental sustainability. *Environmental and Experimental Botany* 223: 105775.

Jiang, Y., Liang, G., Yang, S., and Yu, D. (2014). *Arabidopsis* WRKY57 functions as a node of convergence for jasmonic acid- and auxin-mediated signaling in jasmonic acid-induced leaf senescence. *Plant Cell* 26: 230–245. https://doi.org/10.1105/tpc.113.117838.

Kartal, G., Temel, A., Arican, E., and Gozukirmizi, N. (2009). Effects of brassinosteroids on barley root growth, antioxidant system and cell division. *Plant Growth Regulation* 58: 261–267. https://doi.org/10.1007/s10725-009-9374-z.

Kavi Kishor, P.B., Tiozon, R.N. Jr., Fernie, A.R., and Sreenivasulu, N. (2022). Abscisic acid and its role in the modulation of plant growth, development, and yield stability. *Trends in Plant Science* 27 (12): 1283–1295. https://doi.org/10.1016/j.tplants.2022.08.013.

Kieliszewski M. J., Lamport D. T. A., Tan L., Cannon M. C. (2011). Hydroxyproline-rich glycoproteins: form and function" in *Annual plant reviews: plant polysaccharides, biosynthesis and bioengineering*. Vol. 41. ed. Ulvskov P. (Blackwell), 321–342.

Kim, J.I., Sharkhuu, A., Jin, J.B. et al. (2007). yucca6, a dominant mutation in *Arabidopsis*, affects auxin accumulation and auxin-related phenotypes. *Plant Physiology* 145: 722–735. https://doi.org/10.1104/pp.107.104935.

Kim, J.H., Woo, H.R., Kim, J. et al. (2009). Trifurcate feed-forward regulation of age-dependent cell death involving miR164 in *Arabidopsis*. *Science* 323: 1053–1057. https://doi.org/10.1126/science.1166386.

Kim, Y., Park, S.U., Shin, D.M. et al. (2020). ATBS1-INTERACTING FACTOR 2 negatively regulates dark- and brassinosteroid-induced leaf senescence through interactions with INDUCER OF CBF EXPRESSION 1. *Journal of Experimental Botany* 71: 1475–1490. https://doi.org/10.1093/jxb/erz533.

Komori, R., Amano, Y., Ogawa-Ohnishi, M., and Matsubayashi, Y. (2009). Identification of tyrosylprotein sulfotransferase in Arabidopsis. *Proceedings of the National Academy of Sciences of the USA*. 106: 15067–15072. https://doi.org/10.1073/pnas.0902801106.

Kovaleva, V., Bukhteeva, I., Kit, O.Y., and Nesmelova, I.V. (2020). Plant defensins from a structural perspective. *International Journal of Molecular Sciences* 21 (15): 5307. https://doi.org/10.3390/ijms21155307.

Lalun, V.O. and Butenko, M.A. (2025). Plant peptide ligands as temporal and spatial regulators. *Annual Review of Plant Biology* 76: 20.1–20.25. https://doi.org/10.1146/annurev-arplant-070324-041348.

Lee, G.I. and Howe, G.A. (2003). The tomato mutant *spr1* is defective in systemin perception and the production of a systemic wound signal for defense gene expression. *Plant Journal* 33: 567–576. https://doi.org/10.1046/j.1365-313X.2003.01646.x.

Lee, H.N., Lee, K.H., and Kim, C.S. (2015). Abscisic acid receptor PYRABACTIN RESISTANCE-LIKE 8, PYL8, is involved in glucose response and dark-induced leaf senescence in *Arabidopsis*. *Biochemical and Biophysical Research Communications* 463: 24–28. https://doi.org/10.1016/j.bbrc.2015.05.010.

Li, J., Nagpal, P., Vitart, V. et al. (1996). A role for brassinosteroids in light-dependent development of *Arabidopsis*. *Science* 272: 398–401. https://doi.org/10.1126/science.272.5260.398.

Li, J.R., Yu, K., Wei, J.R. et al. (2010). Gibberellin retards chlorophyll degradation during senescence of *Paris polyphylla*. *Biologia Plantarum* 54: 395–399. https://doi.org/10.1007/s10535-010-0072-5.

Li, Z., Peng, J., Wen, X., and Guo, H. (2013). Ethylene-insensitive3 is a senescence-associated gene that accelerates age-dependent leaf senescence by directly repressing miR164 transcription in *Arabidopsis*. *Plant Cell* 25: 3311–3328. https://doi.org/10.1105/tpc.113.113340.

Li, Z., Zhang, Y., Zou, D. et al. (2020). LSD 3.0: a comprehensive resource for the leaf senescence research community. *Nucleic Acids Research* 48: D1069–D1075.

Li, W., Ye, T., Ye, W. et al. (2024a). S-acylation of a non-secreted peptide controls plant immunity via secreted-peptide signal activation. *EMBO Reports* 25: 489–505. https://doi.org/10.1038/s44319-023-00029-x.

Li, Y., Di, Q., Luo, L., and Yu, L. (2024b). Phytosulfokine peptides, their receptors, and functions. *Frontiers in Plant Science* 14: 1326964. https://doi.org/10.3389/fpls.2023.1326964.

Lim, C., Kang, K., Shim, Y. et al. (2020). Rice ETHYLENE RESPONSE FACTOR 101 promotes leaf senescence through jasmonic acid-mediated regulation of OsNAP and OsMYC2. *Frontiers in Plant Science* 11: 1096. https://doi.org/10.3389/fpls.2020.01096.

Lim, P. O., Woo, H. R., and Nam, H. G. (2003). Molecular genetics of leaf senescence in Arabidopsis. *Trends in Plant Science* 8, 272–278. doi: https://doi.org/10.1016/s1360-1385(03)00103-1

Lin, W., Zhang, H., Huang, D. et al. (2020). Dual-localized WHIRLY1 affects salicylic acid biosynthesis via coordination of ISOCHORISMATE SYNT HASE1, PHENYLALANINE AMMONIA LYASE1 and S-ADENOSYL-L-METHIONINE-DEPENDENT METHYLTRANSFERASE1. *Plant Physiology* 184: 1884–1899.

Liu, J., Novero, M., Charnikhova, T. et al. (2013). Carotenoid cleavage dioxygenase 7 modulates plant growth, reproduction, senescence, and determinate nodulation in the model legume *Lotus japonicus*. *Journal of Experimental Botany* 64: 1967–1981. https://doi.org/10.1093/jxb/ert056.

Lu, S. and Xiao, F. (2024). Small peptides: orchestrators of plant growth and developmental processes. *International Journal of Molecular Sciences* 25 (14): 7627. https://doi.org/10.3390/ijms25147627.

Makandar, R., Nalam, V.J., Chowdhury, Z. et al. (2015). The combined action of enhanced disease susceptibility1, phytoalexin deficient4, and senescence-associated101 promotes salicylic acid-mediated defenses to limit *Fusarium graminearum* infection in *Arabidopsis thaliana*. *Molecular Plant-Microbe Interactions* 28: 943–953. https://doi.org/10.1094/MPMI-04-15-0079-R.

Mao, C., Lu, S., Lv, B. et al. (2017). A rice NAC transcription factor promotes leaf senescence via ABA biosynthesis. *Plant Physiology* 174: 1747–1763. https://doi.org/10.1104/pp.17.00542.

Marmiroli, N. and Maestri, E. (2014). Plant peptides in defense and signaling. *Peptides* 56: 30–44. https://doi.org/10.1016/j.peptides.2014.03.013.

Matsubayashi, Y., Ogawa, M., Morita, A., and Sakagami, Y. (2002). An LRR receptor kinase involved in perception of a peptide plant hormone, phytosulfokine. *Science* 296: 1470–1472. https://doi.org/10.1126/science.1069607.

Matsubayashi, Y., Ogawa, M., Kihara, H. et al. (2006). Disruption and overexpression of Arabidopsis phytosulfokine receptor gene affects cellular longevity and potential for growth. *Plant Physiology* 142: 45–53. https://doi.org/10.1104/pp.106.081109.

Matsubayashi, Y. (2011). Post-translational modifications in secreted peptide hormones in plants. *Plant and Cell Physiology* 52 (1): 5–13. https://doi.org/10.1093/pcp/pcq169.

Matsubayashi, Y. and Sakagami, Y. (2006). Peptide hormones in plants. *Annual Review of Plant Biology* 57: 649–674. https://doi.org/10.1146/annurev.arplant.56.032604.144204.

Mecchia, M. A., Santos-, G., Duss, N. N., Somoza, S. C., Boisson-Dernier, A., Gagliardini, V. Martinez-Bernardini A, Ndinyanka-Fabrice T, Ringli C, Grossniklaus U (2017) RALF4/19 peptides interact with LRX proteins to control pollen tube growth in arabidopsis. *Science*, 358 (6370), 1600–1603. https://doi.org/10.1126/science.aao5467

Miao, Y. and Zentgraf, U. (2007). The antagonist function of *Arabidopsis* WRKY53 and ESR/ESP in leaf senescence is modulated by the jasmonic and salicylic acid equilibrium. *Plant Cell* 19: 819–830. https://doi.org/10.1105/tpc.106.042705.

Morris, K., MacKerness, S.A., Page, T. et al. (2000). Salicylic acid has a role in regulating gene expression during leaf senescence. *The Plant Journal* 23: 677–685. https://doi.org/10.1046/j.1365-313x.2000.00836.x.

Noodén LD. The phenomena of senescence and aging. In: Noodén LD, Leopold AC, eds. Senescence and aging in plants. 1988, *San Diego*, CA: Academic Press, Inc., 1–50.

Ogawa-Ohnishi, M., Matsushita, W., and Matsubayashi, Y. (2013). Identification of three hydroxyproline O-arabinosyltransferases in *Arabidopsis thaliana*. *Nature Chemical Biology* 9: 726–730. https://doi.org/10.1038/nchembio.1351.

Ogawa-Ohnishi, M., Yamashita, T., Kakita, M. et al. (2022). Peptide ligand-mediated trade-off between plant growth and stress response. *Science* 378 (6616): 175–180. https://doi.org/10.1126/science.abq5735.

Oh, S.A., Park, J.H., Lee, G.I. et al. (1997). Identification of three genetic loci controlling leaf senescence in *Arabidopsis thaliana*. *The Plant Journal* 12: 527–535. https://doi.org/10.1111/j.0960-7412.1997.00527.x.

Paál, A. (1919). Über Phototropische Reizleitung. *Jahrbuch für Wissenschaften Botanik* 58: 406–458.

Pandurangi, R., Karwa, A., Sagaram, U.S. et al. (2023). Medicago Sativa Defensin1 as a tumor sensitizer for improving chemotherapy: translation from anti-fungal agent to a potential anti-cancer agent. *Frontiers in Oncology* 13: 1141755. https://doi.org/10.3389/fonc.2023.1141755.

Pearce, G., Strydom, D., Johnson, S., and Ryan, C.A. (1991). A polypeptide from tomato leaves induces wound-inducible proteinase inhibitor proteins. *Science* 253: 895–897. https://doi.org/10.1126/science.253.5022.895.

Pearce, G., Moura, D.S., Stratmann, J., and Ryan, C.A. Jr. (2001). RALF, a 5-kDa ubiquitous polypeptide in plants, arrests root growth and development. *Proceedings of the National Academy of Sciences of the USA* 98: 12843–12847. https://doi.org/10.1073/pnas.201416998.

Pennell, R.I. and Lamb, C. (1997). Programmed cell death in plants. *The Plant Cell* 9: 1157–1168. https://doi.org/10.1105/tpc.9.7.1157.

Piao, W., Kim, S.H., Lee, B.D. et al. (2019). Rice transcription factor OsMYB102 delays leaf senescence by downregulating abscisic acid accumulation and signaling. *Journal of Experimental Botany* 70: 2699–2715. https://doi.org/10.1093/jxb/erz095.

Qi, T., Wang, J., Huang, H. et al. (2015). Regulation of jasmonate-induced leaf senescence by antagonism between bHLH Subgroup IIIe and IIId factors in *Arabidopsis*. *The Plant Cell* 27: 1634–1649. https://doi.org/10.1105/tpc.15.00110.

Qi, T., Yang, Q., Qin, H. et al. (2025). Defensin: the immune system regulatory factor against peripheral nerve disease. *Brain-X* 3: e70022. https://doi.org/10.1002/brx2.70022.

Qiao, H., Liu, Y., Cheng, L. et al. (2021). TaWRKY13-A serves as a mediator of jasmonic acid-related leaf senescence by modulating jasmonic acid biosynthesis. *Frontiers in Plant Science* 12: 717233. https://doi.org/10.3389/fpls.2021.717233.

Qiu, K., Li, Z., Yang, Z. et al. (2015). EIN3 and ORE1 accelerate degreening during ethylene-mediated leaf senescence by directly activating chlorophyll catabolic genes in *Arabidopsis*. *PLoS Genetics* 11: e1005399. https://doi.org/10.1371/journal.pgen.1005399.

Qu, X., Hall, B.P., Gao, Z., and Schaller, G.E. (2007). A strong constitutive ethylene-response phenotype was conferred on *Arabidopsis* plants containing null mutations in the ethylene receptors ETR1 and ERS1. *BMC Plant Biology* 7: 3. https://doi.org/10.1186/1471-2229-7-3.

Raskin, I. (1992). Role of salicylic acid in plants. *Annual Review of Plant Biology* 43: 439–463. https://doi.org/10.1146/annurev.pp.43.060192.002255.

Rautengarten, C., Steinhauser, D., Büssis, D. et al. (2005). Inferring hypotheses on functional relationships of genes: analysis of the Arabidopsis thaliana subtilase gene family. *PLoS Comput. Biol.* 1: e40. https://doi.org/10.1371/journal.pcbi.0010040.

Rehman, M., Saeed, M.S., Fan, X. et al. (2023). The multifaceted role of jasmonic acid in plant stress mitigation: an overview. *Plants* 12 (23): 3982. https://doi.org/10.3390/plants12233982.

Rivas-San Vicente, M. and Plasencia, J. (2011). Salicylic acid beyond defence: its role in plant growth and development. *Journal of Experimental Botany* 62: 3321–3338. https://doi.org/10.1093/jxb/err031.

Sade, N., Del Mar Rubio-Wilhelmi, M., Umnajkitikorn, K., and Blumwald, E. (2018). Stress-induced senescence and plant tolerance to abiotic stress. *Journal of Experimental Botany* 69: 845–853.

Santner, A., Calderon-Villalobos, L., and Estelle, M. (2009). Plant hormones are versatile chemical regulators of plant growth. *Nature Chemical Biology* 5: 301–307. https://doi.org/10.1038/nchembio.165.

Schaller, A. and Ryan, C.A. (1994). Identification of a 50-kDa systemin-binding protein in tomato plasma membranes having Kex2p-like properties. *Proceedings of the National Academy of Sciences of the USA* 91: 11802–11806. https://doi.org/10.1073/pnas.91.25.11802.

Schaller, A. (2004). A cut above the rest : the regulatory function of plant proteases. *Planta* 220: 183–197. https://doi.org/10.1007/s00425-004-1407-2.

Schaller, A., Stintzi, A., Rivas, S. et al. (2018). From structure to function-a family portrait of plant subtilases. *New Phytologist* 218: 901–915. https://doi.org/10.1111/nph.14582.

Schaller, G.E., Street, I.H., and Kieber, J.J. (2014). Cytokinin and the cell cycle. *Current Opinion in Plant Biology* 21: 7–15. https://doi.org/10.1016/j.pbi.2014.05.015.

Schommer, C., Palatnik, J.F., Aggarwal, P. et al. (2008). Control of jasmonate biosynthesis and senescence by miR319 targets. *PLoS Biology* 6: e230. https://doi.org/10.1371/journal.pbio.0060230.

Seltmann, M.A., Stingl, N.E., Lautenschlaeger, J.K. et al. (2010). Differential impact of lipoxygenase 2 and jasmonates on natural and stress-induced senescence in *Arabidopsis*. *Plant Physiology* 152: 1940–1950. https://doi.org/10.1104/pp.110.153114.

Sobieszczuk-Nowicka, E., Paluch-Lubawa, E., Mattoo, A.K. et al. (2019). Polyamines – a new metabolic switch: crosstalk with networks involving senescence, crop improvement, and mammalian cancer therapy. *Frontiers in Plant Science* 10: 859. https://doi.org/10.3389/fpls.2019.00859.

Song, Y., Xiang, F., Zhang, G. et al. (2016). Abscisic acid as an internal integrator of multiple physiological processes modulates leaf senescence onset in *Arabidopsis thaliana*. *Frontiers in Plant Science* 7: 181. https://doi.org/10.3389/fpls.2016.00181.

Starling, E.H. (1905). The chemical correlations of the function of the body. *Lancet* 2: 339–341.

Stirnberg, P., Furner, I.J., and Ottoline Leyser, H.M. (2007). MAX2 participates in an SCF complex which acts locally at the node to suppress shoot branching. *The Plant Journal* 50: 80–94. https://doi.org/10.1111/j.1365-313X.2007.03032.x.

Stührwohldt, N., Dahlke, R.I., Steffens, B. et al. (2011). Phytosulfokine-α controls hypocotyl length and cell expansion in *Arabidopsis thaliana* through phytosulfokine receptor 1. *PLoS One* 6: e21054. https://doi.org/10.1371/journal.pone.0021054.

Stuhrwohldt, N., Dahlke, R.I., Kutschmar, A. et al. (2015). Phytosulfokine peptide signaling controls pollen tube growth and funicular pollen tube guidance in *Arabidopsis thaliana*. *Physiologia Plantarum* 153: 643–653. https://doi.org/10.1111/ppl.12270.

Su, Y., Hu, S., Zhang, B. et al. (2017). Characterization and fine mapping of a new early leaf senescence mutant es3(t) in rice. *Plant Growth Regulation* 81: 419–431. https://doi.org/10.1007/s10725-016-0219-2.

Thomma, B.P., Cammue, B.P., and Thevissen, K. (2002). Plant defensins. *Planta* 216: 193–202. https://doi.org/10.1007/s00425-002-0902-6.

Tian, M.Q., Jiang, K., Takahashi, I., and Li, G.D. (2018). Strigolactone-induced senescence of a bamboo leaf in the dark is alleviated by exogenous sugar. *Journal of Pesticide Science* 43: 173–179. https://doi.org/10.1584/jpestics.D18-003.

Tian, T., Ma, L., Liu, Y. et al. (2020). *Arabidopsis* FAR-RED ELONGATED HYPOCOTYL3 integrates age and light signals to negatively regulate leaf senescence. *The Plant Cell* 32: 1574–1588. https://doi.org/10.1105/tpc.20.00021.

Tsuchisaka, A., Yu, G., Jin, H. et al. (2009). A combinatorial interplay among the 1-aminocyclopropane-1-carboxylate isoforms regulates ethylene biosynthesis in *Arabidopsis thaliana*. *Genetics* 183: 979–1003. https://doi.org/10.1534/genetics.109.107102.

Ueda, H. and Kusaba, M. (2015). Strigolactone regulates leaf senescence in concert with ethylene in *Arabidopsis*. *Plant Physiology* 169: 138–147. https://doi.org/10.1104/pp.15.00325.

van der Graaff, E., Schwacke, R., Schneider, A. et al. (2006). Transcription analysis of *Arabidopsis* membrane transporters and hormone pathways during developmental and induced leaf senescence. *Plant Physiology* 141: 776–792. https://doi.org/10.1104/pp.106.079293.

Vanneste, S. and Friml, J. (2009). Auxin: a trigger for change in plant development. *Cell* 136: 1005–1016. https://doi.org/10.1016/j.cell.2009.03.001.

Vogelmann, K., Drechsel, G., Bergler, J. et al. (2012). Early senescence and cell death in *Arabidopsis* saul1 mutants involve the PAD4-dependent salicylic acid pathway. *Plant Physiology* 159: 1477–1487. https://doi.org/10.1104/pp.112.196220.

Wang, F., Liu, J., Chen, M. et al. (2016). Involvement of abscisic acid in PSII photodamage and D1 protein turnover for light-induced premature senescence of rice flag leaves. *PLoS One* 11: e0161203. https://doi.org/10.1371/journal.pone.0161203.

Wang, Y., Cui, X., Yang, B. et al. (2020). WRKY55 transcription factor positively regulates leaf senescence and the defense response by modulating the transcription of genes implicated in the biosynthesis of reactive oxygen species and salicylic acid in *Arabidopsis*. *Development* 147 (16): https://doi.org/10.1242/dev.189647.

Wang, C., Dai, S., Zhang, Z.L. et al. (2021). Ethylene and salicylic acid synergistically accelerate leaf senescence in *Arabidopsis*. *Journal of Integrative Plant Biology* 63: 828–833. https://doi.org/10.1111/jipb.13075.

Wang, S., Liang, Y., and Jiang, Y. (2025). Plant CLE peptides: functions, challenges, and future prospects. *Plant Hormones* 1 (1): e0006.

Wen, Z., Mei, Y., Zhou, J. et al. (2020). SAUR49 can positively regulate leaf senescence by suppressing sspp in *Arabidopsis*. *Plant & Cell Physiology* 61: 644–658. https://doi.org/10.1093/pcp/pcz231.

Weyers, J.D.B. and Paterson, N.W. (2001). Plant hormones and the control of physiological processes. *The New Phytologist* 152 (3): 375–407. https://doi.org/10.1046/j.0028-646X.2001.00281.x.

Woo, H.R., Kim, H.J., Lim, P.O., and Nam, H.G. (2019). Leaf senescence: systems and dynamics aspects. *Annual Review of Plant Biology* 70: 347–376.

Wu, K., Xu, H., Gao, X., and Fu, X. (2021a). New insights into gibberellin signaling in regulating plant growth-metabolic coordination. *Current Opinion in Plant Biology* 63: 102074: https://doi.org/10.1016/j.pbi.2021.102074.

Wu, W., Du, K., Kang, X., and Wei, H. (2021b). The diverse roles of cytokinins in regulating leaf development. *Horticulture Research* 8: 118. https://doi.org/10.1038/s41438-021-00558-3.

Xiao, X.M., Xu, Y.M., Zeng, Z.X. et al. (2019). Activation of the transcription of BrGA20ox3 by a BrTCP21 transcription factor is associated with gibberellin-delayed leaf senescence in Chinese flowering cabbage during storage. *International Journal of Molecular Sciences* 20: https://doi.org/10.3390/ijms20163860.

Xu, F., Meng, T., Li, P. et al. (2011). A soybean dual specificity kinase, GmSARK, and its *Arabidopsis* homolog, AtSARK, regulate leaf senescence through synergistic actions of auxin and ethylene. *Plant Physiology* 157: 2131–2153. https://doi.org/10.1104/pp.111.182899.

Xu, P., Chen, H., and Cai, W. (2020). Transcription factor CDF4 promotes leaf senescence and floral organ abscission by regulating abscisic acid and reactive oxygen species pathways in *Arabidopsis*. *EMBO Reports* 21: e48967. https://doi.org/10.15252/embr.201948967.

Xu, X., Jibran, R., Wang, Y. et al. (2021). Strigolactones regulate sepal senescence in *Arabidopsis*. *Journal of Experimental Botany* 72: 5462–5477. https://doi.org/10.1093/jxb/erab199.

Yamada, Y. and Umehara, M. (2015). Possible roles of strigolactones during leaf senescence. *Plants (Basel).* 4: 664–677. https://doi.org/10.3390/plants4030664.

Yamakawa, S., Matsubayashi, Y., Sakagami, Y. et al. (1999). Promotive effects of the peptidyl plant growth factor, phytosulfokine-alpha, on the growth and chlorophyll content of *Arabidopsis* seedlings under high night-time temperature conditions. *Bioscience, Biotechnology, and Biochemistry* 63: 2240–2243.

Yan, H., Saika, H., Maekawa, M. et al. (2007). Rice tillering dwarf mutant dwarf3 has increased leaf longevity during darkness-induced senescence or hydrogen peroxide-induced cell death. *Genes & Genetic Systems* 82: 361–366. https://doi.org/10.1266/ggs.82.361.

Yang, J., Worley, E., and Udvardi, M. (2014). A NAP-AAO3 regulatory modüle promotes chlorophyll degradation via ABA biosynthesis in *Arabidopsis* leaves. *The Plant Cell* 26: 4862–4874. https://doi.org/10.1105/tpc.114.133769.

Yang, Z., Wang, C., Qiu, K. et al. (2020). The transcription factor ZmNAC126 accelerates leaf senescence downstream of the ethylene signalling pathway in maize. *Plant, Cell & Environment* 43: 2287–2300. https://doi.org/10.1111/pce.13803.

Yang, Y., Chu, C., Qian, Q., and Tong, H. (2024). Leveraging brassinosteroids towards the next Green Revolution. *Trends in Plant Science* 29 (1): 86–98. https://doi.org/10.1016/j.tplants.2023.09.005.

Yu, J., Zhang, Y., Di, C. et al. (2016). JAZ7 negatively regulates dark-induced leaf senescence in *Arabidopsis*. *Journal of Experimental Botany* 67: 751–762. https://doi.org/10.1093/jxb/erv487.

Yu, Y., Qi, Y., Xu, J. et al. (2021a). *Arabidopsis* WRKY71 regulates ethylene-mediated leaf senescence by directly activating EIN2, ORE1 and ACS2 genes. *The Plant Journal* 107: 1819–1836. https://doi.org/10.1111/tpj.15433.

Yu, X., Xu, Y., and Yan, S. (2021b). Salicylic acid and ethylene coordinately promote leaf senescence. *Journal of Integrative Plant Biology* 63: 823–827. https://doi.org/10.1111/jipb.13074.

Zelman, A.K. and Berkowitz, G.A. (2023). Plant elicitor peptide (Pep) signaling and pathogen defense in tomato. *Plants* 12 (15): 2856. https://doi.org/10.3390/plants12152856.

Zhang, K. and Gan, S.S. (2012). An abscisic acid-AtNAP transcription factor-SAG113 protein phosphatase 2C regulatory chain for controlling dehydration in senescing *Arabidopsis* leaves. *Plant Physiology* 158: 961–969. https://doi.org/10.1104/pp.111.190876.

Zhang, X., Chen, Y., Lin, X. et al. (2013). Adenine phosphoribosyl transferase 1 is a key enzyme catalyzing cytokinin conversion from nucleobases to nucleotides in *Arabidopsis*. *Molecular Plant* 6: 1661–1672. https://doi.org/10.1093/mp/sst071.

Zhang, Y., Zhao, L., Zhao, J. et al. (2017). S5H/DMR6 encodes a salicylic acid 5-hydroxylase that fine-tunes salicylic acid homeostasis. *Plant Physiology* 175: 1082–1093. https://doi.org/10.1104/pp.17.00695.

Zhang, H., Hu, Z., Lei, C. et al. (2018). A plant phytosulfokine peptide initiates auxin-dependent immunity through cytosolic Ca^{2+} signaling in tomato. *The Plant Cell* 30: 652–667. https://doi.org/10.1105/tpc.17.00537.

Zhang, Y., Ji, T.T., Li, T.T. et al. (2020). Jasmonic acid promotes leaf senescence through MYC2-mediated repression of CATALASE2 expression in *Arabidopsis*. *Plant Science* 299: 110604. https://doi.org/10.1016/j.plantsci.2020.110604.

Zhang, W., Peng, K., Cui, F. et al. (2021a). Cytokinin oxidase/dehydrogenase OsCKX11 coordinates source and sink relationship in rice by simultaneous regulation of leaf senescence and grain number. *Plant Biotechnology Journal* 19: 335–350. https://doi.org/10.1111/pbi.13467.

Zhang, Y., Zhang, Y., Sun, Q. et al. (2021b). Citrus transcription factor CsHB5 regulates abscisic acid biosynthetic genes and promotes senescence. *The Plant Journal* 108: 151–168. https://doi.org/10.1111/tpj.15431.

Zhang, D., Zhu, Z., Gao, J. et al. (2021c). The NPR1-WRKY46-WRKY6 signaling cascade mediates probenazole/salicylic acid-elicited leaf senescence in *Arabidopsis thaliana*. *Journal of Integrative Plant Biology* 63: 924–936. https://doi.org/10.1111/jipb.13044.

Zhang, Z., Liu, C., Li, K. et al. (2022). CLE14 functions as a "brake signal" to suppress age-dependent and stress-induced leaf senescence by promoting JUB1-mediated ROS scavenging in *Arabidopsis*. *Molecular Plant* 15: 179–188. https://doi.org/10.1016/j.molp.2021.09.006.

Zhang, Z., Han, H., Zhao, J. et al. (2025a). Peptide hormones in plants. *Molecular Horticulture* 5: 7. https://doi.org/10.1186/s43897-024-00134-y.

Zhang, D., Zhu, Q., Qin, P. et al. (2025b). The peptide-encoding CLE25 gene modulates drought response in cotton. *Agriculture* 15 (11): 1226.

Zhao, Y., Chan, Z., Gao, J. et al. (2016). ABA receptor PYL9 promotes drought resistance and leaf senescence. *Proceedings of the National Academy of Sciences of the United States of America*. 113: 1949–1954. https://doi.org/10.1073/pnas.1522840113.

Zhao, Y., Zhang, Z., Gao, J. et al. (2018). *Arabidopsis* duodecuple mutant of PYL ABA receptors reveals PYL repression of ABA-independent SnRK2 activity. *Cell Reports* 23 (11): 3340–3351. e5. https://doi.org/10.1016/j.celrep.2018.05.044.

Zhao, M.M., Zhang, X.W., Liu, Y.W. et al. (2020). A WRKY transcription factor, TaWRKY42-B, facilitates initiation of leaf senescence by promoting jasmonic acid biosynthesis. *BMC Plant Biology* 20: 444. https://doi.org/10.1186/s12870-020-02650-7.

Zhao, C., Jiang, W., Zayed, O. et al. (2021). The LRXs-RALFs-FER module controls plant growth and salt stress responses by modulating multiple plant hormones. *Natl. Sci. Rev.* 8: nwaa149. https://doi.org/10.1093/nsr/nwaa149.

Zhao, Y., Zhang, Y., Li, S. et al. (2024). Leaf senescence database v5. 0: a comprehensive repository for facilitating plant senescence research. *Journal of Molecular Biology* 436 (17): 168530.

Zhao, W., Sun, Y., Li, J. et al. (2025). Smcle25 coordinates stress resistance and leaf senescence inhibition via transcriptional repression of Rboh-mediated oxidative burst. Available at SSRN 5266931.

Zhong, S., Li, L., Wang, Z. et al. (2022). RALF peptide signaling controls the polytubey block in Arabidopsis. *Science* 375: 290–296. https://doi.org/10.1126/science.abl4683.

Zhu, W., Wang, H., Fujioka, S. et al. (2013). Homeostasis of brassinosteroids regulated by DRL1, a putative acyltransferase in *Arabidopsis*. *Molecular Plant* 6: 546–558. https://doi.org/10.1093/mp/sss144.

Zhu, X., Chen, J., Xie, Z. et al. (2015). Jasmonic acid promotes degreening via MYC2/3/4- and ANAC019/055/072-mediated regulation of major chlorophyll catabolic genes. *The Plant Journal* 84: 597–610. https://doi.org/10.1111/tpj.13030.

Zhu, S., Martínez Pacheco, J., Estevez, J.M., and Yu, F. (2020). Autocrine regulation of root hair size by the RALF-FERONIA-RSL4 signaling pathway. *New Phytologist* 227: 45–49. https://doi.org/10.1111/nph.16497.

Zhuo, M., Sakuraba, Y., and Yanagisawa, S. (2020). A jasmonate-activated MYC2-Dof2.1-MYC2 transcriptional loop promotes leaf senescence in *Arabidopsis*. *The Plant Cell* 32: 242–262. https://doi.org/10.1105/tpc.19.00297.

10

Regulation of Symbiotic Relation in Plants by Phytohormones

Weam El-Aggan and Amel Tammam

Department of Botany and Microbiology, Faculty of Science, Alexandria University, Alexandria, Egypt

10.1 Introduction

Plants play an essential role on the Earth by providing environmental stability and human nutrition. The improvement of plant growth and development is one of the most important issues that must be taken into consideration. Nitrogen is a major macronutrient required for plant growth and productivity. However, despite being the most abundant gas in the atmosphere, nitrogen is not readily available to most plants in a usable form because of the inability of plants and most organisms to enzymatically break down the triple bond in atmospheric N_2 and convert it into forms that plant roots can take up as NO^{3-} and NH_4^+ ions (Kouchi 2011).

The major natural source of N_2 input to the biosphere is biological nitrogen fixation (BNF), adding approximately 50–70 tons of nitrogen globally to agricultural systems (Roy et al. 2020). BNF is the conversion of atmospheric N_2 to NH_3 catalyzed by nitrogenase in nitrogen-fixing organisms (majority of *Cyanobacteria* and a few bacteria), which occurs either in a free-living state or in symbiotic associations with other plants. The symbiotic nitrogen fixation (SNF) associations are formed between different nitrogen-fixing organisms and a variety of plant species; the most well-known are the interactions between legume—rhizobia, nonlegume—*Frankia*, and *Azolla*—cyanobacteria (Kouchi 2011).

The plant family Fabaceae (Legumes) is the third-largest family of flowering plants, consisting of ~19,000 known species, 88% of which form SNF with rhizobia, results in the formation of specialized root organs called nodules (Roy et al. 2020), where atmospheric nitrogen is converted into ammonia. Nodule development is a highly orchestrated and energy-intensive process that involves special chemical and molecular signaling between the symbiont and the host plant. In case of nitrogen deficiency, the legume roots release chemical signaling molecules (flavonoids) that are received by rhizobia and enhance the synthesis of Nod factors (NF), which are the key bacterial signaling molecules causing the plant to undergo physiological and developmental changes that lead to thread infection formation, bacterial entry, and nodule organogenesis (Oldroyd et al. 2011). Regulation of

Peptide Hormones in Plant Development and Stress Responses, First Edition. Edited by Vijay Pratap Singh and Samiksha Singh.

this nodulation process requires a sophisticated signaling network. Among the key regulators of nodule formation are phytohormones (plant hormones) and small signaling peptides, which play pivotal roles in the nodulation process, coordinate nodule initiation and development as well as the balance between nodulation and autoregulation of nodulation (AON) (Ferguson et al. 2014).

Two types of nodules are well studied: determinate and indeterminate nodules. The main difference between them is the position of the cortical root cell layers that give rise to initiate the nodule primordium (Hirsch 1992). Many legumes from the family Leguminosae form nodules with a persistent nodule meristem; indeterminate nodules, which are familiar for inverted repeat-lacking clade (IRLC) legumes, as *Pisum sativum*, *Vicia sativa*, and *Medicago truncatula* (Mamaeva et al. 2025), in which cell divisions exist in the inner part of the cortex and pericycle (Downie and Kondorosi 2021). While in determinate nodules as *Lotus japonicus* and *Glycine max*, cell divisions are limited to the middle or outer cortex, respectively (Hirsch 1992).

Plant hormones are plant growth regulators naturally synthesized within plant cells in minuscule concentrations and participate in various aspects of the plant life cycle. Their action takes place either at the local or at distant sites, which is mediated through many transport mechanisms. They are chemical messengers with specific chemical structures and various chemical properties that exert their effects on plant growth and development through a variety of biochemical processes, including plant responses to biotic and abiotic stress (Lianes et al. 2018). Plant hormones such as auxins (IAA), cytokinins (CKs), gibberellins (GAs), abscisic acid (ABA), ethylene, jasmonic acid (JA), strigolactones (SGLs), and BRs act either positive or negative regulatory role in the nodulation process (Suzaki et al. 2015). For instance, CKs are known to boost nodule initiation by enhancing the division of cortical cells, but ethylene mainly acts as a negative regulator by suppressing the early events of symbiosis. Auxins are required together with CKs to regulate nodule organogenesis, affecting nodule patterning and vascular formation (Plet et al. 2011). The crosstalk and cooperation among these phytohormones permit the host plant to fine-tune nodule formation in response to developmental signals and environmental conditions (Plet et al. 2011).

In addition to the traditional phytohormones, small signaling peptide hormones have been identified as essential regulators in legume–rhizobia interaction. Peptide hormones are small secreted polypeptide-based intercellular communication signal molecules (usually <20 amino acids when mature and rarely >120 amino acids as a full-length precursor), encoded by nuclear genes, and causing changes in gene expression and cellular responses (Zhang et al. 2025). Biosynthesis of these hormones often goes through proteolytic cleavage of preproproteins and posttranslational modification. These small secreted peptides, including CLAVATA3/EMBRYO-SURROUNDING REGION-related (CLE), nodule-specific cysteine rich (NCR), rapid alkalinization factor (RALF), and early nodulin40 (ENOD40) peptides, act as mobile signals that orchestrate nodule formation through systemic pathways (Downie and Kondorosi 2021; Mamaeva et al. 2023; Gultyaev et al. 2023). Notably, CLE and C-terminally encoded peptides (CEPs) have been identified as signals involved in regulating the number of nodules through a mechanism known as AON. In this process, CLE peptides induced in the root in response to rhizobial infection are transported to the shoot, where they are sensed by receptor kinases as HAR1, leading to inhibit further nodulation (Mamaeva et al. 2025). On the other hand, CEP peptides can act to stimulate nodulation under nitrogen-shortage conditions, confirming the complicate balance of local and systemic signals that govern nodulation.

Recent investigations have shed light on the interaction between phytohormones and peptides, providing new insights into the regulatory events that control nodule formation. Understanding these mechanisms is essential for developing biotechnological strategies to stimulate BNF, enhance legume productivity, and diminish dependence on chemical fertilizers.

This chapter aims to provide a comprehensive general review of the roles and regulatory mechanisms by which plant hormones and peptides orchestrate nodule initiation, formation, and autoregulation to sustain symbiotic efficiency, setting the stage for deeper exploration of their molecular interactions and potential applications in agriculture.

10.2 Legume–Rhizobia Interaction: General Considerations

The legume symbiosis process is driven by bacteria that possess the key enzyme nitrogenase, which specifically reduces atmospheric N_2 to ammonia in the symbiotic root nodules. These bacteria belong to the genera *Rhizobium*, *Sinorhizobium*, *Bradyrhizobium*, *Mesorhizobium*, and *Azorhizobium* and collectively are termed rhizobia (Karmakar et al. 2015). These bacteria interact with legume roots, leading to the formation of N_2-fixing nodules.

10.2.1 Legume–Rhizobia Signaling During Interaction

The first stage of symbiosis starts with the secretion of flavonoids and related molecules from the host legume seed coats and roots, which activate the regulatory gene *nodD* in rhizobia (Junior et al. 2015). These flavonoids are diverse and vary among legume hosts. Thus, flavonoid perception is one of the first levels of specificity between hosts and nitrogen-fixing rhizobia. Flavonoids possess chemotactic responses and behave as chemoattractants for rhizobia to induce the nodulation gene (Hirsch and Fujishige 2012).

10.2.1.1 Signaling Expressed by Rhizobia

The nodulation genes are not expressed in free-living bacteria, with the exception of *nodD* (regulatory gene), which is constitutively expressed. The regulatory gene encodes the nodD protein, which activates transcription of other genes required for the synthesis of the Nod factor backbone (Roche et al. 1991). NFs are primary signal molecules produced by bacteria and detected by the plant to induce nodule organogenesis. Upon interacting with a flavonoid, the NodD protein changes its conformation so that it can bind to nod boxes in the promoters of nod genes, leading to their expression and ultimately to the synthesis of Nod factor (Figure 10.1), a lipochitinoligosaccharide (LCO) signal molecule (varying in length from three to six sugar units) with various substitutions on the reducing and non-reducing ends (Denarie et al. 1996).

The nodulation genes (nod) are categorized as common nod genes and host-specific nod genes. The common nod genes *nodABC* (*nodA*, *nodB*, and *nodC*) are found in all rhizobial strains, while the host-specific nod genes such as *nodH*, *nodQ*, and *nodP*, or *nodE*, and *nodF* differ among rhizobial species and determine the host range (Roche et al. 1991; Oldroyd

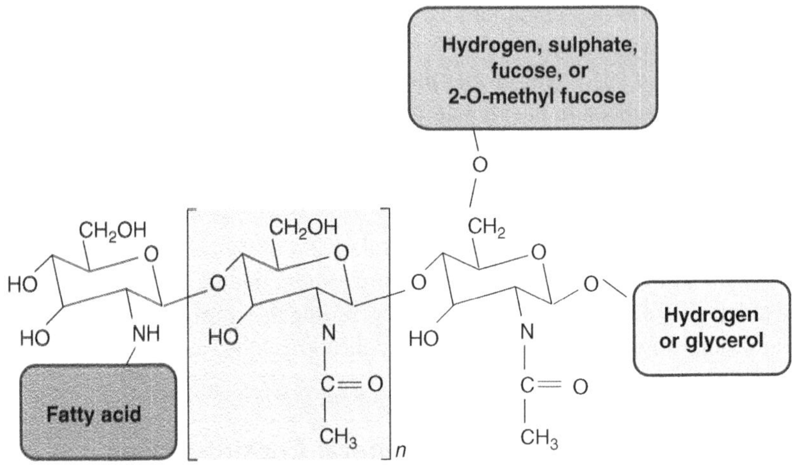

Figure 10.1 Nod factors, the fatty acid chain typically has 16 to 18 carbons. The number of repeated middle sections (*n*) is usually 1 to 4.

2001). The common nod genes are responsible for the production of the *N*-acyl glucosamine oligomer acylated by a default fatty acid, e.g. cis-vaccenic acid, as well as several host specificity nod genes, which are needed for the "molecular decorations" found on both the reducing and nonreducing ends of the molecule (Mus et al. 2016). The bacterial Nod factor perception is mediated by Nod factor receptors (NFRs), which are present in the plant plasma membrane. Hence, NFRs act as host determinants for symbiotic specificity. NFRs are serine/threonine kinases like receptors and contain LysM motifs in their extracellular domains (Arrighi et al. 2006). NFs induce curling of root hairs and allow entry of rhizobia into infection threads.

Host-specific nod genes that vary among rhizobial species are involved in the modification of the fatty acyl chain or the addition of groups important in determining host specificity. NodE and NodF determine the length and degree of saturation of the fatty acyl chain; those of *Rhizobium leguminosarum* and *Rhizobium meliloti* result in the synthesis of an 18:4 and a 16:2 fatty acyl group, respectively (Recall that the number before the colon gives the total number of carbons in the fatty acyl chain, and the number after the colon gives the number of double bonds (Roche et al. 1991). Other enzymes, such as NodL, influence the host specificity of NFs through the addition of specific substitutions at the reducing or nonreducing sugar moieties of the chitin backbone. For example, *Sinorhizobium meliloti* NFs are sulfated on the reducing end of the LCO, enabling these symbionts to nodulate species of *Medicago*, *Melilotus*, and *Trigonella* but not *Pisum*. Proteins that are postulated to function as receptors for Nod factor LysM-RLK/NFP in *M. truncatula* and NFR1/NFR5 in *L. japonicus* have been identified (Jones et al. 2007; Madsen et al. 2010).

A particular legume host responds to a specific Nod factor. The legume receptors for NFs appear to be special lectins (sugar-binding proteins) produced in the root hairs. NFs activate these lectins, increasing their hydrolysis. This lectin activates rhizobia to appropriate hosts and facilitates attachment of the rhizobia to the cell walls of a root hair (Junior et al. 2015). Rhizobia trapped in the root hair curl enter the cell by the degradation of the cell wall and move through the root hair epidermal cell into the cortex by invagination of the plasma membrane, forming an infection thread. The formation of an infection thread requires the expression of ENOD genes and is facilitated by transcription factor signaling in the epidermal root hair cell nucleus (Mus et al. 2016).

10.2.1.2 Signaling Expressed by Legumes

In legume-*Rhizobium* symbioses, the term "nodulin gene" only referred to nodule-specific plant genes. These nodulin genes are grouped according to the time course of their expression into ENODs that are involved in the infection process and root nodule morphogenesis and late nodulins that are expressed following the release of the bacterium from the infection threads, but prior to the commencement of nitrogen fixation (Hirsch et al. 1997).

Amino acid sequence analysis of the ENOD gene (*ENOD2*) has revealed that it encodes the ENOD2 protein, which resembles other plant cell wall proteins. It is likely that ENOD2 is a hydroxyproline-rich glycoprotein. Since the nodule inner cortex has been suggested to be an oxygen diffusion barrier, the presence of the putative cell wall protein ENOD2 may contribute to the morphology of the nodule parenchyma and thus indirectly to the establishment of the oxygen barrier and produced in the invasion zone. Amino acid sequence analysis of other ENODs has suggested that *ENOD3* and *ENOD14* might encode proteins involved in metal transport and that ENOD5 may be a protein in the peribacteroid membrane (Truchet et al. 1991; Hirsch et al. 1989).

Many of the later nodulin genes encode proteins that support some functions, such as (i) Leghemoglobin, the most abundant and important (up to 20%) protein in root nodules. It provides a flux of oxygen to support bacterial respiration at a concentration low enough to prevent the deactivation of nitrogenase. (ii) The ammonia excreted by the bacteroids following the action of nitrogenase is assimilated in the plant cytoplasm by glutamine synthase (GS). Several nodule-specific GS isoenzymes have been identified, providing a rapid removal of the ammonia synthesized by the bacteroids to prevent the inhibition of nitrogenase (Garrocho-Villegas et al. 2007).

10.2.2 Types of Nodules in Legumes

Several types of nodules exist within mature roots of legumes. However, the two principal types are well studied: determinate (spherical nodules lacking a persistent meristem) and indeterminate (containing a persistent meristem resulting in cylindrical nodules) nodules. The main difference between the two types is the position of cortical root cell layers that give rise to initiate the nodule primordium (Hirsch 1992). Many legumes from the family Leguminosae form nodules with a persistent nodule meristem, indeterminate nodules. This nodule type is familiar for IRLC legumes, as *P. sativum*, *V. sativa*, and *M. truncatula* (Figure 10.2; Mamaeva et al. 2025), in which cell divisions exist in the inner part of the cortex and pericycle (Downie and Kondorosi 2021). While in determinate nodules as *L. japonicus* and *G. max*, cell divisions are limited to the middle or outer cortex, respectively (Hirsch 1992). The site of these primary divisions matches the site of auxin signaling in cortical cells, with extra expression in the endodermis and pericycle during initiation of nodules (Mathesius et al. 1998; van Noorden et al. 2007). Both determinate and indeterminate-type nodules contain a central mass of infected cells and peripheral vascular bundles,

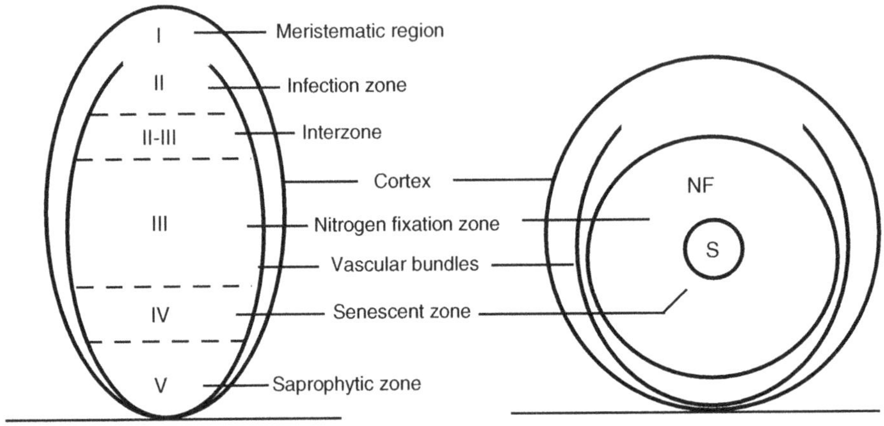

Figure 10.2 Types of nodules in leguminous plants and their structure.

where nitrogen fixation takes place. However, the processes of senescence of nitrogen-fixing cells, nitrogen fixation, and the infection site are separated spatially in indeterminate nodules, whereas this separation does not occur in determinate nodules (Hirsch 1992).

10.3 Roles of Phytohormones During Nodule Formation

Phytohormones play a central role in nodulation by regulating symbiotic interactions between legumes and rhizobia through signaling pathways, modulating root development, and nutrient exchange. The interplay between different hormones ensures the proper establishment and maintenance of these beneficial interactions, allowing plants to adapt to changing environmental conditions and optimize resource acquisition.

Generally, plant hormones have been divided into three categories (Figure 10.3) on the basis of their action on plant growth; group one hormones include auxins, GAs, CKs, and salicylic acid (SA) via their major roles in growth-promoting activities by cell division, elongation, cell differentiation, and enhancing various physiological processes as photosynthesis and nutrient uptake. Hormones in group two are ABA and ethylene; mainly promoting stomatal closure, inducing dormancy, promoting leaf abscission, and flower wilting under stress environmental conditions, such as light intensity, soil salinity, soil water conditions, and nutrient availability (Bari and Jones 2009). The third hormone group comprises brassinosteroids (BRs), SGLs, and jasmonates, which play a role in growth regulation; they inhibit growth under stress but promote plant growth under normal conditions by regulating various physiological processes, such as shoot branching, root development, vascular differentiation, and reproductive development.

Below is an overview of how key plant hormones regulate these symbiotic relationships.

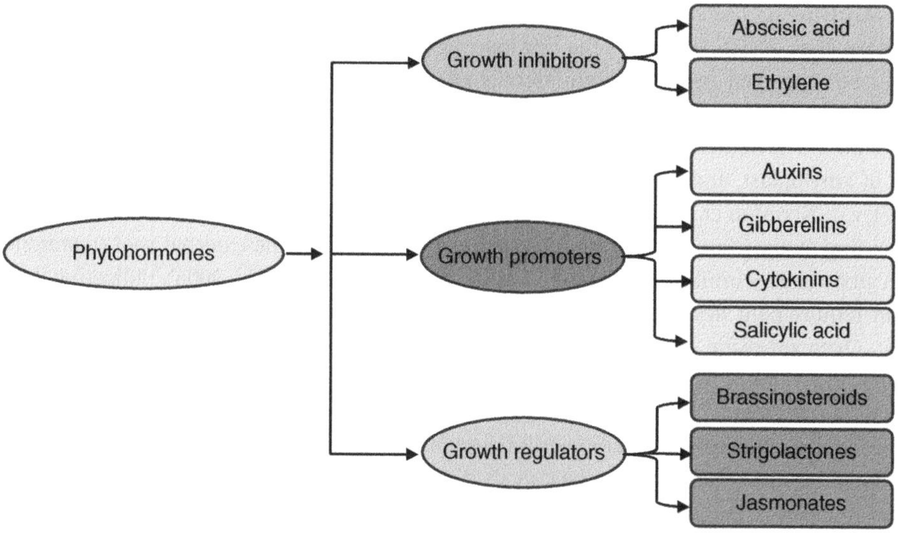

Figure 10.3 Classification of phytohormones that involved in the nodulation process.

10.3.1 Phytohormones Promoting Growth

10.3.1.1 Auxin

Auxin is the first class of main phytohormones discovered to be a major promoter of plant growth and development. The most essential member of this class is indole-3-acetic acid (IAA), which is derived from the phenyl-propanoid biosynthetic pathway in plants. High auxin concentration was detected in cells subjected to cell division, elongation, and differentiation, as well as during the formation of vascular bundles. Therefore, auxin has been hypothesized to play a crucial role during nodule formation (Shimoda et al. 2012). Many studies have focused on shedding light on auxin roles in the development and regulation of root nodule formation. Allen et al. (1953) were the first to show that exogenous addition of auxin polar transport inhibitors to alfalfa roots caused the formation of nodule-like structures in the absence of rhizobia. Mathesius et al. (1998) reported that Nod factor can disturb auxin transport in white clover. Moreover, in *M. truncatula,* the flavonoid deficiency that helps to inhibit auxin flow causes an inhibition in nodule number. These observations proposed an alteration in the auxin flow that affects nodule proliferation, therefore involving auxin in the nodulation process. During spontaneous nodule formation, a localized auxin response is detected by a gain-of-function mutation of the CCaMK; Ca^{2+}/calmodulin independent protein kinase (Suzaki et al. 2013), which is during the process of nodulation, acts to decode Ca^{2+} signals (Madsen et al. 2010; Shimoda et al. 2012).

During nodule formation in *M. truncatula*, three MtLAX proteins are actively expressed in the developing primordia, indicating high auxin import into nodule cells (de Billy et al. 2001). The accumulation of MtPIN proteins (a plant-specific family of transmembrane proteins that transport the plant signal molecule auxin as their substrate) is negatively controlled by MtCRE1-dependent cytokinin signaling. Regulation of the polar localization of certain PIN proteins may be needed for the establishment of localized auxin responses (Plet et al. 2011).

In *Arabidopsis thaliana*, cytokinin inhibits the initiation of lateral root initiator cells (Laplaze et al. 2007). Therefore, the negative regulation of PIN auxin carriers by cytokinin might be conserved in nodule and lateral root development. Many results with different legumes indicated that the exogenous application of cytokinin has a positive effect during the initiation of nodule organogenesis by inducing the division of cortical cells, accumulation of amyloplast, and expression of *ENOD* genes (Mathesius et al. 2000; Tirichine et al. 2007). van Noorden et al. (2007) reported that proteomic analysis of *M. truncatula* showed increasing auxin responses during early nodule primordial, while external root treatment with auxin causes inhibition of nodule formation (van Noorden et al. 2006), indicating that auxin requirement lies within a certain concentration and the stage of nodules development. Overexpression of the gene that is responsible for the auxin receptor did not affect the nodule number in the soybean plant (Turner et al. 2013). In addition, Turner et al. (2013) clarified that auxin requirement does not dependent on the nodule types, overexpression of both *G. max* (determinate nodules) and *M. truncatula* (indeterminate nodules) with *miRNA160* gene, which regulates negatively auxin response factor genes, reduced the number of nodules in both legume plants (Turner et al. 2013; Bustos-Sanmamed et al. 2013).

10.3.1.2 Cytokinins

The plant hormones CKs together with auxins are crucial for regulating diverse aspects of plant cell reproduction and differentiation processes. The two hormones play key roles in controlling root nodule development. Suzaki et al. (2013) reported that cytokinin signaling is essential in initiating nodule primordia. It may play specifically role in nodule organogenesis but not in the rhizobial infection process. After naming the Nod factor as an actual regulator of nodulation (Truchet et al. 1991), many studies identified the similarities between Nod factor, cytokinin, and rhizobial inoculation, regarding their effects on nodulation. Regarding the expression patterns for several *ENOD* genes, the uniformity of the progressing cortical cells induced by Nod factor or rhizobial inoculation appears uniform to those induced by CK treatment in white clover and alfalfa (Fang and Hirsch 1998; Mathesius et al. 2000). For example, the expression of the first gene that induce cortical cell division (ENOD40) in *M. truncatula* is activated by CK (Charon et al. 1997), The early studies which carried out with different species of legume indicated no further progress occur following the enhancing of cortical cell development with cytokinin treatment. Heckmann et al. (2011) indicated the possibility of stimulating bulge formation with the appearance of nodule-like primordia by adding cytokinin to the *L. japonicus* roots under experimental conditions. However, the frequency of these bulges' formation differs among *L.* species, suggesting an interspecies difference in response to cytokinin effect. CKs are involved in regulating cell division during nodulation, in which their response genes are activated during nodule initiation in inner (indeterminate nodule) or outer (determinate nodules) cortex cells (Plet et al. 2011). The cytokinin 5′-monophosphate phosphoribohydrolase, which is involved in cytokinin synthesis, has been shown to be required in *M. truncatula* with a certain expression window for regulating nodule numbers, nodule initiation, and nodule differentiation, suggesting the requirement of the de novo synthesis of cytokinin for nodule formation (Mortier et al. 2014). The observation of CK in the root cortex is essential for nodule initiation. Legumes deformed in cytokinin perception cannot initiate nodule primordia, as in both *L. japonicus* and *M. truncatula* mutants (Murray et al. 2007; Plet et al. 2011), in which delayed nodules with low frequency were produced in both legume mutants, suggesting other cytokinin receptors must be required for nodule initiation. Additional genes in *L. japonicus* LNK mutant that are needed for nodule initiation are expressed during outer and middle cortical cells division (Held et al. 2014). They also illustrated that nodule formation in this mutant is delayed and can be completed by preventing the entry of infected bacteria, suggesting that bacterial entry may provide a certain signal to the cortex that may bypass between the cortex and epidermis, which is mediated by LHK1. Cytokinin is required at early and later stages of nodulation (Plet et al. 2011). It is also involved in shoot-derived cytokinin signaling for AON activation (Mortier et al. 2012). Cytokinin produced by bacterial nodulation mutants has been found to be enough to induce divisions in *Medicago sativa* cortical cells (Cooper and Long 1994). In contrast, there was no relation between the amount and type of cytokinin synthesized by rhizobia species and their capacity to induce nodules (Kisiala et al. 2013). For example, cytokinin production by *Bradyrhizobium* sp. was found to accelerate nodule formation and change nodule size and number in the legume *Aeschynomene*, while deficient mutants of cytokinin were able to form nodules (Podlešáková et al. 2013).

10.3.1.3 Gibberellins

Gibberellins: gibberellic acids (GAs) are involved in a wide range of biological processes, including seed germination, cell elongation, and flowering. GAs are required at various stages of nodule formation (Hayashi et al. 2014). Much evidence shows accumulation of the GA12 precursor in the root epidermis by Nod factor treatment (Jardinaud et al. 2016). Evidence also suggests negative regulation of nodule formation and infection in both nodule types, determinate and indeterminate, by bioactive GAs (Fonouni-Farde et al. 2016). Therefore, Nod-factor-induced GAs may act as a fine-tune Nod factor signaling and rhizobia infection during symbiosis. GA-deficient mutants of pea developed fewer nodules compared to wild-type plants (Ferguson et al. 2011). Applied GA to plants shows either positive or negative effects on nodule numbers. An optimum concentration of GA appears to be applied for successful nodulation which depending on plant species, growth conditions, method of application, and concentration, and the type of GA applied (Ferguson and Mathesius 2003). Transcriptome analysis showed numerous upregulated GAs biosynthesis genes in soybean during early nodulation, among them are *GmGA3ox 1a* and *GmGA20ox a*, which show a high sequence similarity to SrGA20ox1 (Hayashi et al. 2012). The *GmGA20ox* gene was detected in only soybean roots induced to initiate nodules, indicating this gene is nodulation-specific (Hayashi et al. 2012). The upregulation of GAs biosynthesis genes at various stages of nodule formation denotes a temporary requirement for GAs during nodulation, occurring spatially and temporally. The expression pattern is consistent with findings that GAs affect early infection stages (curling of root hair and development of infection thread), as well as the more mature stages of nodulation (nodule primordial formation and the establishment of the correct nodulation zones) (Ferguson et al. 2011).

10.3.1.4 Salicylic Acid

Salicylic acid (SA) is a phytohormone that induces plant defense responses to pathogen infection and systemic acquired resistance (Ferguson and Mathesius 2014). SA is also well-recognized as an endogenous plant growth regulator. Much evidence indicates that

SA plays essential roles in regulating cell division and expansion, the crucial processes that define the final status of the plant. Thus, SA mediates growth regulation by controlling cell division and cell expansion (Li et al. 2022). The role of SA in nodulation processes is not well-defined, with some incompatible results reported. Many works have illustrated that SA highly suppresses nodule formation, leading to reduced nitrogenase activity. Martínez-Abarca et al. (1998) reported that SA produced in *M. sativa* (alfalfa) roots (forms indeterminate-type nodules) when inoculated with a nodC mutant of *S. meliloti,* causing defective in the biosynthesis of Nod factor needed for infection. They also reported that exogenous application of SA resulted in delayed and reduced nodulation of *M. sativa* inoculated with *S. meliloti*. Therefore, they suggested the participation of NFs produced by *S. meliloti* in the suppression defense role of SA in leguminous plants. Van Spronsen et al. (2003) illustrated that the addition of exogenous SA inhibited the formation of indeterminate nodules of *V. sativa* but not the formation of determinate nodules of *L. japonicus.* They explained this response by the fact that *Rhizobium leguminosarum* produces a Nod factor with an 18:4 fatty acid, which may be active in oxylipin signaling, that is defined to be suppressed by SA. While rhizobia species that induce determinate nodules form NFs without polyunsaturated fatty acids, therefore these Nod signals may affect in a different way. In contrast with this theory, Sato et al.(2002) reported that exogenous application of SA to *G. max* (soybean) seedlings (which forms determinate-type nodules) suppressed the early stage of nodule formation with less pronounced inhibition in hypernodulating soybean mutants as compared with wild-type plants, indicating a direct involvement of SA in signal transmission of the autoregulation process. Stacey et al. (2006) also showed an obvious decrease in the level of SA involved with a reduction in the infection thread number of *L. japonicus* and *M. truncatula*, causing a reduction in nodule number. Furthermore, Lian et al. (2000) showed that the addition of low SA concentration has no effect on the nodule formation (determinate-type) in *Phaseolus vulgaris*, *L. japonicus*, *G. max*, and *Glycine soja*, but high SA concentration (4mM SA) significantly reduced nodule number and the dry weight of *G. max.* The above results propose the involvement of endogenous SA levels in regulating nodulation in both indeterminate- and determinate-type nodules in leguminous plants.

10.3.2 Phytohormones Inhibiting Growth

10.3.2.1 Ethylene

Ethylene the gaseous hormone, plays several physiological roles, including promotion of flowering, involving in fruit ripening, abscission of leaf and fruit, germination of seed, morphogenesis of seedling, root emergence, elongation of root hair, and stress response beside its action on plant defense (Prayitno et al. 2006), interacts with ABA (Chan et al. 2013), and involving in auxin transport process (Prayitno et al. 2006). Ethylene acts as a negative nodule regulator and plays a diverse role during nodule development, including the formation of infection thread and regulation of nodule numbers, position, and morphology of the nodules (Guinel and Geil 2002). In numerous legumes, ethylene application inhibits nodule numbers, while ethylene synthesis generally enhances nodule numbers (Ferguson and Mathesius 2003). In different legumes, ethylene-insensitive mutants have provided genetic confirmation of the participation of ethylene signaling in nodule formation (Gresshoff et al.

2009). The nodule numbers are partly affected by ethylene signaling, as insensitive ethylene mutants develop a number of nodules more than wild-type plants (Nukui et al. 2004).

During the process of nodulation, ethylene interferes with calcium signaling and regulates negatively the formation of the infection thread (Oldroyd et al. 2001). Furthermore, all the early plant responses were reduced by ethylene action, including calcium spiking initiation in *M. truncatula*. This may be because ethylene acts at the point of calcium spiking in the pathway of Nod factor signal transduction (Oldroyd et al. 2001). The involvement of ethylene in the regulation of the auxin transport system during nodulation may affect the total nodule numbers (Prayitno et al. 2006). It is possible that ethylene could act through the induction of flavonoids that modulate auxin transport, but this has not been directly indicated during nodule formation (Buer et al. 2006). When ethylene begins to act upstream to regulate auxin transport, there is evidence that ethylene effect is needed for auxin transport change to take place (Rightmyer and Long 2011). It is also observed that ethylene can alter nodule number and morphology of *L. japonicus* (Lohar et al. 2009). Although the number of nodules in the earliest stage of development was increased in *L. japonicus* expressing the CmERS1 gene, the number of nodules in the mature stage was approximately equal to wild-type plants, suggesting that ethylene causes alteration in nodulation number at different stages of nodule formation (Nukui et al. 2004). Aminoethoxyvinyl glycine (AVG), the inhibitor of ethylene synthesis, can stimulate nodule formation in *M. truncatula* roots (Zhang et al. 2022). Ethylene-insensitive mutant of *M. truncatula* has been produced hypernodulation phenotype (Penmetsa and Cook 1997).

Some bacteria can produce 1-aminocyclopropane-1-carboxylic acid (ACC) deaminase that cleaves directly the precursor of ethylene ACC, which is released by plant roots into α-ketobutyrate and NH_3, resulting in decreasing the endogenous ethylene level in plant roots by restricting the amount of ACC available. Consequently, it prevents the inhibition of root growth by ethylene and thus enhances lateral root development (Zhang et al. 2022). It is reported by van Zeijl et al. (2015) that numerous ethylene biosynthesis genes can be induced by Nod factor expression, including ACC synthase (ACS) genes (*MtACS1* and *MtACS2*). Moreover, Larrainzar et al. (2015) indicated three ACS genes with rhizobium are upregulated by Nod factor signaling at 6 hpi. Root nodule formation in *L. japonicus* was inhibited by ACS but was boosted by AVG and Ag+ (Nukui et al. 2004). These results suggest that the ethylene-mediated signaling pathway is involved in the process of nodulation, whatever of the nodule type. Thus, ethylene may not play a significant role in nodule development in all species. The determinate nodules type (soybean plants), the exogenous treatment with ethylene did not suppress nodulation, and application of AVG or Ag^+ did not enhance nodule number (Lee and LaRue 1992).

10.3.2.2 Abscisic Acid

Abscisic acid (ABA) plays a well-characterized role in plant growth, regulating stress responses, enhancing seed and bud dormancy, mediating embryonic cell division, stimulating flowering formation, and mediating stomatal closure during drought conditions (Nagata and Suzuki 2014). ABA has been stated to play negative functions at diverse stages of nodule formation. The treatment of *P. sativum* with ABA repressed nodulation (Xu et al. 2021), *M. truncatula* (Ding et al. 2008), *G. max* (Bano and Harper 2002), *L. japonicus*, and *Trifolium repens* (Suzuki et al. 2004). It is reported that ABA treatment suppresses both determinate (e.g. with no persistent meristem as *L. japonicus*, *G. max*, and *P. vulgaris*) and indeterminate (e.g. with a persistent meristem as *M. truncatula*, *Medicago sativa*, and *P. sativum*) nodules in legumes (Ding et al. 2008). Nevertheless, how the signaling of ABA and nodulation interplays with each other is not clear (Xu et al. 2021). They identified GmbZIP1 that negatively regulates nodule formation in soybean via direct inhibition of GmENOD40–1 expression, while it positively regulates ABA response.

Exogenous treatment of ABA in both *T. repens* (forms indeterminate nodules) and *L. japonicus* (forms determinate nodules) showed a reduction in nodule number (Suzuki et al. 2004). On the other hand, applying a specific biosynthesis inhibitor of ABA (Abamine) enhanced nodule number in many legume species (Liu et al. 2018; Ding et al. 2008). Suzuki et al. (2004) illustrated that, the application of ABA to *T. repens* prevented root hair deformation at the very early stage of rhizobium infection (between swelling stage of root hair and its curling), as well as ABA treatment blocked the formation of infected thread in *M. truncatula* and *L. japonicus* (Ding et al. 2008; Nakatsukasa-Akune et al. 2005) Consistent with these results, Ding et al. (2008) reported an induction of nodulin genes, such as *ENOD11* and RIP (*Rhizobium*-induced peroxidase), was canceled by ABA treatment. They also concluded that ABA inhibited the expression of cytokinin-activated nodulation gene ENOD40. The inhibition of Nod factor-induced calcium spiking by ABA in *M. truncatula* proposed that ABA exerts its action at the very early stage of nodulation (Ding et al. 2008). ABA addition to wild-type soybean and to a super-nodulation mutant (NOD1-3) significantly diminished nodule number as well as decreased the formation of isoflavonoid compounds, which act as the principal material for initiates nodule formation, thereby causing a reduction in rhizobial infection (Cho and Harper 1993).

10.3.3 Phytohormones Regulating Growth

10.3.3.1 Jasmonic Acid

Jasmonic acid (JA) acts as a well-defined role in plant growth, stress regulation, as well as defense against pathogen attack and wound response (Buhian and Bensmihen 2018). Many reports have indicated that JAs can act as either negative or positive regulator role during nodule formation and the nitrogen fixation process. These conflicting reports depend upon the type of JA applied, the legume species, hormone dose, time of application, and where the hormone is applied (Ferguson and Mathesius 2014). In support of the negative role for JAs in nodulation, methyl jasmonate application to *L. japonicus*, shoot strongly suppressed nodulation, infection thread formation, and the expression of the *NIN* gene in both wild-type plants and the Ljhar1 hypernodulation mutant (Nakagawa and Kawaguchi 2006). Similarly, Sun et al. (2006) reported that adding a high concentration of JA to the growth medium of *M. truncatula* plants significantly reduced nodule formation, including the plant's responsiveness to NFs, causing a low number of root hairs that exhibited calcium spiking. Additionally, JA depressed the expression of the early *MtRIP1* and *ENOD11* genes in plants.

In contrast, with support of the positive role for JAs in nodulation, the level of JA in leaves of a *G. max* hypernodulating nts mutant was higher compared with that in wild-type *G. max* under normal conditions. Furthermore, transcription levels of JA-responsive genes augmented in the hypernodulating nts mutant, suggesting the nts mutation causes modification in particular pathways, including JA synthetic metabolism, leading to activation of JA-responsive genes (Seo et al. 2006). Suzuki et al. (2011) reported that treatment of *L. japonicus* with a low level of JA results in boosted nodulation and formation of infection threads when wild-type plants are grown under low red/far-red light conditions. Their results explain that nodule formation is photo-morpho-genetically regulated by sensing the red/far-red ratio via signaling of JA. JA enhanced the expression of nod genes in *Bradyrhizobium japonicum* (Mabood and Smith 2005; they also found that preinoculation of *B. japonicum* with JA enhances nodulation in soybean.

10.3.3.2 Strigolactones

Strigolactones (SGLs) have been identified as phytohormones that have various roles involved in the regulation of auxin transport, defining root architecture, regulating plant shoot branching, and play essential roles as signaling compounds in the symbiotic association between arbuscular mycorrhizal (AM) fungi and the roots of higher plants, stimulating seed germination of parasitic weed plants such as *Orobanche* and *Striga* species (Nagata and Suzuki 2014). SGL boosts nodule formation in *P. sativum*, *L. japonicus*, and *M. sativa* (Foo and Davies 2011; Liu et al. 2013; Soto et al. 2010). Double mutant of *P. sativum* plants of rms1 and the autoregulation mutants, sym29 (Psnark), nod3 (Psrdn1), or sym28 (Psclv2), all show a hypernodulation phenotype, which indicates that the hormone does not act downstream of these genes in nodulation control, and instead likely influences the promotion of nodulation (Foo et al. 2014). Biosynthesis SGL mutant of *P. sativa*, rms1 (Psccd8), formed less nodules than the wild-type plant, while the response SGL mutant rms4 (Psmax2) did not (Foo et al. 2013). Silencing the biosynthesis gene of SGL in *L. japonicus*, CAROTENOID CLEAVAGE DIOXYGENASE7 (LjCCD7), resulting in a slight decrease of nodulation as compared with control plants, these results proposed the positive effect of SGLs on the number of determinate nodules (Liu et al. 2013). Therefore, SGLs play an essential role in elevating nodules in both indeterminate- and determinate-type nodules. Soto et al. (2010) reported that a synthetic SGL analog (GR24) obviously enhanced nodule formation in *M. sativa* (alfalfa). They also explain that treatment of *M. sativa* with GR24 causes enhancement in the metabolism and biosynthesis of the SGLs, consequently, resulting in an increase in the indeterminate nodule formation. They also reported that adding GR24 to the SGL-deficient rms1 mutants of *P. sativum*, which produce lesser nodules than wild-type plants, increased the number of nodules in both wild-type as well as in rms1 mutants of *P. sativum* to reach a level approximately equal to that recorded in untreated wild-type plants, thus endogenous SGLs elevate nodule number in *P. sativum*.

10.3.3.3 Brassinosteroids

Brassinosteroids (BRs) are a set of plant hormones that control a variety of physiological responses, such as seed germination, cell elongation, xylem differentiation, and photomorphogenesis. These plant steroid hormones (BRs) are greatly mobile within the cells of plants, although they exist in very low concentrations (Takatsuto et al. 1983). Ferguson et al. (2005) proposed that BRs play a role in the process of AON on the shoot that participates in the regulation of the root nodule numbers. They also reported that uniconazole-P (the biosynthesis inhibitor of GA3) partially inhibits the biosynthesis of BRs. In *L. japonicus*, the addition of BR combined with uniconazole-P caused a considerably reduction in the infection threads, as compared with uniconazole-P treatment alone, with a less considerable reduction in GAs level. In meanwhile, no great effect on either nodules number or the root and shoot lengths. This suggests that BRs negatively affect root-hair growth, and consequently fewer infection threads develop (Maekawa et al. 2009). BRs are affected in both nodulation and nitrogen fixation processes; their action is dependent mainly on the method of application and the species used, as well as on the type and the dose of BR applied. In this field, several authors have focused on the influence of BR application on nodule numbers. Thus, Shahid et al. (2011) found that imbibition of pea seeds with 24-epibrassinolide directed to enhance the number of nodules, increase in both nodule fresh and dry weights, and boost nitrogenase activity. Applied of epibrassinolide by foliar application to *Arachis hypogaea* (peanut) plants results in increased nodules weight and number as well as the activity of nitrogenase (Vardhini and Rao 1999). The pea (*P. sativum* L.) plant treated with 24-epibrassinosteroid showed higher nodule number, fresh and dry weights, and activity of nitrogenase as compared to those obtained with the control (Shahid et al. 2011). Applied homobrassinolide or epibrassinolide by foliar sprays to French bean plants (*P. vulgaris*) were found to boost nodulation and increase nitrogen fixation activity (Upreti and Murti 2004). Applied BR to the leaves of the supper-nodulation mutant (En6500) of soybean cultivar (Enrei) induced stem elongation as well as inhibited root nodule development, but not in the wild-type (Terakado et al. 2005).

In contrast to the positive effects of BR on nodulation, some authors recorded negative effects for some types of BRs. For example, Hayat and Ahmad (2003) found that various concentrations of 28-homobrassinolide, when soaking with *Lens culinaris* seeds, resulted in decreased root nodule number and length but increased the activity of nitrate reductase. Hunter (2001) reported that applying epibrassinolide to soybean roots resulted in a decrease in both nodule formation and nitrogenase activity. Terakado et al. (2005) also found that root injection or foliar application of brassinolide to a mutant of hypernodulating soybean depressed nodulation and root development, but not in the wild type. On other hand, they reported that foliar treatment of soybean with biosynthesis inhibitor of BR (brassinazole), enhanced the number of nodules with obvious reduction in the length of stem in wild-type, which propose that BR may act to regulate nodule development and that its biosynthesis inhibitor transfers from the culture medium to the shoot and therefore lower the endogenous level of BR in soybean leaves.

It is clear from the above discussion that root nodulation is a complex symbiotic process that occurs between leguminous plants and nitrogen-fixing bacteria (rhizobia). Plant hormones play a crucial role at every step of nodulation, regulating root architecture, bacterial infection, nodule formation, and function. Figure 10.4 shows the steps of nodule formation and the roles of each phytohormone in the different steps.

10.4 Signaling Peptides

Numerous peptides are found to have an essential regulatory role in establishing the symbiotic relationship processes. These peptides are frequently found to act as ligands for receptors to trigger aspects of plant growth and development, like to the role of traditional plant

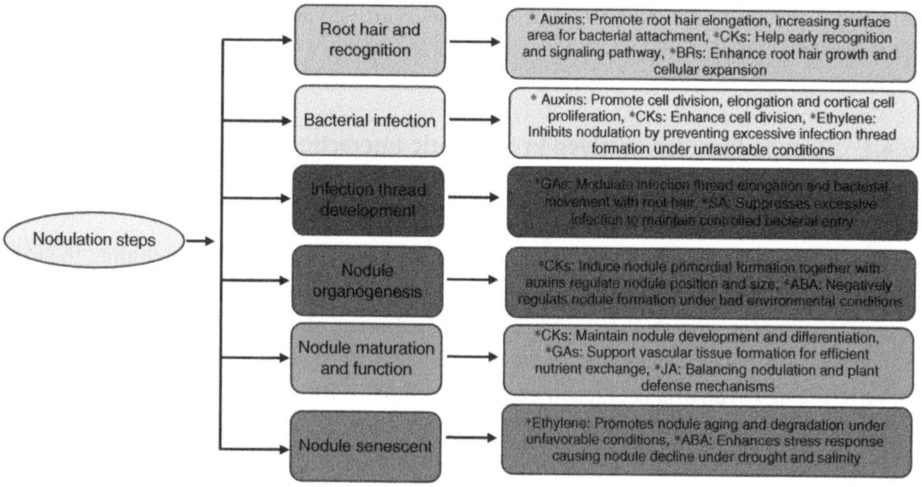

Figure 10.4 The role of different phytohormones in the root nodulation process in leguminous plants.

hormones. Peptides also associate with the pathways of plant hormone signaling to modify development. There are around ten percent of the transcriptomic alterations during rhizobial invasion related to plant regulatory peptides (Kang et al. 2020).

Plant peptides that are involved in the beneficial relationships between rhizobia and legumes can be divided into four classes (Figure 10.5); they are peptide families encoded by PTM peptides (posttranslational modification), cysteine-rich peptides (CRPs), non-CRP and non-PTM, and smORF peptides (small open-reading frames). Each class contains some essential regulatory families that are involved in the nodulation process (Figure 10.5). Plant peptides as signaling molecules have many advantages; they act as regulators for various proccsses because their production costs relatively low energy, their expression can be

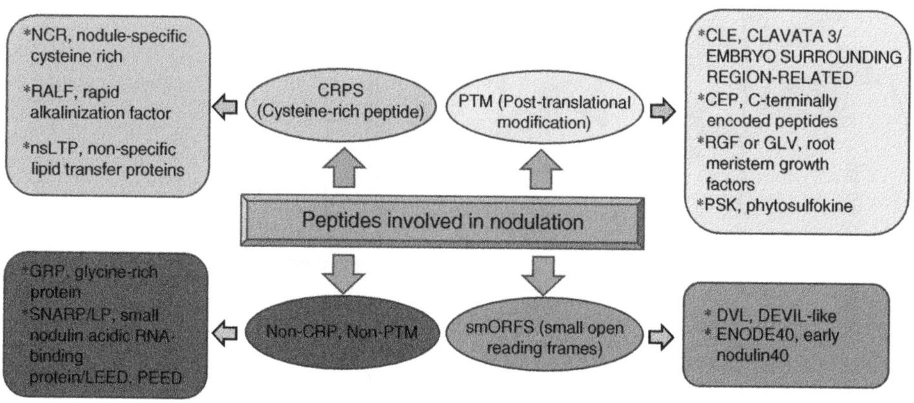

Figure 10.5 Peptide families that involved in the beneficial relationships between rhizobia and legumes.

easily regulated, and their efficiency can be adjusted through posttranslational modification (Mamaeva et al. 2025).

10.4.1 Peptide Families Involved in Symbiotic Interactions

10.4.1.1 Posttranslational Modifications (PTMs) Peptide Families

10.4.1.1.1 CLE Peptides

CLE peptides participate in different processes during the plant life cycle, including the formation of symbiosis with rhizobia species in legumes. CLE peptides are small peptide hormones specific for plants that serve as intermediaries of cell-to-cell communication. Numerous CLE peptide hormones were found to regulate negatively nodule formation, as PvRIC1/2 in *P. vulgaris* (Ferguson et al. 2014), LjCLE-RS1/2 in *L. japonicus* (Okamoto et al. 2009), GmRIC1/2 in soybean (Reid et al. 2011), and MtCLE12/13 in *M. truncatula* (Saur et al. 2011). CLE peptides are normally located near or at the C-terminus of their prepropeptide, containing from 12 to 13 amino acids, and act to control the number of nodules (Reid et al. 2011).

In response to rhizobia infection in legumes, certain *CLE* genes are expressed to enhance AON, a regulatory pathway (Figure 10.6) that depresses the development of excessive nodule numbers (Miyawaki et al. 2013). These genes are expressed at all stages of nodule formation (early, primordial, and developing stages), in both determinate and indeterminate root nodules (Mortier et al. 2012; Nishida et al. 2016). CLE induces AON and boosts TML1/2 expression by mobile miRNA miR2111 and reduces the expression of its repressor (Gautrat et al. 2020; Valmas et al. 2023). The transcriptomic analysis reveals that MtCLE35 in OE

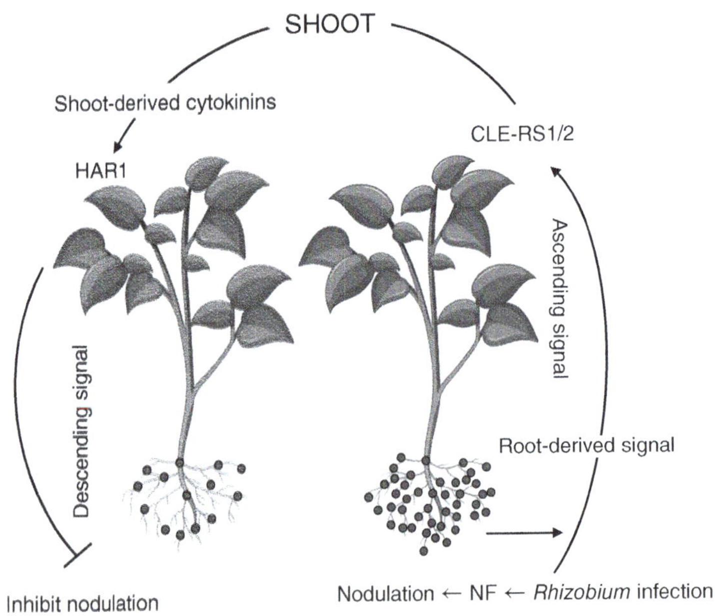

Figure 10.6 Autoregulation of nodulation process by host CLE signaling.

plants prevents the development of new nodules by regulating the rhizobacterial-induced expression (Lebedeva et al. 2024).

Regulation of Nodule Formation by Host CLE Signaling The symbiotic association between legumes and rhizobia is assumed to be pathogenic if the plant fails to regulate the number of nodules. Generally, this symbiotic association has a benefit for the rhizobia because if so, many nodules are formed within the legume roots, which will result in a negative impact on the host vegetative growth and yield (Matsunami et al. 2004). Therefore, legumes use an autoregulation process of nodulation to regulate nodule number by stopping the formation of new nodules (Mortier et al. 2012). This process is thought to involve a signal originating from the root (root-derived signal) that is then transported to the shoot, which, in turn, induces a shoot-derived signal that is transported to the root to inhibit the formation of new nodules. Thus, it is supposed that the autoregulation process is a root-to-shoot-to-root pathway (Suzaki et al. 2015). Reid et al. (2011) considered that CLE peptides act as signal molecules that are transported from the roots to the shoot and are involved in the signal pathway of AON. In *L. japonicus*, AON is mediated by CLE peptide CLE ROOT SIGNAL1/2 (CLE-RS1/2), which is specifically expressed in the root, and directly attaches to HYPERNODULATION ABERRANT ROOT 1 (HAR1) that functions in the shoot (Okamoto et al. 2009). The arabinosylation of CLE-RS1/2 is essential for attachment with HAR1. The CLE-RS1/2 is derived from the CLE domain by a posttranslationally arabinosylated glycopeptide, and Okamoto et al. (2009) reported that, extreme amount of CLE-RS1/2 is enough to inhibit nodule formation. The shoot-derived CKs (Figure 10.6) are transported from the shoot to the root, which could act as the shoot-derived signal to suppress nodule formation. AON may negatively regulate the development of excess root nodule bacteria via controlling auxin responses (Suzaki et al. 2015). Several studies have demonstrated that nodulation is strongly suppressed when the *CLE* genes are constitutively activated (Okamoto et al. 2009; Mortier et al. 2012).

Gretchen Hagen3 (GH3) showed that in white clover, the Nod factor can disturb auxin transport (Mathesius et al. 1998). Furthermore, flavonoid deficiency, which causes inhibition of auxin transport, decreased nodule number in *M. truncatula* (Wasson et al. 2006). These observations indicate that alteration of the auxin transport affects nodule formation, thereby involving auxin in the AON process. On the other hand, Ferguson et al. (2011) have indicated that GAs also act independently in the AON pathway, and the hormone is required as an essential factor in further support for proper nodulation.

In summary, the process of AON in legumes is orchestrated by a mobile interaction between plant hormones and peptide signals. These molecular players confirm that nodulation occurs efficiently and only when beneficial, reflecting a finely tuned system evolved through plant-microbe coevolution.

10.4.1.1.2 C-Terminally Encoded Peptide

C-terminally encoded peptide (CEP) is a member of a multigene peptide family, which occurs in higher plants while being absent from primitive ones. Mature CEPs are around 15 amino acids long, possess a clear role in coordinating root and shoot growth (Roberts et al. 2013). CEPs are synthesized in the roots under conditions of nitrogen starvation and boost nodule formation (Laffont and Frugier 2023). Imin et al. (2013) reported that the addition of MtCEP1 peptide to *M. truncatula* caused lateral root inhibition, increased nodule number, induced peripheral root swelling, and increased auxin response. They also noted that the increase in nodule number is partially tolerant to certain concentrations of nitrate, which usually inhibit nodule formation; these results may indicate that MtCEP1 regulates lateral root and nodulation in *M. truncatula*. Generally, CEP peptides coordinate the root system architecture and ion uptake (Taleski et al. 2024). Both CLE and CEP peptides regulate the development of nodule formation through opposition mechanisms. They depend on the availability of the essential macronutrients and display a fine regulatory system for nodulation, taking into consideration plant needs and resources (Laffont and Frugier 2023). The two peptides regulate miR2111, the mobile miRNA that inhibits the AON component TML1/2: CEP promotes the expression of miR2111 and nodulation, while CLE peptides exhibit the opposite effects (Gautrat et al. 2021). MiR2111 is not the only link between CEPs and CLE. The MtNIN, a transcription factor that regulates interactions between plant and microbe, has been shown to stimulate the expression of MtCLE13 and MtCEP7 (Laffont and Frugier 2023). This orchestrated regulation in the expression of these peptides for controlling nodulation depends on both the nitrogen availability for the plant and the efficiency of the nitrogen-fixing capacity of the symbiont (Laffont and Frugier 2023).

10.4.1.1.3 GLVs or RGFs

Root meristem growth factors (RGFs) or GLV peptides were first reported in *A. thaliana*, but are found to distribute through all land plants and collectively in all plant parts, including shoot and root, and also within all stages of growth and reproduction (Fernandez et al. 2013). However, their expression of various genes is restricted to specific cell types. GLV peptides are involved in the regulation of root gravitropic responses, meristematic activity of primary root meristem, maintain lateral root and root hair development, and shoot formation (Fernandez et al. 2013). The mature bioactive GLV peptides are formed by cleavage of larger precursor proteins that share a common structure, which is predicted to have several 86–163 amino acids. GLV peptides also regulate the processes concerning the beneficial interaction between rhizobia species and different legume plants. About 5 of the 15 RGF peptides that are responsible for nodulation are concerted in controlling the nodulation process in *M. truncatula* (Roy et al. 2024). They also showed two MtRGF peptides in soybean determinate nodules that indicate specific nodule expression. MtRGF3/GLV7 negatively regulates nodule formation in *S. meliloti* and its expression depends on the secretion of Nod factor by this symbiont (Li et al. 2020). Furthermore, MtRGF3/GLV7 suppresses the expression of important nodule formation regulators (Li et al. 2023).

10.4.1.1.4 Phytosulfokines

Phytosulfokines (PSKs) are a family of disulfate pentapeptides and are considered as plant peptide hormones. PSKs are four bioactive compounds via PSK-α, -γ, -δ, and -ϵ. Their precursor genes are widespread in land plants, which play roles in growth and immunity. Furthermore, they are participating in several crucial biological processes, as promoting nodulation, enhancing plant reproduction, regulating somatic embryogenesis, and inducing plant resistance to stresses conditions (Li et al. 2024). Mature PSK peptides are mediated by the plasma membrane localized receptors PSKRs that are related to the leucine-rich repeat receptor-like kinase (LRR-RLK) family (Yu et al. 2022). They identified a novel type of PSK-δ, which is first expressed in the nodules of *M. truncatula*. They detected *PSK*-δ expression in tiny nodules, in root cortical cells undergoing nodule organogenesis, in nodule primordia, as well as in the apical part of mature nodules. Therefore, they suggested that PSK-δ, together with PSK-α peptide, plays a regulatory role during all stages of nodule organogenesis. PSK-ε peptide has been identified in the genomes of *G. max*, *L. japonicus*, and *M. truncatula*. PSK-ε peptide differs from PSK-δ by a single amino acid. It is expressed in the lateral root primordia and root apex. Moreover, its expression is enhanced by bacterial infection and is preserved in the primordial and apical zone during nodule formation. Exogenous application of MtPSK-ε peptide boosts root growth, lateral root formation, and nodule development (Di et al. 2022).

Five PSK genes were recognized in *L. japonicus*, three of which are expressed in nodules (LjPSK1, LjPSK4, and LjPSK5). *LjPSK4* gene is also expressed in spontaneous nodules and *LjPSK1* gene is induced by bacterial infection and is expressed throughout the nodule formation (Wang et al. 2015).

10.4.1.2 Cysteine-Rich Peptide (CRP) Families

10.4.1.2.1 Nodule-Specific Cysteine Rich

Nodule-specific cysteine rich (NCR) is a large class of peptides that regulates the formation of indeterminate-type nodules (Kondorosi et al. 2013) but has not been implicated in determinate-type nodules (Downie and Kondorosi 2021). The mature peptide consists of 35–55 amino acids and contains either 4 or 6 conserved cysteine residues (Lima et al. 2020). The expression patterns of NCR vary according to the stage of nodulation and their localization in the zones of the nodule (Downie and Kondorosi 2021). NCR peptides are similar to defensin peptides (antimicrobial peptides, AMPs); some NCRs, such as MtNCR343, MtNCR211, and MtNCR169, have antimicrobial activity toward free-living rhizobacteria (Zhang et al. 2023; Kim et al. 2015). Thus, NCR peptides can regulate the bacteroid differentiation of *S. meliloti*, the microsymbiont of *M. truncatula* (Wang et al. 2010). NCR peptides as MtNCR343, MtNCR169, and MtNCR211, are crucial for controlling terminal bacterial differentiation and survival (Kim et al. 2015). Saifi et al. (2024) reported that two types of NCR peptides (MtNCR314 and MtNCR086)are required for differentiation of nodules in *M. truncatula*, in which the nodules produced with MtNCR314 and Mt-NCR086 knockout mutants are small, round, without an increase in size, and do not undergo nitrogen fixation or endoreduplication.

The stability and activity of NCR peptides are mainly related to the redox status of the molecule. Thioredoxin s1, which is specific to the nodulation process, reduces the disulfide bonds in NCR335 and NCR247 peptides, which show potent antimicrobial activity in this form; therefore, thioredoxin s1 is essential for terminal bacteroid differentiation (Ribeiro et al. 2017). Iron ions, which are required for nitrogen fixation, were found to induce NCR247; this peptide plays a significant role in regulating the metabolism of metals in symbiotic bacteria (Sankari et al. 2022).

10.4.1.2.2 Rapid Alkalinization Factor

Rapid alkalinization factor (RALF) peptides are cysteine-rich SSPs, a highly protected gene family found all around the plant kingdom, that originate at membranes and are expressed in numerous plant organs, such as roots, shoots, flowers, and leaves (Bedinger et al. 2010). RALF peptides are broadly distributed between various plant taxa from *Physcomitrella* and *Selaginella* to angiosperms (Campbell and Turne 2017). RALF affects nodulation, pollen tube growth, stress resistance, modulates the immune response by either suppressing or stimulating it, and regulates cell expansion (Tena 2016; Mamaeva et al. 2023). Combier et al. (2008) reported that the RALF-encoded gene in *M. truncatula* is upregulated by NF application, while the overexpression (OE) of this gene augmented the infection thread number and consequently the nodules number. In addition, this gene leads to the development of small nodules in size with defective function. They postulated that RALF plays a negative regulatory role in both infection and nodulation processes.

Analysis of the RALF gene family in *G. max* colonized by *B. japonicum* reported that transcription of ten genes was upregulated, while six genes was downregulated (Jia and Li 2023). They suggested the dual functions of RALF peptides with participating in the immune response (downregulated genes) and supporting symbiotic association (upregulated genes). Solis-Miranda et al. (2023) illustrated that the expression of RALF in *P. vulgaris* is induced by bacterial infection and continues in nodule primordia during all stages of nodule formation, but this effect depends on nitrogen status in the growth medium. During nitrogen shortage, the small peptides, PvRALFs, induce nodule formation; however, in sufficient nitrogen, these peptides reduce nodule formation by inducing AON.

10.4.1.2.3 Nonspecific Lipid Transfer Protein

Plant nonspecific lipid transfer proteins (nsLTPs) are normally known as small, basic proteins, with widespread occurrence in many plant taxa (Missaoui et al. 2022). nsLTPs have the capacity to transport various types of hydrophobic molecules *in vitro*, such as glycolipids, phospholipids, fatty acids, and fatty acyl–CoA (de Oliveira Carvalho and Gomes 2007). Structural studies of nsLTP peptides revealed that the mature peptide of nsLTPs contains a preserved eight cysteine motifs, linked by four disulfide bridges, and a hydrophobic internal cavity that houses various ligands, including fatty acids, phospholipids, and acyl-coenzyme A. This structure shows great stability and boosts its function (Missaoui et al. 2022). nsLTP peptides participate in various physiological responses, including plant growth and reproduction, germination process, seed development, cuticle formation, abiotic stress tolerance, and nodulation (Missaoui et al. 2022; Lei et al. 2014). They also play a role in the immune response; nsLTPs accumulate during pathogenic infection (Gao et al. 2015). Furthermore, essential signaling molecules such as methyl jasmonate, ABA, ethylene, and SA are also involved in the expression of nsLTP genes (Safi et al. 2015). In addition, nsLTPs take part in controlling the early stages of symbiotic association in both rhizobia and *Frankia* (Gasser et al. 2023). In the legumes–rhizobia symbiosis, a remarkable percentage of nsLTPs in the genomes are enhanced during nodule development (Gasser et al. 2023). It was observed that nsLTP1 and nsLTPd in legumes are broadly expressed in roots and nodules, supporting the involvement of nsLTP in the nodulation process (Fonseca-Garcia et al. 2021). Notably, subinhibitory doses of AgLTP24 from *Alnus glutinosa* induce transcriptional changes in 142 genes in *Frankia* (Gasser et al. 2023).

10.4.1.3 Non-CRP, Non-PTM Peptide Families

10.4.1.3.1 Glycine-Rich Protein

Glycine-rich protein (GRP) is a large peptide family with high glycine content, which is restricted to the IRLC (Kereszt et al. 2018). GRPs participate in the plant cell growth, cell wall structure, flower development, pollen hydration, and stress responses (Xu et al. 2023). Furthermore, the GRPs play a crucial role in the functioning of the root nodules; they target the symbionts and take an essential part in the bacterial terminal differentiation (Kereszt et al. 2018). In addition, GRPs possess an RNA-binding domain that shows antimicrobial activity (Fu et al. 2007). GRP peptides were first observed in *Medicago* spp. and *Vicia faba* (Alunni et al. 2007). They reported that in *M. truncatula*, the NCR gene family is much larger than the GRP one, with about 30 members. In *M. truncatula*, about 40% of GRP transcripts are present in the infection zone of the nodule, along with only about 18% of NCR transcripts. In this zone, the bacteroid differentiation takes place (Kereszt et al. 2018). This indicates that many GRPs participate in bacteroid differentiation. Both *M. sativa* and *M. truncatula* show GRPs with nodule-specific expression patterns, while these genes were not expressed in green parts of plants, and neither hormones nor stresses induced their expression (Kevei et al. 2002).

10.4.1.3.2 Small Nodulin Acidic RNA-Binding Protein/LP

SNARPs are also identified by LEED. PEEDs form a small family, containing 10 to 13 members, with no more than 70 amino acids in length that are characterized by the presence of one or two protected domains (named LEED or PEED domains) of acidic amino acid residues. SNARPs are also characterized as RNA-binding peptides, as indicated by their full name, the **S**mall **N**odulin **A**cidic **R**NA-binding **P**rotein (Laporte et al. 2010). The 13 members of this family are all transcribed in nodules displaying different expression patterns. Moreover, SNARP/LP expression was discovered in various nodule zones (Laporte et al. 2010). SNARPs were detected in *M. truncatula* as a specific peptide family in legume plants (Graham et al. 2004). MtSNARP2 is an ENOD that reacts to ENOD40 RNA. MtSNARP2/LP11 transcripts were detected in the meristem zone (Trujillo et al. 2014). Silencing *MtSNARP2* leads to a large number of aberrant nodules with a hypertrophied outer cortex, a lack of clear nodule zonation, a brown pigment instead of the pink pigment (leghemoglobin), and inactive rhizobacteria after cell infection (Laporte et al. 2010).

10.4.1.4 Small Open Reading Frames (SmORF) Peptide Families

10.4.1.4.1 Early Nodulin40

Early nodulin40 (ENOD40) is a peptide with many short ORFs; it is an enigmatic signal because it encodes a gene well conserved through the plant kingdom. Although the *ENOD40* gene plays an essential role in nodule formation in legumes, it also exists in non-legume plants and has roles in developmental processes in plants outside nodulation (Gultyaev et al. 2023). Because the production of ENOD40 peptide is not restricted to legumes plant, and is broadly distributed across all plant species and organ, therefore, its expression in other plants plays a crucial role in many vital processes as enhancing mycorrhizal colonization, boosting lateral root development, and functioning widely in tissue and organ formation (Staehelin et al. 2001; Mathesius et al. 2000). *ENOD40 gene* contains two short peptides: peptide A which is 11–13 amino acids long and peptide B is approximately

24–27 amino acids in length (Gultyaev et al. 2023). Both peptides A and B bind the enzyme sucrose synthase, leading to increased sucrose breakdown and consequently enhancing cell carbon supply (Röhrig et al. 2002).

ENOD40 peptides are produced directly from their mRNA, in which they act to form a complex with the ENOD40 mRNA transcript that acts as untranslated signaling molecule (Sousa et al. 2001). Laporte et al. (2010) reported that ENOD40 mRNA induced a nuclear RNA-binding protein in the cytoplasm of *M. truncatula* that interacted with RNA-binding peptides that sustain nodulation during symbiosis. The ENOD40 peptide plays a role during the stage of nodule primordium formation in legumes (Charon et al. 1999). Wan et al. (2007) demonstrated that both *ENOD40-1* and *ENOD40-2*, gene silencing of two *M. truncatula ENOD40* genes, are involved in both stages of nodule initiation and nodule formation.

10.4.1.4.2 DEVIL-like/ROTUNDIFOLIA

DVL/ROT (DEVIL-like/ROTUNDIFOLIA) is a peptide group encoded by smORFs. DVL peptides are about 30 amino acids long and, like RALF, CEP, and CLE peptides, are derived from the C-terminus of their full-length proteins (Valdivia et al. 2013). Although the function of several DVL/RTFL peptides is thought to be nonessential; nevertheless, they exert special control in development through diverse expression patterns (Wen et al. 2004). Yang et al. (2024) reported the involvement of DVLs in the regulation of flower and leaf formation, root growth under nonbiological stress, and the shade response (Wen et al. 2004; Yang et al. 2024). In addition, plants inoculated with *S. meliloti* and overexpressing MtDVL1 showed a reduction in nodule number, with no effect on nodule shape (Combier et al. 2008). Concerning the smORF peptide family, only ENOD40 is well-known to be involved in plant–microbe interactions, whereas the functions of DVL/RTFL peptides are yet to be determined (Mamaeva et al. 2025).

10.5 Conclusion and Future Prospective

Phytohormones and peptides play an orchestral role in the formation and regulation of nodules in legumes. This mechanism ensures a highly complicated signaling network for an effective SNF process. The regulation of nodulation in legumes is a highly harmonized process, in a highly complex manner controlled by phytohormones, including plant promoting growth (IAA, CKs, Gas, and SA), plant inhibiting growth (ethylene and ABA), and plant regulating growth (JA, SGLs, and BRs) as well as small signaling peptides such as, PTM peptides (CLE, CEP, RGF, and PSK), CRP peptides (NCR, RALF, and nsLTP), non-CRP, non-PTM peptides (GRP and SNARP), and smORF peptides (ENOD40 and DVL). These molecules serve as important messengers that facilitate communication between host plants (legumes) and rhizobia (symbiont), ensuring effective and energy-efficient nodule differentiation. Hormonal crosstalk coordinates the balance between stimulation and suppression of nodulation, while peptides serve key roles in autoregulation and nodule formation. Understanding these regulatory mechanisms provides an important insight into improving nitrogen-use efficiency and legume productivity, which is vital for agriculture sustainability. However, additional investigation is needed to completely explain the cross-talk between phytohormones and peptide signaling, as well as their interplay with environmental factors. Therefore, genetic engineering studies are necessary for modifying peptides and hormone pathways which could lead to greater nodulation efficiency with legumes, especially under stress conditions. Using omics methods in integrative studies may serve to identify novel signaling pathways and regulatory nodes. Advances in the tools of gene editing such as CRISPR/Cas9 provide opportunities to activate hormone and peptide pathways to enhance nodulation efficiency, mainly in nutrient-poor soils. A long-term goal is the transfer of SNF to nonleguminous crops. Understanding the hormonal and peptide-based control systems in legumes will be introductory to this aspiring target.

References

Allen, E.K., Allen, O.N., and Newman, A.S. (1953). Pseudo nodulation of leguminous plants induced by 2-bromo-3,5-dichlorobenzoic acid. *American Journal Botany* 40: 429–435.

Alunni, B., Kevei, Z., Redondo-Nieto, M. et al. (2007). Genomic organization and evolutionary insights on GRP and NCR genes, two large nodule-specific gene families in *Medicago truncatula. Moleular Plant-Microbe Interaction* 20: 1138–1148.

Arrighi, J.F., Barre, A., Amor, B.B. et al. (2006). The *Medicago truncatula* lysine motif-receptor-like kinase gene family includes NFP and new nodule-expressed genes. *Plant Physiology* 142: 265–279.

Bano, A. and Harper, J.E. (2002). Plant growth regulators and phloem exudates modulate root nodulation of soybean. *Functional Plant Biology* 29: 1299–1307.

Bari, R. and Jones, J.D. (2009). Role of plant hormones in plant defence responses. *Plant Molecular Biology* 69 (4): 473–488.

Bedinger, P.A., Pearce, G., and Covey, P.A. (2010). RALFs: peptide regulators of plant growth. *Plant Signaling & Behavior* 5: 1342–1346.

Buer, C.S., Sukumar, P., and Muday, G.K. (2006). Ethylene modulates flavonoid accumulation and gravitropic responses in roots of *Arabidopsis. Plant Physiology* 140: 1384–1396.

Buhian, W.P. and Bensmihen, S. (2018). Mini-review: Nod factor regulation of phytohormone signaling and homeostasis during rhizobia-legume symbiosis. *Frontiers in Plant Science* 9: 1247. https://doi.org/10.3389/fpls.2018.01247.

Bustos-Sanmamed, P., Mao, G., Deng, Y. et al. (2013). Overexpression of miR160 affects root growth and nitrogen-fixing nodule number in *Medicago truncatula. Functional Plant Biology* 40: 1208–1220.

Campbell, L. and Turne, S.R.A. (2017). Comprehensive analysis of RALF proteins in green plants suggests there are two distinct functional groups. *Frontiers in Plant Science* 8: 37.

Chan, P.K., Biswas, B., and Gresshoff, P.M. (2013). Classical ethylene insensitive mutants of the *Arabidopsis* EIN2 orthologue lack the expected 'hypernodulation' response in *Lotus japonicus. Journal of Integrative Plant Biology* 55: 395–408.

Charon, C., Johansson, C., Kondorosi, E. et al. (1997). enod40 induces dedifferentiation and division of root cortical cells in legumes. *Proceedings of the National Academy of Sciences of the United States of America* 94: 8901–8906.

Charon, C., Sousa, C., Crespi, M., and Kondorosi, A. (1999). Alteration of enod40 expression modifies *Medicago truncatula* root nodule development induced by *Sinorhizobium meliloti*. *Plant Cell* 11: 1953–1966.

Cho, M.J. and Harper, J.E. (1993). Effect of abscisic acid application on root isoflavonoid concentration and nodulation of wild type and nodulation-mutant soybean plants. *Plant and Soil* 53: 145–149.

Combier, J.P., Kuster, H., Journet, E.P. et al. (2008). Evidence for the involvement in nodulation of the two small putative regulatory peptide-encoding genes MtRALFL1 and MtDVL1. *Molecular Plant-Microbe Interactions* 21: 1118–1127.

Cooper, J.B. and Long, S.R. (1994). Morphogenetic rescue of *Rhizobium meliloti* nodulation mutants by trans-zeatin secretion. *Plant Cell* 6: 215–225.

de Billy, F., Grosjean, C., May, S. et al. (2001). Expression studies on AUX1-like genes in *Medicago truncatula* suggest that auxin is required at two steps in early nodule development. *Molecular Plant-Microbe Interactions* 14: 267–277.

de Oliveira Carvalho, A. and Gomes, V.M. (2007). Role of plant lipid transfer proteins in plant cell physiology—a concise review. *Peptides* 28 (5): 1144–1153.

Denarie, J., Debelle, F., and Prome, J. (1996). Rhizobium lipo-citooligosaccharide nodulation factors: signaling molecules mediating recognition and morphogenesis. *Annual Review of Biochemistry* 65: 503–535.

Di, Q., Li, Y., Zhang, D. et al. (2022). A novel type of phytosulfokine, PSK-ε, positively regulates root elongation and formation of lateral roots and root nodules in *Medicago truncatula*. *Plant Signaling & Behavior* 17: 2134672.

Ding, Y., Kalo, P., Yendrek, C. et al. (2008). Abscisic acid coordinates nod factor and cytokinin signaling during the regulation of nodulation in *Medicago truncatula*. *Plant Cell* 20: 2681–2695.

Downie, J.A. and Kondorosi, E. (2021). Why should nodule cysteine-rich (NCR) peptides be absent from nodules of some goups of legumes but essential for symbiotic N-fixation in others? *Frontiers in Agronomy* 3: 654576.

Fang, Y. and Hirsch, A.M. (1998). Studying early nodulin gene ENOD40 expression and induction by nodulation factor and cytokinin in transgenic alfalfa. *Plant Physiology* 116: 53–68.

Ferguson, B.J. and Mathesius, U. (2003). Signaling interactions during nodule development. *Journal of Plant Growth Regulation* 22: 47–72.

Ferguson, B.J. and Mathesius, U. (2014). Phytohormone regulation of legumerhizobia interactions. *Journal of Chemical Ecology* 40: 770–790.

Ferguson, B.J., Ross, J.J., and Reid, J.B. (2005). Nodulation phenotypes of gibberellin and brassinosteroid mutants of pea. *Plant Physiology* 138: 2396–2405.

Ferguson, B.J., Foo, E., Reid, J.B., and Ross, J.J. (2011). Relationship between gibberellin, ethylene and nodulation in *Pisum sativum*. *New Phytologist* 189: 829–842.

Ferguson, B.J., Li, D., Hastwell, A.H. et al. (2014). The soybean (*Glycine max*) nodulation-suppressive CLE peptide, GmRIC1, functions interpsecifically in common white bean (*Phaseolus vulgaris*), but not in a supernodulating line mutated in PvNARK. *Plant Biotechnology Journal* 12 (8): 1085–1097.

Fernandez, A., Drozdzecki, A., Hoogewijs, K. et al. (2013). Transcriptional and functional classification of the GOLVEN/ROOT GROWTH FACTOR/CLE-like signaling peptides reveals their role in lateral root and hair formation. *Plant Physiology* 161: 954–970.

Fonouni-Farde, C., Tan, S., Baudin, M. et al. (2016). DELLA-mediated gibberellin signaling regulates nod factor signaling and rhizobial infection. *National Communication* 7: 12636.

Fonseca-Garcia, C., Solis-Miranda, J., Pacheco, R., and Quinto, C. (2021). Non-specific lipid transfer proteins in legumes and their participation during root-nodule symbiosis. *Frontiers in Agronomy* 3: 660100.

Foo, E. and Davies, N.W. (2011). Strigolactones promote nodulation in pea. *Planta* 234: 1073–1081.

Foo, E., Yoneyama, K., Hugill, C.J. et al. (2013). Strigolactones and the regulation of pea symbioses in response to nitrate and phosphate deficiency. *Molecular Plant* 6: 76–87.

Foo, E., Ferguson, B.J., and Reid, J.B. (2014). The potential roles of strigolactones and brassinosteroids in the autoregulation of nodulation pathway. *Annual of Botany* 113 (6): 1037–1045.

Fu ZQ, Guo M, Jeong B, Tian F, Elthon TE, Cerny RL, Staiger D, Alfano JRA (2007) Type III effector ADP-ribosylates RNA-binding proteins and quells plant immunity. *Nature*, 447, 284–288.Gao, S.; Guo, W.; Feng, W.; Liu, L.; Song, X.; Chen, J.; Hou, W.; Zhu, H.; Tang, S.; Hu, J.(2015) LTP3 Contributes to Disease Susceptibility in Arabidopsis by Enhancing Abscisic Acid (ABA) Biosynthesis. Mol. Plant Pathol. 17: 412–426.

Garrocho-Villegas, V., Gopalasubramaniam, S.K., and Arredondo-Peter, R. (2007). Plant hemoglobins: what we know six decades after their discovery. *Gene* 398: 78–85.

Gasser, M., Keller, J., Fournier, P. et al. (2023). Identification and evolution of nsLTPs in the root nodule nitrogen fixation clade and molecular response of Frankia to AgLTP24. *Scientific Reports* 13: 16020.

Gautrat, P., Laffont, C., and Frugier, F. (2020). Compact root architecture 2 promotes root competence for nodulation through the miR2111 systemic effector. *Current Biology* 30: 1339–1345.

Gautrat, P., Laffont, C., Frugier, F., and Ruffel, S. (2021). Nitrogen systemic signaling: from symbiotic nodulation to root acquisition. *Trends in Plant Science* 26: 392–406.

Graham, M.A., Silverstein, K.A.T., Cannon, S.B., and VandenBosch, K.A. (2004). Computational identification and characterization of novel genes from legumes. *Plant Physiology* 135: 1179–1197.

Gresshoff, P.M., Lohar, D., Chan, P.K. et al. (2009). Genetic analysis of ethylene regulation of legume nodulation. *Plant Signaling & Behavior* 4: 818–823.

Gao, S., Guo, W., Feng, W. et al. (2015). LTP3 Contributes to Disease Susceptibility in Arabidopsis by Enhancing Abscisic Acid (ABA) Biosynthesis. *Molecular Plant Pathology* 17: 412–426.

Guinel, F.C. and Geil, R.D. (2002). A model for the development of the rhizobial and arbuscular mycorrhizal symbioses in legumes and its use to understand the roles of ethylene in the establishment of these two symbioses. *Canadian Journal of Microbiology* 80: 695–720.

Gultyaev, A.P., Koster, C., van Batenburg, D.C. et al. (2023). Conserved structured domains in plant non-coding RNA Enod40, their evolution and recruitment of sequences from transposable elements. *NAR Genomics and Bioinformatics* 5: lqad091.

Hayashi, S., Reid, D.E., Lorenc, M. et al. (2012). Transient nod factor-dependent gene expression in the nodulation competent zone of soybean (*Glycine max* [L.] Merr.) roots. *Plant Biotechnology Journal* 8: 995–1010.

Hayashi, S., Gresshoff, P.M., and Ferguson, B.J. (2014). Mechanistic action of gibberellins in legume nodulation. *Journal of Integrative Plant Biology* 56 (10): 971–978. https://doi.org/10.1111/jipb.12201.

Hayat, S. and Ahmad, A. (2003). Soaking seeds of *Lens culinaris* with 28-homobrassinolide increased nitrate reductase activity and grain yield in the field in India. *Annals of Applied Biology* 143: 121–124.

Heckmann, A.B., Sandal, N., Bek, A.S. et al. (2011). Cytokinin induction of root nodule primordia in *Lotus japonicus* is regulated by a mechanism operating in the root cortex. *Molecular Plant-Microbe Interactions* 24: 1385–1395.

Held, M., Hou, H., Miri, M. et al. (2014). *Lotus japonicas* cytokinin receptors work partially redundantly to mediate nodule formation. *The Plant Cell* 26 (2): 678–694. https://doi.org/10.1105/tpc.113.119362.

Hirsch, A.M. (1992). Developmental biology of legume nodulation. *New Phytologist* 122: 211–237.

Hirsch, A.M. and Fujishige, N.A. (2012). Molecular signals and receptors: communication between nitrogen-fixing bacteria and their plant hosts. In: *Biocommunication of Plants*, 255–280. Berlin: Springer.

Hirsch, A.M., Bhuvaneswari, T.V., Torrey, J.G., and Bisseling, T. (1989). Early nodulin genes are induced in alfalfa root outgrowths elicited by auxin transport inhibitors. *Proceedings of the National Academy of Sciences of the United States of America* 86: 1244–1248.

Hirsch, A.M., Fang, Y., Asad, S., and Kapulnik, Y. (1997). The role of phytohormones in plant-microbe symbioses. *Plant and Soil* 194: 171–184.

Hunter, W.J. (2001). Influence of root-applied epibrassinolide and carbenoxolone on the nodulation and growth of soybean (*Glycine max* L.) seedlings. *Journal of Agronomy and Crop Science* 186: 217–221.

Imin, N., Mohd-Radzman, N.A., Ogilvie, H.A., and Djordjevic, M.A. (2013). The peptide-encoding CEP1 gene modulates lateral root and nodule numbers in *Medicago truncatula*. *Journal of Experimental Botany* 64: 5395–5409.

Jardinaud, M.F., Boivin, S., Rodde, N. et al. (2016). A laser dissection-RNAseq analysis highlights the activation of cytokinin pathways by Nod factors in the *Medicago* truncatula root epidermis. *Plant Physiology* 171: 2256–2276.

Jia, Y. and Li, Y. (2023). Genome-wide identification and comparative analysis of RALF gene family in legume and non-legume species. *International Journal of Molecular Sciences* 24: 8842.

Jones, K.M., Kobayashi, H., Davies, B.W. et al. (2007). How rhizobial symbionts invade plants: the Sinorhizobium-Medicago model. *Nature Reviews. Microbiology* 5: 619–633.

Junior, L., Nascimento, L.R.S.A., and Fracetto, G.G.M. (2015). Legume-rhizobia signal exchange: promiscuity and environmental effects. *Frontiers in Microbiology* 6: 945.

Kang, W., Jiang, Z., Chen, Y. et al. (2020). Plant transcriptome analysis reveals specific molecular interactions between alfalfa and its rhizobial symbionts below the species level. *BMC Plant Biology* 20: 293.

Karmakar, K., Rana, A., Rajwar, A. et al. (2015). Legume-rhizobia symbiosis under stress. In: *Plant Microbes' Symbiosis: Applied Facets*, 241–258. India: Springer.

Kereszt, A., Mergaert, P., Montiel, J. et al. (2018). Impact of plant peptides on symbiotic nodule development and functioning. *Frontiers in Plant Science* 9: 1026.

Kevei, Z., Vinardell, J.M., Kiss, G.B. et al. (2002). Glycine-rich proteins encoded by a nodule-specific gene family are implicated in different stages of symbiotic nodule development in *Medicago* spp. *Molecular Plant-Microbe Interactions* 15: 922–931.

Kim, M., Chen, Y., Xi, J. et al. (2015). An antimicrobial peptide essential for bacterial survival in the nitrogen-fixing symbiosis. *Proceedings of the National Academy of Sciences of the United States of America* 112: 15238–15243.

Kisiala, A., Laffont, C., Emery, J.R.N., and Frugier, F. (2013). Bioactive cytokinins are selectively secreted by *Sinorhizobium meliloti* nodulating and nonodulating strains. *Molecular Plant-Microbe Interactions* 26: 1225–1231.

Kondorosi, E., Mergaert, P., and Kereszt, A. (2013). A paradigm for endosymbiotic life: cell differentiation of Rhizobium bacteria provoked by host plant factors. *Annual Review of Microbiology* 67: 611–628.

Kouchi, H. (2011). *Symbiotic Nitrogen Fixation*. John Wiley & Sons, Ltd. ISBN: 978-0-470-74703-2.

Laffont, C. and Frugier, F. (2023). Rhizobium symbiotic efficiency meets CEP signaling peptides. *New Phytologist* 241: 24–27.

Laplaze, L., Benkova, E., Casimiro, I. et al. (2007). Cytokinin act directly on lateral root founder cells to inhibit root initiation. *Plant Cell* 19: 3889–3900.

Laporte, P., Satiat-Jeunemaitre, B., Velasco, I. et al. (2010). A novel RNA-binding peptide regulates the establishment of the *Medicago truncatula*–Sinorhizobium meliloti nitrogen fixing symbiosis. *Plant Journal* 62: 24–38.

Larrainzar, E., Riely, B.K., Kim, S.C. et al. (2015). Deep sequencing of the *Medicago truncatula* root transcriptome reveals a massive and early interaction between nodulation factor and ethylene signals. *Plant Physiology* 169: 233–265.

Lebedeva, M.A., Dobychkina, D.A., Bashtovenko, K.A. et al. (2024). MtCLE35 mediates inhibition of rhizobia-induced signaling pathway and upregulation of defense-related genes in rhizobia-inoculated *Medicago truncatula* roots. *Journal of Plant Growth Regulation* 43: 4941–4956.

Lee, K.H. and LaRue, T.A. (1992). Exogenous ethylene inhibits nodulation of *Pisum sativum* L. cv Sparkle. *Plant Physiology* 100: 1759–1763.

Lei, L., Chen, L., Shi, X. et al. (2014). A nodule-specific lipid transfer protein AsE 246 participates in transport of plant-synthesized lipids to symbiosome membrane and is essential for nodule organogenesis in Chinese milk vetch. *Plant Physiology* 164 (2): 1045–1058.

Li, Q., Li, M., Zhang, D. et al. (2020). The peptide-encoding MtRGF3 gene negatively regulates nodulation of *Medicago truncatula*. *Biochemical Biophysical Research Community* 523: 66–71.

Li, A., Sun, X., and Liu, L. (2022). Action of salicylic acid on plant growth. *Frontiers in Plant Science* 13: 878076.

Li, Q., Shan, D., Zheng, W. et al. (2023). MtRGF3 peptide activates defense responses and represses the expressions of nodulation signaling genes in *Medicago truncatula*. *Acta Biochimica et Biophysica Sinica* 55: 1319–1322.

Li, Y., Di, Q., Luo, L., and Yu, L. (2024). Phytosulfokine peptides, their receptors, and functions. *Frontiers in Plant Science* 14: 1326964.

Lian, B., Zhou, X., Miransari, M., and Smith, D.L. (2000). Effects of salicylic acid on the development and root nodulation of soybean seedlings. *Journal of Agronomy and Crop Science* 185: 187–192.

Lianes, A., Reginato, M., Devinar, G., and Luna, V. (2018). What is known about phytohormones in halophytes? A review. *Biologia.* 73 (8): 727–742.

Lima, R.M., Kylarova, S., Mergaert, P., and Kondorosi, É. (2020). Unexplored arsenals of legume peptides with potential for their applications in medicine and agriculture. *Frontiers in Microbiology* 11: 1307.

Liu, J., Novero, M., Charnikhova, T. et al. (2013). CAROTENOID CLEAVAGE DIOXYGENASE 7 modulates plant growth, reproduction, senescence, and determinate nodulation in the model *Lotus japonicus. Journal of Experimental Botany* 64: 1967–1981.

Liu, H., Zhang, C., Yang, J. et al. (2018). Hormone modulation of legumerhizobial symbiosis. *Journal of Integrative Plant Biology* 60: 632–648.

Lohar, D., Stiller, J., Kam, J. et al. (2009). Ethylene insensitivity conferred by the *Arabidopsis* receptor gene alters the nodulation response of transgenic *Lotus japonicus. Annual of Botany* 104: 277–285.

Mabood, F. and Smith, D.L. (2005). Pre-inoculation of *Bradyrhizobium japonicum* with jasmonates accelerates nodulation and nitrogen fixation in soybean (*Glycine max)* at optimal and suboptimal root zone temperatures. *Physiologia Plantarum* 125: 311–323.

Madsen, L.H., Tirichine, L., Jurkiewicz, A. et al. (2010). The molecular network governing nodule organogenesis and infection in the model legume *Lotus japonicus. Nature Communications* 1: 10.

Maekawa, T., Maekawa-Yoshikawa, M., Takeda, N. et al. (2009). Gibberellin controls the nodulation signaling pathway in *Lotus japonicas. Plant Journal* 58: 183–194.

Mamaeva, A., Lyapina, I., Knyazev, A. et al. (2023). RALF peptides modulate immune response in the moss *Physcomitrium patens. Frontiers in Plant Science* 14: 1077301.

Mamaeva, A., Makeeva, A., and Ganaeva, D. (2025). The small key to the treasure chest: endogenous plant peptides involved in symbiotic interactions. *Plants (Basel)* 14: 378.

Martínez-Abarca, F., Herrera-Cervera, J.A., Bueno, P. et al. (1998). Involvement of salicylic acid in the establishment of the *Rhizobium meliloti* alfalfa symbiosis. *Molecular Plant-Microbe Interactions* 11: 153–155.

Mathesius, U., Schlaman, H.R., Spaink, H.P. et al. (1998). Auxin transport inhibition precedes root nodule formation in white clover roots and is regulated by flavonoids and derivatives of chitin oligosaccharides. *Plant Journal* 14: 23–34.

Mathesius, U., Charon, C., Rolfe, B.G. et al. (2000). Temporal and spatial order of events during the induction of cortical cell divisions in white clover by Rhizobium legumin Sarum bv. trifolii inoculation or localized cytokinin addition. *Molecular Plant-Microbe Interactions* 13: 617–628.

Matsunami, T., Kaihatsu, A., Maekawa, T. et al. (2004). Characterization of vegetative growth of a super nodulating soybean genotype, sakukei4. *Plant Production Science* 7: 165–171.

Missaoui, K., Gonzalez-Klein, Z., Pazos-Castro, D. et al. (2022). Plant non-specific lipid transfer proteins: an overview. *Plant Physiology and Biochemistry* 171: 115–127.

Miyawaki, K., Tabata, R., and Sawa, S. (2013). Evolutionarily conserved CLE peptide signaling in plant development, symbiosis, and parasitism. *Current Opinion in Plant Biology* 16: 598–606.

Mortier, V., De Wever, E., Vuylsteke, M., and Goormachtig, H.M. (2012). Nodule numbers are governed by interaction between CLE peptides and cytokinin signaling. *Plant Journal* 70: 367–376.

Mortier, V., Wasson, A., Jaworek, P. et al. (2014). Role of LONELY GUY genes in indeterminate nodulation on *Medicago truncatula. New Phytologist* 202 (2): 582–593. https://doi.org/10.1111/nph.12681.

Murray, J.D., Karas, B.J., Sato, S. et al. (2007). A cytokinin perception mutant colonized by Rhizobium in the absence of nodule organogenesis. *Science* 315: 101–104.

Mus, F., Crook, M.B., Garcia, K. et al. (2016). Symbiotic nitrogen fixation and the challenges to its extension to nonlegumes. *Applied and Environmental Microbiology* 82: 3698–3710.

Nagata, M. and Suzuki, A. (2014). Effects of phytohormones on nodulation and nitrogen fixation in leguminous plants. In: *Advances in Biology and Ecology of Nitrogen Fixation*. IntechOpen https://doi.org/10.5772/57267.

Nakagawa, T. and Kawaguchi, M. (2006). Shoot-applied MeJA suppresses root nodulation in *Lotus japonicus*. *Plant Cell Physiology* 47: 176–180.

Nakatsukasa-Akune, M., Yamashita, K., Shimoda, Y. et al. (2005). Suppression of root nodule formation by artificial expression of the TrEnodDR1 (coat protein of white clover cryptic virus 1) gene in *Lotus japonicus*. *Molecular Plant-Microbe Interactions* 18: 1069–1080.

Nishida, H., Handa, Y., Tanaka, S. et al. (2016). Expression of the CLE-RS3 gene suppresses root nodulation in *Lotus japonicus*. *Journal of Plant Research* 129: 909–919.

Nukui, N., Ezura, H., and Minamisawa, K. (2004). Transgenic *Lotus japonicas* with an ethylene receptor gene Cm-ERS1/H70A enhances formation of infection threads and nodule primordia. *Plant Cell Physiology* 45: 427–435.

Okamoto, S., Ohnishi, E., Sato, S. et al. (2009). Nod factor/nitrate-induced CLE genes that drive HAR1-mediated systemic regulation of nodulation. *Plant Cell Physiology* 50: 67–77.

Oldroyd, G.E. (2001). Dissecting symbiosis: developments in nod factor signal transduction. *Annals of Botany* 87: 709–718.

Oldroyd, G.E., Engstrom, E.M., and Long, S.R. (2001). Ethylene inhibits the Nod factor signal transduction pathway of *Medicago truncatula*. *Plant Cell* 13: 1835–1849.

Oldroyd, G.E., Murray, J.D., Poole, P.S., and Downie, J.A. (2011). The rules of engagement in the legume-rhizobial symbiosis. *Annual Review of Genetics* 45: 119–144.

Penmetsa, R.V. and Cook, D.R. (1997). A legume ethylene-insensitive mutant hyperinfected by its rhizobial symbiont. *Science* 275: 527–530.

Plet, J., Wasson, A., Ariel, F. et al. (2011). MtCRE1-dependent cytokinin signaling integrates bacterial and plant cues to coordinate symbiotic nodule organogenesis in *Medicago truncatula*. *Plant Journal* 65: 622–633.

Podlešáková, K., Fardoux, J., Patrel, D. et al. (2013). Rhizobial synthesized cytokinins contribute to but are not essential for the symbiotic interaction between photosynthetic Bradyrhizobia and Aeschynomene legumes. *Molecular Plant-Microbe Interactions* 26: 1232–1238.

Prayitno, J., Rolfe, B.G., and Mathesius, U. (2006). The ethylene insensitive sickle mutant of Medicago truncatula shows altered auxin transport regulation during nodulation. *Plant Physiology* 142: 168–180.

Reid, D.E., Ferguson, B.J., Hayashi, S. et al. (2011). Molecular mechanisms controlling legume autoregulation of nodulation. *Annals of Botany* 108: 789–795.

Ribeiro, C.W., Baldacci-Cresp, F., Pierre, O. et al. (2017). Regulation of differentiation of nitrogen-fixing bacteria by microsymbiont targeting of pant thioredoxin S1. *Current Biology* 27: 250–256.

Rightmyer, A.P. and Long, S.R. (2011). Pseudonodule formation by wild-type and symbiotic mutant *Medicago truncatula* in response to auxin transport inhibitors. *Molecular Plant-Microbe Interactions* 24: 1372–1384.

Roberts, I., Smith, S., De Rybel, B. et al. (2013). The CEP family in land plants: evolutionary analyses, expression studies, and role in Arabidopsis shoot development. *Journal of Experimental Botany* 64: 5371–5381.

Roche, P., Debellé, F., Maillet, F. et al. (1991). Molecular basis of symbiotic host specificity in *Rhizobium meliloti*: nodH and nodPQ genes encode the sulfation of lipo-oligosaccharide signals. *Cell* 67: 1131–1143.

Röhrig, H., Schmidt, J., Miklashevichs, E. et al. (2002). Soybean *ENOD40* encodes two peptides that bind to sucrose synthase. *Proceedings of the National Academy of Sciences of the United States of America* 99: 1915–1920.

Roy, S., Liu, W., Nandety, R.S. et al. (2020). Celebrating 20 years of genetic discoveries in legume nodulation and symbiotic nitrogen fixation. *Plant Cell* 32 (1): 15–41. https://doi.org/10.1105/tpc.19.00279.

Roy, S., Torres-Jerez, I., Zhang, S. et al. (2024). The peptide GOLVEN10 alters root development and noduletaxis in *Medicago truncatula*. *Plant Journal* 118: 607–625.

Safi, H., Saibi, W., Alaoui, M.M. et al. (2015). A wheat lipid transfer protein (TdLTP4) promotes tolerance to abiotic and biotic stress in *Arabidopsis thaliana*. *Plant Physiology and Biochemistry* 89: 64–75.

Saifi, F., Biro, J.B., Horvath, B. et al. (2024). Two members of a nodule-specific cysteine-rich (NCR) peptide gene cluster are required for differentiation of rhizobia in *Medicago truncatula* nodules. *Plant Journal* 119: 1508–1525.

Sankari, S., Babu, V.M.P., Bian, K. et al. (2022). A haem-sequestering plant peptide promotes iron uptake in symbiotic bacteria. *Nature Microbiology* 7: 1453–1465.

Sato, T., Fujikake, H., Ohtake, N. et al. (2002). Effect of exogenous salicylic acid supply on nodulation formation of hypernodulating mutant and wild type of soybean. *Journal of Soil Science and Plant Nutrition* 48: 413–420;

Saur, I.M., Oakes, M., Djordjevic, M.A., and Imin, N. (2011). Crosstalk between the nodulation signaling pathway and the autoregulation of nodulation in *Medicago truncatula*. *New Phytologist* 190: 865–874.

Seo, H.S., Li, J., Lee, S.Y. et al. (2006). The hypernodulating nts mutation induces jasmonate synthetic pathway in soybean leaves. *Molecular Cell* 24: 185–193.

Shahid, M.A., Pervez, M.A., Balal, R.M. et al. (2011). Brassinosteroid (24-epibrassinolide) enhances growth and alleviates the deleterious effects induced by salt stress in pea (*Pisum sativum L.*). *Australian Journal of Crop Science* 5: 500–510.

Shimoda, Y., Han, L., Yamazaki, T. et al. (2012). Rhizobialand fungal symbioses show different requirements for calmodulin binding to calcium calmodul independent protein kinase in *Lotus japonicus*. *Plant Cell* 24: 304–321.

Solis-Miranda, J., Juarez-Verdayes, M.A., Nava, N. et al. (2023). The *Phaseolus vulgaris* receptor-like kinase PvFER1 and the small peptides PvRALF1 and PvRALF6 regulate nodule number as a function of nitrate availability. *International Journal of Molecular Sciences* 24: 5230.

Soto, M.J., Fernández-Aparicio, M., Castellanos-Morales, V. et al. (2010). First indications for the involvement of strigolactones on nodule formation in alfalfa (*Medicago sativa*). *Soil Biology and Biochemistry* 42: 383–385.

Sousa, C., Johansson, C., Charon, C. et al. (2001). Translational and structural requirements of the early nodulin gene enod40, a short-open reading frame-containing RNA, for elicitation of a cell-specific growth response in the alfalfa root cortex. *Molecular Cell. Biology* 21: 354–366.

Stacey, G., McAlvin, C.B., Kim, S.Y. et al. (2006). Effects of endogenous salicylic acid on nodulation in the model legumes *Lotus japonicus* and *Medicago truncatula*. *Plant Physiology* 141: 1473–1481.

Staehelin, C., Charon, C., Boller, T. et al. (2001). *Medicago truncatula* plants overexpressing the early nodulin gene Enod40 exhibit accelerated mycorrhizal colonization and enhanced formation of arbuscules. *Proceedings of the National Academy of Sciences of the United States of America* 98: 15366–15371.

Sun, J., Cardoza, V., Mitchell, D.M. et al. (2006). Crosstalk between jasmonic acid, ethylene and Nod factor signaling allow integration of diverse inputs forregulation of nodulation. *Plant Journal* 46: 961–970.

Suzaki, T., Ito, M., and Kawaguchi, M. (2013). Genetic basis of cytokinin and auxin functions during root nodule development. *Frontiers in Plant Sience* 4 (42): 1–6.

Suzaki, T., Yoro, E., and Kawaguchi, M. (2015). Leguminous plants: inventors of root nodules to accommodate symbiotic bacteria. *International Review of Cell and Molecular Biology* 316: 111–158.

Suzuki, A., Akune, M., Kogiso, M. et al. (2004). Control of nodule number by the phytohormone abscisic acid in the roots of two leguminous species. *Plant Cell Physiology* 45: 914–922.

Suzuki, A., Suriyagoda, L., Shigeyama, T. et al. (2011). *Lotus japonicus* nodulation is photomorphogenetically controlled by sensing the red/far red (R/FR) ratio through jasmonic acid (JA) signaling. *Proceedings of the National Academy of Sciences of the United States of America* 108: 16837–16842.

Takatsuto, S., Yazawa, N., Ikekawa, N. et al. (1983). Structure activity relationship of brassinosteroids. *Phytochemistry* 22: 2437–2441.

Taleski, M., Jin, M., Chapman, K. et al. (2024). CEP hormones at the nexus of nutrient acquisition and allocation, root development, and plant–microbe interactions. *Journal of Experimental Botany* 75: 538–552.

Tena, G. (2016). Immunity: RALF to the rescue. *Nature Plants* 2: 16067.

Terakado, J., Fujihara, S., Goto, S. et al. (2005). Systemic effect of a brassinosteroid on root nodule formation in soybean as revealed by the application of brassinolide and brassinazole. *Soil Science and Plant Nutrition* 51: 389–395.

Tirichine, L., Sandal, N., Madsen, L.H. et al. (2007). A gain-of-function mutation in a cytokinin receptor triggers spontaneous root nodule organogenesis. *Science* 315: 104–107.

Truchet, G., Roche, P., Lerouge, P. et al. (1991). Sulfated lipo-oligosaccharide signals of *Rhizobium meliloti* elicit root nodule organogenesis in alfalfa. *Nature* 351: 670–673.

Trujillo, D.I., Silverstein, K.A.T., and Young, N.D. (2014). Genomic characterization of the LEED. PEEDs, a gene family unique to the *Medicago lineage. G3: Genes - Genomes - Genetics* 4: 2003–2012.

Turner, M., Nizampatam, N.R., Baron, M. et al. (2013). Ectopic expression of miR160results in auxin hypersensitivity, cytokinin hyposensitivity, and inhibition of symbiotic nodule development in soybean. *Plant Physiology* 162: 2042–2055.

Upreti, K.K. and Murti, G.S.R. (2004). Effects of brassinosteroids on growth, nodulation, phytohormone content and nitrogenase activity in French bean under water stress. *Biologia Plantarum* 48: 407–4111.

Valdivia, E.R., Hertweck, K.L., Cho, S.K., and Walker, J.C. (2013). DVL/RTFL. In: *Handbook of Biologically Active Peptides*, 2e (ed. A.J. Kastin), 15–19. Elsevier Science Publishing, Academic Press.

Valmas, M.I., Sexauer, M., Markmann, K., and Tsikou, D. (2023). Plants recruit peptides and micro RNAs to regulate nutrient acquisition from soil and symbiosis. *Plants (Basel)* 12: 187.

van Noorden, G.E., Ross, J.J., Reid, J.B. et al. (2006). Defective long distance auxin transport regulation in the *Medicago truncatula* super numeric nodulation mutant. *Plant Physiology* 140: 1494–1506.

van Noorden, G.E., Kerim, T., Goffard, N. et al. (2007). Overlap of proteome changes in *Medicago truncatula* in response to auxin and *Sinorhizobium meliloti*. *Plant Physiology* 144: 1115–1131.

van Spronsen, P.C., Tak, T., Rood, A.M.M. et al. (2003). Salicylic acid inhibits indeterminate-type nodulation but not determinate-type nodulation. *Molecular Plant-Microbe Interactions* 16: 83–91.

van Zeijl, A., Op den Camp, R., Deinum, E. et al. (2015). Rhizobium lipo-chitooligosaccharide signaling triggers accumulation of cytokinins in *Medicago truncatula* roots. *Molecular Plant* 8: 1213–1226.

Vardhini, B.V. and Rao, S.S.R. (1999). Effect of brassinosteroids on nodulation and nitrogenase activity in groundnut (*Arachis hypogaea* L.). *Plant Growth Regulation* 28: 165–167.

Wan, X., Hontelez, J., Lillo, A. et al. (2007). *Medicago truncatula* ENOD40-1 and ENOD40-2 are both involved in nodule initiation and bacteroid development. *Journal of Experimental Botany* 58: 2033–2041.

Wang, D., Griffitts, J., Starker, C. et al. (2010). A nodule-specific protein secretory pathway required for nitrogen-fixing symbiosis. *Science* 327: 1126–1129.

Wang, C., Yu, H., Zhang, Z. et al. (2015). Phytosulfokine is involved in positive regulation of *Lotus japonicus* nodulation. *The American Phytopathological Society* 28: 847–855.

Wasson, A.P., Pellerone, F.I., and Mathesius, U. (2006). Silencing the flavonoid pathway in *Medicago truncatula* inhibits root nodule formation and prevents auxin transport regulation by rhizobia. *Plant Cell* 18: 1617–1629.

Wen, J., Lease, K.A., and Walker, J.C.D.V.L. (2004). A novel class of small polypeptides: overexpression alters *Arabidopsis* development. *Plant Journal* 37: 668–677.

Xu, S., Song, S., Dong, X. et al. (2021). GmbZIP1 negatively regulates ABA-induced inhibition of nodulation by targeting GmENOD40–1 in soybean. *BMC Plant Biology* 21: 35.

Xu, P., Li, H., Xu, K. et al. (2023). Genetic variation in the glycine-rich protein gene BnGRP1 contributes to low phosphorus tolerance in *Brassica napus*. *Journal of Experimental Botany* 74: 3531–3543.

Yang, L., Hu, X., Ren, M. et al. (2024). Stem-cell-expressed DEVIL-like small peptides maintain root growth under abiotic stress via abscisic acid signaling. *Plant Physiology* 194: 2372–2386.

Yu, L., Di, Q., Zhang, D. et al. (2022). A legume-specific novel type of phytosulfokine, PSK-δ, promotes nodulation by enhancing nodule organogenesis. *Journal of Experimental Botany* 73: 2698–2713.

Zhang, Q., Yang, J., Zhou, X. et al. (2022). Soil-borne bacterium *Klebsiella pneumoniae* promotes cluster root formation in white lupin through ethylene mediation. *New Phytologist* 237: 1320–1332.

Zhang, R., Shen, Y., He, J. et al. (2023). Nodule-specific cysteine-rich peptide 343 is required for symbiotic nitrogen fixation in *Medicago truncatula*. *Plant Physiology* 193: 1897–1912.

Zhang, Z., Han, H., Zhao, J. et al. (2025). Peptide hormones in plants. *Molecular Horticulture* 5: 7. https://doi.org/10.1186/s43897-024-00134-y.

11

Regulation of Nutrient Assimilation by Plant Hormones and Its Relation to Crop Yields

Rutupurna Das[1], *Gagan Kumar Panigrahi*[1,*], *and Annapurna Sahoo*[2]

[1] *Department of Zoology, School of Applied Sciences, Centurion University of Technology and Management, Bhubaneswar, Odisha, India*
[2] *Department of Zoology, Jatni College, Utkal University, Khordha, Odisha, India*

11.1 Introduction

11.1.1 Nutrient Assimilation in Plants

Nutrient assimilation is the process of acquiring and transporting the essential nutrients from the environment into metabolic pathways. This is a fundamental process for plant growth, development, and reproduction. Nitrogen (N), phosphorus (P), and potassium (K) are macronutrients that serve as building blocks of key biomolecules, including proteins and nucleic acids. Micronutrients such as zinc (Zn), iron (Fe), and boron (B) are also important to enzymatic activity and cellular signaling (Baillie et al. 2024). It is important for agricultural productivity because of effective nutrient assimilation. It indicates the extent to which the plants can use available resources and allocate these resources to the biomass, grain, and fruit production. This process is directly involved in critical functions, including photosynthesis, respiration, and energy storage (Arsova et al. 2018). Nutrient assimilation is vital to maintain plant resilience to biotic and abiotic stresses and is key to sustainable farming in a changing environment. The availability of vital nutrients varies widely with the properties of the soil, the climate, and agricultural practices. Poor growth and yield may occur due to limitations in nutrient supply (Canut et al. 2016). On one hand, applications of fertilizers are rather excessive, leading to such negative environmental impacts as soil degradation and eutrophication of water bodies; on the other hand, the process of fertilizing itself is not very profitable. Improving the efficiency of nutrient assimilation at the physiological and molecular levels is a strategy to reduce these problems while reducing agriculture's ecological footprint and maintaining food security (Chen et al. 2020).

*Corresponding author.

Peptide Hormones in Plant Development and Stress Responses, First Edition. Edited by Vijay Pratap Singh and Samiksha Singh.

11.1.2 Challenges in Agriculture: Nutrient Use Efficiency and Crop Yield Gaps

There are a number of serious challenges that modern agriculture needs to address, and nutrient use efficiency (NUE) is a particularly urgent issue. NUE is the term for plants' efficiencies in absorbing and using nutrients that are added as fertilizers (Corcilius et al. 2017). After the use of nitrogen and phosphorus fertilizers, 80–90% of the nutrients flows or leaches into the environment without being accessed by crops (Dao et al. 2022; Zhang et al. 2024b,c). It is an inefficient system that causes a lot of economic losses for farmers and affects people in different harmful ways with regard to environmental issues (groundwater contamination, greenhouse gas emissions, biodiversity loss). Crop yield gaps are indicative of the difference between potential crop production, or what can be produced under perfect conditions, and what is seen in the field (Fedoreyeva et al. 2022). Contributing to these gaps are poor nutrient uptake, restriction of water supply, pest infestation, and suboptimal management practices. Though technological improvements have increased yields in some areas (for instance, in developing countries), the resource constraints – especially as their soils are deteriorating – may still be a great issue (Hastwell et al. 2014). To address these challenges, therefore, there needs to be a shift toward more resource-efficient farming practices and away from intensive farming. An innovative solution is the hormonal regulation of the nutrient processes (Ito et al. 2006). This leads to plants with higher productivity and reduced dependence on chemical fertilizers by improving the nutrient acquisition and utilization through hormonal pathways. This represents a sustainable solution to close yield gaps while addressing environmental and economic issues associated with inefficient nutrient use (Huang et al. 2022).

11.1.3 Role of Plant Hormones as Key Regulators of Nutrient Processes

Plant hormones, also known as phytohormones, are small signaling molecules that control a wide range of physiological and developmental processes in plants. Root hair formation and root growth control, nutrient transporter expression, and redirection of nutrients within the plant are central functions in nutrient assimilation, and hormones play a central role in these functions. These hormones allow the plant to sense environmental changes and adjust its physiology. For example, auxins are known to induce root elongation as well as lateral root development to boost the surface area for nutrient uptake (Zhong and Qu 2019). Cytokinin regulates nitrogen metabolism and remobilizes the nutrients to actively growing tissues, namely to microtissues. Brassinosteroids provide the basis for the uptake and use of phosphorus, and gibberellins further promote nitrogen assimilation into amino acids and proteins. Plants are helped to optimize nutrient use in adverse conditions, such as drought or limited nutrients by stress-related hormones such as abscisic acid (ABA) and ethylene. Some of the interactions between hormones and nutrient pathways go beyond individual signaling pathways (Ohkubo et al. 2017). There is hormonal cross talk that allows plants to integrate multiple environmental signals to prioritize specific responses depending on resource availability. For example, auxins and cytokinins play antagonistically to control root and shoot development, and ABA and ethylene participate in a cooperative manner to modulate nutrient uptake under stress conditions. Recent advances in science have made the study of hormonal regulation in nutrient processes expand rapidly (Nguyen and Kim

2023). This provides the opportunity to breed crop varieties better able to use nutrients and resist stresses. In addition, hormone-based treatments and biostimulants are practical tools for improving nutrient assimilation in field conditions (Liu et al. 2021). This chapter provides an overview of the relationship between nutrient assimilation and plant hormones, which is important for ensuring agricultural productivity and sustainability.

11.2 Overview of Nutrient Assimilation in Plants

11.2.1 Components and Importance of Nutrient Assimilation

The assimilation of nutrients in plants is referred to as the process by which the plant takes in, metabolizes, and assimilates essential nutrients into plant growth, development, and metabolic processes. It is a complex series of steps that involves absorbing inorganic ions from the soil, transporting them through plant tissues, and then changing them into organic molecules such as amino acids, proteins, and nucleic acids. As plants must synthesize vital molecules to create cellular structure and function, they depend on a wide range of chemical elements for nutrient assimilation (Igarashi et al. 2012). The macronutrients present in these are nitrogen, phosphorus, and potassium, and the micronutrients include iron, manganese, and zinc. Efficient nutrient assimilation supports growth, optimizes disease resistance, and enhances reproductive success. At the same time, it contributes to productivity in agricultural systems and global food security (Hirayama and Mochida 2022).

11.2.2 Macronutrients and Micronutrients: Roles in Plant Growth

The macronutrients are essential components of plants; they are required in large amounts in order for plants to grow, develop, and function metabolically. The primary macronutrients are nitrogen (N), phosphorus (P), and potassium (K), and the secondary macronutrients are calcium (Ca), magnesium (Mg), and sulfur (S). These are very crucial nutrients for plant health and productivity (Guillou et al. 2022). Chlorophyll, vital for photosynthesis, requires nitrogen, and nitrogen is also a base material of amino acids and proteins, which contribute to vegetative growth on the whole. Phosphorus is cited for its role in energy transfer within the plant with the help of compounds like ATP and supporting root development and is also necessary for flower and seed production (Hagelthorn and Fletcher 2023). However, potassium increases water regulation, supports cell wall integrity, and enhances disease and environmental stress resistance. Secondary macronutrients are just as important; calcium helps in maintaining the integrity of the cell wall and in membrane function, magnesium plays a central role in chlorophyll molecules and activates many enzymes, whereas sulfur is needed for the synthesis of specific amino acids and vitamins. These macronutrients, along with other important compounds found in plants, such as organic acids, are the foundation for plant nutrition, which forms the basis for plant life, reproduction, and other challenges of the plants' environment (Kaeothip et al. 2013; Panigrahi and Satapathy 2020a).

Although macronutrients are required by plants in relatively larger amounts as compared with micronutrients, they are equally essential for plant growth, development, and metabolic processes. Among these nutrients are Fe, Mn, Zn, Cu, B, Mo, Cl, and Ni, each of which contributes to different steps in plant physiology. For chlorophyll synthesis, iron is very important and it is a cofactor in many enzymatic reactions. Water splitting and activation of various enzymes by means of the splitting of water molecules are the roles of manganese in photosynthesis. Protein synthesis, hormone regulation, and activation of enzymes require zinc, whereas photosynthetic electron transport requires copper, and zinc is useful for lignin synthesis (Fernandez-Fernandez et al. 2023). Molybdate is an essential regulator of nitrogen fixation and nitrate reduction while boron is required for cell wall formation and sugar transport, and reproduction. It's a known factor that chlorine assists in osmotic regulation, photosynthetic oxygen evolution, and nickel, although needed in very small amounts, activates specific enzymes used in nitrogen metabolism (Furumizu et al. 2021; Panigrahi and Satapathy 2020b). However, even with their minimal concentrations, a deficiency of any of these micronutrients can considerably undermine plant development and productivity and emphasizes their very important position within the nurturing of superior plant health and function. These larger particles allow the soil to drain quickly and hold fewer nutrients (Blackburn et al. 2020). Clay soils, on the other hand, have small particles that trap water and nutrients well, but the speed of drainage can be slower. Nutrient cycling depends on the formation of soil organic matter and the activity of microorganisms. Organic matter releases nutrients in forms that plants can use and digest. The fertilization practices also influence the soil fertility that has to be made to particular nutrient needs of different crops (Boschiero et al. 2019). Temperature and rainfall patterns directly influence the rate of nutrient uptake. For example, nutrient uptake might be limited during drought as water is limited. High temperatures can accelerate nutrient leaching, especially in areas with heavy rainfall (Chaulagain and Frugoli 2021).

11.2.3 Genetic Basis of Nutrient Assimilation Efficiency

A plant's ability to adapt to its environment and increase nutrient uptake and utilization efficiency forms the genetic basis of nutrient assimilation efficiency (NAE). These traits are genetic and also influence nutrient assimilation, as explained in the following:

Variations in root architecture: Root length, density, and branching patterns can play a role in how efficiently roots take up nutrients. Often, plants with deep and extensive roots are better able to access nutrients of the soil (Chapman et al. 2024; Sahoo and Satapathy 2021).

Transporter genes: Some plant species may contain different transporter genes which increase their ability to take up specified nutrients. Thus, plants with improved phosphate transporters are able to efficiently capture phosphorus from low availability soils. Legumes and some other plants have symbiotic relationships with nitrogen-fixing bacteria, from which small pieces of genetic material have been transferred into *Escherichia coli* for scientists to study these genes (Eckardt et al. 2024). These plants, therefore, take nitrogen from the atmosphere, bypassing the need for soil nitrogen, through the genetic regulation of the nitrogen fixation genes (Zhou et al. 2024). Key genes in molecular pathways of nutrient sensing and uptake are relevant to studying nutrient assimilation. Related genes to the regulation of nutrient transporters or nutrient assimilation enzymes, such as nitrate reductase and glutamine synthetase, are included.

11.3 Plant Hormones and Their Roles in Nutrient Assimilation

11.3.1 Overview of Plant Hormones: Diversity and Signaling Mechanisms

Phytohormones are signaling molecules that regulate many physiological processes such as growth, development, stress responses, and nutrient assimilation. They are synthesized in small amounts but act with power on the plant. Plant hormones are divided into five major classes: auxins, cytokinins, gibberellins, ABA, and ethylene (Friso and van Wijk 2015). More recently, recognized hormones that play roles in physiological processes include brassinosteroids (BRs), jasmonates, and salicylic acid (SA) (Furumizu and Aalen 2023). Plant hormones are known to bind to specific receptors and initiate signal transduction pathways, which result in changes in gene expression and cellular behavior. These pathways may be very complex, and many of them involve multiple signal transduction networks (Haruta et al. 2014; Panigrahi and Satapathy 2020c). Hormones usually work in a context-dependent manner, i.e. their action is regulated by interactions between hormones, environmental factors, or cues to developmental events (Hirakawa and Sawa 2019). These signal integrations in plants enable the plants to grow, remain metabolically active, and respond to nutrient availability. Plant hormones are involved centrally in regulating nutrient uptake, distribution, and storage in plant nutrient assimilation (Greifenhagen et al. 2024). Hormonal control also helps plants respond to changing nutrient levels in soil, which aids plants in efficiently using nitrogen. Additionally, auxins and cytokinins, as well as other hormones such as ABA and ethylene in some way, influence root growth and nutrient transporter expression and perform stress responses related to nutrient acquisition (Johns et al. 2021; Je et al. 2018).

11.3.2 Auxins: Role in Root Development and Nutrient Transporter Regulation

Auxins are a group of plant hormones that are mostly linked with controlling growth and development, particularly through cell elongation and cell division (Lebedeva et al. 2020; Matsubayashi 2018). Auxins play one of the key roles in the assimilation of nutrients by influencing root development. Auxins also encourage the plant to seek available nutrients by elongating root cells and stimulating lateral root formation. A well-developed root system increases the amount of surface area for the uptake of nutrients. In fact, auxins are important in regulating nutrient transporters (Matsuzaki et al. 2010; Millar et al. 2019; Ogawa-Ohnishi et al. 2022). Auxins have also been identified to be able to modulate expression of transporters for essential nutrients, such as nitrogen, phosphate, and potassium in other studies. Auxins may adjust the root system architecture and transporter activity to ameliorate the responses to nutrient deficiencies (Okamoto et al. 2022). Auxin signaling pathway interacts with phosphate transporter pathway to ensure that there is enough phosphate for plant when phosphate is scarce. Moreover, auxins are important components in the signaling pathways, permitting the plant to adapt to changing nutrient conditions and ensuring that nutrients are distributed in a balanced way throughout the whole plant. Coordination of root and shoot growth, as well as nutrient uptake (Figure 11.1), is facilitated by the interaction of auxins with other hormones such as cytokinins and gibberellins (Narasimhan and Simon 2022).

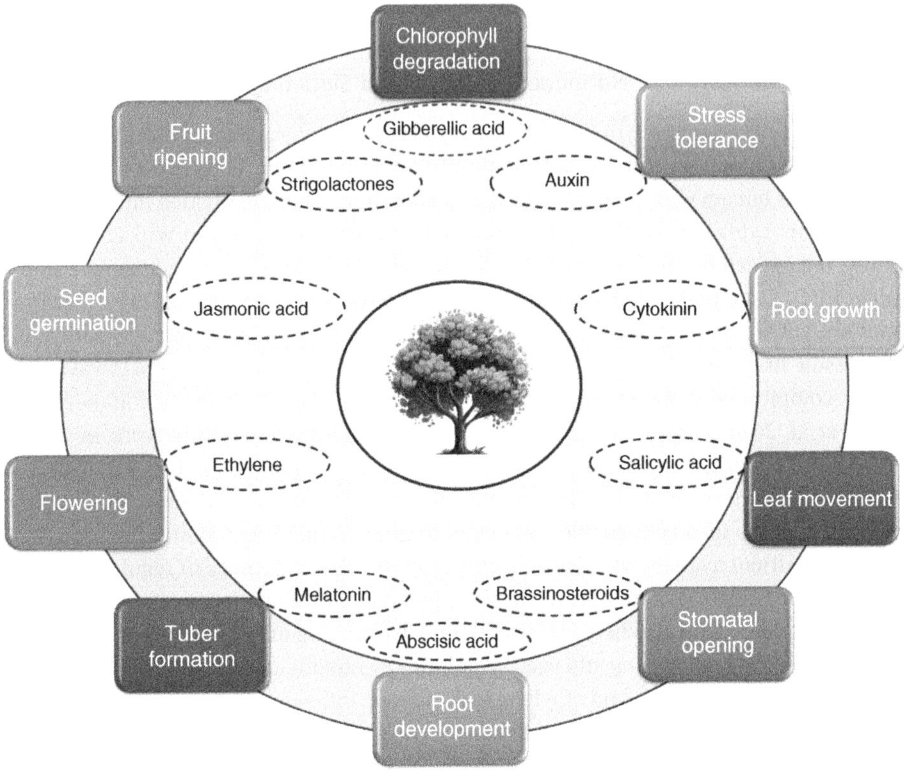

Figure 11.1 Regulation of cellular metabolism by plant hormones. *Source:* available / Adobe Stock Photos.

11.3.3 Cytokinin: Function in Nutrient Redistribution and Photosynthetic Maintenance

Hormones belonging to the class of cytokinins promote cell division, and they are involved in many plant growth aspects, such as root and shoot development. Cytokinin serves an important role in nutrient assimilation and in particular nutrient redistribution (Muller and Harrison 2019; Murphy et al. 2012; Nakagami et al. 2024). Cytokinins can also help shift the movement of dietary nutrients inside the plant so that it transports those nutrients into the tissue where they are most needed. It is particularly important when growth is active and when some nutrients become limiting in the soil (Nakayama et al. 2017; Ohki et al. 2011; Oh et al. 2018). Also, maintaining photosynthetic activity is important for nutrient assimilation, and cytokinins are involved in this process. Cytokinins promote chloroplast development and support photosynthesis to energize plants to provide the required energy for nutrient transport and assimilation (Motomitsu et al. 2015; Pastor-Fernandez et al. 2023). Furthermore, cytokinin can affect the transport of sugars between the roots and shoots as well as other organic compounds, which need to supply energy for the uptake of nutrients (Pecenkova et al. 2024; Phour et al. 2020; Qian et al. 2018). Cytokinins affect the expression of nutrient transporters, predominantly of siRNA of potassium, nitrogen, and phosphate uptake. Such hormones help the plant to maintain nutrient distribution under stress conditions so that the plant can still function properly when the resources are limited (Figure 11.1; Okamoto et al. 2013; Pan et al. 2024; Patel et al. 2018).

11.3.4 Gibberellins (GAs): Influence on Seed Germination and Root-Shoot Growth Balance

Gibberellins (GAs) are a plant hormone known to promote seed germination, stem elongation, and flowering. And in terms of nutrient assimilation, GAs alter the balance of shoot and root growth (Ren et al. 2024; Reichardt et al. 2020; Amano et al. 2007). GAs induce shoot elongation to increase the plant's capacity to intercept sunlight for photosynthesis, but at the same time, in coordination with GAs, the growth of roots is induced to ensure the plant can obtain sufficient nutrients from the soil (Aggarwal et al. 2020; Bashyal et al. 2024; Chapman et al. 2024). During seed germination, GAs keep the seeds breaking dormancy and induce the growth of roots and shoots, helping the early establishment of the plant. On the other hand, for proper nutrient acquisition, a balance between roots and shoots is crucial; a vigorous root system means nutrients are taken up from the soil and vice versa, healthy shoots allow for efficient photosynthesis and the transport of assimilated nutrients to various parts of the plant (Bhalla et al. 2025; Avalos-Rangel et al. 2024; Bari and Jones 2009). Other hormones, like auxins and cytokinins, also interact with gibberellins and other hormones, so there is always a coordinated response to nutrient availability. Gibberellins divert their growth-promoting capacity toward roots at the expense of shoots under nutrient-limiting conditions to promote nutrient acquisition (Abarca et al. 2021; Chang and Xiao 2025; Ding et al. 2023).

11.3.5 Abscisic Acid (ABA): Stress Adaptation and Stomatal Regulation

Most known for its role in stress responses, abscisic acid (ABA) is a plant hormone that mainly regulates water loss, as with other plant hormones, under drought conditions. ABA helps plants adapt to environmental stresses that might limit the nutrient uptake in the metabolic term (Andersen et al. 2018; Carbonnel et al. 2022; de Bang et al. 2017). For example, ABA has been shown to regulate stomatal closure during drought. This action is required to reduce water loss but, in turn, decreases nutrient uptake via limiting the amount of water and dissolved minerals entering the roots. Stress condition imposes a role for ABA besides its role of water regulation, where it is also involved in nutrient allocation (Figure 11.1). ABA modulates the expression of nutrient transporters to ensure that the most essential nutrients will be prioritized when plants are nutrient-limited (Farrokhi et al. 2008; Gancheva et al. 2019). Over and above interaction with other hormones, ABA also senses and responds to changes in the nutrient content of the soil. ABA helps the plant maintain metabolic homeostasis under nutrient stress, and the regulation of metabolism by ABA includes the synthesis of proteins and enzymes involved in nutrient uptake and assimilation (Hsiao and Yamada 2020).

11.3.6 Ethylene: Induction of Root Hairs and Stress-Response Mechanisms

It is a gaseous hormone involved in many physiological processes including senescence, fruit ripening, and stress responses. With an important role in relation to nutrient assimilation in relation to root hair formation, ethylene increases the plant's ability to absorb nutrients from the soil (Kaufmann and Sauter 2019). Root hairs are specialized epidermal cells in the root that are promoted to develop by ethylene and have an increased surface area for absorption of nutrients, primarily in the case of limited resource availability (Kokla et al. 2024; Laffont et al. 2020; Kondo et al. 2011). It is also very important for the plant to respond to abiotic stresses, for example, nutrient deficiencies, salinity, and drought, through ethylene. By influencing the plant's root system architecture under such conditions, ethylene signaling controls root elongation and branching to allow for increased nutrients available to the plant (Lalun et al. 2024; Leasure and He 2012). Moreover, ethylene influences nutrient uptake together with other hormones such as auxins and ABA in coordination with the stress-response mechanisms that integrate growth and nutrient acquisition (Figure 11.1).

11.3.7 Brassinosteroids (BRs): Stress Mitigation and Enhancement of Nutrient Uptake

The brassinosteroids (BRs) are a class of steroid hormones which are critical to their function of increasing plant stress tolerance and increasing plant growth (Kuromori et al. 2022). BRs stimulate rapid cell expansion, elongation, and division, which are essential for root development and nutrient acquisition. Moreover, BRs have been shown to increase the efficiency of nutrient uptake through improving root function. In addition, they can regulate the activity of key enzymes involved in nutrient assimilation, including those involved in nitrate and phosphate metabolism, and boost the plant's capacity to absorb and utilize nutrients (Lyapina et al. 2024; Liu et al. 2022b). Additionally, BRs aid in the plant's resilience to oxidative stress, which can cause nutrient deficiency or other environmental stresses, increasing nutrient acquisition in stressful conditions (Figure 11.1).

11.3.8 Emerging Roles of Jasmonates and Salicylic Acid in Nutrient Signaling

Jasmonates and SA are plant hormones principally involved in defense response; however, growing data demonstrate that they might be equally essential for nutrient signaling. Plant responses to herbivory and pathogen attack are known to be mediated by jasmonates, which also affect nutrient uptake and distribution (Liu et al. 2022a; Mens et al. 2021). They modulate the expression of other types of nutrient transporters, especially of those related to nitrogen and phosphorus uptake. Besides that, SA, traditionally considered involved in plant immune responses, has also been involved in the regulation of nutrient acquisition. Root architecture and the expression of nutrient transporter genes are influenced by SA under conditions of nutrient deficiency. It regulates plants to take up specific nutrients efficiently and promote the overall metabolic activity of the plant (Figure 11.1). Jasmonates and SA interact with other hormones like auxins, cytokinins, and ethylene, creating a complex hormonal network that governs plant responses to nutrient availability, environmental conditions, and biological stressors (Qiu et al. 2022).

11.4 Mechanisms of Hormonal Regulation of Nutrient Assimilation

11.4.1 Hormonal Control of Root System Architecture

Root systems are key components in determining how plants can assimilate nutrients (how a plant explores the soil to access available nutrients) and are determinants of architecture. The development and remodeling of the root system are regulated by plant hormones, particularly in response to environmental cues and nutrient availability (Reid et al. 2011; Ren

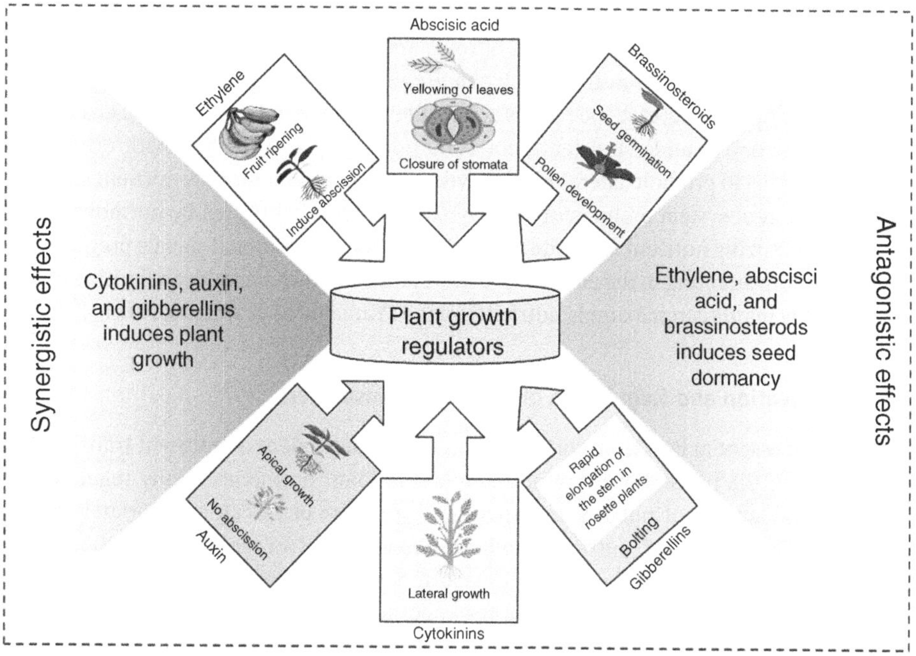

Figure 11.2 Hormonal regulation of plant development through plant growth regulators. *Source:* Muhammad / Adobe Stock Photos.

et al. 2019). All of these plant hormones, namely auxins, cytokinins, ethylene, and gibberellins, regulate root system architecture by regulating processes such as root elongation, lateral root formation, root hair development, and branching pattern. Auxins are essential for overall root architecture. They help to elongate root cells, and their concentration gradient inside the root tips is responsible for the direction of the growth. Also, auxins participate in the priming of lateral root formation (Roman et al. 2022; Panigrahi et al. 2021a,b,c). Auxin signaling stimulates the formation of lateral roots near nutrient-rich regions in the soil when nutrients such as nitrogen or phosphate are limiting, providing access to the nutrients (Figure 11.2). However, auxin levels that are too high can inhibit lateral root formation and reduce the root system under nutrient-poor soils (Yan 2024; Xiao and Zhou 2023).

Cytokinin plays a more complex role in the root system architecture. Promotion of root branching and the development of root meristems is influenced by these hormones. Auxins, to some extent, help regulate root growth in coordination with cytokinin, and they are usually involved in nutrient redistribution (to maintain that the roots are distributed in the soil as optimally as possible, so they get the maximum amount of nutrient acquisition). Cytokinin also encourages the formation of root hairs, which increases root surface area and nutrient absorption (Wulf et al. 2024; Villalobos-Lopez et al. 2022).

In addition, ethylene is also needed to define root system architecture under stress conditions. The most obvious manifestation of ethylene's role is seen during environmental stresses such as low soil compaction, flooding, or nutrient deficiency. In these conditions, ethylene causes the formation of root hairs, aiding in the plant's enhancement of using nutrients (Verma et al. 2016; Vie et al. 2015; Xie et al. 2022). Furthermore, the balance between primary root elongation and lateral root formation is also influenced by ethylene and it allows the root system to acclimate to changing soil conditions (Figure 11.2).

It is well known that GAs affect shoot growth, but they also influence the development of root systems. GAs control the elongation of roots and their effect depends on the nutrient availability in the soil. Under nutrient-rich conditions, GAs may cause plant roots to differentiate so as to further increase the plant's acquisition of water and minerals (Xu et al. 2015; Yang et al. 2023). Yet under nutrient-poor conditions, GAs can interact with other hormones like auxins to promote lateral root growth in order to scout for new nutrient sources. In conclusion, root system architecture is a dynamic process controlled by hormones with the aim of optimizing nutrient acquisition. The integration of hormonal signals prevents the root system from adapting to the environment until nutrients are available in limiting quantities, whereby plants can maximize nutrient uptake (Yamaguchi et al. 2016; Xu et al. 2013).

11.4.2 Activation and Regulation of Nutrient Transporters

The uptake of essential ions from soil into plant cells is dependent on nutrient transporters that are part of the process of nutrient assimilation. Plant hormones tightly regulate the expression and activity of nutrient transporters so that the plant can appropriately adapt to changing availability of nutrients and the environment (Yoshida et al. 2021; Tost et al. 2021). The expression of nutrient transporters for nitrogen, phosphorus, and potassium is specifically affected by auxins. For instance, auxins influence the activity of specific root nitrate and phosphate transporters to ensure that the plant can efficiently acquire these important nutrients when soil levels are minimal (Stintzi and Schaller 2022; Takahashi et al. 2018). The expression of the auxins, besides their roles in different stages of development, also determines transporter expression in different plant tissues, playing a role in regulating the distribution of nutrients. Members of this class of cytokinin promote the allocation of nutrients (and, by extension, the nutrients thus transported) to sites of high metabolic demand (in the case of plants, root and shoot growth from callus) (Takahashi et al. 2020; Taleski et al. 2020). These hormones dictate the expression of nutrient transporters for nitrogen, potassium, and phosphorus, primarily in relation to the stress condition or the nutrient limitation. As shown in Figure 11.2, cytokinin also helps redistribute nutrients between different parts of the plant to ensure that vital organs receive proper amounts of nutrients to drive growth (Royek et al. 2022; Salvi et al. 2021).

It has been established that gibberellins could influence nutrient transporter activity, particularly those concerning the expression of nitrogen uptake. The genes encoding the N transporter genes are regulated by the GAs, and under favorable conditions, GAs play a role in the efficient acquisition of N from the roots (Rzemieniewski et al. 2024; Sauter 2015). The actions of GAs all coordinate with other hormones, such as auxins and cytokinins to adjust the expression of nutrient transporters and root growth in response to nutrient deficiencies. During stress conditions, particularly water stress and nutrient limitations, ABA is a major regulator of nutrient transporter activity (Schardon et al. 2016; Setty et al. 2024). One of the uses for ABA is to reduce expression of some transporters, such as those for nitrate and phosphate uptake to avoid nutrient waste where a lack of water or nutrients occurs. A further function of ABA is that it can affect the transport of nutrients, including potassium and calcium, involved in osmotic regulation and stress tolerance. One of the notable aspects of ABA's role in nutrient transporter regulation is that it helps in the conservation of resources under adverse environmental conditions. Regulation of the expression of nutrient transporters by ethylene also occurs to counteract soil compaction, flooding, and nutrient deficiency. Incorporation of ethylene increases the expression of transporters involved in the uptake of phosphate and nitrogen (Scott et al. 2024; Sang et al. 2025). Furthermore, ethylene regulates the root's capacity to absorb nutrients under low-oxygen conditions, either during waterlogging or when nutrient uptake is hindered because of bad soil aeration (Tabata and Sawa 2014; Suh et al. 2024).

11.4.3 Signal Integration for Environmental Adaptation

Plants are constantly subject to fluctuating environmental conditions, such as nutrient availability, soil moisture, temperature, and light. In order to adapt to these changing conditions, plants need to integrate hormonal signals that help them assimilate nutrients as efficiently as possible and support growth (Tavormina et al. 2015; Tan et al. 2024). Different hormonal signals integrate through complex networks in which several hormones work in concert to coordinate plant responses to the environment.

Signaling pathways: When the soil has low nutrient availability, auxins, cytokinins, and ethylene-signaling pathways affect root architecture and increase in nutrient content are important. For example, auxins and cytokinins interact to promote lateral root growth and the expression of nitrogen transporters under low nitrogen conditions, and ethylene aids the plant in altering its root system to maximize the nutrient access under poor nitrogen conditions (Taleski et al. 2024; Vanyushin et al. 2017). Under drought, ABA has a centralized role in the integration of the signal that controls water conservation mechanisms, including stomatal closure and root elongation, during water stress. Moreover, ABA can also interact with other hormones like cytokinins and ethylene to promote the diversion of plant energy away from nonessential shoot growth, but back into essential things such as root development and nutrient uptake (Taleski et al. 2018; Toruno et al. 2016; Yu et al. 2024a,b). Temperatures and light affect the activity of various plant hormones, which influence nutrient assimilation. While phloem-mobile hormones such as gibberellins and cytokinins are used to adjust plant growth in response to light intensity and temperature changes, auxins and brassinosteroids are responsible for the regulation of root growth in order to effectively uptake nutrients. Root growth may be favored over shoot growth under low light or cool conditions in order to optimize nutrient acquisition and energy production (Yu et al. 2024a,b; Zarsky et al. 2024).

11.4.4 Synergistic and Antagonistic Interactions among Hormones

Nutrient assimilation is not controlled by one hormone, but multiple hormones interact in intricate ways to regulate it. Interactions can be synergistic or antagonistic and they affect nutrient uptake, transport, and utilization in different ways. Many of the hormones engage synergistic interactions, working together to increase nutrient assimilation. For example, auxins and cytokinins function together to stimulate root elongation and lateral root formation under nutrient-limited conditions (Zelman and Berkowitz 2023; Panigrahi et al. 2021a). Likewise, auxins and gibberellins also frequently synergize to help promote root in addition to shoot growth in reaction to environmental cues. Together with these hormones working, plants are able to use the optimal strategy for taking in nutrients and growing roots. Other instances might be where hormones act as antagonists of nutrient assimilation processes.

For example, the relationship between auxins and ethylene is usually antagonistic (Yuan et al. 2021; Zelman and Berkowitz 2023). Auxins promote root elongation and lateral roots; ethylene inhibits primary root elongation and promotes root hair formation. Such antagonistic interaction ensures that root systems respond to more than one environmental condition such as soil compaction or nutrient scarcity. In fact, similar to cytokinin and ABA, cytokinins tend to antagonize ABA, promoting growth under favorable conditions and inhibiting growth under stress conditions. How hormones control nutrient assimilation constitutes molecular mechanisms of activating specific genes that encode nutrient transporters, metabolic enzymes, and stress-response proteins (Zhang et al. 2024a,b,c; Zhou et al. 2022). The production of these genes is controlled by hormonal signaling pathways that orchestrate activity of transcription factors which in turn control their expression of nutrient-responsive genes. Consequently, in order to properly acquire and assimilate nutrients, hormones are integrated for optimal gene activation under appropriate conditions. Environmental cues also influence hormonal control of nutrient-responsive genes. For example, ABA regulates genes related to root growth and nutrient transporter expression under drought conditions to save resources. As is the case with nutrient deficiency, jasmonate and SA help increase gene expression to elevate nutrient uptake (Zhong et al. 2022; Zhang et al. 2022a,b). In summary, a complex network of signaling molecules and transcription factors is molecularly linking hormones to nutrient-responsive genes to permit plants to change their growth and metabolic processes according to changing nutrient availability and environmental conditions (Stewart and Ronald 2022; Zhang et al. 2019, 2024a,b,c).

11.5 Hormonal Regulation and Crop Yields

11.5.1 Relationship Between Nutrient Assimilation Efficiency and Yield Potential

Nutrient assimilation efficiency is a key determinant of plant growth and yield. This is a term used to describe a plant's ability to successfully take up nutrients from the soil, transport them to the desired tissues as well as incorporate them into metabolic processes (Majumder et al. 2025). Efficient assimilation of nutrients guarantees optimal physiological functions, including photosynthesis, energy metabolism, and protein synthesis, and so on which are directly linked with biomass accumulation and yield potential (Simon and Dresselhaus 2015; Song et al. 2022). The role of plant hormones in modulating root architecture, nutrient transporter activity, and nutrient remobilization in the plant is a vital part of the regulation of NAE. For instance, auxins induce root elongation and root branching (lateral root formation) to increase nutrient-uptake surface area. As is the case with cytokinins, nitrate transporter genes are expressed in response to cytokinins to take up and transport nitrate to actively growing tissues (Stewart and Ronald 2022; Zhang et al. 2022a,b). Brassinosteroids increase uptake of essential nutrients, such as phosphorus, and gibberellins increase the activity of enzymes used in nitrogen assimilation. The relationship between NAE and yield potential is highly evident under nutrient-scarcity stress. Plants with higher NAE better support growth and development in conditions of limited nutrient availability, which eventually translates into better yields. Rather, low efficiency in nutrient assimilation causes poor growth, poor grain filling, and reduced yield (Zhang et al. 2023, 2024a,b,c). Hormonal mechanisms can make farmers more productive with less or the same amount of fertilizers. This is significant for sustainable agriculture, since it avoids the negative environmental impacts of nutrient runoff and eutrophication. Knowledge of hormone-evoked NAE molecular mechanisms forms the cornerstone to develop high-yielding, nutrient-efficient crops. This can be done through genetic engineering, the use of marker-assisted breeding or application of hormone-based biostimulants. Increasing NAE, in addition to its positive impacts on yield potential, also provides food security by optimizing resource use (Ryslava et al. 2013; Roy et al. 2018).

11.5.2 Case Studies: Hormone-Mediated Yield Improvements in Major Crops

Having demonstrated that plant hormones can increase yield in major crops (e.g. wheat, rice, maize, and soybean), they are now ready for field use. These crops are used as staple food sources throughout the world and sustenance for food security is dependent on their productivity. Research and farmers have shown yield improvement under different scenarios by modulating hormonal pathways (Mosher et al. 2013; Lin et al. 2023).

Auxins enhance root elongation and branching in wheat roots, thereby increasing nitrogen uptake and improving root growth. Increasing grain filling by regulating sink strength (in particular) has been suggested by cytokinin treatments to enhance grain size and weight. Wheat treated with ABA in the early reproductive stage shows increased water-use efficiency and nutrient remobilization under drought stress, stabilizing yield (Kang et al. 2022; Li et al. 2025).

Ethylene promotes root hair formation in rice, resulting in a significant increase in phosphorus uptake on low-phosphorus soils. Often, GAs are added to rice to speed up panicle development during and prior to harvest and avoid yield reduction. Brassinosteroids also increase photosynthesis and nutrient assimilation, especially by increasing heat-stress tolerance, helping rice plants to become more resilient to changing climatic conditions (Lepetit and Brouquisse 2023).

Brassinosteroids have been proven to improve nitrogen metabolism in maize and increase biomass production and grain yield. Application of ABA under saline conditions helps maize plants retain ionic balance, allowing continued nutrient uptake. Increased potassium uptake associated with ethylene-mediated root development has also been reported in maize, and it is crucial to stress tolerance and yield (Lee et al. 2019; Panigrahi et al. 2021b).

The interaction of auxins and ethylene in soybean is clearly advantageous for nodulation and nitrogen fixation (Hsu et al. 2021). The positive effect of these hormones is that they help the plant to get the nitrogen in low-nitrogen soils and therefore have reduced dependency on chemical fertilizers. Imparting resistance to pathogens under biotic stress has also been attributed to SA treatment in soybean (Gou et al. 2010; Panigrahi et al. 2021c).

11.5.3 Stress-Resilient Crops: Leveraging Hormones Under Abiotic and Biotic Stress

Plants are greatly constrained in crop productivity due to abiotic and biotic stresses, and plant hormones serve as major players in overcoming their damage. Plants are equipped with hormone signals to recognize when they are stressed, to activate survival and

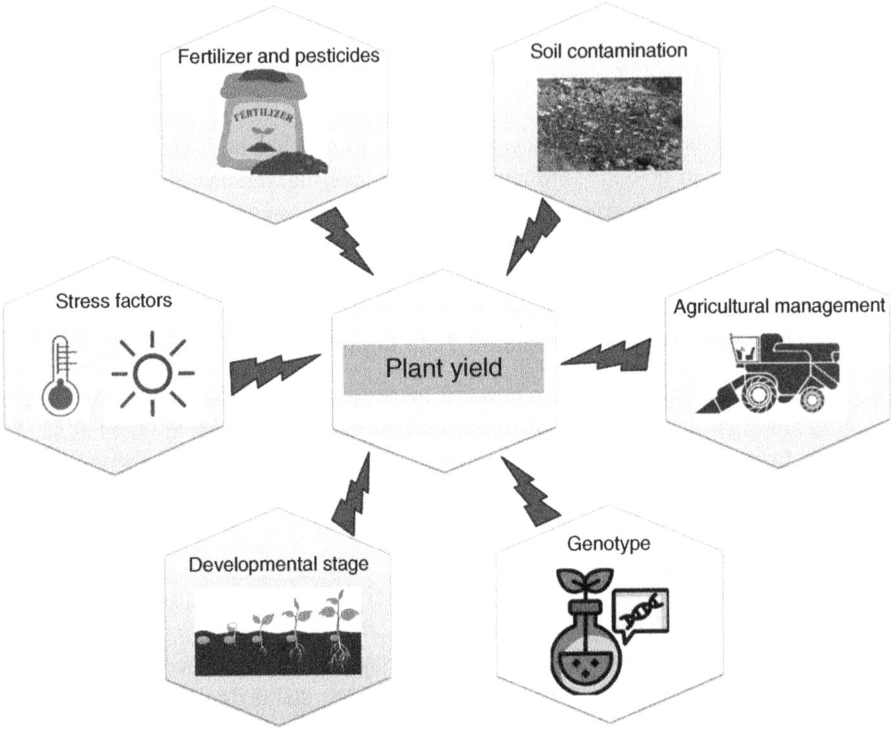

Figure 11.3 Complexities of crop yield: a multidimensional approach to environmental and biological factors. *Source:* Buch & Bee / Adobe Stock Photos; Collab Media / Adobe Stock Photos; Slamlabs / Adobe Stock Photos.

production adaptive responses, and to ensure continued survival and production in an adverse environment (Gorgues et al. 2022; Jung et al. 2020). Plant responses to drought and salinity: ABA is the central component of plant responses to drought and salinity. During drought, ABA accumulation causes stomatal closure to reduce water loss without compromising critical nutrient uptake (Feng et al. 2023; Sahoo et al. 2023). It also controls the expression of stress-responsive genes, which results in increased stress tolerance of the plant. Exposure to extreme temperatures results in oxidative stress in plants, which is mitigated by scavenging of reactive oxygen species (ROS) and stabilization of cellular membranes, through a reaction to forms of brassinosteroids (Fedoreyeva 2023; Patro et al. 2023). Ethylene is often linked to stress-induced senescence, but it promotes root hair development in response to nutrient stress, leading to the increased uptake of nutrients, specifically phosphorus and potassium (Figure 11.3).

Plant defense against pathogens and pests: Hormones such as salicylic acid (SA) and jasmonic acid (JA) are key players in the defense against pathogens and pests. SA is also involved in systemic acquired resistance (SAR), where it is known to activate the plant's defense genes to increase the plant's defense against microbial infections (Gao et al. 2023; Das and Panigrahi 2025a). However, JA controls the response to insect herbivory by inducing the production of secondary metabolites and defensive proteins. The SA and JA pathways interact, allowing the plants to tailor their defense response to the type of biotic stress. Manipulation of hormones to develop stress-resilient crops is of great importance to food security. The yield can be maintained in marginal environments by enhancing the plant's capability of coping with stress, thereby boosting productivity (Ghosh and Roychoudhury 2023; Patro et al. 2024). In the regions prone to climate variability and resource limitations, such strategies are truly valuable.

11.5.4 Role of Hormones in Reducing Fertilizer Dependency and Enhancing Nutrient Use Efficiency (NUE)

However, the use of chemical fertilizers is excessive and causes environmental problems, including soil degradation, water pollution, and greenhouse gas emissions. A sustainable alternative is provided by plant hormones, resulting in increased NUE, allowing crops to be grown at high yields with low fertilizer inputs (Goad et al. 2017). Proper activity of cytokinin is essential to increase NUE by modulating the activities of the nitrate transporters and nitrogen metabolism. For instance, they assist the transformation of absorbed nitrate into amino acids and proteins, a process that can make better use of nitrogen. Just as brassinosteroids activate phosphorus transport genes and increase phosphorus uptake and assimilation, improved root architecture driven by auxins allows plants to sample a larger soil volume, thus improving the uptake of immobile nutrients, such as phosphorus and potassium (Majumder and Panigrahi 2024). ABA regulates NUE under stress conditions, facilitating the remobilization of nutrients to key tissues to enable continued growth and development as a first priority. In a second-stress response mechanism involving ethylene, root hair formation increases nutrient uptake from nutrient-deficient soils (Figure 11.3). Hormonal regulation that decreases the dependence on fertilizer has economic and environmental benefits. It reduces the input costs of farmers, decreases fertilizer-runoff risk, and lowers the carbon footprint associated with agricultural practices (Fedoreyeva 2023; Panigrahi et al. 2024). Advances in hormonal research have led to the development of biostimulants based on the hormones, which are useful in improving the NUE of the crops. Such innovations mirror the goals of sustainable agriculture, achieving high productivity while preserving natural resources.

11.6 Advances in Hormonal Research and Agricultural Applications

11.6.1 Genetic Engineering and Hormonal Manipulation in Crops

Plant hormones are natural products and as such are attractive targets for genetic alteration because they regulate a number of physiological and developmental processes. Biotechnology advances have enabled scientists to manipulate hormonal pathways in crops to increase nutrient assimilation, stress tolerance, and yield potential. For instance, cytokinin biosynthesis gene overexpression has been shown to promote shoot and root growth and consequent nutrient uptake and utilization. Likewise, altering ethylene-signaling pathways has also led to better resistance to abiotic stress factors like drought and salinity. The gibberellin biosynthetic genetic modifications in rice have led to the development of semi-dwarf seed cultivars that have both higher lodging resistance and higher yield. Photosynthetic efficiency and grain production have increased in maize through brassinosteroid-pathway engineering (De Coninck and De Smet 2016; Das and Panigrahi 2025b). The genetic interventions shown here demonstrate the feasibility of the use of hormone manipulation to solve particular agricultural problems and optimize crop performance. However, challenges still exist (Figure 11.3). Due to the complex interactions between hormone pathways and environmental factors, there is a need to achieve consistent results (Combest et al. 2021). Other ethical and regulatory considerations related to GMOs require thorough safety assessments and effective public awareness campaigns.

11.6.2 Hormone-Based Biostimulants: Commercial Applications

Hormone-based biostimulants are eco-friendly alternatives to traditional agrochemicals. They are usually composed of hormones found in plant products extracted from natural sources, such as seaweed extracts, that contain auxins, cytokinins, and brassinosteroids that encourage growth and resilience in the plant. There have been several biostimulants commercially available that have successfully increased germination and root development and nutrient uptake (Sahoo et al. 2024). A auxin stability, for example, containing high auxin content promotes root elongation thereby enhancing the ability of plants to draw water and nutrients from deeper soil layers (Cho et al. 2014). The cytokinin-based products promote cell division and shoot growth, while brassinosteroids ameliorate stress-induced damage by stabilizing membranes of the cells and reducing the ROS levels. Hormone-based biostimulants are applied in such a way that they promote sustainable farming by minimizing chemical fertilizers and pesticides. This, however, has yet to be optimized for application across various crops and environmental conditions (Majumder and Panigrahi 2025).

11.6.3 Marker-Assisted Selection for Hormone-Regulated Nutrient Traits

These limitations should be addressed by advances in precision agriculture and formulation technology and as such, marker-assisted selection (MAS) is expected to become widespread in crop breeding by allowing identification and selection of specific genetic traits determined by hormonal regulation. Breeders can generate nutrient-efficient and resilient crop varieties through the use of molecular markers associated with hormone-mediated nutrient assimilation and stress tolerance. For instance, nodulation-marker genes to cytokinin-regulated nitrate transporters have been used to improve nitrogen use efficiency in wheat. Markers associated with auxin-controlled root architecture traits have helped to breed rice varieties with improved phosphorus uptake (Chen et al. 2023; Singh et al. 2024). They show that in MAS, these advancements demonstrate that the technology can speed up the development of high-performing crops that thrive in difficult environments. However, integrating the MAS with genomic selection and gene editing technologies has the promise for future crop improvement programs. Researchers can combine genetic insights with advanced breeding tools to totally fine-tune the development of hormone-regulated nutrient traits (Behera et al. 2024).

11.6.4 Precision Agriculture: Optimizing Hormone Application for Nutrient Management

Precision farming uses the latest technologies, including sensors, drones, and data analysis, to ensure the optimum input in farming that includes applications of hormones. Using real-time inputs on soil nutrients, the development stages of plants, and external environmental factors, precision farming supports targeted and optimum utilization of hormones (Panigrahi and Satapathy 2023). For example, the use of drones with multispectral cameras helps gauge plant health as well as regions needing hormone-based interventions. Based on these evaluations, hormone-based biostimulants can be applied with precision to induce growth, increase nutrient uptake, and reduce stress (Chen et al. 2024b). The use of artificial intelligence (AI) and machine learning also increases the scope of precision agriculture. Predictive models can suggest the best schedules for hormone application to achieve maximum efficacy and cost-effectiveness. These technologies align with the objectives of sustainable agriculture by reducing resource waste and increasing productivity.

11.7 Policy and Economic Considerations in Hormone-Based Agriculture

To continue implementing hormone-based solutions in agriculture, sophisticated regulatory frameworks need to be in place to guarantee the safety, efficacy, and environmental sustainability of such hormone use (Das et al. 2025). Guidelines should be set for the development, testing, and commercialization of products based on hormones, such as biostimulants and genetically modified crops; in this regard, government bodies as well as international organizations have to act. Global regulations can be harmonized; they can facilitate cross-border trade and adoption of such technologies so that farmers across the world can also benefit from progress made in hormonal research (Chen et al. 2024a). Ethical issues relating to GMOs should also be discussed by regulatory frameworks and should stimulate public awareness for consumer trust. Farmers will continue to face both traditional and emerging climate challenges, and hormonal innovations that could transform the agricultural economies by reducing input costs, increasing yields, and minimizing environmental damage should be taken as seriously as any greenhouse-gas reduction plan. Hormone-based technologies improve nutrient-use efficiency and stress resilience and lower dependence on expensive chemical fertilizers and pesticides thus raising farmers' profitability (Kim et al., 2021; Panigrahi and Satapathy, 2021). Affordable, scalable, hormone-based solutions that can be mass-marketed in developing countries are a critical component of the development. In accelerating the deployment of these technologies, public-private partnerships and government subsidies could be a proximate key to widen the accessibility. Several SDGs overlap with hormone-based agriculture, including Zero Hunger, Responsible Consumption and Production, and Climate Action. These technologies help in food security, minimize greenhouse gas emissions from agricultural practices, and conserve natural resources through promoting sustainable farming (Abarca et al. 2021). Incorporating hormonal innovations in national agricultural development can assist countries in achieving their SDG targets and promote economic growth and rural development. To benefit from these, collaboration among stakeholders, including policymakers, researchers, and farmers, is essential.

11.8 Conclusion

Among plant hormones, those acting on AMT and AMT mimicry transporters, stress tolerance, and yield potential play a central role. Advances in hormonal research have shown the ability of these molecules to revolutionize agriculture by increasing productivity while supporting sustainability. Stakeholders should focus on research in the hormonal innovation aspect that investigates hormone nutrient interactions, affordable biostimulants, and precision agriculture technologies to maximize the benefits of hormonal innovations. Supportive regulatory frameworks and public awareness should be established by policymakers to promote the adoption of this approach. Integrating hormonal pathways with sustainable farming practices is proposed as a new approach to tackling global food security problems. Plant hormones provide the power to build resilient agricultural systems that feed a growing population sustainably while leaving the planet's resources to keep future generations alive and healthy. Much remains to be understood about hormone–nutrient interactions. Both the complexity of hormonal-signaling networks, including their cross talk with environmental factors, and the complexity of nutrient assimilation and stress responses pose some challenges in identifying what role a hormonal-signaling network has in specific processes. Future research should aim to achieve a better understanding of these mechanisms to target interventions. Sustainable practices in farming are a great opportunity to integrate the hormonal pathways and boost crop productivity and resilience. Organic farming, crop rotation, and agroforestry, combined with hormonal innovations, can help produce holistic solutions that are both productive and environmentally friendly. To strengthen hormone-based agriculture, scientists, including plant biologists, agronomists, data scientists, and policymakers, must work together. Innovation is accelerated through collaborative research initiatives, knowledge gaps are bridged and science is translated into practice. As highly integrated into the economy, the policies concerning food security and trade have far-reaching policy implications for hormone-based agriculture. Research, infrastructure, and capacity building should receive government investment to reap the potential of hormonal innovations. Equitable access to these technologies can be ensured through international cooperation and knowledge sharing to safeguard global food security.

References

Abarca, A., Franck, C.M., and Zipfel, C. (2021). Family-wide evaluation of rapid alkalinization factor peptides. *Plant Physiology* 187 (2): 996–1010.

Aggarwal, S., Kumar, A., Jain, M. et al. (2020). C-terminally encoded peptides (CEPs) are potential mediators of abiotic stress response in plants. *Physiology and Molecular Biology of Plants* 26: 2019–2033.

Amano, Y., Tsubouchi, H., Shinohara, H. et al. (2007). Tyrosine-sulfated glycopeptide involved in cellular proliferation and expansion in Arabidopsis. *Proceedings of the National Academy of Sciences of the United States of America* 104: 18333–18338.

Andersen, E.J., Ali, S., Byamukama, E. et al. (2018). Disease resistance mechanisms in plants. *Genes* 9 (7): 339. https://doi.org/10.3390/genes9070339.

Arsova, B., Watt, M., and Usadel, B. (2018). Monitoring of plant protein post-translational modifications using targeted proteomics. *Frontiers in Plant Science* 9: 1168.

Avalos-Rangel, A., Ruiz-Herrera, L.F., Gutierrez-Alanis, D. et al. (2024). CLE14 peptide impairs root tip regeneration and callogenesis in Arabidopsis. *Journal of Plant Growth Regulation* 43 (5): 1458–1465.

Baillie, A.L., Sloan, J., Qu, L.J., and Smith, L.M. (2024). Signalling between the sexes during pollen tube reception. *Trends in Plant Science* 29 (3): 343–354.

Bari, R. and Jones, J.D. (2009). Role of plant hormones in plant defence responses. *Plant Molecular Biology* 69: 473–488.

Bashyal, S., Gautam, C.K., and Muller, L.M. (2024). CLAVATA signaling in plant-environment interactions. *Plant Physiology* 194 (3): 1336–1357. https://doi.org/10.1093/plphys/kiad591.

Behera, A., Panigrahi, G.K., and Sahoo, A. (2024). Nonsense-mediated mRNA decay in human health and diseases: current understanding, regulatory mechanisms and future perspectives. *Molecular Biotechnology* 67 (9): 3374–3390. https://doi.org/10.1007/s12033-024-01267-7.

Bhalla, H., Sudarsanam, K., Srivastava, A., and Sankaranarayanan, S. (2025). Structural insights into the recognition of RALF peptides by FERONIA receptor kinase during Brassicaceae pollination. *Plant Molecular Biology* 115 (1): 1–6.

Blackburn, M.R., Haruta, M., and Moura, D.S. (2020). Twenty years of progress in physiological and biochemical investigation of RALF peptides. *Plant Physiology* 182 (4): 1657–1666.

Boschiero, C., Lundquist, P.K., Roy, S. et al. (2019). Identification and functional investigation of genome-encoded, small, secreted peptides in plants. *Current Protocols in Plant Biology* 4 (3): e20098.

Canut, H., Albenne, C., and Jamet, E. (2016). Post-translational modifications of plant cell wall proteins and peptides: a survey from a proteomics point of view. *Biochimica et Biophysica Acta (BBA)-Proteins and Proteomics* 1864 (8): 983–990.

Carbonnel, S., Falquet, L., and Hazak, O. (2022). Deeper genomic insights into tomato CLE genes repertoire identify new active peptides. *BMC Genomics* 23 (1): 756.

Chang, S. and Xiao, F. (2025). Comprehensive review of plant small signaling peptides: from stress adaptation mechanisms to practical solutions for crop resilience. *International Journal of Biological Macromolecules* 299: 139971.

Chapman, K., Taleski, M., Frank, M., and Djordjevic, M.A. (2024). c-Terminally encoded peptide (cep) and cytokinin hormone signaling intersect to promote shallow lateral root angles. *Journal of Experimental Botany* 75 (2): 631–641.

Chaulagain, D. and Frugoli, J. (2021). The regulation of nodule number in legumes is a balance of three signal transduction pathways. *International Journal of Molecular Sciences* 22 (3): 1117.

Chen, Y.L., Fan, K.T., Hung, S.C., and Chen, Y.R. (2020). The role of peptides cleaved from protein precursors in eliciting plant stress reactions. *The New Phytologist* 225 (6): 2267–2282. https://doi.org/10.1111/nph.16241.

Chen, H., Song, Y., Li, H. et al. (2023). Enhancing the adaptability of tea plants (*Camellia sinensis* L) to high-temperature stress with small peptides and biosurfactants. *Plants (Basel)* 12 (15): 2817. https://doi.org/10.3390/plants12152817.

Chen, C., Buscaill, P., Sanguankiatticha, N. et al. (2024a). Extracellular plant subtilases dampen cold-shock peptide elicitor levels. *Nature Plants* 10 (11): 1749–1760.

Chen, X., van de Sande, J.W., Ritmejeris, J. et al. (2024b). Resolving sulfation PTMs on a plant peptide hormone using nanopore sequencing. bioRxiv.

Cho, H., Ryu, H., Rho, S. et al. (2014). A secreted peptide acts on BIN2-mediated phosphorylation of ARFs to potentiate auxin response during lateral root development. *Nature Cell Biology* 16 (1): 66–76.

Combest, M.M., Moroz, N., Tanaka, K. et al. (2021). St PIP1, a PAMP-induced peptide in potato, elicits plant defenses and is associated with disease symptom severity in a compatible interaction with Potato virus Y. *Journal of Experimental Botany* 72 (12): 4472–4488.

Corcilius, L., Hastwell, A.H., Zhang, M. et al. (2017). Arabinosylation modulates the growth-regulating activity of the peptide hormone CLE40a from soybean. *Cell Chemical Biology* 24 (11): 1347–1355.

Dao, T.Q., Weksler, N., Liu, H.M.H. et al. (2022). Interactive CLV3, CLE16 and CLE17 signaling mediates stem cell homeostasis in the Arabidopsis shoot apical meristem. *Development* 149 (19): dev200787.

Das, R. and Panigrahi, G.K. (2025a). Complexities of salinity stress on zebrafish survivability. *Toxicology International* 32 (1): 57–63. https://doi.org/10.18311/ti/2025/v32i1/45794.

Das, R. and Panigrahi, G.K. (2025b). Messenger RNA surveillance: current understanding, regulatory mechanisms, and future implications. *Molecular Biotechnology* 67 (2): 393–409. https://doi.org/10.1007/s12033-024-01062-4.

Das, R., Singh, S.P., Sahoo, A. et al. (2025). Synergistic stress of zinc oxide nanoparticles and pH triggers gill damage and mortality in zebrafish. *Nanotechnology for Environmental Engineering* 10 (1): 1–11.

de Bang, T.C., Lundquist, P.K., Dai, X. et al. (2017). Genome-wide identification of Medicago peptides involved in macronutrient responses and nodulation. *Plant Physiology* 175 (4): 1669–1689.

De Coninck, B. and De Smet, I. (2016). Plant peptides—taking them to the next level. *Journal of Experimental Botany* 67 (16): 4791–4795. https://doi.org/10.1093/jxb/erw309.

Ding, S., Lv, J., Hu, Z. et al. (2023). Phytosulfokine peptide optimizes plant growth and defense via glutamine synthetase GS2 phosphorylation in tomato. *The EMBO Journal* 42 (6): e111858.

Eckardt, N.A., Avin-Wittenberg, T., Bassham, D.C. et al. (2024). The lowdown on breakdown: open questions in plant proteolysis. *The Plant Cell* 36 (9): 2931–2975.

Farrokhi, N., Whitelegge, J.P., and Brusslan, J.A. (2008). Plant peptides and peptidomics. *Plant Biotechnology Journal* 6 (2): 105–134.

Fedoreyeva, L.I. (2023). Molecular mechanisms of regulation of root development by plant peptides. *Plants (Basel)* 12 (6): 1320. https://doi.org/10.3390/plants12061320.

Fedoreyeva, L.I., Chaban, I.A., and Kononenko, N.V. (2022). Short peptides induce development of root hair *Nicotiana tabacum*. *Plants (Basel)* 11 (7): 852.

Feng, Y.Z., Zhu, Q.F., Xue, J. et al. (2023). Shining in the dark: the big world of small peptides in plants. *aBIOTECH* 4 (3): 238–256. https://doi.org/10.1007/s42994-023-00100-0.

Fernandez-Fernandez, A.D., Stael, S., and Van Breusegem, F. (2023). Mechanisms controlling plant proteases and their substrates. *Cell Death and Differentiation* 30 (4): 1047–1058. https://doi.org/10.1038/s41418-023-01120-5.

Friso, G. and van Wijk, K.J. (2015). Posttranslational protein modifications in plant metabolism. *Plant Physiology* 169 (3): 1469–1487. https://doi.org/10.1104/pp.15.01378.

Furumizu, C. and Aalen, R.B. (2023). Peptide signaling through leucine-rich repeat receptor kinases: insight into land plant evolution. *New Phytologist* 238 (3): 977–982.

Furumizu, C., Krabberod, A.K., Hammerstad, M. et al. (2021). The sequenced genomes of nonflowering land plants reveal the innovative evolutionary history of peptide signaling. *The Plant Cell* 33 (9): 2915–2934.

Gancheva, M.S., Malovichko, Y.V., Poliushkevich, L.O. et al. (2019). Plant peptide hormones. *Russian Journal of Plant Physiology* 66: 171–189.

Gao, Q., Wang, C., Xi, Y. et al. (2023). RALF signaling pathway activates MLO calcium channels to maintain pollen tube integrity. *Cell Research* 33 (1): 71–79.

Ghosh, P. and Roychoudhury, A. (2023). Plant peptides involved in abiotic and biotic stress responses and reactive oxygen species (ROS) signaling. *Journal of Plant Growth Regulation* 43: 1410–1427.

Goad, D.M., Zhu, C., and Kellogg, E.A. (2017). Comprehensive identification and clustering of CLV3/ESR-related (CLE) genes in plants finds groups with potentially shared function. *New Phytologist* 216 (2): 605–616.

Gorgues, L., Li, X., Maurel, C. et al. (2022). Root osmotic sensing from local perception to systemic responses. *Stress Biology* 2 (1): 36.

Gou, X., He, K., Yang, H. et al. (2010). Genome-wide cloning and sequence analysis of leucine-rich repeat receptor-like protein kinase genes in Arabidopsis thaliana. *BMC Genomics* 11: 19. https://doi.org/10.1186/1471-2164-11-19.

Greifenhagen, A., Ruwe, H., Zimmer, V. et al. (2024). The peptide hormone Pj CLE1 stimulates haustorium formation in the parasitic plant *Phtheirospermum japonicum*. *Proceedings of the National Academy of Sciences* 121 (42): e2414582121.

Guillou, M.C., Balliau, T., Vergne, E. et al. (2022). The PROSCOOP10 gene encodes two extracellular hydroxylated peptides and impacts flowering time in Arabidopsis. *Plants (Basel)* 11 (24): 3554.

Hagelthorn, L. and Fletcher, J.C. (2023). The CLAVATA3/ESR-related peptide family in the biofuel crop pennycress. *Frontiers in Plant Science* 14: 1240342.

Haruta, M., Sabat, G., Stecker, K. et al. (2014). A peptide hormone and its receptor protein kinase regulate plant cell expansion. *Science (New York, N.Y.)* 343 (6169): 408–411. https://doi.org/10.1126/science.1244454.

Hastwell, A.H., Gresshoff, P.M., and Ferguson, B.J. (2014). The structure and activity of nodulation-suppressing CLE peptide hormones of legumes. *Functional Plant Biology* 42 (3): 229–238.

Hirakawa, Y. and Sawa, S. (2019). Diverse function of plant peptide hormones in local signaling and development. *Current Opinion in Plant Biology* 51: 81–87. https://doi.org/10.1016/j.pbi.2019.04.005.

Hirayama, T. and Mochida, K. (2022). Plant harmonics: a key tool for deep physiological phenotyping to improve crop productivity. *Plant and Cell Physiology* 63 (12): 1826–1839.

Hsiao, Y.C. and Yamada, M. (2020). The roles of peptide hormones and their receptors during plant root development. *Genes* 12 (1): 22. https://doi.org/10.3390/genes12010022.

Hsu, P.K., Dubeaux, G., Takahashi, Y., and Schroeder, J.I. (2021). Signaling mechanisms in abscisic acid-mediated stomatal closure. *The Plant Journal* 105 (2): 307–321. https://doi.org/10.1016/j.pmpp.2020.101568.

Huang, P., Li, Z., and Guo, H. (2022). New advances in the regulation of leaf senescence by classical and peptide hormones. *Frontiers in Plant Science* 13: 923136.

Igarashi, D., Tsuda, K., and Katagiri, F. (2012). The peptide growth factor, phytosulfokine, attenuates pattern-triggered immunity. *The Plant Journal* 71 (2): 194–204.

Ito, Y., Nakanomyo, I., Motose, H. et al. (2006). Dodeca-CLE peptides as suppressors of plant stem cell differentiation. *Science* 313 (5788): 842–845.

Je, B.I., Xu, F., Wu, Q. et al. (2018). The CLAVATA receptor FASCIATED EAR2 responds to distinct CLE peptides by signaling through two downstream effectors. *eLife* 7: e35673.

Johns, S., Hagihara, T., Toyota, M., and Gilroy, S. (2021). The fast and the furious: rapid long-range signaling in plants. *Plant Physiology* 185 (3): 694–706.

Jung, H.W., Panigrahi, G.K., Jung, G.Y. et al. (2020). PAMP-triggered immunity involves proteolytic degradation of core nonsense-mediated mRNA decay factors during early defense response. *The Plant Cell* 32 (4): 1081–1101. https://doi.org/10.1105/tpc.19.00631.

Kaeothip, S., Ishiwata, A., and Ito, Y. (2013). Stereoselective synthesis of Arabidopsis clavata3 (CLV3) glycopeptide, unique protein post-translational modifications of secreted peptide hormone in plant. *Organic & Biomolecular Chemistry* 11 (35): 5892–5907.

Kang, J., Wang, X., Ishida, T. et al. (2022). A group of CLE peptides regulates de novo shoot regeneration in Arabidopsis thaliana. *New Phytologist* 235 (6): 2300–2312.

Kaufmann, C. and Sauter, M. (2019). Sulfated plant peptide hormones. *Journal of Experimental Botany* 70 (16): 4267–4277.

Kim, J.S., Jeon, B.W., and Kim, J. (2021). Signaling peptides regulating abiotic stress responses in plants. *Frontiers in Plant Science* 12: 704490.

Kokla, A., Leso, M., Simura, J. et al. (2024). A long-distance inhibitory system regulates haustoria numbers in parasitic plants. bioRxiv, 2024-12.

Kondo, T., Yokomine, K., Nakagawa, A., and Sakagami, Y. (2011). Analogs of the CLV3 peptide: synthesis and structure-activity relationships focused on proline residues. *Plant & Cell Physiology* 52 (1): 30–36. https://doi.org/10.1093/pcp/pcq146.

Kuromori, T., Fujita, M., Takahashi, F. et al. (2022). Inter-tissue and inter-organ signaling in drought stress response and phenotyping of drought tolerance. *The Plant Journal* 109 (2): 342–358.

Laffont, C., Ivanovici, A., Gautrat, P. et al. (2020). The NIN transcription factor coordinates CEP and CLE signaling peptides that regulate nodulation antagonistically. *Nature Communications* 11 (1): 3167.

Lalun, V.O., Breiden, M., Galindo-Trigo, S. et al. (2024). A dual function of the IDA peptide in regulating cell separation and modulating plant immunity at the molecular level. *eLife* 12: RP87912.

Leasure, C.D. and He, Z.H. (2012). CLE and RGF family peptide hormone signaling in plant development. *Molecular Plant* 5 (6): 1173–1175.

Lebedeva, M., Azarakhsh, M., Yashenkova, Y., and Lutova, L. (2020). Nitrate-induced CLE peptide systemically inhibits nodulation in *Medicago truncatula*. *Plants (Basel)* 9 (11): 1456.

Lee, Z.H., Hirakawa, T., Yamaguchi, N., and Ito, T. (2019). The roles of plant hormones and their interactions with regulatory genes in determining meristem activity. *International Journal of Molecular Sciences* 20 (16): 4065.

Lepetit, M. and Brouquisse, R. (2023). Control of the rhizobium–legume symbiosis by the plant nitrogen demand is tightly integrated at the whole plant level and requires inter-organ systemic signaling. *Frontiers in Plant Science* 14: 1114840.

Li, J., Huang, Y., Yu, X. et al. (2025). Identification and application of CLE peptides for drought resistance in Solanaceae crops. *Journal of Agricultural and Food Chemistry* 72 (24): 13869–13884.

Lin, Z.B., Zhao, Y.X., and Yang, W.C. (2023). A balanced mechanism between plant growth and stress response. *Advanced Agrochem* 2 (2): 105–106.

Liu, Y., Zuo, T., Qiu, Z. et al. (2021). Genome-wide identification reveals the function of CEP peptide in cucumber root development. *Plant Physiology and Biochemistry* 169: 119–126.

Liu, D., Shen, Z., Zhuang, K. et al. (2022a). Systematic annotation reveals CEP function in tomato root development and abiotic stress response. *Cells* 11 (19): 2935.

Liu, Q., Wu, K., Song, W. et al. (2022b). Improving crop nitrogen use efficiency toward sustainable green revolution. *Annual Review of Plant Biology* 73 (1): 523–551.

Lyapina, I., Ganaeva, D., Rogozhin, E.A. et al. (2024). Comparative analysis of small secreted peptide signaling during defense response: insights from vascular and non-vascular plants. bioRxiv, 2024-05.

Majumder, S. and Panigrahi, G.K. (2024). Molecular docking and ADME evaluation of plant-based bioactive molecules targeting nonsense-mediated mRNA decay pathway factors to modulate tumorigenesis. *Plant Science Today* 11 (4): https://doi.org/10.14719/pst.3737.

Majumder, S. and Panigrahi, G.K. (2025). Advancements in contemporary pharmacological innovation: mechanistic insights and emerging trends in drug discovery and development. *Intelligent Pharmacy* 3 (2): 118. https://doi.org/10.1016/j.ipha.2024.10.001.

Majumder, S., Das, R., and Panigrahi, G.K. (2025). Harnessing phytochemicals to modulate the nonsense-mediated decay pathway for anticancer drug discovery. *Research Journal of Pharmacy and Technology* 18 (2): 489–494. https://doi.org/10.52711/0974-360X.2025.00074.

Matsubayashi, Y. (2018). Exploring peptide hormones in plants: identification of four peptide hormone-receptor pairs and two post-translational modification enzymes. *Proceedings of the Japan Academy. Series B, Physical and Biological Sciences* 94 (2): 59–74. https://doi.org/10.2183/pjab.94.006.

Matsuzaki, Y., Ogawa-Ohnishi, M., Mori, A., and Matsubayashi, Y. (2010). Secreted peptide signals required for maintenance of root stem cell niche in Arabidopsis. *Science* 329: 1065–1067.

Mens, C., Hastwell, A.H., Su, H. et al. (2021). Characterisation of Medicago truncatula CLE34 and CLE35 in nitrate and rhizobia regulation of nodulation. *New Phytologist* 229 (5): 2525–2534.

Millar, A.H., Heazlewood, J.L., Giglione, C. et al. (2019). The scope, functions, and dynamics of posttranslational protein modifications. *Annual Review of Plant Biology* 70 (1): 119–151.

Mosher, S., Seybold, H., Rodriguez, P. et al. (2013). The tyrosine-sulfated peptide receptors PSKR1 and PSY1R modify the immunity of Arabidopsis to biotrophic and necrotrophic pathogens in an antagonistic manner. *The Plant Journal* 73 (3): 469–482.

Motomitsu, A., Sawa, S., and Ishida, T. (2015). Plant peptide hormone signaling. *Essays in Biochemistry* 58: 115–131. https://doi.org/10.1042/bse0580115.

Muller, L.M. and Harrison, M.J. (2019). Phytohormones, miRNAs, and peptide signals integrate plant phosphorus status with arbuscular mycorrhizal symbiosis. *Current Opinion in Plant Biology* 50: 132–139.

Murphy, E., Smith, S., and De Smet, I. (2012). Small signaling peptides in Arabidopsis development: how cells communicate over a short distance. *The Plant Cell* 24 (8): 3198–3217.

Nakagami, S., Kajiwara, T., Tsuda, K., and Sawa, S. (2024). CLE peptide signaling in plant-microbe interactions. *Frontiers in Plant Science* 15: 1481650.

Nakayama, T., Shinohara, H., Tanaka, M. et al. (2017). A peptide hormone required for Casparian strip diffusion barrier formation in Arabidopsis roots. *Science* 355 (6322): 284–286.

Narasimhan, M. and Simon, R. (2022). Spatial range, temporal span, and promiscuity of CLE-RLK signaling. *Frontiers in Plant Science* 13: 906087.

Nguyen, Q.T.C. and Kim, J. (2023). PSY–PSYR peptide–receptor pairs control the trade-off between plant growth and stress response. *Plant Signaling & Behavior* 18 (1): https://doi.org/10.1080/15592324.2023.2260638.

Ogawa-Ohnishi, M., Yamashita, T., Kakita, M. et al. (2022). Peptide ligand-mediated trade-off between plant growth and stress response. *Science* 378 (6616): 175–180.

Oh, E., Seo, P.J., and Kim, J. (2018). Signaling peptides and receptors coordinating plant root development. *Trends in Plant Science* 23 (4): 337–351.

Ohki, S., Takeuchi, M., and Mori, M. (2011). The NMR structure of storage reveals the basis of stomatal density regulation by plant peptide hormones. *Nature Communications* 2: 512. https://doi.org/10.1038/ncomms1520.

Ohkubo, Y., Tanaka, M., Tabata, R. et al. (2017). Shoot-to-root mobile polypeptides involved in systemic regulation of nitrogen acquisition. *Nature Plants* 3 (4): 1–6.

Okamoto, S., Shinohara, H., Mori, T. et al. (2013). Root-derived CLE glycopeptides control nodulation by direct binding to HAR1 receptor kinase. *Nature Communications* 4 (1): 2191.

Okamoto, S., Kawasaki, A., Makino, Y. et al. (2022). Long-distance translocation of CLAVATA3/ESR-related 2 peptide and its positive effect on roots sucrose status. *Plant Physiology* 189 (4): 2357–2367. https://doi.org/10.1093/plphys/kiac227.

Pan, X., Deng, Z., Wu, R. et al. (2024). Identification of CEP peptides encoded by the tobacco (*Nicotiana tabacum*) genome and characterization of their roles in osmotic and salt stress responses. *Plant Physiology and Biochemistry* 209: 108525.

Panigrahi, G.K. and Satapathy, K.B. (2020a). Arabidopsis DCP5, a decapping complex protein interacts with ubiquitin-5 in the processing bodies. *Plant Archives* 20 (1): 2243–2247.

Panigrahi, G.K. and Satapathy, K.B. (2020b). Formation of Arabidopsis poly(A)-specific ribonuclease associated processing bodies in response to pathogenic infection. *Plant Archives* 20 (2): 4907–4912.

Panigrahi, G.K. and Satapathy, K.B. (2020c). Sacrificed surveillance process favours plant defense: a review. *Plant Archives* 20 (1): 2551–2559.

Panigrahi, G.K. and Satapathy, K.B. (2021). Pseudomonas syringae pv. syringae infection orchestrates the fate of the arabidopsis j domain containing cochaperone and decapping protein factor 5. *Physiological and Molecular Plant Pathology* 113 (101598): 1–9. https://doi.org/10.1016/j.pmpp.2020.101598.

Panigrahi, G.K. and Satapathy, K.B. (2023). Biological synthesis of gold nanoparticles by microbes: mechanistic aspects, biomedical applications, and future prospects. In: *Entrepreneurship with Microorganisms*, 371–388. Academic Press: https://doi.org/10.1016/B978-0-443-19049-0.00012-8.

Panigrahi, G.K., Sahoo, A., and Satapathy, K.B. (2021a). Differential expression of selected Arabidopsis resistant genes under abiotic stress conditions. *Plant Science Today* 8 (4): 859–864. https://doi.org/10.14719/pst.2021.8.4.1213.

Panigrahi, G.K., Sahoo, A., and Satapathy, K.B. (2021b). Insights to plant immunity: defense signaling to epigenetics. *Physiological and Molecular Plant Pathology* 113: 1–7.

Panigrahi, G.K., Sahoo, S.K., Sahoo, A. et al. (2021c). Bioactive molecules from plants: a prospective approach to combat SARS-Cov-2. *Advances in Traditional Medicine* 23 (3): 617–630. https://doi.org/10.1007/s13596-021-00599-y.

Panigrahi, G.K., Sahoo, A., and Satapathy, K.B. (2024). The processing body component varicose plays a multiplayer role towards stress management in Arabidopsis. *Plant Physiology Reports* 29: 186–192. https://doi.org/10.1007/s40502-023-00778-w.

Pastor-Fernandez, J., Sanchez-Bel, P., Flors, V. et al. (2023). Small signals lead to big changes: the potential of peptide-induced resistance in plants. *Journal of Fungi* 9 (2): 265.

Patel, N., Mohd-Radzman, N.A., Corcilius, L. et al. (2018). Diverse peptide hormones affecting root growth identified in the *Medicago truncatula* secreted peptidome. *Molecular & Cellular Proteomics* 17 (1): 160–174.

Patro, I., Sahoo, A., Nayak, B.R. et al. (2023). Nonsense-mediated mRNA decay: mechanistic insights and physiological significance. *Molecular Biotechnology* 66 (11): 3077–3091. https://doi.org/10.1007/s12033-023-00927-4.

Patro, A.K., Panigrahi, G.K., Majumder, S. et al. (2024). Nonsense-mediated mRNA decay: physiological significance, mechanistic insights and future implications. *Pathology, Research and Practice* 264: 155677. https://doi.org/10.1016/j.prp.2024.155677.

Pecenkova, T., Potocky, M., and Stegmann, M. (2024). More than meets the eye: knowns and unknowns of the trafficking of small secreted proteins in Arabidopsis. *Journal of Experimental Botany* 75 (12): 3713–3730.

Phour, M., Sehrawat, A., Sindhu, S.S., and Glick, B.R. (2020). Interkingdom signaling in plant-rhizomicrobiome interactions for sustainable agriculture. *Microbiological Research* 241: 126589.

Qian, P., Song, W., Yokoo, T. et al. (2018). The CLE9/10 secretory peptide regulates stomatal and vascular development through distinct receptors. *Nature Plants* 4 (12): 1071–1081.

Qiu, Z., Zhuang, K., Liu, Y. et al. (2022). Functional characterization of C-terminally encoded peptide (CEP) family in Brassica rapa L. *Plant Signaling & Behavior* 17 (1): 2021365.

Reichardt, S., Piepho, H.P., Stintzi, A., and Schaller, A. (2020). Peptide signaling for drought-induced tomato flower drop. *Science* 367 (6485): 1482–1485.

Reid, D.E., Ferguson, B.J., and Gresshoff, P.M. (2011). Inoculation-and nitrate-induced CLE peptides of soybean control NARK-dependent nodule formation. *Molecular Plant-Microbe Interactions* 24 (5): 606–618.

Ren, S.C., Song, X.F., Chen, W.Q. et al. (2019). CLE25 peptide regulates phloem initiation in Arabidopsis through a CLERK-CLV2 receptor complex. *Journal of Integrative Plant Biology* 61 (10): 1043–1061.

Ren, G., Zhang, Y., Chen, Z. et al. (2024). Research progress of small plant peptides on the regulation of plant growth, development, and abiotic stress. *International Journal of Molecular Sciences* 25 (7): 4114. https://doi.org/10.3390/ijms25074114.

Roman, A.O., Jimenez-Sandoval, P., Augustin, S. et al. (2022). HSL1 and BAM1/2 impact epidermal cell development by sensing distinct signaling peptides. *Nature Communications* 13 (1): 876.

Roy, S., Lundquist, P., Udvardi, M., and Scheible, W.R. (2018). Small and mighty: peptide hormones in plant biology. *Plant Cell* 30: tc118tt0718.

Royek, S., Bayer, M., Pfannstiel, J. et al. (2022). Processing of a plant peptide hormone precursor facilitated by posttranslational tyrosine sulfation. *Proceedings of the National Academy of Sciences of the United States of America* 119 (16): e2201195119. https://doi.org/10.1073/pnas.2201195119.

Ryslava, H., Doubnerova, V., Kavan, D., and Vanek, O. (2013). Effect of posttranslational modifications on enzyme function and assembly. *Journal of Proteomics* 92: 80–109. https://doi.org/10.1016/j.jprot.2013.03.025.

Rzemieniewski, J., Leicher, H., Lee, H.K. et al. (2024). CEP signaling coordinates plant immunity with nitrogen status. *Nature Communications* 15 (1): 1–16.

Sahoo, A. and Satapathy, K.B. (2021). Differential expression of Arabidopsis EJC core proteins under short-day and long-day growth conditions. *Plant Science Today* 8 (4): 815–819. https://doi.org/10.14719/pst.2021.8.4.1214.

Sahoo, A., Satapathy, K.B., and Panigrahi, G.K. (2023). Ectopic expression of disease resistance protein promotes resistance against pathogen infection and drought stress in Arabidopsis. *Physiological and Molecular Plant Pathology* 124 (101949): 1–7. https://doi.org/10.1016/j.pmpp.2023.101949.

Sahoo, A., Satapathy, K.B., and Panigrahi, G.K. (2024). Security check: plant immunity under temperature surveillance. *Journal of Plant Biochemistry and Biotechnology* 33: 1–4. https://doi.org/10.1007/s13562-023-00846-0.

Salvi, P., Manna, M., Kaur, H. et al. (2021). Phytohormone signaling and crosstalk in regulating drought stress response in plants. *Plant Cell Reports* 40: 1305–1329.

Sang, T., Zhang, Z., Liu, G., and Wang, P. (2025). Navigating the landscape of plant proteomics. *Journal of Integrative Plant Biology* 67 (3): 740–761.

Sauter, M. (2015). Phytosulfokine peptide signalling. *Journal of Experimental Botany* 66 (17): 5161–5169.

Schardon, K., Hohl, M., Graff, L. et al. (2016). Precursor processing for plant peptide hormone maturation by subtilisin-like serine proteinases. *Science* 354 (6319): 1594–1597.

Scott, T., Correa-Lozano, A., and Foo, E. (2024). Upstairs, Downstairs-conserved and divergent CLAVATA signalling in shoot meristem development and root symbioses. *Annals of Botany* mcae192.

Setty, J., Singh, P., Tantuway, G., and Vijai, P. (2024). Recent advancement on peptide research and their application in eco-agriculture. In: *Plant Elicitor Peptides: New Tool for Sustainable Agriculture*, 269–281. Singapore: Springer Nature Singapore.

Simon, R. and Dresselhaus, T. (2015). Peptides take centre stage in plant signaling. Preface. *Journal of Experimental Botany* 66 (17): 5135–5138. https://doi.org/10.1093/jxb/erv376.

Singh, S.P., Panigrahi, G.K., and Sahoo, A. (2024). Optimizing nanoparticles use for growth promotion in zebrafish: insights on concentration-dependent effects. *Nanotechnology for Environmental Engineering* 9: 757–767. https://doi.org/10.1007/s41204-024-00393-6.

Song, X.F., Hou, X.L., and Liu, C.M. (2022). CLE peptides: critical regulators for stem cell maintenance in plants. *Planta* 255: 1–17.

Stewart, V. and Ronald, P.C. (2022). Sulfotyrosine residues: interaction specificity determinants for extracellular protein–protein interactions. *Journal of Biological Chemistry* 298 (8): 102232.

Stintzi, A. and Schaller, A. (2022). Biogenesis of post-translationally modified peptide signals for plant reproductive development. *Current Opinion in Plant Biology* 69: 102274.

Suh, E.M., Reinhardt, J.K., and Weng, J.K. (2024). The emergence and loss of cyclic peptides in Nicotiana illuminate dynamics and mechanisms of plant metabolic evolution. bioRxiv, 2024-11.

Tabata, R. and Sawa, S. (2014). Maturation processes and structures of small secreted peptides in plants. *Frontiers in Plant Science* 5: 311. https://doi.org/10.3389/fpls.2014.00311.

Takahashi, F., Suzuki, T., Osakabe, Y. et al. (2018). A small peptide modulates stomatal control via abscisic acid in long-distance signaling. *Nature* 556 (7700): 235–238.

Takahashi, F., Kuromori, T., Urano, K. et al. (2020). Drought stress responses and resistance in plants: from cellular responses to long-distance intercellular communication. *Frontiers in Plant Science* 11: 556972.

Taleski, M., Imin, N., and Djordjevic, M.A. (2018). CEP peptide hormones: key players in orchestrating nitrogen-demand signalling, root nodulation, and lateral root development. *Journal of Experimental Botany* 69 (8): 1829–1836.

Taleski, M., Chapman, K., Imin, N. et al. (2020). The peptide hormone receptor CEPR1 functions in the reproductive tissue to control seed size and yield. *Plant Physiology* 183 (2): 620–636.

Taleski, M., Jin, M., Chapman, K. et al. (2024). CEP hormones at the nexus of nutrient acquisition and allocation, root development, and plant–microbe interactions. *Journal of Experimental Botany* 75 (2): 538–552.

Tan, W., Nian, H., Tran, L.S.P. et al. (2024). Small peptides: novel targets for modulating plant–rhizosphere microbe interactions. *Trends in Microbiology* 32 (11): 1072–1083.

Tavormina, P., De Coninck, B., Nikonorova, N. et al. (2015). The plant peptidome: an expanding repertoire of structural features and biological functions. *The Plant Cell* 27 (8): 2095–2118. https://doi.org/10.1105/tpc.15.00440.

Toruno, T.Y., Stergiopoulos, I., and Coaker, G. (2016). Plant-pathogen effectors: cellular probes interfering with plant defenses in spatial and temporal manners. *Annual Review of Phytopathology* 54 (1): 419–441.

Tost, A.S., Kristensen, A., Olsen, L.I. et al. (2021). The PSY peptide family—expression, modification and physiological implications. *Genes* 12 (2): 218. https://doi.org/10.3390/genes12020218.

Vanyushin, B.F., Ashapkin, V.V., and Aleksandrushkina, N.I. (2017). Regulatory peptides in plants. *Biochemistry (Moscow)* 82: 89–94.

Verma, V., Ravindran, P., and Kumar, P.P. (2016). Plant hormone-mediated regulation of stress responses. *BMC Plant Biology* 16: 86. https://doi.org/10.1186/s12870-016-0771-y.

Vie, A.K., Najafi, J., Liu, B. et al. (2015). The IDA/IDA-LIKE and PIP/PIP-LIKE gene families in Arabidopsis: phylogenetic relationship, expression patterns, and transcriptional effect of the PIPL3 peptide. *Journal of Experimental Botany* 66 (17): 5351–5365.

Villalobos-Lopez, M.A., Arroyo-Becerra, A., Quintero-Jimenez, A., and Iturriaga, G. (2022). Biotechnological advances to improve abiotic stress tolerance in crops. *International Journal of Molecular Sciences* 23 (19): 12053. https://doi.org/10.3390/ijms231912053.

Wulf, K., Sun, J., Wang, C. et al. (2024). The role of CLE peptides in the suppression of mycorrhizal colonization of tomato. *Plant and Cell Physiology* 65 (1): 107–119.

Xiao, F. and Zhou, H. (2023). Plant salt response: perception, signaling, and tolerance. *Frontiers in Plant Science* 13: 1053699.

Xie, H., Zhao, W., Li, W. et al. (2022). Small signaling peptides mediate plant adaptions to abiotic environmental stress. *Planta* 255 (4): 72.

Xu, T.T., Song, X.F., Ren, S.C., and Liu, C.M. (2013). The sequence flanking the N-terminus of the CLV3 peptide is critical for its cleavage and activity in stem cell regulation in Arabidopsis. *BMC Plant Biology* 13: 1–12.

Xu, C., Liberatore, K.L., Mac Alister, C.A. et al. (2015). A cascade of arabinosyltransferases controls shoot meristem size in tomato. *Nature Genetics* 47 (7): 784–792.

Yamaguchi, Y.L., Ishida, T., and Sawa, S. (2016). CLE peptides and their signaling pathways in plant development. *Journal of Experimental Botany* 67 (16): 4813–4826.

Yan, Y. (2024). Deciphering the role of the malectin-like RLK and LRR-RLKs in the liverwort Marchantia polymorpha. Doctoral dissertation. Universität zu Köln.

Yang, Z., Gao, C., Zhang, Y. et al. (2023). Recent progression and future perspectives in cotton genomic breeding. *Journal of Integrative Plant Biology* 65 (2): 548–569.

Yoshida, T., Fernie, A.R., Shinozaki, K., and Takahashi, F. (2021). Long-distance stress and developmental signals associated with abscisic acid signaling in environmental responses. *The Plant Journal* 105 (2): 477–488.

Yu, B., Chao, D.Y., and Zhao, Y. (2024a). How plants sense and respond to osmotic stress. *Journal of Integrative Plant Biology* 66 (3): 394–423.

Yu, X., Huang, Z., Cheng, Y. et al. (2024b). Comparative genomics screens identify a novel small secretory peptide, SlSolP12, which activates both local and systemic immune response in tomatoes and exhibits broad-spectrum activity. *Journal of Agricultural and Food Chemistry* 72 (33): 18507–18519.

Yuan, N., Furumizu, C., Zhang, B., and Sawa, S. (2021). Database mining of plant peptide homologues. *Plant Biotechnology* 38 (1): 137–143.

Zarsky, V., Nielsen, M.E., and Blatt, M.R. (2024). Formins, cell wall integrity, ROP guanine exchange factors, secretion regulators, and small secreted peptides in plant cell exocytosis and defence. *Journal of Experimental Botany* 75 (12): 3665–3667.

Zelman, A.K. and Berkowitz, G.A. (2023). Plant elicitor peptide (pep) signaling and pathogen defense in tomato. *Plants (Basel)* 12 (15): 2856.

Zhang, L., Shi, X., Zhang, Y. et al. (2019). CLE9 peptide-induced stomatal closure is mediated by abscisic acid, hydrogen peroxide, and nitric oxide in Arabidopsis thaliana. *Plant, Cell & Environment* 42 (3): 1033–1044.

Zhang, S., Li, C., Si, J. et al. (2022a). Action mechanisms of effectors in plant-pathogen interaction. *International Journal of Molecular Sciences* 23 (12): 6758.

Zhang, Y., Tan, S., Gao, Y. et al. (2022b). CLE42 delays leaf senescence by antagonizing ethylene pathway in Arabidopsis. *New Phytologist* 235 (2): 550–562.

Zhang, Y.M., Ye, D.X., Liu, Y. et al. (2023). Peptides, new tools for plant protection in eco-agriculture. *Advanced Agrochem* 2 (1): 58–78.

Zhang, C., Xie, Y., He, P., and Shan, L. (2024a). Unlocking nature's defense: plant pattern recognition receptors as guardians against pathogenic threats. *Molecular Plant-Microbe Interactions* 37 (2): 73–83.

Zhang, P., Zhao, J., Zhang, W. et al. (2024b). Sulfated peptides: key players in plant development, growth, and stress responses. *Frontiers in Plant Science* 15: 1474111. https://doi.org/10.3389/fpls.2024.1474111.

Zhang, P., Gao, W., Guo, L. et al. (2024c). Functional characterization of plant peptide-containing sulfated tyrosine (PSY) family in wheat (*Triticum aestivum* L). *International Journal of Molecular Sciences* 25 (23): 12663.

Zhong, S. and Qu, L.J. (2019). Peptide/receptor-like kinase-mediated signaling involved in male-female interactions. *Current Opinion in Plant Biology* 51: 7–14.

Zhong, S., Li, L., Wang, Z. et al. (2022). RALF peptide signaling controls the polytube block in Arabidopsis. *Science* 375 (6578): 290–296.

Zhou, H., Xiao, F., Zheng, Y. et al. (2022). Pamp-induced secreted peptide 3 modulates salt tolerance through receptor-like kinase 7 in plants. *The Plant Cell* 34 (2): 927–944.

Zhou, Y., Zheng, J., Wu, H. et al. (2024). A novel toolbox to record CLE peptide signaling. *Frontiers in Plant Science* 15: 1468763.

12

Peptide Hormone Signaling in Photoprotection: Integrating Developmental and Stress Pathways to Enhance Photosynthetic Resilience

Kanagaraj Muthu-Pandian Chanthini[1] *and Sengottayan Senthil-Nathan*[2]

[1] *Department of Pharmacology, Saveetha Institute of Medical and Technical Sciences (SIMATS), Chennai, Tamil Nadu, India*
[2] *Division of Bio-pesticides and Environmental Toxicology, Sri Paramakalyani Centre for Excellence in Environmental Sciences, Manonmaniam Sundaranar University, Tirunelveli, Tamil Nadu, India*

12.1 Introduction

Peptide hormones are small, mobile signaling molecules composed of short chains of amino acids (Stührwohldt and Schaller 2019). Unlike traditional plant hormones like auxins or gibberellins, peptide hormones function primarily through cell-to-cell communication, often acting in a highly specific and localized manner (Pandita et al. 2023). They are synthesized as precursor proteins that undergo posttranslational modifications (PTMs) such as proteolytic cleavage and sometimes sulfation or hydroxylation to become biologically active. These active peptides are then perceived by plasma-membrane-localized receptor-like kinases (RLKs), triggering intracellular signaling cascades that regulate gene expression, development, and responses to environmental stimuli (Zhu et al. 2023).

In recent years, peptide hormones have gained recognition for their pivotal roles in regulating diverse physiological processes in plants, including root and shoot meristem maintenance, vascular differentiation, and defense against pathogens (Zhang et al. 2025). Families such as CLAVATA3/EMBRYO SURROUNDING REGION-related (CLE), rapid alkalinization factor (RALF), and Inflorescence deficient in abscission (IDA) represent well-studied examples of plant peptides with specific regulatory roles (Goyal and Singh 2025). Beyond development, growing evidence suggests that peptide hormones are also integral in mediating stress responses, including those triggered by abiotic factors such as drought, salinity, and light stress, thus positioning them as key modulators in the plant's adaptive toolkit (Kaya 2024).

Photosynthesis is the fundamental process by which plants convert light energy into chemical energy, producing glucose and releasing oxygen as a by-product. It occurs primarily in the chloroplasts of leaf cells and is driven by the absorption of light by chlorophyll and other pigments (Nair et al. 2021). Through a series of light-dependent and light-independent reactions, plants synthesize carbohydrates that serve as essential energy sources for cellular

Peptide Hormones in Plant Development and Stress Responses, First Edition. Edited by Vijay Pratap Singh and Samiksha Singh.

functions, growth, and reproduction. Photosynthesis also underpins the global food chain, making it the most critical biological process for life on Earth (Uchiyama and Orio 2024).

From an agricultural perspective, the efficiency of photosynthesis directly impacts plant productivity and crop yield. Enhancing photosynthetic capacity is a major goal in crop improvement efforts, particularly in the face of rising global food demands and climate change stressors (Croce et al. 2024). However, photosynthesis is highly sensitive to environmental fluctuations such as light intensity, temperature, water availability, and nutrient status. Thus, understanding how plants regulate photosynthetic processes under variable conditions is vital for developing strategies to sustain productivity and ensure food security (Smith et al. 2023).

Photoprotection refers to the suite of physiological and biochemical mechanisms that plants employ to mitigate the damaging effects of excessive light energy (Simkin et al. 2022). Under natural conditions, plants are often exposed to fluctuating light intensities, and when the absorbed light exceeds the capacity of photosynthesis, it can lead to the formation of reactive oxygen species (ROS), protein damage, and impaired photosystem function—particularly in Photosystem II (Sohrabi et al. 2024). To counteract this, plants activate nonphotochemical quenching (NPQ) pathways, adjust chloroplast positioning, and upregulate antioxidant enzymes to dissipate excess energy and detoxify harmful compounds (Sharma et al. 2023a, b).

This photoprotective response is not merely a passive defense but a dynamic regulatory system that maintains photosynthetic efficiency and prevents long-term photodamage (Shuang et al. 2022). Key components of photoprotection include the xanthophyll cycle, which facilitates the harmless dissipation of excess light energy as heat, and proteins such as PsbS that sense and respond to proton gradients in the thylakoid lumen (Levin and Schuster 2023). Photoprotection is especially crucial under abiotic stress conditions like drought or nutrient limitation, which impair carbon fixation and increase the risk of photooxidative damage. Understanding these mechanisms is vital for engineering crops that can maintain high productivity even under environmental stress (Muhammad et al. 2021).

Understanding the intricate signaling networks that govern plant responses to their environment is key to advancing both basic and applied plant sciences. Among these, peptide hormones have emerged as critical players in coordinating growth, development, and stress adaptation. Simultaneously, photosynthesis and its protective mechanisms are central to plant survival under varying light conditions. In the sections that follow, we explore the roles of peptide hormones, the fundamentals of photosynthesis, and the sophisticated strategies plants use to safeguard this vital process.

12.2 Peptide Hormones in Plants

Plant peptide hormones are small, secreted proteins that are usually 5–50 amino acids in length and serve as signaling molecules to control a broad array of physiological processes (Stührwohldt and Schaller 2019). In contrast to traditional plant hormones that are low-molecular-weight organic compounds, plant peptide hormones are gene-encoded and undergo specific posttranslational processing to become biologically active (Zhang et al.

2025). They also serve as messengers during cell-to-cell communication, controlling development, immune reactions, and environmental adaptation. The majority of peptide hormones are synthesized in a tissue-specific manner and exert their action through very specific receptor interactions, allowing for localized and targeted transduction of signals. Their fine-tuned regulation and fast signaling proficiency enable them to play a crucial part in plant homeostasis and stress resilience (Abid et al. 2021).

12.2.1 Biosynthesis

The biosynthesis of peptide hormones starts from the translation of precursor proteins, commonly termed as prepropeptides, which have signal peptides to target them to the endoplasmic reticulum (ER) for secretion (Figure 12.1).

The precursor proteins usually have a signal peptide, a pro-region, and the mature peptide sequence (Malandrino and Smith 2017). Upon entry into the secretory pathway, the signal peptide is removed and the propeptide is processed further to liberate the mature, active

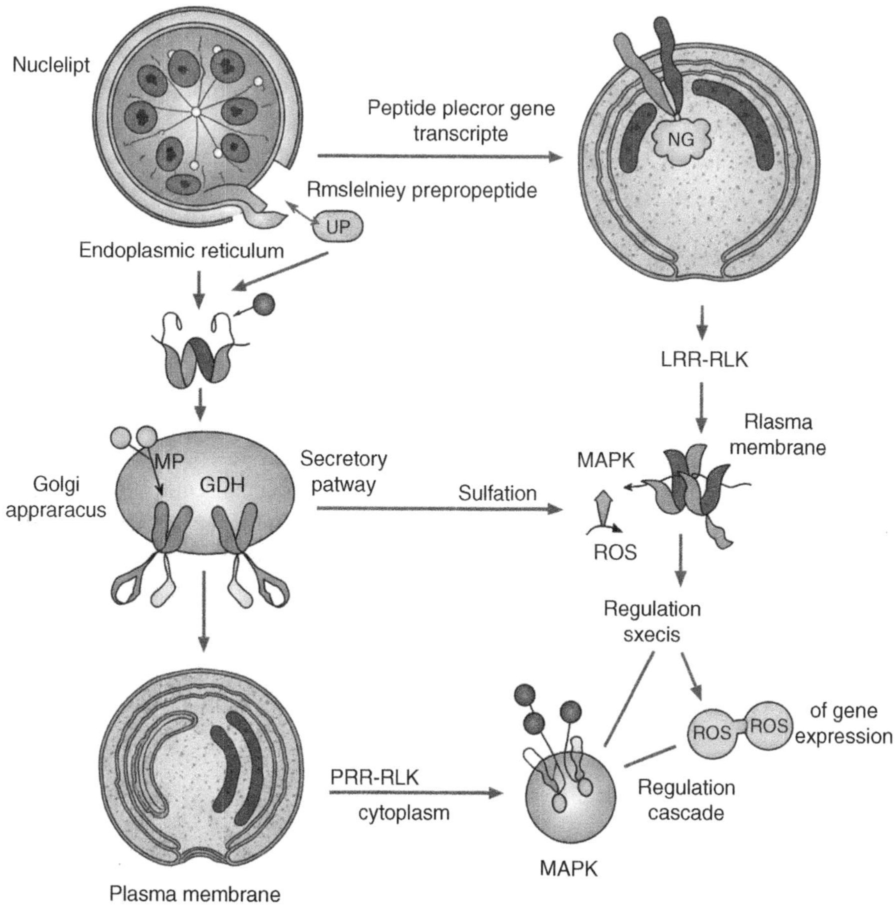

Figure 12.1 General model of peptide hormone biosynthesis and signaling pathway.

peptide. This process is under strict control to maintain the spatial and temporal regulation of peptide hormone levels. Certain peptides are produced as part of multigene families so that redundancy in or diversification of signaling function across tissues or developmental states is facilitated (Eggelkraut-Gottanka and Beck-Sickinger 2004).

12.2.1.1 Posttranslational Modifications

Posttranslational modifications (PTMs) are vital for peptide hormone maturation and activity. General PTMs include proteolytic cleavage, hydroxylation, glycosylation, and tyrosine sulfation, and these profoundly affect the stability, activity, and receptor-binding capacity of the peptides (Stührwohldt and Schaller 2019). Phytosulfokines (PSKs), for example, need to undergo tyrosine sulfation for their biological activity, whereas CLE peptides usually undergo specific proteolytic cleavage to yield the active signaling domain. Such alterations are catalyzed by specific enzymes found in the secretory pathway, and PTM defects can impair peptide function and cause developmental disabilities (Sauter 2015). PTMs guarantee that peptide hormones are properly processed and biologically active to participate in accurate intercellular signaling (Stührwohldt and Schaller 2019).

12.2.1.2 Receptor-Mediated Signaling Mechanisms

Peptide hormones act by binding to particular RLKs on the plasma membrane of target cells (Cui et al. 2022). Such receptors are usually members of the leucine-rich repeat receptor-like kinase (LRR-RLK) family, which have extracellular domains for ligand binding, one transmembrane domain, and an intracellular kinase domain for signal transduction (Chakraborty et al. 2019). Once bound by the peptide, the receptor changes its conformation, which in turn activates the kinase domain, promoting phosphorylation cascades to convey the signal inside the cell (Hohmann et al. 2017). The receptor-ligand specificity results in only cells with the correct receptor responding to a certain peptide, thereby enabling plant growth and environmental responses to be finely controlled (Martin-Ramirez et al. 2025).

In most instances, peptide signaling is further mediated by coreceptors like SOMATIC EMBRYOGENESIS RECEPTOR KINASEs (SERKs) that increase receptor sensitivity and downstream signaling (Becraft 2002). Upon activation of the receptor complex, it triggers intracellular signaling cascades such as mitogen-activated protein kinase (MAPK) cascades, calcium signaling, or alterations in transcriptional regulation. These cascades regulate the expression of particular target genes controlling development, stress responses, or immune activation (Gou and Li 2020). Negative regulators such as protein phosphatases are also brought in to restore the signaling apparatus and ensure homeostasis in other systems. Receptor spatial expression and dynamic regulation of these signaling modules also make plant peptide hormone action specific and plastic (Kim et al. 2021).

12.2.2 Major Classes of Peptide Hormones

12.2.2.1 CLE Peptides

CLAVATA3/EMBRYO SURROUNDING REGION-related (CLE) peptides are perhaps the most investigated families of plant peptide hormones (Yamaguchi et al. 2016). They are usually implicated in the regulation of meristem maintenance, stem cell differentiation, and vascular growth. CLE peptides act via specific LRR-RLK receptors like CLAVATA1 (CLV1) in the shoot apical meristem (Willoughby and Nimchuk 2021). For example, the CLV3 peptide limits the pool size of the stem cells through signaling via CLV1 to maintain a balance between cell proliferation and differentiation. CLE peptides encode small gene families that exhibit unique expression patterns in different tissues to achieve functional diversification, as well as precise developmental regulation (Hirakawa 2021).

12.2.2.2 Rapid Alkalinization Factors

RALFs are small, cysteine-rich peptides first identified for their ability to induce rapid alkalinization of the culture medium in tobacco cell cultures. They are involved in diverse processes such as cell expansion, pollen tube integrity, and immune responses (Sharma et al. 2016). RALF peptides signal through RLKs such as FERONIA (FER), which modulates cellular pH, calcium fluxes, and ROS production. This peptide family is central to the integration of abiotic and biotic stress signals with growth control (Lin et al. 2022). Because of their ability to respond to environmental stimuli, RALFs are identified as mediators at the center of the growth-defense trade-off in plants (Murphy and De Smet 2014).

12.2.2.3 Phytosulfokine

Phytosulfokines (PSKs) are sulfated pentapeptides that need tyrosine sulfation for complete biological function. They are most famous for their role in inducing cell proliferation, tissue regeneration, and somatic embryogenesis (Wu et al. 2025). PSKs interact with LRR-RLKs referred to as PSK receptors (PSKRs), thereby initiating downstream processes responsible for cellular growth and enlargement (Kaufmann et al. 2021). They are most active when responding to wounding or tissue injury, in which they reactivate growth and repair mechanisms. Furthermore, PSKs are also found to be involved in regulating immune responses, which renders them multifaceted actors in developmental as well as defense-related processes (Liu et al. 2024a).

12.2.2.4 IDA and Related Peptides

INFLORESCENCE DEFICIENT IN ABSCISSION (IDA) and associated peptides are essential regulators of abscission in organs, especially flowers and leaves (Aalen et al. 2013). These peptides act via RLKs like HAESA (HAE) and HAESA-LIKE2 (HSL2) to trigger cell separation activities in the abscission zones (Kumpf et al. 2013). The IDA signaling module is responsible for the timely detachment of organs required for reproduction and stress adjustment. IDA-related peptides have also been indicated to play a role in lateral root emergence and abscission of leaves due to drought stress by recent studies, broadening their function beyond reproductive growth. Their narrowly regulated expression highlights their role in spatially and temporally controlled developmental processes (Lalun et al. 2024).

12.2.3 Role in Growth, Development, and Stress Responses

Peptide hormones play a central role in orchestrating plant growth by regulating cell division, elongation, and differentiation (Table 12.1). For instance, CLE peptides modulate meristem activity, maintaining the delicate balance between stem cell maintenance and differentiation, while PSKs promote cell proliferation and tissue regeneration (Ali and

Table 12.1 Functional roles of major peptide hormone families in plants (Gancheva et al. 2019; Yadav 2024).

Peptide hormone family	Receptor(s)	Primary function(s)	Developmental/stress role
CLE (CLAVATA3/ ESR-related)	CLV1, BAM1/2, TDR/PXY	Regulates stem cell maintenance, vascular patterning	Shoot/root meristem control, response to nutrient stress
RALF (rapid alkalinization factor)	FERONIA (FER), ANX1/2	Modulates cell expansion, pH, ROS signaling	Pollen tube integrity, defense, abiotic stress response
PSK (phytosulfokine)	PSKR1, PSKR2	Promotes cell division, tissue regeneration	Wound healing, somatic embryogenesis
IDA (inflorescence deficient in abscission)	HAE, HSL2	Triggers organ abscission, lateral root emergence	Flower and leaf drop, drought-induced leaf shedding

Baloch 2020). These peptides often act in concert with classical plant hormones like auxin and cytokinin, fine-tuning developmental outcomes. The spatial and temporal precision of peptide hormone signaling enables plants to form complex organs and adapt their architecture in response to environmental and internal cues (Wong et al. 2023).

In developmental processes, peptide hormones guide key transitions such as root and shoot patterning, vascular formation, and organ abscission (Hsiao and Yamada 2020). IDA peptides ensure the proper timing of floral and leaf drop, an essential process for reproduction and stress escape (Lalun et al. 2024). RALF peptides influence pollen tube guidance and cell wall dynamics, facilitating successful fertilization. These signaling molecules often act in short-range signaling contexts, allowing individual cells or small groups of cells to influence their immediate neighbors, thereby contributing to the patterned development of plant tissues (Qu et al. 2015).

Beyond growth and development, peptide hormones are increasingly recognized for their roles in plant defense and abiotic stress responses. Under conditions such as high light intensity, drought, or pathogen attack, specific peptides modulate signaling pathways that activate stress tolerance genes, regulate ROS levels, and interact with hormonal pathways like abscisic acid (ABA) or jasmonic acid (JA) (Chowdhary and Songachan 2025). For example, RALF-FER signaling can modulate the oxidative burst during pathogen recognition, while CLE peptides have been linked to drought response regulation (Ren et al. 2024). Thus, peptide hormones function as integrative hubs linking developmental cues with environmental adaptation.

Peptide hormones are pivotal in coordinating plant growth through the regulation of cell division, elongation, and differentiation (Table 12.1). For example, CLE peptides regulate meristem activity, sustaining the fine balance between stem cell maintenance and differentiation, while PSKs drive cell proliferation and tissue regeneration (Ali and Baloch 2020). These peptides tend to act in combination with traditional plant hormones such as auxin and cytokinin, modulating developmental outcomes. The spatial and temporal accuracy of peptide hormone signaling allows plants to build elaborate organs and modify their structure in line with environmental and endogenous signals (Wong et al. 2023).

During developmental processes, peptide hormones control important transitions including root and shoot patterning, vascular development, and organ abscission (Hsiao and Yamada 2020). IDA peptides coordinate proper timing of flower and leaf drop, a critical reproductive and stress escape process (Lalun et al. 2024). RALF peptides mediate pollen tube guidance and cell wall dynamics, ensuring effective fertilization. These signaling molecules tend to function in short-range signaling contexts, and thus individual cells or small sets of cells have the ability to influence their immediate neighbors and contribute to the patterned development of plant tissues (Qu et al. 2015).

In addition to growth and development, peptide hormones are gaining recognition for their involvement in plant defense and abiotic stress responses. Under stressful conditions like elevated light intensity, drought, or pathogen attack, certain peptides regulate signaling pathways activating stress-tolerance genes, managing ROS, and interfacing with hormonal pathways like ABA or JA (Chowdhary and Songachan 2025). For instance, RALF-FER signaling may regulate the oxidative burst during pathogen recognition, and drought response regulation has been associated with CLE peptides (Ren et al. 2024). Therefore, peptide hormones act as integration centers connecting developmental signals and environmental adaptation.

12.3 Photosynthesis–Overview

12.3.1 Basic Light and Dark Reactions

Photosynthesis is split between light reactions and dark reactions (or the Calvin-Benson cycle). Light reactions take place in thylakoid membranes within chloroplasts, where the Sun's energy is absorbed by chlorophyll and other pigments. The light is utilized to break apart water molecules, releasing oxygen and creating the high-energy molecules ATP and NADPH (Hossain et al. 2022).

Overall equation of the light reactions:

$$2H_2O + 2NADP^+ + 3ADP + 3Pi + \text{light energy} \rightarrow O_2 + 2NADPH + 3ATP$$

Key processes include:

Photolysis of water (in photosystem II):

$$2H_2O \rightarrow 4H^+ + 4e^- + O^2$$

$NADP^+$ reduction (in photosystem I):

$$2NADP^+ + 4e^- + 4H^+ \rightarrow 2NADPH + 2H^+$$

ATP formation via photophosphorylation:

$$3ADP + 3Pi \rightarrow 3ATP$$

The energy carriers, ATP and NADPH, are subsequently used in the dark reactions, which occur within the stroma of chloroplasts. The dark reactions are responsible for fixing atmospheric CO_2 into organic molecules such as glucose with the energy supplied by ATP and NADPH from the light reactions (Kumar and pal, 2022).

Overall equation of the Calvin cycle (to form one glucose molecule):

$$6CO_2 + 18ATP + 12NADPH + 12H_2O \rightarrow C_6H_{12}O_6 + 18ADP + 18Pi + 12NADP^+ + 6O_2$$

This equation assumes synthesis of one glucose over two full cycles; often simplified per CO_2 fixed as:

$$CO_2 + 2NADPH + 3ATP \rightarrow (CH_2O) + 2NADP^+ + 3ADP + 3Pi$$

12.3.2 Photosystems I and II

Photosystems I (PSI) and II (PSII) are thylakoid membrane–embedded multiprotein pigment complexes responsible for the light reactions of photosynthesis. PSII captures light and starts water molecule splitting into oxygen, protons, and electrons—the process of photolysis (Schwarz et al. 2018). The electrons are passed through an electron transport chain (ETC) to PSI, which reenergizes them via further light energy. PSI then drives the reduction of $NADP^+$ to NADPH. These two photosystems act together in what is referred to as the "Z-scheme," transforming light energy to chemical energy with good efficiency and facilitating ATP and NADPH synthesis (Wang et al. 2022a, b).

12.3.3 Electron Transport Chain and ATP/NADPH Synthesis

The photosynthetic electron transport chain (ETC) consists of a series of protein complexes and mobile carriers that are inserted in the thylakoid membrane. Electrons excited in PSII are transferred along the ETC via plastoquinone (PQ), the cytochrome b6f complex, and plastocyanin to PSI (Ustynyuk and Tikhonov 2022). During the passage of electrons through the chain, protons are pumped from the stroma to the thylakoid lumen and a proton gradient is formed. ATP synthase utilizes this gradient to synthesize ATP from ADP, a process referred to as photophosphorylation (Figure 12.2). Concurrently, PSI donates electrons to $NADP^+$, which is reduced to NADPH. ATP and NADPH, together, supply the energy and reducing power required for carbon fixation in the dark reactions (Tikhonov 2024).

12.3.4 Carbon Fixation: Calvin–Benson Cycle

The Calvin–Benson cycle, taking place in the chloroplast stroma, is the major pathway of carbon fixation in plants. It employs the ATP and NADPH produced from the light reactions to fix carbon dioxide into organic compounds (Gurrieri et al. 2021). The cycle starts with the enzyme RuBisCO catalyzing the carboxylation of ribulose-1,5-bisphosphate (RuBP) to form two molecules of 3-phosphoglycerate. By a series of reduction and regeneration steps, glyceraldehyde-3-phosphate (G3P) is generated, part of which is directed toward the synthesis of glucose and other carbohydrates, and the remaining amount is utilized to regenerate RuBP (Meloni et al. 2023). The cycle not only maintains plant growth but also forms the basis of the global carbon cycle.

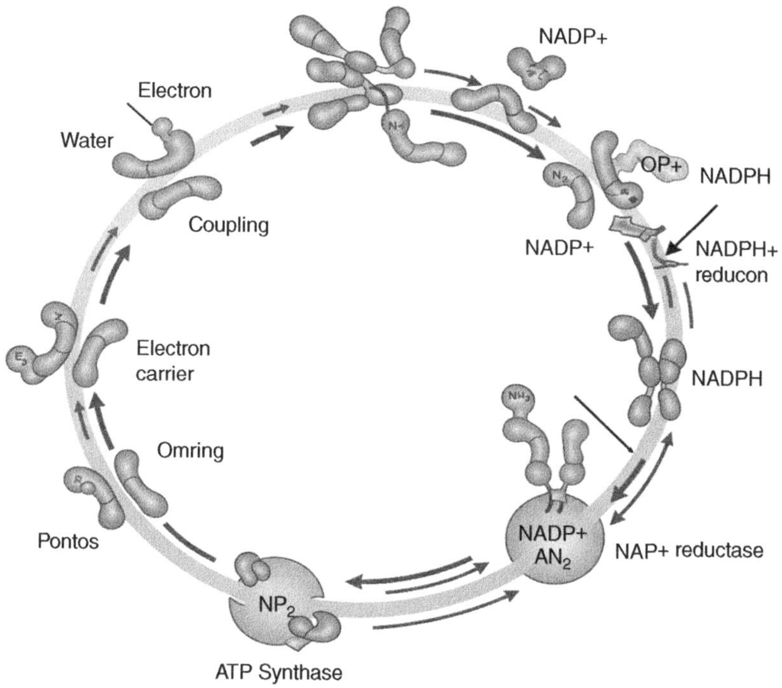

Figure 12.2 Electron transport chain and ATP/NADPH synthesis.

12.4 Photodamage and the Requirement for Photoprotection

12.4.1 Consequences of Excess Light: Reactive Oxygen Species (ROS)

When plants receive excess light more than they require for photosynthesis, the excess energy results in the creation of ROS, like superoxide radicals, hydrogen peroxide, and singlet oxygen (Bassi and Dall'Osto 2021). ROS are mostly produced in chloroplasts, particularly at Photosystem II (PSII), due to the over-reduction of the ETC, causing an excess of electrons to be passed on to oxygen (Foyer 2018). ROS are very reactive and cause oxidative damage to proteins, lipids, and nucleic acids, disturbing cellular structures as well as signaling pathways. ROS accumulation is one of the major signs of photodamage and can induce programmed cell death during extreme stress conditions (Juan et al. 2021).

To counteract ROS toxicity, plants utilize enzymatic (e.g. superoxide dismutase, ascorbate peroxidase [APX], catalase [CAT]) and nonenzymatic (e.g. carotenoids, tocopherols, ascorbate) defense mechanisms (Ahmad et al. 2009). Yet, under chronic or severe light stress, these defense systems can be overburdened, leading to photoinhibition and a reduction in photosynthetic performance. This supports the significance of photoprotection mechanisms such as NPQ that release surplus light energy as heat before it causes ROS generation (Didaran et al. 2024).

12.4.2 Effects of Excess Light: D1 Protein Damage

The D1 protein is a central constituent of the PSII reaction center and is a key player in the initial photochemistry of photosynthesis (Gupta et al. 2017). During conditions of high light, the D1 protein is particularly susceptible to damage because of its involvement in charge separation and water-splitting reactions. Photodamage of D1 inhibits electron transport, decreases the PSII efficiency, and restricts the capacity of the plant to convert light energy into chemical energy (Wang and Domen 2019).

Plants have a repair system referred to as the PSII repair cycle, where selectively damaged D1 proteins are degraded and replaced by newly synthesized versions (Theis and Schroda 2016). Yet, the level of damage under excess light frequently surpasses this repair system's capacity, particularly under integrated stress conditions. Under conditions where D1 degradation is faster than synthesis, there is constant photoinhibition, which emphasizes the importance of efficient photoprotective processes, such as regulation of light gathering and enhanced thermal energy dissipation (Didaran et al. 2024).

12.4.3 Environmental Triggers

12.4.3.1 UV Radiation

UV radiation, especially UV-B (280–315 nm), is a major hazard for photosynthetic life. It enters plant tissues and causes direct DNA damage, protein degradation, and lipid peroxidation (Kumari et al. 2023). UV-B also increases ROS production in chloroplasts, intensifying oxidative stress and photosynthetic component damage such as PSII and RuBisCO. Plants respond by producing UV-absorbing compounds such as flavonoids and expressing UV-resistance genes, but too much still causes stunted growth and impaired photosynthesis (Mahdavian 2024).

12.4.3.2 High Light Intensity

Natural sunlight varies very quickly, and bright light at noon or outdoors in open spaces can exceed the photosynthetic saturation value, causing overexcitation of chlorophyll molecules (Gu et al. 2017). Overexcitation creates more ROS production and photodamage to the D1 protein. Although plants utilize dynamic photoprotection through xanthophyll cycle pigments and NPQ, prolonged high light with no relief may nevertheless induce irreversible damage to PSII, confining carbon fixation and overall growth (Simkin et al. 2022).

12.4.3.3 Drought Stress

Drought closes stomata, which lowers CO_2 availability for the Calvin cycle, leading to an imbalance between light capture and its utilization. This deficit worsens the production of ROS within the chloroplast because excess light energy has no direction to go (Lemonnier and Lawson 2024). Coupled with water restriction, the ability to thermal regulation and antioxidant defense is likewise compromised, augmenting the photodamage and precipitating a drastic reduction in photosynthetic efficiency (Semwal and Khanna-Chopra 2020).

12.4.4 Effect on Photosynthetic Efficiency and Crop Yield

Photosynthetic efficiency in plants is highly sensitive to environmental fluctuations. Stress conditions such as UV radiation, high light intensity, and drought can disrupt the delicate balance between light capture and energy utilization. These triggers often lead to excess ROS production, causing photodamage and impairing photosynthesis (Figure 12.3).

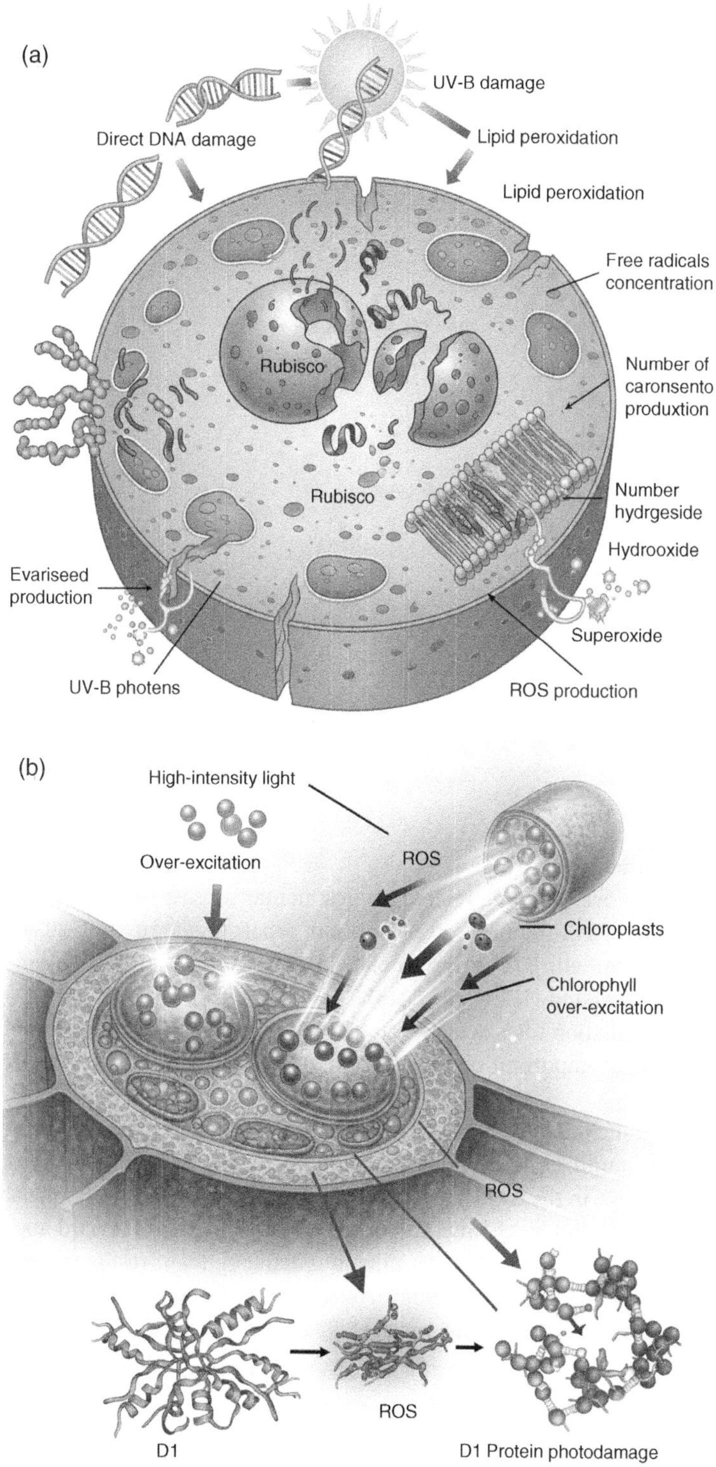

Figure 12.3 Environmental triggers of photodamage in plants: (a) UV-B radiation; (b) high light intensity; (c) drought stress.

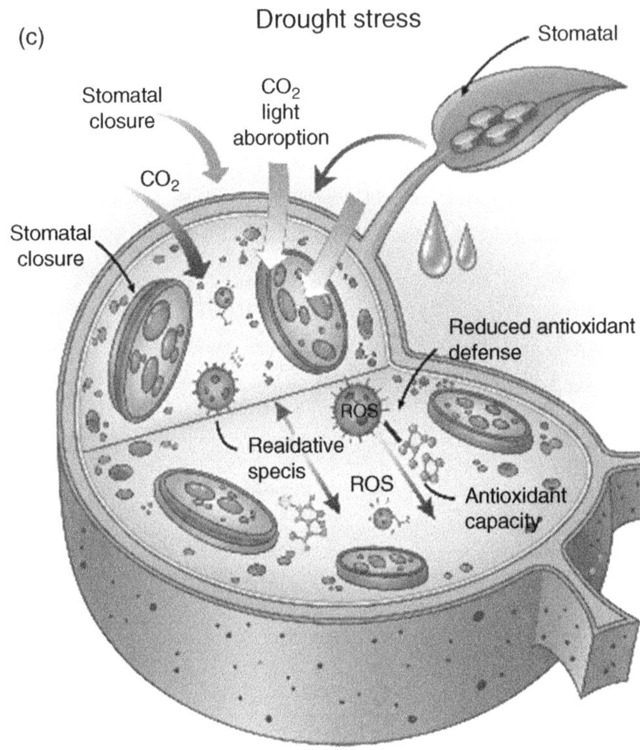

Figure 12.3 (Continued)

12.4.4.1 Photosynthetic Efficiency

Photodamage incurred due to overlight and related environmental stresses triggers photoinhibition, in which the rate of PSII damage exceeds that of repair (Didaran et al. 2024). Directly damaging the photosynthetic ETC, this restricts ATP and NADPH synthesis, essential for the Calvin cycle (Murata and Nishiyama 2018).

Consequently, carbon assimilation rates decrease, and the overall photosynthetic performance weakens, particularly at important phases of plant growth like flowering or grain filling (Table 12.2).

Photoprotective processes such as NPQ and state transitions ensure a balance between energy intake and output, sustaining photosynthetic efficiency in changing light conditions (Derks et al. 2015). Nevertheless, in the case of extended or severe stress, these mechanisms prove to be inadequate. Genetic and biochemical alterations that are directed toward augmenting photoprotection have had encouraging results in enhancing plant resilience and thus indicate possible avenues of future breeding or engineering crops with enhanced photosynthetic stability under stress (KhokharVoytas et al. 2023).

12.4.4.2 Crop Yield

The yield effect of photodamage is notable, especially for crops that develop in conditions of high irradiance, drought, or UV stress (Alexieva et al. 2001). Yield loss is due to decreased

Table 12.2 Photodamage impacts, photoprotection, and effects on crop yield.

Crop/species	Stress factor	Photodamage/ stress	Photoprotection mechanism	Yield/ performance Impact	References
Wheat *(Triticum aestivum)*	UV-B radiation	ROS accumulation, PSII damage	Flavonoid synthesis, antioxidant enzymes	10–15% reduction in biomass	Zhao et al. (2003)
Maize *(Zea mays)*	Drought + high light	ROS, PSII photoinhibition	Antioxidant enzymes, heat shock proteins	15–30% lower yield under combined stress	Long et al. (2006)
Soybean *(Glycine max)*	High light	Singlet oxygen, PSII inactivation	NPQ, chloroplast movement	~12% loss in seed yield	Baker (2008)
Arabidopsis (model)	UV + drought	DNA damage, oxidative stress	UVR8 pathway activation, antioxidants	Experimental growth reduction	Hideg et al. (2013)
Rice (*Oryza sativa*)	High light	D1 protein degradation	Xanthophyll cycle, NPQ	~20% yield loss during grain filling	Kromdijk et al. (2016)
Rice (Transgenic *O. sativa*)	High light (field)	Photoinhibition, ROS increase	Overexpression of GLK → ↑ NPQ, pigments	30–40% boost in yield and biomass	Fang et al. (2021)
Maize hybrids (e.g. DH618)	General light stress	Imbalanced antenna/core pigments	Lower Chl a/b ratio → improved NPQ efficiency	Higher yield potential	Liu et al. (2021)
Rice and wheat	Midday heat + high light	D1 damage, ROS accumulation	Carotenoids, regulated chlorophyll fluorescence	Maintained PSII performance under stress	Kumar et al. (2022)
Leymus chinensis	Drought	ROS, H_2O_2 accumulation	Nitrate fertilization reduced oxidative stress	Improved physiological resilience (nonyield)	Zhang et al. (2024)

biomass production, disrupted reproductive development, and shorter growth cycles. In rice, maize, and wheat, experiments have demonstrated that brief exposure to high light stress during vulnerable growth phases such as anthesis or grain filling reduces yields by 10–30% (Bhattacharya and Bhattacharya 2021).

Enhancing photoprotection—either by agronomic means (e.g. shading, irrigation) or genetic approaches (e.g. overexpression of antioxidant enzymes or xanthophyll cycle pigment components)—can insulate plants against these losses (Simkin et al. 2022). Incorporating these approaches into crop breeding programs is essential to preserve food security in the face of the mounting intensity and unpredictability of climate-related stresses.

12.5 Mechanisms of Photoprotection

Plants have developed a variety of photoprotective processes to avoid light-induced damage and keep photosynthetic efficiency intact in changing and excessive light conditions (Khan et al. 2025). They possess physical, biochemical, and molecular mechanisms to balance energy absorption and dissipation. At the core of these are the capacity to dissipate excess excitation energy and quench ROS (Bassi and Dall'Osto 2021).

12.5.1 Nonphotochemical Quenching (NPQ)

NPQ is a critical mechanism that enables plants to safely dissipate excess light absorbed energy as heat and thereby guard the photosystems against photodamage. NPQ is dynamically regulated and activated upon changes in light intensity (Zuo 2025).

12.5.1.1 Xanthophyll Cycle

The xanthophyll cycle is the reversible conversion of the pigments violaxanthin to antheraxanthin and subsequently to zeaxanthin under conditions of high light. This conversion improves NPQ by enabling the thermal loss of excess energy in the light-harvesting complexes (Demmig-Adains and Adams 2017). Zeaxanthin is at the heart of relaxing excitation pressure on PSII, especially during stress. Upon decreasing light intensity, zeaxanthin is recycled to violaxanthin, permitting effective light capturing to be resumed (Tang et al. 2021). This dynamic process is an important part of the plant's capacity to adjust its light-harvesting capability in real time.

12.5.1.2 Role of PsbS Protein

The PsbS protein is a pH-sensing subunit of the PSII antenna complex and plays a crucial role in eliciting NPQ. In high light, thylakoid lumen acidification occurs through enhanced proton accumulation following increased electron transport (Nicol and Croce 2021). This acidification triggers PsbS activation, inducing conformational changes in the light-harvesting complexes that favor energy dissipation as heat. PsbS-deficient mutants have compromised NPQ capacity and increased susceptibility to photoinhibition, highlighting its key regulatory function in photoprotection (Pinnola 2019).

12.5.2 Antioxidant Systems

Antioxidant defense systems safeguard the plant by detoxifying ROS produced through photosynthesis. Both enzymatic and nonenzymatic components are present in these systems that act together to restrict oxidative damage (Saed-Moucheshi et al. 2014; Ighodaro et al. 2018).

12.5.2.1 Ascorbate-Glutathione Cycle

The ascorbate-glutathione (AsA-GSH) cycle is an important antioxidant pathway that functions within chloroplasts, mitochondria, and peroxisomes. It is the reduction of hydrogen peroxide (H_2O_2) to water by the action of ascorbate (vitamin C) as an electron donor. During this cycle, APX reduces H_2O_2 to water while ascorbate is oxidized to monodehydroascorbate (MDHA). MDHA is subsequently reduced back to ascorbate by monodehydroascorbate reductase (MDHAR) or to dehydroascorbate (DHA), which is reduced by glutathione (GSH) through dehydroascorbate reductase (DHAR). Glutathione reductase (GR) re-reduces GSH with NADPH, ensuring redox homeostasis. This process is crucial in regulating ROS and safeguarding chloroplast operation (Rajput et al. 2021).

12.5.2.2 Superoxide Dismutase (SOD)

SOD is the first line of enzymatic defense against ROS, catalyzing the dismutation of the superoxide radical (O_2^-) to hydrogen peroxide (H_2O_2) and molecular oxygen (O_2) (Saibi and Brini 2018). There are various isoforms of SOD in chloroplasts (Cu/Zn-SOD), mitochondria (Mn-SOD), and cytosol, providing broad-spectrum protection. Its activity decreases the level of superoxide radicals and inhibits their damaging interaction with cellular macromolecules (Rahman et al. 2024).

12.5.2.3 Catalase (CAT)

Catalase is an enzyme found primarily in peroxisomes, where it breaks down hydrogen peroxide (H_2O_2) into oxygen and water (Lismont et al. 2019). CAT, unlike APX, does not need a reducing substrate and works effectively under high H_2O_2 concentrations. CAT plays a crucial role during photorespiration, where extensive quantities of H_2O_2 are generated, thereby avoiding oxidative damage in peroxisomal compartments (Anjum et al. 2016).

12.5.2.4 Ascorbate Peroxidase (APX)

APX is a major enzyme in the AsA-GSH cycle that employs ascorbate to convert hydrogen peroxide into water (Hasanuzzaman et al. 2019). It occurs in various cell compartments such as chloroplasts (thylakoid and stroma), cytosol, and mitochondria. APX cooperates with SOD to kill ROS produced during photosynthesis. It is particularly important in chloroplasts, where it avoids the accumulation of H_2O_2 that could otherwise damage photosynthetic proteins and membranes (Foyer and Hanke 2022).

12.5.3 Chloroplast Movements

Chloroplasts are able to move within cells in a plant in response to light intensity to maximize the absorption of light and minimize photodamage. Chloroplasts extend along the cell surface to maximize the capture of light when there is low light (accumulation response) (Shi et al. 2022). When there is high light, they redistribute to the cell periphery to reduce exposure (avoidance response). These movements, controlled by actin filaments and photoreceptors such as phototropins, assist in regulating internal levels of light and minimizing ROS generation (Bhatla and Lal 2023).

12.5.4 Leaf Positioning

Plants modulate leaf angles and orientation to control light interception. Under high light conditions, the leaves can assume steeper angles or show paraheliotropism (leaf avoidance of direct sunlight) to minimize excessive radiation (Nilsen and Forseth 2018). These movements help avoid overheating and the potential for photodamage, particularly at midday when irradiance is highest. Leaf movement is frequently synchronized with stomatal responses and chloroplast positioning for holistic photoprotection (Karabourniotis et al. 2021).

12.5.5 Cyclic Electron Flow Around PSI

Cyclic electron flow (CEF) around photosystem I represents a different electron-transport route that plays a role in ATP synthesis without NADPH production. This cycle returns electrons from ferredoxin to the pool of PQ, further increasing proton-gradient generation across the thylakoid membrane (Leister 2020). CEF is also crucial in maintaining the ATP/NADPH ratio needed for the Calvin cycle and enhances thermal dissipation through NPQ. CEF also shields PSI and PSII against fluctuating light by reducing over-reduction of electron carriers and stabilizing the photosynthetic machinery (Roach and Krieger-Liszkay 2019).

12.6 Cross Talk Between Peptide Hormones and Photoprotection

Peptide hormones in plants, initially thought to be largely developmental regulators, have emerged as modulators of stress signaling pathways of special importance. Studies have established that different classes of peptides—such as RALFs, CLEs, and PSKs—are expressed differentially in response to environmental stresses like high light, drought, and salinity (Lu and Xiao 2024). These peptides signal via RLKs at the plasma membrane, stimulating intracellular signaling cascades that converge with known stress response pathways. For instance, peptide–RLK interactions may induce MAPK cascades, calcium signaling, and ROS production—all important to the perception and alleviation of abiotic stress, including photodamage (Zhong and Qu 2019).

Additionally, peptide hormone signaling tends to act in parallel or an additive fashion with conventional stress hormones like ABA and salicylic acid (SA), implying a complex integration of stress signals (Parwez et al. 2022). Under conditions of light stress, peptide pathway activation has been associated with increased systemic acquired acclimation (SAA), a defense response where light stress on one leaf triggers resistance in the remote tissues (Szechyńska-Hebda et al. 2017). The spatiotemporal precision of peptide action enables refined response to localized or dynamic stress, thus maximizing photoprotective strategy without wasteful energy expenditure. Such findings decisively attest to the participation of peptide hormones as active players in the plant's extended stress signaling network (Xiao et al. 2025).

12.6.1 Modulation of Gene Expression Associated with Light Stress

Peptide hormones regulate gene expression by affecting transcriptional networks that control stress responses. By binding to their receptors, peptides have the ability to initiate downstream phosphorylation reactions and activation of transcription factors (TFs), thus changing the expression of the genes implicated in photoprotection, such as those coding for antioxidant enzymes, photoreceptors, and chloroplast-localized proteins (Xie et al. 2019). For example, application or overexpression of particular peptides can induce upregulation of APX, SOD, and ELIP (early light-inducible proteins), which are important for coping with photooxidative stress (Valledor et al. 2012) (Figure 12.4).

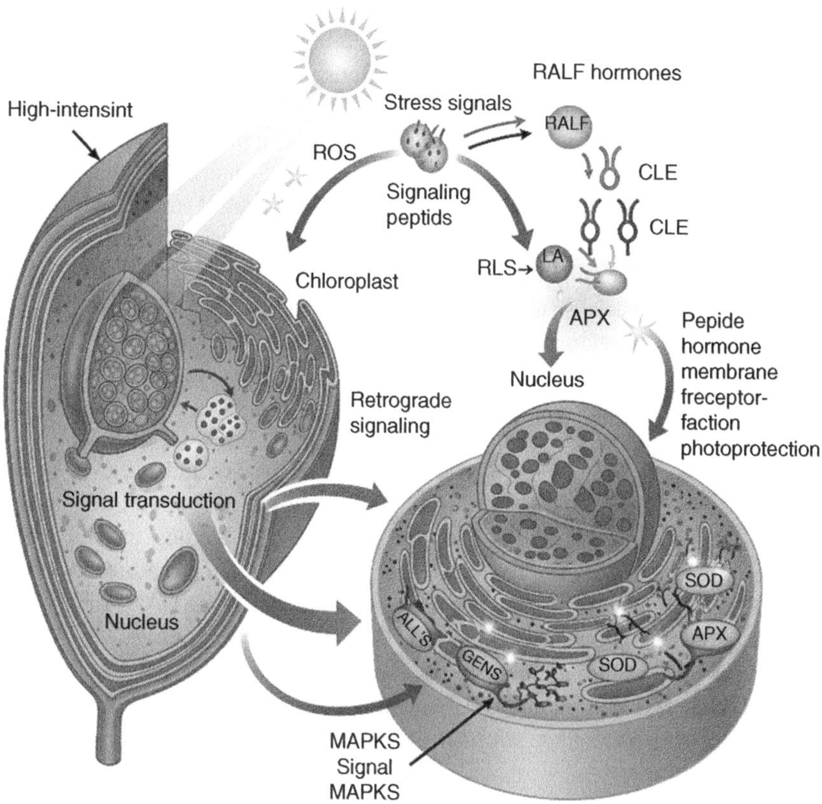

Figure 12.4 Modulation of gene expression related to light stress.

Furthermore, transcriptomic data have indicated that light stress induces expression of peptide hormone genes themselves, indicating a feedback regulatory loop. This two-way interaction allows for speedy and context-dependent regulation of the transcriptome. General cis-regulatory elements that are sensitive to both light and peptide signals have been found in the promoters of target genes, suggesting a molecular mechanism of signal integration (Marand et al. 2023). This gene control leads to photoprotective results such as regulated stomatal aperture, chlorophyll breakdown, and repair of photosystem damage, enabling plants to strike a balance between growth and stress tolerance (Dsilva and Galande 2024).

12.6.2 Hormonal Priming for Oxidative and Photodamage Tolerance

Peptide hormones have been found to be very effective agents in hormonal priming, where preexposure to a low-level stress or signaling molecule makes a plant more tolerant to a future stress incident (Hutchinson et al. 2017). In photoprotection, peptides such as RALFs and CLEs have been found to condition antioxidant responses by regulating basal ROS-scavenging systems. Being in this condition state entails not just biochemical alterations (e.g. increased glutathione content) but epigenetic modifications that make important gene promoters receptive to subsequent activation. This priming helps plants respond more quickly and robustly against high light or UV stress (Ghosh and Roychoudhury 2024).

In addition, peptide-induced priming may induce long-lasting photoprotection at various developmental phases. For instance, preexposure during seedling phase with certain peptides may increase adult plant resistance to photooxidative stress (Stintzi and Schaller 2022). This implies that peptide hormones could modulate plant "memory" of stress, perhaps via stable chromatin remodeling or hormonally regulated changes in sensitivity in some manner (Sharma et al. 2023a, b). This highlights the application potential of peptides for enhancing stress acclimation, especially for crops cultivated under more fluctuating light conditions because of climate change.

12.6.2.1 RALF Peptides and ROS Modulation

RALF peptides, originally known for their involvement in cell growth and pH level maintenance, have also been shown to be involved in ROS signaling during stress. RALFs are capable of modulating ROS levels in the apoplast through their binding to FERONIA (FER), a receptor kinase that plays roles in mechanosensing and immune response (Chowdhary and Songachan 2025). RALF binding causes FER to regulate NADPH oxidases (RBOHs), resulting in a regulated level of ROS that serves as a secondary messenger. This short-lived burst of ROS is essential for triggering protective gene expression and reorientation of stomatal responses in light stress (Ghosh and Roychoudhury 2024). The RALF-FER module is a case of balancing act: stimulating sufficient ROS to trigger stress without becoming cytotoxic.

The RALF peptide or FER signaling-deficient mutants typically display overaccumulation of ROS and hypersensitivity to light damage (Zhang et al. 2022). Moreover, RALFs have also been associated with calcium channel modulation, influencing stress signal transmission further. Their function in ROS fine-tuning positions them as important photoprotection players, connecting growth-regulatory pathways to environmental responsiveness (Ravi et al. 2023).

12.6.2.2 CLE Peptides in Abiotic Stress Contexts

CLE peptides, once linked to vascular differentiation and meristem maintenance, are gaining attention for their function in adapting to abiotic stress. A number of CLE genes are induced under drought, high light, and nutrient stress (Xie et al. 2022). CLE25 has specifically been highlighted for its systemic function in stress signaling—being translocated from roots to shoots to modulate ABA biosynthesis under drought, having an indirect impact on stomatal closure and light-driven water loss (Fletcher 2020).

In photoprotection, CLE peptides potentially function as long-distance messengers orchestrating shoot-root responses. CLE signaling by their receptors (e.g. CLAVATA1, BAM receptors) can regulate gene networks related to cell wall integrity, scavenging ROS, and hormonal communication that are critical under conditions of photodamage (Yamaguchi et al. 2016). Their bifunctional developmental and stress-related roles demonstrate how plants repurpose constitutive regulatory molecules for distinct purposes with maximal resource utilization and upkeep of defense capacity (Whitewoods 2021).

12.6.3 Interfacing with Classical Hormones (ABA, SA, JA)

Peptide hormones often interact with classical phytohormones such as ABA, SA, and JA to establish a multilayered defense system. For instance, under light-induced oxidative stress, levels of ABA usually increase to facilitate stomatal closure and limit water loss (Huang et al. 2022). RALF and CLE peptides may modulate ABA sensitivity or act upon its biosynthesis by controlling specific enzymes like NCED3. Such coordination is crucial to ensure that light-protective processes such as control of transpiration are accurately timed and spatially organized (Datta et al. 2024). In the same way, cross talk with the SA and JA pathways coalesces peptide signaling with defense and repair processes.

SA is associated with systemic acquired resistance (SAR), yet also plays a part in responses to light stress through redox homeostasis regulation (Liu et al. 2024a, b). Peptides such as PLANT ELICITOR PEPTIDES (PIPs) have the ability to enhance SA-mediated signaling, increasing stress perception and response. At the same time, JA combines peptide signals to achieve a balance between growth inhibition and protective secondary metabolism in photodamage (Figure 12.5). This hormonal coordination increases stress tolerance by

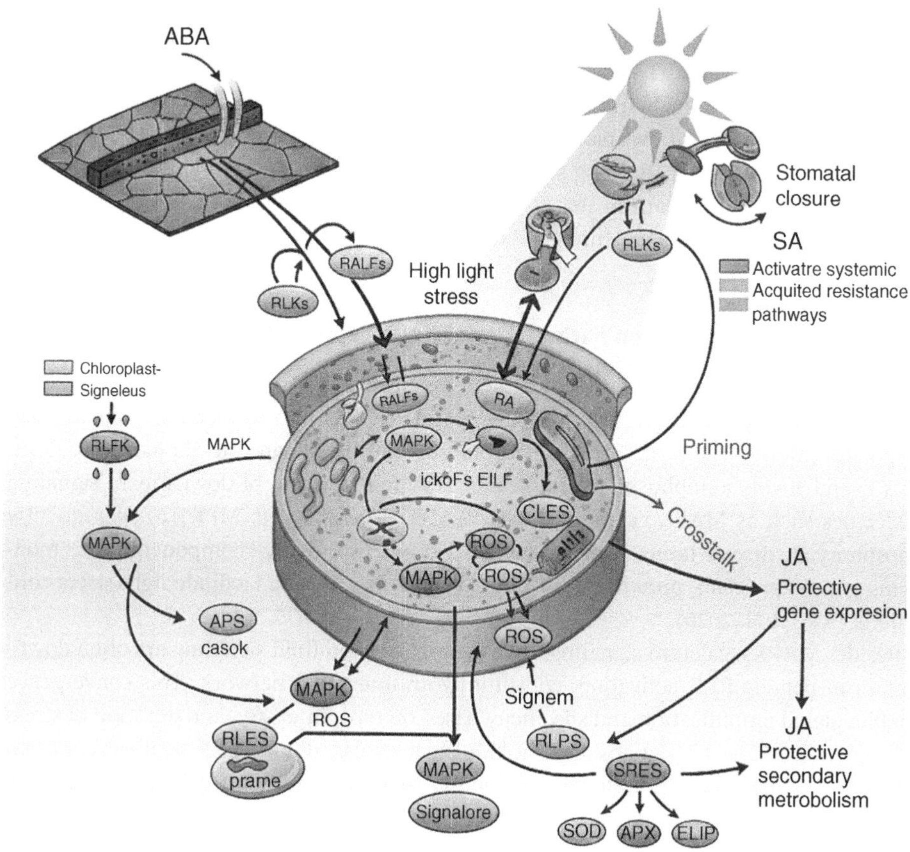

Figure 12.5 Integration of peptide hormones with classical plant hormones (ABA, SA, JA) during oxidative and light stress.

redundancy and synergy, promoting robustness under fluctuating environmental conditions (Pastor-Fernández et al. 2023).

12.7 Transcription Factors Regulated by Peptide Hormones Affecting Photosynthetic Genes

Peptide hormones control individual TFs that have an effect on photosynthesis-associated gene expression. In response to perception by RLKs localized in the plasma membrane, peptides like RALFs, CLEs, and PSKs trigger intracellular signaling pathways that result in transcriptional reprogramming (Jiang et al. 2022). For example, RALF-FERONIA signaling controls WRKY, NAC, and MYB transcription factors—previously documented regulators of stress-responsive and photosynthetic genes. These TFs, in their activated state, regulate the expression of genes that encode photosystem II component proteins, chlorophyll-binding proteins, and Calvin cycle enzymes, thereby directly linking perception of peptides to photosynthetic performance under stress (Huang et al. 2022).

Furthermore, this regulation is context-dependent, frequently requiring combinatorial control with traditional hormones such as ABA or environmental stimuli such as high light or oxidative stress. Some TFs regulated by peptides also serve as redox signal integrators—converting oxidative signals from the chloroplast into nuclear changes in gene expression (Jiang et al. 2022). This allows plants to favor photoprotective outputs such as NPQ or xanthophyll cycle activation. In doing so, peptide hormones feed into a highly regulated network of transcriptional control that is able to preserve photosynthetic performance and avoid light-induced damage (Murchie and Ruban 2020).

12.7.1 Signal Transduction Pathways (MAPKs, Receptor-Like Kinases)

Signaling by peptide hormones is initiated in the cell surface through RLKs, which have an important role in converting extracellular peptide signals into intracellular biochemical responses (Motomitsu et al. 2015). Upon binding with the ligand, RLKs like FERONIA, CLV1, and BAMs autophosphorylate, leading to the activation of downstream signaling molecules such as MAPK cascades. These MAPK pathways (e.g. MPK3/6) subsequently phosphorylate diverse target proteins, such as TFs and cytoskeletal components, thus modifying gene expression, protein stability, and cell architecture to facilitate light-stress conditions (Jose et al. 2020).

Besides MAPKs, calcium signaling, ROS generation, and lipid signaling are often downstream of peptide-RLK activation, constituting an integrated network. This convergence enables signal amplification and specificity based on type, intensity, and duration of stress (Wang et al. 2022a, b). For instance, RALF peptides can induce FER-mediated calcium influx, activating calmodulin-dependent kinases that cross talk with light-stress signaling modules. This cross-communication ensures peptide signals get contextually interpreted and combined with other stress signaling pathways, facilitating a holistic and adaptive photoprotective response (Hamann 2015).

12.7.2 Chloroplast–Nucleus Communication–Retrograde Signaling

Chloroplast-to-nucleus retrograde signaling is crucial for photoprotection as it regulates nuclear gene expression in response to chloroplast status. Peptide hormones may affect or be integrated into these retrograde pathways (Ge et al. 2025). For example, peptides can control the generation or signaling of retrograde signals like Mg-protoporphyrin IX, ROS, and 3′-phosphoadenosine 5′-phosphate (PAP). These molecules are produced in response to light stress and are transported to the nucleus to regulate the expression of photosynthetic and stress genes (Zheng et al. 2020). Peptide-controlled TFs or kinases can also modulate retrograde signaling sensitivity or output, thus adjusting the cellular response to photodamage (Figure 12.6).

Notably, peptide signaling and retrograde communication frequently have reciprocal interactions. Signals derived from chloroplasts are able to modify the expression or activity

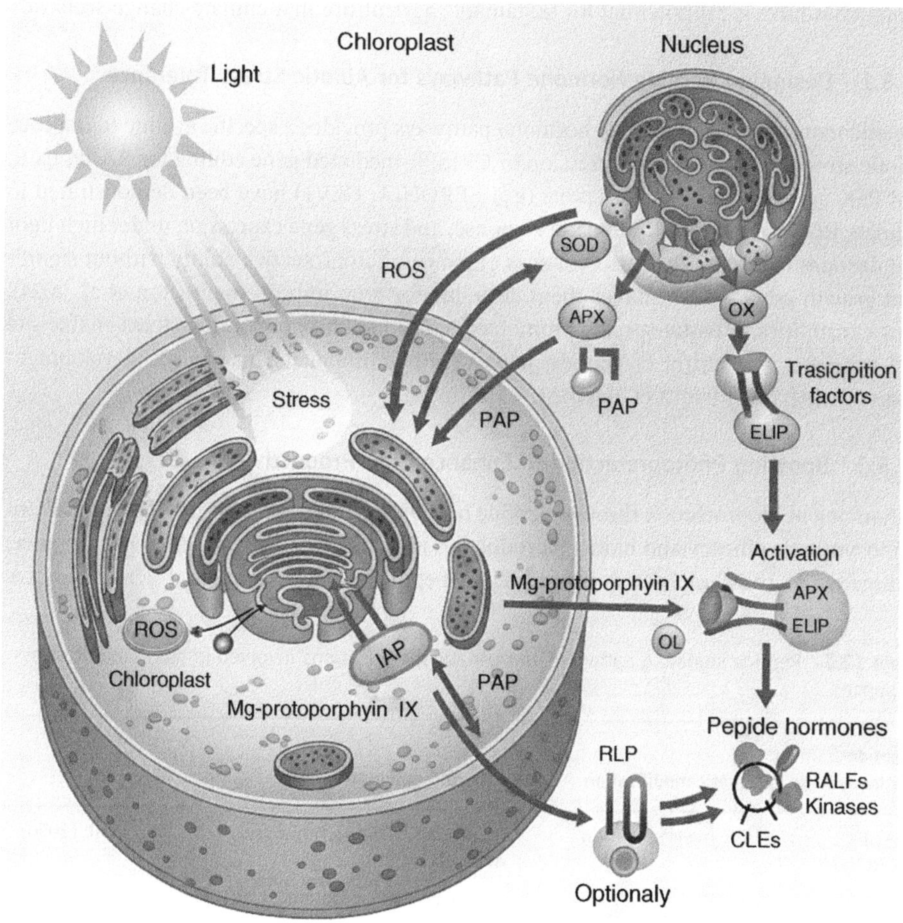

Figure 12.6 Light stress triggers retrograde signaling from chloroplast to nucleus, regulating protective gene expression.

of peptide hormones and their receptors. Excessive light, for instance, may overexpress genes for CLE or RALF peptides, embedding peptide signaling in the retrograde communication network (Xie et al. 2022). This bidirectional control enables the chloroplast and the rest of the cell to respond coordinately, maximizing the use of light energy and reducing oxidative damage through tightly regulated gene expression modulation.

12.8 Biotechnological and Agricultural Implications

Peptide hormones hold great promise for engineering light and oxidative stress tolerance in plants. Their combination with traditional stress pathways increases resilience and yield stability (Raza et al. 2022). Recent advances in synthetic biology facilitate accurate modulation of peptide signaling networks. Exogenous application of synthetic peptides provides new opportunities for field-level stress alleviation (Dutta and Phani 2023). These approaches have great potential for sustainable agriculture in a climate-change scenario.

12.8.1 Designing Peptide Hormone Pathways for Abiotic Stress Tolerance

Genetic manipulation of peptide hormone pathways provides a specific avenue to improve abiotic stress tolerance. Overexpression or CRISPR-mediated gene editing for RALF, CLE, and PSK peptides and their receptors (e.g. FERONIA, CLV1) have been demonstrated to regulate ROS accumulation, stomatal response, and stress gene expression under high light and drought (Table 12.3). These changes enhance photoprotective ability without significant growth costs, which makes them desirable for crop improvement (Ren et al. 2024). New capabilities in tissue-specific promoters and inducible expression systems enable site and time-specific control of peptide activity with minimal defense versus development trade-offs (Allahverdiyeva et al. 2015).

12.8.2 Boosting Photoprotection to Enhance Crop Productivity

Enhancing photoprotection through peptide hormone signaling can result in more resilient photosynthetic efficacy and biomass production in fluctuating light regimes. For instance, transgenic plants that bear stress-inducible peptides exhibit decreased photooxidative

Table 12.3 Peptide signaling pathways and genetic modifications involved in plant abiotic stress responses.

Peptide pathway	Genetic modification	Reported outcome	Reference
RALF-FERONIA	FER overexpression	Enhanced ROS tolerance in *Arabidopsis*	Zhu et al. (2020)
CLE25	CLE25 upregulation	Improved ABA-mediated drought response	Takahashi et al. (2018)
PSK-PSKR1	PSK receptor knockout	Reduced stress signaling and sensitivity	Mosher et al. (2013)

Table 12.4 Peptide-based strategies for enhancing stress resilience and yield in crop models.

Crop model	Strategy	Result	Reference
Rice (*Oryza sativa*)	CLE peptide overexpression	Increased grain yield under light stress	Zhang et al. (2021)
Tomato (*Solanum lycopersicum*)	RALF pathway modulation	Delayed leaf senescence, improved fruit set	Liu et al. (2019)
Wheat (*Triticum aestivum*)	Synthetic PSK application	Enhanced chlorophyll retention, yield stability	Kumar et al. (2022)

Table 12.5 Exogenous application of synthetic peptide analogs and their impact on crop stress resilience.

Synthetic peptide	Application method	Observed benefit	Reference
RALF analog	Foliar spray	Improved ROS detox, stress priming in maize	Torres et al. (2023)
CLE mimic	Seed treatment	Faster germination under abiotic stress	Yamada et al. (2021)
PSK analog	Soil drench	Enhanced root growth and drought resilience	Wang et al. (2020)

injury, increased chlorophyll maintenance, and enhanced yield in high light and heat stress (Simkin et al. 2022). These enhancements are converted into enhanced canopy performance and resource utilization efficiency in field conditions, especially in climates with harsh sunlight or drought-related seasons (Table 12.4).

12.8.3 Synthetic Peptides as Agricultural Treatments

Synthetic peptide analogs provide a convenient means of stress tolerance enhancement in a non-genomodified manner. Peptides may be used as foliar sprays or seed coatings to precondition plant defense against subsequent stress (Pérez-Pizá et al. 2024). Field applications of RALF-, CLE-, and PSK-mimicking peptides have demonstrated better tolerance to oxidative stress, water loss reduction, and yield stability (Table 12.5). Their stability, minimal dosage requirement, and specificity render them a promising addition to sustainable agriculture practices.

12.9 Future Directions

Peptide hormones constitute a fast-developing area of research in plant biology with immense untapped potential. While their developmental roles are well established, their role in abiotic stress and photoprotection is not yet fully understood. New tools in genomics and molecular signaling provide new avenues for exploration. Through the development of these avenues, our ability to make crops resilient to future environmental stresses will be advanced.

12.9.1 Unexplored Peptide Signaling Pathways in Photoprotection

Numerous plant peptides are still uncharacterized in relation to light stress, even though transcriptomic data demonstrated their induction under high light, UV, and drought. New families like C-terminally encoded peptides (CEPs), INFLORESCENCE DEFICIENT IN ABSCISSION-LIKE (IDLs), and ROOT GROWTH FACTORS (RGFs) could potentially have untapped functions in protecting chloroplasts or systemic signaling under photodamage (Ren et al. 2024). Functional characterization by mutant screening, peptide-receptor mapping, and stress assays would uncover new members of the photoprotective network. These new peptides would work in concert with traditional signaling modules, providing new handles on stress adaptation (Ren et al. 2024).

12.9.2 Multiomics Strategies to Unravel Peptide-Hormone–Photosynthesis Interactions

Combining transcriptomics, proteomics, metabolomics, and phosphoproteomics will be key to understanding how peptide hormones control photosynthesis during stress. These layers of data can uncover regulatory centers, posttranslational adjustments, and negative feedback loops for peptide perception and chloroplast function (Satrio et al. 2024). For instance, phosphoproteomics can be used to determine downstream MAPK targets for peptide receptors, and coexpression networks can connect peptide-responsive genes to photosynthetic pathways (Zhang et al. 2017). Spatially resolved omics (e.g. single-cell RNA-seq, spatial metabolomics) will further clarify how these interactions function in different tissues or developmental phases (Xie et al. 2023).

12.9.3 Relevance of Climate Change and Breeding Programs

Since climate change exacerbates light and oxidative stresses, the inclusion of peptide signaling in crop breeding assumes strategic importance. Peptides provide tissue-specific, transient, and energy-effective regulation and thus qualify as excellent tools for climate-resilient agriculture (Sulis et al. 2025). Marker-assisted selection or genome editing of peptide genes and receptors can be incorporated into current breeding pipelines (Kumar et al. 2024). Moreover, deciphering natural variation in peptide-regulated photoprotection among landraces or wild relatives may reveal new alleles for stress tolerance (Yang et al. 2024). This will ensure food security by making molecular findings concomitant with practical breeding results.

12.10 Conclusion

Peptide hormones are becoming central actors in the intricate cross talk between environmental signaling and photosynthesis. By modulating ROS homeostasis, activating TFs, and interacting with traditional hormones (ABA, SA, JA), they design rapid and localized signals in response to light-induced stress. Their capacity to manipulate gene expression, signal transduction, and chloroplast–nucleus communication indicates their diversity in regulating photoprotection and oxidative stress tolerance.

Synthesis of peptide signaling research with photosynthetic biology provides a strong platform for crop enhancement. Manipulation of peptide pathways on a targeted basis—by genetic modification, synthetic peptides, or precision breeding—is extremely promising for deriving robust, high-yielding crops under variable light and climate regimes. Integration between these fields will be essential for maximizing energy utilization, enhancing yield stability, and guaranteeing long-term agricultural resilience.

References

Aalen, R.B., Wildhagen, M., Stø, I.M., and Butenko, M.A. (2013). IDA: a peptide ligand regulating cell separation processes in Arabidopsis. *Journal of Experimental Botany* 64: 5253–5261.

Abid, M.S., Mousavi, S., and Checco, J.W. (2021). Identifying receptors for neuropeptides and peptide hormones: challenges and recent progress. *ACS Chemical Biology* 16: 251–263.

Ahmad, P., Jaleel, C.A., Azooz, M.M., and Nabi, G. (2009). Generation of ROS and non-enzymatic antioxidants during abiotic stress in plants. *Botany Research International* 2: 11–20.

Alexieva, V., Sergiev, I., Mapelli, S., and Karanov, E. (2001). The effect of drought and ultraviolet radiation on growth and stress markers in pea and wheat. *Plant, Cell & Environment* 24: 1337–1344.

Ali, S. and Baloch, A.M. (2020). Overview of sustainable plant growth and differentiation and the role of hormones in controlling growth and development of plants under various stresses. *Recent Patents on Food, Nutrition & Agriculture* 11: 105–114.

Allahverdiyeva, Y., Battchikova, N., Brosche, M. et al. (2015). Integration of photosynthesis, development and stress as an opportunity for plant biology. *The New Phytologist* 208: 647–655.

Anjum, N.A., Sharma, P., Gill, S.S. et al. (2016). Catalase and ascorbate peroxidase—representative H_2O_2-detoxifying heme enzymes in plants. *Environmental Science and Pollution Research* 23: 19002–19029.

Baker, N.R. (2008). Chlorophyll fluorescence: a probe of photosynthesis in vivo. *Journal of Experimental Botany* 59: 89–113.

Bassi, R. and Dall'Osto, L. (2021). Dissipation of light energy absorbed in excess: the molecular mechanisms. *Annual Review of Plant Biology* 72: 47–76.

Becraft, P.W. (2002). Receptor kinase signaling in plant development. *Annual Review of Cell and Developmental Biology* 18: 163–192.

Bhatla, S.C. and Lal, M.A. (2023). Light perception and transduction. In: *Plant Physiology, Development and Metabolism*, 363–390. Singapore: Springer Nature.

Bhattacharya, A. and Bhattacharya, A. (2021). Dry matter production, partitioning, and seed yield under soil water deficit: a review. In: *Soil Water Deficit and Physiological Issues in Plants*, 585–702. Singapore: Springer.

Chakraborty, S., Nguyen, B., Wasti, S.D., and Xu, G. (2019). Plant leucine-rich repeat receptor kinase (LRR-RK): structure, ligand perception, and activation mechanism. *Molecules* 24: 3081.

Chowdhary, N.A. and Songachan, L.S. (2025). Plant peptides involved in ROS signalling and biotic and abiotic stress responses. *International Journal of Peptide Research and Therapeutics* 31: 53.

Croce, R., Carmo-Silva, E., Cho, Y.B. et al. (2024). Perspectives on improving photosynthesis to increase crop yield. *Plant Cell* 36: 3944–3973.

Cui, Y., Lu, X., and Gou, X. (2022). Receptor-like protein kinases in plant reproduction: current understanding and future perspectives. *Plant Communications* 3: 100265.

Datta, T., Kumar, R.S., Sinha, H., and Trivedi, P.K. (2024). Small but mighty: peptides regulating abiotic stress responses in plants. *Plant, Cell & Environment* 47: 1207–1223.

Demmig-Adains, B. and Adams, W.W. (2017). The xanthophyll cycle. In: *Antioxidants in Higher Plants*, 91–110. CRC Press.

Derks, A., Schaven, K., and Bruce, D. (2015). Diverse mechanisms for photoprotection in photosynthesis. *Biochimica et Biophysica Acta – Bioenergetics* 1847: 468–485.

Didaran, F., Kordrostami, M., Ghasemi-Soloklui, A.A. et al. (2024). The mechanisms of photoinhibition and repair in plants under high light conditions and interplay with abiotic stressors. *Journal of Photochemistry and Photobiology. B* 259: 113004.

Dsilva, G.J. and Galande, S. (2024). From sequence to consequence: deciphering the complex cis-regulatory landscape. *Journal of Biosciences* 49: 46.

Eggelkraut-Gottanka, R.V. and Beck-Sickinger, A.G. (2004). Biosynthesis of peptide hormones derived from precursor sequences. *Current Medicinal Chemistry* 11: 2651–2665.

Fang, Z., Chen, L., Zhang, Y. et al. (2021). Overexpression of GLK transcription factors in rice enhances chloroplast development and photosynthetic efficiency under field conditions. *Cells* 10: 1010.

Fletcher, J.C. (2020). Recent advances in Arabidopsis CLE peptide signaling. *Trends in Plant Science* 25: 1005–1016.

Foyer, C.H. (2018). Reactive oxygen species, oxidative signaling and the regulation of photosynthesis. *Environmental and Experimental Botany* 154: 134–142.

Foyer, C.H. and Hanke, G. (2022). ROS production and signalling in chloroplasts: cornerstones and evolving concepts. *The Plant Journal* 111: 642–661.

Gancheva, M.S., Malovichko, Y.V., Poliushkevich, L.O. et al. (2019). Plant peptide hormones. *Russian Journal of Plant Physiology* 66: 171–189.

Ge, Z., Jing, Y., Zhu, J. et al. (2025). APE1 localizes to chloroplast stromules and interacts with ATI1 in Arabidopsis. *Plant Signaling & Behavior* 20: 2511830.

Ghosh, P. and Roychoudhury, A. (2024). Plant peptides involved in abiotic and biotic stress responses and reactive oxygen species (ROS) signaling. *Journal of Plant Growth Regulation* 43: 1410–1427.

Gou, X. and Li, J. (2020). Paired receptor and coreceptor kinases perceive extracellular signals to control plant development. *Plant Physiology* 182: 1667–1681.

Goyal, N. and Singh, K. (2025). Unveiling recent advances in plant peptides and their impact on agriculture and biotechnology. *Nucleus* 68: 1–21.

Gu, J., Zhou, Z., Li, Z. et al. (2017). Photosynthetic properties and potentials for improvement of photosynthesis in pale green leaf rice under high light conditions. *Frontiers in Plant Science* 8: 1082.

Gupta, S.K., Sharma, M., Deeba, F., and Pandey, V. (2017). Role of reactive oxygen species in photophosphorylation and damage to D1 protein: past and present. In: *Reactive Oxygen Species in Plants: Boon or Bane—Revisiting the Role of ROS* (ed. V.P. Singh, S. Singh, D.K. Tripathi, et al.), 165–186. India: Wiley.

Gurrieri, L., Fermani, S., Zaffagnini, M. et al. (2021). Calvin–Benson cycle regulation is getting complex. *Trends in Plant Science* 26: 898–912.

Hamann, T. (2015). The plant cell wall integrity maintenance mechanism–a case study of a cell wall plasma membrane signaling network. *Phytochemistry* 112: 100–109.

Hasanuzzaman, M., Bhuyan, M.B., Anee, T.I. et al. (2019). Regulation of ascorbate-glutathione pathway in mitigating oxidative damage in plants under abiotic stress. *Antioxidants* 8: 384.

Hideg, É., Jansen, M.A.K., and Strid, Å. (2013). UV-B exposure, ROS, and stress: inseparable companions or loosely linked associates? *Environmental and Experimental Botany* 94: 64–70.

Hirakawa, Y. (2021). CLAVATA3, a plant peptide controlling stem cell fate in the meristem. *Peptides* 142: 170579.

Hohmann, U., Lau, K., and Hothorn, M. (2017). The structural basis of ligand perception and signal activation by receptor kinases. *Annual Review of Plant Biology* 68: 109–137.

Hossain, M.A., Polash, M.A., Karim, M.M. et al. (2022). Photosynthesis: fundamentals and advances. *International Journal of Biosciences* 21: 443–463.

Hsiao, Y.C. and Yamada, M. (2020). The roles of peptide hormones and their receptors during plant root development. *Genes* 12: 22.

Huang, P., Li, Z., and Guo, H. (2022). New advances in the regulation of leaf senescence by classical and peptide hormones. *Frontiers in Plant Science* 13: 923136.

Hutchinson, J.A., Burholt, S., and Hamley, I.W. (2017). Peptide hormones and lipopeptides: from self-assembly to therapeutic applications. *Journal of Peptide Science* 23: 82–94.

Ighodaro, O.M. and Akinloye, O.A. (2018). First line defence antioxidants—superoxide dismutase (SOD), catalase (CAT) and glutathione peroxidase (GPX): their fundamental role in the entire antioxidant defence grid. *Alexandria Journal of Medicine* 54: 287–293.

Jiang, Y. et al. (2022). Photosynthetic physiology performance and expression of transcription factors in soybean of water use efficiency difference. *Russian Journal of Plant Physiology* 69: 9.

Jose, J., Ghantasala, S., and Roy Choudhury, S. (2020). Arabidopsis transmembrane receptor-like kinases (RLKs): a bridge between extracellular signal and intracellular regulatory machinery. *International Journal of Molecular Sciences* 21: 4000.

Juan, C.A., Pérez de la Lastra, J.M., Plou, F.J., and Pérez-Lebeña, E. (2021). The chemistry of reactive oxygen species (ROS) revisited: outlining their role in biological macromolecules (DNA, lipids and proteins) and induced pathologies. *International Journal of Molecular Sciences* 22: 4642.

Karabourniotis, G., Liakopoulos, G., Bresta, P., and Nikolopoulos, D. (2021). The optical properties of leaf structural elements and their contribution to photosynthetic performance and photoprotection. *Plants* 10: 1455.

Kaufmann, C., Stührwohldt, N., and Sauter, M. (2021). Tyrosylprotein sulfotransferase-dependent and -independent regulation of root development and signaling by PSK LRR receptor kinases in Arabidopsis. *Journal of Experimental Botany* 72: 5508–5521.

Kaya, C. (2024). Microbial modulation of hormone signaling, proteomic dynamics, and metabolomics in plant drought adaptation. *Food and Energy Security* 13: e513.

Khan, I. et al. (2025). Adaptive responses of plants to light stress: mechanisms of photoprotection and acclimation. A review. *Frontiers in Plant Science* 16: 1550125.

KhokharVoytas, A. et al. (2023). Genetic modification strategies for enhancing plant resilience to abiotic stresses in the context of climate change. *Functional & Integrative Genomics* 23: 283.

Kim, M.J. et al. (2021). Peptide signaling during plant reproduction. *Trends in Plant Science* 26: 822–835.

Kromdijk, J., Glowacka, K., Leonelli, L. et al. (2016). Improving photosynthesis and crop productivity by accelerating recovery from photoprotection. *Science* 354 (6314): 857–861.

Kumar, R., Das, S.P., Choudhury, B.U. et al. (2024). Advances in genomic tools for plant breeding: harnessing DNA molecular markers, genomic selection, and genome editing. *Biological Research* 57 (1): 80.

Kumar, A., Meena, S.K., Verma, V. et al. (2022). Exogenous application of synthetic phytosulfokine (PSK) improves drought tolerance and yield in wheat (*Triticum aestivum*). *Plant Signaling & Behavior* 17 (1): e2001015.

Kumar, M. and Pal, M. (2022). Midday high-light and heat stress effects on photosystem II stability in rice and wheat: role of carotenoids and regulated chlorophyll fluorescence. *Vegetos* 35 (2): 470–480.

Kumari, P., Thakur, R., Singh, N. et al. (2023). Relationships of oxidative stress and ultraviolet-B radiation in plants. In: *UV-B Radiation and Crop Growth*, 277–291. Singapore: Springer Nature Singapore.

Kumpf, R.P., Shi, C.L., Larrieu, A. et al. (2013). Floral organ abscission peptide IDA and its HAE/HSL2 receptors control cell separation during lateral root emergence. *Proceedings of the National Academy of Sciences of the United States of America* 110 (13): 5235–5240.

Lalun, V.O., Breiden, M., Galindo-Trigo, S. et al. (2024). A dual function of the IDA peptide in regulating cell separation and modulating plant immunity at the molecular level. *eLife* 12: RP87912.

Leister, D. (2020). Alternative electron pathways in photosynthesis. *The New Phytologist* 228 (4): 1166–1168.

Lemonnier, P. and Lawson, T. (2024). Calvin cycle and guard cell metabolism impact stomatal function. In: *Seminars in Cell and Developmental Biology*, vol. 155 (ed. Y. Mao), 59–70.

Levin, G. and Schuster, G. (2023). LHC-like proteins: the guardians of photosynthesis. *International Journal of Molecular Sciences* 24 (3): 2503.

Lin, H., Han, X., Feng, X. et al. (2022). Molecular traits and functional analysis of Rapid Alkalinization Factors (RALFs) in four *Gossypium* species. *International Journal of Biological Macromolecules* 194: 84–99.

Lismont, C., Revenco, I., and Fransen, M. (2019). Peroxisomal hydrogen peroxide metabolism and signaling in health and disease. *International Journal of Molecular Sciences* 20 (15): 3673.

Liu, R., Gao, J., Song, Y. et al. (2021). Regulation of chlorophyll a/b ratio improves light-use efficiency and photosynthetic capacity in maize hybrids under stress. *Photosynthesis Research* 149 (2): 181–194.

Liu, D., Jelenska, J., Morgan, J.M., and Greenberg, J.T. (2024b). Phytosulfokine downregulates defense-related WRKY transcription factors and attenuates pathogen-associated molecular pattern-triggered immunity. *The Plant Journal* 120 (6): 2367–2384.

Liu, C., Liu, Q., and Mou, Z. (2024a). Redox signaling and oxidative stress in systemic acquired resistance. *Journal of Experimental Botany* 75 (15): 4535–4548.

Liu, Y., Wang, H., Zhang, S. et al. (2019). Modulation of the RALF signaling pathway delays senescence and enhances fruit set in tomato (*Solanum lycopersicum*). *Plant Physiology and Biochemistry* 139: 1–10.

Long, S.P., Zhu, X.G., Naidu, S.L., and Ort, D.R. (2006). Can improvement in photosynthesis increase crop yields? *Plant, Cell & Environment* 29 (3): 315–330.

Lu, S. and Xiao, F. (2024). Small peptides: orchestrators of plant growth and developmental processes. *International Journal of Molecular Sciences* 25 (14): 7627.

Mahdavian, K. (2024). Effects of ultraviolet radiation on plants and their protective mechanisms. *Russian Journal of Plant Physiology* 71 (6): 184.

Malandrino, N. and Smith, R.J. (2017). Synthesis, secretion and transport of peptide hormones. In: *Principles of Endocrinology and Hormone Action*, 1–14. Cham: Springer.

Marand, A.P., Eveland, A.L., Kaufmann, K., and Springer, N.M. (2023). cis-Regulatory elements in plant development, adaptation, and evolution. *Annual Review of Plant Biology* 74 (1): 111–137.

Martin-Ramirez, S., Stouthamer, J., and Smakowska-Luzan, E. (2025). More questions than answers: insights into potential cysteine-rich receptor-like kinases redox signalling in *Arabidopsis*. *The Plant Journal* 122 (2): e70176.

Meloni, M., Gurrieri, L., Fermani, S. et al. (2023). Ribulose-1, 5-bisphosphate regeneration in the Calvin-Benson-Bassham cycle: focus on the last three enzymatic steps that allow the formation of Rubisco substrate. *Frontiers in Plant Science* 14: 1130430.

Mosher, S., Seybold, H., Rodríguez, P. et al. (2013). The tyrosine-sulfated peptide receptors PSKR1 and PSY1R modify the growth-defense balance in *Arabidopsis thaliana*. *The New Phytologist* 200 (3): 911–922.

Motomitsu, A., Sawa, S., and Ishida, T. (2015). Plant peptide hormone signalling. *Essays in Biochemistry* 58: 115–131.

Muhammad, I., Shalmani, A., Ali, M. et al. (2021). Mechanisms regulating the dynamics of photosynthesis under abiotic stresses. *Frontiers in Plant Science* 11: 615942.

Murata, N. and Nishiyama, Y. (2018). ATP is a driving force in the repair of photosystem II during photoinhibition. *Plant, Cell & Environment* 41 (2): 285–299.

Murchie, E.H. and Ruban, A.V. (2020). Dynamic non-photochemical quenching in plants: from molecular mechanism to productivity. *The Plant Journal* 101 (4): 885–896.

Murphy, E. and De Smet, I. (2014). Understanding the RALF family: a tale of many species. *Trends in Plant Science* 19 (10): 664–671.

Nair, P.R., Kumar, B.M., and Nair, V.D. (2021). General principles of plant productivity. In: *An Introduction to Agroforestry: Four Decades of Scientific Developments* (ed. B.M. Kumar and V.D. Nair), 263–280.

Nicol, L. and Croce, R. (2021). The PsbS protein and low pH are necessary and sufficient to induce quenching in the light-harvesting complex of plants LHCII. *Scientific Reports* 11 (1): 7415.

Nilsen, E.T. and Forseth, I.N. Jr. (2018). The role of leaf movements for optimizing photosynthesis in relation to environmental variation. In: *The Leaf: A Platform for Performing Photosynthesis*, 401–423. Cham: Springer.

Pandita, D., Bhat, J.A., Wani, S.H. et al. (2023). Mobile signaling peptides: secret molecular messengers with a mighty role in plant life. *Journal of Plant Growth Regulation* 42 (10): 6801–6834.

Parwez, R., Aftab, T., Gill, S.S., and Naeem, M. (2022). Abscisic acid signaling and crosstalk with phytohormones in regulation of environmental stress responses. *Environmental and Experimental Botany* 199: 104885.

Pastor-Fernández, J., Sánchez-Bel, P., Flors, V. et al. (2023). Small signals lead to big changes: the potential of peptide-induced resistance in plants. *Journal of Fungi* 9 (2): 265.

Pérez-Pizá, M.C., Sautua, F.J., Szparaga, A. et al. (2024). New tools for the management of fungal pathogens in extensive cropping systems for friendly environments. *Critical Reviews in Plant Sciences* 43 (2): 63–93.

Pinnola, A. (2019). The rise and fall of Light-Harvesting Complex Stress-Related proteins as photoprotection agents during evolution. *Journal of Experimental Botany* 70 (20): 5527–5535.

Qu, L.J., Li, L., Lan, Z., and Dresselhaus, T. (2015). Peptide signalling during the pollen tube journey and double fertilization. *Journal of Experimental Botany* 66 (17): 5139–5150.

Rahman, M.M., Ghosh, P.K., Akter, M. et al. (2024). Green vanguards: Harnessing the power of plant antioxidants, signal catalysts, and genetic engineering to combat reactive oxygen species under multiple abiotic stresses. *Plant Stress* 13: 100547.

Rajput, V.D., Harish, S.R.K., Verma, K.K. et al. (2021). Recent developments in enzymatic antioxidant defence mechanism in plants with special reference to abiotic stress. *Biology* 10 (4): 267.

Ravi, B., Foyer, C.H., and Pandey, G.K. (2023). The integration of reactive oxygen species (ROS) and calcium signalling in abiotic stress responses. *Plant, Cell & Environment* 46 (7): 1985–2006.

Raza, A., Salehi, H., Rahman, M.A. et al. (2022). Plant hormones and neurotransmitter interactions mediate antioxidant defenses under induced oxidative stress in plants. *Frontiers in Plant Science* 13: 961872.

Ren, G., Zhang, Y., Chen, Z. et al. (2024). Research progress of small plant peptides on the regulation of plant growth, development, and abiotic stress. *International Journal of Molecular Sciences* 25 (7): 4114.

Roach, T. and Krieger-Liszkay, A. (2019). Photosynthetic regulatory mechanisms for efficiency and prevention of photo-oxidative stress. *Annual Review of Plant Biology* 2 (1): 1–34.

Saed-Moucheshi, A., Shekoofa, A., and Pessarakli, M. (2014). Reactive oxygen species (ROS) generation and detoxifying in plants. *Journal of Plant Nutrition* 37 (10): 1573–1585.

Saibi, W. and Brini, F. (2018). Superoxide dismutase (SOD) and abiotic stress tolerance in plants: an overview. In: *Superoxide Dismutase: Structure, Synthesis and Applications* (ed. S. Magliozzi), 101–142. Nova Science Publishers, Inc.

Satrio, R.D., Fendiyanto, M.H., and Miftahudin, M. (2024). Tools and techniques used at global scale through genomics, transcriptomics, proteomics, and metabolomics to investigate plant stress responses at the molecular level. In: *Molecular Dynamics of Plant Stress and its Management*, 555–607. Singapore: Springer.

Sauter, M. (2015). Phytosulfokine peptide signalling. *Journal of Experimental Botany* 66 (17): 5161–5169.

Schwarz, E.M., Tietz, S., and Froehlich, J.E. (2018). Photosystem I-LHCII megacomplexes respond to high light and aging in plants. *Photosynthesis Research* 136: 107–124.

Semwal, V.K. and Khanna-Chopra, R. (2020). Enhanced oxidative stress, damage and inadequate antioxidant defense contributes towards insufficient recovery in water deficit stress and heat stress combination compared to either stress alone in *Chenopodium album* (Bathua). *Physiology and Molecular Biology of Plants* 26 (7): 1331–1339.

Sharma, A., Hussain, A., Mun, B.G. et al. (2016). Comprehensive analysis of plant rapid alkalization factor (RALF) genes. *Plant Physiology and Biochemistry* 106: 82–90.

Sharma, N., Nagar, S., Thakur, M. et al. (2023b). Photosystems under high light stress: throwing light on mechanism and adaptation. *Photosynthetica* 61 (2): 250.

Sharma, M., Sidhu, A.K., Samota, M.K. et al. (2023a). Post-translational modifications in histones and their role in abiotic stress tolerance in plants. *Proteomes* 11 (4): 38.

Shi, Y., Ke, X., Yang, X. et al. (2022). Plants response to light stress. *Journal of Genetics and Genomics* 49 (8): 735–747.

Shuang, S.P., Zhang, J.Y., Cun, Z. et al. (2022). A comparison of photoprotective mechanism in different light-demanding plants under dynamic light conditions. *Frontiers in Plant Science* 13: 819843.

Simkin, A.J., Kapoor, L., Doss, C.G. et al. (2022). The role of photosynthesis related pigments in light harvesting, photoprotection and enhancement of photosynthetic yield in planta. *Photosynthesis Research* 152 (1): 23–42.

Smith, E.N., van Aalst, M., Tosens, T. et al. (2023). Improving photosynthetic efficiency toward food security: strategies, advances, and perspectives. *Molecular Plant* 16: 1547–1563.

Sohrabi, F., Saed-Moucheshi, A., Saed-Moucheshi, Z., and Pessarakli, M. (2024). Effects of abiotic stresses on the photosynthesis apparatus in plants. *Energy* 680: 700.

Stintzi, A. and Schaller, A. (2022). Biogenesis of post-translationally modified peptide signals for plant reproductive development. *Current Opinion in Plant Biology* 69: 102274.

Stührwohldt, N. and Schaller, A. (2019). Regulation of plant peptide hormones and growth factors by post-translational modification. *Plant Biology* 21: 49–63.

Sulis, D.B., Lavoine, N., Sederoff, H. et al. (2025). Advances in lignocellulosic feedstocks for bioenergy and bioproducts. *Nature Communications* 16: 1244.

Szechyńska-Hebda, M., Lewandowska, M., and Karpiński, S. (2017). Electrical signaling, photosynthesis and systemic acquired acclimation. *Frontiers in Physiology* 8: 684.

Takahashi, F., Suzuki, T., Osakabe, Y. et al. (2018). A small peptide modulates stomatal control via abscisic acid in long-distance signalling. *Nature* 556: 235–238.

Tang, C., Xie, J., Lv, J. et al. (2021). Alleviating damage of photosystem and oxidative stress from chilling stress with exogenous zeaxanthin in pepper (*Capsicum annuum* L.) seedlings. *Plant Physiology and Biochemistry* 162: 395–409.

Theis, J. and Schroda, M. (2016). Revisiting the photosystem II repair cycle. *Plant Signaling & Behavior* 11: e1218587.

Tikhonov, A.N. (2024). The cytochrome b6f complex: plastoquinol oxidation and regulation of electron transport in chloroplasts. *Photosynthesis Research* 159: 203–227.

Torres, M.A., Rivera, M., Campos, M.L. et al. (2023). Foliar application of a RALF analog primes ROS detoxification and stress tolerance in maize (*Zea mays*). *Plant Physiology and Biochemistry* 199: 113054.

Uchiyama, N. and Orio, M. (2024). The power of photosynthesis: exploring the biological, chemical, and ecological dimensions of earth's primary energy source. *Asian American Research Letters Journal* 1: 3.

Ustynyuk, L.Y. and Tikhonov, A.N. (2022). Plastoquinol oxidation: rate-limiting stage in the electron transport chain of chloroplasts. *Biochemistry (Mosc)* 87: 1084–1097.

Valledor, L., Cañal, M.J., Pascual, J. et al. (2012). Early induced protein 1 (PrELIP1) and other photosynthetic, stress and epigenetic regulation genes are involved in *Pinus radiata* D. don UV-B radiation response. *Physiologia Plantarum* 146: 308–320.

Wang, Q. and Domen, K. (2019). Particulate photocatalysts for light-driven water splitting: mechanisms, challenges, and design strategies. *Chemical Reviews* 120: 919–985.

Wang, X., Hou, L., Shen, L. et al. (2020). Application of synthetic phytosulfokine (PSK) analog improves root architecture and drought tolerance in wheat (*Triticum aestivum*). *Plant Growth Regulation* 91: 187–196.

Wang, Z., Hu, Y., Zhang, S., and Sun, Y. (2022b). Artificial photosynthesis systems for solar energy conversion and storage: platforms and their realities. *Chemical Society Reviews* 51: 6704–6737.

Wang, G., Zhao, Z., Zheng, X. et al. (2022a). How a single receptor-like kinase exerts diverse roles: lessons from FERONIA. *Molecular Horticulture* 2: 25.

Whitewoods, C.D. (2021). Evolution of CLE peptide signalling. *Seminars in Cell & Developmental Biology* 109: 12–19.

Willoughby, A.C. and Nimchuk, Z.L. (2021). WOX going on: CLE peptides in plant development. *Current Opinion in Plant Biology* 63: 102056.

Wong, C., Alabadí, D., and Blázquez, M.A. (2023). Spatial regulation of plant hormone action. *Journal of Experimental Botany* 74: 6089–6103.

Wu, C., Munir, R., Li, F. et al. (2025). From biosynthesis to signaling: unveiling the multifaceted roles of phytosulfokine peptide in plants. *Journal of Experimental Botany (in press)* 76 (19): 5682–5697.

Xiao, F., Zhou, H., and Lin, H. (2025). Decoding small peptides: regulators of plant growth and stress resilience. *Journal of Integrative Plant Biology* 67: 596–631.

Xie, B., Gao, D., Zhou, B. et al. (2023). New discoveries in the field of metabolism by applying single-cell and spatial omics. *Journal of Pharmaceutical Analysis* 13: 711–725.

Xie, Z., Nolan, T.M., Jiang, H., and Yin, Y. (2019). AP2/ERF transcription factor regulatory networks in hormone and abiotic stress responses in Arabidopsis. *Frontiers in Plant Science* 10: 228.

Xie, H., Zhao, W., Li, W. et al. (2022). Small signaling peptides mediate plant adaptations to abiotic environmental stress. *Planta* 255: 72.

Yadav, V. (2024). Endogenous peptides involved in plant growth and development. In: *Plant Elicitor Peptides: New Tool for Sustainable Agriculture*, 113–130. Singapore: Springer Nature Singapore.

Yamada, M., Ito, Y., Niwa, M. et al. (2021). A synthetic CLE peptide promotes seed germination and early seedling development under abiotic stress in Arabidopsis and rice. *Journal of Plant Physiology* 258–259: 153361.

Yamaguchi, Y.L., Ishida, T., and Sawa, S. (2016). CLE peptides and their signaling pathways in plant development. *Journal of Experimental Botany* 67: 4813–4826.

Yang, T., Geng, F., Tang, X. et al. (2024). UV radiation-induced peptides in frog skin confer protection against cutaneous photodamage through suppressing MAPK signaling. *MedComm* 5: e625.

Zhang, H., Han, Z., Song, W. et al. (2021). Overexpression of CLE peptide enhances grain yield and stress tolerance in rice (*Oryza sativa*). *Plant Biotechnology Journal* 19: 101–103.

Zhang, Z., Han, H., Zhao, J. et al. (2025). Peptide hormones in plants. *Molecular Horticulture* 5: 7.

Zhang, T., Schneider, J.D., Zhu, N., and Chen, S. (2017). Identification of MAPK substrates using quantitative phosphoproteomics. In: *Plant Pattern Recognition Receptors: Methods and Protocols*, 133–142. Springer.

Zhang, Y., Wang, J., Sun, L. et al. (2024). Nitrate fertilization mitigates drought-induced oxidative stress in Leymus chinensis through regulation of antioxidant enzymes. *Frontiers in Plant Science* 15: 1234567.

Zhang, H., Zhu, J., Gong, Z., and Zhu, J.K. (2022). Abiotic stress responses in plants. *Nature Reviews. Genetics* 23: 104–119.

Zhao, D., Reddy, K.R., Kakani, V.G., and Sailaja, K. (2003). Effects of ultraviolet-B radiation on photosynthesis, growth and fluorescence characteristics of two cultivars of cotton. *Plant Physiology and Biochemistry* 41: 807–815.

Zheng, Q., Chen, Y., Jia, X. et al. (2020). MicroRNA156 (miR156) negatively impacts mg-Protoporphyrin IX (mg-proto IX) biosynthesis and its plastid-nucleus retrograde signaling in apple. *Plants* 9: 653.

Zhong, S. and Qu, L.J. (2019). Peptide/receptor-like kinase-mediated signaling involved in male–female interactions. *Current Opinion in Plant Biology* 51: 7–14.

Zhu, Q., Feng, Y., Xue, J. et al. (2023). Advances in receptor-like protein kinases in balancing plant growth and stress responses. *Plants* 12: 427.

Zhu, S., Martínez Pacheco, J., Estevez, J.M., and Yu, F. (2020). Autocrine regulation of root hair size by the RALF-FERONIA-RSL4 signaling pathway. *The New Phytologist* 227: 45–49.

Zuo, G. (2025). Non-photochemical quenching (NPQ) in photoprotection: insights into NPQ levels required to avoid photoinactivation and photoinhibition. *The New Phytologist* 246: 1967–1974.

13

Peptide Hormones and Regulation of Sources–Sink Regulation

Somayeh Rastegar[1] *and Emad Hamdy Khedr*[2]

[1] *Department of Horticultural Sciences, Faculty of Agriculture and Natural Resources, University of Hormozgan, Bandar Abbas, Iran*
[2] *Department of Pomology, Faculty of Agriculture, Cairo University, Giza, Egypt*

13.1 Introduction

Plant growth and development are regulated by a complex network of signaling molecules, among which peptide hormones have emerged as important regulators alongside traditional plant hormones. For decades, the classical phytohormones—auxin, gibberellins, cytokinins, abscisic acid (ABA), ethylene, brassinosteroids, jasmonic acid, salicylic acid, and the more recently discovered strigolactones have been recognized as primary regulators of plant growth and development (Mochida and Shinozaki 2011; Cheng et al. 2013; Asami and Nakagawa 2018). However, research over the past few decades has revealed that peptide hormones also play significant roles in plant signaling networks (De Coninck and De Smet 2016).

Plant peptide hormones are small signaling molecules typically shorter than 120 amino acids, with their biologically active forms often comprise fewer than 20 amino acids (Ren et al. 2024). These peptides are encoded by nuclear genes and generally undergo proteolytic processing of preproproteins and various posttranslational modifications (PTMs). The first functional plant peptide identified was tomato systemin, an 18-amino-acid polypeptide involved in defense responses (Pearce et al. 1991). Shortly after, phytosulfokine (PSK), a disulfated pentapeptide, was isolated as a potent mitogenic factor (Fukuda and Higashiyama 2011).

Most peptide hormones function as secreted signaling molecules that mediate intercellular communication. They are typically released from cells and act on neighboring cells in a non-cell-autonomous manner by binding to membrane-associated receptors (Zhang et al. 2025; Yan et al. 2025). Many of these receptors belong to the receptor-like kinase (RLK) family, of which the *Arabidopsis* genome contains more than 600 genes (Fukuda and Higashiyama 2011). This receptor–ligand interaction initiates downstream signaling cascades, ultimately leading to changes in gene expression and cellular responses (Zhang et al. 2025). They are produced through a complex process involving proteolytic cleavage of preprohormone precursors. This process begins with the removal of signal peptides by signal peptidase complexes, followed by cleavage at specific sites by prohormone convertases or furin (Mirabeau et al. 2007). Many peptide hormones undergo PTMs, with more than

Peptide Hormones in Plant Development and Stress Responses, First Edition. Edited by Vijay Pratap Singh and Samiksha Singh.

50% becoming amidated at their C-terminus. These mature peptides are then released into the extracellular space, where they bind to specific cell surface receptors to modulate cellular functions (Mirabeau et al. 2007).

Plant peptide hormones are now recognized as essential components in regulating various aspects of plant growth and development, including meristem maintenance, vascular development, stomatal patterning, and reproductive processes (Fukuda and Higashiyama 2011; Matsubayashi and Sakagami 2006). Additionally, they play important roles in plant responses to environmental stresses, including drought, salinity, temperature extremes, and pathogen attacks (Ren et al. 2024; Kapale et al. 2020). The discovery of these peptide signaling molecules has expanded our understanding of plant hormones beyond the traditional classes, establishing peptides as a new and important class of plant hormones (Matsubayashi and Sakagami 2006; Yan et al. 2025).

The growth and productivity of plants are highly dependent on the efficient coordination between source tissues, typically mature, photosynthetically active leaves—and sink tissues, such as roots, developing fruits, seeds, and young leaves. This coordination, known as the source–sink relationship, governs the allocation of assimilates, nutrients, and signaling molecules throughout the plant. The balance between source supply and sink demand is dynamically regulated and plays a critical role in plant development, stress responses, and yield formation (Fan et al. 2021; Al-Khayri et al. 2024). Traditionally, plant hormones such as auxins, cytokinins, gibberellins, and ABA have been well-known for their involvement in modulating source–sink dynamics (Patrick 2013). However, in recent years, peptide hormones have emerged as a crucial class of signaling molecules in plants. These small peptides, usually composed of 5–50 amino acids, function as mobile signals that influence various physiological processes, including organ development, nutrient allocation, and intercellular communication (Ren et al. 2024).

Peptide hormones such as CLAVATA3/EMBRYO SURROUNDING REGION-related (CLE), C-terminally encoded peptides (CEP), PSK, and INFLORESCENCE DEFICIENT IN ABSCISSION (IDA) have been shown to participate in the fine-tuning of source–sink interactions. Their precise roles in regulating carbohydrate transport, phloem unloading, and sink strength are being increasingly recognized through molecular and genetic studies.

This chapter explores the emerging roles of peptide hormones in regulating source–sink dynamics in plants. It begins with an overview of key peptide hormone types, along with their biosynthesis and signaling pathways. We then delve into the mechanisms by which these peptides modulate source–sink communication under both optimal and stress conditions. Finally, we discuss the biotechnological potential of targeting peptide hormone pathways to enhance crop yield and resource use efficiency. A deeper understanding of these regulatory systems offers promising avenues for sustainable improvements in plant growth and agricultural productivity.

13.2 Classification and Main Characteristics of Peptide Hormones

Plant peptide hormones can be broadly classified into two main categories based on their biosynthesis and structural characteristics: secreted and nonsecreted peptides (Guo et al. 2015; Dodueva et al. 2021). This classification highlights key differences in their production,

PTM, and mechanisms of action. Secreted peptide hormones are defined by the presence of an N-terminal signal peptide that directs them through the secretory pathway (Guo et al. 2015). They are initially synthesized as biologically inactive prepropeptides, which are processed by signal peptidases into propeptides (Matsubayashi 2010; Matsubayashi 2018; Khedr and Khedr 2024). These peptides can be further divided into two major subclasses. The first group includes posttranslationally modified peptides, which are typically synthesized as large precursor proteins, often consisting of several hundred amino acids. These precursors are processed into smaller bioactive peptides of approximately 5–20 amino acids and undergo various modifications, such as hydroxylation and glycosylation of conserved proline residues or sulfation of tyrosine residues (Dodueva et al. 2021). Representative examples of this group include CLE, PSK, and plant peptide–containing sulfated tyrosine (PSY) peptides (Bajguz and Piotrowska-Niczyporuk 2023).

The second subclass consists of cysteine-rich peptides, which contain multiple conserved cysteine residues, typically ranging from 2 to 16, that form intramolecular disulfide bonds essential for their structural stability and function (Dodueva et al. 2021; Matsubayashi 2018). Some members of this group require proteolytic processing, while others function in their original form (Fedoreyeva 2023). Rapid alkalinization factor (RALF) peptides are a well-known example.

In contrast, nonsecreted peptide hormones lack signal peptides and are not directed through the secretory pathway (Guo et al. 2015; Dodueva et al. 2021). These peptides are generally retained in the cytoplasm rather than being secreted to the apoplast (Guo et al. 2015). Examples include systemin, early nodulin 40 (Enod40), POLARIS (PLS), ROT FOUR LIKE (RTFL), and plant elicitor peptides (PEPs) (Guo et al. 2015). The genome of *Arabidopsis* harbors a large number of genes potentially encoding peptide hormones. An in-silico analysis identified 979 putative secreted peptide genes with open reading frames ranging between 50 and 150 amino acids (Matsubayashi 2010). A significant portion of these remains functionally uncharacterized, suggesting that the diversity of plant peptide hormones is likely greater than currently understood (Fedoreyeva 2023). Plant peptide hormones represent a diverse and expanding class of signaling molecules that complement classical phytohormones in regulating various aspects of plant growth, development, and responses to environmental stimuli. Since the discovery of the first plant peptide hormone, systemin, in tomato in 1991 (Pearce et al. 1991), more than 20 distinct types have been identified (Hirakawa et al. 2017). The following sections outline the major types of plant peptide hormones and their key biological functions.

Plant peptide hormones represent a structurally diverse group of signaling molecules involved in various aspects of plant development, stress response, and immunity. These peptides range from sulfated and hydroxylated small peptides like CLE and PSK to cysteine-rich types like RALF and EPF, each with specialized functions and receptor-mediated signaling pathways. While CLE and RGF peptides primarily regulate meristem activity and root patterning, peptides such as PEP and SCOOP play pivotal roles in plant defense. Others, including IDA and PSY, mediate organ abscission and wound responses. Nonsecreted peptides like PLS and RTFL also contribute significantly to developmental processes through hormonal cross-talk. Together, these peptides underscore the complexity and specificity of peptide-based signaling networks in plant physiology (Table 13.1).

Table 13.1 Overview of key plant peptide hormones.

Peptide group	Structure	Function	Additional notes	Reference
CLE (CLAVATA3/ ESR-related)	Small peptides (12–13 amino acids); undergo proline hydroxylation and hydroxyproline glycosylation	Regulate shoot and root apical meristem maintenance	Interact with CLV1-type receptor-like kinases	Ito et al. (2006); Fletcher et al. (1999); Kinoshita et al. (2007); Jun et al. (2010); Hazak and Hardtke (2016)
PSK (phytosulfokine)	Disulfated pentapeptides with tyrosine sulfation	Promote cellular proliferation and expansion	First isolated from *Asparagus officinalis* cell cultures	Yang et al. (2001); Kaufmann and Sauter (2019)
RALF (rapid alkalinization factor)	Cysteine-rich peptides	Regulate root growth and cell expansion	Active throughout the plant kingdom	Pearce et al. (2001); Fiori (2010)
EPF (epidermal patterning factor)	Small peptides with conserved cysteine residues	Regulate stomatal development and patterning	Depend on TOO MANY MOUTHS and ERECTA receptors	Hara et al. (2007)
RGF/CLEL/ GLV	Tyrosine-sulfated peptides	Maintain root stem cell niche; guide root and lateral root growth	Important for directional root development	Matsuzaki et al. (2010a, b); Meng et al. (2012); Fernandez et al. (2012)
PEP (plant elicitor peptides)	Derived from PROPEP precursor proteins	Involved in immune responses and pathogen resistance	Function as damage-associated molecular patterns	Huffaker and Ryan (2007); Wang et al. (2022)
Systemin	18-amino acid polypeptide	Activates defense genes in response to wounding	First discovered plant peptide hormone	Pearce et al. (1991); Pearce and Ryan (2003)
CEP (C-terminally encoded peptide)	Secreted peptides with hydroxyproline	Modulate root architecture in response to environment	Responsive to external nutrient status	Patel et al. (2017); Taleski et al. (2022)
PSY (plant peptide with sulfated tyrosine)	18-amino-acid sulfated glycopeptide	Promotes proliferation and wound response	Contains tyrosine sulfation	Amano et al. (2007)

Table 13.1 (Continued)

Peptide group	Structure	Function	Additional notes	Reference
IDA and IDL (IDA-like)	Small proteins with N-terminal signal peptides	Regulate floral organ abscission	Key in organ separation processes	Butenko et al. (2003); Cho et al. (2008); Stenvik et al. (2008)
SCOOP (serine-rich endogenous peptide)	Brassicaceae-specific; 14 members in *Arabidopsis*	Involved in immune signaling	Recognized by MIK2 receptor kinase	Stahl et al. (2022)
PLS and AREP	Nonsecreted peptides from small ORFs	Regulate root growth and vascular development	Interact with auxin and cytokinin	Casson et al. (2002); Yang et al. (2014)
RTFL/DVL (ROT FOUR LIKE/DEVIL)	Small peptides with RTF domain	Control leaf and fruit development via cell proliferation	Involved in morphogenetic patterning	Narita et al. (2004); Wen et al. (2004)

13.3 Biosynthesis and Posttranslational Modifications

Plant peptide hormones are encoded by nuclear genes and typically produced as larger inactive precursor proteins that require multiple processing steps to become biologically active (Zhang et al. 2025). The biosynthesis of these signaling molecules follows a complex pathway that includes proteolytic processing of preproproteins and various PTMs essential for their function (Kaufmann and Sauter 2019).

The initial step in the biosynthesis of secreted peptide hormones involves the translation of genes into prepropeptides (Matsubayashi 2010). These prepropeptides contain an N-terminal signal peptide that directs them through the secretory pathway. The signal peptide is subsequently removed by a signal peptidase to form propeptides, which then undergo further processing (Matsubayashi 2010). The sites of cleavage in prepropeptides can be predicted with high accuracy using computational tools such as SignalP software (Matsubayashi 2010). PTMs are critical for the bioactivity of many peptide hormones (Kaufmann and Sauter 2019). Several important modifications include:

- **Tyrosine sulfation**: This modification was first discovered in PSK, a disulfated pentapeptide isolated from cultured Asparagus officinalis cells (Kaufmann and Sauter 2019; Moore 2009). Tyrosine sulfation is catalyzed by tyrosylprotein sulfotransferase (TPST), a 500-amino acid type I transmembrane protein in *Arabidopsis* that is structurally distinct from animal TPSTs, suggesting convergent evolution of this enzyme in plants (Pandita et al. 2023). Peptide families that undergo tyrosine sulfation include PSK, PSY, root growth factor (RGF), and Casparian strip integrity factor (CIF) (Pandita et al. 2023).

- **Proline hydroxylation:** This modification occurs in several peptide families, including CLE, CEP, and RGF peptides (Pandita et al. 2023). Hydroxylated proline residues are crucial for receptor recognition and downstream signaling (Ito et al. 2006).
- **Hydroxyproline glycosylation (*O*-arabinosylation):** This modification is found in peptides such as CLV3 and CLE2, where it enhances peptide stability and activity (Pandita et al. 2023). For instance, a dodecapeptide with two hydroxyproline residues derived from the CLE family was shown to be bioactive at concentrations as low as 10^{-11} M (Ito et al. 2006).

Unlike modifications in metabolite hormones such as indole-3-acetic acid or ABA, which often lead to hormone inactivation, PTMs of peptide hormones typically result in their activation (Kaufmann and Sauter 2019). These modifications are essential for proper receptor binding and signal transduction (Pandita et al. 2023).

For some peptide hormones, proteolytic processing by specific enzymes is a key part of their biosynthesis. For example, PEPs are cleaved from their precursors by metacaspases during calcium influx in the cytosol as a rapid response to wounding or pathogen attack (Lyapina et al. 2021; Hander et al. 2019). This proteolytic activation occurs within 30 seconds after injury and peaks around five minutes, allowing for rapid defense responses (Hander et al. 2019).

The biosynthesis of plant peptide hormones involves intricate PTMs, making it difficult to accurately predict their mature forms based solely on primary amino acid sequences (Matsubayashi 2010). To overcome this challenge, advanced peptidomics techniques have been developed to directly identify and characterize endogenous peptide hormones in planta. For example, o-chlorophenol extraction combined with liquid chromatography–mass spectrometry (LC-MS) enables the efficient detection of peptides secreted into the culture medium of submerged whole-plant systems (Ohyama et al. 2008). Additionally, integrative mass spectrometry and bioinformatics approaches have allowed for comprehensive profiling of the plant secretome, uncovering mature peptides with unanticipated N- and C-terminal extensions and diverse PTMs, such as hydroxylation and sulfation (Patel et al. 2017). These methodologies have significantly advanced our understanding of peptide hormone maturation and functional diversity in plant signaling.

In silico analysis has identified a large number of genes potentially encoding peptide hormones. In *Arabidopsis* alone, 979 putative secreted peptide genes with open reading frames between 50 and 150 amino acids have been identified (Matsubayashi 2010). Many of these genes remain functionally uncharacterized, suggesting that the diversity of plant peptide hormones is much greater than currently recognized (Matsubayashi 2010).

13.4 Signaling Pathways of Plant Peptide Hormones

Plant peptide hormones primarily function as signaling molecules that mediate cell-to-cell communication through specific receptor interactions. Most plant peptide hormones are perceived by leucine-rich repeat receptor kinases (LRR-RKs) that are localized in the plasma membranes of neighboring cells (Boller and Felix 2009). To date, receptors for members of at least 10 plant peptide families have been identified (Lyapina et al. 2021).

The signaling mechanism typically follows a conserved pattern where the peptide hormone binds to its specific receptor, triggering downstream signaling cascades. Upon binding of the peptide ligand, the receptor typically undergoes a conformational change that activates its intracellular kinase domain, initiating a phosphorylation cascade that ultimately leads to changes in gene expression and cellular responses (Yan et al. 2025).

Many peptide receptors form complexes with coreceptors to facilitate signal transduction. The best-characterized example is the flagellin receptor FLS2, which forms a heteromeric complex with BRI1-associated kinase 1 (BAK1) upon ligand binding (Boller and Felix 2009). Interestingly, BAK1 also serves as a coreceptor for the brassinosteroid receptor BRI1, suggesting common signaling components between peptide hormone and traditional phytohormone pathways (Boller and Felix 2009).

Calcium signaling plays a crucial role in peptide hormone pathways, particularly in defense responses. For instance, PEPs are released from their precursors by metacaspases during calcium influx in the cytosol as a rapid response to wounding or pathogen attack (Lyapina et al. 2021; Hander et al. 2019). This proteolytic activation occurs within 30 seconds after injury and peaks around five minutes, allowing for rapid defense responses (Hander et al. 2019).

Plant peptide hormone signaling pathways often interact with traditional phytohormone pathways, creating a complex network of signaling cross talk. For example, the peptide hormone systemin works in combination with jasmonate and ethylene to activate effective defense responses (Witzany 2010). Similarly, PEP activates the jasmonate/ethylene signaling pathway, demonstrating how peptide and traditional hormone signaling pathways can converge (Huffaker et al. 2006; Al-Khayri and Khedr 2024).

Peptide hormone signaling can also influence the production, transport, or response to other hormones. This cross talk can occur at the level of gene expression, where hormones may regulate synthesis (hormone levels), sensitivity (hormone response), and transport (hormone distributions) of other hormones (Cheng et al. 2013). Such coordination enables plants to respond appropriately to environmental factors such as temperature, shading, day length, and nutrient availability (Cheng et al. 2013).

Recent research has demonstrated that many plant peptide hormones, once believed to function solely in developmental regulation, also play integral roles in defense signaling, underscoring their multifunctional nature (Hu et al. 2018). This dual functionality enables plants to coordinate growth and immune responses through shared signaling components, thereby optimizing resource allocation under varying environmental conditions (Hsiao and Yamada 2020). The temporal dynamics of peptide signaling are equally critical. Some responses, such as those triggered by PEPs, are remarkably rapid, initiating within seconds of tissue damage to activate defense pathways (Hander et al. 2019). In contrast, other peptides mediate long-term developmental processes that require sustained and finely regulated signaling cascades (Stührwohldt and Schaller 2018). As our understanding of peptide hormone signaling deepens, it is becoming evident that these networks are highly complex, involving multiple layers of regulation and extensive cross talk with classical hormone pathways (Chen et al. 2020). Deciphering these intricate signaling interactions is essential for elucidating how plants integrate developmental cues with environmental signals to maintain homeostasis and resilience (Farrokhi et al. 2008).

Plant peptide hormones achieve intercellular communication through sophisticated signaling mechanisms that begin with their synthesis and secretion. Most plant peptide hormones are synthesized as precursor proteins and undergo proteolytic cleavage to become active signaling molecules. Unlike systemin, which localizes in the cytosol, most plant peptide hormones contain N-terminal signal peptides that direct their secretion into the apoplastic space (cell wall space) (Tameshige et al. 2015; Pearce et al. 1991).

Once secreted, these peptide hormones function primarily through two main signaling modes: paracrine and autocrine. In paracrine signaling, peptides released from one cell affect nearby target cells. A well-studied example is the CLAVATA3 (CLV3) peptide, which is expressed in the central zone of the shoot apical meristem and influences surrounding cells within a short range. In contrast, autocrine signaling involves peptides like PSK, where both the peptide ligand and its receptor are expressed by the same cell (Chakraborty et al. 2019; Hartmann et al. 2013).

Beyond local communication, peptide hormones also facilitate systemic signaling between different plant organs. These peptides can be transported over long distances through vascular tissues, enabling coordination between distant parts of the plant (Khavinson et al. 2024). This long-distance mobility makes peptide hormones important systemic coordinators of plant growth and development in response to developmental cues and environmental stimuli (Valmas et al. 2023). An emerging aspect of peptide hormone signaling is their cross talk with traditional plant hormones. For example, the PAMP-INDUCED PEPTIDE 2 (PIP2) in *Arabidopsis* is an auxin-responsive gene that regulates both root and hypocotyl elongation, demonstrating how peptide hormones and classical phytohormones can work together to coordinate plant growth and development (Shishkova et al. 2023).

13.5 Function of Plant Peptide Hormones

Plant peptide hormones have emerged as pivotal regulators of numerous cellular and physiological processes, complementing the roles of classical phytohormones in orchestrating plant growth, development, and defense responses (Dharmasiri et al. 2013; Hirakawa et al. 2017). Since the discovery of the first plant peptide hormone, systemin, over three decades ago, more than 20 distinct peptide hormone families have been identified, each demonstrating diverse and essential functions in plant biology. Below is a summary of the major functions of plant peptide hormones.

13.5.1 Development and Plant Peptides

Plant peptides act as highly specialized regulators of plant growth and development, functioning as signaling molecules that orchestrate intricate cellular communication networks. These short chains of amino acids interact with specific plasma membrane-localized receptors to initiate downstream transcriptional and metabolic responses that determine cell fate, division rate, and patterning. Unlike classical hormones that often have systemic effects, peptides tend to act locally, ensuring that developmental processes occur with spatial precision and temporal coordination. They influence both vegetative and reproductive stages of the plant life cycle, fine-tuning processes such as meristem activity, root and vascular differentiation, leaf expansion, fruit morphogenesis, stomatal distribution, and reproductive transitions. By integrating environmental cues with intrinsic developmental programs, plant peptides play a central role in optimizing plant architecture and productivity, making them valuable targets for crop improvement strategies.

13.5.1.1 Meristem Maintenance

In both shoot and root apical meristems, the maintenance of a delicate balance between stem cell proliferation and differentiation is crucial for continuous organogenesis. CLE peptides, particularly CLV3, are fundamental in maintaining this balance. Acting through the CLAVATA receptor kinase complex, CLV3 restricts stem cell proliferation by negatively regulating the expression of the WUSCHEL transcription factor, which promotes stem cell identity. This negative feedback loop ensures a stable pool of undifferentiated cells while preventing excessive meristem enlargement, thereby sustaining organized growth throughout the plant's lifespan (Kinoshita et al. 2007; Jun et al. 2010).

13.5.1.2 Root Development

Root system architecture, which determines a plant's ability to absorb water and nutrients, is under precise control by several peptide families. POLARIS (PLS) and AUXIN-RESPONSIVE ENDOGENOUS POLYPEPTIDE 1 (AREP1) influence primary root elongation and responsiveness to hormonal signals such as auxin and ethylene (Casson et al. 2002; Yang et al. 2014). ROOT GROWTH FACTOR/CLE-LIKE/GOLVEN (RGF/CLEL/GLV) peptides regulate the root stem cell niche, contributing to proper root zonation and gravitropic responses, while also promoting lateral root initiation for enhanced soil exploration (Matsuzaki et al. 2010a, b; Meng et al. 2012; Fernandez et al. 2012). Additionally, CEPs act as systemic signals that integrate nutrient status with root development, enabling plants to dynamically adjust root architecture in response to environmental nutrient availability (Taleski et al. 2022).

13.5.2 Vascular Development

The efficient transport of water, minerals, and photosynthates depends on a well-differentiated vascular system. Peptides such as PLS not only affect root elongation but also influence vascular tissue differentiation, coordinating the formation of xylem and phloem elements. By modulating cell division and differentiation in vascular initials, these peptides ensure optimal connectivity between roots and shoots, directly affecting growth efficiency and resource allocation (Casson et al. 2002).

13.5.2.1 Leaf and Fruit Development

Peptides such as DEVIL1 (DVL1) and ROTUNDIFOLIA4 (ROT4) regulate the polar proliferation of cells, which determines the final size and shape of leaves and fruits. By influencing cell division orientation and expansion rates, these peptides contribute to organ symmetry and overall morphology, ultimately impacting photosynthetic performance in leaves and commercial quality traits in fruits (Wen et al. 2004).

13.5.2.2 Stomatal Development

The distribution and density of stomata directly affect a plant's gas exchange capacity and water-use efficiency. EPIDERMAL PATTERNING FACTORs (EPFs) are key regulators of stomatal patterning, functioning through signaling pathways that restrict stomatal formation in adjacent cells to maintain optimal spacing. This control prevents excessive water loss while ensuring sufficient CO_2 uptake for photosynthesis, thereby balancing productivity with stress resilience (Hara et al. 2007).

13.5.2.3 Reproductive Processes

Peptides also coordinate developmental transitions in reproductive organs. INFLORESCENCE DEFICIENT IN ABSCISSION LIKE (IDL) peptides regulate the abscission process, enabling the precise shedding of floral organs postfertilization. This process is crucial for seed dispersal and preventing the retention of senescent floral tissues that could serve as entry points for pathogens (Butenko et al. 2003; Cho et al. 2008; Stenvik et al. 2008).

13.5.3 Plant Peptides and Stress Responses

In addition to shaping plant growth, peptides function as rapid-response messengers in stress perception and signaling pathways. These bioactive molecules act at the frontline of plant defense, detecting and translating external biotic and abiotic stress signals into coordinated physiological and biochemical responses. By modulating hormone signaling, transcriptional reprogramming, and metabolic adjustments, stress-related peptides help plants balance growth with survival. They can serve as damage-associated molecular patterns (DAMPs), initiate systemic defense signaling, and orchestrate protective adaptations that enhance resilience under challenging conditions. This dual role—supporting growth under favorable conditions while swiftly activating defense under stress—makes peptides central to plant fitness and agricultural sustainability.

13.5.3.1 Pathogen Defense

Plant elicitor peptides are among the most well-characterized DAMPs, generated upon proteolytic cleavage of precursor proteins following pathogen attack or tissue damage. Once released, PEPs bind to specific RLKs, triggering a calcium influx and activating mitogen-activated protein kinase (MAPK) cascades. This leads to the rapid induction of defense genes, production of reactive oxygen species (ROS), and strengthening of cell walls, effectively limiting pathogen spread (Wang et al. 2022; Huffaker et al. 2006; Hander et al. 2019).

13.5.3.2 Wounding Response

Systemin, the first plant peptide hormone identified, plays a central role in wound-induced defense in tomato. Produced in response to tissue injury, it triggers a jasmonate-mediated signaling pathway that activates protease inhibitor genes, deterring herbivores, and preventing further damage. This localized signal can also propagate systemically, preparing undamaged tissues for potential attack and ensuring whole-plant readiness (Pearce et al. 1991; Guo et al. 2015).

13.5.3.3 Abiotic Stress Adaptation

Small peptides also participate in the modulation of abiotic stress responses, functioning as early warning signals that prime plants for environmental challenges such as salinity, drought, alkalinity, extreme temperatures, and cold stress. These peptides regulate stomatal behavior, osmolyte accumulation, and antioxidant enzyme activities, thereby mitigating oxidative damage and maintaining cellular homeostasis under stress. Through integration with hormonal networks such as ABA and ethylene signaling, they fine-tune plant responses to achieve a balance between stress tolerance and energy conservation (Ren et al. 2024).

13.5.3.4 Cellular Processes

Peptide hormones play a pivotal role in the regulation of fundamental cellular processes, ensuring that growth, differentiation, and morphogenesis proceed in a coordinated and adaptive manner. Their activity often integrates intrinsic developmental programs with environmental signals, enabling plants to fine-tune cell behavior for optimal tissue and organ formation.

13.5.3.5 Cell Proliferation and Expansion

Phytosulfokine (PSK) and plant peptide–containing sulfated tyrosine (PSY) are notable examples of peptides that directly stimulate cell proliferation and expansion. By binding to their respective plasma membrane receptor kinases, these peptides initiate signaling cascades that promote cell cycle progression and enhance cell wall extensibility. Such actions are essential for organ growth, tissue regeneration after injury, and maintenance of meristematic activity. Beyond primary growth, these peptides also contribute to wound healing processes and may participate in callus formation during tissue culture, making them valuable for both in vivo plant development and in vitro propagation systems (Feng et al. 2024).

13.5.3.6 Cell-to-Cell Communication

A defining feature of plant peptides is their function as mediators of intercellular communication. Acting as short-range signaling molecules, they enable adjacent or nearby cells to coordinate their developmental trajectories. This is critical for processes such as tissue patterning, boundary formation, and organ polarity establishment. By facilitating the exchange of positional information, peptide hormones help synchronize morphogenetic events across diverse tissues, ensuring that structural and functional integration is maintained. For example, peptides can influence epidermal and mesophyll interactions during leaf development or regulate cell layer formation in roots, illustrating their versatile role in spatial coordination (Kapale et al. 2020; Wen et al. 2004).

13.5.4 Hormone Interaction and Regulation

Plant peptides do not operate in isolation but instead form part of a complex and dynamic network of signaling pathways that includes classical plant hormones. This interconnectedness allows peptides to fine-tune plant physiology and development by modulating, enhancing, or attenuating hormonal responses according to internal status and environmental conditions.

13.5.4.1 Senescence Regulation

Emerging research highlights the involvement of peptide hormones in regulating leaf senescence, a highly controlled process that reallocates nutrients from aging tissues to developing organs or storage reserves. By influencing hormonal pathways such as ethylene, cytokinins, and ABA, peptides can either delay or accelerate senescence depending on developmental or stress-related cues. This regulatory capacity not only impacts plant longevity and reproductive success but also has significant agricultural implications for extending the shelf life of leafy vegetables and maintaining photosynthetic capacity in crops (Huang et al. 2022).

13.5.4.2 Hormone Cross Talk

The interplay between peptide hormones and classical hormones—such as auxins, cytokinins, gibberellins, jasmonates, and ABA—creates intricate signaling networks that orchestrate plant responses to a variety of stimuli. Through such cross talk, peptides can modulate hormone biosynthesis, signal transduction, and receptor sensitivity, thereby influencing processes ranging from root initiation to stress acclimation. For instance, certain peptides enhance auxin transport during lateral root formation, while others modulate jasmonate-mediated defense responses following herbivore attack (Lyapina et al. 2021; Stührwohldt and Schaller 2018). The remarkable diversity of plant peptide hormones, combined with their ability to function in both local and systemic signaling contexts, underscores their indispensable role in plant adaptation and development (Chen et al. 2020; Ren et al. 2024). The evolutionary conservation of some peptide families—such as RALF, which is found in taxa ranging from early-diverging mosses to advanced flowering plants—attests to their fundamental role in plant biology (Fiori 2010). As research advances, decoding the complexity of peptide signaling networks continues to open new opportunities for enhancing crop productivity, resilience, and resource efficiency, with potential applications in precision agriculture and sustainable food production systems (Huffaker et al. 2006).

13.5.5 Sink Source Relationships in Plant

The source–sink relationship refers to the movement of photosynthates (mainly sucrose) and nutrients from source organs (mature leaves) to sink organs (developing fruits, roots, tubers, or young leaves). This process is regulated by:

1) **Photosynthetic activity and carbon fixation in source leaves**
2) **Vascular transport through the phloem**
3) **Sink demand, metabolism, and developmental stage**

13.6 Hormonal Regulation, Including Peptide Hormones

Traditional plant hormones like auxins, cytokinins, gibberellins, and ABA have been widely studied in source–sink regulation. However, recent studies indicate that peptide hormones also play a significant role in coordinating source–sink interactions, acting as systemic signals that regulate carbon partitioning, phloem loading, and sink organ development (Tegeder and Hammes 2018).

In higher plants, the "source–sink" relationship forms the foundation of resource allocation that ultimately determines plant development and productivity. Sources are photosynthesizing organs, primarily mature leaves, that export assimilates, while sinks are organs that consume or accumulate these assimilates, such as fruits, seeds, and growing tissues (Fan et al. 2021). The photoassimilates are typically transported from sources to sinks in the form of simple sugars, particularly sucrose (Fan et al. 2021).

The communication between sources and sinks is a critical determinant of carbohydrate partitioning in plants, which directly impacts crop yield and quality (Fan et al. 2021; White et al. 2016; Khedr et al. 2023). This source–sink balance is regulated by a complex signaling network that integrates information from sugars, hormones, and environmental factors (Fan et al. 2021). Understanding the mechanisms controlling this balance is fundamental to plant evolution, ecology, and crop science, as it helps identify growth-limiting factors and potential targets for yield improvement (White et al. 2016).

Recent research has highlighted that the source–sink ratio, which indicates the competition for sugars within the plant, is a key component of developmental regulation, particularly evident in branching control in annual species (Schneider et al. 2019). Sugars produced in source organs, mainly photosynthetic leaves, are transported through the phloem to sink organs such as growing shoot apices and root zones. High sugar availability near dormant buds, resulting from a high source-to-sink ratio, promotes bud outgrowth and branching (Schneider et al. 2019).

Plant hormones play a crucial role in this source–sink regulation. They can communicate sink demand and regulate plasmodesmal conductance that controls the movement of resources through phloem unloading pathways (Patrick 2013). Evidence suggests that hormones like cytokinins are particularly important in this process. For example, increased cytokinin levels in developing reproductive structures have been shown to enhance meristem activity, resulting in more inflorescences, seeds, fruits, and tubers per plant (Patrick 2013; Radchuk et al. 2012).

Furthermore, there is significant interplay between sugar signaling and hormonal networks in regulating sink strength. Studies have shown that trehalose 6-phosphate (Tre6P), a signal of sucrose availability, increases in axillary buds following decapitation, coinciding with bud outgrowth (Schneider et al. 2019; Fichtner et al. 2017). This suggests that sugar signaling works in conjunction with hormonal networks, particularly by antagonizing the inhibitory effects of auxin through inhibition of strigolactone signaling (Schneider et al. 2019). This intricate relationship between source–sink dynamics and hormonal regulation provides plants with an efficient mechanism to adjust their development according to resource availability, ensuring new resource-demanding structures like branches are only created when the plant's resource status is sufficient to support their growth (Schneider et al. 2019; Barbier et al. 2015).

13.7 Plant Hormones in Sink–Source Regulation

Plant hormones function as crucial signaling molecules that orchestrate the intricate regulation of source–sink relationships in plants. This regulation involves the concerted action of multiple phytohormones, each with specific roles in the process. Understanding these hormonal regulators is essential for comprehending how plants allocate resources and maintain developmental equilibrium. In plants, source–sink relationships describe the dynamic process of assimilate production in photosynthetically active tissues (sources) and their subsequent translocation to growing or storage tissues (sinks). Peptide hormones, alongside other phytohormones, serve as key regulators of these relationships through complex signaling networks.

Cytokinin acts as a primary regulator by stimulating RuBisCO activity in leaves, influencing the expression of photosynthetic genes, and delaying leaf senescence, thereby enhancing source capacity (Qin et al. 2023; Cortleven and Schmülling 2015). By extending the photosynthetically active period of leaves, this hormone helps sustain a steady supply of assimilates to developing sink tissues. The significant role of cytokinins in delaying senescence can markedly alter source–sink dynamics, as observed in P(SARK); IPT plants under water stress, where modified hormonal cross talk resulted in increased sink capacity (Peleg et al. 2011).

Auxin represents another critical hormone governing source–sink dynamics. Recent investigations have identified an auxin signaling cascade involving IAA-OsARF18-OsARF2 that regulates sucrose allocation between sources and sinks, particularly impacting the development of reproductive organs in rice (Qin et al. 2023; Zhao et al. 2023). Furthermore, the exogenous application of indoleacetic acid (IAA) enhances the activity of key enzymes related to sucrose metabolism in developing seeds, thereby facilitating their growth (Qin et al. 2023). This highlights how peptide hormones can directly influence carbohydrate partitioning between photosynthetic and reproductive tissues.

Gibberellins play an important role in regulating plant growth and development, as well as stress responses. The interaction between the growth-regulating factor 4 (GRF4) transcription factor and DELLA proteins synergistically modulates growth and carbon and nitrogen metabolism, coordinating source–sink relationships (Qin et al. 2023).

Other hormones also significantly contribute to source–sink regulation. ABA typically promotes leaf senescence, while ethylene affects the grain filling process in rice and enhances assimilate transport in cotton (Qin et al. 2023). In contrast to ABA, cytokinins and auxins generally inhibit senescence, underscoring the antagonistic functions of different hormones in regulating leaf senescence and source capability (Lv et al. 2020). ABA also plays a role in source–sink regulation, particularly under stress conditions. In salt-tolerant plants, elevated ABA concentrations in young leaves can help maintain coordinated source–sink relationships by controlling photosynthetic organ growth and preserving sucrose synthesis in source leaves (Salah et al. 2018). This hormone-mediated adaptation mechanism emphasizes the importance of peptide hormones in stress responses that affect nutrient allocation.

The interplay of these hormones establishes a complex regulatory network that mediates various aspects of source–sink relationships. These hormonal signals play significant roles in all source–sink regulation, including patterns of seed maturation during development (Nguyen et al. 2016). Moreover, hormonal cross talk enables plants to respond appropriately to environmental stimuli and challenging conditions such as nutrient deprivation and various stresses by mediating growth, development, nutrient allocation, and source–sink transitions (Saeed et al. 2017).

Ultimately, optimizing the balance of these hormonal signals holds the potential to enhance seed development to its full capacity (Nguyen et al. 2016). However, it is crucial to recognize that the relationship between source–sink dynamics and hormonal regulation is not simple model but rather a complex network of interactions (Nguyen et al. 2016; Lv et al. 2020).

The interaction between peptide hormones and other signaling molecules forms a complex regulatory network. The GRF4 transcription factor and DELLA proteins exemplify this complexity, acting antagonistically to regulate growth, nitrogen assimilation, and carbon fixation (Qin et al. 2023). The GRF4-DELLA-NGR5 module has been shown to improve nitrogen use efficiency, coordinate source–sink relationships, and increase crop yield, offering potential strategies for agricultural sustainability (Qin et al. 2023).

Peptide phytohormones also interact with traditional plant hormones to regulate various developmental processes. RGF peptides regulate polar auxin transport in roots, while PSK peptides stimulate cell division only in the presence of auxin and cytokinin (Kapale et al. 2020). Additionally, CLE peptides can inhibit genes activated by cytokinin signaling, demonstrating the intricate cross-talk between different hormone signaling pathways (Kapale et al. 2020).

In economically important crops such as sugarcane, the regulation of internodal development, source–sink relationships, and yield is influenced by traditional hormones (auxin, gibberellin, cytokinin, ABA, and ethylene) as well as recently identified regulators including peptide hormones (Khan et al. 2023; Gray 2004). These hormone-mediated processes are critical for determining crop productivity and quality. The regulation of leaf senescence, a process that facilitates nutrient translocation from senescing tissues to developing organs, is also heavily influenced by hormone cross talk (Lv et al. 2020). While ABA induces senescence, cytokinins and auxin inhibit this process, creating a balance that determines the timing and rate of nutrient remobilization (Lv et al. 2020). This hormonal regulation of senescence directly impacts source–sink dynamics by controlling the transition of leaves from sources to nutrient-exporting senescing tissues.

13.8 The Role of Peptide Hormones in Regulating Source–Sink Relationships

Peptide hormones play a crucial role in regulating source–sink dynamics in plants by influencing phloem loading, nutrient partitioning, carbohydrate allocation, regulating leaf senescence, and sink strength. These signaling peptides fine-tune physiological responses to optimize resource distribution between photosynthetic (source) and nonphotosynthetic (sink) organs, especially under fluctuating environmental conditions (Figure 13.1).

13.8.1 Peptide Hormones Regulating Phloem Transport and Sugar Partitioning

While the intricate network of signals regulating phloem transport and sugar partitioning is still being unraveled, several peptide hormones have emerged as key players in these processes. These small signaling molecules, often posttranslationally modified, interact with RLKs to trigger downstream signaling pathways that influence vascular development

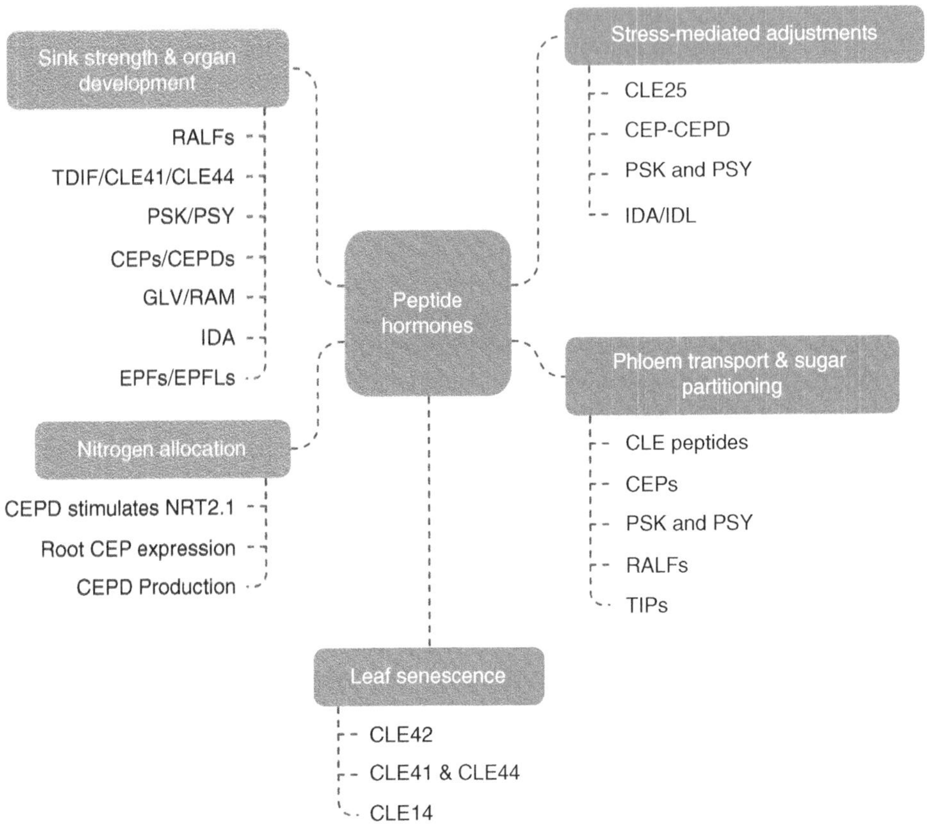

Figure 13.1 The regulatory influence of peptide hormones on source–sink relationships.

and resource allocation. Here are some examples of peptide hormones implicated in phloem transport and sugar partitioning:

- **CLAVATA3/EMBRYO surrounding region-related (CLE) peptides**: This large family of peptides is involved in various developmental processes, including vascular tissue development. Some CLE peptides, like CLE25, have been shown to act as long-distance signals in response to water stress, potentially influencing resource allocation. Phloem-derived CLE peptides such as TDIF/CLE41/CLE44 regulate the vascular cambium, which is crucial for secondary phloem development. Interestingly, other CLE peptides, including CLE25, CLE26, CLE33, and CLE45, can inhibit phloem formation, highlighting the complex regulatory roles within this family. (Yoshida et al. 2020) (Takahashi et al. 2018)
- **C-terminally encoded peptides (CEPs):** These root-derived peptides are involved in long-distance signaling related to nitrogen demand. CEPs interact with receptors in the shoot vasculature and trigger the production of phloem-mobile signals that regulate nitrate transporter expression in roots. This suggests a role in coordinating nutrient uptake with shoot demand, which is linked to sugar partitioning. (Ohkubo et al. 2017) (Kumawat et al. 2024)

- **Phytosulfokine (PSK) and plant peptide-containing sulfated tyrosine (PSY):** These sulfated peptides are known to promote cell proliferation and transdifferentiation, processes that are integral to vascular tissue development and function. While their direct role in phloem transport is still under investigation, their influence on cell growth and differentiation within vascular tissues suggests a potential indirect impact on phloem capacity and sink strength. (Li et al. 2023a, b)
- **Rapid alkalinization factors (RALFs):** This family of peptides is involved in various processes, including cell wall modifications and stress responses. Given the importance of cell wall structure and integrity in phloem function, RALFs might indirectly influence transport processes. (Shumilina et al. 2023) (Zhang et al. 2023)
- **Touch-induced peptides (TIPs):** While primarily known for their role in thigmomorphogenesis (plant response to touch), some evidence suggests TIPs might also be involved in vascular development.
- **Other peptide hormones:** Emerging research continues to identify novel peptide hormones and their roles in plant development. For instance, ROOT MERISTEM GROWTH FACTORs (RGFs) are crucial for root meristem development, and their influence on root sink strength could indirectly affect sugar partitioning. (Shinohara 2021) (Ou et al. 2022)

It is important to note that the precise mechanisms by which these peptide hormones regulate phloem transport and sugar partitioning are still being actively investigated. Their involvement often entails complex interactions with other signaling molecules, including classical plant hormones like auxin and cytokinin, as well as intricate downstream signaling cascades.

13.8.2 CEP Peptides and Nitrogen Allocation

CEPs play a crucial role in systemic nitrogen (N) demand signaling and the allocation of nitrogen resources within the plant. Here's a breakdown of their involvement: (Roy et al. 2022; Rzemieniewski et al. 2024)

When nitrogen levels in the soil are deficient, plant roots sense this limitation and respond by upregulating the expression of genes encoding CEPs. These peptides serve as molecular signals produced specifically in the roots under nitrogen-deficient conditions.

Once synthesized, CEP peptides are transported from the roots to the shoots via the xylem, functioning as long-distance signaling molecules. In the shoots, these peptides are recognized by specific receptors, CEP receptor 1 (CEPR1) and CEP receptor 2 (CEPR2), located in the vascular tissue. The recognition of CEPs by these receptors activates downstream signaling cascades, including the production of CEP DOWNSTREAM (CEPD) polypeptides.

These CEPD polypeptides, unlike CEPs, are phloem-mobile, meaning they can travel from the shoot back to the root system. Once in the roots, they act to upregulate the expression of key nitrate transporter genes such as *NRT2.1*, significantly enhancing the plant's capacity for nitrate uptake under low nitrogen conditions.

In addition to regulating nitrate absorption, CEP signaling also modulates root system architecture. Under nitrogen-limited conditions, CEPs can promote lateral root development to increase the exploratory capacity of the root system. Interestingly, this effect is modulated by carbon availability, as CEP signaling can inhibit lateral root growth in response to sucrose, indicating a sophisticated interplay between nitrogen and carbon signaling pathways.

Moreover, CEP signaling interacts with other hormonal signaling networks, particularly auxin and cytokinin. This cross-talk allows plants to finely tune their growth responses. For example, CEP and cytokinin signaling converge at CEPD-glutaredoxins to inhibit primary root growth under certain environmental conditions, demonstrating an integrated regulatory network that balances growth and nutrient acquisition.

In leguminous plants, CEP peptides also play a vital role in regulating nodule development, essential for symbiotic nitrogen fixation. By promoting nodule formation and coordinating it with nitrate transporter gene expression, CEP signaling bridges symbiotic and direct pathways of nitrogen acquisition, ensuring the plant can adapt flexibly to changing nitrogen sources.

Overall, CEP peptides serve as central messengers that communicate the nitrogen status from roots to shoots and back again. Through this dynamic, bidirectional signaling, plants can coordinate systemic responses that enhance nitrogen uptake and allocation, ultimately promoting growth and resilience under nutrient-limited conditions (Luo et al. 2024).

13.8.3 Peptide Hormone Regulation of Leaf Senescence

Leaf senescence represents the final stage of leaf development, characterized by a series of coordinated physiological and biochemical changes rather than a passive degenerative process. During senescence, plants undergo controlled degradation of intracellular organelles and hydrolysis of macromolecules, including the dismantling of chloroplasts, decline in photosynthetic activity, and decreased cellular metabolism (Yang et al. 2021). This process serves a critical adaptive function in plants, as it enables the relocation of nutrients and energy from older tissues (sources) to newly developing tissues or storage organs (sinks) (Ricachenevsky et al. 2013; Thomas 2013; El-Beltagi et al. 2025a).

Unlike senescence in animals, leaf senescence in plants has significant implications for reproductive success by facilitating nutrient recycling and redistribution (Liang and Chu 2015). As leaves senesce, their nutrients are transferred to developing organs such as new buds, young leaves, flowers, and seeds, ultimately resulting in the death of the senescing leaves (Ritonga et al. 2023). This source–sink relationship is regulated by multiple factors, including sugar levels, which can induce senescence-related degradation of chlorophyll and proteins when present in high concentrations (Ricachenevsky et al. 2013; Rolland et al. 2006; EL-Beltagi et al. 2025b).

The source–sink signaling mechanisms vary depending on species and environmental conditions. Experimental evidence shows that senescence in source leaves can be delayed by removing strong sink tissues, demonstrating the dynamic nature of this relationship (Ricachenevsky et al. 2013). This intricate balance between source and sink tissues is crucial for plant development and is highly regulated by various hormonal signals, particularly cytokinins, which play a significant role in delaying senescence (Ricachenevsky et al. 2013).

Recent research has expanded our understanding of leaf senescence regulation beyond classical phytohormones to include peptide hormones, which form a diverse family of signaling molecules affecting plant growth, development, and stress responses (Huang et al.

2022). Among these, members of the CLE family, PSK, and IDA peptides have been identified as important regulators of leaf senescence (Li et al. 2023a, b).

CLE peptides have garnered significant attention for their pivotal role in regulating leaf senescence, a process integral to plant aging and nutrient remobilization. Among the CLE peptides, CLE42 has been shown to function as a negative regulator of leaf senescence in *Arabidopsis*. In particular, the cle42 mutant exhibits accelerated senescence phenotypes, whereas overexpression of CLE42 can delay both age-dependent and dark-induced leaf senescence (Sakuraba 2022; Zhang et al. 2023). Intriguingly, the application of synthesized CLE42 peptide (CLE42p) alone is sufficient to delay leaf senescence under both natural and dark conditions, underscoring the direct role of this peptide hormone in the regulation of leaf aging (Sakuraba 2022).

At the molecular level, CLE42 exerts its senescence-delaying effect by suppressing the ethylene pathway, which is a key regulator of senescence. RNA sequencing analysis has shown that overexpression of CLE42 inhibits ethylene biosynthesis, which leads to an increase in the accumulation of ethylene-insensitive 3 (EIN3)–binding F-box protein (EBF). This, in turn, reduces the activity of the ethylene-related transcription factor EIN3 (Ren et al. 2024). The importance of this mechanism was confirmed when mutations in EIN3/EIL1 or overexpression of EBF1 suppressed the accelerated senescence phenotypes seen in cle42 mutants (Zhang et al. 2023). These findings highlight the critical role of ethylene signaling in the regulation of senescence and the regulatory potential of CLE42 in this context.

However, CLE42 does not act in isolation when it comes to senescence regulation. It shares functional redundancy with CLE41 and CLE44, with the cle41 cle42 cle44 triple mutant exhibiting more pronounced early senescence phenotypes than any of the single mutants. This underscores the collaborative nature of these peptides in regulating leaf senescence (Sakuraba 2022). This redundancy complicates efforts to understand the distinct roles of each peptide, but it also suggests that CLE peptides work together to finely tune senescence processes, potentially offering a system of checks and balances that ensures proper timing of senescence in response to environmental and developmental cues (Ren et al. 2024; Czyzewicz et al. 2013).

Another member of the CLE family, CLE14, plays a role in senescence regulation but through a distinct mechanism. Treatment with CLE14 peptide has been shown to reduce senescence in detached *Arabidopsis* leaves by targeting JUNGBRUNNEN1 (JUB1), a key regulator of ROS management. This action enhances the expression of genes involved in ROS removal, thereby mitigating both stress-induced and age-induced leaf senescence (Ren et al. 2024). By regulating oxidative stress, CLE14 helps protect leaves from premature aging, further highlighting the complexity of peptide hormones in controlling senescence through multiple pathways.

The involvement of peptide hormones in leaf senescence regulation has profound implications for the source–sink relationship in plants. Leaf senescence, the final stage of leaf development, is crucial for the translocation of nutrients from source tissues (leaves) to sink tissues (such as roots, seeds, and developing fruits). This nutrient remobilization is essential for overall plant fitness, particularly during stressful or resource-limited conditions (Huang et al. 2022). Peptide hormones, through their regulation of senescence, add another layer of control to this process, ensuring that nutrient allocation aligns with the plant's developmental needs and environmental conditions (Li et al. 2023a, b; Guo et al. 2021).

Additionally, the discovery that CLE peptides can function as long-distance signaling molecules between roots and leaves further emphasizes their role in coordinating plant responses to both developmental transitions and environmental stresses. These peptides' ability to relay signals from one part of the plant to another enhances the plant's ability to respond to changes in nutrient availability or other stress factors, offering a sophisticated means of regulating the source–sink relationship across the entire plant (Ren et al. 2024). This long-distance signaling capability adds another layer of complexity to our understanding of how plants integrate environmental and developmental cues to manage resource allocation during critical growth stages.

13.8.4 Peptide Hormones Enhancing Sink Strength and Organ Development

Sink strength, defined as the ability of an organ to attract and utilize photoassimilates, is a major determinant of crop yield. Several peptide hormones play crucial roles in enhancing sink strength and promoting organ development by influencing resource allocation, vascular development, and cell proliferation in sink tissues.

13.8.4.1 RALF Peptides

Rapid alkalinization factor peptides regulate cell expansion and metabolism in sink tissues like tubers and seeds. RALF signaling affects sucrose metabolism in developing seeds, influencing seed weight and nutrient accumulation (Campbell and Turner 2017). Overexpression of RALF genes in soybean and maize has been linked to enhanced seed size and oil content, indicating their role in improving sink efficiency (Zhang et al. 2020).

13.8.4.2 TDIF (Tracheary Element Differentiation Inhibitory Factor)/CLE41/CLE44

While initially identified for its role in inhibiting tracheary element differentiation, TDIF and related CLE peptides are vital for maintaining the vascular cambium. This meristematic layer produces both xylem and phloem, the conduits for resource transport. By promoting cambial activity, TDIF ensures a continuous supply of vascular tissues to developing organs, thereby enhancing their sink strength by facilitating efficient nutrient import. (Yuan and Wang 2021)

13.8.4.3 PSK (Phytosulfokine) and PSY (Plant Peptide–Containing Sulfated Tyrosine)

These sulfated peptides are potent promoters of cell proliferation and expansion. In developing organs (sinks), PSK and PSY can stimulate cell division and growth, leading to increased sink size and metabolic activity. This enhanced cellular activity directly contributes to a stronger demand for photoassimilates and nutrients, thus increasing sink strength. Furthermore, their role in cell differentiation can contribute to the proper development of vascular connections within the sink organ, further enhancing its ability to import resources. PSK is a peptide hormone that enhances sink strength by promoting cell division and expansion in young developing tissues, such as fruit and tubers. PSK receptor (PSKR1) signaling stimulates fruit expansion and carbohydrate accumulation, particularly in crops such as tomato and grape (Matsubayashi and Sakagami 2006). Exogenous application of PSK has been shown to increase fruit set, seed filling, and starch deposition, demonstrating its role in strengthening sinks (Kutschmar et al. 2009).

13.8.4.4 CEPs (C-Terminally Encoded Peptides) and CEPDs (CEP Downstream)

As discussed earlier, the CEP-CEPD signaling pathway plays a critical role in systemic nitrogen signaling. By optimizing nitrogen uptake in roots and its subsequent allocation, this system indirectly enhances sink strength. Adequate nitrogen supply is essential for the synthesis of proteins, nucleic acids, and other vital compounds required for organ growth and development. Therefore, efficient nitrogen allocation mediated by CEPs supports the metabolic demands of developing sinks.

13.8.4.5 GOLVEN (GLV)/Root Apical Meristem Regulatory Peptide (RAM)

This family of peptides is involved in regulating meristem size and activity in both roots and shoots. In shoot apical meristems, GLV signaling influences the development of aerial organs, including leaves and flowers, which act as significant sinks during their development. By controlling the rate and pattern of cell division and differentiation in these meristems, GLVs contribute to the establishment of sink tissues (Fernandez et al. 2013).

13.8.4.6 Inflorescence Deficient in Abscission (IDA)

While primarily known for its role in abscission (organ shedding), IDA and its receptor HAE/HSL2 have also been implicated in vascular development and cell separation processes during organogenesis. Proper vascular connections are crucial for supplying developing organs with resources, and IDA signaling may contribute to establishing or maintaining these connections (Kumpf et al. 2013).

13.8.4.7 Epidermal Patterning Factors (EPFs)/Epidermal Patterning Factor-Likes (EPFLs)

These peptides are mainly known for regulating stomatal development. However, some EPFLs have also been shown to influence vascular development. Proper vascularization is essential for efficient resource delivery to developing organs, indirectly contributing to their sink strength (Xiong et al. 2022) (Shimada et al. 2011).

It is important to recognize that the roles of these peptide hormones are often context-dependent and involve intricate cross-talk with other signaling pathways, including classical plant hormones like auxin, cytokinin, and gibberellins. These interactions fine-tune the balance between source and sink activities, ensuring proper organ development and resource allocation throughout the plant life cycle. In essence, these peptide hormones act as signaling molecules that coordinate cellular activities, vascular development, and resource allocation to actively growing organs, thereby enhancing their sink strength and supporting their proper development. Their precise roles and interactions are areas of ongoing research, promising further insights into the complex regulatory networks governing plant growth and development.

13.8.5 Peptide Hormones and Stress-Mediated Source–Sink Adjustments

Environmental stresses such as drought, salinity, and nutrient deficiency alter source–sink relationships by affecting phloem loading and sink development. Peptide hormones play a crucial role in regulating the dynamic adjustments between source and sink tissues in plants, particularly under stress conditions. One of the key ways in which they mediate these adjustments is by altering vascular development and function. For example, peptide hormones such as TDIF, and potentially some members of the CLE and RALF families, can influence the development and maintenance of xylem and phloem tissues. This regulation ensures efficient transport of water and nutrients, which is essential for maintaining proper resource allocation, especially when environmental stresses like drought or nutrient deficiency threaten the plant's stability (Zhang et al. 2023).

In addition to vascular regulation, peptide hormones can also modulate stomatal conductance, a critical process for balancing water loss and carbon dioxide (CO_2) uptake. CLE25, for instance, is involved in ABA signaling and stomatal closure, which directly impacts transpiration rates and CO_2 uptake. This regulation, in turn, affects the plant's photosynthetic capacity—a critical source activity—and controls water availability for sink tissues. By fine-tuning stomatal conductance, peptide hormones help maintain efficient resource use under water-limited conditions, ensuring that both source and sink functions are adequately supported.

The CEP-CEPD signaling pathway further exemplifies the ability of peptide hormones to regulate nutrient uptake and allocation in plants. This pathway plays a pivotal role in the systemic regulation of nutrient acquisition, particularly nitrogen, and its delivery to sink tissues. By coordinating nutrient acquisition at the source and directing it to sinks, peptide hormones help optimize nutrient use, ensuring that developing tissues, such as roots and fruits, receive the necessary resources even under stressful conditions.

In sink tissues, peptide hormones can also influence cell proliferation and growth to counteract the inhibitory effects of stress. For instance, PSK and PSY peptides promote cell division and expansion in sink tissues, helping maintain sink strength and ensuring that critical tissues continue to develop despite stress. This regulation is essential for ensuring the plant's overall productivity, particularly in times of environmental stress, where growth in sink tissues can be hindered.

In more severe stress conditions, some peptide hormones, such as IDA/IDLs, play a role in promoting senescence and nutrient remobilization from source tissues to more critical sinks. These peptides facilitate the controlled breakdown of source tissues, ensuring that nutrients are salvaged and redirected to areas of the plant that require them most. This process is particularly important during prolonged stress periods, where the plant must prioritize nutrient distribution for survival and continued development (Estornell et al. 2015).

Peptide hormones do not operate in isolation, and their effects are often integrated with other stress signaling pathways. They interact with classical stress hormones like ABA, jasmonates, and ethylene, forming an intricate network that fine-tunes the plant's response to stress. By collaborating with these other hormones, peptide signaling ensures that source–sink adjustments are optimized, balancing the plant's survival and defense needs with its growth and development.

In conclusion, peptide hormones are essential components of the plant's stress response machinery, playing diverse roles in regulating the dynamic adjustments between source and sink tissues. By influencing processes like vascular development, stomatal conductance, nutrient allocation, and sink tissue growth, peptide hormones help optimize resource allocation during periods of stress. Their interactions with other stress signaling pathways are crucial for maintaining plant resilience, ensuring that resources are efficiently distributed to prioritize survival, defense, and continued growth. The complexity and coordination of these peptide signals highlight their importance in plant stress adaptation and their potential for enhancing crop resilience in challenging environmental conditions.

13.9 Molecular Mechanisms of Peptide Hormone Action in Source–Sink Regulation

Peptide hormones exert their effects on source–sink relationships through complex **molecular signaling networks**, involving ligand–receptor recognition, downstream signal transduction cascades, transcriptional reprogramming, and physiological changes in source or sink organs. These mechanisms ensure that resource allocation is dynamically adjusted in response to developmental and environmental cues.

13.9.1 Ligand–Receptor Interactions: The First Step in Peptide Hormone Signaling

Most plant peptide hormones function via ligand–receptor binding at the plasma membrane of target cells. Peptide hormones are perceived by RLKs or receptor-like proteins (RLPs), often containing leucine-rich repeat (LRR) domains for specific ligand recognition.

- CLE peptides bind to receptors such as CLV1, CLV2, BAM1/2, and PXY, initiating vascular differentiation and meristem regulation (Nimchuk et al. 2015).
- CEP peptides are recognized by CEPR1 and CEPR2, leading to transcriptional changes involved in nitrogen metabolism and root architecture (Tabata et al. 2014).
- PSK peptides bind to PSKR1 and PSKR2, which are LRR-RLKs that promote cell proliferation and expansion (Ladwig et al. 2015).
- The IDA/IDL peptides interact with HAE/HSL2 receptors, mediating organ abscission and tissue remodeling (Shi et al., 2019).

The receptor-ligand complex often recruits coreceptors such as BAK1/SERK proteins, which enhance signal transduction through phosphorylation cascades.

13.9.2 Signal Transduction Pathways and Secondary Messengers

Upon ligand binding, RLKs are activated and initiate phosphorylation cascades that amplify the signal and lead to the activation or repression of transcription factors.

- PSKR1 signaling, for example, leads to activation of cyclic GMP-dependent protein kinases (PKG-like kinases) and Ca^{2+} fluxes, influencing gene expression related to growth and sugar transport (Matsubayashi and Sakagami 2006).
- CEP signaling activates NIN-like transcription factors (NLPs), which are central to nitrogen-responsive gene expression (Nakagami et al., 2024).
- CLE-PXY interaction induces the WOX4 transcription factor, which promotes vascular stem cell proliferation, essential for building an efficient transport system (Etchells et al. 2013).

Other secondary messengers involved include:

- ROS
- Inositol triphosphate (IP_3)
- MAPK pathways

These molecules integrate peptide signals with other hormonal and environmental cues, allowing for coordinated regulation of source–sink function.

13.9.3 Transcriptional Regulation and Functional Gene Targets

The ultimate outcomes of peptide hormone signaling are the modulation of gene expression in source or sink tissues. Several transcription factors have been identified as downstream targets of peptide signaling:

- WOX4 (regulated by CLE41–PXY) controls cambial activity and vascular development, crucial for source–sink transport capacity (Etchells et al. 2013).
- NLP7 (regulated by CEP–CEPR1 signaling) modulates nitrogen uptake and assimilation pathways (Tabata et al. 2014).
- MYB and bHLH transcription factors, regulated downstream of PSK and RALF peptides, control cell wall biosynthesis, sugar metabolism, and cell expansion in sink organs (Zhang et al. 2022).

Additionally, the transcription of sugar transporter genes (e.g. SWEETs, SUTs) and nitrogen transporter genes (*NRT1.1*, AMTs) is often upregulated, reinforcing sink strength and transport efficiency (Tegeder and Masclaux-Daubresse 2018).

13.9.4 Cross Talk with Classical Hormones in Source–Sink Regulation

Peptide hormone signaling often operates in **synergy or antagonism** with traditional plant hormones such as auxin, cytokinin, gibberellin, and ABA. This hormonal cross talk allows plants to fine-tune source–sink regulation based on internal developmental programs and external stimuli.

Examples include:

- **CLE41 signaling and auxin flow:** CLE41-PXY-WOX4 pathways intersect with auxin-mediated vascular patterning, controlling cambial division, and phloem loading (Fukuda and Hardtke 2019).
- **CEP and cytokinin interactions:** CEP signals from roots enhance shoot cytokinin responses, modulating source activity and nitrogen assimilation (Ohkubo et al. 2017).
- **PSK and brassinosteroids:** Coregulate cell expansion and sugar accumulation in sink tissues (Kutschmar et al. 2009).

Such interactions contribute to developmental plasticity and homeostasis in carbohydrate and nutrient distribution.

13.9.5 Spatial-Temporal Expression and Tissue-Specific Signaling

Peptide hormone effects are determined not only by the presence of ligands and receptors but also by their spatial and temporal expression patterns.

- CLE41 is expressed in phloem cells, while PXY is localized in adjacent procambial cells, forming a non–cell-autonomous signaling loop (Etchells et al. 2013).
- CEP genes are expressed in N-starved root tissues, but their action takes place in shoot tissues via long-distance signaling.
- PSK precursor genes (*PSK1–5*) show tissue-specific expression in expanding sink organs like young leaves and fruits.

This precise localization ensures targeted modulation of source–sink components, reducing resource waste and enhancing metabolic efficiency.

13.10 Peptide Hormones and Environmental Modulation of Source–Sink Dynamics

Environmental conditions such as nutrient availability, water stress, temperature fluctuations, and biotic interactions significantly influence source–sink relationships. Recent findings highlight that peptide hormones act as mobile signals, integrating these external cues to modulate plant development and resource allocation.

13.10.1 Nutrient Availability and Root-to-Shoot Peptide Signals

Peptide hormones play pivotal roles in nutrient sensing and signaling pathways, particularly in the regulation of nitrogen, phosphorus, and sulfur homeostasis. Among these, the CEP family is activated under nitrogen-deficient conditions in roots and functions as a long-distance signal to coordinate shoot-to-root communication for nitrogen demand. As demonstrated by Tabata et al. (2014), CEPs synthesized in N-starved roots travel via the xylem to the shoot, where they are recognized by the CEPR1/2 receptors. This interaction initiates a systemic signaling cascade that promotes lateral root development in nitrogen-rich zones, enhancing nutrient acquisition efficiency. Similarly, CLE peptides, such as CLE14 and CLE45, are upregulated during phosphorus deficiency and contribute to the suppression of lateral root growth, thereby prioritizing resource conservation and redistribution under low P conditions (Gutiérrez-Alanís et al. 2017). In the case of sulfur and boron, CIF peptides are involved in root barrier formation by responding to these nutrient levels, thereby influencing water and nutrient fluxes across source–sink tissues (Doblas et al. 2017). Through these mechanisms, peptide hormones serve as integrative signals that fine-tune nutrient allocation and developmental plasticity in response to fluctuating environmental conditions.

13.10.2 Drought, Salinity, and Osmotic Stress Responses

Peptide hormones play pivotal roles in enabling plants to withstand abiotic stresses that disturb water and solute balance, thereby impacting source–sink dynamics. RALF peptides, for instance, are key regulators of cell wall remodeling and stomatal closure, processes vital for maintaining osmotic stress tolerance. These peptides interact with FERONIA (FER) receptors to trigger calcium influx and the production of ROS, initiating protective responses across both source tissues, such as leaves, and sink tissues, including roots and seeds (Pearce et al. 2001). Another group, the PSY peptides, is implicated in drought-responsive root growth regulation. They enhance sink capacity under water-deficient conditions by promoting cell expansion and mediating hormone cross talk (Matsuzaki et al. 2010a, b). In rice, the expression of OsCEP6.1 increases in response to drought and nutrient stress, where it influences nitrogen uptake and biomass partitioning, further supporting adaptive adjustments in resource allocation (Mohammadi et al. 2014). Collectively, these peptides integrate environmental signals to orchestrate physiological responses that maintain growth and survival under stress.

13.10.3 Temperature Stress and Peptide Signal Integration

Temperature fluctuations significantly influence carbon assimilation in source tissues and the assimilate demand in sink organs, ultimately impacting source–sink dynamics. Certain peptide hormones help modulate plant developmental transitions and stress acclimation mechanisms in response to heat or cold stress. For example, CLE25 functions as a mobile signal that travels from roots to shoots under dehydration stress, promoting stomatal closure by enhancing ABA biosynthesis. This mechanism allows the plant to integrate both temperature and water availability signals to regulate source activity, as reported by Takahashi et al. (2018). Additionally, temperature-induced IDA peptide signaling plays a crucial role in processes such as flower abscission and fruit development, which are particularly sensitive to changes in the source–sink relationship (Patharkar and Walker 2016). These peptide-mediated responses ensure that plants maintain physiological balance and reproductive success under fluctuating environmental conditions.

13.10.4 Biotic Stress and Immune-Associated Peptide Hormones

Peptide hormones are increasingly recognized as integral components of the plant immune system, playing key roles in mediating source–sink redistribution during pathogen attacks. One classical example is systemin, a peptide hormone identified in Solanaceae species. Upon herbivory or mechanical wounding, systemin is synthesized at the site of damage and transported to distant tissues, where it activates jasmonic acid (JA)-mediated defense responses (Ryan and Pearce 2003). This defensive signaling often results in the reallocation of resources—reducing carbon export from source leaves and enhancing assimilate flow toward roots or developing tissues less exposed to stress. Similarly, PEPs act as DAMPs and are perceived by PEPR receptors, which trigger the expression of resistance genes and modulate sugar transporter activity (Huffaker et al. 2006). This leads to the redirection of sugars and amino acids away from infected tissues toward healthy sinks, promoting survival and maintaining systemic defense readiness.

13.10.5 Cross Talk with Hormonal and Environmental Signaling Networks

Peptide hormone signaling does not function in isolation; rather, it operates in concert with classical stress hormones such as ABA, jasmonates, ethylene, and salicylic acid to fine-tune plant responses to environmental stress and coordinate source–sink dynamics. For example, ABA and CLE25 work synergistically to mediate drought-induced stomatal closure while also influencing phloem transport, ensuring efficient water use and carbohydrate distribution during stress (Takahashi et al. 2018). Similarly, CEP peptides interact with cytokinin signaling pathways to modulate nitrogen-related shoot growth, allowing plants to optimize nutrient allocation in response to fluctuating environmental conditions (Ohkubo et al. 2017). In addition, RALF peptides, through their interaction with FER receptors, integrate mechanical and salt stress signals to regulate cell expansion and sugar distribution, thus reinforcing the plant's structural and metabolic resilience under abiotic stress (Stegmann et al. 2017). These intricate hormonal cross-talks enable plants to dynamically adjust assimilate partitioning and maintain both growth and survival when faced with suboptimal conditions.

13.11 Potential Applications of Peptide Hormones in Crop Improvement and Yield Optimization

Peptide hormones have emerged as promising tools for enhancing crop performance, resource allocation, and stress resilience due to their inherent specificity, mobility, and regulatory capabilities. Unlike traditional phytohormones, peptides offer the advantage of tissue- and condition-specific signaling, positioning them as ideal candidates for precision agriculture and biotechnological applications aimed at revolutionizing crop improvement and boosting yields.

13.11.1 Enhancing Nutrient Use Efficiency

- **Optimizing nitrogen acquisition:** Manipulating the CEP-CEPD signaling pathway holds the potential to develop crops with enhanced efficiency in nitrogen uptake from the soil. This could significantly reduce the necessity for excessive nitrogen fertilization and its associated environmental consequences. Strategies include developing varieties with naturally higher CEP or CEPR activity under nitrogen-limiting conditions, applying synthetic CEP analogs or small molecules that enhance CEP signaling, and engineering root systems with altered CEP production or perception for improved nitrogen foraging. Furthermore, CEP- and CLE-based interventions can directly contribute to reducing the dependency on synthetic nitrogen fertilizers (Luo et al. 2024).
- **Improving phosphate uptake:** While the direct involvement of specific peptide hormones in phosphate uptake mechanisms is still under investigation, identifying such peptides and manipulating their signaling pathways could lead to the development of crops with enhanced phosphorus acquisition, thereby reducing the reliance on phosphate fertilizers (Zhang et al. 2025; Ren et al. 2024).

13.11.2 Enhancing Sink Strength and Yield Components

- **Boosting seed/fruit development:** Peptides such as PSK and PSY, known for their role in promoting cell proliferation, could be strategically employed to enhance the size and number of cells within developing fruits, grains, and tubers, directly contributing to increased yield. This could involve targeted application of synthetic PSK/PSY analogs during critical developmental stages or the development of transgenic crops with increased endogenous PSK/PSY production or enhanced receptor sensitivity in sink tissues (Zhang et al. 2025).
- **Improving vascular development:** Enhancing the activity of peptides like TDIF could lead to the development of more robust vascular systems within plants. This improved vascular network would ensure more efficient transport of photoassimilates to developing sinks, consequently increasing overall yield potential. Strategies might include selecting for varieties with naturally high TDIF activity in vascular tissues or developing methods to transiently enhance TDIF signaling during key growth phases (Yuan and Wang 2021; Fukuda and Hardtke 2019).

13.11.3 Enhancing Stress Tolerance and Yield Stability

- **Improving drought tolerance:** Understanding and manipulating the function of CLE peptides, such as CLE25, and their role in stomatal regulation could lead to the development of drought-tolerant crops capable of maintaining photosynthesis and yield under water-limited conditions. This may involve identifying natural variations in CLE25 signaling pathways that confer drought tolerance or developing targeted breeding or genetic engineering strategies to enhance drought-responsive CLE signaling (Cui et al. 2018)
- **Enhancing nutrient stress tolerance:** Modulating CEP signaling pathways can assist crops in maintaining growth and yield even under nutrient-poor conditions by optimizing both nutrient uptake and allocation within the plant.
- **Improving disease resistance:** Certain CLE peptides are implicated in plant defense responses. Enhancing their activity or manipulating their signaling pathways could lead to the development of crops with improved resistance to various pathogens, thereby reducing yield losses attributable to disease. Additionally, PEPs and systemins have demonstrated efficacy in inducing defense responses in crops, enhancing innate immunity, and potentially decreasing the need for chemical pesticides. Altering RALF-FERONIA signaling pathways has also shown promise in improving cell wall integrity and pathogen resistance, indirectly stabilizing sink development during stress (Wang and Fiers 2010).

13.11.4 Modulating Plant Architecture for Optimal Resource Capture

- **Optimizing root architecture:** Manipulating CEP and other peptides that regulate root meristem activity could lead to the development of root systems that are better adapted to acquire water and essential nutrients from specific soil environments, ultimately positively impacting overall plant health and yield.

- **Controlling shoot branching and leaf development:** Peptide hormones like GLVs play a significant role in shoot meristem activity. Fine-tuning their signaling pathways could lead to optimized plant architecture that facilitates efficient light capture and resource allocation to reproductive organs. Exogenous application of CLE peptides can also be used to modulate shoot apical meristem activity, improving plant architecture and sink development, particularly in cereal crops (Zhao et al. 2021).

13.11.5 Integration in Smart Agriculture and Stress-Responsive Systems

Peptide signals can be effectively integrated into biosensors or synthetic gene circuits to facilitate smart agriculture practices. Stress-inducible peptide promoters can be utilized to trigger the targeted expression of protective genes or mobilize sink resources precisely when specific environmental stress conditions are detected. Furthermore, recent advancements in synthetic biology are focused on developing "switchable" peptides that become active only in response to specific cues such as drought, heat, or pathogen attack. The vision is to engineer crops with self-adjusting source–sink balance mechanisms through engineered peptide circuits, leading to enhanced resilience, resource efficiency, and adaptability (Ogawa-Ohnishi et al. 2022).

13.11.6 Role in Sustainable Agriculture and Reduced Agrochemical Use

Given their natural occurrence in plants and the feasibility of synthesizing them in small quantities, peptide hormones present an environmentally friendly alternative to traditional chemical fertilizers and pesticides. For instance, CEP- and CLE-based interventions can help reduce dependence on synthetic nitrogen fertilizers by improving nutrient uptake and utilization efficiency, while PEPs and systemins can strengthen plants' innate immune systems, thereby lowering the need for chemical pesticide applications. Such approaches align directly with global objectives for sustainable agriculture, low-input farming systems, and climate-smart crop strategies. To translate these benefits into practical agricultural solutions, several delivery strategies are under active exploration. One approach involves the development of cost-effective synthetic peptide applications, delivered through methods such as foliar sprays or seed treatments; however, enhancing peptide stability and uptake efficiency remains a critical challenge. Another promising avenue is genetic engineering, where the expression levels of endogenous peptide hormone genes, their receptors, or downstream signaling components are modified—using advanced tools such as CRISPR/Cas9—to achieve long-term, targeted effects. Additionally, marker-assisted selection enables breeders to identify and utilize natural genetic variations in peptide hormone genes and their regulatory elements to improve traits such as nutrient use efficiency, yield potential, and stress tolerance. A more recent and innovative method involves microbial delivery systems, where beneficial microbes are engineered or selected to produce and release peptide hormones in the rhizosphere, offering a potentially sustainable and self-sustaining mechanism for crop enhancement (Gancheva et al. 2019) (Figure 13.2).

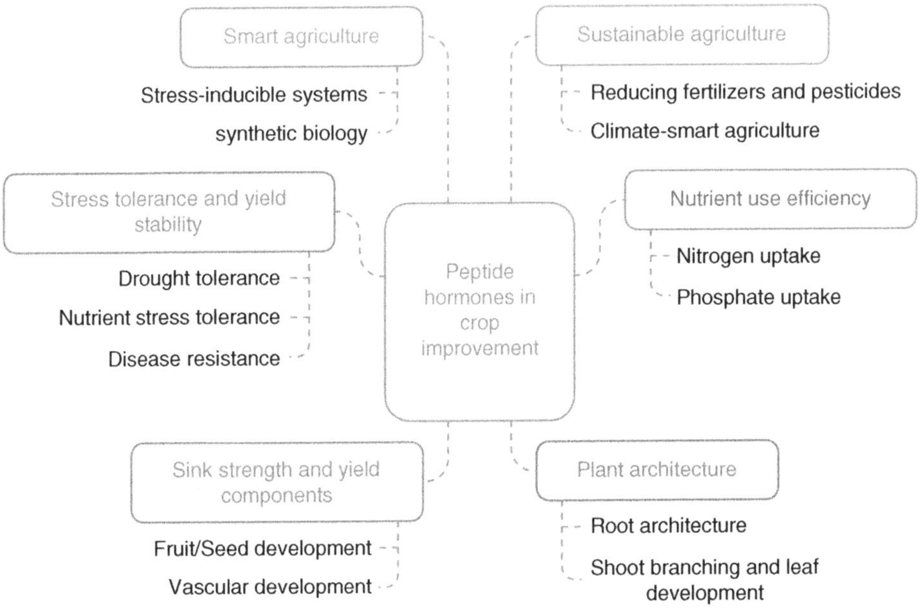

Figure 13.2 Potential applications of peptide hormones in crop improvement and yield optimization.

13.12 Biotechnological Applications of Peptide Hormones in Agriculture

Given their regulatory roles in source–sink relationships, peptide hormones present promising avenues for biotechnological applications in crop improvement.

13.12.1 Genetic Engineering for Enhanced Endogenous Peptide Hormone Function

Several promising strategies are emerging to harness peptide hormones for crop improvement through genetic engineering. One approach involves overexpressing peptide hormone genes by introducing stronger promoters or increasing gene copy numbers, which can elevate endogenous levels of peptides that enhance sink strength (e.g. PSK, PSY) or improve nutrient uptake via the CEP pathway (Sui et al. 2016; Jun et al. 2010). Another strategy is to enhance receptor sensitivity by engineering plants with upregulated or hyperactive receptors, thereby amplifying responses to endogenous signals and promoting more efficient resource allocation or stress tolerance (Araya et al. 2014). Additionally, since many peptide hormones require PTMs such as sulfation or glycosylation to become active, manipulating the enzymes responsible for these modifications could increase the production of bioactive peptide forms (Matsubayashi 2014). Tissue-specific expression using organ-targeted promoters offers a refined approach, enabling the manipulation of specific developmental processes or stress responses—for instance, enhancing sink strength in fruits or improving nutrient uptake in roots. Finally, gene editing technologies such as CRISPR-Cas9 offer powerful tools for precisely modifying peptide hormone genes, their regulatory elements, or associated receptors and processing enzymes. These technologies not only allow fine-tuning of hormone levels and signaling cascades but also open avenues for designing novel peptide variants with improved or tailored functions.

13.12.2 Development and Application of Synthetic Peptide Analogs

Exogenous application: Synthesizing and applying peptide hormone analogs directly to crops presents a targeted strategy to modulate specific physiological processes during critical growth stages. This could involve foliar sprays for enhanced photosynthesis, stress tolerance, or fruit development; seed treatments by coating seeds with peptide analogs to promote early seedling vigor and nutrient uptake; and soil drenching with peptide solutions to influence root development and nutrient acquisition.

Designing stable and bioavailable analogs: A significant challenge lies in the stability and bioavailability of synthetic peptides. Biotechnological approaches can address this by engineering peptide analogs with increased resistance to degradation by plant enzymes and improved uptake by plant tissues. Strategies may include amino acid modifications, substituting natural amino acids with nonnatural ones to enhance stability; conjugation with carrier molecules to facilitate uptake and translocation within the plant; and encapsulation technologies, encasing peptides in protective matrices for controlled release and enhanced stability.

13.12.3 Exploiting Microbial Production of Peptide Hormones

Engineering beneficial microbes: Genetically modifying rhizosphere bacteria or endophytic fungi to produce and secrete beneficial plant peptide hormones could provide a sustainable and localized delivery system. These microbes could colonize plant roots or tissues and continuously supply the plant with the desired signaling molecules. Developing Microbial Consortia: Creating consortia of engineered microbes that produce different peptide hormones with complementary effects (e.g. one enhancing nutrient uptake, another boosting stress tolerance) could lead to synergistic benefits for crop growth and yield.

13.12.4 Utilizing Peptide Hormones as Diagnostic Tools

Developing biosensors: Peptide hormones can serve as highly specific biomarkers for plant physiological status or stress responses. Developing biosensors that can detect and quantify specific peptide hormones in plant tissues or soil solutions could provide valuable information for precision agriculture, allowing for timely interventions like targeted fertilization or irrigation.

13.13 Challenges and Future Prospects

Despite significant progress, identifying novel peptide hormones remains a formidable challenge in plant biology. The plant kingdom likely harbors many undiscovered peptide hormones that are characterized by low abundance, structural diversity, and a lack of universally applicable prediction methods. Their identification is technically challenging, requiring high-throughput "omics" approaches, such as proteomics, transcriptomics, and peptidomics, combined with advanced bioinformatics tools. While these technologies have shown promise, further refinement is needed to reliably discover new peptides across diverse plant species. Once peptides are identified, understanding their precise roles and how they interact within complex signaling networks remains a major hurdle. Many peptide families exhibit redundancy, and they often interact with other hormonal pathways, adding layers of complexity to their functional characterization. This necessitates sophisticated genetic, biochemical, and physiological studies to uncover their specific roles in regulating plant growth, development, and stress responses. Another significant challenge is identifying the specific receptors for many of these peptides. The RLK families in plants are vast and complex, and the possibility of multiple receptors binding to a single peptide complicates the identification of ligand–receptor pairs. This makes it difficult to fully understand the downstream signaling events that occur upon peptide binding. Confirming these ligand–receptor interactions is crucial for a comprehensive understanding of peptide signaling pathways. Moreover, deciphering the downstream signaling mechanisms activated by peptide–receptor interactions presents another significant challenge. This includes identifying key intermediate signaling molecules, second messengers, and transcription factors involved in mediating cellular responses. Understanding these pathways in detail is essential for translating peptide hormone research into practical applications. For agricultural applications, efficient delivery systems for synthetic or recombinant peptide hormones must be developed. Peptides are often prone to degradation by plant enzymes, limiting their stability and bioavailability. Developing cost-effective, stable, and targeted delivery methods will be crucial to ensuring the successful application of peptides in crop improvement and stress management. Additionally, when manipulating peptide hormone signaling, ensuring specificity to targeted processes or tissues while minimizing off-target effects on other developmental pathways is critical.

Peptide hormones rarely function in isolation, and their interactions with other hormones and environmental signals add further complexity. Understanding how these peptides act synergistically or antagonistically with classical hormones like ABA, jasmonates, and ethylene is essential for developing precise applications. Additionally, regulatory hurdles for the deployment of genetically modified crops, particularly those with altered peptide hormone pathways, will require thorough risk assessments and safety evaluations before they can be widely adopted. Despite these challenges, the future of peptide hormone research holds great promise. Tailoring peptide hormone signaling pathways can lead to crops with enhanced yield, optimized flowering times, improved seed or fruit development, and better overall plant architecture. Peptides involved in nutrient uptake, such as components of the CEP pathway, could help in developing crops that require fewer fertilizers, promoting sustainable agriculture. Additionally, engineering peptide hormone pathways related to stress tolerance could result in more resilient crops, capable of withstanding drought, salinity, and temperature stresses.

Furthermore, peptide hormones could open avenues for the development of novel agrochemicals, including bio-based stimulants and biopesticides. Synthetic peptide analogs with enhanced stability and uptake could act as targeted biostimulants, improving plant growth and yield. Peptides with antimicrobial or insecticidal properties could provide environmentally friendly alternatives to traditional pesticides. Advancements in basic plant biology will also benefit from continued research into peptide hormones. These studies will provide deeper insights into plant development, physiology, and their adaptation to changing environments. Peptide hormones and their receptors could serve as valuable tools for dissecting complex signaling networks and developmental processes, contributing to a more holistic understanding of plant biology. In the realm of precision agriculture, peptide-based biosensors could be developed to monitor plant physiological status in real-time, enabling more precise and timely interventions. Additionally, comparative studies of peptide hormone families across different plant species could reveal evolutionary insights into their diversification and functions, shedding light on their role in plant adaptation and survival.

References

Al-Khayri, J.M., Abdel-Haleem, M., and Khedr, E.H. (2024). Harnessing GABA pathways to improve plant resilience against salt stress. *Horticulturae* 10 (12): 1296. https://doi.org/10.3390/horticulturae10121296.

Al-Khayri, J.M. and Khedr, E.H. (2024). Augmenting apricot fruit quality and prolonging storage life via modulating ethylene metabolism and antioxidants with 1-(3-phenyl-propyl) cyclopropene and psyllium treatments. *Journal of Horticultural Science and Biotechnology* 99 (6): 696–712. https://doi.org/10.1080/14620316.2024.2338130.

Amano, Y., Tsubouchi, H., Shinohara, H. et al. (2007). Tyrosine-sulfated glycopeptide involved in cellular proliferation and expansion in Arabidopsis. *Proceedings of the National Academy of Sciences* 104 (48): 19132–19137. https://doi.org/10.1073/pnas.0706164104.

Araya, T., Miyamoto, M., Wibowo, J. et al. (2014). CLE-CLAVATA1 peptide-receptor signaling module regulates the expansion of plant root systems in a nitrogen-dependent manner. *Proceedings of the National Academy of Sciences* 111 (5): 2029–2034.

Asami, T. and Nakagawa, Y. (2018). Preface to the special issue: brief review of plant hormones and their utilization in agriculture. *Journal of Pesticide Science* 43 (3): 154–158.

Bajguz, A. and Piotrowska-Niczyporuk, A. (2023). Biosynthetic pathways of hormones in plants. *Metabolites* 13 (2): 267. https://doi.org/10.3390/metabo13020267.

Barbier, F., Péron, T., Lecerf, M. et al. (2015). Sucrose is an early modulator of the key hormonal mechanisms controlling bud outgrowth in Rosa hybrida. *Plant Physiology* 167 (4): 1319–1333. https://doi.org/10.1104/pp.114.253148.

Boller, T. and Felix, G. (2009). A renaissance of elicitors: perception of microbe-associated molecular patterns and danger signals by pattern-recognition receptors. *Annual Review of Plant Biology* 60 (1): 379–406. https://doi.org/10.1146/annurev.arplant.58.032806.103950.

Butenko, M., Patterson, S.E., Grini, P.E. et al. (2003). INFLORESCENCE DEFICIENT IN ABSCISSION controls floral organ abscission in Arabidopsis and identifies a novel family of putative ligands in plants. *The Plant Cell* 15 (10): 2296–2307. https://doi.org/10.1105/tpc.014365.

Campbell, L. and Turner, S.R. (2017). A Rapid Alkalinization factor gene regulates secondary cell wall patterning in Arabidopsis. *Nature Communications* 8: 14705.

Casson, S., Chilley, P.M., Topping, J.F. et al. (2002). The POLARIS gene of Arabidopsis encodes a predicted peptide required for correct root growth and leaf vascular patterning. *The Plant Cell* 14 (8): 1705–1721. https://doi.org/10.1105/tpc.002618.

Chakraborty, S., Nguyen, B., Wasti, S.D., and Xu, G. (2019). Plant Leucine-rich repeat receptor kinase (LRR-RK): structure, ligand perception, and activation mechanism. *Molecules* 24 (16): 2934. https://doi.org/10.3390/molecules24162934.

Chen, Y.-L., Fan, K.-T., Hung, S.-C., and Chen, Y.-R. (2020). The role of peptides cleaved from protein precursors in eliciting plant stress reactions. *New Phytologist* 227 (6): 1684–1698. https://doi.org/10.1111/nph.16644.

Cheng, X., Ruyter-Spira, C., and Bouwmeester, H. (2013). The interaction between strigolactones and other plant hormones in the regulation of plant development. *Frontiers in Plant Science* 4: 199.

Cho, S., Larue, C.T., Chevalier, D. et al. (2008). Regulation of floral organ abscission in Arabidopsis thaliana. *Proceedings of the National Academy of Sciences* 105 (36): 13570–13575.

Cortleven, A. and Schmülling, T. (2015). Regulation of chloroplast development and function by cytokinin. *Journal of Experimental Botany* 66 (16): 4979–4992. https://doi.org/10.1093/jxb/erv259.

Cui, Y., Li, M., Yin, X. et al. (2018). OsDSSR1, a novel small peptide, enhances drought tolerance in transgenic rice. *Plant Science* 270: 85–96.

Czyzewicz, N., Yue, K., Beeckman, T., and Smet, I.D. (2013). Message in a bottle: small signalling peptide outputs during growth and development. *Journal of Experimental Botany* 64 (17): 5281–5296.

De Coninck, B. and De Smet, I. (2016). Plant peptides–taking them to the next level. *Journal of Experimental Botany* 67 (16): 4791–4795.

Dharmasiri, S., Jayaweera, T., and Dharmasiri, N. (2013). Plant hormone signalling: current perspectives on perception and mechanisms of action. *Ceylon Journal of Science (Biological Sciences)* 42 (1).

Doblas, V.G., Smakowska-Luzan, E., Fujita, S. et al. (2017). Root diffusion barrier control by a vasculature-derived peptide binding to the SGN3 receptor. *Science* 355 (6322): 280–284.

Dodueva, I.E., Lebedeva, M.A., and Lutova, L.A. (2021). Dialog between kingdoms: enemies, allies and peptide phytohormones. *Plants* 10 (2): 315. https://doi.org/10.3390/plants10020315.

El-Beltagi, H.S., Abdel-Haleem, M., Al Saikhan, M.S., and Khedr, E.H. (2025a). Optimization of progesterone treatment via response surface methodology to maintain quality and mitigate chilling injury in *Prunus persica* L. *New Zealand Journal of Crop and Horticultural Science* 1–19: https://doi.org/10.1080/01140671.2025.2458819.

EL-Beltagi, H.S., Abdel-haleem, M., EL-Mogy, M.M., and Khedr, E.H. (2025b). Resilience of forests in a changing climate: responses, mechanisms, and adaptations. *Notulae Botanicae Horti Agrobotanici Cluj-Napoca* 53 (1): 14360. https://doi.org/10.15835/nbha53114360.

Estornell, L.H., Wildhagen, M., Pérez-Amador, M.A. et al. (2015). The IDA peptide controls abscission in Arabidopsis and Citrus. *Frontiers in plant science* 6: 1003.

Etchells, J.P., Provost, C.M., Mishra, L., and Turner, S.R. (2013). WOX4 and WOX14 act downstream of the PXY receptor kinase to regulate plant vascular proliferation independently of any role in vascular organisation. *Development* 140 (10): 2224–2234.

Fan, S., Wang, D., Xie, H. et al. (2021). Sugar transport, metabolism and signaling in fruit development of *Litchi chinensis* Sonn: a review. *International Journal of Molecular Sciences* 22 (1): 12. https://doi.org/10.3390/ijms22010012.

Farrokhi, N., Whitelegge, J., and Brusslan, J. (2008). Plant peptides and peptidomics. *Plant Biotechnology Journal* 6 (8): 731–747. https://doi.org/10.1111/j.1467-7652.2008.00365.x.

Fedoreyeva, L. (2023). Molecular mechanisms of regulation of root development by plant peptides. *Plants* 12 (15): 2748.

Feng, S., Liu, Z., Chen, H. et al. (2024). PHGD: an integrative and user-friendly database for plant hormone-related genes. *iMeta* 3 (1): e195. https://doi.org/10.1002/imt2.195.

Fernandez, A., Drozdzecki, A., Hoogewijs, K. et al. (2013). Transcriptional and functional classification of the GOLVEN/ROOT GROWTH FACTOR/CLE-Like signaling peptides reveals their role in lateral root and hair formation. *Plant Physiology* 161: 954–970. https://doi.org/10.1104/pp.112.206029.

Fernandez, A., Drozdzecki, A., Hoogewijs, K. et al. (2012). Transcriptional and functional classification of the GOLVEN/ROOT GROWTH FACTOR/CLE-like signaling peptides reveals their role in lateral root and hair formation1[W][OA]. *Plant Physiology* 160 (2): 922–937. https://doi.org/10.1104/pp.112.203516.

Fichtner, F., Barbier, F., Feil, R. et al. (2017). Trehalose 6-phosphate is involved in triggering axillary bud outgrowth in garden pea (*Pisum sativum* L.). *The Plant Journal* 90 (1): 146–157. https://doi.org/10.1111/tpj.13489.

Fiori, C. S. (2010). Identificação e caracterização de componentes da via de transdução de sinais do peptídeo hormonal RALF (Doctoral dissertation).

Fletcher, J.C., Brand, U., Running, M.P. et al. (1999). Signaling of cell fate decisions by CLAVATA3 in Arabidopsis shoot meristems. *Science* 283 (5409): 1911–1914. https://doi.org/10.1126/science.283.5409.1911.

Fukuda, H. and Hardtke, C. (2019). Peptide signaling pathways in vascular differentiation [OPEN]. *Plant Physiology* 180 (1): 7–17. https://doi.org/10.1104/pp.18.01355.

Fukuda, H. and Higashiyama, T. (2011). Diverse functions of plant peptides: entering a new phase. *Plant and Cell Physiology* 52 (11): 1831–1839. https://doi.org/10.1093/pcp/pcr133.

Gancheva, M.S., Malovichko, Y.V., Poliushkevich, L.O. et al. (2019). Plant peptide hormones. *Russian Journal of Plant Physiology* 66: 171–189.

Gray, W. (2004). Hormonal Regulation of Plant Growth and Development. In: *Advances in Agricultural Biotechnology*, 43–67. Springer https://doi.org/10.1007/978-3-642-18745-1_3.

Guo, Y., Ren, G., Zhang, K. et al. (2021). Leaf senescence: progression, regulation, and application. *Molecular Horticulture* 1: 1–25.

Guo, H., Zhang, W., Tian, H. et al. (2015). An auxin responsive CLE gene regulates shoot apical meristem development in Arabidopsis. *Frontiers in Plant Science* 6: 759. https://doi.org/10.3389/fpls.2015.00759.

Gutiérrez-Alanís, D., Yong-Villalobos, L., Jiménez-Sandoval, P. et al. (2017). Phosphate starvation-dependent iron mobilization induces CLE14 expression to trigger root meristem differentiation through CLV2/PEPR2 signaling. *Developmental Cell* 41 (5): 555–570.

Hander, T., Fernández-Fernández, Á., Kumpf, R.P. et al. (2019). Damage on plants activates Ca^{2+}-dependent metacaspases for release of immunomodulatory peptides. *Science* 363 (6427): 646–649. https://doi.org/10.1126/science.aau0745.

Hara, K., Kajita, R., Torii, K.U. et al. (2007). The secretory peptide gene EPF1 enforces the stomatal one-cell-spacing rule. *Genes & Development* 21 (14): 1720–1725. https://doi.org/10.1101/gad.1565807.

Hartmann, J., Stührwohldt, N., Dahlke, R., and Sauter, M. (2013). Phytosulfokine control of growth occurs in the epidermis, is likely to be non-cell autonomous and is dependent on brassinosteroids. *The Plant Journal* 76 (3): 517–526. https://doi.org/10.1111/tpj.12310.

Hazak, O. and Hardtke, C.S. (2016). CLAVATA 1-type receptors in plant development. *Journal of Experimental Botany* 67 (2): 513–526. https://doi.org/10.1093/jxb/erv480.

Hirakawa, Y., Torii, K.U., and Uchida, N. (2017). Mechanisms and strategies shaping plant peptide hormones. *Plant and Cell Physiology* 58 (1): 34–42. https://doi.org/10.1093/pcp/pcw194.

Hsiao, Y.-C. and Yamada, M. (2020). The roles of peptide hormones and their receptors during plant root development. *Genes* 11 (11): 1334. https://doi.org/10.3390/genes11111334.

Hu, Z., Zhang, H., and Shi, K. (2018). Plant peptides in plant defense responses. *Plant Signaling & Behavior* 13 (11): e1535160. https://doi.org/10.1080/15592324.2018.1535160.

Huang, P., Li, Z., and Guo, H. (2022). New advances in the regulation of leaf senescence by classical and peptide hormones. *Frontiers in Plant Science* 13: 842494. https://doi.org/10.3389/fpls.2022.842494.

Huffaker, A., Pearce, G., and Ryan, C.A. (2006). An endogenous peptide signal in Arabidopsis activates components of the innate immune response. *Proceedings of the National Academy of Sciences of the United States of America* 103 (26): 10098–10103. https://doi.org/10.1073/pnas.0603950103.

Huffaker, A. and Ryan, C.A. (2007). Endogenous peptide defense signals in Arabidopsis differentially amplify signaling for the innate immune response. *Proceedings of the National Academy of Sciences* 104 (25): 10732–10737. https://doi.org/10.1073/pnas.0703944104.

Ito, Y., Nakanomyo, I., Motose, H. et al. (2006). Dodeca-CLE peptides as suppressors of plant stem cell differentiation. *Science* 313 (5793): 1516–1519. https://doi.org/10.1126/science.1128446.

Jun, J., Fiume, E., Roeder, A.H.K. et al. (2010). Comprehensive analysis of CLE polypeptide signaling gene expression and overexpression activity in Arabidopsis1[C][W][OA]. *Plant Physiology* 154 (4): 1721–1736. https://doi.org/10.1104/pp.110.165095.

Kapale, V., Agrawal, S., Patel, C. et al. (2020). Small polypeptide hormones in plant. *International Journal of Chemical Studies* 8 (4): 2415–2421.

Kaufmann, C. and Sauter, M. (2019). Sulfated plant peptide hormones. *Journal of Experimental Botany* 70 (3): 751–761. https://doi.org/10.1093/jxb/ery388.

Khan, Q., Qin, Y.-Z., Guo, D.-J. et al. (2023). A review of the diverse genes and molecules involved in sucrose metabolism and innovative approaches to improve sucrose content in sugarcane. *Agronomy* 13 (2): 421. https://doi.org/10.3390/agronomy13020421.

Khavinson, V., Linkova, N., Diatlova, A., and Dudkov, A. (2024). Peptide regulation of plant cells differentiation and growth. *BIO Web of Conferences* 94: 01003. https://doi.org/10.1051/bioconf/20249401003.

Khedr, E.H. and Khedr, N. (2024). Enhancing productivity, modifying biochemical parameters, and regulating the phenylpropanoid pathway in 'Le-Conte' pears through optimal protocatechuic acid treatments. *BMC Plant Biology* 24 (1): 50. https://doi.org/10.1186/s12870-023-04715-9.

Khedr, E.H., Khedr, N., and Abdel-Haleem, M. (2023). Harnessing the metabolic modulatory and antioxidant power of 1-(3-Phenyl-Propyl) cyclopropane and melatonin in maintaining mango fruit quality and prolongation storage life. *BMC Plant Biology* 23 (1): 464. https://doi.org/10.1186/s12870-023-04485-4.

Kinoshita, A., Nakamura, Y., Sasaki, E. et al. (2007). Gain-of-function phenotypes of chemically synthetic CLAVATA3/ESR-related (CLE) peptides in Arabidopsis thaliana and *Oryza sativa*. *Plant and Cell Physiology* 48 (12): 1821–1829. https://doi.org/10.1093/pcp/pcm163.

Kumawat, G., Cao, D., Park, C., and Xu, D. (2024). C-terminally encoded peptide-like genes are associated with the development of primary root at *qRL16.1* in soybean. *Frontiers in Plant Science* https://doi.org/10.3389/fpls.2024.1387954.

Kumpf, R.P., Shi, C.-L., Larrieu, A. et al. (2013). Floral organ abscission peptide IDA and its HAE/HSL2 receptors control cell separation during lateral root emergence. *Proceedings of the National Academy of Sciences of the United States of America* 110: 5235–5240. https://doi.org/10.1073/pnas.1210835110.

Kutschmar, A., Rzewuski, G., Stührwohldt, N. et al. (2009). PSK peptide signaling controls root growth by affecting S-phase progression of the cell cycle. *The Plant Journal* 60 (4): 512–522.

Ladwig, F., Dahlke, R.I., Stührwohldt, N. et al. (2015). Phytosulfokine regulates growth in Arabidopsis through a response module at the plasma membrane that includes CYCLIC NUCLEOTIDE-GATED CHANNEL17, H+-ATPase, and BAK1. *The Plant Cell* 27 (6): 1718–1729.

Li, Y., Di, Q., Luo, L., and Yu, L. (2023b). Phytosulfokine peptides, their receptors, and functions. *Frontiers in Plant Science* https://doi.org/10.3389/fpls.2023.1326964.

Li, Z., Oelmüller, R., Guo, H., and Miao, Y. (2023a). Signal transduction of plant organ senescence and cell death. *Frontiers in Plant Science* 14: 1172373.

Liang, C. and Chu, C. (2015). Towards understanding abscisic acid-mediated leaf senescence. *Science China. Life Sciences* 58 (5): 506.

Luo L, Yu L, Yang J, Wang E. Peptide signals regulate nitrogen deficiency adaptation of dicotyledonous model plants. *Plant, Cell & Environment* 2024 Oct 17. doi:https://doi.org/10.1111/pce.15203. Epub ahead of print. PMID: 39420598.

Lv, X., Zhang, Y., Zhang, Y. et al. (2020). Source-sink modifications affect leaf senescence and grain mass in wheat as revealed by proteomic analysis. *BMC Plant Biology* 20 (1): 547. https://doi.org/10.1186/s12870-020-02720-7.

Lyapina, I., Ivanov, V., and Fesenko, I. (2021). Peptidome: Chaos or inevitability. *International Journal of Molecular Sciences* 22 (3): 1428. https://doi.org/10.3390/ijms22031428.

Matsubayashi, Y. (2010). Post-translational modifications in secreted peptide hormones in plants. *Plant and Cell Physiology* 51 (1): 5–14. https://doi.org/10.1093/pcp/pcp167.

Matsubayashi, Y. (2014). Posttranslationally modified small-peptide signals in plants. *Annual Review of Plant Biology* 65: 357–377. https://doi.org/10.1146/annurev-arplant-050213-040258.

Matsubayashi, Y. (2018). Exploring peptide hormones in plants: identification of four peptide hormone-receptor pairs and two post-translational modification enzymes. *Proceedings of the Japan Academy. Series B, Physical and Biological Sciences* 94 (2): 39–53. https://doi.org/10.2183/pjab.94.003.

Matsubayashi, Y. and Sakagami, Y. (2006). Peptide hormones in plants. *Annual Review of Plant Biology* 57: 649–679. https://doi.org/10.1146/annurev.arplant.57.032905.105304.

Matsuzaki, Y., Ogawa-Ohnishi, M., Mori, A., and Matsubayashi, Y. (2010b). Secreted peptide signals required for maintenance of root stem cell Niche in Arabidopsis. *Science* 329 (5996): 1065–1068. https://doi.org/10.1126/science.1192353.

Matsuzaki, Y. et al. (2010a). Phytosulfokine stimulates cellular proliferation in the Arabidopsis root meristem. *Plant Physiology* 153 (4): 1545–1555.

Meng, L., Buchanan, B.B., Feldman, L.J., and Luan, S. (2012). CLE-like (CLEL) peptides control the pattern of root growth and lateral root development in Arabidopsis. *Proceedings of the National Academy of Sciences* 109 (43): 17632–17637. https://doi.org/10.1073/pnas.1214137109.

Mirabeau, O., Perlas, E., Severini, C. et al. (2007). Identification of novel peptide hormones in the human proteome by hidden Markov model screening. *Genome Research* 17 (3): 320–327. https://doi.org/10.1101/gr.5646607.

Mochida, K. and Shinozaki, K. (2011). Advances in omics and bioinformatics tools for systems analyses of plant functions. *Plant and Cell Physiology* 52 (12): 2017–2038.

Mohammadi, A., Kav, N.N., and Deyholos, M.K. (2014). Transcriptome characterization of a model halophyte plant, *Eutrema salsugineum*, during salinity stress. *BMC Genomics* 15 (1): 1–17.

Moore, K.L. (2009). Protein tyrosine sulfation: a critical posttranslational modification in plants and animals. *Proceedings of the National Academy of Sciences of the United States of America* 106 (25): 10149–10150. https://doi.org/10.1073/pnas.0904897106.

Narita, N.N., Moore, S., Horiguchi, G. et al. (2004). Overexpression of a novel small peptide ROTUNDIFOLIA4 decreases cell proliferation and alters leaf shape in Arabidopsis thaliana. *The Plant Journal* 38 (4): 699–713. https://doi.org/10.1111/j.1365-313X.2004.02073.x.

Nakagami, S., Kajiwara, T., Tsuda, K., and Sawa, S. (2024). CLE peptide signaling in plant-microbe interactions. *Frontiers in Plant Science* 15: 1481650.

Nguyen, Q.T., Kisiala, A., Andreas, P. et al. (2016). Soybean seed development: fatty acid and phytohormone metabolism and their interactions. *Current Genomics* 17 (1): 51–67. https://doi.org/10.2174/1389202917666151215221408.

Nimchuk, Z.L., Zhou, Y., Tarr, P.T. et al. (2015). Plant stem cell maintenance by transcriptional cross-regulation of related receptor kinases. *Developmental Cell* 33 (1): 103–113.

Ogawa-Ohnishi, M., Yamashita, T., Kakita, M. et al. (2022). Peptide ligand-mediated trade-off between plant growth and stress response. *Science* 378 (6616): 175–180.

Ohkubo, Y., Tanaka, M., Tabata, R. et al. (2017). Shoot-to-root mobile polypeptides involved in systemic regulation of nitrogen acquisition. *Nature Plants* 3: 17029.

Ohyama, K., Ogawa, M., and Matsubayashi, Y. (2008). Identification of a biologically active, small, secreted peptide in Arabidopsis by in silico gene screening, followed by LC-MS-based structure analysis. *The Plant Journal* 55 (4): 633–642. https://doi.org/10.1111/j.1365-313X.2008.03530.x.

Ou, Y., Tao, B., Wu, Y. et al. (2022). Essential roles of SERKs in the ROOT MERISTEM GROWTH FACTOR-mediated signaling pathway. *Plant Physiology* 189: 165–177. https://doi.org/10.1093/plphys/kiac036.

Pandita, D., Bhat, J., Wani, S.H. et al. (2023). Mobile signaling peptides: secret molecular messengers with a mighty role in plant life. *Journal of Plant Growth Regulation* 42 (10): 4311–4328. https://doi.org/10.1007/s00344-022-10914-y.

Patel, N., Mohd-Radzman, N.A., Corcilius, L. et al. (2017). Diverse peptide hormones affecting root growth identified in the *Medicago truncatula* secreted peptidome∗. *Molecular & Cellular Proteomics* 16 (12): 2181–2195. https://doi.org/10.1074/mcp.RA117.000226.

Patharkar, O.R. and Walker, J.C. (2016). Leaf shedding in Arabidopsis is regulated by the IDA-HAE-HSL2 pathway. *Plant Cell* 28 (10): 2431–2446.

Patrick, J. (2013). Does Don Fisher's high-pressure manifold model account for phloem transport and resource partitioning? *Frontiers in Plant Science* 4: 37. https://doi.org/10.3389/fpls.2013.00037.

Pearce, G., Moura, D.S., Stratmann, J., and Ryan, C.A. (2001). RALF, a ubiquitous polypeptide, arrests root growth and development. *PNAS* 98 (22): 12843–12847.

Pearce, G. and Ryan, C.A. (2003). Systemic signaling in tomato plants for defense against herbivores. *Journal of Biological Chemistry* 278 (32): 30065–30072. https://doi.org/10.1074/jbc.M302245200.

Pearce, G., Strydom, D., Johnson, S., and Ryan, C.A. (1991). A polypeptide from tomato leaves induces wound-inducible proteinase inhibitor proteins. *Science* 253 (5022): 895–898. https://doi.org/10.1126/science.1887166.

Peleg, Z., Reguera, M., Tumimbang, E., Walia, H., & Blumwald, E. (2011). Cytokinin-mediated source/sink modifications improve drought tolerance and increase grain yield in rice under water-stress. *Plant Biotechnology Journal*, 9(7), 747–758

Qin, A.-G., Aluko, O., Liu, Z. et al. (2023). Improved cotton yield: can we achieve this goal by regulating the coordination of source and sink? *Frontiers in Plant Science* 14: 1109991. https://doi.org/10.3389/fpls.2023.1109991.

Radchuk, V., Radchuk, R., Pirko, Y. et al. (2012). A somaclonal line SE7 of finger millet (*Eleusine coracana*) exhibits modified cytokinin homeostasis and increased grain yield. *Journal of Experimental Botany* 63 (1): 357–368. https://doi.org/10.1093/jxb/err291.

Ren, G., Zhang, Y., Chen, Z. et al. (2024). Research progress of small plant peptides on the regulation of plant growth, development, and abiotic stress. *International Journal of Molecular Sciences* 25 (7): 4114.

Ricachenevsky, F.K., Menguer, P.K., and Sperotto, R.A. (2013). kNACking on heaven's door: how important are NAC transcription factors for leaf senescence and Fe/Zn remobilization to seeds? *Frontiers in Plant Science* 4: 226.

Ritonga, F.N., Zhou, D., Zhang, Y. et al. (2023). The roles of gibberellins in regulating leaf development. *Plants* 12 (6): 1243.

Rolland, F., Baena-Gonzalez, E., and Sheen, J. (2006). Sugar sensing and signaling in plants: conserved and novel mechanisms. *Annual Review of Plant Biology* 57 (1): 675–709.

Roy, S., Griffiths, M., Torres-Jerez, I. et al. (2022). Application of synthetic peptide CEP1 increases nutrient uptake rates along plant roots. *Frontiers in Plant Science* 13: https://doi.org/10.3389/fpls.2021.793145.

Ryan, C.A. and Pearce, G. (2003). Systemins: a functionally defined family of peptide signals that regulate defensive genes in Solanaceae species. *PNAS* 100 (Suppl 2): 14577–14580.

Rzemieniewski, J., Leicher, H., Lee, H.K. et al. (2024). CEP signaling coordinates plant immunity with nitrogen status. *Nature Communications* 15: 10686. https://doi.org/10.1038/s41467-024-55194-x.

Saeed, W., Naseem, S., and Ali, Z. (2017). Strigolactones biosynthesis and their role in abiotic stress resilience in plants: a critical review. *Frontiers in Plant Science* 8: 1487.

Sakuraba, Y. (2022). Molecular basis of nitrogen starvation-induced leaf senescence. *Frontiers in Plant Science* 13: 1013304.

Salah, I.B., Jelali, N., Slatni, T. et al. (2018). Involvement of source-sink relationship and hormonal control in the response of *Medicago Ciliaris — Sinorhizobium Medicae* symbiosis to salt stress. *Nature* 553 (7689): 521–525.

Schneider, A., Godin, C., Boudon, F. et al. (2019). Light regulation of axillary bud outgrowth along plant axes: an overview of the roles of sugars and hormones. *Frontiers in Plant Science* 10: 11. https://doi.org/10.3389/fpls.2019.00118.

Shimada, T., Sugano, S.S., and Hara-Nishimura, I. (2011). Positive and negative peptide signals control stomatal density. *Cellular and Molecular Life Sciences* 68: 2081–2088. https://doi.org/10.1007/s00018-011-0685-7.

Shi, C.L., Alling, R.M., Hammerstad, M., and Aalen, R.B. (2019). Control of organ abscission and other cell separation processes by evolutionary conserved peptide signaling. *Plants* 8 (7): 225.

Shinohara, H. (2021). Root meristem growth factor RGF, a sulfated peptide hormone in plants. *Peptides* 142: 170556. https://doi.org/10.1016/j.peptides.2021.170556.

Shishkova, S., Huang, L., Rodriguez, R. et al. (2023). Editorial: root development: towards understanding regulatory networks and complex interactions between cell populations. *Frontiers in Plant Science* 14: 1296202. https://doi.org/10.3389/fpls.2023.1296202.

Shumilina, J., Kiryushkin, A.S., Frolova, N. et al. (2023). Integrative proteomics and metabolomics analysis reveals the role of small signaling peptide Rapid Alkalinization factor 34 (RALF34) in cucumber roots. *International Journal of Molecular Sciences.* https://doi.org/10.3390/ijms24087654.

Stahl, E., Fernandez Martin, A., Glauser, G. et al. (2022). The MIK2/SCOOP signaling system contributes to Arabidopsis resistance against herbivory by modulating Jasmonate and Indole Glucosinolate biosynthesis. *Frontiers in Plant Science* 13: 838802. https://doi.org/10.3389/fpls.2022.838802.

Stegmann, M., Monaghan, J., Smakowska-Luzan, E. et al. (2017). The receptor kinase FER is a RALF-regulated scaffold controlling plant immune signaling. *Science* 355 (6322): 287–289.

Stenvik, G.E., Tandstad, N.M., Guo, Y. et al. (2008). The EPIP peptide of INFLORESCENCE DEFICIENT IN ABSCISSION is sufficient to induce abscission in Arabidopsis through the receptor-like kinases HAESA and HAESA-LIKE2. *Plant Cell* 20 (7): 1805–1817.

Stührwohldt, N. and Schaller, A. (2018). Regulation of plant peptide hormones and growth factors by post-translational modification. *Plant Biology* 20 (S1): 7–14. https://doi.org/10.1111/plb.12645.

Sui, Z., Wang, T., Li, H. et al. (2016). Overexpression of peptide-encoding OsCEP6. 1 results in pleiotropic effects on growth in rice (O. sativa). *Frontiers in Plant Science* 7: 228.

Tabata, R., Sumida, K., Yoshii, T. et al. (2014). Perception of root-derived peptides by shoot LRR-RKs mediates systemic N-demand signaling. *Science* 346 (6207): 343–346.

Takahashi, F., Suzuki, T., Osakabe, Y. et al. (2018). A small peptide modulates stomatal control via abscisic acid in long-distance signaling. *Nature* 556: 235–238.

Taleski, M., Chapman, K., Novák, O., Schmülling, T., Frank, M., & Djordjevic, M. A. (2022). CEP peptide and cytokinin pathways converge on CEPD glutaredoxins to inhibit root growth. https://doi.org/10.1101/2022.07.21.500969

Tameshige, T., Hirakawa, Y., Torii, K., and Uchida, N. (2015). Cell walls as a stage for intercellular communication regulating shoot meristem development. *Frontiers in Plant Science* 6: 1166. https://doi.org/10.3389/fpls.2015.01166.

Tegeder, M. and Masclaux-Daubresse, C. (2018). Source and sink mechanisms of nitrogen transport and use. *New Phytologist* 217 (1): 35–53.

Tegeder, M. and Hammes, U.Z. (2018). The way out and in: phloem loading and unloading of amino acids. *Current Opinion in Plant Biology* 43: 16–21.

Thomas, H. (2013). Senescence, ageing and death of the whole plant. *New Phytologist* 197 (3): 696–711.

Valmas, M.I., Sexauer, M., Markmann, K., and Tsikou, D. (2023). Plants recruit peptides and micro RNAs to regulate nutrient acquisition from soil and symbiosis. *Plants* 12 (11): 2180. https://doi.org/10.3390/plants12112180.

Wang, G. and Fiers, M. (2010). CLE peptide signaling during plant development. *Protoplasma* 240: 33–43.

Wang, A., Guo, J.-A., Wang, S. et al. (2022). BoPEP4, a C-terminally encoded plant elicitor peptide from broccoli, plays a role in salinity stress tolerance. *International Journal of Molecular Sciences* 23 (11): 6194. https://doi.org/10.3390/ijms23116194.

Wen, J., Lease, K.A., and Walker, J.C. (2004). DVL, a novel class of small polypeptides: overexpression alters Arabidopsis development. *The Plant Journal* 37 (5): 775–786. https://doi.org/10.1046/j.1365-313X.2003.01985.x.

White, A., Rogers, A., Rees, M., and Osborne, C. (2016). How can we make plants grow faster? A source-sink perspective on growth rate. *Journal of Experimental Botany* 67 (1): 1–17. https://doi.org/10.1093/jxb/erv447.

Witzany, G. (2010). Uniform categorization of biocommunication in bacteria, fungi and plants. *World Journal of Biological Chemistry* 1 (4): 183–189. https://doi.org/10.4331/wjbc.v1.i4.183.

Xiong, L., Huang, Y., Liu, Z. et al. (2022). Small EPIDERMAL PATTERNING FACTOR-LIKE2 peptides regulate awn development in rice. *Plant Physiology* 190: 516–531. https://doi.org/10.1093/plphys/kiac278.

Yan, C.-L., Guan, K.-X., Lin, H. et al. (2025). Peptides in plant reproduction—small yet powerful. *Frontiers in Plant Science* 16: 1454891. https://doi.org/10.3389/fpls.2024.1454891.

Yang, K., An, J.P., Li, C.Y. et al. (2021). The apple C2H2-type zinc finger transcription factor MdZAT10 positively regulates JA-induced leaf senescence by interacting with MdBT2. *Horticulture Research* 8.

Yang, H., Matsubayashi, Y., Nakamura, K., and Sakagami, Y. (2001). Diversity of Arabidopsis genes encoding precursors for phytosulfokine, a peptide growth factor. *Plant Physiology* 125 (3): 1345–1358. https://doi.org/10.1104/pp.125.3.1345.

Yang, F., Song, Y., Yang, H. et al. (2014). An auxin-responsive endogenous peptide regulates root development in Arabidopsis. *Journal of Integrative Plant Biology* 56 (1): 59–70. https://doi.org/10.1111/jipb.12118.

Yoshida, T., Fernie, A.R., Shinozaki, K., and Takahashi, F. (2020). Long-distance stress and developmental signals associated with abscisic acid signaling in environmental responses. *Plant Journal* 105: 477–488. https://doi.org/10.1111/tpj.15101.

Yuan, B. and Wang, H. (2021). Peptide signaling pathways regulate plant vascular development. *Frontiers in Plant Science* 12: https://doi.org/10.3389/fpls.2021.719606.

Zhang, Z., Han, H., Zhao, J. et al. (2025). Peptide hormones in plants. *Molecular Horticulture* 5: 7. https://doi.org/10.1186/s43897-024-00134-y.

Zhang, Y., Xu, Y., Huang, D. et al. (2022). Research progress on the MYB transcription factors in tropical fruit. *Tropical Plants* 1 (1): 1–15.

Zhang, R., Shi, P.-T., Zhou, M. et al. (2023). Rapid Alkalinization factor: function, regulation, and potential applications in agriculture. *Stress biology*. https://doi.org/10.1007/s44154-023-00093-2.

Zhang, H., Jing, X., Chen, Y. et al. (2020). The genome-wide analysis of RALF-like genes in strawberry (wild and cultivated) and five other plant species (Rosaceae). *Genes* 11 (2): 174.

Zhao, L., Li, Y., Ding, Q. et al. (2021). CD36 senses dietary lipids and regulates lipids homeostasis in the intestine. *Frontiers in Physiology* 12: 713859. https://doi.org/10.3389/fphys.2021.713859.

Zhao, Z., Wang, C., Yu, X. et al. (2023). Auxin regulates source-sink carbohydrate partitioning and reproductive organ development in rice. *Plant Physiology* 191 (1): 708–724. https://doi.org/10.1093/plphys/kiac498.

14

Priming, Plant Defense Against Insects, and Peptide Hormones

Bimal Kumar Sahoo[1], *Ansh Raj*[2], *Ipsita Samal*[2], *Deepak Kumar Mahanta*[3], *Tanmaya Kumar Bhoi*[4], *Mohammad Ikram*[1], *and Iqra Ali*[1]

[1] *Department of Agricultural Entomology, Centre for Plant Protection Studies, Tamil Nadu Agricultural University, Coimbatore, Tamil Nadu, India*

[2] *ICAR-National Research Centre on Litchi, Muzaffarpur, Bihar, India*

[3] *Forest Entomology Discipline, Forest Protection Division, Indian Council of Forestry Research and Education-Forest Research Institute (ICFRE-FRI), Uttarakhand, India*

[4] *Forest Protection Division, ICFRE-Arid Forest Research Institute (ICFRE-AFRI), Jodhpur, Rajasthan, India*

14.1 Introduction

Plants remain stationary because of their immobile nature thus they cannot avoid various abiotic and biotic stresses (Mauch-Mani et al. 2017). Plants can endure these stresses through their inducible defense system and innate immune system, which includes cellular barriers such as cell wall and membrane function and the production of secondary metabolites (Jones and Dangl 2006). Plants possess the ability to develop adaptive immunity in addition to their innate immunity systems through which they acquire better defense capabilities from previous encounters (Conrath et al. 2015). Insects function as the most influential group of biotic stressors that activate plant immune responses. Plant pests activate two different types of stimuli, which include physical contact and biological signals. Leaf trichomes contact herbivores as they move, which triggers physical stimuli (Howe and Jander 2008; Peiffer et al. 2009). The biological stimuli consist of oral secretions together with ovipositional cues and damage-associated molecular patterns (DAMPs) or insect-associated molecular patterns (IAMPs) (Hilker and Fatouros 2015; Hilker and Meiners 2010).

The defense responses of plants become active through compounds found in insect oral secretions, including amino acids and peptides, and fatty acid conjugates, when herbivores encounter these compounds (Alborn et al. 1997; Howe and Jander 2008). These secretions function as essential elements for plant immunity because they enable the system to distinguish between mechanical damage and herbivore feeding. Plant defense pathways can be activated through ovipositional cues, insect stylet penetration, and trichome-based mechanical sensing, according to Peiffer et al. (2009) and Pérez-Hedo et al. (2015). The plant defense system can be triggered against potential threats through herbivore-induced plant volatiles (HIPVs) released by neighboring plants. The volatile signals known as terpenoids function as primers to protect maize plants against *Spodoptera littoralis* (Ton et al. 2007; Frost et al. 2008).

Peptide Hormones in Plant Development and Stress Responses, First Edition. Edited by Vijay Pratap Singh and Samiksha Singh.

Plants activate their immune responses when exposed to various molecular patterns such as PAMPs, MAMPs, DAMPs, and xenobiotic compounds, including fungicides and insecticides (Conrath 2011; Návarová et al. 2012). When plants encounter stress factors, these stressors trigger multiple physiological, transcriptional, epigenetic, and metabolic changes that enable the plants to endure these stresses during the priming phase (Mauch-Mani et al. 2017). Plants exhibit defense responses through priming as an inactive state until subsequent stress exposure activates these responses to become stronger than unprimed native plants (Slaughter et al. 2012; Kim and Felton 2013). Plants maintain durable priming effects that last through their entire lifespan and may transmit these effects to future generations, thus demonstrating plant immunological memory. The primed state of *Arabidopsis thaliana* proved heritable because offspring from primed plants displayed superior pathogen resistance to descendants of unprimed plants, according to Slaughter et al. (2012).

The priming process leads to increased signaling intermediate accumulation that includes transcription factors, which stay inactive until stressors appear (Hilker and Schmülling 2019). Plants use four different priming-based resistance mechanisms, which include systemic acquired resistance (SAR), β-aminobutyric acid-induced resistance (BABA-IR), rhizobacteria-mediated induced systemic resistance (ISR), and volatile organic compound-induced resistance (VOC-IR). The activation of salicylic acid (SA)-regulated genes occurs through SAR and BABA-IR responses, but jasmonic acid (JA)-mediated defense system activation mainly relies on ISR and VOC-IR mechanisms (Ton et al. 2009). The priming responses do not heavily rely on the signaling mechanisms that use abscisic acid (ABA) and phosphoinositide (PI) signaling pathways. The primed state can be initiated through biological interactions as well as by using various natural and synthetic elicitors such as JA, SA, and BABA (Worrall et al. 2012). The seeds of tomato plants treated with JA or BABA developed enhanced resistance to both insect herbivores, including spider mites, caterpillars, and aphids, and fungal pathogens (Worrall et al. 2012).

Peptide hormones are small, secreted polypeptides that act as intercellular signaling molecules, playing significant roles in plant communication (Lease and Walker 2006). Nuclear genes encode these hormones, which are then subjected to posttranslational changes and proteolytic processing of preproproteins. After being processed, the majority of peptide hormones are released from the cell to interact with nearby cell-membrane-bound receptors, starting signal transduction pathways that change gene expression and cause biochemical reactions (Zhang et al. 2025). There are two types of peptide hormones: secretory and nonsecretory. Secretory peptides have an N-terminal signal peptide that guides them through the secretion pathway, whereas the nonsecretory peptides lack this. Secretory peptide hormones include CLAVATA3 (CLV3)/EMBRYO SURROUNDING REGION-RELATED peptides (CLEs), C-TERMINALLY ENCODED PEPTIDES (CEPs), PHYTOSULFOKINES (PSKs), EPIDERMAL PATTERNING FACTORS (EPFs), and RAPID ALKALINIZATION FACTORS (RALFs). Nonsecretory types include SYSTEMINS (SYS), PLANT ELICITOR PEPTIDES (PEPs), *ZEA MAYS* IMMUNE SIGNALING PEPTIDE 1 (ZIP1), and GRIM REAPER (GRI) PEPTIDE. Among these peptide hormones, systemin, PEPs, and RALFs are essential in plant immune responses, particularly in initiating priming defenses against herbivores by triggering DAMPs (Pearce et al. 2001). For example, systemin activates JA biosynthesis through its binding to SYR1 receptor to trigger proteinase inhibitor genes for insect defense (Wang et al. 2018). AtPep1, a peptide hormone in *A.*

thaliana activates defense mechanisms that use ethylene and JA (Hou et al. 2019). During a subsequent insect attack, PEPs activate PEPR receptors, which produce a stronger immune response that functions as an effective priming agent (Zelman and Berkowitz 2024). The peptide hormones play a role in stress priming of plants through enhanced defense gene expression and secondary metabolite production. The chapter provides detailed research on priming processes that strengthen plant resistance against insect herbivores through the management of peptide hormones as defense controllers.

14.2 Mechanism of Plant Priming

Priming, formerly known as sensitization, is a defense process that allows organisms to build sensitivity to biotic or abiotic stimuli in order to respond faster and better in future events. The scientific community now recognizes that this adaptive immunological preparation process happens throughout kingdoms, including plants and insects. The perception of any challenge increases metabolic costs during priming, yet the process remains efficient because defenses are activated only after difficulties are recognized (Conrath et al. 2015).

14.2.1 Types of Priming

14.2.1.1 Chemical Priming in Plants

Chemical priming changes plants' immune systems while increasing their response capacity. β-aminobutyric acid (BABA), a nonprotein amino acid, is a well-studied priming factor that causes plants to respond more quickly and effectively to attacks. The BABA-IR defense mechanism, which operates via the SA route, causes callose deposition and activates the transcription factor activity pathway (Cohen 2002). The significant chemical inducers comprise three substances: azelaic acid (AZA), pipecolic acid (PA), and benzothiadiazole (a functional analogue of SA, which causes SAR). Plants develop systemic protection against future pathogens after localized infections cause the accumulation of endogenous metabolites, including AZA and PA (Conrath et al. 2015).

14.2.1.2 Biological Priming in Plants

Living organisms activate biological priming through beneficial microbes or herbivores. The best-understood case of biological priming involves ISR that starts when beneficial root-associated microbes such as *Pseudomonas fluorescens* and *Trichoderma asperellum* activate this process. The JA, together with ethylene (ET) signaling pathways, drive ISR activation but SAR does not involve pathogenesis-related (PR) protein accumulation. Plants become primed through signals from herbivorous animals that include both insect oviposition and exposure to HIPVs. The signals from actual herbivore damage activate defense mechanisms in both unattacked plant regions and neighboring plants with increased speed and effectiveness (Pangesti et al. 2016).

14.2.1.3 Transgenerational Immune Priming in Insects

Transgenerational immune priming (TGIP) is a unique type of priming seen in insects. In TGIP, parental exposure to pathogens leads to stronger immunity in their offspring. This can happen through paternal epigenetic processes, such as DNA methylation and small RNAs in sperm, or maternal transfer of microbial fragments into developing eggs (Vilcinskas 2021; Tetreau et al. 2019). While paternal TGIP offers more extensive protection, maternal TGIP is well-documented and is highly specific. Both types are known to prepare the offspring for any upcoming infections they have not yet experienced. The highly specific and persistent nature of these effects is supported by several experiments from species such as *Apis mellifera*, *Tribolium castaneum*, and *Galleria mellonella* (Mukherjee et al. 2017; Harwood et al. 2021).

Environmental factors like heat stress can drastically lower the efficacy of TGIP. For example, elevated temperatures have been shown to disrupt immune memory passed on to offspring, emphasizing the ecological context of this phenomenon (Schneider et al. 2024; Lang et al. 2022).

14.2.2 Molecular Basis of Priming

14.2.2.1 Mitogen-Activated Protein Kinases (MAPKs) in Plants

MAPKs are essential signaling proteins in plant immunity. MPK3 and MPK6 are present in an inactive form during priming and become rapidly activated upon stress perception. These kinases promote a quicker and better-coordinated immune response by inducing the transcription of downstream defense genes like PR1 and PAL1 (Conrath et al. 2015). In particular, MPK3 is closely associated with SAR, highlighting its critical function in primed defense signaling (Shoresh et al. 2006).

14.2.2.2 Epigenetic and Transgenerational Mechanisms

Epigenetic modifications provide a molecular framework for both intra- and intergenerational priming. In plants, modifications such as histone acetylation (H3K4me3, H3K9ac) and H4 acetylation help the promoter regions of primary defense genes to remain open, in a transcriptionally permissive state. This allows rapid transcription when a stress signal is detected. Genes like *WRKY29*, *WRKY6*, and *WRKY53* display these modifications in a primed state (Conrath et al. 2015; Jaskiewicz et al. 2011).

Similar epigenetic modifications and the participation of regulatory molecules such as microRNAs (miRNAs) maintain TGIP in insects. These small RNAs have the capacity to alter gene expression posttranscriptionally and are becoming more widely recognized as significant participants in immune memory (Sułek et al. 2021). The cohesive effect of all these changes allows the inheritance of immune traits without altering the DNA sequence.

14.2.3 Signaling Pathways Involved in Priming

14.2.3.1 Salicylic Acid Pathway

The salicylic acid (SA) pathway is one of the main components of SAR in plants. The accumulation of SA and activation of SA-responsive genes are increased by priming agents such as BABA, AZA, and PA. Furthermore, chromatin modifications and transcription factor dynamics are closely associated with SA signaling. SA can lead to the primed state by upregulating WRKY transcription factors, which in turn, govern a group of downstream defense genes (Conrath et al. 2015).

14.2.3.2 Jasmonic Acid and Ethylene Pathways

Jasmonic acid (JA) and ethylene (ET) are the fundamental participants in ISR, predominantly triggered by necrotrophic pathogens and herbivory. While the beneficial rhizobacteria activate the ORA59 branch, the MYC2 branch of JA/ET signaling is activated by feeding by herbivorous insects, thereby enabling the plants to fine-tune their responses based on the type of threat perceived (Pangesti et al. 2016; Jung et al. 2024). Recent studies suggest that ISR can also include aspects of SA signaling, underscoring the interlinked nature of defense pathways. This flexibility allows plants to modify their responses in accordance with the environmental challenges.

14.2.4 Rhizobacteria-Induced Systemic Resistance (ISR)

ISR is representative of one of the most vital examples of priming in plants. When the plant-growth-promoting rhizobacteria (PGPR) colonize the roots, it leads to priming of the systemic tissues to fend off a variety of attackers. This is further achieved via the activation of defense-related secondary metabolites and hormonal changes. ISR, in particular, promotes the synthesis of aliphatic glucosinolates (GLS) and camalexin, compounds which are known to reduce herbivory (Table 14.1) (Pangesti et al. 2016). An intriguing extension of ISR is the concept of "soil legacy effects," where herbivore-induced changes in root-associated microbial communities endure and impact the defense potential of subsequent generations or plants nearby.

14.2.5 Ecological Relevance and Costs of Priming

Priming is energetically more efficient compared to constitutive defense but it still carries ecological costs. The maintenance of such a high level of receptiveness is at the expense of diversion of resources from development or reproduction. However, it gives a survival advantage in environments that are prone to herbivores or are pathogen-rich (Contreras-Garduño et al. 2016).

Table 14.1 Summary of gene expression and chemical changes during ISR.

Treatment condition	Induced pathway	Major defensive compounds affected	Herbivore impact	Reference
Herbivory alone	JA-MYC2 branch	VSP2, Camalexin, Indolic GLS	Moderate feeding suppression	Pangesti et al. (2016)
Rhizobacteria + herbivory	JA/ET-ORA59 branch	PDF1.2, Aliphatic GLS, Camalexin	Strong suppression	
Rhizobacteria alone	JA/ET-ORA59 branch	Camalexin, Aliphatic GLS	Basal priming	

Octadecanoid-Responsive Arabidopsis 59 (ORA59).
Vegetative Storage Protein 2 (VSP2).
Plant Defensin 1.2 (PDF 1.2).
Glucosinolates (GLS).

In insects, TGIP has shown a biphasic pattern of immunity, wherein the immune markers first decline after priming but are reactivated more strongly after a subsequent challenge is encountered. However, environmental stressors such as temperature fluctuations have a negative impact on such mechanisms, necessitating the ecological settings to be in compliance with the normal environmental conditions required for the development of plants (Schneider et al. 2024).

14.2.6 Comparative Themes in Insect and Plant Priming

Despite the phylogenetic distance among plants and insects, they exhibit striking similarities in their priming mechanisms (Table 14.2):

- **Signal readiness:** Pattern recognition receptors (PRRs) in insects, hormone signaling in plants
- **Epigenetic regulation:** Histone modifications, DNA methylation, and miRNAs
- **Systemic/inherited preparedness:** ISR in plants, TGIP in insects
- **Cost–benefit trade-offs:** Energy use compared to survival benefits

Such similarities are reflective of evolutionary convergence. Additionally, utilizing such priming-based approaches like ISR in agriculture can improve crop resilience and reduce the dependency on pesticides (Jung et al. 2024).

14.3 Plant Defense Responses to Insect Herbivory

14.3.1 Constitutive and Induced Defense

Plants initially defend against insects using static or constitutive mechanisms, where they pre-produce and store defensive compounds during normal growth. These include physical barriers such as thorns and waxy cuticles, and chemical compounds like tannins and

Table 14.2 Shared features of immune priming in insects and plants.

Feature	Insect immune priming	Plant-induced systemic resistance (ISR)	References
Primary trigger	Pathogen or parasite exposure	Beneficial rhizobacteria colonization	Tetreau et al. (2019)
Memory mechanism	Epigenetics, microRNA, PRRs	Hormonal reprogramming, signaling cross talk	Vilcinskas (2021)
Response speed	Faster AMP expression	Faster production of defensive compounds	Conrath et al. (2015)
Inheritance potential	Yes (transgenerational)	No (but long-term systemic readiness)	Schneider et al. (2024)
Key regulators	Hemocytes, miRNAs, histone marks	JA, ET, MYC2/ORA59 pathways	Jung et al. (2024)

alkaloids synthesized via secondary metabolism, which deter or kill herbivores upon (Gatehouse 2002; Balakrishnan et al. 2024). In contrast, induced defenses are triggered by herbivore or pathogen attack and involve the activation of protease inhibitors, enhanced production of phenolics, and the release of plant volatiles and signaling molecules like JA. Some defense traits, such as trichomes, alkaloids, and phenolics, can function as both constitutive and induced defenses, depending on whether they are pre-formed or produced in greater amounts in response to stress (War et al. 2012).

14.3.2 Morphological Defense Traits

Structural defense, such as spinescence (spines and thorns), pubescence (trichomes), sclerophylly (hardened leaves), hardening of plant tissues through mineral deposits, and growth of complex, widely spread branches, acts as physical defense against herbivory (Hanley et al. 2007). Trichomes can be straight, spiral, hooked, branched, or unbranched and glandular or nonglandular, which can negatively affect the movement of the insects and ovipositional behavior of insect pest. Flavonoids, terpenoids, and alkaloids act as a chemical defense secreted by the glandular trichomes (Sharma et al. 2009). Black mustard plants increase their trichome density and glucosinolate concentrations after the attack by *Pieris rapae* (L.) (Traw 2002). Longer and denser trichomes reduce the population of nymphs and adults of BPH (Sandhu and Sarao 2021). *Solanum carolinense* and *Solanum atropurpureum* have higher spine densities that act as defensive structures against *Manduca sexta* caterpillars (Kariyat et al. 2017). Bernaola et al. (2021) reported that wild-type rice plants with normal epicuticular wax layers showed reduced insect infestation of rice water weevil, *Lissorhoptrus oryzophilus,* and fall armyworm, *Spodoptera frugiperda,* compared to the mutant rice plants.

14.3.3 Plant Secondary Metabolites

Plant produces secondary metabolites to defend against herbivores by making the plant tissues less palatable. These compounds are either always present in an inactive state (phytoanticipins), which include GLS or formed after an attack by pests or pathogens (phytoalexins), which include isoflavonoids, terpenoids, alkaloids, etc. (Howe and Jander 2008). Secondary metabolites like C-glycosyl flavone maysin [2″-*O*-a-L-rhamnosyl-6-*C*-(6-deoxy-*xylo*-hexos-4-ulosyl) luteolin] and the phenylpropanoid product, chlorogenic acid have been identified as key factors in the resistance of maize with host plant resistance (HPR) to the corn earworm, *Helicoverpa zea* (Manachini et al. 2007). Plants produce and accumulate phenols throughout their normal growth, making them consistently present in the plant structures (War et al. 2011). Phenols contribute to plant defense as a physical barrier integrated into cell walls with lignin and suberin, and also act as chemical defenses by deterring herbivores or directly harming herbivores through toxicity (Walling 2000). Phenolic compounds like SA act as antifeedants, but their main role is as phytohormones that help *Salix* leaves activate defense responses against *Operophtera brumata* (Simmonds 2003). Flavonoids and isoflavonoids are toxic to insect cells, bind to various enzymes, and protect plants by disturbing insect growth and behavior, and by neutralizing harmful reactive oxygen species through metal binding (Harborne and

Grayer 2017; Gautam et al. 2023). Bentivenha et al. (2018) reported that the flavonoids rutin and genistein in soybean genotypes PI 274453, PI 274454, and IAC 100 contributed to resistance against *Piezodorus guildinii*. Tannins are polyphenolic in nature, known for their astringent taste. They reduce the nutritive value of plants consumed by herbivores by binding to the proteins and digestive enzymes, making them harder to digest and also bind to the metal ions, making these essential nutrients less available to the herbivores (Barbehenn and Constabel 2011). Condensed tannins, also called proanthocyanidins, are oligomeric or polymeric flavonoids that play a defensive role in plants (War et al. 2012). Higher condensed tannin levels in aspen leaves reduced aphid reproduction and caused feeding difficulties (Rodríguez et al. 2022).

14.3.4 Plant Defensive Proteins and Enzymes

Plant proteins play a crucial defensive role by disrupting insect digestion and physiology when ingested. Some proteins remain stable in the insect gut and even cross into the hemolymph, influencing internal processes. The use of protease inhibitors enhances the stability and toxicity of these proteins, strengthening plant resistance (Napoleão et al. 2019). Storage organs such as seeds and tubers contain 1–10% of defensive proteinase inhibitors as part of their total protein content. These inhibitors inhibit various digestive enzymes in insect herbivores, contributing to the plant's defense (Dunse et al. 2010). Plants such as *Sorghum bicolor*, tomato, *Gossypium hirsutum*, *Solanum nigrum*, *Nicotiana attenuata*, and transgenic lines of *Arabidopsis*, oilseed rape, and tobacco possess proteinase inhibitors that serve as a plant defense against insect herbivores (War et al. 2012). Hartl et al. (2010) reported that *S. nigrum* produces serine protease inhibitors that serve as a plant defense against *M. sexta*. Enzymes like peroxidases (PODs), polyphenol oxidases (PPOs), and lipoxygenases (LOXs) interfere with insect nutrients by oxidizing phenols into reactive electrophilic compounds like o-quinones, which disrupt digestion (Gill et al. 2010; Ajaharuddin et al. 2024). Monomeric hemoproteins known as PODs play a key role in regulating the plant defense processes such as lignification, wound healing, and somatic embryogenesis, with their expression influenced by tissue, development, and environmental factors (Gulsen et al. 2010). According to Stout et al. (2009), PODs found in rice act as a defensive mechanism against *S. frugiperda*. PPOs regulate insect feeding and growth by producing quinones, which reduce plant nutritional quality and create oxidative stress. These quinones crosslink proteins, thereby limiting the availability of essential nutrients for herbivores (Taranto et al. 2017). Plant species such as tomato and buffalo grass produce PPOs that serve as a plant defense against herbivorous insects, including *M. sexta*, *Blissus occiduus*, *S. frugiperda*, and *Helicoverpa armigera* (War et al. 2012). The octadecanoid pathway in plants is activated by LOXs, antioxidative enzymes that play a key role in defense against various stresses (Avanci et al. 2010). Zhao et al. (2009) reported that wheat produces LOXs that serve as a plant defense against the herbivorous insect *Sitobion avenae*. Lectins are glycoproteins found in nature that bind to carbohydrates, play a defensive role against various pests. *Allium sativum* leaf lectin expressed in chickpea has been shown to provide defense against *Aphis craccivora* (Chakraborti et al. 2009).

14.4 Herbivore-Induced Plant Volatiles as Defense Signals: Role of Insect Oral Secretions

Plants emit a unique blend of plant volatiles after the attack of herbivores that act as chemical signals to deter pests or attract predators, known as herbivore-induced plant volatiles (HIPVs). The emission of volatiles depends on the plant and herbivore species, as well as plant health and developmental stage. HIPVs activate specific defense pathways in plants, facilitating interactions with herbivores, natural enemies, and neighboring plants (Penaflor and Bento 2013). Kutty and Mishra (2023) categorized HIPVs as green leaf volatiles (GLVs), terpenes, methyl salicylate, and other VOCs. Plant volatiles such as methyl salicylate induce defensive signals to attract beneficial predators like the green lacewing, *Chrysopa nigricornis* (James 2003). Methyl salicylate, present in leaf and floral scents, is used in sticky traps for natural enemy monitoring. Soybean plants release methyl salicylate after aphid infestation, attracting the predator *Coccinella septempunctata* (Zhu and Park 2005). GLVs are stress-induced volatiles derived from the hydroperoxide lyase (HPL) branch of the oxylipin pathway and play a key role in plant defense. They attract natural enemies of herbivores and include C6-aldehydes and related compounds (Matsui and Engelberth 2022). These defense volatiles come from several pathways, such as lipoxygenase, shikimic acid, and terpenoid pathways (Gill et al. 2010). Ghosh et al. (2023) reported that shoot and fruit borers avoid oviposition on eggplants that release geraniol, a defensive monoterpene volatile. According to Moreira et al. (2018bb), female *Baccharis salicifolia* plants showed defense responses to VOCs from neighboring plants of the same sex, with distinct VOCs being released when attacked by *Uroleucon macolai* or *Aphis gossypii*, and neighboring plants only responding after damage by the same aphid species (Moreira et al. 2018).

In intercropping, companion plants act as "sentinels," releasing VOCs that alert neighboring cash crops to biotic threats. This VOC-mediated signaling triggers a process known as priming, enhancing the plants' defensive readiness (Brilli et al. 2019). Volatile priming in plants, such as with GLVs, helps to enhance defense potential by boosting JA and sesquiterpene production upon herbivore cues, such as beet armyworm, *Spodoptera exigua* caterpillar regurgitant. Engelberth et al. (2004) and Muroi et al. (2011) reported that (E)-β-ocimene–emitting transgenic tobacco plants induce the resistance of neighboring bean and maize against the herbivores, *Tetranychus urticae* and *Mythimna separata*. According to Cascone et al. (2015), tomato defense responses are enhanced by the volatiles released by the tobacco plant. It worked through direct suppression of the aphid, *Macrosiphum euphorbiae* and indirect parasitoid, *Aphidius ervi* attraction.

Chewing herbivores, such as caterpillars and borers, damage plant tissues, triggering changes in the plant's transcriptomes, proteomes, and metabolomes through chemical signals like oral secretions, regurgitant, saliva, ventral eversible gland secretions, waste products, and oviposition fluids. These signals, known as effectors, activate defense responses, including the emission of volatiles to repel herbivores and attract natural predators (Kant et al. 2009). Tomato plants exposed to glandular secretions from *S. exigua* caterpillars show increased release of defense volatiles such as GLVs, monoterpenes, sesquiterpenes, and methyl salicylate (Zebelo et al. 2014). Fatty acid-amino acid conjugates (FACs), such as volicitin and N-linolenoyl-glu, are found in the oral secretions of insects

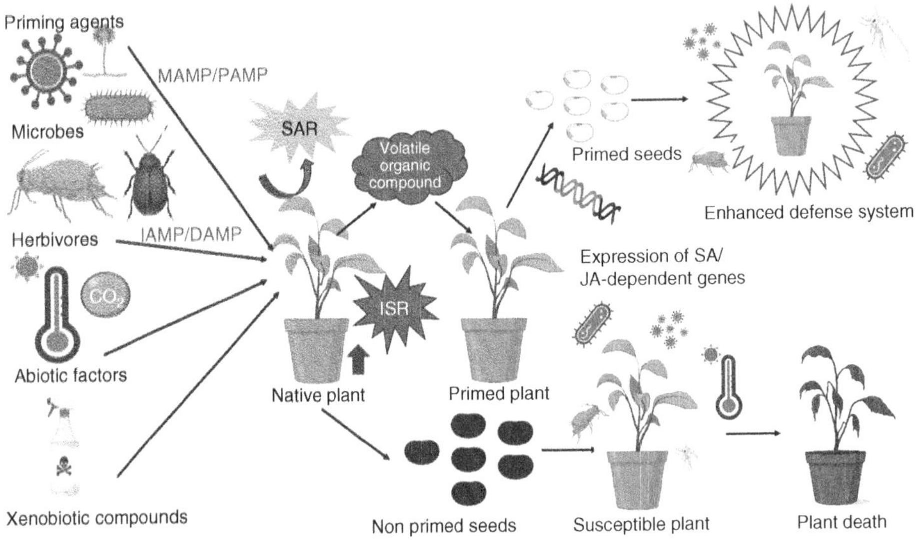

Figure 14.1 Effect of priming on plant defense against biotic and abiotic stress.

like the beet armyworm and tobacco hornworm. When applied to plants like *Zea mays* and tobacco, these FACs trigger the emission of volatile compounds that attract natural predators of the herbivores. FACs induce plant defense against herbivore populations (Alborn et al. 1997; Halitschke et al. 2001) (Figure 14.1).

14.5 Phytohormones as Induced Plant Defenses

Plants activate defense pathways through JA, SA, and ethylene in response to herbivore attacks. Herbivores and pathogens activate the JA and SA signaling pathways, which mediate plant defense (Zhang et al. 2018). JA is triggered by necrotrophic pathogens and chewing herbivores, whereas SA is activated by biotrophic pathogens and phloem-feeding insects (Rowen and Kaplan 2016). JA signaling pathways are triggered by the VOCs (Z)-3-hexenal, (Z)-3-hexen-1-ol, and (Z)-3-hexenyl acetate, which primes plants to defend against herbivores and pathogens (Ameye et al. 2015; Engelberth and Engelberth 2019). Phytohormones serve as both intra- and interplant signaling during insect-induced stress by regulating growth, stress responses, and defense gene expression (Lazebnik et al. 2014). JA serves as both a direct and indirect defense hormone produced from linolenic acid through the octadecanoid pathway against herbivores (Schaller 2001). Jasmonates act as a part of defense response promoting the formation of protein inhibitors (PIs), VOCs, alkaloids, trichomes, and the secretion of extrafloral nectar (EFN) in plants (War et al. 2012). JA plays a key role in defense against biotic and abiotic stresses through signal transduction by influencing the expression and activity of calcium-dependent protein kinases (CDPKs) in potato plants (Ulloa et al. 2002). (Z)-3-hexen-1-ol compound influences ethylene, auxin, and JA signaling in tea, initiating both direct and indirect defense mechanisms against *Ectropis obliqua* (Xin et al. 2016). These defenses were characterized by enhanced PPO activity and volatile emissions. The accumulation of SA, a benzoic acid derivative, triggers the NPR1 gene through redox pathways, modulates plant defense via transcription factors, and offers stronger resistance against piercing and sucking insects than chewing pests (Zhao et al. 2009; Roychowdhury et al. 2024). Peng et al. (2004) reported that the SA pathway induces ROS, contributing to enhanced pest resistance in crops like tomato against *H. armigera*. Ethylene interacts with JA in either a supportive or opposing manner to regulate plant defense. In tomatoes, their combined action promotes the expression of proteinase inhibitors, enhancing resistance to herbivores (Kachroo and Kachroo 2007). According to Tscharntke et al. (2001), infestation of *Alnus glutinosa* by *Alnus alni* triggered the release of ethylene along with several volatiles such as monoterpenes, sesquiterpenes, and homoterpenes. Ethylene response factor 1 (ERF1), a key transcription factor, enhances defense responses in *Arabidopsis* by integrating signals from both JA and ET pathways (Lorenzo et al. 2003).

14.6 Peptide Hormones in Plant Defense

Peptide hormones are small, secreted signaling molecules derived from nuclear genes that undergo posttranslational processing to become biologically active. They are secreted via conventional (endoplasmic reticulum–Golgi) or unconventional pathways (e.g. vesicles, vacuoles, exocyst-positive organelles) and bind to receptors on nearby cells, activating signal transduction cascades that regulate gene expression and plant responses (Del Corpo et al. 2024). Different families of peptide hormones in plants are defined by their molecular sequences, modes of secretion, functional roles, and physiological origins (Zhang et al. 2025). The discovery that proteinase inhibitors accumulate in both damaged and undamaged leaves of Solanaceous plants after Colorado potato beetle larvae infestations marked the beginning of studies on systemic plant defense against herbivores (Green and Ryan 1972). Plants activate defense genes both locally at the wound site and systemically in undamaged leaves in response to insect attack. Systemin and jasmonates are key signals that mediate these systemic defense responses (Green and Ryan 1972; Ryan 2000). Prosystemin undergoes cleavage to produce systemin, a peptide consisting of 18 amino acids involved in immune modulation (Pearce et al. 1991). Wang et al. (2018) reported that systemin receptor SYR1 enhances tomato resistance to insect herbivores like *S. littoralis* by mediating systemin perception. Larvae feeding on SYR1-functional plants showed reduced weight gain, highlighting SYR1's role in defense. CLV3/embryo surrounding region peptides contribute significantly to stem cell regulation in multiple meristematic areas, including those responsible for shoot, root, and vascular development. (Selby and Jones 2023). CLE peptides, when synthetically applied or overexpressed, disrupt root development by terminating the root apical meristem, leading to stunted primary roots and abnormal lateral roots, hence being recognized as root-active CLE peptides (Strabala et al. 2006; Betsuyaku et al. 2011). CEP peptides found in various plant species are vital for processes like root development, nitrogen uptake, and stress response. Their expression is influenced by environmental stresses such as drought, salinity, and pathogens, indicating their role in stress mediation. The overexpression or application of CEP peptides inhibits root elongation, and

CEP3 expression is notably elevated under salt stress, with *cep3* mutants exhibiting longer primary roots under various stress conditions (Delay et al. 2013; Roberts et al. 2013, 2016).

RALF peptides have a major impact on cellular alkalization and root growth inhibition, making them important for plant development and adaptation to different environmental conditions (Du et al. 2016). According to Zhu et al. (2020), the ralf1 knockout mutant displays larger root cells, increased number of lateral roots, and longer roots. In *A. thaliana*, receptors like RGF1 INSENSITIVE (RGI) or RGF RECEPTOR (RGFR), which belong to the leucine-rich repeat receptor-like kinase family, sense the root meristem growth factor (RGF), golven (GLV), and CLLEL peptides. These genes exhibit variable expression, particularly in the leaf tissues and root meristem (Fernandez et al. 2013). In *Medicago truncatula*, MtRGF3 knockdown improved nodule formation, whereas its overexpression or peptide application decreased nodules (Li et al. 2020). Overexpression of inflorescence deficient in abscission (IDA) and IDA-like (IDL) causes premature abscission in a variety of tissues, while *ida* mutants delay the flower drop and IDA peptides control the shedding of plant organs (Butenko et al. 2003; Wang et al. 2023). Nakayama et al. (2017) and Zhang et al. (2023) indicate that the formation of the Casparian strip, pollen wall, and embryonic cuticle is influenced by the Casparian strip integrity factor (CIF) peptides with mutations in rice and *Arabidopsis* damage the endodermal barriers in the root. EPF/stomagen peptides like EPFL4, EPFL5, and EPFL6 influence a number of biological processes, including stamen filament elongation for self-pollination in both cold and normal conditions, while EPFL2 and EPFL9 coordinate ovule initiation, influencing seed number and fruit size (Kawamoto et al. 2020; Negoro et al. 2023; He et al. 2023). The sulfated pentapeptides called phytosulfokines (PSKs) regulate cell division via the PSK RECEPTOR-1 (PSKR1) and PSKR2 receptors (Matsubayashi et al. 2002). They are essential in controlling tomato abscission under drought stress (Reichardt et al. 2020). By altering the phosphorylation of tomato glutamine synthetase GS2, PSK peptide enhances plant defense and development (Ding et al. 2023). Plant elicitor peptides (PEPs) are activated by wounding, JA, and pathogen attack (Zelman and Berkowitz 2023). Wang et al. (2022) reported that *pep7* mutants in *A. thaliana* showed a decline in lateral root primordia across developmental stages, while external application of PEP7 stimulated their development at those stages. According to Hou et al. (2014), exogenous PAMPs such as flagellin and chitin strongly induce the expression of PAMP-induced secreted peptides (PIP). These peptides, encoded by eleven PIP/PIP-LIKE (PIPL) genes in *A. thaliana*, are vital for plant growth and tolerance to both biotic and abiotic stresses. Hussain et al. (2021) observed that transgenic plants with increased expression of *PIP2* and *PIP3* have shorter primary roots, whereas *pip2* mutants do not show notable root abnormalities, and *pip3* or *pip2 pip3* double mutants have stunted primary roots.

14.7 Top of Form

14.7.1 Bottom of Form

14.7.1.1 Enhancing Crop Resistance via Priming

Arbuscular mycorrhizal fungi (AMF) can be effectively used as a priming agent for the plant to induce immune responses against herbivores. For example, priming the tomato plant with *Glomus mosseae* upregulated defense-related genes, such as lipoxygenase D, allene oxide cyclase, and serine protease inhibitors, which resulted in reduced larval weight of *H. armigera* by 62.3% and 78.2% in wild-type and JA-overexpressing tomato genotypes, respectively (Song et al. 2013). The biosynthesis of JA is initiated by lipoxygenase-driven oxygenation of α-linolenic acid at position 13 (Christensen et al. 2013); however, allene oxide cyclase is involved in the conversion of fatty acid derivatives (Schaller et al. 2004). Xu et al. (2022) also reported that rice plants treated with microbial metabolites like decoyinine and angustmycin A enhanced the resistance against *Nilaparvata lugens* by decreasing fecundity, longevity, oviposition, honeydew secretions, and insect body weight. JAs can also be used as priming agents in enhancing defense activity against herbivores in *Arabidopsis* and tomato plants; however, the transgenerational effect was not observed in mutants. Similarly, Liu et al. (2017) revealed that ABA treatment induced resistance to *N. lugens* in rice by decreasing fecundity, while gibberellin application also resulted in reduced reproduction in *N. lugens* (Wang et al. 2021a, b).

Apart from the direct priming in plants, seed priming is an best alternative for inducing plant defense against herbivores. For instance, priming the rice seeds of the susceptible TN1 rice variety with decoyinine significantly enhanced resistance against *N. lugens* (Ma et al. 2022). Additionally, this priming resulted in suppression of ovary development, and fecundity might be due to reduced vitellogenin levels. Demographic parameters like intrinsic and net reproductive rates, as well as finite rate of increase and population size, were also altered by this priming. Similarly, the feeding acquisition period was also altered due to higher frequency and duration of electrical penetration graph (EPG) waveforms (np, N2, and N3), suggesting greater difficulty in intracellular stylet penetration (Ma et al. 2022). Similarly, tomato seeds primed with JA enhanced resistance against several insect pests like *M. sexta*, *Myzus persicae*, and *T. urticae* (Worrall et al. 2012), against the pine weevil *Hylobius abietis* in spruce (Berglund et al. 2016). In line with this, Wu et al. (2021) observed that BPH females showed reduced preference for rice plants grown from triflumezopyrim-treated seeds, along with decreased honeydew excretion. Similarly, wheat seeds primed with calcium chloride resulted in increased calcium concentration in leaves and callose deposition, which further reinforced the cell wall, ultimately hampered feeding by the greenbug aphid *Schizaphis graminum* (Wang et al. 2021a, b)

Xenobiotic compounds can also be used in priming the plants to enhance their resistance against insect pests. Shah et al. (2022) also reported that rice variety Wuyujing 3 pretreated with decoyinine resulted in the reduction of fecundity of the small brown planthopper (*Laodelphax striatellus*) by altering the defense-related enzymes in rice plants. Additionally, ingestion of phloem content as well as female fecundity was significantly hampered in *N. lugens* when the rice plants were treated with a xenobiotic compound, triflumezopyrim, conferring resistance against rice (Wu et al. 2021; Xi et al. 2022). Similarly, exogenous application of flavonols kaempferol and quercetin to TN1 rice seedlings improved survival of rice seedlings even when attacked by white-backed planthopper (*Sogatella furcifera*) (Jan et al. 2020). Priming with the xenobiotic compound synthetic elicitor 4-fluorophenoxyacetic acid impaired the phloem ingestion of *S. furcifera* by prolonging the nonprobing and stylet pathway activities (Wang et al. 2020).

Silica plays an important role in plant resistance against insect pests by developing a physical barrier in plant tissue or altering the feeding habits, like chewing, penetration, and digestion (Kvedaras and Keeping 2007; Reynolds et al. 2009). In addition, Si is also involved in the regulation of defense-related genes, plant hormones, and volatile compounds signaling by several biochemical processes (Ye et al. 2013; Nazaralian et al. 2017). Based on these facts, Si can be used as a potential priming agent in developing resistance against herbivores. Lin et al. (2022) revealed that Si treatment in rice plants increased the accumulation of SA as well as upregulated the activities of defense enzymes like POD, superoxide dismutase (SOD), protease, PPO, and phenylalanine ammonia-lyase (PAL) which ultimately enhanced resistance against *N. lugens* by affecting the several life-table parameters like oviposition behavior, hatch rates, and honeydew secretion. Similarly, Yang et al. (2018) demonstrated that silicon supplementation in rice enhanced *N. lugens* resistance through reinforcement of physical barriers, disruption of phloem sap ingestion, and extension of the nonprobing and stylet pathway phases.

14.8 Reducing Chemical Inputs in Pest Management

Chemical control is often considered a quick method in pest control in modern agriculture. Similarly, seed treatment with chemical protectants has been used for several years as a quick method for pest management to protect emerging seedlings from soilborne pathogens as well as seedborne insect pests (Munkvold et al. 2014). However, treatment with chemical protectants like insecticides and fungicides offers protection for a limited duration with severe ecological and environmental drawbacks like insecticide resistance (Vojvodić and Bažok 2021), pest resurgence (Wu et al. 2020), and secondary pest outbreaks (Malagnoux et al. 2015). Moreover, residual chemicals infiltrate the soil and water bodies and can also impact other beneficial arthropods like pollinators, predators, nematodes, and entomopathogenic microbes, ultimately threatening the natural equilibrium as well as natural pest control processes (Okon and Antia 2023). As an alternative to chemical protectants, microbes like bacteria and fungi can be used as bioprotectants for inducing plant defense in the progenies. By the process of myco-seed priming, seeds are inoculated with entomopathogenic fungi such as *Trichoderma* spp. or *Metarhizium anisopliae* have emerged as a new technology in pest control called biopriming (Anjum et al. 2020; Chowdhury et al. 2024). This method also has several other benefits, like enhancement of plant morphogenesis, systemic defense, growth, and yield (Ahsan et al. 2024; Batool et al. 2022; Di Lelio et al. 2021; Zhang et al. 2021). These fungi can also form a microbial shield by colonizing the rhizosphere and releasing secondary metabolites that will deter the soil-inhabiting pathogens as well as herbivores (Hu and Bidochka 2021). For example, maize seeds bioprimed with fungal agents like *Beauveria bassiana* and *T. asperellum* significantly hampered the survival of *Ostrinia furnacalis* larvae. Apart from this, biopriming with these fungal consortia enhanced enzymatic activities, including SOD, POD, PPO, and proline (Batool et al. 2022). Similarly, *Phaseolus vulgaris* seeds treated with *Metarhizium robertsii* and *B. bassiana* decreased the spider mite population of *T. urticae*, without affecting the predation capability of *Phytoseiulus persimilis*, a key predator of spider mites. Beyond fungi, bacteria can also be used as a potential biopriming agent. According to the findings of Disi et al. (2018), the feeding behavior and host selection preference of *Diabrotica virgifera virgifera* were altered when maize seeds were primed with *Bacillus pumilus*. Similarly, rice seed priming with *Serratia marcescens and Bacillus velezensis* increased the seed vigor and resistance against

N. lugens (Niu et al. 2022; Harun-Or-Rashid et al. 2018). Overall, myco-priming can be used as an alternative to chemical pest control methods with increased plant morphogenesis, growth, and yield as sustainable agricultural practices.

14.9 Existing Gap in Knowledge and Future Directions

Despite significant advancements in seed priming techniques and plant defense systems, the exact mechanism and function of peptide hormones in enhancing plant resistance against herbivores are still unclear. Although priming with biological agents like beneficial microbes and xenobiotic compounds can enhance the systemic resistance of plants against herbivores, the downstream signaling pathways, especially the involvement of peptide hormones like systemin, PSK, and RALFs, are still poorly understood. The current knowledge has mostly concentrated on the hormonal cross talk involving SA, JA, and ethylene, however, there is a significant void regarding the pathways through which the peptide hormones play an intricate role in insect lethality. Furthermore, the geographical and temporal dynamics of peptide hormones in the plant after priming effects are still unclear. Future research should focus on omics-based approaches, like transcriptomic, metabolomic, and proteomic techniques, to find out the molecular mechanisms responsible for the activation of peptide hormone-mediated defense activities by priming effects. It may also be feasible to identify these relationships and develop more precise and durable pest control strategies by employing endogenous signaling peptides as bioindicators or targets for crop development.

References

Ahsan, S.M., Injamum-Ul-Hoque, M., Das, A.K. et al. (2024). Plant–entomopathogenic fungi interaction: recent progress and future prospects on endophytism-mediated growth promotion and biocontrol. *Plants* 13 (10): 1420.

Ajaharuddin, S.M., Das, K.K., Kar, P. et al. (2024). Insect-plant-pathogens: toxicity, dependence, and defense dynamics. In: *Biocontrol Agents for Improved Agriculture*, 385–411. Elsevier.

Alborn, H.T., Turlings, T.C.J., Jones, T. et al. (1997). An elicitor of plant volatiles from beet armyworm oral secretion. *Science* 276 (5314): 945–949.

Ameye, M., Audenaert, K., De Zutter, N. et al. (2015). Priming of wheat with the green leaf volatile Z-3-hexenyl acetate enhances defense against *Fusarium graminearum* but boosts deoxynivalenol production. *Plant Physiology* 167 (4): 1671–1684.

Anjum, Z.A., Hayat, S., Ghazanfar, M.U. et al. (2020). Does seed priming with Trichoderma isolates have any impact on germination and seedling vigor of wheat. *International Journal of Botany Studies* 5 (2): 65–68.

Avanci, N.C., Luche, D.D., Goldman, G.H., and Goldman, M.H.D.S. (2010). Jasmonates are phytohormones with multiple functions, including plant defense and reproduction. *Genetics and Molecular Research* 9 (1): 484–505.

Balakrishnan, D., Bateman, N., and Kariyat, R.R. (2024). Rice physical defenses and their role against insect herbivores. *Planta* 259 (5): 110.

Barbehenn, R.V. and Constabel, C.P. (2011). Tannins in plant–herbivore interactions. *Phytochemistry* 72 (13): 1551–1565.

Batool, R., Umer, M.J., Wang, Y. et al. (2022). Myco-synergism boosts herbivory-induced maize defense by triggering antioxidants and phytohormone signaling. *Frontiers in Plant Science* 13: 790504.

Bentivenha, J.P., Canassa, V.F., Baldin, E.L. et al. (2018). Role of the rutin and genistein flavonoids in soybean resistance to *Piezodorus guildinii* (Hemiptera: Pentatomidae). *Arthropod-Plant Interactions* 12: 311–320.

Berglund, T., Lindström, A., Aghelpasand, H. et al. (2016). Protection of spruce seedlings against pine weevil attacks by treatment of seeds or seedlings with nicotinamide, nicotinic acid and jasmonic acid. *Forestry* 89 (2): 127–135.

Bernaola, L., Butterfield, T.S., Tai, T.H., and Stout, M.J. (2021). Epicuticular wax rice mutants show reduced resistance to rice water weevil (Coleoptera: Curculionidae) and fall armyworm (Lepidoptera: Noctuidae). *Environmental Entomology* 50 (4): 948–957.

Betsuyaku, S., Sawa, S., and Yamada, M. (2011). The function of the CLE peptides in plant development and plant-microbe interactions. *The Arabidopsis Book/American Society of Plant Biologists* 9: e0149.

Brilli, F., Loreto, F., and Baccelli, I. (2019). Exploiting plant volatile organic compounds (VOCs) in agriculture to improve sustainable defense strategies and productivity of crops. *Frontiers in Plant Science* 10: 264.

Butenko, M.A., Patterson, S.E., Grini, P.E. et al. (2003). Inflorescence deficient in abscission controls floral organ abscission in Arabidopsis and identifies a novel family of putative ligands in plants. *The Plant Cell* 15 (10): 2296–2307.

Cascone, P., Iodice, L., Maffei, M.E. et al. (2015). Tobacco overexpressing β-ocimene induces direct and indirect responses against aphids in receiver tomato plants. *Journal of Plant Physiology* 173: 28–32.

Chakraborti, D., Sarkar, A., Mondal, H.A., and Das, S. (2009). Tissue specific expression of potent insecticidal, *Allium sativum* leaf agglutinin (ASAL) in important pulse crop, chickpea (Cicer arietinum L.) to resist the phloem feeding *Aphis craccivora*. *Transgenic Research* 18: 529–544.

Chowdhury, M.Z.H., Mim, M.F., Siddique, S.S. et al. (2024). Seed priming with *Metarhizium anisopliae* (MetA1) improves physiology, growth and yield of wheat. *Heliyon* 10 (17).

Christensen, S.A., Nemchenko, A., Borrego, E. et al. (2013). The maize lipoxygenase, Zm LOX 10, mediates green leaf volatile, jasmonate and herbivore-induced plant volatile production for defense against insect attack. *The Plant Journal* 74 (1): 59–73.

Cohen, Y.R. (2002). β-aminobutyric acid-induced resistance against plant pathogens. *Plant Disease* 86 (5): 448–457.

Conrath, U. (2011). Molecular aspects of defence priming. *Trends in Plant Science* 16 (10): 524–531.

Conrath, U., Beckers, G.J., Langenbach, C.J., and Jaskiewicz, M.R. (2015). Priming for enhanced defense. *Annual Review of Phytopathology* 53 (1): 97–119.

Contreras-Garduño, J., Lanz-Mendoza, H., Franco, B. et al. (2016). Insect immune priming: ecology and experimental evidences. *Ecological Entomology* 41 (4): 351–366.

Del Corpo, D., Coculo, D., Greco, M. et al. (2024). Pull the fuzes: processing protein precursors to generate apoplastic danger signals for triggering plant immunity. *Plant Communications* 5 (8): 100931.

Delay, C., Imin, N., and Djordjevic, M.A. (2013). CEP genes regulate root and shoot development in response to environmental cues and are specific to seed plants. *Journal of Experimental Botany* 64 (17): 5383–5394.

Di Lelio, I., Coppola, M., Comite, E. et al. (2021). Temperature differentially influences the capacity of Trichoderma species to induce plant defense responses in tomato against insect pests. *Frontiers in Plant Science* 12: 678830.

Ding, S., Lv, J., Hu, Z. et al. (2023). Phytosulfokine peptide optimizes plant growth and defense via glutamine synthetase GS2 phosphorylation in tomato. *The EMBO Journal* 42 (6): e111858.

Disi, J.O., Kloepper, J.W., and Fadamiro, H.Y. (2018). Seed treatment of maize with *Bacillus pumilus* strain INR-7 affects host location and feeding by Western corn rootworm, *Diabrotica virgifera*. *Journal of Pest Science* 91: 515–522.

Du, C., Li, X., Chen, J. et al. (2016). Receptor kinase complex transmits RALF peptide signal to inhibit root growth in Arabidopsis. *Proceedings of the National Academy of Sciences* 113 (51): E8326–E8334.

Dunse, K.M., Stevens, J.A., Lay, F.T. et al. (2010). Coexpression of potato type I and II proteinase inhibitors gives cotton plants protection against insect damage in the field. *Proceedings of the National Academy of Sciences* 107 (34): 15011–15015.

Engelberth, J., Alborn, H.T., Schmelz, E.A., and Tumlinson, J.H. (2004). Airborne signals prime plants against insect herbivore attack. *Proceedings of the National Academy of Sciences* 101 (6): 1781–1785.

Engelberth, J. and Engelberth, M. (2019). The costs of green leaf volatile-induced defense priming: temporal diversity in growth responses to mechanical wounding and insect herbivory. *Plants* 8 (1): 23.

Fernandez, A., Drozdzecki, A., Hoogewijs, K. et al. (2013). Transcriptional and functional classification of the GOLVEN/ROOT GROWTH FACTOR/CLE-like signaling peptides reveals their role in lateral root and hair formation. *Plant Physiology* 161 (2): 954–970.

Frost, C.J., Mescher, M.C., Carlson, J.E., and De Moraes, C.M. (2008). Plant defense priming against herbivores: getting ready for a different battle. *Plant Physiology* 146 (3): 818–824.

Gatehouse, J.A. (2002). Plant resistance towards insect herbivores: a dynamic interaction. *New Phytologist* 156 (2): 145–169.

Gautam, H., Sharma, A., and Trivedi, P.K. (2023). The role of flavonols in insect resistance and stress response. *Current Opinion in Plant Biology* 73: 102353.

Ghosh, R., Metze, D., Sant, S. et al. (2023). Chemical ecology of Himalayan eggplant variety's antixenosis: identification of geraniol as an oviposition deterrent against the eggplant shoot and fruit borer. *New Phytologist* 240 (3): 1259–1274.

Gill, R., Gupta, A., Taggar, G., and Taggar, M. (2010). Role of oxidative enzymes in plant defenses against insect herbivory. *Acta Phytopathologica et Entomologica Hungarica* 45 (2): 277–290.

Green, T.R. and Ryan, C.A. (1972). Wound-induced proteinase inhibitor in plant leaves: a possible defense mechanism against insects. *Science* 175 (4023): 776–777.

Gulsen, O., Eickhoff, T., Heng-Moss, T. et al. (2010). Characterization of peroxidase changes in resistant and susceptible warm-season turfgrasses challenged by *Blissus occiduus*. *Arthropod-Plant Interactions* 4: 45–55.

Halitschke, R., Schittko, U., Pohnert, G. et al. (2001). Molecular interactions between the specialist herbivore *Manduca sexta* (Lepidoptera, Sphingidae) and its natural host *Nicotiana attenuata*. III. Fatty acid-amino acid conjugates in herbivore oral secretions are necessary and sufficient for herbivore-specific plant responses. *Plant Physiology* 125 (2): 711–717.

Hanley, M.E., Lamont, B.B., Fairbanks, M.M., and Rafferty, C.M. (2007). Plant structural traits and their role in anti-herbivore defence. *Perspectives in Plant Ecology, Evolution and Systematics* 8 (4): 157–178.

Harborne, J.B. and Grayer, R.J. (2017). Flavonoids and insects. In: *The Flavonoids Advances in Research Since 1986* (ed. J.B. Harborne), 589–618. Routledge.

Hartl, M., Giri, A.P., Kaur, H., and Baldwin, I.T. (2010). Serine protease inhibitors specifically defend *Solanum nigrum* against generalist herbivores but do not influence plant growth and development. *The Plant Cell* 22 (12): 4158–4175.

Harun-Or-Rashid, M., Kim, H.J., Yeom, S.I. et al. (2018). Bacillus velezensis YC7010 enhances plant defenses against brown planthopper through transcriptomic and metabolic changes in rice. *Frontiers in Plant Science* 9: 1904.

Harwood, G., Salmela, H., Freitak, D., and Amdam, G. (2021). Social immunity in honey bees: royal jelly as a vehicle in transferring bacterial pathogen fragments between nestmates. *Journal of Experimental Biology* 224 (7): jeb231076.

He, Y., He, X., Wang, X. et al. (2023). An EPFL peptide signaling pathway promotes stamen elongation via enhancing filament cell proliferation to ensure successful self-pollination in *Arabidopsis thaliana*. *New Phytologist* 238 (3): 1045–1058.

Hilker, M. and Fatouros, N.E. (2015). Plant responses to insect egg deposition. *Annual Review of Entomology* 60 (1): 493–515.

Hilker, M. and Meiners, T. (2010). How do plants "notice" attack by herbivorous arthropods? *Biological Reviews* 85 (2): 267–280.

Hilker, M. and Schmülling, T. (2019). Stress priming, memory, and signalling in plants. *Plant, Cell & Environment* 42 (3): 753–761.

Hou, S., Liu, Z., Shen, H., and Wu, D. (2019). Damage-associated molecular pattern-triggered immunity in plants. *Frontiers in Plant Science* 10: 646.

Hou, S., Wang, X., Chen, D. et al. (2014). The secreted peptide PIP1 amplifies immunity through receptor-like kinase 7. *PLoS Pathogens* 10 (9): e1004331.

Howe, G.A. and Jander, G. (2008). Plant immunity to insect herbivores. *Annual Review of Plant Biology* 59 (1): 41–66.

Hu, S. and Bidochka, M.J. (2021). Root colonization by endophytic insect-pathogenic fungi. *Journal of Applied Microbiology* 130 (2): 570–581.

Hussain, S., Wang, W., Ahmed, S. et al. (2021). PIP2, an auxin induced plant peptide hormone regulates root and hypocotyl elongation in Arabidopsis. *Frontiers in Plant Science* 12: 646736.

James, D.G. (2003). Field evaluation of herbivore-induced plant volatiles as attractants for beneficial insects: methyl salicylate and the green lacewing, *Chrysopa nigricornis*. *Journal of Chemical Ecology* 29: 1601–1609.

Jan, R., Khan, M.A., Asaf, S. et al. (2020). Overexpression of OsF 3 H modulates WBPH stress by alteration of phenylpropanoid pathway at a transcriptomic and metabolomic level in *Oryza sativa*. *Scientific Reports* 10 (1): 14685.

Jaskiewicz, M., Conrath, U., and Peterhänsel, C. (2011). Chromatin modification acts as a memory for systemic acquired resistance in the plant stress response. *EMBO Reports* 12 (1): 50–55.

Jones, J.D. and Dangl, J.L. (2006). The plant immune system. *Nature* 444 (7117): 323–329.

Jung, J., Ahn, S., Kim, D.H., and Riu, M. (2024). Triple interactions for induced systemic resistance in plants. *Frontiers in Plant Science* 15: 1464710.

Kachroo, A. and Kachroo, P. (2007). Salicylic acid-, jasmonic acid-and ethylenemediated regulation of plant defense signaling. In: *Genetic Engineering: Principles and Methods*, vol. 28 (ed. J.K. Setlow), 55–83. Boston, MA: Springer.

Kant, M.R., Bleeker, P.M., Van Wijk, M. et al. (2009). Plant volatiles in defence. *Advances in Botanical Research* 51: 613–666.

Kariyat, R.R., Hardison, S.B., De Moraes, C.M., and Mescher, M.C. (2017). Plant spines deter herbivory by restricting caterpillar movement. *Biology Letters* 13 (5): 20170176.

Kawamoto, N., Del Carpio, D.P., Hofmann, A. et al. (2020). A peptide pair coordinates regular ovule initiation patterns with seed number and fruit size. *Current Biology* 30 (22): 4352–4361.e4.

Kim, J. and Felton, G.W. (2013). Priming of antiherbivore defensive responses in plants. *Insect Science* 20 (3): 273–285.

Kutty, N.N. and Mishra, M. (2023). Dynamic distress calls: volatile info chemicals induce and regulate defense responses during herbivory. *Frontiers in Plant Science* 14: 1135000.

Kvedaras, O.L. and Keeping, M.G. (2007). Silicon impedes stalk penetration by the borer *Eldana saccharina* in sugarcane. *Entomologia Experimentalis et Applicata* 125 (1): 103–110.

Lang, S., Simone-Finstrom, M., and Healy, K. (2022). Context-dependent viral transgenerational immune priming in honey bees (Hymenoptera: Apidae). *Journal of Insect Science* 22 (1): 19.

Lazebnik, J., Frago, E., Dicke, M., and Van Loon, J.J. (2014). Phytohormone mediation of interactions between herbivores and plant pathogens. *Journal of Chemical Ecology* 40: 730–741.

Lease, K.A. and Walker, J.C. (2006). The Arabidopsis unannotated secreted peptide database, a resource for plant peptidomics. *Plant Physiology* 142 (3): 831–838.

Li, Q., Li, M., Zhang, D. et al. (2020). The peptide-encoding MtRGF3 gene negatively regulates nodulation of *Medicago truncatula*. *Biochemical and Biophysical Research Communications* 523 (1): 66–71.

Lin, Y., Lin, X., Ding, C. et al. (2022). Priming of rice defense against a sap-sucking insect pest brown planthopper by silicon. *Journal of Pest Science* 1–15.

Liu, J., Du, H., Ding, X. et al. (2017). Mechanisms of callose deposition in rice regulated by exogenous abscisic acid and its involvement in rice resistance to *Nilaparvata lugens* Stål (Hemiptera: Delphacidae). *Pest Management Science* 73 (12): 2559–2568.

Lorenzo, O., Piqueras, R., Sánchez-Serrano, J.J., and Solano, R. (2003). Ethylene response factor1 integrates signals from ethylene and jasmonate pathways in plant defense. *Plant Cell* 15: 165–178.

Ma, C., Zhang, Y., Gui, W. et al. (2022). Priming of rice seed with decoyinine enhances resistance against the brown planthopper *Nilparvata lugens*. *Crop Protection* 157: 105970.

Malagnoux, L., Capowiez, Y., and Rault, M. (2015). Impact of insecticide exposure on the predation activity of the European earwig *Forficula auricularia*. *Environmental Science and Pollution Research* 22: 14116–14126.

Manachini, B., Casati, P., Cinanni, L., and Bianco, P. (2007). Role of *Myzus persicae* (Hemiptera: Aphididae) and its secondary hosts in plum pox virus propagation. *Journal of Economic Entomology* 100 (4): 1047–1052.

Matsubayashi, Y., Ogawa, M., Morita, A., and Sakagami, Y. (2002). An LRR receptor kinase involved in perception of a peptide plant hormone, phytosulfokine. *Science* 296 (5572): 1470–1472.

Matsui, K. and Engelberth, J. (2022). Green leaf volatiles—the forefront of plant responses against biotic attack. *Plant and Cell Physiology* 63 (10): 1378–1390.

Mauch-Mani, B., Baccelli, I., Luna, E., and Flors, V. (2017). Defense priming: an adaptive part of induced resistance. *Annual Review of Plant Biology* 68 (1): 485–512.

Moreira, X., Nell, C.S., Katsanis, A. et al. (2018a). Herbivore specificity and the chemical basis of plant–plant communication in *Baccharis salicifolia* (A steraceae). *New Phytologist* 220 (3): 703–713.

Moreira, X., Nell, C.S., Meza-Lopez, M.M. et al. (2018b). Specificity of plant–plant communication for *Baccharis salicifolia* sexes but not genotypes. *Ecology* 99 (12): 2731–2739.

Mukherjee, K., Grizanova, E., Chertkova, E. et al. (2017). Experimental evolution of resistance against Bacillus thuringiensis in the insect model host *Galleria mellonella* results in epigenetic modifications. *Virulence* 8 (8): 1618–1630.

Munkvold, G.P., Watrin, C., Scheller, M. et al. (2014). Benefits of chemical seed treatments on crop yield and quality. In: *Global Perspectives on the Health of Seeds and Plant Propagation Material* (ed. M. Gullino and G. Munkvold), 89–103. Dordrecht: Springer Netherlands.

Muroi, A., Ramadan, A., Nishihara, M. et al. (2011). The composite effect of transgenic plant volatiles for acquired immunity to herbivory caused by inter-plant communications. *PLoS One* 6 (10): e24594.

Nakayama, T., Shinohara, H., Tanaka, M. et al. (2017). A peptide hormone required for Casparian strip diffusion barrier formation in Arabidopsis roots. *Science* 355 (6322): 284–286.

Napoleão, T.H., Albuquerque, L.P., Santos, N.D. et al. (2019). Insect midgut structures and molecules as targets of plant-derived protease inhibitors and lectins. *Pest Management Science* 75 (5): 1212–1222.

Návarová, H., Bernsdorff, F., Döring, A.C., and Zeier, J. (2012). Pipecolic acid, an endogenous mediator of defense amplification and priming, is a critical regulator of inducible plant immunity. *The Plant Cell* 24 (12): 5123–5141.

Nazaralian, S., Majd, A., Irian, S. et al. (2017). Comparison of silicon nanoparticles and silicate treatments in fenugreek. *Plant Physiology and Biochemistry* 115: 25–33.

Negoro, S., Hirabayashi, T., Iwasaki, R. et al. (2023). EPFL peptide signalling ensures robust self-pollination success under cool temperature stress by aligning the length of the stamen and pistil. *Plant, Cell & Environment* 46 (2): 451–463.

Niu, H., Sun, Y., Zhang, Z. et al. (2022). The endophytic bacterial entomopathogen *Serratia marcescens* promotes plant growth and improves resistance against *Nilaparvata lugens* in rice. *Microbiological Research* 256: 126956.

Okon, O.G. and Antia, U.E. (2023). Plants and soil microbiota health implications of agrochemicals: potential alternatives for the safe propagation of food crops. In: *One Health Implications of Agrochemicals and their Sustainable Alternatives* (ed. M.C. Ogwu and S. Chibueze Izah), 441–460. Singapore: Springer Nature.

Pangesti, N., Reichelt, M., van de Mortel, J.E. et al. (2016). Jasmonic acid and ethylene signaling pathways regulate glucosinolate levels in plants during rhizobacteria-induced systemic resistance against a leaf-chewing herbivore. *Journal of Chemical Ecology* 42: 1212–1225.

Pearce, G., Moura, D.S., Stratmann, J., and Ryan, C.A. Jr. (2001). RALF, a 5-kDa ubiquitous polypeptide in plants, arrests root growth and development. *Proceedings of the National Academy of Sciences* 98 (22): 12843–12847.

Pearce, G., Strydom, D., Johnson, S., and Ryan, C.A. (1991). A polypeptide from tomato leaves induces wound-inducible proteinase inhibitor proteins. *Science* 253 (5022): 895–897.

Peiffer, M., Tooker, J.F., Luthe, D.S., and Felton, G.W. (2009). Plants on early alert: glandular trichomes as sensors for insect herbivores. *New Phytologist* 184 (3): 644–656.

Penaflor, M.F.G.V. and Bento, J.M.S. (2013). Herbivore-induced plant volatiles to enhance biological control in agriculture. *Neotropical Entomology* 42: 331–343.

Peng, J., Deng, X., Huang, J., Jia, S., Miao, X., & Huang, Y. (2004). Role of salicylic acid in tomato defense against cotton bollworm, *Helicoverpa armigera* Hubner.

Pérez-Hedo, M., Urbaneja-Bernat, P., Jaques, J.A. et al. (2015). Defensive plant responses induced by *Nesidiocoris tenuis* (Hemiptera: Miridae) on tomato plants. *Journal of Pest Science* 88: 543–554.

Reichardt, S., Piepho, H.P., Stintzi, A., and Schaller, A. (2020). Peptide signaling for drought-induced tomato flower drop. *Science* 367 (6485): 1482–1485.

Reynolds, O.L., Keeping, M.G., and Meyer, J.H. (2009). Silicon-augmented resistance of plants to herbivorous insects: a review. *Annals of Applied Biology* 155 (2): 171–186.

Roberts, I., Smith, S., De Rybel, B. et al. (2013). The CEP family in land plants: evolutionary analyses, expression studies, and role in Arabidopsis shoot development. *Journal of Experimental Botany* 64 (17): 5371–5381.

Roberts, I., Smith, S., Stes, E. et al. (2016). CEP5 and XIP1/CEPR1 regulate lateral root initiation in Arabidopsis. *Journal of Experimental Botany* 67 (16): 4889–4899.

Rodríguez, B.D., Kloth, K.J., and Albrectsen, B.R. (2022). Effects of condensed tannins on behavior and performance of a specialist aphid on aspen. *Ecology and Evolution* 12 (8): e9229.

Rowen, E. and Kaplan, I. (2016). Eco-evolutionary factors drive induced plant volatiles: a meta-analysis. *New Phytologist* 210 (1): 284–294.

Roychowdhury, R., Mishra, S., Anand, G. et al. (2024). Decoding the molecular mechanism underlying salicylic acid (SA)-mediated plant immunity: an integrated overview from its biosynthesis to the mode of action. *Physiologia Plantarum* 176 (3): e14399.

Ryan, C.A. (2000). The systemin signaling pathway: differential activation of plant defensive genes. *Biochimica et Biophysica Acta (BBA)-Protein Structure and Molecular Enzymology* 1477 (1–2): 112–121.

Sandhu, R.K. and Sarao, P.S. (2021). Evaluation of antixenosis resistance in wild rice accessions against brown planthopper, *Nilaparvata lugens* (Stål). *International Journal of Tropical Insect Science* 41 (1): 65–73.

Schaller, F. (2001). Enzymes of the biosynthesis of octadecanoid-derived signalling molecules. *Journal of Experimental Botany* 52 (354): 11–23.

Schaller, F., Schaller, A., and Stintzi, A. (2004). Biosynthesis and metabolism of jasmonates. *Journal of Plant Growth Regulation* 23: 179–199.

Schneider, R.F., Dubin, A., Marten, S.M., and Roth, O. (2024). Parent-specific transgenerational immune priming enhances offspring defense—unless heat stress negates it all. *Ecology and Evolution* 14 (11): e70552.

Selby, R. and Jones, D.S. (2023). Complex peptide hormone signaling in plant stem cells. *Current Opinion in Plant Biology* 75: 102442.

Shah, A.Z., Ma, C., Zhang, Y. et al. (2022). Decoyinine induced resistance in rice against small brown planthopper *Laodelphax striatellus*. *Insects* 13 (1): 104.

Sharma, H.C., Sujana, G., and Manohar Rao, D. (2009). Morphological and chemical components of resistance to pod borer, *Helicoverpa armigera* in wild relatives of pigeonpea. *Arthropod-Plant Interactions* 3: 151–161.

Shoresh, M., Gal-On, A., Leibman, D., and Chet, I. (2006). Characterization of a mitogen-activated protein kinase gene from cucumber required for Trichoderma-conferred plant resistance. *Plant Physiology* 142 (3): 1169–1179.

Simmonds, M.S. (2003). Flavonoid–insect interactions: recent advances in our knowledge. *Phytochemistry* 64 (1): 21–30.

Slaughter, A., Daniel, X., Flors, V. et al. (2012). Descendants of primed Arabidopsis plants exhibit resistance to biotic stress. *Plant Physiology* 158 (2): 835–843.

Song, Y.Y., Ye, M., Li, C.Y. et al. (2013). Priming of anti-herbivore defense in tomato by arbuscular mycorrhizal fungus and involvement of the jasmonate pathway. *Journal of Chemical Ecology* 39: 1036–1044.

Stout, M.J., Riggio, M.R., and Yang, Y. (2009). Direct induced resistance in *Oryza sativa* to *Spodoptera frugiperda*. *Environmental Entomology* 38 (4): 1174–1181.

Strabala, T.J., O'Donnell, P.J., Smit, A.M. et al. (2006). Gain-of-function phenotypes of many CLAVATA3/ESR genes, including four new family members, correlate with tandem variations in the conserved CLAVATA3/ESR domain. *Plant Physiology* 140 (4): 1331–1344.

Sułek, M., Kordaczuk, J., and Wojda, I. (2021). Current understanding of immune priming phenomena in insects. *Journal of Invertebrate Pathology* 185: 107656.

Taranto, F., Pasqualone, A., Mangini, G. et al. (2017). Polyphenol oxidases in crops: biochemical, physiological and genetic aspects. *International Journal of Molecular Sciences* 18 (2): 377.

Tetreau, G., Dhinaut, J., Gourbal, B., and Moret, Y. (2019). Trans-generational immune priming in invertebrates: current knowledge and future prospects. *Frontiers in Immunology* 10: 1938.

Ton, J., D'Alessandro, M., Jourdie, V. et al. (2007). Priming by airborne signals boosts direct and indirect resistance in maize. *The Plant Journal* 49 (1): 16–26.

Ton, J., Ent, S.V.D., Van Hulten, M.H.A. et al. (2009). Priming as a mechanism behind induced resistance against pathogens, insects and abiotic stress. *IOBC/wprs Bulletin* 44: 3–13.

Traw, M.B. (2002). Is induction response negatively correlated with constitutive resistance in black mustard? *Evolution* 56 (11): 2196–2205.

Tscharntke, T., Thiessen, S., Dolch, R., and Boland, W. (2001). Herbivory, induced resistance, and interplant signal transfer in *Alnus glutinosa*. *Biochemical Systematics and Ecology* 29 (10): 1025–1047.

Ulloa, R.M., Raíces, M., MacIntosh, G.C. et al. (2002). Jasmonic acid affects plant morphology and calcium-dependent protein kinase expression and activity in *Solanum tuberosum*. *Physiologia Plantarum* 115 (3): 417–427.

Vilcinskas, A. (2021). Mechanisms of transgenerational immune priming in insects. *Developmental & Comparative Immunology* 124: 104205.

Vojvodić, M. and Bažok, R. (2021). Future of insecticide seed treatment. *Sustainability* 13 (16): 8792.

Walling, L.L. (2000). The myriad plant responses to herbivores. *Journal of Plant Growth Regulation* 19: 195–216.

Wang, L., Einig, E., Almeida-Trapp, M. et al. (2018). The systemin receptor SYR1 enhances resistance of tomato against herbivorous insects. *Nature Plants* 4 (3): 152–156.

Wang, W., Jin, N., Mo, X. et al. (2021b). Exogenous gibberellin GA 3 enhances defense responses in rice to the brown planthopper *Nilaparvata lugens* (Stål). *Journal of Plant Biology* 64: 379–387.

Wang, J., Song, J., Wu, X.B. et al. (2021a). Seed priming with calcium chloride enhances wheat resistance against wheat aphid *Schizaphis graminum* Rondani. *Pest Management Science* 77 (10): 4709–4718.

Wang, P., Wu, T., Jiang, C. et al. (2023). Brt9SIDA/IDALs as peptide signals mediate diverse biological pathways in plants. *Plant Science* 330: 111642.

Wang, J., Xi, L., Wu, X.N. et al. (2022). PEP7 acts as a peptide ligand for the receptor kinase SIRK1 to regulate aquaporin-mediated water influx and lateral root growth. *Molecular Plant* 15 (10): 1615–1631.

Wang, W., Zhou, P., Mo, X. et al. (2020). Induction of defense in cereals by 4-fluorophenoxyacetic acid suppresses insect pest populations and increases crop yields in the field. *Proceedings of the National Academy of Sciences* 117 (22): 12017–12028.

War, A.R., Paulraj, M.G., Ahmad, T. et al. (2012). Mechanisms of plant defense against insect herbivores. *Plant Signaling & Behavior* 7 (10): 1306–1320.

War, A.R., Paulraj, M.G., War, M.Y., and Ignacimuthu, S. (2011). Herbivore-and elicitor-induced resistance in groundnut to Asian armyworm, *Spodoptera litura* (Fab.) (Lepidoptera: Noctuidae). *Plant Signaling & Behavior* 6 (11): 1769–1777.

Worrall, D., Holroyd, G.H., Moore, J.P. et al. (2012). Treating seeds with activators of plant defence generates long-lasting priming of resistance to pests and pathogens. *New Phytologist* 193 (3): 770–778.

Wu, J., Ge, L., Liu, F. et al. (2020). Pesticide-induced planthopper population resurgence in rice cropping systems. *Annual Review of Entomology* 65 (1): 409–429.

Wu, Q., Zhang, G., Chen, Y. et al. (2021). Seed dressing with triflumezopyrim controls brown planthopper populations by inhibiting feeding behavior, fecundity and enhancing rice plant resistance. *Pest Management Science* 77 (6): 2870–2886.

Xi, C., Ahmad, S., Yu, J. et al. (2022). Seed coating with triflumezopyrim induces the rice plant's defense and inhibits the brown planthopper's feeding behavior. *Agronomy* 12 (5): 1202.

Xin, Z., Li, X., Li, J. et al. (2016). Application of chemical elicitor (Z)-3-hexenol enhances direct and indirect plant defenses against tea geometrid *Ectropis obliqua*. *BioControl* 61: 1–12.

Xu, G., She, S., Gui, W. et al. (2022). Seed priming of rice varieties with decoyinine improve their resistance against the brown planthopper *Nilaparvata lugens*. *Agronomy* 13 (1): 72.

Ye, M., Song, Y., Long, J. et al. (2013). Priming of jasmonate-mediated antiherbivore defense responses in rice by silicon. *Proceedings of the National Academy of Sciences* 110 (38): E3631–E3639.

Zebelo, S., Piorkowski, J., Disi, J., and Fadamiro, H. (2014). Secretions from the ventral eversible gland of *Spodoptera exigua* caterpillars activate defense-related genes and induce emission of volatile organic compounds in tomato, *Solanum lycopersicum*. *BMC Plant Biology* 14: 1–12.

Zelman, A.K. and Berkowitz, G.A. (2023). Plant elicitor peptide (Pep) signaling and pathogen defense in tomato. *Plants* 12 (15): 2856.

Zelman, A.K. and Berkowitz, G.A. (2024). Peps, pathogens, and pests: challenges and opportunities for usage of pep signaling in sustainable farming. In: *Plant Elicitor Peptides: New Tool for Sustainable Agriculture*, 335–351. Springer Nature.

Zhang, Z., Han, H., Zhao, J. et al. (2025). Peptide hormones in plants. *Molecular Horticulture* 5 (1): 7.

Zhang, B., Xin, B., Sun, X. et al. (2023). Small peptide signaling via OsCIF1/2 mediates Casparian strip formation at the root endodermal and nonendodermal cell layers in rice. *The Plant Cell* 36 (2): 383.

Zhang, W., Zhao, F., Jiang, L. et al. (2018). Different pathogen defense strategies in Arabidopsis: more than pathogen recognition. *Cells* 7 (12): 252.

Zhang, Y., Zhou, X., Dong, Y. et al. (2021). Seed priming with melatonin improves salt tolerance in cotton through regulating photosynthesis, scavenging reactive oxygen species and coordinating with phytohormone signal pathways. *Industrial Crops and Products* 169: 113671.

Zhao, L.Y., Chen, J.L., Cheng, D.F. et al. (2009). Biochemical and molecular characterizations of *Sitobion avenae*-induced wheat defense responses. *Crop Protection* 28 (5): 435–442.

Zhu, S., Estevez, J.M., Liao, H. et al. (2020). The RALF1–FERONIA complex phosphorylates eIF4E1 to promote protein synthesis and polar root hair growth. *Molecular Plant* 13 (5): 698–716.

Zhu, J. and Park, K.C. (2005). Methyl salicylate, a soybean aphid-induced plant volatile attractive to the predator *Coccinella septempunctata*. *Journal of Chemical Ecology* 31: 1733–1746.

15

Regulation of Abiotic Stress Tolerance by Peptide Hormones

Sumera Javad[1], Aina Inam[1], Sadia Afzal[1], and Ajit Singh[2]

[1] *Department of Botany, Lahore College for Women University, Lahore, Pakistan*
[2] *School of Biological and Environmental Sciences, University of Nottingham Malaysia, Semenyih, Selangor, Malaysia*

15.1 Introduction

Plant stress refers to any unfavorable environmental condition that negatively impacts a plant's growth, development, or productivity. Many stressors play their role in this stress and activate a wide array of physiological and molecular responses, including changes in gene expression, alterations in cellular metabolism, and modifications in growth patterns and yield outcomes. Typically, stress arises from abrupt changes in the surrounding environment. However, certain stress-tolerant species of plants can acclimate over time, gradually adjusting their internal systems to better withstand specific stress conditions (Verma et al. 2013).

Plant stress is generally classified into two broad categories: **abiotic stress** and **biotic stress**. Abiotic stress originates from nonliving factors such as extreme temperatures, drought, salinity, and exposure to harmful chemicals, while biotic stress is caused by living organisms like pests, pathogens, or competing weeds. When plants are exposed to stress, it can lead to significant physiological disruptions. Mild or short-term stress may only cause temporary damage, allowing plants to recover once favorable conditions return. In contrast, prolonged or intense stress can severely impair essential processes like flowering and seed development, potentially triggering premature aging or even plant death. Plants that cannot recover from such stress are considered stress susceptible.

Nevertheless, some species, such as desert **ephemerals**, have evolved strategies to **avoid** stress altogether. These plants complete their life cycles rapidly, escaping harsh environmental conditions rather than enduring them. This ability to either tolerate or avoid stress is crucial for survival and has important implications for agriculture, especially in the context of climate variability (Sah et al. 2016).

Abiotic stresses such as drought, waterlogging, extreme temperatures (including heat, cold, and frost), salinity, and mineral toxicity have detrimental effects on plant growth, development, seed quality, and overall yield. With projections indicating a rise in freshwater

Peptide Hormones in Plant Development and Stress Responses, First Edition. Edited by Vijay Pratap Singh and Samiksha Singh.

scarcity, the frequency and severity of abiotic stresses are expected to intensify. This highlights the urgent need to develop stress-resilient crop varieties to ensure future food security and safety. The root system serves as a plant's first defense against abiotic stress, and plants growing in healthy, biologically rich soils have a better chance of surviving adverse conditions. One of the most important reactions to high salinity stress is the disturbance of the balance of potassium (K^+) and sodium (Na^+) in cytoplasm. Plants also respond to stress conditions by creating reactive oxygen species (ROS), phytohormones, and activating signal transduction cascades. Peptide hormones play a key role in the activation of these signal transduction cascades as a response to combat the effects of environmental stresses. Peptide hormones also interact with usual phytohormones, such as abscisic acid (ABA). Such an interaction plays a central function in plant regulation of adaptation to diverse environmental stresses, such as drought, salinity, low temperature, and mechanical injury (Gull et al. 2019).

15.1.1 Importance of Stress Tolerance Mechanisms

The series of physiological, biochemical, and molecular mechanisms enabling organisms to resist stress conditions caused by factors like salinity, drought, heavy metals, high temperature, nutrient starvation, and infection by pathogens is referred to as stress tolerance mechanisms. These intrinsic mechanisms also serve as a defense against all living organisms, including plants. For instance, plants respond to environmental stress by triggering antioxidative defense mechanisms, molecular chaperones, cell membrane stabilization, stomatal regulation, osmolyte accumulation, synthesis of secondary metabolites, phytohormone biosynthesis, and signaling transduction pathways (Hasanuzzaman et al. 2011; Sachdev et al. 2021).

Stress tolerance mechanisms are crucial for the survival, growth, and yield of organisms, especially in response to environmental stresses in the form of drought, salinity, high temperature, and oxidative stress. Stress tolerance mechanisms allow cells to maintain cellular homeostasis and physiological processes even under stressful conditions. In plants, for example, stress tolerance is controlled by a range of physiological, biochemical, and molecular processes, including the accumulation of osmoprotectants such as proline and glycine betaine, activation of antioxidant defense mechanisms, and expression of stress-responsive genes (Zhu 2016). The action of heat shock proteins (HSPs) and late embryogenesis abundant (LEA) proteins is also well described in inducing stress resilience through protein and membrane stabilization during dehydration and heat stress (Wang et al. 2004). In animals and humans, stress tolerance mechanisms also include the activation of the hypothalamic–pituitary–adrenal (HPA) axis, leading to the release of cortisol, which helps to modulate metabolic and immune responses to stress (Charmandari et al. 2005).

Oxidative stress, a prevalent outcome of environmental and physiological stressors, is neutralized by enzymatic and nonenzymatic antioxidants like superoxide dismutase (SOD) and glutathione, which avoid cell damage (Foyer and Noctor 2005). Knowledge of the same is critical for the production of stress-resistant plants and the optimization of human stress management techniques, which have major outcomes in agriculture, medicine, and general health.

15.1.2 Role of Signaling Molecules in Stress Adaptation

Effective communication within and between cells, tissues, and organs is critical for the plant to properly respond to abiotic stress, and this will facilitate a coordinated movement toward an adaptive response to these harsh conditions. This interconnected signaling mechanism is complex, with the plant relying on numerous molecules, including classical phytohormones, namely abscisic acid (ABA), ethylene, auxins, cytokinins, gibberellins, and brassinosteroids, each of which is at least partially responsible for modifying gene expression and metabolic processes that help the plant minimize damage from stress (Cutler et al. 2010; Verma et al. 2016). ABA is vital in the response to drought and salinity stress, as it leads to stomatal closure, regulates the expression of stress-responsive genes, and initiates osmoprotectant accumulation for maintaining homeostasis (Zhu 2016). Ethylene acts as another important plant stress hormone that facilitates changes in plant growth, including senescence, directs processes such as root elongation and leaf abscission, as well as regulates antioxidant enzymes in response to stress (Iqbal et al. 2017).

In more recent research, peptide hormones have also been found to be essential regulators of stress response, and along with the classical phytohormones, they will adapt stress responses. These peptide hormones are small proteins like CLAVATA3/embryo surrounding region-related (CLE) peptides, rapid alkalinization factor (RALF), and phytosulfokine (PSK), that work in either an autocrine or paracrine fashion and provide localized control of stress perception and responses (Matsubayashi 2014; Delay et al. 2013a, b). Peptide hormones can control stress pathways more rapidly and in a more localized manner than classical hormones, which move in the vascular system and rarely move site-specifically (to be fair, very few would rely on classical hormones for local site-specific response). For example, CLE peptides can modulate root architecture in response to drought stress, while PSK can enhance oxidative stress resistance and cell survival. Furthermore, peptide signaling works with ROS and calcium (Ca^{2+}) dependent signaling, which enhances resilience in plants by increasing the efficiency of signal transduction and selectivity of responses. Understanding how classical phytohormones and the newly found peptide hormones interact provides opportunities to enhance the abiotic stress tolerance of crops with a biotechnological approach, ultimately helping to create a sustainable method of agriculture in the face of climate change (Tavormina et al. 2015).

15.1.3 Brief Introduction to Peptide Hormones and Their Significance in Plant Biology

Peptide hormones are short chains of amino acids that are produced from larger precursor proteins, through the processes of proteolytic processing and posttranslational modifications, such as hydroxylation, sulfation, and glycosylation, that are required for maturation and biological function (Matsubayashi 2014). Peptide hormones are not like traditional plant hormones that act at long distances and in a wide array of tissues, because they are secreted locally; they enter the apoplast at the site of secretion, and once secreted, peptide hormones will have a strong specificity due to binding to membrane-localized leucine-rich repeat receptor-like kinases (LRR-RLKs) that trigger downstream intracellular signaling pathways (Tavormina et al. 2015). This ligand–receptor binding mechanism will activate a suite of downstream signaling mechanisms, including mitogen-activated protein kinases (MAPKs), calcium fluxes, and transcriptional regulators, allowing plants to respond quickly and more accurately to both developmental cues and environmental changes (Olsson et al. 2019).

Peptide hormones have been implicated in several plant physiological processes, including meristem maintenance (i.e. CLAVATA3 [CLV3]), vascular differentiation (i.e. TDIF/Tracheary element Differentiation Inhibitory Factor), and stomatal patterning (i.e. EPF1/EPF2), suggesting the fundamental role of peptide hormones in plant development (Delay et al. 2013a, b). Specifically, in addition to their role in development, recent work has revealed that many classes of peptides are also involved in responses to abiotic stress.

For example, RALF peptides can modulate pH and ion homeostasis under salt stress, and the peptide CLE25 plays a role in signaling ABA over long distances during dehydration stress, connecting root-derived signals to leaf stomata (Takahashi et al. 2018). These examples highlight the adaptive advantage of signaling via peptides, nature's highly specific and redundant method of helping organisms' size and process information about their environment to develop responses that are spatially and temporally precise. The extreme specificity and redundancy within peptide ligands and their receptors offer a flexible regulatory framework in which plants can use redundant signaling to fine-tune their responses to stressors without compromising developmental signaling, all while maintaining flexibility in their responses by retaining the ability to distinguish between developmental signaling and stress signaling. Hence, the study of peptide hormone signaling offers a potentially advantageous approach to engineering stress-resilient crops, given the context of a changing climate and the increasingly variable environment (Haruta et al. 2014) (Table 15.1).

15.2 Peptide Hormones: Structure and Functions

The molecular features of peptide hormones are indispensable for their abilities to act as precise and flexible signaling molecules during plant responses to abiotic stress. These tiny peptides fall into categories or families based on their sequences, structures, and functions, e.g. CLE peptides, C-terminally encoded peptides (CEPs), PSKs, and RALFs, and each performs unique physiological roles (Tavormina et al. 2015).

Peptide hormones are synthesized as long precursor proteins and require regulated levels of proteolytic processing, and in some cases, a slew of posttranslational modifications like hydroxylation, glycosylation, and sulfation to achieve maturation, stability, and/or receptor recognition (Matsubayashi 2014). For example, sulfonated tyrosine residues of PSK peptides are imperative for the biological activity and receptor binding (Olsson et al. 2019).

Peptide receptors are proteins found mainly as integral membrane proteins in the plasma membrane that either sense binding of specific signals, e.g. damage-inducing molecules such as stress-related peptides (PEPs) or pathogens. When proteins are processed into mature peptides, they are secreted into the apoplast and may interact with specific receptor-like kinases (RLKs) that usually reside in the plasma membrane. Examples of RLKs for peptides include CLAVATA1 (CLV1) and phytosulfokine receptor1 (PSKR1) that contain extracellular leucine-rich repeat (LRR) domains for the acceptance of peptides and intracellular kinase domains, which, when activated, initiate downstream signaling pathways

Table 15.1 Role of peptide hormones in abiotic stress mitigation.

Family	Peptide name	Full name	Role in stress mitigation	Model organism	Reference
OSIPs	OSIP108	Oxidative stress-induced peptide 108	Oxidative stress mitigation (H_2O_2)	*Arabidopsis thaliana*	De Coninck et al. (2013)
LEA proteins	*Mf*LEA3	*Medicago falcata* Late embryogenesis abundant-3	Drought, light, cold	Tobacco	Shi et al. (2020)
P5CS	P5CS2	Δ-1-pyrroline-5-carboxylate synthase-2	Salt tolerance	*Arabidopsis thaliana*	Funck et al. (2020)
NAC	TwNAC01	*Triticale wittmack* NAC01	Drought tolerance	*Arabidopsis thaliana*	Wang et al. (2022)
RACK	GmRACK1	*Glycine max* Receptor for activated C kinase 1	Drought tolerance	Soybean	Li et al. (2018)
HSP	CsHSP70-derived peptides	*Camelina sativa* Heat shock protein 70	Thermal salinity	*Camelina sativa*	Heidari et al. (2024)
NAC	SbNAC1	*Sorghum bicolor* NAC-2	Drought tolerance	sorghum	Jin et al. (2021)
S33	AtPAP1	*Arabidopsis* prolyl aminopeptidase-1	Drought salt	*Arabidopsis thaliana*	Sun et al. (2013)
	OsDSSR1	*Oryza sativa* DSSR-1	Drought tolerance Salinity	Rice	Cui et al. (2018)

(Butenko et al. 2009). These signaling pathways will include the action of secondary messengers Ca^{2+} ions, ROS, and MAPK pathways leading to transcriptional reprogramming and physiological tolerance to stress (Takahashi et al. 2018). Peptide hormones are qualitatively different from classic phytohormones such as ABA, auxins, or cytokinins, because classic phytohormones act through long-distance transport and general systemic regulation. Peptide hormones operate in a localized, cell-to-cell manner, facilitating precise spatial and temporal regulation of stress responses and signaling (Matsubayashi 2014).

This distinction highlights the differing roles of classical and peptide hormones within the context of plant stress adaptation. While classical hormones act as global regulators that coordinate developmental and stress responses, peptide hormones provide the level of specificity and fine-tuning that allows for context-sensitive adjustments to rapid and reversible environmental stimuli. With our burgeoning understanding of peptide hormone biosynthesis and signaling mechanisms, we are also starting to develop novel biotechnological approaches to harness peptide-mediated stress signaling networks to improve crop stress resilience caused by climate change-related stressors.

15.2.1 Classification of Plant Peptide Hormones

Peptide hormones in plants are a structurally and functionally heterogeneous class of signaling molecules that have essential functions in regulating plant development and environmental responses, especially in the environment of abiotic stress. Peptide hormones arise from larger precursor proteins and are subject to posttranslational modifications such as hydroxylation, sulfation, and proteolytic cleavage (the mechanism by which proteins are inactivated and/or short-lived proteins are degraded during the protein turnover) necessary for stability, receptor binding, and biological activity (Matsubayashi 2014; Olsson et al. 2019). Peptide hormones can broadly be grouped into families based on their structure or functional mechanisms of action. Each family can activate different receptors and activate different physiological events.

One of the best-understood families is the **CLE peptides**, which are normally 12–13 amino acids long and work on a short distance communicating within plant cells. CLE peptides govern multiple processes of meristem activity, vascular differentiation, and root development, hallmarks of developmental plasticity and stress responses (Fletcher et al. 1999; Kondo et al. 2006). Under abiotic stress situations, alterations in CLE signaling and meristem activity can fine-tune the response of plants to abiotic or biotic stress states, allowing greater adjustment in growth and developmental processes, including a willingness to sacrifice growth or reproduction potential to ensure survival.

The PSK family, or phytosulfokines, comprises tyrosine-sulfated pentapeptides that stimulate cell proliferation, differentiation, and immune responses. PSKs have also been implicated in providing plants with resilience against abiotic stress, particularly drought tolerance, whereby they stimulate cellular expansion and growth in an environment with limited water availability (Matsubayashi 2014). This growth-promoting effect, along with stress-responsive signaling, establishes two competing features under which the life of PSK operates.

RALF peptides form one of the eminent classes capable of inducing both rapid alkalinization of the extracellular matrix and modifying the wall pH and rigidity. These peptides confer potently inhibitory effects on root growth and pollen tube integrity, while also possibly being involved in stress responses such as salinity and mechanical perturbations (Pearce et al. 2001; Haruta et al. 2014). RALF peptides affect ion flux and cell wall remodeling, two crucial aspects of tolerating environmental and chemical stressors, through modifying extracellular pH.

The **CEP** family has recently become increasingly prominent for its involvement in nutrient sensing and root system architecture. CEPs are expressed in nitrogen-deficient conditions, and they mediate systemic signaling that adjusts both lateral root formation and primary root elongation to increase foraging ability under conditions of nutrient stress (Delay et al. 2013a, b; Ohkubo et al. 2017). Adjusting root development is a straightforward physiological mechanism through which plants can modulate their behavior to changing environmental conditions.

The structural diversity of these peptide families allows them to interact with a wide variety of RLKs, with each RLK providing a unique pairing for specific ligand-receptor pairs, leading to downstream signaling activities that are distinct from one another. This specificity has enabled peptide hormones to evoke very specific physiological responses from plants, allowing them to adapt better to various abiotic stressors, and as the research into these molecules continues, their potential to impact biotechnology increases for instance around genetically engineering crops with enhanced stress resilience.

15.2.2 Biosynthesis, Processing, and Signaling Pathways

The path from genes to active peptide hormone is a tightly regulated and orchestrated process to ensure space and time accuracy in plant signaling, and especially under abiotic stress conditions. This often multistep process originates at the genome level with the transcription and translation of the peptide hormone genes. The peptide hormone genes are first transcribed as messenger RNA (mRNA) in the nucleus, then translated in the cytoplasm into larger proteins known as precursor proteins or prepropeptides that vary across the plant. These peptide hormone precursors usually contain an N-terminal signal peptide sequence to deliver them to the endoplasmic reticulum (ER) and begin their entry to the secretory pathway (Matsubayashi 2014; Olsson et al. 2019). A lot of this happens to ensure the peptide hormones are directed to the correct cellular compartments to process and eventually secrete into the extracellular environment.

The precursor proteins are now in the ER, where they will undergo **posttranslational processing.** This irreversible modification implies that they are committed to producing a functional mature protein. Proteases act to cleave the propeptides at conserved sites to release the bioactive core peptide. Additional modifications occur as well, including hydroxylation (often of proline), *O*-glycosylation, and tyrosine sulfation. While these modifications can seem to be decorative at times, they are, in fact, critical for determining the binding stability to the receptors, stability to proteolytic cleavage, and ultimately the biological activity of the peptides. For instance, tyrosine sulfation of **PSK** at the hands of tyrosyl protein sulfotransferases is required for receptor binding and signaling activity. Similarly, hydroxylation of proline in **CLV3** peptides is an essential configuration for maintaining proper three-dimensional structure and functional specificity (Matsubayashi 2014).

Once processing is complete, the **secretion and perception phase** starts, in which the mature peptide is transported from the Golgi apparatus to the plasma membrane and secreted into the apoplast. In the extracellular space, the peptide encounters its cognate receptor (normally an **RLK** located on the plasma membrane). These receptors typically exhibit extracellular LRR domains involved in ligand binding and contain cytoplasmic kinase domains activated to relay intracellular signals (Tavormina et al. 2015). The binding of the peptide to the receptor causes conformational changes to the receptor, often leading to receptor dimerization or the formation of a complex with coreceptors like Somatic Embryogenesis Receptor Kinases (SERKs) that are essential for full activation (Gish and Clark 2011).

Lastly, **signal transduction** is initiated when the receptor kinase is activated, producing secondary messengers such as calcium (Ca^{2+}), ROS, or **MAPKs** that phosphorylate other substrates. These secondary messengers transmit the signal and produce a variety of transcriptional changes in the nucleus that can translate stress into a wide variety of physiological changes (such as stomatal closure, roots changing architecture, or osmoprotectant generation) (Takahashi et al. 2018). Importantly, the signaling pathways are under tight regulation to facilitate appropriate perception and response to the level and time-course of stress to avoid maladaptive responses or developmental trade-offs.

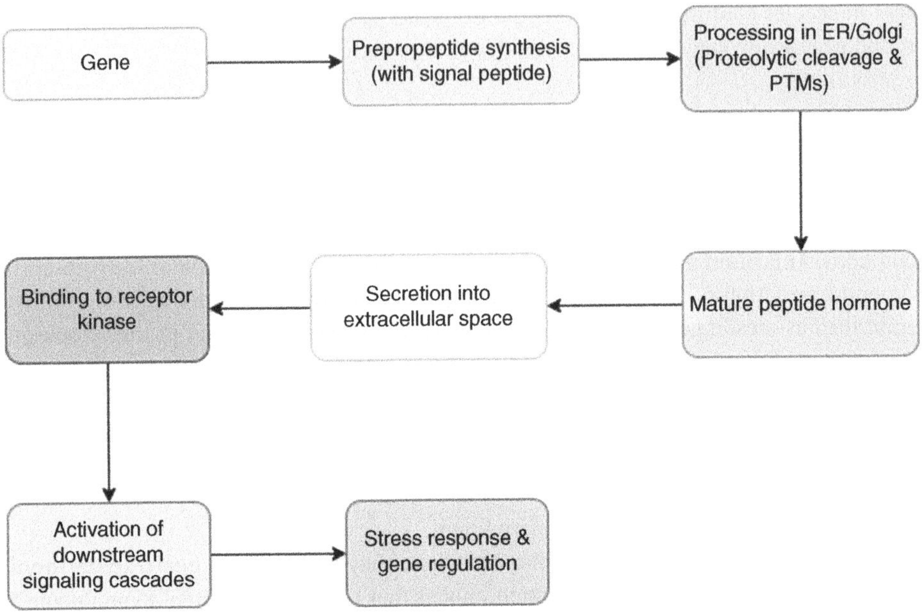

Figure 15.1 A flowchart summarizes the biosynthesis and signaling pathway of peptide hormones.

Collectively, these actions create a dynamic, modular signaling apparatus that gives plants the precision to perceive and respond to many kinds of environmental inputs. Likewise, as we learn more about peptide hormone biosynthesis and signaling, the opportunities for exploitation of these molecules to develop stress-tolerant crop varieties could become even more realistic, whether through genetic engineering or application of synthetic peptides (Figure 15.1).

15.2.3 Comparison with Classical Plant Hormones in Stress Responses

Abscisic acid, auxins, cytokinins, ethylene, and jasmonate are some examples of classical plant hormones and have long been known to be key regulators of plant growth and development as well as responses to stress. These hormones are synthesized in set tissues and move long distances via the vascular system of the plant, which allows coordinated responses of physiological processes across the plant (Santner et al. 2009). On the other hand, peptide hormones are an emerging but increasingly important class of signaling molecules that typically act locally or in a tissue-specific manner. Peptide hormones often have very specific actions, which allow precise responses to internal developmental states, as well as external environmental stimuli (Matsubayashi 2014; Olsson et al. 2019; Xiao et al. 2025a).

One of the main differences between classical (or nonpeptide) hormones and peptide hormones is their **spatial and temporal specificity**. For example, classical hormones such as **ABA** are often made in one part of the plant and are instead transported a distance to another part, where they induce physiological changes over a relatively longer timeframe. ABA is made in plant roots when the plant experiences drought stress to transport it to leaves to induce stomatal closure to reduce water loss from the plant (Cutler et al. 2010). Thus, ABA operates over a larger distance and timeframe to induce physiological change. In contrast, many **peptide hormones,** such as CLE peptides, are prime examples of peptide hormones functioning in a more localized, typically cell-to-cell manner. CLE peptides interact with receptors in adjacent cells to regulate shoot and root meristem function (Fletcher et al. 1999; Takahashi et al. 2018). Functionally, it is possible to reduce the region of spatial growth when the plant experiences the stress of either nutrient deficiency or limited water. This localized nature of signal transduction creates a plasticity of development with limited physiological changes when compared to the global physiological changes that can occur as a result of classical hormone signaling during stress.

An additional characteristic difference is **signal integration and crosstalk**. Classical hormone pathways, for example, auxin and cytokinin, have been demonstrated to utilize complex feedback loops and cross-regulatory systems to maintain plant development and responses to environmental signals (Kieber and Schaller 2018). Although characterized more recently, peptide hormones are beginning to be understood as having crosstalk through both synergistic and antagonistic interactions within classical hormone pathways. Some CLE peptides have been demonstrated to modulate ABA signaling in dehydration stress, for instance, increasing the plant's ability to respond to dehydration signals by affecting stomatal flow and root development (Takahashi et al. 2018; Delay et al. 2013a, b). The ability to modify classical hormone signals provides an additional mechanism for **fine-tuning** the ability of plants to respond to changing conditions.

Further differentiation exists in the **speed of response**. They are relatively small molecules, rapidly processed, and secreted locally; therefore, they can initiate signal transduction within minutes of perception by binding directly to membrane-localized receptor kinases (Tavormina et al. 2015). This speed of response is particularly surprising given that some classical hormone pathways take significantly longer to activate, as they may involve transcriptional regulation and long-distance transport, not to mention some will also include more complicated feedback mechanisms. In those cases where immediate changes are required, as in salt shock, pathogen attack, and sudden drought onset, the rapidity at which peptide signaling can be activated will serve a distinct survival advantage by facilitating a timely physiological response (Haruta et al. 2014).

In summary, **peptide hormones provide a different function to classical hormones in that they have precision, locality, and speed of response in extreme abiotic conditions**. The emerging nature of peptide hormones in stress signaling not only indicates they serve an important function, but they can be included in crop improvement and stress tolerance strategies in a climate-altered growing conditions (Xiao et al. 2025b) (Figure 15.2).

15.3 Peptide Hormone Signaling Mechanisms

15.3.1 Receptor Kinases and Downstream Signal Transduction

Receptor kinases are membrane proteins with vital roles in cellular signaling. These transmembrane proteins receive extracellular signals of signaling molecules like peptide hormones and transmit those signals to the intracellular environment. At the outer surface

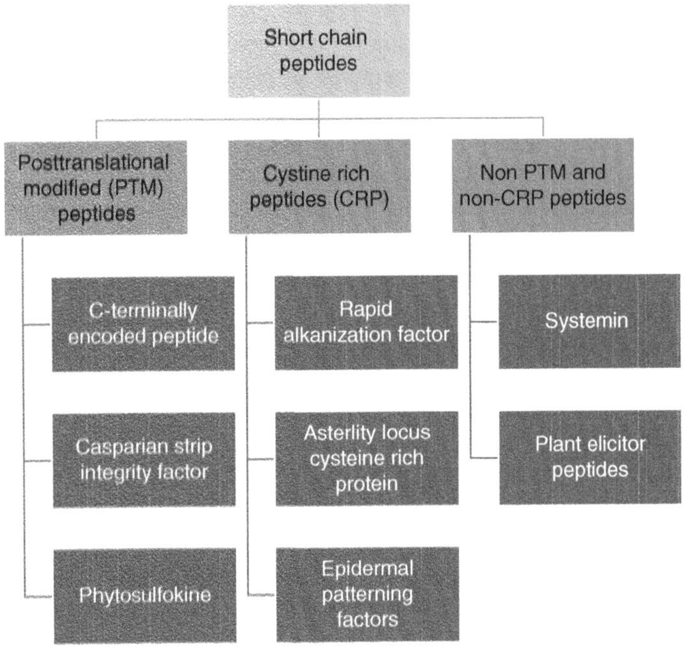

Figure 15.2 Classification of short-chain peptides.

of their target cells, the signaling molecules (peptide hormones) bind to receptor sites of protein kinases, to form a receptor-ligand (peptide hormone) complex. This complex causes configurational changes on receptor kinases, which leads to activation and phosphorylation of their intracellular kinase domains. The phosphorylation activity occurs by the addition of phosphate groups to tyrosine residues of the receptor's protein or other cellular proteins. Phosphorylation activates the receptor kinase activity to switch the signaling cascade, which involves proteins and secondary messengers like calcium ions (Ca^{2+}), cyclic AMPs, or phosphoinositides to propagate the signals and regulate transcription factors. Transcription factors are amino acid sequences that bind at specific DNA sites to trigger the expression of target genes for stress responses. The activity of the signaling cascade is regulated by a negative feedback mechanism to avoid over-response. Likewise, under excessive or prolonged signaling conditions, the receptor can control its kinase activity by autosignaling (Zhang et al. 2025).

15.3.2 Molecular Networks Involving Transcription Factors, Kinases, and Secondary Messengers

One of the molecular networks involving kinase transduction is the **CLV** signaling cascade, responsible for regulating plant growth and development. CLV cascade acts as a checkpoint to ensure a balanced growth of the shoot apical meristem (SAM). It involves RLK, CLAVATA1 (CLV1, the main receptor of the network), CLAVATA2 (CLV2, the coreceptor of CLV1), CLV3 (signaling peptide), and WUS (WUSCHEL; transcriptional factor) (Wu et al. 2022). In this cascade, the **CLV3** peptide binds to CLV1 and CLV2 receptors of

RLK (receptor-like kinase). CLV3-receptor complex and causes configurational change on CLV1, leading to its autophosphorylation. This autophosphorylation switches the kinase activity of CLV1, which initiates downstream signaling through a cascade of MAPK. **MAPK** cascade suppresses the activity of transcriptional factors (WUS) for the expression of the *CLV3* gene. So, CLV cascade acts in a negative-feedback manner to regulate the activity of meristematic stem cells (Li et al. 2022; Zhang et al. 2025).

Another example of a signal transduction cascade is **PSKs,** peptide hormones. PSKs are a class of small, sulfated peptides having a sulfate group on their tyrosine residue. The tyrosine residue binds to the PSK receptor site on the **LRR-RLK** receptor. Once the PSK-protein complex is formed, it causes autophosphorylation of the receptor's intracellular domain that switches its kinase activity. This kinase activity triggers a signaling mechanism by the involvement of the **MAPK** cascade and secondary messengers like $\mathbf{Ca^{2+}}$ ions. MAPK protein kinases and Ca^{2+} ions help to amplify and propagate signals through a transduction cascade to transcription factors. Transcription factors trigger the expression of target genes to regulate growth, embryogenesis, development, and stress responses. This cascade helps plants to endure challenging environments like drought, salinity, and thermal variability (Takahashi et al. 2018) (Figure 15.3).

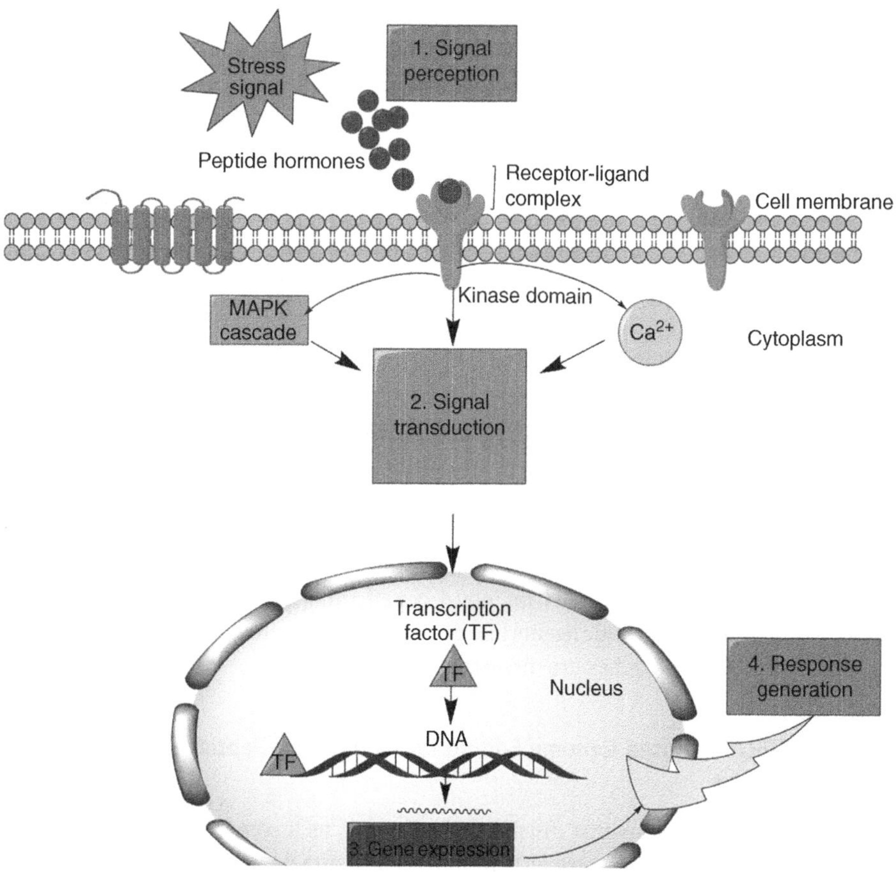

Figure 15.3 Stress response mechanism by peptide hormones.

15.3.3 Crosstalk with Phytohormones like ABA and Auxins

Phytohormones like ABA, IAA, gibberellins, and cytokinins are involved in coping with abiotic stresses. IAA and cytokinin help to combat water loss during drought stress. **CLV** cascade interacts directly or indirectly with these phytohormones by regulating their production. Moreover, plants reduce their growth under stress conditions of drought, salinity, and high temperature. In plants, the CLV cascade facilitates the growth reduction activity to conserve energy for their survival (Wu et al. 2022). Likewise, **CEP-5** peptide also negatively interacts with IAA to cope with stress. Under normal conditions, IAA facilitates overall plant growth by promoting cell elongation in roots, shoots, and leaves in the presence of optimal water and nutrients. The cellular activity of IAA is controlled by the AUX/IAA protein. Normally, AUX/IAA remains inactive in cells. Its activity is dependent on CEP-5. CEP-5 is a small peptide hormone. Under drought and salinity stress, CEP-5 peptide binds to CEPR-1 and CEPR-2 receptors to trigger the signal transduction pathway. This leads to the activation of AUX/IAA. The activated version of AUX/IAA inhibits the production of IAA to reduce the loss of water and nutrients in plant growth. So, CEP-5 and AUX/IAA act as checkpoints of IAA to divert plant activity from the utility of water in growth to the conservation of water for survival (Cao et al. 2025; Smith et al. 2020).

ABA is a major phytohormone that plays a role in tolerating drought and salt stress. In rice, higher levels under stress conditions acts as a signal for the expression of the *OsNAC2* gene. OsNAC2 helps improve rice yield at the maturation phase under water scarcity and salt stress in the ABA-dependent pathway. (Shen et al. 2017). Likewise, **CLE-25** is a peptide hormone found in root cells of *Arabidopsis thaliana*. It is involved in closing stomata in response to drought stress. The pathway of stomatal closure is ABA-dependent. And CLE-25 interacts with the *NCED-3* gene for ABA activation, which then translocates to leaves and causes stomatal closure (Takahashi et al. 2018). Although signal transduction-dependent activity of phytohormones helps plants to better adapt to environmental stresses, but detailed mechanisms of interactions and chemistry of receptor sites need further exploration.

15.4 Genetic and Biotechnological Approaches

Recently, abiotic stresses, including drought, salinity, heat, and nutrient deficiency, have become growing challenges in the agriculture sector. These stresses make crops vulnerable and impact overall yield, while the use of chemical agents to enhance crop yield also poses environmental threats. So, the need for novel genetic and biotechnological approaches arises for the modulation of biotic agents like peptide hormones in making climate-resilient crops. Some of these approaches are discussed below.

15.4.1 CRISPR/Cas-Based Genome Editing for Modulating Peptide Hormone Pathways

The CRISPR/Cas system was first discovered in bacteria as a component of the immune system, where it helps to protect bacteria from pathogen attack. CRISPR/Cas is comprised of two basic components: the **CRISPR sequence** and the **Cas protein**. The CRISPR sequence is a DNA segment with genetic information to prevent infections. In contrast, Cas is a CRISPR-linked protein with the enzymatic properties of molecular scissors. This protein can make precise cuts on specific DNA sequences. Due to this ability, the CRISPR/Cas system has gained popularity and is a widely used tool to manipulate genomes of various organisms, including plants (Rehman et al. 2022).

In plants, the CRISPR/Cas system with **gRNA** can be used to specifically target certain genes that are correlated with hormonal and peptide signaling cascades. The activity of a peptide signaling cascade could be optimized by CRISPR/Cas-mediated manipulation of peptide-encoding genes (Li et al. 2025). For example, the manipulation of CLE family genes in corn, tomato, and *Arabidopsis* could not only improve the ability of plants to tolerate stress but also crop yield. Similarly, the editing of *RALF* genes could be used to modulate the RALF signaling cascade to resist drought in plants (Ji et al. 2025).

15.4.2 Overexpression and Knockout Studies in Model and Crop Plants

"Overexpression" is a genetic and biotechnological term used to refer to levels of expression of target genes that are higher than normal for the subject of interest. "Knockout" refers to manipulating target DNA by silencing or deleting part of the DNA sequence(s) that code for target genes. Both overexpression and knockout studies facilitate understanding how plant systems respond to stressed environments. For example,

- The overexpression of the *CEP-3* **(C terminally encoded peptide-3)** gene leads to inhibited root growth in a model system (we used *Arabidopsis thaliana*). In contrast, the knockout of *CEP-3* created a mutant CEP-3-1a organism. The knockout CEP-3-1a mutant in *Arabidopsis thaliana* facilitated its survival by promoting root and shoot growth under nitrogen deficit and salt stress (Delay et al. 2013a, b).
- Conversely, overexpression of **AtPep-3** promotes the ability of *Arabidopsis thaliana* to tolerate salinity stress. But the knockdown (partial silencing of genes by RNAi or siRNA) process of AtPep-3 enables it to interact with the immune system for triggering the immune-related pathways of *A. thaliana*. This interaction of the knockdown AtPep-3 peptide and the immune system enable the plant to adapt to higher salinity levels (Ghosh and Roychoudhury 2024; Nakaminami et al. 2018).
- Expression of **AtPROPEP-3** is helpful in tolerating salt stress in seedlings. But its knockdown activity helps to make seedlings immune to salt stress (Ghosh and Roychoudhury 2024).
- In *Oryza sativa*, overexpression of *OsDSSR-1* genes encodes a peptide that facilitates drought and salt stress tolerance by accumulation of osmolytes, reduction of ABA sensitivity, and stimulating antioxidant enzymes (SOD and APX). The knockdown process of OsDSSR1 reduced the stress tolerance ability of the plant (Cui et al. 2018).
- Conversely, overexpression of the *OsNAC-2* gene in *Oryza sativa* hampered stress tolerance ability. This makes plants more vulnerable to stress responses. OsNAC-2 transcription factor has direct interactions with promoter sequences of other stress-responsive genes (e.g. OsSAPK-1 and OsLEA-3). The binding of OsNAC-2 transcription factor at the promoter sequences of other genes causes their downregulation. In this way stress regulation mechanism slows down and causes yield reduction in plants. In contrast,

the knockdown process of OsNAC-2 with the help of RNAi facilitates the upregulation of stress-tolerant mechanisms. This increases drought and salinity tolerance ability and makes plants resilient to climatic changes (Shen et al. 2017).

15.4.3 Application of Synthetic Peptides for Stress Mitigation

Synthetic peptides are artificially made protein-like fragments produced as an exact copy of a certain protein or with some modulations. These modulations increase their stability and shelf life to generate stress responses in a better way than conventional chemicals (Groß et al. 2016). Synthetic peptides can be applied to plants in the form of nanoparticles, emulsions, formulations, or agrochemicals. The mode of application may include foliar spray, root irrigation, or transfection, depending upon need. Due to their specificity, synthetic peptides bind to receptor sites of their target cells to trigger transduction pathways for stress responses. It helps to address various types of stresses, including drought, salinity, heavy metal toxicity, nutrient deficiency, and thermal stress. For instance, salinity stress hampers seedling growth by bleaching of chlorophyll. This issue can be resolved by exogenous applications of synthetic **AtPROPEP-3** because it helps plants to tolerate salt stress by triggering the signal transduction pathway (Ghosh and Roychoudhury 2024).

Synthetic peptides like **OSIP108** (De Coninck et al. 2013) and **CLE14** have a great ability to cope with the effects of droughts, salinity, heavy metals, thermal, and nutrient stress. In case of drought and nutrient deficiency, CLE14 peptide interacts with ROS and nitric oxide (NO) producing genes. Because ROS and NO act as stimulants for increasing plant root area by root hair production and elongation. Increased root area facilitates plants for water and nutrient absorption. But a prolonged and excessive amount of ROS is damaging to plant tissues. So, the CLE14 peptide also has the ability to switch genes that can scavenge ROS to prevent their toxicity. Likewise, the application of synthetic **BrCEP-3** can induce drought stress response by H_2O_2 and Ca^{2+} mediated signal transduction cascade. BrCEP-3 peptide acts as a signaling molecule of receptor CEPR present on plasma membrane. This signal activates the kinase activity of receptor and triggers the transduction cascade with the involvement of H_2O_2 and Ca^{2+}. It causes root elongation to absorb more water under drought conditions. Although it seems a good approach to address challenges of abiotic stress but knowledge about detailed mechanisms and interactions of synthetic peptides with receptor active sites is limited (Qiu et al. 2022).

15.5 Future Prospects and Challenges

15.5.1 Potential of Peptide Hormones in Climate-Resilient Crops

The population is increasing day by day, posing a potential threat to environmental resources and health. It leads to changing climatic conditions and global warming. This directly affects the agricultural sector by reducing the overall crop yield. Conversely, the growing population also needs more food for its survival. So, it is a future need to produce climate-resilient crops. In this regard, peptide hormones offer a significant approach to increasing the stress tolerance ability of plants by switching signal transduction pathways and by interacting with other phytohormones, including ABA, IAA, gibberellins, and ethylene. Moreover, overexpression and knockout approaches for gene expression regulation can also be incorporated into global food security. Overexpression of many peptide hormones **(CLV3, AtPep-3, and OsDSSR1)** helps plants to withstand drought, salinity, heavy metals, heat, and cold stresses. Knockout or knockdown approaches can produce mutant phenotypes **(CEP-3-1a)** that can help to increase plants' adaptability in a challenging environment (Delay et al. 2013a, b). So, further research on overexpression and knockdown studies of peptide hormones will help to explore new peptide families that can respond to multiple stresses simultaneously.

15.5.2 Integrating Peptide Hormone Research with Precision Agriculture

Precision agriculture is a modern technique to optimize crop yield with the use of technological tools like sensors, drones, GPS, and data analysis. These tools help to detect real-time data about plant health and needs (watering, nutrients, and harvesting). Its major goal is to provide optimum amounts of water, nutrients, and agrochemicals at the right time. For instance, sensors can detect air moisture, heat intensity, nutrients, and water levels in soil. These sensors can also detect the amount of phytohormones (peptide hormones, IAA, ABA, etc.), metabolites, and ROS (Bukhamsin et al. 2025). The real-time data of these biomolecules helps to estimate the actual plant needs. Higher or lower levels of phytohormones are associated with stress responses in plants. For instance, the presence of CLE-25 and ABA in root and leaf cells of *Arabidopsis thaliana* represents the stress response toward drought stress (Takahashi et al. 2018). This sensor-based data can help farmers to decide about the amount and time of watering and other agrochemicals to overcome stress. Moreover, recent technological advances have made it possible to visualize 3D structures, molecular interactions, protein-ligand interactions, and drug likeness of peptides. Knowledge of these characteristics is helpful to make environment-friendly peptide-based formulations and agrochemicals (Zhang et al. 2023, 2025). Sensors can detect the crop needs of peptide-based formulations under stress conditions, and drone technology can be used to deliver these formulations. However, knowledge about the integration of peptide-based hormones in precision agriculture is limited and needs to be explored. Further research to explore new peptide formulations and their integration with precision agriculture will help to produce stress-resilient crops.

15.5.3 Challenges in Practical Applications

Although peptide hormones have a significant role in stress mitigation in plants, certain challenges exist regarding their use in agriculture. Some of these challenges are given below.

- Physicochemical properties like stability, receptor specificity, translocation, and mode of action of different peptide hormones vary according to their nature. A complete understanding of all these aspects is needed before their practical applications, especially in the case of synthetic peptides. Understanding structures and modes of action for each peptide hormone will also help to select their most appropriate type compatible with the plant and its environment. Poor handling and knowledge may cause unintended effects on the entire ecosystem.

- Synthetic peptide hormones are less explored, and a high budget is needed for their production. So, synthetic peptide-based formulations are unaffordable for farmers in low-income countries. Conversely, natural peptide-based agrochemicals are more reliable and cost-effective, but issues related to their stability exist. Natural peptides need optimum conditions for their functionality. Minor fluctuations of the surrounding environment, like temperature, pH, and solvents, can denature and hamper their activity (Zhang et al. 2023).
- Peptide hormones also interact with conventional phytohormones (IAA, ABA, gibberellin, and ethylene, etc.). In contrast, the mechanisms underlying these interactions under stress conditions are less understood. So, the production of peptide-based formulations that can better interact with conventional phytohormones is a big challenge.

15.6 Conclusion

Abiotic stressors such as extreme temperatures, drought, salinity, waterlogging, and exposure to harmful chemicals are potential threats to the growth, development, and productivity of plants. Plants respond to these environmental stressors by activating antioxidative defense mechanisms, molecular chaperones, cell membrane stabilization, stomatal regulation, osmolyte accumulation, synthesis of secondary metabolites, phytohormone biosynthesis, and signal transduction cascades. Amongst these, peptide hormone-based signal transduction cascades are major contributors to stress tolerance. Peptide hormones can control stress pathways more rapidly and in a more localized manner than traditional hormones (ABA, IAA, gibberellin, and ethylene). Peptide signaling involves ROS, calcium (Ca^{2+}) ions, MAPK-dependent signaling, and transcription factors to propagate molecular signals for the expression of target genes to normalize the plant growth under stress conditions. Some peptide hormones also interact with traditional phytohormones to switch the signal transduction cascade for stress response against drought, salinity, and heat stress. For example, **CLE-25** is a peptide hormone found in root cells of *Arabidopsis thaliana*. It is involved in closing stomata in response to drought stress. The pathway of stomatal closure is ABA-dependent. Activity of peptide-encoding genes could be manipulated by CRISPR/Cas system. Overexpression and knockdown/knockout techniques can play a significant role in increasing drought and salinity tolerance of plants and making them more resilient to climatic changes. Nowadays, synthetic peptides are also gaining popularity due to their longer stability and shelf life than natural peptides. The extreme specificity and redundancy within peptide ligands and their receptors offer a flexible regulatory framework in which plants can use redundant signaling to fine-tune their responses to stressors without compromising developmental signaling. However, knowledge about the integration of peptide-based hormones in precision agriculture is limited. Further research to explore new peptide formulations and mechanisms behind the interactions of these formulations with target receptor sites and other traditional hormones will help to produce stress-resilient crops.

References

Bukhamsin, A., Kosel, J., Blilou, I., and Salama, K.N. (2025). Accelerating adoption of species-agnostic plant sensors for precision farming. *Nature Reviews Electrical Engineering* 1–13.

Butenko, M.A., Vie, A.K., Brembu, T. et al. (2009). Plant peptides in signalling: looking for new partners. *Trends in Plant Science* 14 (5): 255–263.

Cao, Y., Yang, P., and Li, M. (2025). Research progress of peptides discovery and function in resistance to abiotic stress in plant. *Stress Biology* 5 (1): 1–12.

Charmandari, E., Tsigos, C., and Chrousos, G. (2005). Endocrinology of the stress response. *Annual Review of Physiology* 67: 259–284.

Cui, Y., Li, M., Yin, X. et al. (2018). OsDSSR1, a novel small peptide, enhances drought tolerance in transgenic rice. *Plant Science* 270: 85–96.

Cutler, S.R., Rodriguez, P.L., Finkelstein, R.R., and Abrams, S.R. (2010). Abscisic acid: emergence of a core signaling network. *Annual Review of Plant Biology* 61 (1): 651–679.

De Coninck, B., Carron, D., Tavormina, P. et al. (2013). Mining the genome of Arabidopsis thaliana as a basis for the identification of novel bioactive peptides involved in oxidative stress tolerance. *Journal of Experimental Botany* 64 (17): 5297–5307. https://doi.org/10.1093/jxb/ert295.

Delay, C., Imin, N., and Djordjevic, M.A. (2013a). Regulation of Arabidopsis root development by small signaling peptides. *Frontiers in Plant Science* 4: 352.

Delay, C., Imin, N., and Djordjevic, M.A. (2013b). CEP genes regulate root and shoot development in response to environmental cues and are specific to seed plants. *Journal of Experimental Botany* 64 (17): 5383–5394.

Fletcher, J.C., Brand, U., Running, M.P. et al. (1999). Signaling of cell fate decisions by CLAVATA3 in Arabidopsis shoot meristems. *Science* 283 (5409): 1911–1914.

Foyer, C.H. and Noctor, G. (2005). Redox homeostasis and antioxidant signaling: a metabolic interface between stress perception and physiological responses. *The Plant Cell* 17 (7): 1866–1875.

Funck, D., Baumgarten, L., Stift, M. et al. (2020). Differential contribution of P5CS isoforms to stress tolerance in Arabidopsis [Original research]. *Frontiers in Plant Science* 11: https://doi.org/10.3389/fpls.2020.565134.

Ghosh, P. and Roychoudhury, A. (2024). Plant peptides involved in abiotic and biotic stress responses and reactive oxygen species (ROS) signaling. *Journal of Plant Growth Regulation* 43 (5): 1410–1427.

Gish, L.A. and Clark, S.E. (2011). The RLK/Pelle family of kinases. *The Plant Journal* 66 (1): 117–127.

Groß, A., Hashimoto, C., Sticht, H., and Eichler, J. (2016). Synthetic peptides as protein mimics [Review]. *Frontiers in Bioengineering and Biotechnology* 3: https://doi.org/10.3389/fbioe.2015.00211.

Gull, A., Lone, A.A., and Wani, N.U.I. (2019). *Biotic and Abiotic Stresses in Plants*. IntechOpen.

Haruta, M., Sabat, G., Stecker, K. et al. (2014). A peptide hormone and its receptor protein kinase regulate plant cell expansion. *Science* 343 (6169): 408–411.

Hasanuzzaman, M., Hossain, M.A., da Silva, J.A.T., and Fujita, M. (2011). Plant response and tolerance to abiotic oxidative stress: antioxidant defense is a key factor. In: *Crop Stress and Its Management: Perspectives and Strategies*, 261–315. Dordrecht: Springer Netherlands.

Heidari, P., Rezaee, S., Hosseini Pouya, H.S., and Mora-Poblete, F. (2024). Insights into the heat shock protein 70 (Hsp70) family in *Camelina sativa* and its roles in response to salt stress. *Plants* 13 (23): 3410. https://www.mdpi.com/2223-7747/13/23/3410.

Iqbal, N., Umar, S., Per, T.S., and Khan, N.A. (2017). Ethephon increases photosynthetic-nitrogen use efficiency, proline and antioxidant metabolism to alleviate decrease in photosynthesis under salinity stress in mustard. *Plant Signaling & Behavior* 12 (5): e1297000.

Ji, C., Li, H., Zhang, Z. et al. (2025). The power of small signaling peptides in crop and horticultural plants. *The Crop Journal* 13 (3): 656–667.

Jin, X., Long, Y., Xiong, S. et al. (2021). SbNAC2 enhances abiotic stress tolerance by upregulating ROS scavenging activities and inducing stress-response genes in sorghum. *Environmental and Experimental Botany* 192: 104664.

Kieber, J.J. and Schaller, G.E. (2018). Cytokinin signaling in plant development. *Development (Cambridge, England)* 145 (4): dev149344.

Kondo, T., Sawa, S., Kinoshita, A. et al. (2006). A plant peptide encoded by CLV3 identified by in situ MALDI-TOF MS analysis. *Science* 313 (5788): 845–848.

Li, D.-H., Chen, F.-J., Li, H.-Y. et al. (2018). The soybean *GmRACK1* gene plays a role in drought tolerance at vegetative stages. *Russian Journal of Plant Physiology* 65: 541–552.

Li, S., Meng, S., Weng, J., and Wu, Q. (2022). Fine-tuning shoot meristem size to feed the world. *Trends in Plant Science* 27 (4): 355–363.

Li, G., Zhou, X., Zhu, G. et al. (2025). Macrobrachium rosenbergii Genome Editing Breeding with CRISPR–Cas Nucleases, Base Editors, and Prime Editors. *Animals* 15 (15): 2161.

Matsubayashi, Y. (2014). Posttranslationally modified small-peptide signals in plants. *Annual Review of Plant Biology* 65 (1): 385–413.

Nakaminami, K., Okamoto, M., Higuchi-Takeuchi, M. et al. (2018). AtPep3 is a hormone-like peptide that plays a role in the salinity stress tolerance of plants. *Proceedings of the National Academy of Sciences* 115 (22): 5810–5815.

Ohkubo, Y., Tanaka, M., Tabata, R. et al. (2017). Shoot-to-root mobile polypeptides involved in systemic regulation of nitrogen acquisition. *Nature Plants* 3 (4): 1–6.

Olsson, V., Joos, L., Zhu, S. et al. (2019). Look closely, the beautiful may be small: precursor-derived peptides in plants. *Annual Review of Plant Biology* 70 (1): 153–186.

Pearce, G., Moura, D.S., Stratmann, J., and Ryan, C.A. Jr. (2001). RALF, a 5-kDa ubiquitous polypeptide in plants, arrests root growth and development. *Proceedings of the National Academy of Sciences* 98 (22): 12843–12847.

Qiu, Z., Zhuang, K., Liu, Y. et al. (2022). Functional characterization of C-terminally encoded peptide (CEP) family in *Brassica rapa* L. *Plant Signaling & Behavior* 17 (1): 2021365.

Rehman, I., Shaukat, F., Anwar, Z. et al. (2022). Applications of CRISPR/Cas system in plants. *International Journal of Agriculture and Biosciences* 11 (4): 231–237. https://doi.org/10.47278/journal.ijab/2022.031.

Sachdev, S., Ansari, S.A., Ansari, M.I. et al. (2021). Abiotic stress and reactive oxygen species: generation, signaling, and defense mechanisms. *Antioxidants* 10 (2): 277.

Sah, S.K., Reddy, K.R., and Li, J. (2016). Abscisic acid and abiotic stress tolerance in crop plants. *Frontiers in Plant Science* 7: 571.

Santner, A., Calderon-Villalobos, L.I.A., and Estelle, M. (2009). Plant hormones are versatile chemical regulators of plant growth. *Nature Chemical Biology* 5 (5): 301–307.

Shen, J., Lv, B., Luo, L. et al. (2017). The NAC-type transcription factor OsNAC2 regulates ABA-dependent genes and abiotic stress tolerance in rice. *Scientific Reports* 7 (1): 40641.

Shi, H., He, X., Zhao, Y. et al. (2020). Constitutive expression of a group 3 LEA protein from Medicago falcata (MfLEA3) increases cold and drought tolerance in transgenic tobacco. *Plant Cell Reports* 39: 851–860.

Smith, S., Zhu, S., Joos, L. et al. (2020). The CEP5 peptide promotes abiotic stress tolerance, as revealed by quantitative proteomics, and attenuates the AUX/IAA equilibrium in Arabidopsis. *Molecular & Cellular Proteomics* 19 (8): 1248–1262.

Sun, X., Wang, F., Cai, H. et al. (2013). Functional characterization of an Arabidopsis prolyl aminopeptidase AtPAP1 in response to salt and drought stresses. *Plant Cell, Tissue and Organ Culture* 114: 325–338.

Takahashi, F., Suzuki, T., Osakabe, Y. et al. (2018). A small peptide modulates stomatal control via abscisic acid in long-distance signalling. *Nature* 556 (7700): 235–238.

Tavormina, P., De Coninck, B., Nikonorova, N. et al. (2015). The plant peptidome: an expanding repertoire of structural features and biological functions. *The Plant Cell* 27 (8): 2095–2118.

Verma, S., Nizam, S., and Verma, P.K. (2013). Biotic and abiotic stress signaling in plants. *Stress Signaling in Plants: Genomics and Proteomics Perspective* 1: 25–49.

Verma, V., Ravindran, P., and Kumar, P.P. (2016). Plant hormone-mediated regulation of stress responses. *BMC Plant Biology* 16: 1–10.

Wang, M., Ren, L.-T., Wei, X.-Y. et al. (2022). NAC transcription factor TwNAC01 positively regulates drought stress responses in Arabidopsis and Triticale [Original Research]. *Frontiers in Plant Science* 13: https://doi.org/10.3389/fpls.2022.877016.

Wang, W., Vinocur, B., Shoseyov, O., and Altman, A. (2004). Role of plant heat-shock proteins and molecular chaperones in the abiotic stress response. *Trends in Plant Science* 9 (5): 244–252.

Wu, Q., Schmidt, W., Aalen, R.B. et al. (2022). Editorial: peptide signaling in plants [editorial]. *Frontiers in Plant Science* 13: https://doi.org/10.3389/fpls.2022.843918.

Xiao, W., Jiang, W., Chen, Z. et al. (2025a). Advance in peptide-based drug development: delivery platforms, therapeutics and vaccines. *Signal Transduction and Targeted Therapy* 10 (1): 74.

Xiao, F., Zhou, H., and Lin, H. (2025b). Decoding small peptides: Regulators of plant growth and stress resilience. *Journal of Integrative Plant Biology* 67 (3): 596–631.

Zhang, Z., Han, H., Zhao, J. et al. (2025). Peptide hormones in plants. *Molecular Horticulture* 5 (1): 7.

Zhang, Y.-M., Ye, D.-X., Liu, Y. et al. (2023). Peptides, new tools for plant protection in eco-agriculture. *Advanced Agrochem* 2 (1): 58–78.

Zhu, J.K. (2016). Abiotic stress signaling and responses in plants. *Cell* 167 (2): 313–324.

16

Peptide Hormone Receptors and Plant Defense

Farhad Nazarian-Firouzabadi

Production Engineering and Plant Genetics Department, Faculty of Agriculture, Lorestan University, Khorramabad, Iran

16.1 Introduction

Plants, especially crops, are continuously exposed to a wide array of biotic stresses, including bacteria, fungi, and viruses (Gull et al. 2019). To counter these persistent threats, in turn, plants have evolved some complex defense mechanisms, among which plant peptide hormones (PPHs) play a pivotal role (Stührwohldt and Schaller 2019). These small but effective signaling molecules are entangled in regulating different cellular physiological processes, including growth, development, and immunity. In particular, PPHs have emerged as key modulators of plant resistance to disease by influencing the activation of defense responses, the regulation of stress tolerance, and the coordination of signaling pathways that enhance pathogen resistance (Denancé et al. 2013).

Peptide hormones in plants are typically 5–50 amino acids in length and are synthesized as precursors that are processed into their active forms (Table 16.1). One of the best-known families of PPHs involved in disease resistance is the *systemin* family, which triggers defense responses upon pathogen attack (Lindsey 2001). Systemin, an 18-amino-acid peptide, was first discovered in *tomato* (*Solanum lycopersicum L.*) and is involved in systemic acquired resistance (SAR) and the induction of defense-related genes (Pearce et al. 1991). Other peptide hormones, such as *flg22*, derived from the flagellin protein of bacteria, act as pathogen-associated molecular patterns (PAMPs) that trigger the plant's immune response through the activation of pattern recognition receptors (PRRs), a component of the plant's innate immunity (Chisholm et al. 2006). Additionally, *pep1*, a peptide hormone in *Arabidopsis*, is also involved in triggering immune responses against various pathogens by activating defense-related pathways (Huffaker et al. 2006).

A critical aspect of plant immunity involves the complex interplay between peptide hormones and various signaling pathways, such as those mediated by ethylene, Salicylic Acid (SA), and Jasmonic Acid (JA) hormones (Yang et al. 2015). These phytohormones work together to modulate the expression of defense genes and promote resistance to biotic stressors. Recent studies have demonstrated that peptide hormones can act as crucial

Peptide Hormones in Plant Development and Stress Responses, First Edition. Edited by Vijay Pratap Singh and Samiksha Singh.

Table 16.1 A list of characterized plant peptide hormone receptors, their functions, and roles in plant defense.

Peptide hormone	Receptor(s)	Activity	Role in defense	Key references
Systemin	SR160 (BRI1-associated kinase)	Mediates response to wounding and herbivore attack.	Induces defense-related gene expression and production of protease inhibitors to deter herbivores.	Ryan (2000); Yamaguchi et al. (2013)
AtPeps (e.g. AtPep1)	PEPR1/PEPR2	Amplifies pathogen-triggered immune responses.	Enhances pathogen resistance by inducing reactive oxygen species (ROS) production and defense gene expression.	Huffaker et al. (2006); Yamaguchi et al. (2006)
HypSys peptides	Unknown	Trigger systemic signaling in response to wounding.	Induce jasmonic acid-mediated defenses against herbivores.	Pearce and Ryan (2003)
RALF peptides	FERONIA (FER)	Regulates cell wall integrity and ion fluxes.	Maintains cell wall strength, which can contribute to defense against pathogen invasion.	Haruta et al. (2014)
CLAVATA3/ESR-related peptides (CLEs)	CLV1/CLV2/CRN Complex	Regulate stem cell homeostasis in meristems.	Some CLE peptides are induced during pathogen attack and may contribute to immune responses.	Kondo et al. (2006); Wang and Fiers (2010)
PSK (phytosulfokine)	PSKR1/PSKR2	Promotes cell growth and proliferation.	Contributes to wound healing and may modulate immune responses during stress conditions.	Mosher (2013)
CEP (C-terminally encoded peptides)	CEPR1/CEPR2	Regulate nitrogen uptake and root architecture.	Potential role in immune responses by modulating root interactions with soil pathogens.	Tabata et al. (2014)
ZIP peptides	RLKs (unknown specific receptor)	Influence immune system priming and defense.	Priming defense-related pathways against future pathogen attacks.	Hou et al. (2021); Ziemann et al. (2018)
GOLVEN peptides	RGI (RGF/GLV/INFLORESCENCE)	Regulate root development.	Role in defense not fully established but implicated in stress signaling pathways.	Fernandez et al. (2013); Whitford et al. (2012)

intermediaries in the cross talk between these hormones, fine-tuning the plant's immune response and providing a more coordinated defense against diverse pathogens (Derksen et al. 2013). For example, the activation of peptide hormones like C-terminally encoded peptide 1 (CEP1) has been shown to interact with the SA and JA signaling networks, leading to enhanced resistance against fungal and bacterial infections (Cipollini et al. 2004).

Furthermore, peptide hormones also contribute to resistance by modulating cell wall integrity and the formation of physical barriers that limit pathogen entry. For instance, rapid alkalinization factor (RALF) peptides can promote pH changes in the plant's extracellular matrix, creating an environment unfavorable for pathogen growth (Matos et al. 2008). These peptides also influence the deposition of callose and lignin, strengthening plant cell walls and enhancing pathogen resistance (Blackburn et al. 2020). Another notable peptide hormone, AtPeps, can induce resistance to the necrotrophic fungus *Botrytis cinerea* in *Arabidopsis* by activating defense genes associated with oxidative stress and cell wall fortification (Cui et al. 2024).

According to genome-wide *in silico* analysis, plant genomes contain at least 10 times more predicted genes encoding peptide receptors or transporters than those of animals. For instance, it is estimated that the *Arabidopsis thaliana* genome contains over 1000 genes encoding potential secretory peptides (Lease and Walker 2006), suggesting the wide range of such peptides in plants' different pathways and mechanisms.

The involvement of PPHs in disease resistance opens up new avenues for crop protection. Given their relatively small size and potent biological effects, these hormones or their analogs have the potential to be used as novel biotechnological tools to enhance plant immunity, reduce pesticide dependence, and improve agricultural sustainability. As research continues to elucidate the diverse roles of peptide hormones in plant disease resistance, these molecules may become central players in the development of next-generation disease-resistant crops.

16.2 Role of Peptide Hormones in Stress Responses

Peptide hormones play a significant role in regulating both biotic and abiotic stress responses by acting as chemical messengers that regulate physiological and behavioral adaptations to stress. Such small signaling peptides mostly interact with their specific protein receptors to activate downstream signaling pathways, enabling plants to maintain growth and survive adverse environmental conditions. Until recently, the significance of plant peptides in regulating and responding to environmental stresses was largely overlooked. However, over the past two decades, research has revealed their involvement in defense response mechanisms and established them as a distinct class of hormone molecules (Segonzac and Monaghan 2019). Such important peptides can act both systemically and locally at remarkably small concentrations (nM to pM), playing a crucial role in regulating plant development and stress responses (Wang and Irving 2011).

16.2.1 Abiotic Stress Responses

Plants have evolved various physiological, biochemical, and molecular adaptations to cope with drought and water stress. These strategies help maintain water balance, reduce water loss, and ensure survival under limited water availability. Among plants, stress-responsive genes have evolved encoding a wide range of peptides and proteins to regulate responses to diverse abiotic stresses such as drought and water deficit. Upregulation of drought-responsive genes, such as those encoding late embryogenesis abundant (LEA) proteins, aquaporins, and transcription factors like dehydration-responsive element-binding proteins (DREB), plays a pivotal role in coping with water stress in plants.

CLE genes encoding a group of small peptides (CLAVATA3/EMBRYO SURROUNDING REGION-RELATED) are essential plant signaling molecules that regulate cell differentiation and maintain meristem activity, playing a significant role in plant growth, development, and water uptake under drought stress. Derived from precursor proteins, the active CLE peptides consist of a conserved 12–13 amino acid domain and act in shoot apical meristems, root apical meristems, and vascular tissues. By binding to receptor-like kinases (RLKs; e.g. CLAVATA1), CLE peptides influence the balance between stem cell maintenance and differentiation. For example, CLAVATA3 restricts stem cell proliferation in the SAM by signaling through CLV1 and interacting antagonistically with the transcription factor WUSCHEL, whereas CLE40 regulates root stem cell niche activity in the RAM. Beyond development, CLE peptides also control meristem size, regulate vascular differentiation (e.g. CLE41/CLE44 in procambium activity), and adapt meristem function under stress to prioritize survival. These peptides hold significant potential in agriculture, with applications ranging from improving crop yields and stress resilience to advancing tissue engineering techniques. Their multifaceted roles underscore their importance in maintaining the dynamic balance of growth and environmental adaptation in plants.

16.2.2 Biotic Stress Responses

PPHs, such as systemin and phytosulfokine (PSK), play a crucial role in mediating plant responses to biotic stress, including pathogen attack. These peptides function as signaling molecules, activating defense pathways like the hypersensitive response (HR) and SAR, which help to limit pathogen spread (Bari and Jones 2009). Systemin, for instance, is involved in the induction of JA-mediated defenses against herbivores and pathogens (Turner et al. 2002). Additionally, peptides such as AtPep1 and its receptors trigger immunity by enhancing the plant's ability to perceive PAMPs, initiating innate immune responses (Boller and Felix 2009). Together, these PPHs constitute an essential component of the plant's immune system, boosting resistance to diseases through both local and systemic defense mechanisms.

16.3 Plant Peptide Receptors

In plants, peptide hormones exert their effects through specific receptor proteins located on the surface of plant cells. These receptors are essential for converting extracellular signals or stimuli into intracellular responses, enabling plants to adapt and respond to their dynamic environment (de Azevedo et al. 2021). In principle, PPRs are specialized proteins that recognize and bind to small bioactive peptides through specific motifs. Such receptors are typically membrane-bound proteins and belong to diverse families of RLKs or G-protein coupled receptors (GPCRs) (Pandey 2020). Once the peptide binds to the receptor, it triggers a series of signaling events within the cell, leading to changes in gene expression, ion fluxes, and cellular activities that guide growth, differentiation, defense, and environmental adaptation.

Genome-wide *in silico* analysis involving many different plant species has identified different classes of cell surface membrane-bound receptor proteins. Table 16.2 summarizes RLKs that have been identified by using genome-wide analysis in a number of model plants. The number of RLKs in plants is by far higher than any other gene family, suggesting their crucial role in different plant cell processes and pathways.

Typically, most RLKs consist of three main domains (Figure 16.1): an extracellular domain that detects various ligands, a transmembrane domain, and a cytoplasmic kinase

Table 16.2 Some of the well-characterized RLKs, their subfamilies, functions, and associated references.

RLK name	Subfamily	Plant species	Function	Key references
FLS2	LRR-RLK	*Arabidopsis*	Flagellin perception, triggers pattern-triggered immunity (PTI)	Gómez-Gómez and Boller (2000), Chinchilla et al. (2007)
EFR	LRR-RLK	*Arabidopsis*	Perception of bacterial EF-Tu (elf18), activates PTI	Chinchilla et al. (2007)
Xa21	LRR-RLK	Rice	Resistance to *Xanthomonas oryzae*, pathogen recognition	Song et al. (1995)
CERK1	LysM-RLK	*Arabidopsis*, rice	Chitin perception, fungal resistance, symbiosis signaling	Song et al. (1995); Miya et al. (2007)
PERK1	Proline-rich RLK	Rapeseed, Wheat	Senses cell wall damage, involved in growth and stress responses	Babu et al. (2008); Shumayla et al. (2022)
PR5K	Thaumatin-like RLK	*Arabidopsis*, Wheat	Pathogen resistance, chitinase activity, stress response	Sharma et al. (2020)
FER	CrRLK1L	*Arabidopsis*, Soybean	Cell wall integrity, growth, pathogen resistance	Wang et al. (2022)
LYK4, LYK5	LysM-RLK	*Arabidopsis*	Recognizes peptidoglycan/chitin	Willmann et al. (2011)
LYK7	LysM-RLK	Tomato	Recognizes peptidoglycan/chitin	Zeng et al. 2012
Ve1	LRR-RLK	Tomato	Recognizes *Verticillium* effector	Fradin et al. (2009)
BAK1	LRR-RLK	*Arabidopsis*		Li et al. (2002)
PEPR1, PEPR2	LRR-RLK	*Arabidopsis*	Perception of pathogen peptides	Yamaguchi et al. (2006)

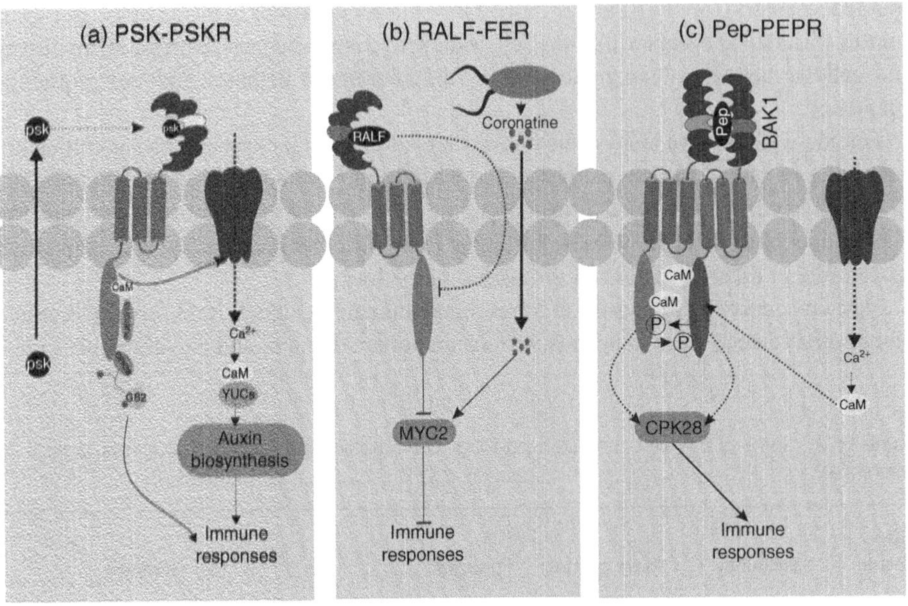

Figure 16.1 A schematic representation of receptor-like kinases with plant peptide hormones (PSK, RALF, and Pep). (a) A working model for PSKR (LRR-RLK) and PSK peptide signaling. The mature PSK is secreted to apoplasts, where it is perceived by the PSKRs' extracellular LRRs containing domain. The Ser/Thr kinase cytoplasmic domain is regulated by CaM binding. Next, the elevated cytosolic Ca^{2+} signal is further transduced to CaM. The CaM molecules bind to auxin biosynthetic proteins such as YUCs, eventually leading to auxin-mediated immunity against pathogens. The PSKR receptor binds and interacts with the CPK28 protein, which in turn phosphorylates the GS2 protein to inversely regulate cell immunity against plant pathogens. Furthermore, the PUB12/13 protein can interact with the PSKR receptor, resulting in ubiquitylation and further degradation of the receptor, maintaining a tight control of PSKR abundance at the posttranslational level. The kinase activity of PSKRs and/or the generated Ca^{2+} signaling predictably regulate downstream elements to confer biological functions, including regulation of disease resistance. (b) A working model for FER (LRR-RLK) and RALF peptide regulating MYC2 protein during bacterial infection. FER effectively phosphorylates and inhibits MYC2 to positively contribute to plant defense. The processing of RALF peptide is increased following bacterial infection, which functions to inhibit FER receptor signaling, hence negatively contributing to bacterial defense. (c) A working model showing the cooperation of the PEPR receptor with the BAK1 protein. Increasing Ca^{2+} efflux and further transduced to CaM, facilitate the autophosphorylation and/or trans-phosphorylation of PEPR and BAK1 proteins. Furthermore, the PEPR-BAK1 complex changes the CPK28 expression, eventually leading to a bunch of immune responses.

domain (Greeff et al. 2012). The cytoplasmic kinase domain serves as the core of the signaling cascade, where it induces the transmission of external signals to downstream protein components. The cytoplasmic kinase domain possesses catalytic activity by phosphorylating and dephosphorylating downstream proteins in the signaling pathways, leading to downstream gene expression. In most cases, the transmembrane domain of RLKs, composed of ~22–28 amino acids, connects the cytoplasmic kinase domain with that of the extracellular ligand-binding domain. The extracellular ligand-binding domain is situated at the N-terminal of RLKs and it is involved in signal perception. This domain exhibits significant diversity across different RLKs, contributing to the functional versatility of RLKs in plants. Most RLKs feature an N-terminal transit peptide. In plants, the RLKs superfamily is basically categorized into three major groups according to the presence or absence of the receptor domain and kinase activity domain (Torii 2009). RLKs can be further divided into more than 44 subgroups, based on the divergence of extracellular domains, i.e. leucine-rich repeat (LRR) RLKs, S-domain RLKs, and so on (Gish and Clark 2011). Among all these subgroups, the LRR-RLK group is by far the largest subgroup in plants. LRR-RLK members contain multiple repeats, each approximately 24 amino acids long, usually with conserved leucine (Leu) residues in their extracellular regions (Zhang 1998).

16.3.1 Phytosulfokine Receptors (PSK)

PSK peptides, members of plant peptide growth factors, are disulfated pentapeptides that are encoded by precursor genes mainly found in higher plants with universal occurrence. To gain full functionality, PSK preproproteins undergo sulfonylation by a tyrosylprotein sulfotransferase in the trans-Golgi apparatus, followed by a proteolytic cleavage in the apoplast, eventually producing PSK-α, -γ, -δ, and -ε full functional peptides (Sauter 2015; Li et al. 2024). PSK receptors (PSKR1 and PSKR2) are member of RLKs in plants. They play a significant role in plant innate immunity and disease resistance via activating specific signaling pathways that enhance plant responses to pathogen infections.

The fully functional PSK peptides are perceived by corresponding plasma membrane-localized RLK receptors. As can be seen from Figure 16.1a, the PSK-specific RLKs possess a transmembrane domain, a cytoplasmic domain, an extracellular domain containing ~21 LRRs and an island domain to which PSKs bind. Upon binding mature PSK peptides, the cytoplasmic domain of RLK receptors initiates phosphorylation activities by aid of their Ser/Thr kinase catalytic domain (Wong and Gehring 2013). Through Ca^{2+} accumulation and calmodulin (CaM) binding, these cascades lead to auxin-mediated immunity against pathogens (Zhang et al. 2018). The kinase activity of RLKs and/or the generated Ca^{2+} signaling regulates downstream proteins to confer biological functions, including resistance to biotic stress.

The activation of RLK receptors by PSK peptides can also lead to systemic immunity in plants, resulting in greater resistance to pathogenic infections. Therefore, PSK receptors are essential for the regulation of both basal immunity and enhanced resistance in plants. PSK receptors assist plants in responding effectively to pathogen attacks by activating a wide array of immune pathways and hormone signaling.

16.3.2 Rapid Alkalinization Factor Receptors (RALF)

RALF polypeptides are a family of small ~5-kDa peptides that are mainly composed of ~50 amino acids, which were originally isolated from tobacco leaves (Pearce et al. 2001). Reports describe on the role of RALF peptides in the growth of roots, activating mitogen-activated protein kinases (MAPKs), alkalinizing the cell and plant growth (Pearce et al. 2001; Yamaguchi and Huffaker 2011; Germain et al. 2005; Scheer et al. 2005). In 2013, Atkinson et al. (2013) showed that RALF peptides play a significant role in plant defense responses (Atkinson et al. 2013).

The receptor kinase FERONIA (FER), belonging to the CrRLK1L family of RLKs, forms a critical signaling hub in plants' diverse biological functionalities such as hormone signaling, biotic and abiotic stress responses, plant growth, and development. The interaction of RALF and FER which leads to signaling can either activate or suppress plants' innate immune responses (Guo et al. 2018). Depending on context and peptide type, RALF peptides primarily bind to FER receptors and other associated CrRLK1L family members, eventually triggering downstream signaling pathways, including auxin, JA, and ABA signaling, production of ROS, pathogen defense regulation, and calcium signaling. Coronatine (COR), a phytotoxin produced by some bacterial pathogens, structurally and functionally resembles the active form of JA known as JA-isoleucine (JA-Ile). It has been documented that COR can hijack the host plant cells' JA-signaling pathway. Leading to host disease susceptibility (Xin and He 2013). To do this, COR employs the MYC2 transcription factor as one of the main regulators of JA signaling, to activate NAC transcription factors, which in turn inhibit the accumulation of SA and thus compromise host immunity (Zheng et al. 2012). In *A. thaliana*, FER activity phosphorylates and consequently inhibits MYC2 to positively contribute to plant defense. The processing of RALF23 peptide (Figure 16.1b) is increased by bacterial infection, which functions to inhibit FER receptor signaling, hence negatively contributes to bacterial defense (Guo et al. 2018).

16.3.3 Plant Elicitor Peptide Receptors

Plant elicitor peptides (Peps) are small endogenous peptides of approximately 23–27 amino acids that serve as potent signals to activate plant innate immunity following pathogen attack. Peps are, indeed, damage-associated molecular patterns (DAMPs), small molecules that are released from plant cells when they are damaged, either by mechanical injury or pathogen attack. In other words, Peps are produced by the plant cells as a way to "sense" internal damages. They are usually derived from precursor proteins called PROcessed Elicitor Peptide (PROPEPs), which are cleaved to release the active Pep peptides. So far, a number of such genes encoding Peps have been identified in *Arabidopsis*, maize (ZmPEP1), tomato (SlPeps), and rice (OsPep) (Trivilin et al. 2014; Huffaker et al. 2011; Shen et al. 2022). The PROPEP family in *Arabidopsis thaliana* (AtPeps) plays a pivotal role in plant innate immunity and stress responses. Genome-wide *in silico* analysis showed that *Arabidopsis* genome comprises eight AtPeps genes (PROPEP1–8). AtPeps modulate JA and SA pathways, and interact with cell wall integrity and calcium signaling to enhance plant resilience. Recent research has highlighted diverse expression patterns, receptor binding, and proteolytic release mechanisms of AtPeps (Yamaguchi et al. 2010). They are rapidly induced by pathogen attack, wounding, and abiotic stress, eventually triggering defense responses through their corresponding receptors PEPR1 and PEPR2 (Yamaguchi et al. 2010).

Crystal structure analysis, mutagenesis studies, and co-immunoprecipitation analysis confirm that PEPR receptors are members of LRR-RLKs. Upon stress stimuli such as pathogen attack, AtPep1 binds and fits into a conserved groove on the extracellular LRR domain of PEPR1 receptor (Figure 16.1c), involving Tyr21, Leu6, Gly17 residues via a key and lock mechanism (Tang et al. 2015). To stabilize the binding to PEPR1, AtPep1 recruits BRI1-ASSOCIATED KINASE 1 (BAK1), a coreceptor kinase (Claus et al. 2025; Wang et al. 2025). Once PEPR1 is bound to AtPep1, the expression of defense genes like *PDF1.2*, *WRKYs*, and *PR1* is upregulated, callose is produced and deposited, and ROS bursts occur, collectively resulting in pathogen resistance (Ranatnuge and Hasara 2024).

16.4 Implication of Peptide Hormone Receptors in Plant Breeding

The breeding implications of peptide hormones and their association with corresponding receptors can be used for marker-assisted breeding, transgenic/genome-edited crops, and functional genomics studies. Due to the role of interaction between PPHs and corresponding receptors, they can initiate pattern-triggered immunity (PTI) as well as effector-triggered immunity (ETI) by activating downstream signaling cascades, producing defense-related hormones (e.g. SA, JA, ethylene), strengthening the cell wall, and inducing reactive oxygen species (ROS) production. Overexpression of these receptors often leads to enhanced resistance to biotic and abiotic stresses, especially through the activation of PTI, SAR, and ROS production pathways. Many studies have highlighted the importance of receptors like PEPRs, the tyrosine-sulfated peptide receptors (PSY1R), and CLAVATA-like receptors in enhancing plant cells' defense responses.

By targeting peptide hormone receptors such as LRR-RLKs, breeders can modulate plant architecture and improve resistance to pathogens. Some peptide hormones, such as Peps and RALFs, interact with receptors like PEPR1/2 and FERONIA. Therefore, manipulating these receptors by genetic engineering tools can lead to pathogen-resistant cultivars without compromising growth, a common challenge in classical breeding. Peptide-receptor signaling helps plants respond to mechanical wounding and pathogen attacks. Since peptide hormones interact with other hormonal pathways such as auxin, cytokinin, and ethylene biosynthesis, understanding receptor-level signaling allows for more precise hormone manipulation and balanced growth-immunity trade-offs. Such understanding enables plant breeding methods for multitrait optimization. Recently, breeders have been able to use gene editing tools like CRISPR genome editing to change the receptor sensitivity or expression of synthetic peptides to control development without genetic modification. Furthermore, peptide receptors can serve as biomarkers for selecting elite genotypes in breeding programs.

References

Atkinson, N.J., Lilley, C.J., and Urwin, P.E. (2013). Identification of genes involved in the response of Arabidopsis to simultaneous biotic and abiotic stresses. *Plant Physiology* 162 (4): 2028–2041.

de Azevedo, E., Manhães, A.M., Ortiz-Morea, F.A. et al. (2021). Plant plasma membrane-resident receptors: surveillance for infections and coordination for growth and development. *Journal of Integrative Plant Biology* 63 (1): 79–101.

Babu, C.S., Babar, S.M.E., Song, E.J. et al. (2008). *Kinetic Analysis of the MAPK and PI3K/Akt Signaling Pathways*, vol. 25. Elsevier.

Bari, R. and Jones, J.D. (2009). Role of plant hormones in plant defence responses. *Plant Molecular Biology* 69: 473–488.

Blackburn, M.R., Haruta, M., and Moura, D.S. (2020). Twenty years of progress in physiological and biochemical investigation of RALF peptides. *Plant Physiology* 182 (4): 1657–1666.

Boller, T. and Felix, G. (2009). A renaissance of elicitors: perception of microbe-associated molecular patterns and danger signals by pattern-recognition receptors. *Annual Review of Plant Biology* 60 (1): 379–406.

Chinchilla, D., Zipfel, C., Robatzek, S. et al. (2007). A flagellin-induced complex of the receptor FLS2 and BAK1 initiates plant defence. *Nature* 448 (7152): 497–500.

Chisholm, S.T., Coaker, G., Day, B., and Staskawicz, B.J. (2006). Host-microbe interactions: shaping the evolution of the plant immune response. *Cell* 124 (4): 803–814.

Cipollini, D., Enright, S., Traw, M., and Bergelson, J. (2004). Salicylic acid inhibits jasmonic acid-induced resistance of *Arabidopsis thaliana* to *Spodoptera exigua*. *Molecular Ecology* 13 (6): 1643–1653.

Claus, L.A.N., Ortiz-Morea, F.A., Yang, S.-L. et al. (2025). Endocytosis of the damage-associated molecular pattern receptor PEPR1 is BAK1-dependent. *bioRxiv:2025.2003. 2015.643409.*

Cui, J., Sa, E., Wei, J. et al. (2024). The truncated peptide AtPEP1 (9–23) has the same function as AtPEP1 (1–23) in inhibiting primary root growth and triggering of ROS burst. *Antioxidants* 13 (5): 549.

Denancé, N., Sánchez-Vallet, A., Goffner, D., and Molina, A. (2013). Disease resistance or growth: the role of plant hormones in balancing immune responses and fitness costs. *Frontiers in Plant Science* 4: 155.

Derksen, H., Rampitsch, C., and Daayf, F. (2013). Signaling cross-talk in plant disease resistance. *Plant Science* 207: 79–87.

Fernandez, A., Hilson, P., and Beeckman, T. (2013). GOLVEN peptides as important regulatory signalling molecules of plant development. *Journal of Experimental Botany* 64 (17): 5263–5268.

Fradin, E.F., Zhang, Z., Juarez Ayala, J.C. et al. (2009). Genetic dissection of *Verticillium* wilt resistance mediated by tomato Ve1. *Plant Physiology* 150 (1): 320–332.

Germain, H., Chevalier, E., Caron, S., and Matton, D.P. (2005). Characterization of five RALF-like genes from *Solanum chacoense* provides support for a developmental role in plants. *Planta* 220: 447–454.

Gish, L.A. and Clark, S.E. (2011). The RLK/Pelle family of kinases. *The Plant Journal* 66 (1): 117–127.

Gómez-Gómez, L. and Boller, T. (2000). FLS2: an LRR receptor–like kinase involved in the perception of the bacterial elicitor flagellin in Arabidopsis. *Molecular Cell* 5 (6): 1003–1011.

Greeff, C., Roux, M., Mundy, J., and Petersen, M. (2012). Receptor-like kinase complexes in plant innate immunity. *Frontiers in Plant Science* 3: 209.

Gull, A., Lone, A.A., and Wani, N.U.I. (2019). Biotic and abiotic stresses in plants. *Abiotic and Biotic Stress in Plants* 7: 1–9.

Guo, H., Nolan, T.M., Song, G. et al. (2018). FERONIA receptor kinase contributes to plant immunity by suppressing jasmonic acid signaling in Arabidopsis thaliana. *Current Biology* 28 (20): 3316–3324. e3316.

Haruta, M., Sabat, G., Stecker, K. et al. (2014). A peptide hormone and its receptor protein kinase regulate plant cell expansion. *Science* 343 (6169): 408–411.

Hou, S., Liu, D., and He, P. (2021). Phytocytokines function as immunological modulators of plant immunity. *Stress Biology* 1: 1–14.

Huffaker, A., Dafoe, N.J., and Schmelz, E.A. (2011). ZmPep1, an ortholog of Arabidopsis elicitor peptide 1, regulates maize innate immunity and enhances disease resistance. *Plant Physiology* 155 (3): 1325–1338.

Huffaker, A., Pearce, G., and Ryan, C.A. (2006). An endogenous peptide signal in Arabidopsis activates components of the innate immune response. *Proceedings of the National Academy of Sciences* 103 (26): 10098–10103.

Kondo, T., Sawa, S., Kinoshita, A. et al. (2006). A plant peptide encoded by CLV3 identified by in situ MALDI-TOF MS analysis. *Science* 313 (5788): 845–848.

Lease, K.A. and Walker, J.C. (2006). The Arabidopsis unannotated secreted peptide database, a resource for plant peptidomics. *Plant Physiology* 142 (3): 831–838.

Li, Y., Di, Q., Luo, L., and Yu, L. (2024). Phytosulfokine peptides, their receptors, and functions. *Frontiers in Plant Science* 14: 1326964.

Li, J., Wen, J., Lease, K.A. et al. (2002). BAK1, an Arabidopsis LRR receptor-like protein kinase, interacts with BRI1 and modulates brassinosteroid signaling. *Cell* 110 (2): 213–222.

Lindsey, K. (2001). Plant peptide hormones: the long and the short of it. *Current Biology* 11 (18): R741–R743.

Matos, J.L., Fiori, C.S., Silva-Filho, M.C., and Moura, D.S. (2008). A conserved dibasic site is essential for correct processing of the peptide hormone AtRALF1 in Arabidopsis thaliana. *FEBS Letters* 582 (23–24): 3343–3347.

Miya, A., Albert, P., Shinya, T. et al. (2007). CERK1, a LysM receptor kinase, is essential for chitin elicitor signaling in Arabidopsis. *Proceedings of the National Academy of Sciences* 104 (49): 19613–19618.

Mosher, S. (2013). *The Tyrosine-Sulfated Peptide Receptors PSKR1 and PSY1R Modulate Arabidopsis Immune Responses*. Universität Tübingen.

Pandey, S. (2020). Plant receptor-like kinase signaling through heterotrimeric G-proteins. *Journal of Experimental Botany* 71 (5): 1742–1751.

Pearce, G., Moura, D.S., Stratmann, J., and Ryan, C.A. Jr. (2001). RALF, a 5-kDa ubiquitous polypeptide in plants, arrests root growth and development. *Proceedings of the National Academy of Sciences* 98 (22): 12843–12847.

Pearce, G. and Ryan, C.A. (2003). Systemic signaling in tomato plants for defense against herbivores: isolation and characterization of three novel defense-signaling glycopeptide hormones coded in a single precursor gene. *Journal of Biological Chemistry* 278 (32): 30044–30050.

Pearce, G., Strydom, D., Johnson, S., and Ryan, C.A. (1991). A polypeptide from tomato leaves induces wound-inducible proteinase inhibitor proteins. *Science* 253 (5022): 895–897.

Ranatnuge, N.P. and Hasara, W.T. (2024). Plant elicitor peptides as amplifiers of immune responses against biotic stressors. In: *Plant Elicitor Peptides: New Tool for Sustainable Agriculture* (ed. S. Singh and R. Mehrotra), 1–18. Springer.

Ryan, C.A. (2000). The systemin signaling pathway: differential activation of plant defensive genes. *Biochimica et Biophysica Acta (BBA)-Protein Structure and Molecular Enzymology* 1477 (1–2): 112–121.

Sauter, M. (2015). Phytosulfokine peptide signalling. *Journal of Experimental Botany* 66 (17): 5161–5169.

Scheer, J.M., Pearce, G., and Ryan, C.A. (2005). Le RALF, a plant peptide that regulates root growth and development, specifically binds to 25 and 120 kDa cell surface membrane proteins of Lycopersicon peruvianum. *Planta* 221 (5): 667–674.

Segonzac, C. and Monaghan, J. (2019). Modulation of plant innate immune signaling by small peptides. *Current Opinion in Plant Biology* 51: 22–28.

Sharma, A., Tyagi, S., Alok, A. et al. (2020). Thaumatin-like protein kinases: molecular characterization and transcriptional profiling in five cereal crops. *Plant Science* 290: 110317.

Shen, W., Zhang, X., Liu, J. et al. (2022). Plant elicitor peptide signalling confers rice resistance to piercing-sucking insect herbivores and pathogens. *Plant Biotechnology Journal* 20 (5): 991–1005.

Shumayla, M.V., Singh, K., and Upadhyay, S.K. (2022). Insight into the roles of proline-rich extensin-like receptor protein kinases of bread wheat (*Triticum aestivum* L.). *Life* 12 (7): 941.

Song, W.-Y., Wang, G.-L., Chen, L.-L. et al. (1995). A receptor kinase-like protein encoded by the rice disease resistance gene, Xa21. *Science* 270 (5243): 1804–1806.

Stührwohldt, N. and Schaller, A. (2019). Regulation of plant peptide hormones and growth factors by post-translational modification. *Plant Biology* 21: 49–63.

Tabata, R., Sumida, K., Yoshii, T. et al. (2014). Perception of root-derived peptides by shoot LRR-RKs mediates systemic N-demand signaling. *Science* 346 (6207): 343–346.

Tang, J., Han, Z., Sun, Y. et al. (2015). Structural basis for recognition of an endogenous peptide by the plant receptor kinase PEPR1. *Cell Research* 25 (1): 110–120.

Torii, K.U. (2009). Transmembrane receptors in plants: receptor kinases and their ligands. *Annual Plant Reviews* 33: 1–29.

Trivilin, A., Hartke, S., and Moraes, M. (2014). Components of different signalling pathways regulated by a new orthologue of A t PROPEP 1 in tomato following infection by pathogens. *Plant Pathology* 63 (5): 1110–1118.

Turner, J.G., Ellis, C., and Devoto, A. (2002). The jasmonate signal pathway. *The Plant Cell* 14 (suppl_1): S153–S164.

Wang, P., Clark, N.M., Nolan, T.M. et al. (2022). Integrated omics reveal novel functions and underlying mechanisms of the receptor kinase FERONIA in Arabidopsis thaliana. *The Plant Cell* 34 (7): 2594–2614.

Wang, G. and Fiers, M. (2010). CLE peptide signaling during plant development. *Protoplasma* 240: 33–43.

Wang, Y.H. and Irving, H.R. (2011). Developing a model of plant hormone interactions. *Plant Signaling & Behavior* 6 (4): 494–500.

Wang, X., Yu, W., Yuan, Q. et al. (2025). The pathogen-induced peptide CEP14 is perceived by the receptor-like kinase CEPR2 to promote systemic disease resistance in Arabidopsis. *Plant Physiology* 197 (1): kiae549.

Whitford, R., Fernandez, A., Tejos, R. et al. (2012). GOLVEN secretory peptides regulate auxin carrier turnover during plant gravitropic responses. *Developmental Cell* 22 (3): 678–685.

Willmann, R., Lajunen, H.M., Erbs, G. et al. (2011). Arabidopsis lysin-motif proteins LYM1 LYM3 CERK1 mediate bacterial peptidoglycan sensing and immunity to bacterial infection. *Proceedings of the National Academy of Sciences* 108 (49): 19824–19829.

Wong, A. and Gehring, C. (2013). The Arabidopsis thaliana proteome harbors undiscovered multi-domain molecules with functional guanylyl cyclase catalytic centers. *Cell Communication and Signaling* 11: 1–6.

Xin, X.-F. and He, S.Y. (2013). Pseudomonas syringae pv. tomato DC3000: a model pathogen for probing disease susceptibility and hormone signaling in plants. *Annual Review of Phytopathology* 51 (1): 473–498.

Yamaguchi, Y. and Huffaker, A. (2011). Endogenous peptide elicitors in higher plants. *Current Opinion in Plant Biology* 14 (4): 351–357.

Yamaguchi, Y., Huffaker, A., Bryan, A.C. et al. (2010). PEPR2 is a second receptor for the Pep1 and Pep2 peptides and contributes to defense responses in Arabidopsis. *The Plant Cell* 22 (2): 508–522.

Yamaguchi, Y., Pearce, G., and Huffaker, A. (2013). Defense signal peptides. In: *Handbook of Biologically Active Peptides* (ed. A.J. Kastin), 5–14. Elsevier.

Yamaguchi, Y., Pearce, G., and Ryan, C.A. (2006). The cell surface leucine-rich repeat receptor for At Pep1, an endogenous peptide elicitor in Arabidopsis, is functional in transgenic tobacco cells. *Proceedings of the National Academy of Sciences* 103 (26): 10104–10109.

Yang, Y.-X., Ahammed, G.J., Wu, C. et al. (2015). Crosstalk among jasmonate, salicylate and ethylene signaling pathways in plant disease and immune responses. *Current Protein and Peptide Science* 16 (5): 450–461.

Zeng, L., Velásquez, A.C., Munkvold, K.R. et al. (2012). A tomato LysM receptor-like kinase promotes immunity and its kinase activity is inhibited by AvrPtoB. *The Plant Journal* 69 (1): 92–103.

Zhang, X. (1998). Leucine-rich repeat receptor-like kinases in plants. *Plant Molecular Biology Reporter* 16: 301–311.

Zhang, H., Hu, Z., Lei, C. et al. (2018). A plant phytosulfokine peptide initiates auxin-dependent immunity through cytosolic Ca2+ signaling in tomato. *The Plant Cell* 30 (3): 652–667.

Zheng, X.-Y., Spivey, N.W., Zeng, W. et al. (2012). Coronatine promotes *Pseudomonas syringae* virulence in plants by activating a signaling cascade that inhibits salicylic acid accumulation. *Cell Host & Microbe* 11 (6): 587–596.

Ziemann, S., Van Der, K., Lahrmann, U. et al. (2018). An apoplastic peptide activates salicylic acid signalling in maize. *Nature Plants* 4: 172–180.

17

Plant Peptide Hormones in Salt and Drought Stress Tolerance

Tülay Alp Öztürk and Melike Bor

Faculty of Science, Department of Biology, Ege University, Bornova-Izmir, Türkiye

17.1 Introduction

In multicellular organisms, intercellular signaling is essential for the coordinated regulation of growth and development. Plants, like mammals, use signaling pathways through small peptides and their receptors to control and regulate growth, development, and adaptation to environmental constraints (Czyzewicz et al. 2013). Peptide hormones are defined as short- or long-range secreted or nonsecreted signaling molecules with small polypeptide structures. Since the first plant peptide hormone, "systemin," was identified in tomato in 1991, a wide variety of plant peptide hormones with different functions ranging from growth and development to stress coping mechanisms have been identified. There are studies showing that plant peptide hormones play a critical role in various processes such as cell differentiation, organogenesis, embryo development, fertilization, pollination, biotic and abiotic stress responses (Zhang et al. 2025). The effects of plant peptide hormones on growth, development, and stress responses have been shown to be associated with transcription factors, but the steps and autonomy of this control mechanism have not been clearly defined (Czyzewicz et al. 2013). Accordingly, membrane-localized receptor kinases are important components of peptide hormone signaling and recent studies put forward the importance of conserved motifs for receptor binding and activation mechanisms (Zhang et al. 2016). The extracellular domains of receptor kinases are able to recognize small signaling peptides and binding of peptide hormones induces homo- or heterodimerization of these receptors that lead to alterations in the transcriptional programming (Han et al. 2014; Zhang et al. 2016). Therefore, modulation of receptor kinases' activity by structural changes through peptide hormones may be critical not only in growth- and development-related processes but also for stress tolerance.

Plant peptide hormones usually contain 2–100 amino acids and according to the latest categorization they are divided into two major groups: **conventional** (precursor protein derived) and **nonconventional** (nonprecursor protein derived) peptides (Datta et al. 2024; Xiao et al. 2025) (Figure 17.1). The conventional group is derived from functional

Peptide Hormones in Plant Development and Stress Responses, First Edition. Edited by Vijay Pratap Singh and Samiksha Singh.

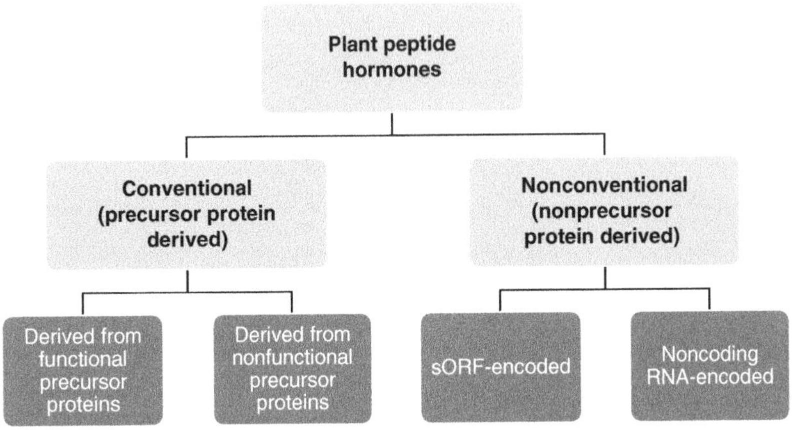

Figure 17.1 Recent classification of plant peptide hormones. *Source:* Adapted from Datta et al. (2024); Xiao et al. (2025).

or nonfunctional precursor proteins. The functional precursor-derived peptides have a signal peptide at their N-terminal and carry out their biological roles after being secreted within the plant, while the nonfunctional precursor peptides contain posttranslational modifications (PTMs), cysteine-rich peptides (CRPs), and non–cysteine-rich/non-PTM peptides (Lu and Xiao 2024; Chang and Xiao 2025; Xiao et al. 2025). Among the secreted peptide hormones, CLAVATA3/EMBRYO SURROUNDING REGION-RELATED peptides (CLEs), RAPID ALKALIZATION FACTORS (RALFs), EPIDERMAL PATTERNING FACTORS (EPFs), C-TERMINALLY ENCODED PEPTIDES (CEPs), PHYTOSULFOKINES (PSKs), PLANT DEFENSINS (PDFs), THIONIN-LIKE (THL), and NODULE-SPECIFIC CYSTEINE-RICH PEPTIDE (NCR) can be given, which all have different functions not only in *Arabidopsis* but also in other plants including crop species as well (Lu and Xiao 2024; Chang and Xiao 2025; Xiao et al. 2025). The mature CLE peptides derived from the conserved CLE domain are all composed of 12 amino acids and contain hydroxyproline residues (Hirakawa and Sawa 2019). Among the best characterized CLE peptides, CLV3 is an important signal for shoot apical meristem growth, while its functional counterpart CLE40 serves the same role in the root apical meristem in *Arabidopsis* (Clark et al. 1995; Fiers et al. 2005; Czyzewicz et al. 2013). The first identified sulfated peptide hormone, PSK, was isolated from *Asparagus* mesophyll cells (Matsubayashi and Sakagami 1996; Ding et al. 2023). PSK has two sulfated tyrosine residues and is recognized by leucine-rich repeat receptor-like kinase (LRR-RLK) receptors localized on membranes. The signal transduction cascade activated by PSK is reported to regulate the balance between growth and defense against Pst DC3000 in tomato (Ding et al. 2023). RALFs affect growth by controlling cell expansion not only in *Arabidopsis* but also in *S. lycopersicum* and *N. attenuate* (Wu et al. 2007; Srivastava et al. 2009; Covey et al. 2010; Czyzewicz et al. 2013). The EPF2 peptide regulates the formation of stomata by controlling the spacing and the number of cells that will develop into stomata (Hunt and Gray 2009; Czyzewicz et al. 2013).

The nonsecreted and nonfunctional precursor derived peptide hormones are; SYSTEMIN (SYS), EARLY NODULIN 40 (ENOD40), KISS OF DEATH (KOD), *ZEA MAYS* IMMUNE

SIGNALING PEPTIDE 1 (ZIP1), GRIM REAPER PEPTIDE (GRI), and PLANT ELICITOR PEPTIDES (PEPs) which also have diverse functions in plants (Lu and Xiao 2024; Chang and Xiao 2025; Xiao et al. 2025). Although they are nonsecreted, these peptides are also recognized by receptors and activate cellular signaling processes. For instance, the first identified peptide, tomato systemin, is transferred to the apoplast when plant cells are wounded and perceived by the systemin receptor (SYR), a leucine-rich repeat receptor kinase (Hirakawa and Sawa 2019). Processing of peptide hormones by PTMs is important for their functionality; therefore, a better understanding of PTMs will be important for the manipulation of signaling networks via peptide hormones. Undergoing PTM is critical for the biological activity of peptide hormones and these modifications include tyrosine sulfation, proline hydroxylation, and arabinosylation. To date, several PTM patterns have been reported to be associated with the maturation process of plant peptide hormones and sometimes the same peptide hormone may be processed by different PTMs according to the physiological conditions. PTMs also determine the stability and specificity of peptide-receptor interaction and enable regulation of their biological functions in growth and stress responses (Chang and Xiao 2025; Xiao et al. 2025). Besides PTMs, proteolytic processing of plant peptide hormones for their maturation is also important for nonsecreted peptide hormones. Hander et al. (2019) found that wounding induced Ca^{2+} accumulation activated metacaspase 4 (MC4) for the release of nonsecreted peptide (PEP1) from its tonoplast-localized precursor protein.

The second class, nonconventional peptide hormones, is grouped into small open reading frame (sORF)-encoded and noncoding RNA (long noncoding RNAs—lncRNAs, circular RNAs—circRNAs, and mitochondrial RNAs—mtRNAs) originated (Xiao et al. 2025) (Figure 17.1). Example of the sORF-encoded ones is the conserved peptide upstream open reading frame (CPuORF), while FAst-growing MOSS (FAMOSS), ELVd-ORF1 (E-ORF1), and mitochondrial-derived peptides (MDPs) are examples of the ones encoded by noncoding RNAs (Xiao et al. 2025). With the aid of advances in sequencing (RNA-seq, Ribo-seq, etc.) and bioinformatics, several nonconventional peptide hormones have been identified (Yu et al. 2025; Xiao et al. 2025). The ncRNA-encoded peptides (ncPEPs) are defined as critical regulatory components of gene expression during plant development and stress responses. In light of this, a specific database for presenting ncRNA and ncPEP data across different plant species has been launched (https://bis.zju.edu.cn/ncPlantDB) (Liu et al. 2025). This database provides information on ncRNA expression patterns, tissue-specific ncRNAs, potential coding capacity of ncRNAs, and potential ncPEPs, interactions, and networks, which all will improve our understanding of plant ncPEPs (Liu et al. 2025). Although these peptide hormones' biosynthesis and structural properties are quite different from the conventional ones, they are reported to have similar regulatory functions in growth, development, and stress responses in plants (Chang and Xiao 2025). For instance, Yu et al. (2025) reported a unique 31-amino acid containing nonconventional peptide hormone; microRPG1 regulates kernel dehydration during the kernel maturation process through ethylene signaling in maize. Accordingly, the exogenous microRPG1 treatment had the same effect in *Arabidopsis*, which indicated a plausible conserved response for dehydration control among plants (Yu et al. 2025).

Another interesting group of nonconventional peptide hormones is the microRNA-encoded peptides (miPEPs). These peptides are translated from sORFs of the primary miRNA transcripts and they regulate their corresponding miRNAs by positive feedback (Lauressergues et al. 2022; Xiao et al. 2025). Recent studies showed that miRNA regulation through miPEPs might also play an important role in plant growth, development, and stress responses. In *Medicago truncatula* miPEP171b reported to enhance the accumulation of mature miR171b, which is associated with root development (Lauressergues et al. 2015) and, miPEP858a in *Arabidopsis* and miPEP172c in *Glycine max*, regulate flavonoid biosynthesis and nodulation, respectively (Couzigou et al. 2016; Sharma et al. 2020). miPEPs are proposed to be internalized to plant cells by endocytosis or passive diffusion rather than by ligand-receptor association (Ormancey et al. 2020; Ren et al. 2021). Ormancey et al. (2020) reported that miPEP165a was transported into the whole root system within 24 hours, while the inhibition of endocytosis affected this peptide's uptake in *Arabidopsis*. Their detailed regulatory mechanisms and mode of action are yet to be defined. However, direct interaction of miPEPs with proteins or enzymes was proposed to be a way to explain their function in growth, development, and stress responses. miPEPs may bind or interact with enzymes, including polymerase II for regulating transcription, similar to other small peptides such as Early nodulin 40 (ENOD40) in soybean that functions in root nodule organogenesis (Röhrig et al. 2002; Ren et al. 2021). *ENOD40* genes contain only short ORFs and 5′-located conserved region I of soybean *ENOD40* encodes a small peptide that binds explicitly to sucrose synthase's nodulin 100 subunit. It was proposed to be responsible for the control of sucrose consumption in nitrogen-fixing nodules (Röhrig et al. 2002; Ren et al. 2021).

Under environmental stress conditions, one of the biggest constraints is the maintenance of the growth-defense balance in plants. Therefore, plant peptide hormones provide a handful of opportunities for developing new strategies for enabling this balance. To date, different functions have been attributed to plant peptide hormones, and most of them have been reported to be conserved among different plant species. Integration of this knowledge into breeding strategies will improve crop plants in terms of yield and stress resilience. In the coming sections of this chapter, we will present and discuss the recent information on plant peptide hormones' function in salt and drought stress tolerance. Plant peptide hormones' potential to increase crop performance and yield is important not only for sustainable agriculture practices but also for adaptation to global climate change. Although most of the information depends on the experiments done in the model plant *Arabidopsis thaliana*, homologs of peptide hormone encoding genes, processing mechanisms, and similar receptor interactions are also starting to be defined in agricultural plants, which is promising for improved abiotic and biotic stress tolerance.

17.2 Plant Peptide Hormones Mediate Salt and Drought Stress Responses

The biosynthesis, maturation (proteolytic cleavage, PTMs), transport, recognition, perception, and various functions of plant peptide hormones are being characterized in detail. These peptides form complexes and modules with various molecules to initiate and carry out their functions. Understanding how these processes are triggered by environmental factors is important not only for increasing crop performance and production but also for developing plants resistant to abiotic and biotic stress. Plant peptide hormones are likely to have dual functions as sensors or mobile signals for stress tolerance. These peptides can be "**sensors**" because most of them are derived from precursors and mature through PTMs. Sensing/recognition of stressors activates all of these processes; hence, peptide hormones may trigger them. They can be "**mobile signals**" for intracellular and/or root-to-shoot communication since several reports indicated most of the defined peptides are synthesized in roots and transported to shoots for inducing several defense processes, including stomatal movements in the leaves or activating the MAPK signaling cascade for transcriptional reprogramming. In literature, there is a vast amount of information on salt and drought stress responses and tolerance mechanisms; however, we still lack exact information on regulatory molecules that we can use to increase agricultural yield under stress conditions, which can be easily activated after the detection of the stressor. In this sense, plant peptide hormones are promising candidates.

17.2.1 Salt Stress

Salt stress affects sustainable agricultural practices all over the world and it is among the most studied stress factors in plants. Growth and development of plants are affected to a higher extent by salt stress because of its osmotic and ionic effects. Sensing and signaling processes are important components of defense against salt stress and peptide hormones are good candidates for these processes (Figure 17.2). Recent studies on RALF22, RALF23, CAP-derived peptide (CAPE), Pep3, CEP, and PAMP-induced secreted peptide3 (PIP3) defined how these peptides contribute to salt stress tolerance in *Arabidopsis* and other plants (Xiao et al. 2025). However, as stated in these studies, further clarification is needed regarding biosynthesis and maturation processes, recognition patterns, receptor interactions, and the mode of action of plant peptide hormones.

FERONIA (FER) internalization within plant cells leads to increased cell wall integrity, inhibition of ABA signaling, and preventing of reactive oxygen species (ROS) accumulation for enhancing salt stress tolerance. RALF22 and RALF23 peptides play a critical role in FER activation. Without stress conditions, two leucine-rich repeat extensins, LRX3 and LRX4, bind to RALF22 and RALF23 peptides to prevent their interaction with FER. These LRX proteins play an important role in cell wall-plasma membrane communication through their N- and C-terminal extension domains (Fabrice et al. 2018). Upon exposure to salt stress, disruption of the cell wall leads to the activation of S1P protease, which enables the release of RALF22 and RALF23 from LRX proteins for facilitating their interaction with FER. On the other hand, RALF1 also associates with FER and modulates ABA signaling under salt stress (Yu and Assmann 2018; Li et al. 2025). There may be other regulatory processes controlled by the RALF1-FER module by means of salt stress tolerance that need further elucidation, such as photorespiration. Jiang et al. (2024) recently reported that FER-mediated stabilization of shikimate oxidoreductase homolog 1 (SHM1) regulates the photorespiratory pathway, which contributes to salt stress tolerance.

CAPE and subtilase peptide (SUBPEP) are functional precursors involved in salt tolerance (Lu and Xiao 2024). CAPE1 is a negative regulator of salt tolerance in *Arabidopsis*

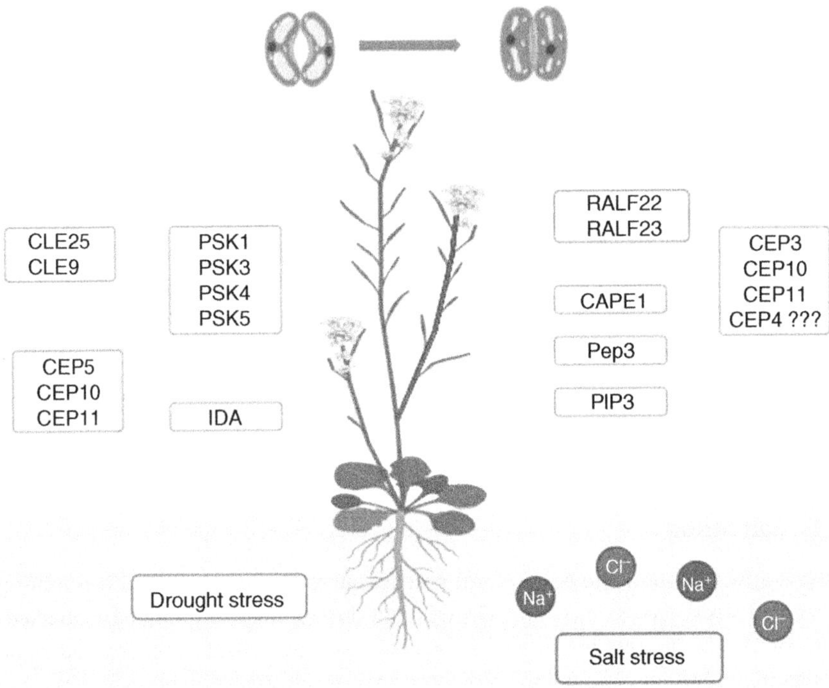

Figure 17.2 Summary of peptide hormones involved in salt and drought stress tolerance in *Arabidopsis thaliana*. Their effects on stomatal closure to prevent excessive water loss under salt and drought stress have been well characterized. **CAPE1**, CYSTEINE-RICH SECRETORY PROTEIN, ANTIGEN 5 AND PR-1 (CAP)-DERIVED PEPTIDE 1; **CEP**, C-TERMINALLY ENCODED PEPTIDE; **CLE**, CLAVATA3 EMBRYO SURROUNDING REGION-RELATED PEPTIDE; **IDA**, INFLORESCENCE DEFICIENT IN ABSCISSION; **Pep3**, PLANT ELICITOR PEPTIDE3; **PIP3**, PAMP-INDUCED SECRETED PEPTIDE3; **PSK**, PHYTOSULFOKINE; **RALF**, RAPID ALKALIZATION FACTOR. *Source:* Created in BioRender.

(Chien et al. 2015). CAPE1 is derived from the C-terminus of the 172-amino-acid PROCAPE1 and has 11 amino acids in its sequence (Chen et al. 2020). The maturation process, receptor, and recognition pattern of CAPE1 are not clearly defined. Since salt stress affects CAPE1 levels in *Arabidopsis*, PROCAPE1 expression and cleavage of PROCAPE1 are rapidly reduced upon exposure to salt stress, but this is quickly recovered. CAPE1 was reported to inhibit osmolyte production by regulating the expression of DELTA1-PYRROLINE-5-CARBOXYLATE SYNTHASE 1 (P5CS1), GALACTINOL SYNTHASE 2 (GolS2), and ALDEHYDE DEHYDROGENASE 7B4 (ALDH7B4) encoding genes, which are all important for the osmotic phase of the salt stress (Chien et al. 2015; Chen et al. 2020). On the other hand, Chien et al. (2015) reported that CAPE1 repressed ABA-responsive gene, AREB1, under salt stress, and ABA did not induce PROCAPE1. Although the gene encoding PROCAPE1 is not highly transcribed in the shoot and the transport process from root to shoot has not yet been elucidated, it is predicted that CAPE1 may also function as a long-distance signal under salinity in *Arabidopsis* (Chen et al. 2020).

Pep peptides form a conserved family and besides Pep3, all others are reported to regulate plant immunity. Pep3 peptide was first defined to function in plant immunity (Huffaker et al. 2006) and its contribution to salinity tolerance was recently discovered (Nakaminami et al. 2018). Pep3 was reported to be recognized by PEPR1; however, its mode of action under salt stress needs to be elucidated. Nakaminami et al. (2018) reported that, in *Arabidopsis,* Pep3 and its other forms have an ameliorative effect on chlorophyll bleaching under salinity. In *Arabidopsis*, propep3 knockdown mutant hypersensitivity to salt stress was reported, which can be recovered by exogenous Pep3 treatment. Accordingly, Pep3 treatment and PROPEP3 overexpression enhanced salt stress tolerance via PEPRECEPTOR 1 (Nakaminami et al. 2018). WRKY33 expression is induced by Pep3, which is known to inhibit the ABA biosynthetic genes' expression (Liu et al. 2015). Therefore, Pep3 or other yet to be defined Pep peptides may act on an ABA-independent signaling route for salt stress tolerance (Ma et al. 2012; Chen et al. 2020). Although these data present a significant function for Pep3 in salt stress responses, we still lack precise information on the perception, localization, processing, and translocation of these peptides for understanding their complete biological function. It is plausible that common receptors may perceive Pep peptides and their precursors are differentially localized within plant cells. Therefore, it would be interesting to discover how the differentiation between biotic and abiotic stress-responsive Pep peptides is achieved and do they have any differences by means of precursors and their processing mechanisms.

CEP peptides have a conserved 15-amino-acid-region at their C-terminus, are synthesized in the root, and serve as a mobile signal between root and shoot. Although recent studies provide information on their function in root development and stress tolerance, the complete mode of action for this diverse family of peptides is still lacking. To date, 15 CEP-encoding genes have been defined in *Arabidopsis* and mutational analysis has put forward relevant information on their regulatory functions (Mei et al. 2024; Keerthana et al. 2025). Upregulation of CEP3 enhanced salt stress tolerance, while CEP3 mutants have increased primary roots under salt stress in *Arabidopsis* (Delay et al. 2013; Aggarwal et al. 2020; Keerthana et al. 2025). Not only in *Arabidopsis* but also in crop plants, CEP peptides may play a regulatory role in salt stress tolerance. Xu et al. (2023) reported that transcript levels of CEP10 and CEP11 encoding SlCEP10 and SlCEP11 were slightly increased by 100 μM ABA treatment, while they were significantly induced by 200 mM NaCl treatment in tomato plants. On the other hand, activation of downstream defense through Ca^{2+} signaling for abiotic and biotic stress tolerance through CEP peptides may be a common regulatory step. CEP4 peptide is perceived by the RECEPTOR-LIKE KINASE 7 (RLK7) and this interaction induces Ca^{2+} influx in response to *Pseudomonas syringae* Pto infection in *Arabidopsis*. Moreover, CEP4 peptide is also involved in ethylene biosynthesis (Rzemieniewski et al. 2022; Ma et al. 2012). Both mechanisms are also contributing to salt stress tolerance, and to date CEP4's contribution to salt stress tolerance has not been defined. RLK7 activation by salt stress positively regulates MAPK3/MAPK6 signaling for salt stress responses (Zhou et al. 2022; Ma et al. 2022) in *Arabidopsis*. Moreover, under high salinity, increased ethylene accumulation aided by increased 1-aminocyclopropane-1-carboxylate synthase (ACS) activity for salt stress adaptation was reported (Zhu et al. 2016; Ma et al. 2022).

CASPARIAN STRIP INTEGRITY FACTORs (CIFs) are sulfated peptides as PSKs and along with CIF-like peptide, TWISTED SEED 1 (TWS1), they are reported to function in the development of Casparian strip, pollen cell wall, and embryo epidermal layer (Doblas et al. 2017; Nakayama et al. 2017; Fujita et al. 2020; Zhang et al. 2024; He et al. 2025). In plant roots, the Casparian strip is located in the endodermis and is responsible for the control of water and nutrient diffusion. Two LRR-RLKs, GASSHO 1 (GSO1) and GSO2, play a critical role in the development of the Casparian strip and CIFs can directly bind to these receptors (Tsuwamoto et al. 2008; Doblas et al. 2017; Nakayama et al. 2017; Fujita et al. 2020; Zhang et al. 2024; He et al. 2025). CIF1 and CIF2 are perceived by the receptor GSO1, which is located on the newly formed Casparian strip and upon entry of these peptides development of the Casparian strip is completed by the events regulated through cytoplasmic receptor kinase, SCHENGEN 1 (SGN1) signaling. GSO1 was also found to interact and activate an important component of salt stress tolerance; SALT OVERLY SENSITIVE 2 via phosphorylation (Chen et al. 2023) therefore, it is plausible that CIF peptides contribute to salt stress tolerance via the maintenance of the Casparian strip integrity for avoiding Na^+ diffusion to vascular system through this mechanism (Chen et al. 2023; He et al. 2025). It would be worth defining in detail how CIFs and GSO1/GSO2 interaction contribute to salt stress tolerance.

On the other hand, one of the PIPs, PIP3, was reported to be involved in salt stress tolerance. PIP peptides were first discovered when their accumulation was highly induced by pathogen infection and elicitor treatment, such as flagellin and chitin (Hou et al. 2014). To date, several PIP/PIP-like peptide encoding genes have been characterized in *Arabidopsis* in relation to their functions in growth, development, abiotic and biotic stress responses (Hou et al. 2014; Vie et al. 2015). PrePIP3 encoding gene's expression is induced by salt stress and upon biosynthesis, matured PIP3 interacts with RLK7 receptor to activate MAPK3/6-signaling cascade for the maintenance of Na^+ homeostasis (Zhou et al. 2022). Accordingly, other PIPs may also have a role in salt stress tolerance through the regulation of growth. Hussain et al. (2021) reported that PIP2 regulates cell division and elongation for root and hypocotyl growth along with auxin signaling in *Arabidopsis*. Mutational analysis provided information on the importance of PIP signaling for primary and lateral root phenotype along with hypocotyl phenotype, which also may be significant in salt stress tolerance as well (Hussain et al. 2021). On the other hand, all of the PIPs including PIP3 regulate plant immunity through salicylate and jasmonate signaling (Najafi et al. 2020). These peptides may be common stress-responsive elements for activating defense processes against biotic and abiotic stresses.

17.2.2 Drought Stress

Drought stress affects yield and crop performance adversely, and this impact has been increased nowadays due to increased atmospheric and soil temperature, decreased water resources and reduced water-use efficiency. Regulation of stomatal closure, deep-going root morphology, and biosynthesis of compatible osmolytes are among the well-known drought stress coping mechanisms. Recent studies on plant peptide hormones provided information especially for their function in controlling and regulating stomatal movements and root growth, which are both critical for keeping growth-defense balance under drought stress conditions. To date, CLEs, PSKs, INFLORESCENCE DEFICIENT IN ABSCISSION (IDA), EPF and EPF-LIKE (EPFL), and CEP peptides have been found to be associated with drought stress responses and tolerance (Figure 17.2). Moreover, information on these peptides can also be integrated into the studies of salt stress tolerance because their osmotic effects are similar to drought stress.

As indicated before, CLE peptides are among the first defined and functionally characterized plant peptide hormones, which have different regulatory functions in several physiological processes in plants. Among the CLE peptides, CLE25 and CLE9 are reported to have ameliorative effects under drought stress in *A. thaliana* (Takahashi et al. 2018; Zhang et al. 2019; Kim et al. 2021). In response to water deficit, CLE25 plays an important function as a root-to-shoot signaling molecule for the induction of stomatal closure through abscisic acid (ABA) biosynthesis. CLE25 is derived from the C-terminal region of the CLE25 precursor protein and has two hydroxylated proline residues within its 12-amino-acid sequence (Chen et al. 2020). CLE25 controls stomatal movements through ABA signaling and reducing water loss is advantageous both in drought and salt stress (Takahashi et al. 2018; Chen et al. 2020). Under water deficiency, CLE25 peptides are synthesized in roots and transported to leaves to induce the expression of NINE CIS-EPOXYCAROTENOID DIOXYGENASE3 (NCED3) encoding gene for ABA biosynthesis (Takahashi et al. 2018; Kim et al. 2021). This process is facilitated by the formation of BARELY ANY MERISTEM 1 and 3 (BAM1/3) and CLE25 module to trigger ABA biosynthesis through NCED3. Besides CLE25, another CLE peptide, CLE9, is also involved in drought stress tolerance through an onsite modulation of stomatal closure (Zhang et al. 2019). Zhang et al. (2019) reported that CLE9 induces stomatal closure in an ABA-dependent MAPK3/6 signaling cascade along with increased H_2O_2 and NO production for drought stress resilience. Unlike root-expressed and transported CLE25, CLE9 is expressed in guard cells. It acts as a local signal by interacting with a protein kinase, OPEN STOMATA 1 (OST1), and an anion channel protein, SLOW ANION CHANNEL ASSOCIATED 1 (SLAC1) (Zhang et al. 2019). CLE9 may act as a sensor for drought stress and upon interaction with the unidentified receptor, CLE9-triggered downstream events take place to prevent water loss via stomatal closure (Zhang et al. 2019).

PSK promotes primary root growth in *Arabidopsis thaliana* and is associated with drought-induced flower drop in *Lycopersicon esculentum* (Kutschmar et al. 2009; Reichardt et al. 2020; Ding et al. 2023). SBTs process proPSK1 for PSK peptide formation, and their interaction contributes to drought stress tolerance as indicated by different studies (Kim et al. 2021). Upregulation of PSK1, PSK3, PSK4, PSK5, SBT1.4, SBT3.7, and SBT3.8 encoding genes' expression was detected in response to osmotic stress (Stührwohldt et al. 2021). Exogenous PSK peptide treatment had an ameliorative effect on osmotic stress-sensitive sbt3.8 mutants. Accordingly, in the proPSK1-overexpressing mutants, enhanced root and hypocotyl growth was found in relation to osmotic stress tolerance (Kim et al. 2021). Reichardt et al. (2020) reported that phytosulfokine signaling is associated with drought-induced flower drop. Accordingly, mutants of proPSK cleaving enzyme phytaspase 2 (phyt2) had different effects on flower drop in tomato plants. Overexpressing lines enhanced flower abscission while knockdowns had reduced levels (Reichardt et al. 2020). Detailed analysis put forward that under drought stress conditions, SlPhyt2 produces bioactive PSK, which activates cell wall hydrolases for abscission.

The IDA peptide has 20 amino acids and the C-terminus has a Pro-rich motif It functions in lateral root formation and floral organ abscission (Motomitsu et al. 2015; Kim et al. 2021). ABA, ethylene, auxin, and abiotic stress induce abscission. Under drought conditions, IDA encoding gene's expression is induced at leaf abscission zones and this leads to cell-wall breakdown and separation, facilitating the organ abscission (Patharkar and Walker 2016; Zhu et al. 2019; Takahashi et al. 2019; Kim et al. 2021). IDA's precursor peptide has 77 amino acids, and it is processed by subtilisin-like serine proteinases (SBTs). HAESA and HAESA-LIKE (HSL)2 receptor-like kinases recognize the processed IDA peptides in the floral abscission zone (Cho et al. 2008; Takahashi et al. 2019). Besides IDA, IDA-LIKE (IDL)1 is also recognized by the HSL2 receptor. Other than floral abscission, both peptides are reported to regulate abiotic and biotic stress responses and root morphogenesis through auxin signaling (Takahashi et al. 2019).

Among the CEP peptides, CEP5 has been reported to function in root development and osmotic stress tolerance in *A. thaliana* (Roberts et al. 2016; Smith et al. 2020). CEP5 alters several abiotic stress-responsive proteins, such as osmotic stress-inducible transcription factors, and mutant analysis indicated its positive impact on drought stress responses and recovery (Kim et al. 2021). Although the mode of action is not clearly defined yet, CEP5 negatively regulates AUXIN RESPONSE FACTORs for AUX/IAA stabilization, leading to drought stress tolerance (Smith et al. 2020). In *S. lycopersicum*, several abiotic stress-responsive cis- and hormone-binding elements were detected on CEP encoding promoters (Xu et al. 2023). Dehydration-responsive element (DRE/MBS); antioxidant responsive element (ARE); abscisic acid-responsive element (ABRE); MeJA-responsive element; ethylene-responsive element (ERE); low temperature-responsive element (LTR); and stress-responsive TC-rich repeats can be given (Xu et al. 2023). Among these elements, ABRE and DRE were two important drought stress-responsive ones that are usually associated with drought stress tolerance (Sakuma et al. 2006; Kim et al. 2011; Xu et al. 2023). Accordingly, Xu et al. (2023) reported that exogenous SlCEP10 and SlCEP11 peptide treatments increased drought tolerance in tomato.

The EPF and EPF-LIKE (EPFL) peptides along with their interaction with receptor kinases regulate several developmental processes, including stomatal patterning and density, which all have critical importance by means of water use efficiency and drought stress adaptation (Lee et al. 2012; He et al. 2025). In *Arabidopsis*, eleven EPF and EPFL peptides were characterized to have different regulatory functions in xylem and epidermal differentiation, stem elongation, and inflorescence growth (He et al. 2025). Among them, EPF1, EPF2, EPFL6, and EPFL9 (STOMAGEN) are the regulators of stomatal development via controlling cell fate transitions and asymmetric cell division. Their mode of action involves ERECTA, ERL1, and ERL2 receptors and MAPK signaling. AtEPF1 and AtEPF2 are expressed in the epidermis for the initiation of asymmetric cell division in epidermal differentiation for proper stomatal orientation (Hara et al. 2007; Hunt et al. 2010). EPFL9 (STOMAGEN) regulates further development of stomata by repressing EPF1 and EPF2 in the mesophyll cells (Hunt et al. 2010). Mutational and transcriptomic analyses revealed the function of EPF peptides in drought stress tolerance by regulating stomatal development not only in *Arabidopsis* but also in *Oryza sativa*, *Triticum aestivum*, *Zea mays*, *Solanum lycopersicum*, *Sorghum bicolor*, and *Gossypium hirsutum* (Hunt et al. 2010; Lee et al. 2012; Yin et al. 2017; Lu et al. 2019; Jiang et al. 2019; Dunn et al. 2019; He et al. 2023; Jiao et al. 2023; Xia et al. 2024; Liu et al. 2024; He et al. 2025). Since this regulation process is conserved both in monocotyledons and dicotyledons, it is plausible that EPF peptides are among the first evolved regulatory components in plant cells for developing efficient water use strategies under limited water conditions.

17.3 Peptide Hormones Requiring Further Characterization for Their Salt and Drought Stress Effects

Some of the peptides identified to be associated with salt and drought stress resistance have been identified only by bioinformatics tools and transcriptomic analyses; therefore, further functional characterization, particularly in terms of their biosynthesis, maturation processes, receptors, and activated downstream signaling pathways, is required. One of them is *Oryza sativa* drought and salt stress response-1 (OsDSSR1) peptide. In rice, the transcript levels of the OsDSSR1 peptide-encoding gene were highly induced by ABA and hydrogen peroxide (H_2O_2) treatments, salt and drought stresses (Cui et al. 2018). Rice plants overexpressing OsDSSR1 were also reported to have increased drought stress tolerance with higher osmoprotectant contents and increased expression of ROS scavenging-related genes (Cui et al. 2018). Another example is a cysteine-rich peptide, rice small peptide drought tolerance 11 (OsDT11) with 88 amino acids (Li et al. 2017). Plants have increased OsDT11 expression under polyethylene glycol (PEG) treatment and overexpressing lines displayed high ABA concentrations along with reduced water loss and stomatal density, which all led to enhanced drought stress tolerance (Li et al. 2017). The drought tolerance effect of the OsDT11 peptide may be related to the ABA signaling pathway. Small paraquat resistance protein (SPQ) is a peptide with 69 amino acids and an N-terminal secretion signal that was isolated from the halophytic *Lepidium crassifolium*. Overexpression of LcSPQ resulted in increased drought resistance in *Arabidopsis* (Farago et al. 2022; Ren et al. 2024). This peptide may be a unique regulatory molecule by means of salt stress tolerance since it has no similarity to any other known peptide (Farago et al. 2022; Ren et al. 2024). cysteine-rich transmembrane module (CYSTM) peptides are novel cysteine-rich transmembrane molecules that function in different abiotic stresses (Venancio and Aravind 2010). Particularly, mitochondria-localized CYSTM3 adversely affects Na^+ efflux and ROS levels, indicating its negative regulatory function for salt stress tolerance in *Arabidopsis* (Xu et al. 2018, 2019).

The functions of some peptides have not been linked to salt or drought stress, but their mode of action, identified so far, may also play a role in stress tolerance. For example, ROOT MERISTEM GROWTH FACTORs (RGFs) were defined as sulfated peptides other than PSKs, PSYs, and CIFS as the regulators of primary root growth (Kaufmann and Sauter 2019; Zhang et al. 2024). RGF precursors are composed of 86–163 amino acids with a conserved C-terminal region and N-terminal signal. The mature peptides vary between 13 and 18 amino acids and can be modified by hydroxylation and sulfonation (Meng et al. 2012; Shinohara 2021; Zhang et al. 2024). To date, three RGF receptor LRR-RLKs (RGFR1, RGFR2, and RGFR3) have been defined; however, the downstream signaling, which is most likely to be through the MAPK cascade, is not well understood. On the other hand, RGF1-INDUCIBLE TRANSCRIPTION FACTOR 1 (RITF1) that controls the accumulation of superoxide anions in the meristematic zone for the fate of root tip cells, whether to propagate or elongate, is also regulated by RGF signaling (Yamada et al. 2020; Wang et al. 2022; Zhang et al. 2024). RGF signaling appears to be tightly controlled in plant cells, and its mechanisms remain poorly defined. Furthermore, ROS accumulation, particularly during salt and drought stress, may also be important for maintaining the balance between growth and defense associated with RGF signaling.

Previous studies indicated that overexpression of wound-signaling related peptide, prosystemin (ProSys) encoding genes also enhanced salt stress tolerance in tomato (Coppola et al. 2010, 2015, 2019; Cirillo et al. 2022). ProSys has 200 amino acid residues, and upon proteolytic cleavage, Sys is formed, and activation of defense genes in response to biotic stress is induced in tomato plants. Exogenous Sys treatment increased salt stress tolerance in tomato through the upregulation of the sodium transporters; SOS1, NHX1, and HKT1.1, enhanced redox status, and protein stability (Cirillo et al. 2022). Cirillo et al. (2022) proposed that Sys may be a common regulator for enhancing plant responses to both abiotic and biotic stresses via shared tolerance mechanisms.

17.4 Conclusion

We presented some of the studies that confirmed the regulatory functions of plant peptide hormones not only in growth and development but also in salt and drought stress tolerance. Although more information about the peptides derived from functional and nonfunctional precursors is available, there are still missing points in understanding their maturation processes, receptors, cell-type–specific interactions, interactions with other macromolecules, and signaling components. On the other hand, less is known about the noncoding RNA (ncRNA) part of the story, although ncRNA-encoded peptides (ncPEPs) are the hot topics of research on regulatory processes. Besides functional characterization of ncPEPs, bioinformatic approaches such as ncPlantDB will also improve our understanding of how plant ncRNAs contribute to peptide signaling and complex networks for stress resilience. Among well-studied peptide hormones, CLEs, Pep3, CAPEs, RALFs, and PIPs share similar signaling pathways, including cytosolic Ca^{2+} accumulation, ABA, ROS, and MAPK signaling. Stimulation of these pathways leads to distinct stress responses following recognition of these peptides by pattern recognition receptors (PRRs), such as leucine-rich repeat receptor-like kinases. There are different aspects regarding the regulatory functions of peptide hormones on stress responses. Both positive and negative regulatory mechanisms are defined, CLE25 and Pep3 are examples of the former while CAPE1 is an example of the latter. Especially the negative regulator ones may be important for fine-tuning and modulation of growth and development under salt and drought stress conditions. On the other hand, the aforementioned studies put forward the role of several peptide hormone signaling processes in physiological processes such as water use efficiency, ion homeostasis, protein biosynthesis and stability, ROS scavenging, and cell wall and membrane integrity. The modules formed by peptide hormones and their receptors are reported to regulate these processes efficiently, indicating their importance in the stress tolerance of plants. One of the issues that should be resolved is deciphering the counterparts of such interactions and regulatory processes in crop plants, since most of the presented data was found in *Arabidopsis*. Integrating plant peptide hormone regulatory mechanisms into selection strategies will improve plant breeding programs for salt and drought stress tolerance. Several properties of plant peptide hormones stand out in this regard. First, their ability to exist in different conformations (different peptide hormones can be synthesized from the same precursor as a result of maturation processes). Second, they utilize known receptors to activate various signaling pathways related to salt and drought stress. Third, a large portion of the identified peptides play roles in controlling various growth and developmental processes, such as those affecting roots and reproductive organs. As previously mentioned, these properties are particularly important for maintaining the growth-defense balance under salt and drought stress conditions. In order to integrate peptide hormones into agricultural practices, further studies are needed on their mode of action, receptors, maturation processes, interactions with other signaling pathways, their common and specific functions, and most importantly, their presence in agricultural plants.

References

Aggarwal, S., Kumar, A., Jain, M. et al. (2020). C-terminally encoded peptides (CEPs) are potential mediators of abiotic stress response in plants. *Physiology and Molecular Biology of Plants* 26 (10): 2019–2033.

Chang, S. and Xiao, F. (2025). Comprehensive review of plant small signaling peptides: From stress adaptation mechanisms to practical solutions for crop resilience. *International Journal of Biological Macromolecules* 299: 139971.

Chen, Y.L., Fan, K.T., Hung, S.C., and Chen, Y.R. (2020). The role of peptides cleaved from protein precursors in eliciting plant stress reactions. *The New Phytologist* 225: 2267–2282.

Chen, C., He, G., Li, J. et al. (2023). A salt stress activated GSO1-SOS2-SOS1 module protects the *Arabidopsis* root stem cell niche by enhancing sodium ion extrusion. *The EMBO Journal* 42: e113004.

Chien, P.S., Nam, H.G., and Chen, Y.R. (2015). A salt-regulated peptide derived from the CAP superfamily protein negatively regulates salt-stress tolerance in *Arabidopsis*. *Journal of Experimental Botany* 66: 5301–5313.

Cho, S.K., Larue, C.T., Chevalier, D. et al. (2008). Regulation of floral organ abscission in *Arabidopsis thaliana*. *PNAS* 105: 15629–15634.

Cirillo, V., Molisso, D., Aprile, A.M. et al. (2022). Systemin peptide application improves tomato salt stress tolerance and reveals common adaptation mechanisms to biotic and abiotic stress in plants. *Environmental and Experimental Botany* 199: 104865.

Clark, S.E., Running, M.P., and Meyerowitz, E.M. (1995). CLAVATA3 is a specific regulator of shoot and floral meristem development affecting the same processes as CLAVATA1. *Development* 121: 2057–2067.

Coppola, M., Corrado, G., Coppola, V. et al. (2015). Prosystemin overexpression in tomato enhances resistance to different biotic stresses by activating genes of multiple signalling pathways. *Plant Molecular Biology Reporter* 33: 1270–1285.

Coppola, M., Di Lelio, I., Romanelli, A. et al. (2019). Tomato plants treated with systemin peptide show enhanced levels of direct and indirect defense associated with increased expression of defense-related genes. *Plants* 8 (10): 395.

Coppola, M., Ruocco, M., Digilio, M.C. et al. (2010). Prosystemin coordinates multiple defense responses in tomato. *Minerva Biotecnology* 22: 20–22.

Couzigou, J.M., André, O., Guillotin, B. et al. (2016). Use of microRNA encoded peptide miPEP172c to stimulate nodulation in soybean. *The New Phytologist* 211: 379–381.

Covey, P.A., Subbaiah, C.C., Parsons, R.L. et al. (2010). A pollen-specific RALF from tomato that regulates pollen tube elongation. *Plant Physiology* 153: 703–715.

Cui, Y.C., Li, M.J., Yin, X.M. et al. (2018). OsDSSR1, a novel small peptide, enhances drought tolerance in transgenic rice. *Plant Science* 270: 85–96.

Czyzewicz, N., Yue, K., Beeckman, T., and De Smet, I. (2013). Message in a bottle: small signalling peptide outputs during growth and development. *Journal of Experimental Botany* 64 (17): 5281–5296.

Datta, T., Kumar, R.S., Sinha, H., and Prabodh, K.T. (2024). Small but mighty: peptides regulating abiotic stress responses in plants. *Plant, Cell & Environment* 47: 1207–1223.

Delay, C., Imin, N., and Djordjevic (2013). MA CEP genes regulate root and shoot development in response to environmental cues and are specific to seed plants. *Journal of Experimental Botany* 64 (17): 5383–5394.

Ding, S., Jianrong, L., Hu, Z. et al. (2023). Phytosulfokine peptide optimizes plant growth and defense via glutamine synthetase GS2 phosphorylation in tomato. *The EMBO Journal* 42 (e111858): 1–17.

Doblas, V.G., Smakowska-Luzan, E., Fujita, S. et al. (2017). Root diffusion barrier control by a vasculature-derived peptide binding to the SGN3 receptor. *The Sciences* 355: 280–284.

Dunn, L.H., Afsharinafar, M., Meselmani, M.A. et al. (2019). Reduced stomatal density in bread wheat leads to increased water-use efficiency. *Journal of Experimental Botany* 70 (18): 4737–4748.

Fabrice, T.N., Vogler, H., Draeger, C. et al. (2018). LRX proteins play a crucial role in pollen grain and pollen tube cell wall development. *Plant Physiology* 176: 1981–1992.

Farago, D., Zsigmond, L., Benyo, D. et al. (2022). Small paraquat resistance proteins modulate paraquat and ABA responses and confer drought tolerance to overexpressing Arabidopsis plants. *Plant, Cell & Environment* 7: 1985–2003.

Fiers, M., Golemiec, E., Xu, J. et al. (2005). The 14-amino acid CLV3, CLE19, and CLE40 peptides trigger consumption of the root meristem in Arabidopsis through a CLAVATA2-dependent pathway. *The Plant Cell* 17: 2542–2553.

Fujita, S., De Bellis, D., Edel, K.H. et al. (2020). SCHENGEN receptor module drives localized ROS production and lignification in plant roots. *The EMBO Journal* 39: e103894.

Han, Z., Sun, Y., and Chai, J. (2014). Structural insight into the activation of plant receptor kinases. *Current Opinion in Plant Biology* 20: 55–63.

Hander, T., Fernandez-Fernandez, A.D., Kumpf, R.P. et al. (2019). Damage on plants activates Ca^{2+}-dependent metacaspases for release of immunomodulatory peptides. *Science* 363: eaar7486.

Hara, K., Kajita, R., Torii, K.U. et al. (2007). The secretory peptide gene EPF1 enforces the stomatal one-cell-spacing rule. *Genes & Development* 21 (14): 1720–1725.

He, Y., He, X., Wang, X. et al. (2023). An EPFL peptide signaling pathway promotes stamen elongation via enhancing filament cell proliferation to ensure successful self-pollination in *Arabidopsis thaliana*. *The New Phytologist* 238 (3): 1045–1058.

He, S., Sun, H., Chen, Q. et al. (2025). Comprehensive identification of cotton EPF/EPFL receptors and functional characterization of the GhEPFL1-1-GhER1 module in drought tolerance. *BMC Plant Biology* 25: 901.

Hirakawa, Y. and Sawa, S. (2019). Diverse function of plant peptide hormones in local signaling and development. *Current Opinion in Plant Biology* 51: 81–87.

Hou, S., Wang, X., Chen, D. et al. (2014). The secreted peptide PIP1 amplifies immunity through receptor-like kinase 7. *PLoS Pathogens* 10: e1004331.

Huffaker, A., Pearce, G., and Ryan, C.A. (2006). An endogenous peptide signal in *Arabidopsis* activates components of the innate immune response. *PNAS* 103: 10098–10103.

Hunt, L., Bailey, K.J., and Gray, J.E. (2010). The signalling peptide EPFL9 is a positive regulator of stomatal development. *The New Phytologist* 186 (3): 609–614.

Hunt, L. and Gray, J.E. (2009). The signaling peptide EPF2 controls asymmetric cell divisions during stomatal development. *Current Biology* 19: 864–869.

Hussain, S., Wang, W., Ahmed, S. et al. (2021). PIP2, an auxin induced plant peptide hormone regulates root and hypocotyl elongation in Arabidopsis. *Frontiers in Plant Science* 12: 646736.

Jiang, W., Wang, Z., Li, Y. et al. (2024). FERONIA regulates salt tolerance in *Arabidopsis* by controlling photorespiratory flux. *Plant Cell* 36 (11): 4732–4751.

Jiang, Q., Yang, J., Wang, Q. et al. (2019). Overexpression of MdEPF2 improves water use efficiency and reduces oxidative stress in tomato. *Environmental and Experimental Botany* 162: 321–332.

Jiao, Z., Wang, J., Shi, Y. et al. (2023). Genome-wide identification and analysis of the EPF gene family in *Sorghum bicolor* (L.) Moench. *Plants* 12 (22): 3912.

Kaufmann, C. and Sauter, M. (2019). Sulfated plant peptide hormones. *Journal of Experimental Botany* 70: 4267–4277.

Keerthana, K., Ramakrishnan, M., Ahmad, Z. et al. (2025). Root-derived small peptides: Key regulators of plant development, stress resilience, and nutrient acquisition. *Plant Science* 354 (112433): 1–17.

Kim, J.S., Jeon, B.W., and Kim, J. (2021). Signaling peptides regulating abiotic stress responses in plants. *Frontiers in Plant Science* 12: 704490.

Kim, J.S., Mizoi, J., Yoshida, T. et al. (2011). An ABRE promoter sequence is involved in osmotic stress-responsive expression of the DREB2A gene, which encodes a transcription factor regulating drought-inducible genes in *Arabidopsis*. *Plant & Cell Physiology* 52: 2136–2146.

Kutschmar, A., Rzewuski, G., Stuhrwohldt, N. et al. (2009). PSK-α promotes root growth in *Arabidopsis*. *The New Phytologist* 181: 820–831.

Lauressergues, D., Couzigou, J.M., Clemente, H.S. et al. (2015). Primary transcripts of microRNAs encode regulatory peptides. *Nature* 520: 90–93.

Lauressergues, D., Ormancey, M., Guillotin, B. et al. (2022). Characterization of plant microRNA-encoded peptides (miPEPs) reveals molecular mechanisms from the translation to activity and specificity. *Cell Reports* 38: 110339.

Lee, J.S., Kuroha, T., Hnilova, M. et al. (2012). Direct interaction of ligand–receptor pairs specifying stomatal patterning. *Genes & Development* 26 (2): 126–136.

Li, X., Han, H., Chen, M. et al. (2017). Overexpression of OsDT11, which encodes a novel cysteine-rich peptide, enhances drought tolerance and increases ABA concentration in rice. *Plant Molecular Biology* 93: 21–34.

Li, X., Zhu, L., Wang, H. et al. (2025). Peptide hormone-mediated regulation of plant development and environmental adaptability. *Advancement of Science* 12 (34): e06590.

Liu, S., Kracher, B., Ziegler, J. et al. (2015). Negative regulation of ABA signaling by WRKY33 is critical for *Arabidopsis* immunity towards *Botrytis cinerea* 2100. *eLife* 4: e07295.

Liu, L., Liu, E., Hu, Y. et al. (2025). ncPlantDB: a plant ncRNA database with potential ncPEP information and cell type-specific interaction. *Nucleic Acids Research* 53: 1587–1594.

Liu, R., Xu, K., Li, Y. et al. (2024). Investigation on the potential functions of ZmEPF/EPFL family members in response to abiotic stress in maize. *International Journal of Molecular Sciences* 25 (13): 7196.

Lu, J., He, J., Zhou, X. et al. (2019). Homologous genes of epidermal patterning factor regulate stomatal development in rice. *Journal of Plant Physiology* 234–235.

Lu, S. and Xiao, F. (2024). Small peptides: orchestrators of plant growth and developmental processes. *International Journal of Molecular Sciences* 25: 7627.

Ma, L., Liu, X., Lv, W., and Yang, Y. (2022). Molecular mechanisms of plant responses to salt stress. *Frontiers in Plant Science* 13: 934877.

Ma, Y., Walker, R.K., Zhao, Y., and Berkowitz, G.A. (2012). Linking ligand perception by PEPR pattern recognition receptors to cytosolic Ca^{2+} elevation and downstream immune signaling in plants. *PNAS* 109 (48): 19852–19857.

Matsubayashi, Y. and Sakagami, Y. (1996). Phytosulfokine, sulfated peptides that induce the proliferation of single mesophyll cells of *Asparagus officinalis* L. *PNAS* 93: 7623–7627.

Mei, Z., Li, B., Zhu, S. et al. (2024). A genome-wide analysis of the CEP gene family in cotton and a functional study of GhCEP46-D05 in plant development. *International Journal of Molecular Sciences* 25 (8): 4231.

Meng, L., Buchanan, B.B., Feldman, L.J., and Luan, S. (2012). A putative nuclear CLE-like (CLEL) peptide precursor regulates root growth in *Arabidopsis*. *Molecular Plant* 5: 955–957.

Motomitsu, A., Sawa, S., and Ishida, T. (2015). Plant peptide hormone signaling. *Essays in Biochemistry* 58: 115–131.

Najafi, J., Brembu, T., Vie, A.K. et al. (2020). PAMP-INDUCED SECRETED PEPTIDE 3 modulates immunity in Arabidopsis. *Journal of Experimental Botany* 71: 850–864.

Nakaminami, K., Okamoto, M., Higuchi-Takeuchi, M. et al. (2018). AtPep3 is a hormone-like peptide that plays a role in the salinity stress tolerance of plants. *PNAS* 115: 5810–5815.

Nakayama, T., Shinohara, H., Tanaka, M. et al. (2017). A peptide hormone required for Casparian strip diffusion barrier formation in Arabidopsis roots. *The Sciences* 355: 284–286.

Ormancey, M., Le Ru, A., Duboe, C. et al. (2020). Internalization of miPEP165a into Arabidopsis roots depends on both passive diffusion and endocytosis-associated processes. *International Journal of Molecular Sciences* 21: 2266.

Patharkar, O.R. and Walker, J.C. (2016). Core mechanisms regulating developmentally timed and environmentally triggered abscission. *Plant Physiology* 172: 510–520.

Reichardt, S., Piepho, H.P., Stintzi, A., and Schaller, A. (2020). Peptide signaling for drought-induced tomato flower drop. *Science* 367: 1482–1485.

Ren, Y., Song, Y., Zhang, L. et al. (2021). Coding of non-coding RNA: insights into the regulatory functions of Pri-MicroRNA-encoded peptides in plants. *Frontiers in Plant Science* 12: 641351.

Ren, G., Zhang, Y., Chen, Z. et al. (2024). Research progress of small plant peptides on the regulation of plant growth, development, and abiotic stress. *International Journal of Molecular Sciences* 25: 4114.

Roberts, I., Smith, S., Stes, E. et al. (2016). CEP5 and XIP1/CEPR1 regulate lateral root initiation in *Arabidopsis*. *Journal of Experimental Botany* 67: 4889–4899.

Röhrig, H., Schmidt, J., Miklashevichs, E. et al. (2002). Soybean ENOD40 encodes two peptides that bind to sucrose synthase. *PNAS* 99: 1915–1920.

Rzemieniewski, J., Leicher, H., Lee, H. et al. (2022). Phytocytokine signaling integrates cell surface immunity and nitrogen limitation. *BioRxiv* 12 (20): 521212.

Sakuma, Y., Maruyama, K., Qin, F. et al. (2006). Dual function of an Arabidopsis transcription factor DREB2A in water-stress-responsive and heat-stress-responsive gene expression. *PNAS* 103: 18822–18827.

Sharma, A., Badola, P.K., Bhatia, C. et al. (2020). Primary transcript of miR858 encodes regulatory peptide and controls flavonoid biosynthesis and development in *Arabidopsis*. *Nature Plants* 6: 1262–1274.

Shinohara, H. (2021). Root meristem growth factor RGF, a sulfated peptide hormone in plants. *Peptides* 142: 170556.

Smith, S., Zhu, S., Joos, L. et al. (2020). The CEP5 peptide promotes abiotic stress tolerance, as revealed by quantitative proteomics, and attenuates the AUX/IAA equilibrium in *Arabidopsis*. *Molecular & Cellular Proteomics* 19: 1248–1262.

Srivastava, R., Liu, J.X., Guo, H.Q. et al. (2009). Regulation and processing of a plant peptide hormone, AtRALF23, in *Arabidopsis*. *The Plant Journal* 59: 930–939.

Stührwohldt, N., Bühler, E., Sauter, M., and Schaller, A. (2021). Phytosulfokine (PSK) precursor processing by subtilase SBT3.8 and PSK signaling improve drought stress tolerance in *Arabidopsis*. *Journal of Experimental Botany* 72: 3427–3440.

Takahashi, F., Hanada, K., Kondo, T., and Shinozaki, K. (2019). Hormone-like peptides and small coding genes in plant stress signalling and development. *Current Opinion in Plant Biology* 51: 88–95.

Takahashi, F., Suzuki, T., Osakabe, Y. et al. (2018). A small peptide modulates stomatal control via abscisic acid in long-distance signaling. *Nature* 556: 235–238.

Tsuwamoto, R., Fukuoka, H., and Takahata, Y. (2008). GASSHO1 and GASSHO2 encoding a putative leucine-rich repeat trans membrane type receptor kinases are essential for the normal development of the epidermal surface in *Arabidopsis* embryos. *The Plant Journal* 54: 30–42.

Venancio, T.M. and Aravind, L. (2010). CYSTM, a novel cysteine-rich transmembrane module with a role in stress tolerance across eukaryotes. *Bioinformatics* 2: 149–152.

Vie, A.K., Najafi, J., Liu, B. et al. (2015). The IDA/IDA-LIKE and PIP/PIP-LIKE gene families in Arabidopsis: phylogenetic relationship, expression patterns, and transcriptional effect of the PIPL3 peptide. *Journal of Experimental Botany* 66: 5351–5365.

Wang, Y., Chen, W., Ou, Y. et al. (2022). Arabidopsis ROOT ELONGATION RECEPTOR KINASES negatively regulate root growth putatively via altering cell wall remodeling gene expression. *Journal of Integrative Plant Biology* 64: 1502–1513.

Wu, J., Kurten, E.L., Monshausen, G. et al. (2007). NaRALF, a peptide signal essential for the regulation of root hair tip apoplastic pH in *Nicotiana attenuata*, is required for root hair development and plant growth in native soils. *The Plant Journal* 52: 877–890.

Xia, H., Wang, Q., Chen, Z. et al. (2024). Identification and functional analysis of the EPF/EPFL gene family in maize (*Zea mays* L.): implications for drought stress response. *Agronomy* 14 (8): 1734.

Xiao, F., Zhou, H., and Lin, H. (2025). Decoding small peptides: regulators of plant growth and stress resilience. *Journal of Integrative Plant Biology* 67: 596–631.

Xu, K., Tian, D., Wang, T.J. et al. (2023). Small secreted peptides (SSPs) in tomato and their potential roles in drought stress response. *Molecular Horticulture* 3: 17.

Xu, Y., Yu, Z.P., Zhang, D. et al. (2018). CYSTM, a novel non-secreted cysteine-rich peptide family, involved in environmental stresses in *Arabidopsis thaliana*. *Plant & Cell Physiology* 2: 423–438.

Xu, Y., Yu, Z.P., Zhang, S.Z. et al. (2019). CYSTM3 negatively regulates salt stress tolerance in *Arabidopsis*. *Plant Molecular Biology* 99: 395–406.

Yamada, M., Han, X., and Benfey, P.N. (2020). RGF1 controls root meristem size through ROS signalling. *Nature* 577: 85–88.

Yin, X., Biswal, A.K., Dionora, J. et al. (2017). CRISPR-Cas9 and CRISPR-Cpf1 mediated targeting of a stomatal developmental gene EPFL9 in rice. *Plant Cell Reports* 36 (5): 745–757.

Yu, Y. and Assmann, S.M. (2018). Inter-relationships between the heterotrimeric Gβ subunit AGB1, the receptor-like kinase FERONIA, and RALF1 in salinity response. *Plant, Cell & Environment* 41: 2475–2489.

Yu, Y., Li, W., Liu, Y. et al. (2025). A Zea genus-specific micro peptide controls kernel dehydration in maize. *Cell* 188 (44–59): e21.

Zhang, H., Han, Z., Song, W., and Chai, J. (2016). Structural insight into recognition of plant peptide hormones by receptors. *Molecular Plant* 9: 1454–1463.

Zhang, Z., Han, H., Zhao, J. et al. (2025). Peptide hormones in plants. *Molecular Horticulture* 5 (7): 1–50.

Zhang, L., Shi, X., Zhang, Y. et al. (2019). CLE9 peptide-induced stomatal closure is mediated by abscisic acid, hydrogen peroxide, and nitric oxide in *Arabidopsis* thaliana. *Plant, Cell & Environment* 42: 1033–1044.

Zhang, B., Xin, B., Sun, X. et al. (2024). Small peptide signaling via OsCIF1/2 mediates Casparian strip formation at the root endodermal and non-endodermal cell layers in rice. *Plant Cell* 36: 383–403.

Zhou, H., Xiao, F., Zheng, Y. et al. (2022). Pamp induced secreted peptide 3 modulates salt tolerance through receptor-like kinase 7 in plants. *Plant Cell* 34: 927–944.

Zhu, T., Deng, X., Zhou, X. et al. (2016). Ethylene and hydrogen peroxide are involved in brassinosteroid induced salt tolerance in tomato. *Scientific Reports* 6: 35392.

Zhu, Q., Shao, Y., Ge, S. et al. (2019). A MAPK cascade downstream of IDA-HAE/HSL2 ligand-receptor pair in lateral root emergence. *Nature Plants* 5: 414–423.

18

The Role of Peptide Hormones in Plant Immunity

Muhammad Imran[1,*], *Zhongke Sun*[1], *and Kamal A. M. Abo-Elyousr*[2,3,*]

[1] *School of Biological Engineering, Henan University of Technology, Zhengzhou, Henan, China*
[2] *Department of Agriculture, Faculty of Environmental Sciences, King Abdulaziz University, Jeddah, Saudi Arabia*
[3] *Department of Plant Pathology, Faculty of Agriculture, University of Assiut, Assiut, Egypt*

18.1 Introduction and Overview to Plant Immunity and Peptide Hormones

Plant immunity is a sophisticated and dynamic defense mechanism that has evolved to safeguard plants from an array of pathogenic organisms, including bacteria, fungi, viruses, and nematodes. Unlike animals, plants lack specialized mobile immune cells and do not possess an adaptive immune system. Instead, they rely entirely on an intricate innate immune system that enables early recognition of potential threats and initiation of effective defense responses (Jones and Dangl 2006; Bentham et al. 2020). This immune system is composed of both preformed barriers and inducible responses that act together to limit pathogen invasion and proliferation. Preformed or structural defenses include physical components such as the cell wall, cuticle, and stomatal structures, as well as chemical compounds like phytoalexins, saponins, and antimicrobial proteins that inhibit pathogen establishment (Bentham et al. 2020). However, when these barriers are breached, plants initiate inducible defense responses triggered by the perception of pathogen-derived signals. Plasma membrane-localized pattern recognition receptors (PRRs), including receptor-like kinases (RLKs) and receptor-like proteins (RLPs), recognize conserved microbial signatures known as pathogen-associated molecular patterns (PAMPs), or host-derived signals referred to as damage-associated molecular patterns (DAMPs) (Fang and Gu 2021). The recognition of these signals activates a defense strategy termed pattern-triggered immunity (PTI), which represents the first line of inducible immune response. In parallel, plants have evolved intracellular immune receptors, primarily nucleotide-binding leucine-rich repeat (NB-NLR) proteins, that recognize specific pathogen effector molecules delivered into the host cytoplasm. This triggers effector-triggered immunity (ETI), a second layer of defense often

*Corresponding: Kamal A. M Abo-Elyousr. Email: kaaboelyousr@agr.au.edu.eg; Muhammad Imran. Email: Imranpathologist@cau.edu.cn

Peptide Hormones in Plant Development and Stress Responses, First Edition. Edited by Vijay Pratap Singh and Samiksha Singh.

characterized by a rapid and localized hypersensitive response (HR), which involves programmed cell death (PCD) to contain the infection (Jones and Dangl 2006; Bentham et al. 2020).

Both PTI and ETI responses initiate complex downstream signaling cascades. These include the rapid generation of reactive oxygen species (ROS), an increase in cytosolic calcium levels, activation of mitogen-activated protein kinase (MAPK) pathways, and extensive transcriptional reprogramming that leads to the accumulation of defense-related metabolites and pathogenesis-related (PR) proteins (Bentham et al. 2020; Fang and Gu 2021). These local responses are further reinforced by systemic signals such as salicylic acid (SA), jasmonic acid (JA), and ethylene (ET), which coordinate systemic acquired resistance (SAR) and induced systemic resistance (ISR) in distal tissues, preparing the plant for future attacks (Fang and Gu 2021). Interestingly, peptide hormones have emerged as critical regulators within this immune signaling network. These small peptides, typically comprising short chains of amino acids, are known to regulate a range of physiological processes, including growth, development, and stress responses. In the context of immunity, peptide hormones serve as signaling molecules that contribute to immune receptor activation, intercellular communication, and the regulation of defense gene expression. Notable among these is systemin, the first peptide hormone discovered in plants, which plays a vital role in long-distance wound signaling and defense activation (Hu et al. 2018). Peptide hormones function analogously to classical plant hormones but are distinct in their mode of action. They are often produced as precursor proteins that undergo posttranslational processing to generate biologically active peptides. These peptides then bind to specific receptors on target cells, activating downstream defense responses. In addition to systemin, other peptide hormones such as phytocytokines and CLAVATA3/endosperm surrounding region-related (CLE) peptides are known to operate either locally at infection sites or systemically to orchestrate immune responses (Hu et al. 2018). The innate immune architecture of plants, despite lacking an adaptive counterpart, is highly flexible and responsive. It integrates input from both external signals and internal physiological status to mount a well-coordinated response to environmental cues and pathogen invasion (Bajguz and Hayat 2009; Han 2019). The two-branched innate immune system comprising PTI and ETI is not only interconnected but also modulated by a variety of signaling molecules, including both classical phytohormones (e.g. SA, JA, ET) and peptide hormones (Reimer-Michalski and Conrath 2016; Hirakawa and Sawa 2019).

The interplay between peptide hormones and classical phytohormones is essential for immune fine-tuning. For example, peptide hormones can amplify or suppress specific hormone signaling pathways to adjust the strength and specificity of immune responses. This intricate cross talk ensures optimal resource allocation between growth and defense, allowing plants to adapt to rapidly changing environments (Kim et al. 2021a, 2021b; Datta et al. 2024). Furthermore, peptide hormones have been implicated in the regulation of cell division, differentiation, and chemical defenses, highlighting their multifunctionality in plant development and immunity (Kim et al. 2021a, 2021b; Xie et al. 2022; Datta et al. 2024). Recent studies have also identified antimicrobial peptides (AMPs) as important components of the plant defense system. These peptides can exhibit direct antimicrobial activity by disrupting pathogen membranes or interfering with essential microbial processes, offering an additional layer of protection beyond signaling functions (Xie et al. 2022).

Collectively, plant immunity comprises a highly coordinated and multilayered network where physical barriers, immune receptors, and hormonal signals work synergistically to detect, respond to, and neutralize pathogen threats. Peptide hormones serve as crucial mediators in this system, bridging communication between cells and modulating the immune landscape. Their interactions with other signaling molecules further enhance the adaptability and efficiency of the plant immune system.

This chapter will delve into the diverse roles of peptide hormones in plant immunity. We will explore their specific involvement in PTI and ETI, their functional interaction with classical hormone signaling pathways, and their emerging roles as both signaling molecules and antimicrobial agents. By understanding how peptide hormones shape immune responses, we can gain deeper insights into the complex molecular orchestration that underlies plant defense strategies.

18.1.1 Pattern-Triggered Immunity (PTI)

Pattern-triggered immunity constitutes the first layer of inducible plant defense, initiated upon recognition of conserved molecular signatures, PAMPs or DAMPs, by cell-surface localized PRRs (Nguyen et al. 2021). These PRRs are typically classified into two main types: RLKs and RLPs (Yamaguchi et al. 2013; Nguyen et al. 2021; Zhang et al. 2024a, 2024b). Both classes are anchored in the plasma membrane via transmembrane domains, enabling direct interaction with extracellular signals. Upon ligand binding, PRRs activate a cascade of immune responses aimed at halting pathogen establishment and proliferation. These responses include a rapid influx of calcium ions into the cytosol, the activation of MAPK signaling pathways, and the production of ROS, all of which contribute to the transcriptional activation of defense-related genes (Nguyen et al. 2021). This signal transduction is further supported by various secondary messengers and defense hormones, reinforcing the immune barrier at the site of infection. The extracellular domains of RLKs and RLPs play a critical role in microbe-associated molecular pattern (MAMP) recognition. These domains vary structurally and include leucine-rich repeats (LRRs), lysin motifs (LysM), lectin domains, malectin-like motifs, and epidermal growth factor (EGF)-like repeats, which determine ligand specificity (Nguyen et al. 2021; Zhang et al. 2024a, 2024b) RLPs, lacking an intracellular kinase domain, must associate with kinase-active RLKs to transduce signals across the plasma membrane. This PRR complex formation is not only essential for ligand perception but also for the phosphorylation events required to initiate downstream signaling (TA and Zipfel 2021).

Notably, the PTI signaling network is not isolated. It interconnects closely with ETI, forming a continuum of immune responses. PTI can promote the generation of DAMPs, which further enhance defense responses through a mechanism termed DAMP-triggered immunity (DTI) (TA and Zipfel 2021; Nguyen et al. 2021). DTI amplifies the PTI signaling output, and this amplification appears to potentiate ETI-associated responses, creating a synergistic feedback loop essential for a robust immune defense. Recent insights underscore the modulatory role of peptide hormones in PTI. These small signaling peptides can either amplify or suppress PTI responses, depending on their nature and the context of pathogen attack (Igarashi et al. 2012; Rzemieniewski and Stegmann 2022; Stegmann et al. 2022). For instance, phytosulfokine (PSK), a peptide known to promote growth, has been shown to negatively regulate PTI in *Arabidopsis thaliana*. Its perception through the PSKR1 receptor dampens immune signaling, likely to mitigate the fitness cost associated with sustained defense activation (Igarashi et al. 2012). Mutants deficient in PSKR1 exhibit heightened expression of defense genes and enhanced sensitivity to MAMPs, indicating that PSK signaling acts as a negative regulator of immunity (Igarashi et al. 2012). In contrast, certain peptides such as GOLVEN (GLV) peptides have been shown to positively regulate PRR signaling and improve resistance against bacterial pathogens by enhancing the amplitude of PTI responses (Stegmann et al. 2022). Similarly, PAMP-induced peptides (PIPs) are secreted in response to MAMP detection and can reinforce immune signaling, further integrating peptide hormone action within the PTI framework (Hou et al. 2014; Chen et al. 2020). The plant must continuously balance these processes to maintain optimal fitness under fluctuating environmental and pathogenic pressures. The integration of peptide hormone signaling into the PTI network underscores their significance as modulators of immune homeostasis, fine-tuning the intensity and duration of defense responses while coordinating developmental cues (Chen et al. 2020; Igarashi et al. 2012; Rzemieniewski and Stegmann 2022). Thus, PTI serves as a foundational immune strategy in plants, mediated by PRRs that recognize conserved microbial and endogenous patterns to initiate downstream defense signaling. Peptide hormones, acting as both positive and negative regulators, intricately modulate this response, enabling plants to prioritize defense without compromising growth. The convergence of these signaling pathways is central to the plant's ability to dynamically adapt its immune responses, a theme that will be further explored in the subsequent sections.

18.1.2 Effector-Triggered Immunity (ETI)

Effector-triggered immunity represents a robust defense mechanism activated when plants recognize specific pathogen-derived effector proteins (Nguyen et al. 2021). This recognition is mediated by resistance (R) proteins encoded by *R* genes, primarily intracellular NB-LRR proteins (Deslandes et al. 2002; Gu et al. 2005; Wu et al. 2014). These proteins can detect effectors either directly or indirectly through guard or decoy models. Upon effector recognition, NB-LRR proteins undergo a conformational change involving the exchange of ADP for ATP within their central nucleotide-binding (NB) domain, initiating downstream defense signaling (Van Doorn 2011; Wu et al. 2014). NB-LRRs are subclassified based on their N-terminal domains: Coiled-Coil (CC)-NB-LRR and Toll/Interleukin-1 Receptor-like (TIR)-NB-LRR (Lukasik and Takken 2009; Wu et al. 2014). The activation of these R proteins leads to a cascade of immune responses, including the production of ROS, accumulation of SA, transcriptional activation of defense-related genes, and HR, a localized PCD that restricts pathogen spread (Van Doorn 2011; Wu et al. 2014).

The activation of ETI requires specific cofactors: ENHANCED DISEASE SUSCEPTIBILITY1 (EDS1) is crucial for TIR-NB-LRR-mediated signaling, while NON-RACE-SPECIFIC DISEASE RESISTANCE1 (NDR1) facilitates CC-NB-LRR-dependent immunity (Aarts et al. 1998; Takken et al. 2006). These pathways converge into a complex signaling network that coordinates broad-spectrum resistance responses and large-scale transcriptional reprogramming (Van Doorn 2011). Recent studies also underscore a strong synergy between ETI and PTI, whereby ETI amplifies PTI by enhancing the abundance and activity of its signaling components, ultimately leading to more durable and robust immunity (Van Doorn 2011; Tena 2021). Peptide hormones play a regulatory role in ETI by modulating ROS production and interacting with defense-related signaling cascades (Hsiao and Yamada 2020). These small signaling molecules often exert their function via leucine-rich repeat receptor-like kinases (LRR-RLKs), a family of transmembrane receptors that are pivotal in plant perception of biotic and abiotic stimuli (Man et al. 2020; Nie et al. 2022; Soltabayeva et al. 2022). LRR-RLKs possess an extracellular LRR domain for ligand binding and a cytoplasmic kinase domain for transmitting signals downstream (Man et al. 2023; Zhiqi et al. 2024). Upon effector recognition or stress perception, LRR-RLKs form receptor complexes and trigger signaling cascades involving calcium influx, ROS burst, and MAPK activation, leading to defense gene expression and immunity enhancement (Dixon et al. 1996; Tang et al. 2017; Soltabayeva et al. 2022; Huang and Joosten 2024). Well-known LRR-RLKs such as FLAGELLIN SENSING 2 (FLS2) and EF-TU RECEPTOR (EFR) in *Arabidopsis* detect bacterial PAMPs like flagellin and elongation factor Tu, demonstrating their role in bridging innate immunity and receptor-mediated signaling (Soltabayeva et al. 2022). LRR-RLKs also interact with receptor-like cytoplasmic kinases (RLCKs) to fine-tune defense responses and establish resistance (Huang and Joosten 2024).

Beyond immunity, LRR-RLKs participate in plant development, hormonal signaling, and responses to abiotic stresses like temperature and salinity, reflecting their multifaceted roles (Kobe and Kajava 2001; Liu et al. 2017; Mubassir 2019; Mu et al. 2024). Notably, peptide hormone-mediated signaling through LRR-RLKs, such as ROOT GROWTH FACTOR1 (RGF1) and its receptor RGFR, regulates ROS distribution and impacts downstream transcription factors like PLETHORA2 (PLT2), integrating developmental cues with immune responses (Hsiao and Yamada 2020). ROS are central to ETI, functioning both as direct antimicrobial agents and secondary messengers in signal transduction (Mittler 2017; Dumanović et al. 2021). The oxidative burst, a hallmark of ETI, generates high ROS concentrations at infection sites, contributing to HR and defense signaling (Dumanović et al. 2021). Hydrogen peroxide (H_2O_2), in particular, can modify cysteine residues of target proteins, altering their function to mediate specific stress responses (Hasanuzzaman et al. 2020). The physiological role of ROS is concentration-dependent: low levels promote signaling and adaptation, while excessive accumulation induces cytotoxicity and cell death (Wu et al. 2023).

Together, the coordination between ETI signaling components, peptide hormone activity, LRR-RLKs, and ROS homeostasis allows plants to mount a finely tuned immune response. This balance between developmental regulation and stress signaling ensures optimal survival during pathogen attack (Hsiao and Yamada 2020).

18.2 Peptide Hormones in Plant Signaling

Peptide hormones are critical regulators in plant signaling, functioning as mobile intercellular messengers that integrate growth, development, and stress responses. These small, secreted polypeptides modulate various physiological processes, including cell proliferation, organ formation, and adaptation to both biotic and abiotic stressors (Motomitsu et al. 2015; Hirakawa et al. 2017). Their action is initiated by binding to membrane-localized

LRR-RLKs, which subsequently trigger downstream intracellular signaling cascades (Hirakawa et al. 2017). Importantly, peptide hormones do not act in isolation but engage in intricate cross talk with classical phytohormones such as auxins, brassinosteroids (BRs), ET, JA, and SA. This interaction occurs through shared downstream elements, transcriptional regulation, and synergistic or antagonistic functional relationships (Hsiao and Yamada 2020; Dodueva et al. 2021; Zhang et al. 2024a, b). For instance, sulfated peptides like PSK can coordinate with BRs to regulate cell division in the root quiescent center and modulate ET pathways involved in floral organ abscission (Zhang et al. 2024a, b). Similarly, CLE peptides interact with auxin signaling pathways to fine-tune procambium development and root architecture (Wang et al. 2016). These cross-regulatory networks are essential for enabling plants to optimize their physiological responses under changing environmental conditions (Wang et al. 2016).

Beyond development, peptide hormones also serve as pivotal elements in plant immune responses. Several peptides, such as systemin, function as signaling molecules that orchestrate local and systemic defense mechanisms upon wounding or pathogen attack. Systemin, one of the earliest identified defense-related peptides, activates chemical defense by inducing protease inhibitor production to deter herbivores (Hu et al. 2018). In addition, members of the CLE peptide family have been linked to various stress responses and are believed to modulate immune signaling in conjunction with other hormone pathways (Hsiao and Yamada 2020; Kim et al. 2021a, b). Many peptide hormones also act as DAMPs, which are recognized by PRRs to activate PTI (Zelman and Berkowitz 2023). Upon recognition, these DAMPs stimulate defense signaling pathways involving ROS production and transcriptional activation of defense-related genes, thereby strengthening plant resistance to microbial pathogens. In defense signaling, peptide hormones operate synergistically with key phytohormones such as SA, JA, and ET. SA is central to resistance against biotrophic pathogens and plays a key role in SAR, regulating the expression of numerous PR genes (Klessig et al. 2000; Lu 2009; War et al. 2011). JA mediates responses to wounding and necrotrophic pathogens and often functions in coordination with ET to balance defense and growth processes (Yang et al. 2019; Ruan et al. 2019; Li et al. 2022a, b). ET contributes to resistance against necrotrophic pathogens and acts in concert with both JA and SA, modulating a broad spectrum of stress responses (Yang et al. 2019). Systemin, integrating with these hormonal pathways, reinforces plant immunity through the activation of protease inhibitors and other defense responses (Schaller and Ryan 1996; Pearce 2011; Molisso et al. 2022). Collectively, peptide hormones form a central hub in the signaling network of plants, interfacing with multiple phytohormones and immune components to orchestrate a dynamic and robust response to developmental cues and environmental challenges.

18.2.1 Salicylic Acid (SA)

Salicylic acid is a phenolic compound that functions as a crucial signaling molecule in plant immunity and regulates a broad spectrum of physiological processes critical for plant defense (Seyfferth and Tsuda 2014). As a key player in plant innate immunity, SA plays an essential role in providing both local and systemic resistance against biotic stresses, particularly microbial pathogens. Its involvement in plant defense is well established, with SA's ability to activate various immune responses in the affected tissues and induce protective mechanisms in distant tissues, preparing the plant for subsequent attacks (Li et al. 2019; Ding and Ding 2020). The model plant *A. thaliana* has been instrumental in advancing our understanding of SA's multifaceted roles in immunity. Genetic analyses have demonstrated that mutations in genes such as *SID2* and *EDS5*, which are critical for SA biosynthesis, impair the plant's ability to mount an effective immune response, particularly through SAR (Zhang et al. 2010). In *Arabidopsis*, SID2 encodes isochorismate synthase 1 (ICS1), a pivotal enzyme involved in the isochorismate pathway, which is one of the primary routes for pathogen-induced SA biosynthesis. This pathway works alongside the phenylpropanoid pathway, both of which converge to synthesize SA under stress conditions (Zhang et al. 2010). These two biosynthetic branches ensure the rapid accumulation of SA in response to pathogen attack, enabling the plant to trigger various defense mechanisms. Once produced, SA activates defense responses by binding to receptors, including Nonexpresser of Pathogenesis-Related (NPR) proteins primarily NPR1, NPR3, and NPR4 in *Arabidopsis*. These proteins serve as key mediators for SA perception and initiate downstream transcriptional reprogramming, which is critical for inducing immune-related genes (Seyfferth and Tsuda 2014). Notably, NPR1 functions as a transcriptional coactivator, and mutations in NPR1 lead to impaired SA-mediated transcriptional changes, which significantly hinder the plant's immune responses (Seyfferth and Tsuda 2014).

The role of SA extends beyond simple pathogen defense. It also regulates a complex array of physiological processes in plants, particularly in response to biotic stress. One of the most critical roles of SA is in the activation of SAR, which provides long-lasting protection against future pathogen attacks. By priming the plant's immune system, SAR enhances the speed and magnitude of subsequent immune responses, ensuring the plant is better prepared for future infections (Kim and Lim 2023). SA also plays a key role in HR, a form of PCD at the site of infection that limits pathogen spread and prevents further damage to the plant. This local cell death is typically accompanied by the accumulation of ROS and other signaling molecules, which further amplify the immune response (Radojičić et al. 2018). Beyond its direct involvement in immunity, SA engages in extensive cross talk with other plant hormones to fine-tune the plant's defense responses. These interactions allow the plant to modulate its immune strategy based on the type of stress encountered. For example, SA and JA often exhibit antagonistic interactions (Spoel and Dong 2024). SA is primarily involved in defenses against biotrophic pathogens, which rely on living host tissue, while JA is more effective in combating necrotrophic pathogens, which thrive on dead or dying tissue, as well as herbivores (Dempsey and Klessig 2017; Pokotylo et al. 2022). In addition, ET often acts synergistically with JA to enhance defense responses, particularly in the context of necrotrophic pathogens and herbivory. Together, these hormonal interactions allow plants to activate defense mechanisms that are tailored to the specific type of stress they are experiencing, ensuring an efficient use of energy and resources in response to pathogen attacks.

Other hormones, such as abscisic acid (ABA), auxin, gibberellic acid (GA), and cytokinin, also interact with SA to regulate plant immunity. For instance, ABA, which is often associated with stress responses such as drought and salinity, can modulate SA signaling to maintain a balance between stress adaptation and immune responses (Tan et al. 2020; Ali et al. 2020). Similarly, auxin and GA can influence SA-mediated defenses by regulating plant growth and developmental processes, which are closely linked to the plant's ability to allocate resources between growth and defense (Weiss et al. 2005). This intricate network of hormonal cross talk ensures that plants can respond to a variety of environmental conditions, adjusting their immune responses as needed. Moreover, SA's influence extends beyond transcriptional regulation to include posttranslational modifications of key signaling components. These modifications are critical for amplifying the immune response and enabling rapid signaling in response to pathogen attack. One of the mechanisms by which SA enhances signaling is through the formation of biomolecular condensates, which serve as dynamic hubs for cellular signaling. These condensates enable the efficient spatial and temporal organization of signaling components, ensuring that the plant's defense machinery is activated in a coordinated manner across multiple cellular compartments (Ding and Ding 2020). By promoting such condensate formation, SA facilitates the integration of various signaling pathways, enhancing the robustness of the plant's immune responses. Thus, SA is a central regulator of plant immunity, orchestrating a complex network of defense responses through both local and systemic mechanisms. Its interactions with other hormones further fine-tune these responses, allowing plants to balance growth and defense in a rapidly changing environment. The multifunctional role of SA, coupled with its ability to integrate multiple signaling pathways, underscores its importance in plant immune systems and highlights its potential as a target for improving disease resistance in crops.

18.2.2 Jasmonic Acid (JA)

Jasmonic acid is a crucial plant hormone involved in regulating chemical defense mechanisms, contributing to resistance against a broad spectrum of biotic stressors, including pathogens and herbivores (Campos et al. 2014). JA's role in plant immunity is complex, encompassing its biosynthesis, signaling pathways, and interaction with other plant hormones (Roychowdhury et al. 2025). This multilayered signaling network ensures that the plant can mount an appropriate immune response to diverse environmental challenges. The mechanism behind jasmonic acid-triggered immunity (JATI) is conceptualized as a multistage signal transduction cascade. This cascade begins with the recognition of danger signals by PRRs, followed by the activation of a conserved JA signaling module. This ultimately leads to the expression of defense-related genes orchestrated by transcription factors that regulate immune responses (Campos et al. 2014). Furthermore, negative feedback loops play a critical role in modulating the duration and intensity of the defense responses, preventing excessive activation and minimizing potential fitness costs (Campos et al. 2014). JA biosynthesis is initiated in the chloroplasts and completed in the peroxisomes (Wasternack and Hause 2019). The process begins with the release of α-linolenic acid (α-LeA) from galactolipids in the chloroplast membranes by phospholipase A1 (PLA1), a key enzyme in the pathway (Wasternack and Song 2017). The production of JA is induced upon pathogen or herbivore attack, triggering the synthesis of jasmonates (JAs), which are essential for activating defense mechanisms (Caarls et al. 2017). JA and its derivatives are lipid-derived phytohormones that regulate plant defense, especially against pathogens and herbivorous insects (Okada et al. 2015).

JA signaling acts as a central component in the plant's response to both abiotic and biotic stresses. It interacts with other phytohormones such as ABA, ET, and SA, forming a complex network that modulates plant immunity (Wang et al. 2021). For example, JA induces the expression of late-defense genes, while SA is primarily involved in initiating early defense responses (Yang et al. 2019). ET, together with JA, plays a pivotal role in defending against necrotrophic pathogens and inducing systemic resistance (Van Wees et al. 2000; Stintzi et al. 2001). The role of JA in plant immunity extends beyond pathogen defense to also include responses to herbivores and abiotic stress factors. In addition to its function in defense, JA enhances plant tolerance to abiotic stresses such as drought, salinity, temperature extremes, and heavy metal toxicity by modulating antioxidant systems and other stress-related pathways (Wang et al. 2021). The interplay between JA and other hormones like SA and ABA is crucial for fine-tuning the plant's immune response, ensuring an optimal balance between defense and growth (Wang et al. 2021). A defining feature of JATI is the recognition of pathogen-, herbivore-, and damage-associated molecular patterns (PAMPs, HAMPs, and DAMPs) by PRRs at the surface of plants. PRRs are essential for initiating the signaling cascade that leads to the synthesis of bioactive JA (Campos et al. 2014). This step is fundamental, linking the perception of danger signals to downstream immune activation.

The JATI signaling cascade is regulated by a conserved module involving the degradation of jasmonate ZIM domain (JAZ) proteins. Under nonstress conditions, JAZ proteins suppress transcription factors (TFs) in the nucleus. Upon stress, the elevated levels of JA bind to JAZ proteins, leading to their degradation. This degradation releases transcription factors such as MYC2, MYC3, and MYC4, which then activate the expression of JA-responsive genes. These genes contribute to various defense mechanisms, including the production of defense compounds and morphological changes that aid in the plant's survival (Campos et al. 2014). Negative feedback mechanisms within this signaling network help prevent excessive defense activation, ensuring that the plant's immune response remains balanced and does not incur unnecessary energy costs. PRRs are the primary sensors of danger signals, initiating the defense response by activating JA synthesis and subsequent immune reactions. They are responsible for detecting molecular patterns from pathogens and herbivores, triggering the downstream signaling that leads to JA-mediated defense responses (Campos et al. 2014). The versatility of PRRs allows plants to recognize a broad range of molecular patterns, enabling the activation of targeted defense mechanisms against specific threats (Campos et al. 2014). In summary, JA plays an essential role in plant immunity, regulating both local and systemic defense mechanisms. Through its complex signaling pathways and interactions with other phytohormones, JA enables plants to mount effective defense responses against pathogens, herbivores, and abiotic stresses. The tightly regulated signaling processes, including PRR activation, JAZ degradation, and transcription factor activation, ensure that JA-mediated immune responses are balanced, promoting plant survival while minimizing fitness costs.

18.2.3 Ethylene (ET)

Ethylene is a critical phytohormone that regulates various physiological processes in plants, such as growth, development, and responses to both biotic and abiotic stresses. A key aspect of ET's role is its involvement in plant-microbe interactions, where it mediates plant defense responses against pathogens and plays a regulatory role in interactions with beneficial microbes (Shekhawat et al. 2022). In response to MAMPs, plants increase ET production as part of their defense strategy (Shekhawat et al. 2022). Additionally, ET influences plant interactions with beneficial microbes, especially under abiotic stresses like heat and salt stress, where it modulates ethylene-responsive factor (ERF) genes. These ERFs regulate the plant's transcriptome, Na^+/K^+ homeostasis, and antioxidant systems, including those involved in the scavenging of ROS (Kacperska and Kanellis 1997; Groen et al. 2013; Peng et al. 2014; Keunen et al. 2016). The biosynthesis of ET begins with the conversion of methionine to S-adenosylmethionine (SAM) by SAM synthetase. SAM is then converted to 1-aminocyclopropane-1-carboxylic acid (ACC) by ACC synthase, and ACC is subsequently converted into ET by ACC oxidase (Schellingen et al. 2014; Ravanbakhsh et al. 2018). ET perception occurs at the endoplasmic reticulum membrane, where receptors such as ETHYLENE RESPONSE SENSOR1 (ERS1), ERS2, ETHYLENE RESPONSE1 (ETR1), ETR2, and ETHYLENE INSENSITIVE4 (EIN4) negatively regulate the ET signaling pathway (Groen et al. 2013; Scarpeci et al. 2017). Upon pathogen or herbivore attack, ET signaling is activated, initiating calcium fluxes and MAPK cascades that lead to increased ET production. The quantity and timing of ET synthesis are tightly regulated in response to different stresses, optimizing the plant's defense response (Groen and Whiteman 2014; Rehrig et al. 2014).

The role of ET in plant defense varies during different stages of plant development. For instance, in *Nicotiana attenuata*, the emission of ET in response to herbivore damage, such as feeding by *Manduca sexta* caterpillars, decreases with the age of the leaf and plant (Diezel et al. 2011). ET not only promotes local defense responses at the site of infection or injury but also triggers systemic immunity. This includes pathogen-induced SAR, herbivore-induced systemic resistance (ISR), and resistance induced by beneficial microbes. These systemic defenses provide protection to the uninfected parts of the plant, ensuring the survival and future resistance of the entire organism (Pieterse et al. 2014). Interestingly, some plant-growth-promoting bacteria (PGPB) degrade the ET precursor ACC to enhance their own growth, while ET can inhibit the growth of certain mycorrhizal fungi (Shekhawat et al. 2022). For example, ET reduces nodulation in legume-rhizobia symbioses in *Medicago truncatula* and restricts the growth of *Glomus aggregatum*, an arbuscular mycorrhizal fungus in *Pisum sativum* (Guinel 2015). This suggests that ET concentrations must be tightly controlled, as lower ET levels can enhance mycorrhizal colonization, while higher levels may inhibit it. Furthermore, ET interacts with other hormones, such as ABA, which is essential for the proper formation of arbuscules in mycorrhizal fungi. In ABA-deficient plants, ET accumulation leads to a reduced abundance of arbuscules (Guinel 2015). The involvement of ET in ISR has been well established, especially in *Arabidopsis thaliana*, where the JA and ET-dependent ISR pathway operates independently of SA. Certain strains of *Pseudomonas* induce systemic resistance in a variety of plants, including carnation, cucumber, radish, tobacco, and *Arabidopsis*, enhancing their defense capacity upon pathogen challenge (Choudhary et al. 2007; Reimer-Michalski and Conrath 2016). While ET's role in SAR and herbivore-induced systemic resistance is less pronounced, its essential function in ISR is widely recognized (Fu and Dong 2013; Pieterse et al. 2014). Additionally, ET is implicated in volatile signaling, either as a volatile itself or by regulating the production of other volatile compounds that serve as signals to distant plant tissues or neighboring plants. This volatile signaling can function as an indirect defense mechanism, repelling herbivores or attracting their natural enemies (Scala et al. 2013).

Ethylene also plays a dual role in regulating plant-microbial interactions, depending on whether the microbe is beneficial or pathogenic. In response to beneficial microbes, plants increase ET biosynthesis, signaling the need to optimize microbial colonization. Furthermore, ET contributes to inter-plant communication, alerting nearby plants to potential threats. The effects of ET on microbial colonization are context-dependent, with some microbial strains promoting growth in the presence of ET, while others may be inhibited. Therefore, the precise regulation of ET levels is critical for maintaining a balance between defense responses and the establishment of beneficial plant-microbe relationships. In summary, ET plays a crucial role in plant defense mechanisms, regulating both local and systemic responses to biotic and abiotic stresses. Its interaction with other hormones, such as ABA and JA, further fine-tunes the plant's immune responses. The regulation of ET synthesis and perception is essential for balancing defense against pathogens and fostering beneficial plant-microbial associations. Future research should aim to better understand the interplay between ET and other signaling networks to optimize plant resistance strategies and improve agricultural practices.

18.2.4 Systemin

Systemin is an 18-amino-acid peptide plant hormone that plays a central role in coordinating both local and systemic immune responses in plants (Ryan and Pearce 1998). It acts as a key signal in plant defense, particularly in response to herbivore attacks, by activating defense genes and triggering the production of defensive chemicals. Even at femtomolar concentrations, systemin can induce the activation of defense-related genes in tomato leaves (Zhang et al. 2020). This hormone is considered a primary signal for systemic defense, essential for both initiating local immune responses and enhancing defense across the entire plant (Ryan and Pearce 1998; Zhang et al. 2020). The mechanism through which systemin exerts its effects is primarily based on lipid-based signaling pathways. Upon activation, systemin triggers the release of linolenic acid from plant membranes, which is subsequently converted into JA, a critical activator of defense gene expression (Ryan 2000; Ryan and Pearce 1998; Zhang et al. 2020). The production of JA, in turn, leads to the synthesis of defensive proteins, such as proteinase inhibitors (PIs), and enhances the plant's ability to resist future attacks, thereby inducing SAR. Studies have shown that systemin enhances plant protection when combined with *Trichoderma afroharzianum* (strain T22), a beneficial microorganism, suggesting that systemin actively contributes to plant defense activation (Aprile et al. 2022). Genetic analysis has revealed that systemin and its precursor, prosystemin, are key components in a wound-induced intercellular signaling pathway, indicating that systemin is an upstream regulator of plant responses to physical damage (Li et al. 2002). Notably, systemin works synergistically with JA in the same signaling pathway to trigger the expression of various defense genes (Li et al. 2002; Ryan and Pearce 1998; War et al. 2011). Grafting experiments further support this interaction, showing that JA, regulated by systemin, serves as the systemic wound signal that amplifies defense responses across the plant (Sun et al. 2011).

Systemin-primed plants exhibit increased resistance to herbivorous pests, attract more parasitoid insects, and demonstrate a heightened response to wounding (Coppola et al.

2017). This effect is attributed to the ability of systemin to stimulate the production of proteolytic enzymes through its interaction with JA, which induces the production of additional prosystemin and other defense-related enzymes (Ryan and Pearce 1998). This coordinated action between systemin and JA is crucial for robust plant defense, especially in tomato plants (Li et al. 2002; Sun et al. 2011). Moreover, systemin is involved in the modulation of other phytohormones such as SA and ET, which also play significant roles in plant defense. Systemin can induce metabolic changes that prime neighboring plants for enhanced defense, including downregulation of SA-responsive genes, which are typically associated with resistance to biotrophic pathogens (Coppola et al. 2017). While systemin mainly interacts with the JA pathway, ET, alongside JA, is known to mediate the activation of various defense responses, particularly against certain pathogens (Klessig et al. 2000). Although the precise relationship between SA, ET, and JA pathways remains complex and not fully defined, systemin's role in activating JA suggests that ET might influence the JA response triggered by systemin, providing an additional layer of regulation in the plant defense network (Klessig et al. 2000; Coppola et al. 2017). Systemin's role in plant defense extends beyond its direct interaction with JA (Sun et al. 2011). It has been found to influence other plant responses through interactions with additional signaling pathways (Singh 2022). A key example of this is its involvement with ROS, which are typically produced during pathogen attacks or environmental stressors (Dumanović et al. 2021). Systemin regulates ROS levels, acting as an early signaling molecule that prepares the plant for subsequent oxidative bursts (Zhou et al. 2024; Wang et al. 2024a, b). These bursts of ROS are essential for strengthening the plant's defense systems, particularly in response to herbivore feeding and pathogen infection (Lee et al. 2020; Sahu et al. 2022). Additionally, systemin has been shown to play a role in maintaining the plant's cell wall integrity by inducing the expression of enzymes that modify the cell wall, thus enhancing the physical barrier against pathogens (Silva-Sanzana et al. 2020). This dual role, activating systemic defense responses while modulating oxidative stress and structural changes, underscores the complexity of systemin's function in plant immunity (Prasad et al. 2023; Wang et al. 2024a, b). As a central signaling molecule, systemin coordinates both local and systemic immune responses, preparing the plant to confront a range of biotic and abiotic challenges (Buonanno et al. 2018; Lee et al. 2020).

18.3 Interplay between Peptide Hormones and Immune Responses

The interaction between peptide hormones and plant immune responses plays a pivotal role in helping plants balance their growth, development, and defense mechanisms against pathogens and environmental stresses (Gilroy and Breen 2022). This cross talk enables plants to prioritize defense responses during stressful interactions, effectively reducing the fitness costs that often accompany sustained immune activation. Plant peptide hormones act as signaling molecules, orchestrating cell-to-cell communication that is essential for both plant growth and the regulation of defense processes (Hu et al. 2018). In this way, these hormones not only guide the plant's normal physiological processes but also allow for a timely and appropriate response to biotic and abiotic stresses. Hormonal signaling cross talk is especially crucial for plant defense against a diverse array of pathogens, pests, and other environmental stressors, including wounding (Derksen et al. 2013; Ding et al. 2022). To effectively defend against pathogen invasion, plants have evolved both passive and active resistance mechanisms. Passive defenses are typically preexisting, such as physical barriers or chemical defenses, while active defense responses are rapidly triggered after pathogen detection. These include the release of ROS, activation of the HR, and the synthesis of phytoalexins (Almagro et al. 2009; Bacete et al. 2018). These active defense mechanisms are tightly regulated and initiated upon recognition of PAMPs and effector proteins (Tian et al. 2019; Ding et al. 2022).

Both PTI and ETI are modulated by key plant hormones, including SA, JA, and ET (Ding et al. 2022). These hormones act as primary signals that mediate various resistance responses related to basal immunity (Ding et al. 2022; Rzemieniewski et al. 2022; Pastor-Fernández et al. 2023). Generally, the JA signaling pathway cooperates synergistically with the ET pathway, while the SA and JA pathways tend to have an antagonistic relationship (Aerts et al. 2021; Val-Torregrosa et al. 2021). This dynamic interaction between hormone signaling pathways allows plants to fine-tune their defense responses, providing a robust and adaptable immune system capable of responding to a wide range of biotic threats (Pieterse et al. 2009; Ding et al. 2022).

The interplay between plant hormone signaling pathways also ensures specificity in defense responses. Synergistic, antagonistic, and additive effects among different hormone pathways help manage the plant's defense strategies under diverse conditions. For instance, synergistic interactions between ROS, calcium ions (Ca^{2+}), and phytohormones activate downstream defense genes, amplifying the immune response (Yu et al. 2024). On the other hand, antagonistic interactions, such as those between the SA and JA pathways, provide robustness to the plant's immune system. In the event that one signaling pathway is compromised, the other can compensate, ensuring continued protection (Aerts et al. 2021). The most well-studied example of antagonism is the reciprocal interaction between SA and JA, where damage caused by one pathogen or insect can trigger a shift in the relative activity of these pathways (Moreira et al. 2018; Aerts et al. 2021). Plants respond to pathogen or herbivore attacks by synthesizing specific blends of SA, JA, and ET, which direct the plant's immune system to target particular types of threats. SA-mediated defenses are typically more effective against biotrophic pathogens, while JA and ET signaling pathways are more effective in responding to necrotrophic pathogens and herbivorous insects (Stout et al. 2006; Pieterse and Dicke 2007). This distinction highlights the specialization of plant immune responses, ensuring that the plant can deploy the most appropriate defense strategy based on the nature of the invader. The SA and JA/ET signaling pathways are considered central to plant immunity, providing a solid foundation for resistance against biotic invaders (Dombrecht et al. 2007; Koornneef and Pieterse 2008). By integrating these hormonal signals, plants are able to mount an effective and highly specific immune response, balancing growth with defense to optimize overall fitness.

18.3.1 Insights into Peptide Hormones and Their Role in Immune Modulation

To further elaborate on the interplay between peptide hormones and immune responses, it is essential to highlight how this cross talk helps plants modulate their defenses against various environmental challenges. Peptide hormones, including systemin, rapidly trigger complex signaling networks that allow plants to adapt their immune responses based on the type and intensity of the stress encountered. For example, the interaction between systemin and JA plays a crucial role in activating defense mechanisms against herbivore attacks. This dynamic signaling process is not only confined to defense but also has a broader impact on the plant's overall immune readiness, allowing plants to maintain a balance between immune activation and growth. The integration of multiple signaling pathways provides a sophisticated system that enables plants to fine-tune their responses to different stress factors. Peptide hormones like systemin, together with other plant hormones like SA, ET, and ABA, influence plant immune responses by regulating gene expression, protein synthesis, and the production of secondary metabolites. This coordination of peptide hormone action supports the plant's ability to resist pathogens while ensuring growth and development are not overly compromised by immune activation. This dual functionality of peptide hormones is particularly evident in their role in regulating both local and systemic defense responses.

Moreover, the regulatory feedback between peptide hormones and plant immune responses extends to the modulation of metabolic processes, such as those involved in ROS production and cellular signaling. For example, ROS serve as both signaling molecules and direct effectors of defense responses. They play a pivotal role in activating defense genes by promoting systemic signaling through cell-to-cell communication networks. The balance between ROS production and scavenging is fine-tuned by peptide hormones, ensuring that oxidative stress does not overwhelm the plant, while still providing the necessary signals for defense activation. In addition, the cross talk between peptide hormones and immune responses is essential for fine-tuning the plant's reaction to both biotic and abiotic stresses. For instance, under drought or salt stress, peptide hormones can modulate the plant's response to pathogen attack, ensuring that defense mechanisms are maintained even in unfavorable conditions. This interplay underscores the importance of peptide hormones in regulating not only defense but also stress adaptation, offering plants a versatile tool to thrive in fluctuating environmental conditions. Through this network of signaling interactions, plants can maximize their immune potential while maintaining overall growth and survival, demonstrating the critical role of peptide hormones in plant defense strategies.

18.3.2 Cross-Talk Mechanisms

The cross talk between peptide hormones plays a pivotal role in regulating plant immunity, enabling plants to fine-tune their defense responses against a wide range of biotic and abiotic stresses (Verhage et al. 2010; Yang et al. 2015). This sophisticated communication network allows for both synergistic and antagonistic interactions between different signaling pathways, influencing the activation of specific defense mechanisms, such as the production of AMPs (Vos et al. 2013; Broekgaarden et al. 2015; Bolouri Moghaddam et al. 2016). Understanding these cross-talk mechanisms is essential for decoding the intricate regulation of plant immunity and for developing strategies aimed at improving crop protection. Hormonal cross talk is integral to balancing plant growth, immunity, and stress tolerance (Coolen et al. 2016; Shigenaga et al. 2017; Davila Olivas et al. 2017). A classic example of hormonal interaction is the interplay between SA, JA, and ET, which orchestrate plant defenses against pathogens and herbivores. While SA-induced defenses are particularly effective against biotrophic pathogens, JA and ET are crucial for combating necrotrophic pathogens and herbivorous insects (Bolouri Moghaddam et al. 2016). However, these pathways are not always mutually supportive. Evidence shows that the SA-mediated defense response can antagonize JA signaling. Specifically, SA inhibits the transcription of JA-inducible genes by activating transcriptional regulators that downregulate JA-responsive genes (Caarls et al. 2015). This antagonism is a critical aspect of plant immune regulation, as it allows the plant to prioritize defense against specific types of threats.

The interaction between peptide hormones and transcription factors forms a central hub in plant immune responses. For example, AMPs play a crucial role in this network, binding to various proteins, such as COP9 signalosome subunits, MAPK2, and thioredoxins, which are involved in regulating plant immunity (Bolouri Moghaddam et al. 2016). These interactions help activate specific AMPs and other defense-related genes, ensuring an appropriate immune response depending on the pathogen or environmental stress encountered. Systemin, a well-characterized peptide hormone, contributes significantly to modulating plant immune responses. Upon binding to its receptor, systemin activates a signaling cascade that enhances the production of jasmonates, which are essential for initiating long-distance signaling in plant defense (Greifenhagen et al. 2024; Zhang et al. 2025). This pathway links systemin to other hormonal signaling networks, strengthening the plant's overall immune response. Recent studies have also shown that peptides like PIP3 can regulate plant immunity by modulating the cross talk between ABA and the RALF peptide (Zhou et al. 2024). These interactions further highlight the complexity of hormonal regulation in plant immunity.

Targeted engineering of phytohormone cross talk represents a promising strategy for enhancing both pathogen resistance and reproductive growth in plants (Johnston et al. 2024). By manipulating these hormonal interactions, it may be possible to optimize defense responses according to the specific type of pathogen or environmental condition. The interplay between AMPs and other immune system components, as well as their corresponding signaling pathways, is crucial for a robust and effective immune response. Additionally, this cross talk extends to the plant's response to abiotic stresses, such as drought, high salinity, and temperature extremes, demonstrating the broad scope of its importance in plant survival and resilience (Bolouri Moghaddam et al. 2016). A deeper understanding of how peptide hormones, AMPs, and other immune system components cooperate could lead to the development of more effective crop protection strategies. Ongoing research into emerging peptide hormones and their roles in plant immunity will provide valuable insights into plant defense mechanisms and potential targets for future crop improvement (Bolouri Moghaddam et al. 2016; Tanaka and Heil 2021).

18.4 Role of Peptide Hormones in Plant Defense Against Pathogens

There is no doubt that one of the survival strategies of plants is a response to pathogen attack. It is beginning to be understood that peptide hormones play an important role in the interaction between plants and various pathogens (Tan et al. 2024; Tian et al. 2024). In many cases, pathogen-induced secretion of peptides by plants is the initial response to attack and may lead to the formation of immunity. Various pathogens also secrete peptides that either inhibit the defense of the host plant or, conversely, mimic the peptide hormones of plants, activating their immune responses or other defense pathways (Roychoudhury et al. 2024; Ghosh and Roychoudhury 2024). Peptide hormones are important weapons in the arsenal of a plant that increase the underlying resistance and resistance of a plant to pathogens, and this is true for many types of pathogens (Dodueva et al. 2021).

There are many different types of peptides related to peptide hormones involved in defense against pathogens. Such peptides include several defensins, oligomeric peptides and immune peptides, as well as the recently discovered family of "nonphyto" peptide hormones, expression of which is induced by pathogen attack in vascular plants (Ghosh and Roychoudhury 2024; Chowdhary and Songachan 2025; Lyapina et al. 2025) The importance of peptide hormones in both plant immune signaling and attacks on pathogens is supported by abundant evidence. The effect of peptide hormones on some aspects of immunity continues to be studied, but there is absolutely no doubt that it is an integral part of plant immunity (Ghosh and Roychoudhury 2024;; Ali et al. 2024; Lyapina et al. 2025). The principle of the role of peptides as immunity signals or invigorators in the plant-pathogen interaction has been confirmed for the plant orientation response, extracellular poisons, phytoalexins, defense genes, the so-called "defense—hypersensitivity" reaction (DHR), etc. Broad-spectrum resistance to pathogens is a consequence of the effect of peptides, as it is acquired after exogenous treatment with synthetic analogs of such peptides or the herbivory of plants that secrete them (Dodds et al. 2024; Simas et al. 2025; Yang et al. 2025).

18.4.1 Antimicrobial Peptides

Plants depend on a network of sophisticated defense mechanisms to defend against biotic stresses such as pathogen invasion (Du et al. 2024). PAMPs, such as lipopolysaccharides (LPS), flagellin, chitin, and EF-Tu, are actively recognized by the PRRs of the plants, thus initiating immune responses, namely PTI (Du et al. 2024; Li et al. 2024; Snoeck et al. 2025). PTI is usually sufficient to largely inhibit continued invader adaptation (CIA) in juvenile plants, which often exist in a relatively pathogen-free environment. Nonetheless, some virulent pathogens evolve virulence factors (effectors) that are xenohormones, phosphatases, proteases, or type III secretion system (T3SS) that interfere with positive regulators of PTI, thereby inducing effector-triggered susceptibility (ETS) (Dar et al. 2024; Hückelhoven and Schouten 2024). Consequently, most adult plants inevitably suffer from biotic stress and must invest enormous energy to mount another defense system: ETI (Ali et al. 2024). The race between arm and counter-arm controls evolution, and in the co-occurrence of ancient effector genes and receptors, ancestral effector genes are bitterly lost even at the species level, while receptor genes are duplicated and diversified (Parisi et al. 2024).

However, as a result of the activation of defense mechanisms, plants become not only sick but also starved, as most defense responses are pro-apoptotic and/or booby-trapped at the cost of considerable nutrients (Dutta et al. 2024; Li et al. 2024). Overall, many plants eventually overcome biotic stress, albeit at the cost of considerable energy investments. The success of late-stage defense relies on seven key inflammatory defense responses, which are inevitable corollaries of ETI signaling (Munns and Millar 2023; Du et al. 2024). Whether the initiation of ETI signaling is sufficient for the activation of late-stage defense is yet-unreported. Here, a synthetic default type of ETI signaling module (DTESM) was test-constructed (Wang et al. 2025). The potential of DTESM to induce late-stage inflammatory defense responses such as the repression of chloroplast reactivity (CR), cell death (CD), and the inhibition of root growth (IRG) was successfully tested and evaluated in the absence of the effector, providing suggestive evidence for a small fraction of components actively repressing the initiation of late-stage defense responses in noninfected plants (Wang et al. 2022; König et al. 2024; Tian et al. 2024).

18.5 Peptide Hormones and Abiotic Stress Responses

Plant hormones are signaling molecules that regulate various physiological processes. Plant hormones include different natural compounds such as peptide hormones, which influence cell division and elongation by regulating gene expression, ensuring biomass increase and hence higher crop yields (Dodueva et al. 2021). Phytohormones produced by biotrophic and hemibiotrophic pathogens, symbionts, and microbes interacting with plants are abnormal but effective parts of the molecular interface of these organisms with a host plant. These include indole-3-acetic acid, gibberellins, cytokinin, strigolactones, and peptide phytohormones (Ali et al. 2024; Roychoudhury et al. 2024; Chen et al. 2023). They activate growth responses in plants via classical and nonclassical signaling pathways, ensuring plant coexistence with these kinds of parasites in certain conditions. Much less data is available on the origin of nonplant peptide phytohormones (Bao et al. 2024; Wei et al. 2024). Nonetheless, it is a very exciting area of research, with different hypotheses made for ethylene gas, benzoic acid, JA, SA, and peptide phytohormones. The origin of peptide phytohormones is especially intriguing, as many different peptides, of a diverse origin, were discovered in different taxa (Bach et al. 2022; Potgieter et al. 2023; Ali et al. 2024). Hence, their origin is still uncertain. Different peptides could arise either as a result of the coevolution of a pathogen/symbiont and a host plant or as a result of horizontal gene transfer from other organisms. Peptide hormones produced by plant pathogens, symbionts, and microbes interacting with plants are part of the molecular interface that ensures their coexistence with the host plant (Osborne et al. 2023; Rehneke and Schäfer 2024; Tan et al. 2024). Over the last 25 years, a lot of data on this issue have been accumulated, both on the functions of plant peptide hormones and on the identified homologs of peptide phytohormones outside the plant kingdom. New discoveries in this area should help to resolve the issue of the origin of these unique peptides (Fan et al. 2023; Lalun and Butenko 2025).

18.5.1 Mechanisms of Action

The activity of PAMPs is limited by highly specific receptors that are either membrane-bound RLKs or RLPs, similar to the concept of self/other discrimination used for cell recognition (Bender and Zipfel 2023; Reyes Zamora et al. 2024; Cai et al. 2024). The earliest event of recognition of microbial PAMPs is the phosphorylation of Ser/Thr residues in the carboxyl-terminal cytoplasmic domains of RLKs. Phosphorylation of activated RLKs leads to a cascade of phosphorylation through a series of kinases, resulting in the activation of MAPK cascades (Gandhi and Oelmüller 2023; Reyes Zamora et al. 2024; Liu et al. 2024).

The transduction leads to a plant defense response, which is often accompanied by ROS bursts, localized cell death, the expression of PR genes, and defense-related hormones (Dodueva et al. 2021). To counteract the immune system of the host plant, pathogens produce virulence factors or effector proteins. These effectors are recognized by receptor proteins in the host plant, resulting in ETI (Qin et al. 2022; Xiao et al. 2024). ETI is generally more powerful but slower than PAMP-Triggered Immunity (PaTI) and is often associated with the generation of an HR (Yang et al. 2022; Cai et al. 2024). HR results in rapid ion flux changes, production of various defense-related phytohormones, generation of ROS, activation of transcription factors and downstream PR genes, and production of SAR, making plants give amplified immunity to the next attack (Zhang et al. 2021a, b; Gogoi et al. 2024; Iqbal and Poór 2025).

Some plant hormones and their receptors are perceived by hemibiotrophic pathogens as susceptible effectors. Then, reprogramming of the host immune system is started. A large number of studies that provide evidence for this situation are considered along with the current understanding of the wonders of plant immunity (Liao et al. 2022; Todd et al. 2022; Zhou et al. 2023a, b). Flg22 is the 22-amino-acid peptide derived from the flagellin of pathogenic bacteria that acts as a microbial PAMP. Perception of flg22 mainly through the transmembrane receptor FLAGELLIN-SENSING 2 (FLS2) triggers rapid and localized immunity (Cheng et al. 2021; Sands et al. 2022; Murakami et al. 2022). FLS2 has a unique ligand-induced dimerization mode that counterpoises high signal-on and low signal-off rates. Ligand-induced receptor dimerization activates the receptor-associated cytoplasmic kinase BRI1-ASSOCIATED RECEPTOR KINASE 1 (BAK1) that interacts with the inactive FLS2 monomer (Jeon et al. 2024; Xi et al. 2024). On the FLS2 dimerization model, BAK1 participates together with the FLS2 dimer to regulate downstream immune responses (Zhou et al. 2023a, b; Ehinger 2024).

18.5.2 Signal Transduction Pathways

In plants, pathogen response mechanisms involve the rapid release of a vast array of signaling molecules (S) in order to detect and signal the pathogen invasion and activate tolerance and defense responses to protect themselves against pathogens (Mittler et al. 2022; Roychowdhury et al. 2025). The current understanding of signaling elements involved in plant biotic stress signaling has come almost entirely from studies of well-characterized pea and *Arabidopsis* (Yaschenko et al. 2024; Sarma et al. 2025). Given the evolutionary conservation of plant S and family members, it is likely that similar mechanisms exist in signal processing and signaling cascade activation (Slavik and Kranzusch 2023; Hong et al. 2024). However, compositional data overriding the aforementioned model are emerging in several species like wheat and grapevine for PAMP signaling (Sarwar and Khan 2023; Gallucci et al. 2024). Also, additional S, receptors, and components of the related pathways might exist, as suggested by mutational analyses, proteomic studies, and recent discoveries of novel players in biotic stress signaling pathways (Dodueva et al. 2021). PAMPs, as a class, may not control responses to all microbes, and potentially untapped signal processing levels must remain to be discovered. Despite ongoing studies of the relationships among them in individual types of signals, coordinated cascades of signals still await being unmasked (Yu et al. 2022; Meddya et al. 2023). Alternative means of defense signal production and processing in plants, like vesicle trafficking or calcium wave propagation, need to be elucidated in terms of pathway participants, as continued compartmentalization of information from the source to its deciphering point might be integrated over the plant body (Kaur et al. 2022; Roychowdhury et al. 2025). Multiple S that could be produced upon specific or more general cues have not been investigated at the mechanistic level and their interactive effects remain to be explored. Induced pathogen defense responses, transcriptional changes, and accumulation of antimicrobial compounds may also need to be better related regarding their timescale and how plants process pathways and layers of information (Kaur et al. 2022; Nguyen et al. 2022). Integration points of the various stress and growth signals passed via such general common pathways as MAPK cascades or ROS production have only been speculated on in terms of effects. Each plant's size may impose specific additional regulation of a parameter (Das et al. 2025; Xia et al. 2025). Whether this affects the ability of biotic stress signaling mechanisms to form microdomain-like compartments and the processing route still needs research. The capacity of the defenses leading to nonhost responses may relate to each organism's complexity (Kuzminsky and Ghanim 2025; Wang et al. 2025). Also unresolved is how much each plant contributes to effective processing from perception to response (Novoplansky et al. 2024). Reliably establishing input–output relations for a broad range of combinations of input and degree of response from behavioral readouts remains an unattained goal (Subeshan et al. 2024; Shanks and Byblow 2025). Exploring signal processing stages via experimentally tractable biological materials could help clarify such questions as differential maintenance of specificity, affinity, dynamics, and duration of responses or multiple routes to the same response, such as peaked versus dampened patterns (Delplace et al. 2022; Roychowdhury et al. 2025). Other approaches to developmentally more tractable and amenable materials are needed to continue data acquisition and address questions such as spatiotemporal integration and programming of input succession and interacting cross talk and feedback (Levin 2023; Paul Friedman et al. 2025).

18.5.3 Gene Expression Modulation

Peptide hormones are moderately sized polypeptides of 5 to 50 amino acids, which are encoded by their pre-pro-precursor forms in the genome. The pre-pro-region directs the hormone synthesis in the endoplasmic reticulum and proteolytic cleavage removes it to convert it into the bioactive form (Zhang et al. 2021a, b; Rouaud 2024). These forms are posttranslationally modified to yield the active molecules. Some amino acids and their derivatives and their amino acid-like compounds act as signaling molecules in plants; one such category of signaling molecules is a peptide hormone (Dodueva et al. 2021; Huang et al. 2022; Ali et al. 2024). Plant peptide hormones, comprised of 5 to 50 amino acid residues, are a long-standing topic of study in plant biology. A variety of small polypeptides, with three subfamilies in eukaryotes, translocate from one cell to the other and modulate gene expression in the target cells over the last two decades (Huang et al. 2022; Chowdhary and Songachan 2025).

Few peptide hormones have been directly associated with protection against biotic stress and their downstream manipulation would have implications for the understanding of resistance mechanisms (Kumar et al. 2023; Ali et al. 2024; Ghosh and Roychoudhury

2024). Moreover, there is a wide range of peptide hormones still unstudied in relation to plant defense, including ribosome-induced small peptides with potential defensin activity, Jack bean α-amylase inhibitor, lycorine, and SCLY (residues 13–20) have never been analyzed in the context of plant-pathogen interactions (Kozyulina et al. 2023; Ghosh et al. 2024). While most peptide hormones are exclusive to plants, there are some exceptions with homologs outside the plant kingdom. The latter can be exploited for both molecular and biochemical studies (Chekan et al. 2024; Zada et al. 2024). Additionally, the technology for systematic characterization of the hormonal output at the organismal scale using phage display has the potential to yield data on dozens of new peptide hormones (Wang et al. 2024a, b; Pierzynowska et al. 2024). Furthermore, the application of this technology can improve tonalities to be used and nonredundant peptide hormone molecules identified within libraries (Georgiades et al. 2023; Geuna et al. 2024; Li et al. 2025).

This review will summarize the currently identified and best-studied small-peptide hormones produced in plants, focusing on the several synthetic strategies established in vitro to generate their active forms (Li et al. 2021; Montesinos et al. 2021; Shen et al. 2022; Li et al. 2025). The focus will also be clamped on recent advances that are being made in the understanding of their mode of action and considering the receptors responsible for mediating the downstream hormonal responses (Akbarian et al. 2022; Latic and Erben 2022; Malécange et al. 2023). It will also highlight the emergent perception of their involvement in cross talk and cross-regulation between several plant signaling pathways, which include other classes of hormones and environmental stresses such as drought and salinity (Zhao et al. 2021; Xie et al. 2022; Waadt et al. 2022).

18.6 Future Directions in Research

Research on signaling peptides in defense responses against pathogens is an actively developing field. Modern studies focus, first, on the formation of the peptide signal pool; second, on the recognition of the widest array of peptides; and third, on deciphering the intracellular signaling chains that result in effective defense responses (Fernandes and Ghag 2022; Dou et al. 2023; Peláez-Vico et al. 2024). Research using high-throughput methods recently demonstrated that the proteome of a functional protein may contain information hidden in the form of cryptic peptides that are bioactive in various cell compartments (Abril et al. 2022; Othoum and Maher 2023; Coelho et al. 2024). New biologically active peptides involved in plant immunity, as well as in the regulation of the immune responses of sensitive and resistant species to pathogens, have been identified (Ali et al. 2024). On the one hand, the development of next-generation sequencing technologies and the advancement of search strategies available in bioinformatics now make it possible to conduct high-throughput analyses of peptide pools (Vanella et al. 2022; Satam et al. 2023). On the other hand, the improvement of sensitive methods of mass-spectrometric analysis allows for comprehensive studies of proteomes and peptidomes (Li et al. 2023; Fan et al. 2023).

Although there are still many unexplored peptide-based signaling pathways, this topic will expand into systems biology approaches. Thus, future studies will be aimed at mapping a number of previously unknown, similar to those involved in immune responses, peptide-based signaling pathways that ensure regulation of plant ontogenesis, organogenesis, and seed development (Pandita et al. 2023; Xiao et al. 2025; Ghosh and Roychoudhury 2024). The recognition by one receptor of peptides with different structures and, vice versa, with the same structure recognized by various receptors, is manifested by the differential modulation of the receptors' signaling activity (Snoeck et al. 2022; Snoeck et al. 2025). The specificity of peptide signals is maintained by overall signal topology, dose-dependent activation, cellular location, and desensitization of the receptors (Lyapina et al. 2025). However, extensive data on signaling were obtained for only a handful of receptors and peptide ligands. It is extremely important to elucidate mechanisms, templates, and general principles of perception governing specificity and pleiotropy (Kim et al. 2023; Huang et al. 2025). These studies will be conducted with the sister pairs of receptors and peptides on the one hand, and with symmetrical receptors on the other hand (Thompson et al. 2024; Kawada et al. 2025; Korsching 2025). The complexity of receptor-peptide interactions strengthens the need not only for mathematical modeling but also for deeper exploration of emergent properties of peptide signaling networks (Jefferson et al. 2023; Nikitovic et al. 2024; Yuan et al. 2025).

18.7 Emerging Peptide Hormones in Plant Immunity

Plant hormones, or phytohormones, are organic chemical substances with a relatively low molecular weight produced by plants that significantly affect their growth, development, morphogenesis, and other vital processes (EL Sabagh et al. 2022; Asif et al. 2022). Plant hormones are divided into five major groups according to their functions, chemical structure, and sources of production: auxins, gibberellins, cytokinins, ET, and ABA (Vaishnav and Chowdhury 2023; Sosnowski et al. 2023). Groups of hormones playing similar roles and having similar chemical structure are often classified into "families," for example, strigolactones, BRs, and SA (Song et al. 2024). The latest research has revealed that numerous peptides consisting of 5–60 amino acids and acting in a paracrine or autocrine manner, either alone or with other hormones, also perform the functions of hormones in plants (Sun et al. 2024; Guo et al. 2021). Many phytohormones have been well-characterized, including their structure/ligand-receptor-binding sites that mediate signaling pathways. In addition, the perception or receptor families, along with downstream signaling pathways, have been characterized for some peptide hormones (Iqbal et al. 2022; Shafqat et al. 2024). On the contrary, the research on peptide hormones in plants is in its infancy (Ghosh and Roychoudhury 2024). The role of peptide hormones in interkingdom interactions between plants and some known prokaryotic and eukaryotic pathogens that cause plant diseases is emerging (Chiquito-Contreras et al. 2024). Recently, some effector proteins of phytopathogenic bacteria, fungi, symbiotic microbes, and parasitic plants as mimic peptide phytohormones were revealed with a specific binding region close to plant peptide hormone receptor domains and perturbation of the plant hormone metabolism through hijacking the peptide hormone signaling pathways and inhibiting plant growth and immunity responses (Roychoudhury et al. 2024; Duhan and Pasrija 2025; Shukla et al. 2025). However, there are numerous prokaryotic and eukaryotic plant pathogens having phylogenetically distant homologs of plant peptide hormones. These unrelated protein families harbor similar topologies and partially conserved binding or catalytic residues, leading to the assumption that they might act as peptide hormone mimics (Ma et al. 2022; Zhou et al. 2023a, b). Consequently, it is crucial to develop methodologies to predict the functional implications of such "rando-gnomes." This chapter highlights the involvement of peptide hormones of different origins in the establishment of host plants and nonhost interactions (Rys et al. 2024; Novoplansky et al. 2024; Wang et al. 2025). It also considers induction mechanisms, evolution, and strategies of hijacking plant peptide hormones by pathogens, although most of the interactions that have been characterized for peptide phytohormones and effector proteins involve only local defense in the developmental programs of plants (Huang et al. 2022; Roychoudhury et al. 2024; Ali et al. 2024; Lalun and Butenko 2025).

18.8 Conclusion

While relatively small in size, peptide hormones are ubiquitous factors of intracellular information exchange in organisms of all domains of life. Peptides play a wide range of roles in living organisms. They act as the most specific and evolutionary conserved signal able to evoke a response in the target cells. Because of their positioning at the tip of the signaling cascade, peptide hormones are established key players in regulating almost any process in plant life, including growth and development, resistance to abiotic and biotic factors.

Because of the importance of peptide hormones in plant life, these signals were evolved by plant pathogens as a means of suppressing host plant immunity. This question remains a challenge for plant biologists and evolutionists and still attracts wide interest. Questions concerning the origin of nonplant peptide phytohormones, distribution in different biological kingdoms, and the pattern of their evolution are still put forward. Plants received innovations in the peptide signaling networks from two main sources: other multicellular eukaryotes, in particular fungi and chlorophytes, and two domains of prokaryotes that include pathogenic bacteria of the Proteobacteria division, as well as the Archaea group.

In recent years, a lot of data have been accumulated both on the functions of plant peptide hormones and on the identified homologs of peptide phytohormones outside the plant kingdom. There are comprehensive studies on the evolution of peptide signaling networks in fungi and green algae, as well as a huge database on the peptide hormones in the most versatile taxa of bacteria and archaea. All of this permitted the revelation of plant-injected peptide phytohormones and the establishment of a taxonomical group of algae-plants and peptidome in the Kingdom Plantae.

References

Aarts, N., Metz, M., Holub, E. et al. (1998). Different requirements for EDS1 and NDR1 by disease resistance genes define at least two R gene-mediated signaling pathways in Arabidopsis. *Proceedings of the National Academy of Sciences of the United States of America* 95: 10306–10311.

Abril, A.G., Pazos, M., Villa, T.G. et al. (2022). Proteomics characterization of food-derived bioactive peptides with anti-allergic and anti-inflammatory properties. *Nutrients* 14: 4400.

Aerts, N., Pereira Mendes, M., and Van Wees, S.C.M. (2021). Multiple levels of crosstalk in hormone networks regulating plant defense. *The Plant Journal* 105: 489–504.

Akbarian, M., Khani, A., Eghbalpour, S., and Uversky, V.N. (2022). Bioactive peptides: synthesis, sources, applications, and proposed mechanisms of action. *International Journal of Molecular Sciences* 23: 1445.

Ali, S., Khan, N., and Xie, L. (2020). Molecular and hormonal regulation of leaf morphogenesis in Arabidopsis. *International Journal of Molecular Sciences* 21: 5132.

Ali, J., Mukarram, M., Ojo, J. et al. (2024). Harnessing phytohormones: advancing plant growth and defence strategies for sustainable agriculture. *Physiologia Plantarum* 176: e14307.

Almagro, L., Gómez Ros, L.V., Belchi-Navarro, S. et al. (2009). Class III peroxidases in plant defence reactions. *Journal of Experimental Botany* 60: 377–390.

Aprile, A.M., Coppola, M., Turrà, D. et al. (2022). Combination of the systemin peptide with the beneficial fungus *Trichoderma afroharzianum* T22 improves plant defense responses against pests and diseases. *Journal of Plant Interactions* 17: 569–579.

Asif, R., Yasmin, R., Mustafa, M. et al. (2022). Phytohormones as plant growth regulators and safe protectors against biotic and abiotic stress. In: *Plant Hormones-Recent Advances, New Perspectives and Applications* (ed. C. Hano). IntechOpen.

Bacete, L., Mélida, H., Miedes, E., and Molina, A. (2018). Plant cell wall-mediated immunity: Cell wall changes trigger disease resistance responses. *The Plant Journal* 93: 614–636.

Bach, E., Passaglia, L.M.P., Jiao, J., and Gross, H. (2022). Burkholderia in the genomic era: from taxonomy to the discovery of new antimicrobial secondary metabolites. *Critical Reviews in Microbiology* 48: 121–160.

Bajguz, A. and Hayat, S. (2009). Effects of brassinosteroids on the plant responses to environmental stresses. *Plant Physiology and Biochemistry* 47: 1–8.

Bao, D., Chang, S., Li, X., and Qi, Y. (2024). Advances in the study of auxin early response genes: Aux/IAA, GH3, and SAUR. *The Crop Journal* (in press).

Bender, K.W. and Zipfel, C. (2023). Paradigms of receptor kinase signaling in plants. *The Biochemical Journal* 480: 835–854.

Bentham, A.R., De la Concepcion, J.C., Mukhi, N. et al. (2020). A molecular roadmap to the plant immune system. *The Journal of Biological Chemistry* 295: 14916–14935.

Bolouri Moghaddam, M.R., Vilcinskas, A., and Rahnamaeian, M. (2016). Cooperative interaction of antimicrobial peptides with the interrelated immune pathways in plants. *Molecular Plant Pathology* 17: 464–471.

Broekgaarden, C., Caarls, L., Vos, I.A. et al. (2015). Ethylene: traffic controller on hormonal crossroads to defense. *Plant Physiology* 169: 2371–2379.

Buonanno, M., Coppola, M., Di Lelio, I. et al. (2018). Prosystemin, a prohormone that modulates plant defense barriers, is an intrinsically disordered protein. *Protein Science* 27: 620–632.

Caarls, L., Elberse, J., Awwanah, M. et al. (2017). Arabidopsis JASMONATE-INDUCED OXYGENASES down-regulate plant immunity by hydroxylation and inactivation of the hormone jasmonic acid. *Proceedings of the National Academy of Sciences of the United States of America* 114: 6388–6393.

Caarls, L., Pieterse, C.M., and Van Wees, S.C. (2015). How salicylic acid takes transcriptional control over jasmonic acid signaling. *Frontiers in Plant Science* 6: 170.

Cai, M., Yu, H., Sun, E., and Zuo, C. (2024). Receptor-like proteins: decision-makers of plant immunity. *Phytopathology Research* 6: 58.

Campos, M.L., Kang, J.H., and Howe, G.A. (2014). Jasmonate-triggered plant immunity. *Journal of Chemical Ecology* 40 (7): 657–675.

Chekan, J.R., Mydy, L.S., Pasquale, M.A., and Kersten, R.D. (2024). Plant peptides – redefining an area of ribosomally synthesized and post-translationally modified peptides. *Natural Product Reports* 41: 1020–1059.

Chen, Y.L., Fan, K.T., Hung, S.C., and Chen, Y.R. (2020). The role of peptides cleaved from protein precursors in eliciting plant stress reactions. *The New Phytologist* 225 (6): 2267–2282.

Chen, Y., Wang, Y., Liang, X. et al. (2023). Mass spectrometric exploration of phytohormone profiles and signaling networks. *Trends in Plant Science* 28 (4): 399–414.

Cheng, J.H., Bredow, M., Monaghan, J., and DiCenzo, G.C. (2021). Proteobacteria contain diverse flg22 epitopes that elicit varying immune responses in Arabidopsis thaliana. *Molecular Plant-Microbe Interactions* 34 (5): 504–510.

Chiquito-Contreras, C.J., Meza-Menchaca, T., Guzmán-López, O. et al. (2024). Molecular insights into plant-microbe interactions: a comprehensive review of key mechanisms. *Frontiers in Bioscience-Elite* 16 (1): 9.

Choudhary, D.K., Prakash, A., and Johri, B.N. (2007). Induced systemic resistance (ISR) in plants: mechanism of action. *Indian Journal of Microbiology* 47 (4): 289–297.

Chowdhary, N.A. and Songachan, L.S. (2025). Plant peptides involved in ROS signalling and biotic and abiotic stress responses. *International Journal of Peptide Research and Therapeutics* 31 (3): 53.

Coelho, L.P., Santos-Júnior, C.D., and de la Fuente-Nunez, C. (2024). Challenges in computational discovery of bioactive peptides in 'omics data. *Proteomics* 24 (12–13): 2300105.

Coolen, S., Proietti, S., Hickman, R. et al. (2016). Transcriptome dynamics of Arabidopsis during sequential biotic and abiotic stresses. *The Plant Journal* 86: 249–267.

Coppola, M., Cascone, P., Madonna, V. et al. (2017). Plant-to-plant communication triggered by systemin primes anti-herbivore resistance in tomato. *Scientific Reports* 7 (1): 15522.

Dar, S.A., Hasan, W., Devi, Y.K. et al. (2024). Enzyme-mediated adaptation of herbivorous insects to host phytochemicals. *Phytochemistry Reviews* 23 (5): 1–24.

Das, S., Shil, S., Rime, J. et al. (2025). Phytohormonal signaling in plant resilience: advances and strategies for enhancing abiotic stress tolerance. *Plant Growth Regulation* 1–32.

Datta, T., Kumar, R.S., Sinha, H., and Trivedi, P.K. (2024). Small but mighty: Peptides regulating abiotic stress responses in plants. *Plant, Cell & Environment* 47 (4): 1207–1223.

Davila Olivas, N.H., Kruijer, W., Gort, G. et al. (2017). Genome-wide association analysis reveals distinct genetic architectures for single and combined stress responses in Arabidopsis thaliana. *The New Phytologist* 213: 838–851.

Delplace, F., Huard-Chauveau, C., Berthomé, R., and Roby, D. (2022). Network organization of the plant immune system: from pathogen perception to robust defense induction. *The Plant Journal* 109 (2): 447–470.

Dempsey, D.M.A. and Klessig, D.F. (2017). How does the multifaceted plant hormone salicylic acid combat disease in plants and are similar mechanisms utilized in humans? *BMC Biology* 15: 1–11.

Derksen, H., Rampitsch, C., and Daayf, F. (2013). Signaling cross-talk in plant disease resistance. *Plant Science* 207: 79–87.

Deslandes, L., Olivier, J., Theulières, F. et al. (2002). Resistance to Ralstonia solanacearum in Arabidopsis thaliana is conferred by the recessive RRS1-R gene, a member of a novel family of resistance genes. *Proceedings of the National Academy of Sciences of the United States of America* 99 (4): 2404–2409.

Diezel, C., Allmann, S., and Baldwin, I.T. (2011). Mechanisms of optimal defense patterns in Nicotiana attenuata: flowering attenuates herbivory-elicited ethylene and jasmonate signaling. *Journal of Integrative Plant Biology* 53: 971–983.

Ding, P. and Ding, Y. (2020). Stories of salicylic acid: a plant defense hormone. *Trends in Plant Science* 25 (6): 549–565.

Ding, L.N., Li, Y.T., Wu, Y.Z. et al. (2022). Plant disease resistance-related signaling pathways: recent progress and future prospects. *International Journal of Molecular Sciences* 23 (24): 16200.

Dixon, M.S., Jones, D.A., Keddie, J.S. et al. (1996). The tomato Cf-2 disease resistance locus comprises two functional genes encoding leucine-rich repeat proteins. *Cell* 84 (3): 451–459.

Dodds, P.N., Chen, J., and Outram, M.A. (2024). Pathogen perception and signaling in plant immunity. *Plant Cell* 36 (5): 1465–1481.

Dodueva, I., Lebedeva, M., and Lutova, L. (2021). Dialog between kingdoms: enemies, allies and peptide phytohormones. *Plants (Basel)* 10 (11): 2243.

Dombrecht, B., Xue, G.P., Sprague, S.J. et al. (2007). MYC2 differentially modulates diverse jasmonate-dependent functions in Arabidopsis. *Plant Cell* 19: 2225–2245.

Dou, X., Yan, D., Liu, S. et al. (2023). Host defense peptides in nutrition and diseases: a contributor of immunology modulation. *Journal of Agricultural and Food Chemistry* 71 (7): 3125–3140.

Du, B., Haensch, R., Alfarraj, S., and Rennenberg, H. (2024). Strategies of plants to overcome abiotic and biotic stresses. *Biological Reviews* 99 (4): 1524–1536.

Duhan, L. and Pasrija, R. (2025). Unveiling exogenous potential of phytohormones as sustainable arsenals against plant pathogens: molecular signaling and crosstalk insights. *Molecular Biology Reports* 52 (1): 98.

Dumanović, J., Nepovimova, E., Natić, M. et al. (2021). The significance of reactive oxygen species and antioxidant defense system in plants: a concise overview. *Frontiers in Plant Science* 11: 552969.

Dutta, A., Dracatos, P.M., and Khan, G.A. (2024). Balancing act: the dynamic relationship between nutrient availability and plant defence. *The Plant Journal* 120 (5): 1724–1734.

Ehinger, A.S. (2024). Analysis of the spatiotemporal organization of immune receptors in plant plasma membrane by sptPALM and the impact of HIR2 as putative organizing factor. Doctoral dissertation, Universität Tübingen.

EL Sabagh, A., Islam, M.S., Hossain, A. et al. (2022). Phytohormones as growth regulators during abiotic stress tolerance in plants. *Frontiers in Agronomy* 4: 765068.

Fan, K.T., Hsu, C.W., and Chen, Y.R. (2023). Mass spectrometry in the discovery of peptides involved in intercellular communication: from targeted to untargeted peptidomics approaches. *Mass Spectrometry Reviews* 42 (6): 2404–2425.

Fang, Y. and Gu, Y. (2021). Regulation of plant immunity by nuclear membrane-associated mechanisms. *Frontiers in Immunology* 12: 771065.

Fernandes, L.B. and Ghag, S.B. (2022). Molecular insights into the jasmonate signaling and associated defense responses against wilt caused by *Fusarium oxysporum*. *Plant Physiology and Biochemistry* 174: 22–34.

Fu, Z.Q. and Dong, X. (2013). Systemic acquired resistance: turning local infection into global defense. *Annual Review of Plant Biology* 64: 839–863.

Gallucci, A., Giordano, D., Facchiano, A. et al. (2024). Transmembrane proteins in grape immunity: current knowledge and methodological advances. *Frontiers in Plant Science* 15: 1515163.

Gandhi, A. and Oelmüller, R. (2023). Emerging roles of receptor-like protein kinases in plant response to abiotic stresses. *International Journal of Molecular Sciences* 24 (19): 14762.

Georgiades, M., Alampounti, A., Somers, J. et al. (2023). Hearing of malaria mosquitoes is modulated by a beta-adrenergic-like octopamine receptor which serves as insecticide target. *Nature Communications* 14 (1): 4338.

Geuna, F., Pensotti, A., Vecchione, R., and Germano, R. (2024). Hints of biological activity of xerosydryle: preliminary evidence on the early stages of seedling development. *International Journal of Molecular Sciences* 25 (16): 8717.

Ghosh, P. and Roychoudhury, A. (2024). Plant peptides involved in abiotic and biotic stress responses and reactive oxygen species (ROS) signaling. *Journal of Plant Growth Regulation* 43 (5): 1410–1427.

Ghosh, E., van Nouhuys, S., and Ode, P.J. (2024). Anthropogenic effects on the eco-immunology of herbivorous insects. *Current Opinion in Insect Science* 66: 101285.

Gilroy, E. and Breen, S. (2022). Interplay between phytohormone signalling pathways in plant defence – other than salicylic acid and jasmonic acid. *Essays in Biochemistry* 66 (5): 657–671.

Gogoi, K., Gogoi, H., Borgohain, M. et al. (2024). The molecular dynamics between reactive oxygen species (ROS), reactive nitrogen species (RNS) and phytohormones in plant's response to biotic stress. *Plant Cell Reports* 43 (11): 1–25.

Greifenhagen, A., Ruwe, H., Zimmer, V. et al. (2024). The peptide hormone Pj CLE1 stimulates haustorium formation in the parasitic plant *Phtheirospermum japonicum*. *Proceedings of the National Academy of Sciences of the United States of America* 121 (42): e2414582121.

Groen, S.C. and Whiteman, N.K. (2014). The evolution of ethylene signaling in plant chemical ecology. *Journal of Chemical Ecology* 40: 700–716.

Groen, S.C., Whiteman, N.K., Bahrami, A.K. et al. (2013). Pathogen-triggered ethylene signaling mediates systemic-induced susceptibility to herbivory in Arabidopsis. *Plant Cell* 25: 4755–4766.

Gu, K., Yang, B., Tian, D. et al. (2005). R gene expression induced by a type-III effector triggers disease resistance in rice. *Nature* 435: 1122–1125.

Guinel, F.C. (2015). Ethylene, a hormone at the center-stage of nodulation. *Frontiers in Plant Science* 6: 1121.

Guo, N., Zhang, S., Gu, M., and Xu, G. (2021). Function, transport, and regulation of amino acids: what is missing in rice? *Crop Journal* 9 (3): 530–542.

Han, G.Z. (2019). Origin and evolution of the plant immune system. *The New Phytologist* 222 (1): 70–83.

Hasanuzzaman, M., Bhuyan, M.B., Zulfiqar, F. et al. (2020). Reactive oxygen species and antioxidant defense in plants under abiotic stress: revisiting the crucial role of a universal defense regulator. *Antioxidants* 9 (8): 681.

Hirakawa, Y. and Sawa, S. (2019). Diverse function of plant peptide hormones in local signaling and development. *Current Opinion in Plant Biology* 51: 81–87.

Hirakawa, Y., Torii, K.U., and Uchida, N. (2017). Mechanisms and strategies shaping plant peptide hormones. *Plant & Cell Physiology* 58 (8): 1313–1318.

Hong, Y., Boiti, A., Vallone, D., and Foulkes, N.S. (2024). Reactive oxygen species signaling and oxidative stress: transcriptional regulation and evolution. *Antioxidants* 13 (3): 312.

Hou, S., Wang, X., Chen, D. et al. (2014). The secreted peptide PIP1 amplifies immunity through receptor-like kinase 7. *PLoS Pathogens* 10: e1004331.

Hsiao, Y.C. and Yamada, M. (2020). The roles of peptide hormones and their receptors during plant root development. *Genes* 12 (1): 22.

Hu, Z., Zhang, H., and Shi, K. (2018). Plant peptides in plant defense responses. *Plant Signaling & Behavior* 13 (8): e1475175.

Huang, P., Gao, W., Fu, C. et al. (2025). Clinical functional proteomics of intercellular signalling in pancreatic cancer. *Nature* 637 (8046): 726–735.

Huang, W.R. and Joosten, M.H. (2024). Immune signaling: receptor-like proteins make the difference. *Trends in Plant Science* 30 (1): 54–68.

Huang, P., Li, Z., and Guo, H. (2022). New advances in the regulation of leaf senescence by classical and peptide hormones. *Frontiers in Plant Science* 13: 923136.

Hückelhoven, R. and Schouten, A. (2024). Plant immunity and plant defense. In: *Agrios' Plant Pathology* (ed. G.N. Agrios), 161–210. Academic Press.

Igarashi, D., Tsuda, K., and Katagiri, F. (2012). The peptide growth factor, phytosulfokine, attenuates pattern-triggered immunity. *The Plant Journal* 71 (2): 194–204.

Iqbal, N. and Poór, P. (2025). Plant protection by tannins depends on defence-related phytohormones. *Journal of Plant Growth Regulation* 44 (1): 22–39.

Iqbal, S., Wang, X., Mubeen, I. et al. (2022). Phytohormones trigger drought tolerance in crop plants: outlook and future perspectives. *Frontiers in Plant Science* 12: 799318.

Jefferson, R.E., Oggier, A., Füglistaler, A. et al. (2023). Computational design of dynamic receptor-peptide signaling complexes applied to chemotaxis. *Nature Communications* 14 (1): 2875.

Jeon, B.W., Kang, N.Y., Park, W.Y. et al. (2024). Control of lateral root formation by rapamycin-induced dimerization of engineered RGI/BAK1 and by BIR3 chimera. *Physiologia Plantarum* 176 (1): e14155.

Johnston, G.A., Berry, H.M., Kojima, M. et al. (2024). Concerted plant growth and defense through targeted phytohormone crosstalk modification. *bioRxiv: 2024-04.*

Jones, J.D.G. and Dangl, J.L. (2006). The plant immune system. *Nature* 444: 323–329.

Kacperska, A. and Kanellis, A.K. (1997). Ethylene synthesis and a role in plant responses to different stressors. In: *Biology and biotechnology of the plant hormone ethylene* (ed. A.K. Kanellis), 207–216. Dordrecht: Springer.

Kaur, S., Samota, M.K., Choudhary, M. et al. (2022). How do plants defend themselves against pathogens – biochemical mechanisms and genetic interventions. *Physiology and Molecular Biology of Plants* 28 (2): 485–504.

Kawada, T., Aoyama, M., Matsubara, S. et al. (2025). Ovarian tachykinin signaling system induces the growth of secondary follicles during the gonadotropin-independent process. *The Journal of Biological Chemistry* 301 (4): 108375.

Keunen, E., Schellingen, K., Vangronsveld, J., and Cuypers, A. (2016). Ethylene and metal stress: small molecule, big impact. *Frontiers in Plant Science* 7: 23.

Kim, J.S., Jeon, B.W., and Kim, J. (2021a). Signaling peptides regulating abiotic stress responses in plants. *Frontiers in Plant Science* 12: 704490.

Kim, M.J., Jeon, B.W., Oh, E. et al. (2021b). Peptide signaling during plant reproduction. *Trends in Plant Science* 26: 822–835.

Kim, K.K., Kim, M., Pyun, K. et al. (2023). A substrate-less nanomesh receptor with meta-learning for rapid hand task recognition. *Nature Electronics* 6: 64–75.

Kim, T.J. and Lim, G.H. (2023). Salicylic acid and mobile regulators of systemic immunity in plants: transport and metabolism. *Plants* 12: 1013.

Klessig, D.F., Durner, J., Noad, R. et al. (2000). Nitric oxide and salicylic acid signaling in plant defense. *Proceedings of the National Academy of Sciences of the United States of America* 97: 8849–8855.

Kobe, B. and Kajava, A.V. (2001). The leucine-rich repeat as a protein recognition motif. *Current Opinion in Structural Biology* 11: 725–732.

König, L.E., Rodriguez, S., Hug, C. et al. (2024). TYK2 as a novel therapeutic target in Alzheimer's disease with TDP-43 inclusions. *bioRxiv*.

Koornneef, A. and Pieterse, C.M. (2008). Cross talk in defense signaling. *Plant Physiology* 146: 839–844.

Korsching, S.I. (2025). Evolution of vertebrate olfactory receptor repertoires and their function. *Current Opinion in Behavioral Sciences* 61: 101483.

Kozyulina, P.Y., Pavlova, O.A., Kantsurova, E.S. et al. (2023). Transcriptomic analysis of pea plant responses to chitooligosaccharides' treatment revealed stimulation of mitogen-activated protein kinase cascade. *Frontiers in Plant Science* 14: 1092013.

Kumar, S., Korra, T., Thakur, R. et al. (2023). Role of plant secondary metabolites in defence and transcriptional regulation in response to biotic stress. *Plant Stress* 8: 100154.

Kuzminsky, I. and Ghanim, M. (2025). Immunity responses as checkpoints for efficient transmission of begomoviruses by whiteflies. *Virology* 605: 110462.

Lalun, V.O. and Butenko, M.A. (2025). Plant peptide ligands as temporal and spatial regulators. *Annual Review of Plant Biology* 76: 229–253.

Latic, N. and Erben, R.G. (2022). Interaction of vitamin D with peptide hormones with emphasis on parathyroid hormone, FGF23, and the renin-angiotensin-aldosterone system. *Nutrients* 14: 5186.

Lee, D., Lal, N.K., Lin, Z.J.D. et al. (2020). Regulation of reactive oxygen species during plant immunity through phosphorylation and ubiquitination of RBOHD. *Nature Communications* 11: 1838.

Levin, M. (2023). Darwin's agential materials: evolutionary implications of multiscale competency in developmental biology. *Cellular and Molecular Life Sciences* 80: 142.

Li, N., Han, X., Feng, D. et al. (2019). Signaling crosstalk between salicylic acid and ethylene/jasmonate in plant defense: do we understand what they are whispering? *International Journal of Molecular Sciences* 20: 671.

Li, J., Hu, S., Jian, W. et al. (2021). Plant antimicrobial peptides: structures, functions, and applications. *Botanical Studies* 62: 5.

Li, L., Li, C., Lee, G.I., and Howe, G.A. (2002). Distinct roles for jasmonate synthesis and action in the systemic wound response of tomato. *Proceedings of the National Academy of Sciences of the United States of America* 99: 6416–6421.

Li, T., Moreno-Pérez, A., and Coaker, G. (2024). Plant pattern recognition receptors: exploring their evolution, diversification, and spatiotemporal regulation. *Current Opinion in Plant Biology* 82: 102631.

Li, L., Simopoulos, C.M., Mayne, J. et al. (2025). Systematic metaproteomics mapping reveals functional and ecological landscapes of human gut microbiota responses to therapeutic drugs. *bioRxiv*.

Li, A., Sun, X., and Liu, L. (2022b). Action of salicylic acid on plant growth. *Frontiers in Plant Science* 13: 878076.

Li, L., Wu, J., Lyon, C.J. et al. (2023). Clinical peptidomics: advances in instrumentation, analyses, and applications. *BME Frontiers* 4: 0019.

Li, C., Xu, M., Cai, X. et al. (2022a). Jasmonate signaling pathway modulates plant defense, growth, and their trade-offs. *International Journal of Molecular Sciences* 23: 3945.

Liao, C.J., Hailemariam, S., Sharon, A., and Mengiste, T. (2022). Pathogenic strategies and immune mechanisms to necrotrophs: differences and similarities to biotrophs and hemibiotrophs. *Current Opinion in Plant Biology* 69: 102291.

Liu, P.L., Du, L., Huang, Y. et al. (2017). Origin and diversification of leucine-rich repeat receptor-like protein kinase (LRR-RLK) genes in plants. *BMC Evolutionary Biology* 17: 1–16.

Liu, J., Li, W., Wu, G., and Ali, K. (2024). An update on evolutionary, structural, and functional studies of receptor-like kinases in plants. *Frontiers in Plant Science* 15: 1305599.

Lu, H. (2009). Dissection of salicylic acid-mediated defense signaling networks. *Plant Signaling & Behavior* 4: 713–717.

Lukasik, E. and Takken, F.L. (2009). Standing strong, resistance proteins instigators of plant defence. *Current Opinion in Plant Biology* 12: 427–436.

Lyapina, I., Ganaeva, D., Rogozhin, E.A. et al. (2025). Comparative analysis of small secreted peptide signaling during defense response: insights from vascular and non-vascular plants. *Physiologia Plantarum* 177: e70147.

Ma, L., Meng, Q., Shi, F. et al. (2022). Genome-wide analysis of maize PR-1 gene family and expression profiles induced by plant hormones and fungal phytopathogens. *American Journal of Translational Research* 14: 8315.

Malécange, M., Sergheraert, R., Teulat, B. et al. (2023). Biostimulant properties of protein hydrolysates: recent advances and future challenges. *International Journal of Molecular Sciences* 24: 9714.

Man, J., Gallagher, J.P., and Bartlett, M. (2020). Structural evolution drives diversification of the large LRR-RLK gene family. *The New Phytologist* 226: 1492–1505.

Man, J., Harrington, T.A., Lally, K., and Bartlett, M.E. (2023). Asymmetric evolution of protein domains in the leucine-rich repeat receptor-like kinase family of plant signaling proteins. *Molecular Biology and Evolution* 40: msad220.

Meddya, S., Meshram, S., Sarkar, D. et al. (2023). Plant stomata: an unrealized possibility in plant defense against invading pathogens and stress tolerance. *Plants* 12: 3380.

Mittler, R. (2017). ROS are good. *Trends in Plant Science* 22: 11–19.

Mittler, R., Zandalinas, S.I., Fichman, Y., and Van Breusegem, F. (2022). Reactive oxygen species signalling in plant stress responses. *Nature Reviews. Molecular Cell Biology* 23: 663–679.

Molisso, D., Coppola, M., Buonanno, M. et al. (2022). Not only systemin: prosystemin harbors other active regions able to protect tomato plants. *Frontiers in Plant Science* 13: 887674.

Montesinos, L., Gascón, B., Ruz, L. et al. (2021). A bifunctional synthetic peptide with antimicrobial and plant elicitation properties that protect tomato plants from bacterial and fungal infections. *Frontiers in Plant Science* 12: 756357.

Moreira, X., Abdala-Roberts, L., and Castagneyrol, B. (2018). Interactions between plant defence signalling pathways: evidence from bioassays with insect herbivores and plant pathogens. *Journal of Ecology* 106: 2353–2364.

Motomitsu, A., Sawa, S., and Ishida, T. (2015). Plant peptide hormone signalling. *Essays in Biochemistry* 58: 115–131.

Mu, Z., Xu, M., Manda, T. et al. (2024). Characterization, evolution, and abiotic stress responses of leucine-rich repeat receptor-like protein kinases (LRR-RLK) in *Liriodendron chinense*. *BMC Genomics* 25: 748.

Mubassir, M. (2019). A synopsis of different plant LRR-RLKs structures and functionality. *American Journal of Biomedical Science & Research* 1: 2.

Munns, R. and Millar, A.H. (2023). Seven plant capacities to adapt to abiotic stress. *Journal of Experimental Botany* 74: 4308–4323.

Murakami, T., Katsuragi, Y., Hirai, H. et al. (2022). Distribution of flagellin CD2-1, flg22, and flgII-28 recognition systems in plant species and regulation of plant immune responses through these recognition systems. *Bioscience, Biotechnology, and Biochemistry* 86: 490–501.

Nguyen, Q.M., Iswanto, A.B.B., Son, G.H., and Kim, S.H. (2021). Recent advances in effector-triggered immunity in plants: new pieces in the puzzle create a different paradigm. *International Journal of Molecular Sciences* 22: 4709.

Nguyen, N.H., Trotel-Aziz, P., Clément, C. et al. (2022). Camalexin accumulation as a component of plant immunity during interactions with pathogens and beneficial microbes. *Planta* 255: 116.

Nie, Z., Li, W., Deng, L. et al. (2022). Comprehensive analysis of LRR-RLKs and key gene identification in *Pinus massoniana* resistant to pine wood nematode. *Frontiers in Plant Science* 13: 1043261.

Nikitovic, D., Kukovyakina, E., Berdiaki, A. et al. (2024). Enhancing tumor targeted therapy: the role of iRGD peptide in advanced drug delivery systems. *Cancers* 16: 3768.

Novoplansky, A., Souza, G.M., Brenner, E.D. et al. (2024). Exploring the complex information processes underlying plant behavior. *Plant Signaling & Behavior* 19: 2411913.

Okada, K., Abe, H., and Arimura, G.I. (2015). Jasmonates induce both defense responses and communication in monocotyledonous and dicotyledonous plants. *Plant & Cell Physiology* 56: 16–27.

Osborne, R., Rehneke, L., Lehmann, S. et al. (2023). Symbiont-host interactome mapping reveals effector-targeted modulation of hormone networks and activation of growth promotion. *Nature Communications* 14: 4065.

Othoum, G. and Maher, C.A. (2023). CrypticProteinDB: an integrated database of proteome and immunopeptidome derived non-canonical cancer proteins. *NAR Cancer* 5: zcad024.

Pandita, D., Bhat, J.A., Wani, S.H. et al. (2023). Mobile signaling peptides: secret molecular messengers with a mighty role in plant life. *Journal of Plant Growth Regulation* 42: 6801–6834.

Parisi, M.G., Ozón, B., Vera González, S.M. et al. (2024). Plant protease inhibitors as emerging antimicrobial peptide agents: a comprehensive review. *Pharmaceutics* 16: 582.

Pastor-Fernández, J., Sánchez-Bel, P., Flors, V. et al. (2023). Small signals lead to big changes: the potential of peptide-induced resistance in plants. *Journal of Fungi (Basel, Switzerland)* 9: 265.

Paul Friedman, K., Thomas, R.S., Wambaugh, J.F. et al. (2025). Integration of new approach methods for the assessment of data-poor chemicals. *Toxicological Sciences* 205: 74, kfaf019–105.

Pearce, G. (2011). Systemin, hydroxyproline-rich systemin and the induction of protease inhibitors. *Current Protein & Peptide Science* 12: 399–408.

Peláez-Vico, M.Á., Zandalinas, S.I., Devireddy, A.R. et al. (2024). Systemic stomatal responses in plants: coordinating development, stress, and pathogen defense under a changing climate. *Plant, Cell & Environment* 47: 1171–1184.

Peng, J., Li, Z., Wen, X. et al. (2014). Salt-induced stabilization of EIN3/EIL1 confers salinity tolerance by deterring ROS accumulation in Arabidopsis. *PLoS Genetics* 10: e1004664.

Pierzynowska, K., Morcinek-Orłowska, J., Gaffke, L. et al. (2024). Applications of the phage display technology in molecular biology, biotechnology and medicine. *Critical Reviews in Microbiology* 50: 450–490.

Pieterse, C.M.J. and Dicke, M. (2007). Plant interactions with microbes and insects: from molecular mechanisms to ecology. *Trends in Plant Science* 12: 564–569.

Pieterse, C.M., Leon-Reyes, A., Van der Ent, S., and Van Wees, S.C. (2009). Networking by small-molecule hormones in plant immunity. *Nature Chemical Biology* 5: 308–316.

Pieterse, C.M.J., Zamioudis, C., Berendsen, R.L. et al. (2014). Induced systemic resistance by beneficial microbes. *Annual Review of Phytopathology* 52: 347–375.

Pokotylo, I., Hodges, M., Kravets, V., and Ruelland, E. (2022). A ménage à trois: salicylic acid, growth inhibition, and immunity. *Trends in Plant Science* 27: 460–471.

Potgieter, M.G., Nel, A.J., Fortuin, S. et al. (2023). MetaNovo: an open-source pipeline for probabilistic peptide discovery in complex metaproteomic datasets. *PLoS Computational Biology* 19: e1011163.

Prasad, A., Sharma, S., and Prasad, M. (2023). Suppressing plant defense: scavenge the ROS! *Physiologia Plantarum* 175: e13942.

Qin, S., Xiao, W., Zhou, C. et al. (2022). Pseudomonas aeruginosa: pathogenesis, virulence factors, antibiotic resistance, interaction with host, technology advances and emerging therapeutics. *Signal Transduction and Targeted Therapy* 7: 199.

Radojičić, A., Li, X., and Zhang, Y. (2018). Salicylic acid: a double-edged sword for programed cell death in plants. *Frontiers in Plant Science* 9: 1133.

Ravanbakhsh, M., Sasidharan, R., Voesenek, L.A. et al. (2018). Microbial modulation of plant ethylene signaling: ecological and evolutionary consequences. *Microbiome* 6: 52.

Rehneke, L. and Schäfer, P. (2024). Symbiont effector-guided mapping of proteins in plant networks to improve crop climate stress resilience. *BioEssays* 46: e2300172.

Rehrig, E.M., Appel, H.M., Jones, A.D., and Schultz, J.C. (2014). Roles for jasmonate- and ethylene-induced transcription factors in the ability of Arabidopsis to respond differentially to damage caused by two insect herbivores. *Frontiers in Plant Science* 5: 407.

Reimer-Michalski, E.M. and Conrath, U. (2016). Innate immune memory in plants. *Seminars in Immunology* 28: 319–327.

Reyes Zamora, O., Troncoso-Rojas, R., Báez-Flores, M.E. et al. (2024). Signaling of plant defense mediated by receptor-like kinases, receptor-like cytoplasmic protein kinases and MAPKs triggered by fungal chitin in horticultural crops. *Horticulturae* 10: 361.

Rouaud, L. (2024). Tregs independent GARP functions in the tumor microenvironment. Doctoral dissertation, Universite de Liege (Belgium).

Roychoudhury, A., Manna, S., and Banerjee, D. (2024). Plant elicitor peptide mediated signalling cascades during plant-pathogen interaction. In: *Plant Elicitor Peptides: New Tool for Sustainable Agriculture*, 49–98. Singapore: Springer Nature.

Roychowdhury, R., Hada, A., Biswas, S. et al. (2025). Jasmonic acid (JA) in plant immune response: unravelling complex molecular mechanisms and networking of defence signalling against pathogens. *Journal of Plant Growth Regulation* 44: 89–114.

Ruan, J., Zhou, Y., Zhou, M. et al. (2019). Jasmonic acid signaling pathway in plants. *International Journal of Molecular Sciences* 20 (10): 2479.

Ryan, C.A. (2000). The systemin signaling pathway: differential activation of plant defensive genes. *Biochimica et Biophysica Acta* 1477 (1–2): 112–121.

Ryan, C.A. and Pearce, G. (1998). Systemin: a polypeptide signal for plant defensive genes. *Annual Review of Cell and Developmental Biology* 14: 1–17.

Rys, M., Saja-Garbarz, D., Fodor, J. et al. (2024). Heat pre-treatment modified host and non-host interactions of powdery mildew with barley brassinosteroid mutants and wild types. *Life* 14 (1): 160.

Rzemieniewski, J., Leicher, H., Lee, H. et al. (2022). Phytocytokine signaling integrates cell surface immunity and nitrogen limitation. *bioRxiv*.

Rzemieniewski, J. and Stegmann, M. (2022). Regulation of pattern-triggered immunity and growth by phytocytokines. *Current Opinion in Plant Biology* 68: 102230.

Sahu, P.K., Jayalakshmi, K., Tilgam, J. et al. (2022). ROS generated from biotic stress: effects on plants and alleviation by endophytic microbes. *Frontiers in Plant Science* 13: 1042936.

Sands, L.B., Cheek, T., Reynolds, J. et al. (2022). Effects of harpin and Flg22 on growth enhancement and pathogen defense in *Cannabis sativa* seedlings. *Plants* 11 (9): 1178.

Sarma, H., Ramchiary, C., Sen, B. et al. (2025). Integration of nitric oxide signaling in plant stress responses: unveiling its role in enhancing crop tolerance to abiotic stress. *Discover Plants* 2 (1): 1–19.

Sarwar, A. and Khan, M.U. (2023). Technologies for crop water stress monitoring. In: *Encyclopedia of Digital Agricultural Technologies*, 1456–1470. Cham: Springer International Publishing.

Satam, H., Joshi, K., Mangrolia, U. et al. (2023). Next-generation sequencing technology: current trends and advancements. *Biology* 12 (7): 997.

Scala, A., Allmann, S., Mirabella, R. et al. (2013). Green leaf volatiles: a plant's multifunctional weapon against herbivores and pathogens. *International Journal of Molecular Sciences* 14: 17781–17811.

Scarpeci, T.E., Frea, V.S., Zanor, M.I., and Valle, E.M. (2017). Overexpression of AtERF019 delays plant growth and senescence, and improves drought tolerance in Arabidopsis. *Journal of Experimental Botany* 68: 673–685.

Schaller, A. and Ryan, C.A. (1996). Systemin—a polypeptide defense signal in plants. *BioEssays* 18 (1): 27–33.

Schellingen, K., Van Der Straeten, D., Vandenbussche, F. et al. (2014). Cadmium-induced ethylene production and responses in Arabidopsis thaliana rely on ACS2 and ACS6 gene expression. *BMC Plant Biology* 14: 214.

Seyfferth, C. and Tsuda, K. (2014). Salicylic acid signal transduction: the initiation of biosynthesis, perception and transcriptional reprogramming. *Frontiers in Plant Science* 5: 697.

Shafqat, A., Abbas, S., Ambreen, M. et al. (2024). Exploring the vital role of phytohormones and plant growth regulators in orchestrating plant immunity. *Physiological and Molecular Plant Pathology* 133: 102359.

Shanks, M.J. and Byblow, W.D. (2025). Corticomotor pathway function and recovery after stroke: a look back and a way forward. *The Journal of Physiology* 603 (3): 651–662.

Shekhawat, K., Fröhlich, K., García-Ramírez, G.X. et al. (2022). Ethylene: a master regulator of plant-microbe interactions under abiotic stresses. *Cells* 12 (1): 31.

Shen, W., Zhang, X., Liu, J. et al. (2022). Plant elicitor peptide signalling confers rice resistance to piercing-sucking insect herbivores and pathogens. *Plant Biotechnology Journal* 20 (5): 991–1005.

Shigenaga, A.M., Berens, M.L., Tsuda, K., and Argueso, C.T. (2017). Towards engineering of hormonal crosstalk in plant immunity. *Current Opinion in Plant Biology* 38: 164–172.

Shukla, K., Nikita, A.A., Noorani, M.S., and Gupta, R. (2025). Phytohormones and emerging plant growth regulators in tailoring plant immunity against viral infections. *Physiologia Plantarum* 177 (2): e70171.

Silva-Sanzana, C., Estevez, J.M., and Blanco-Herrera, F. (2020). Influence of cell wall polymers and their modifying enzymes during plant-aphid interactions. *Journal of Experimental Botany* 71 (13): 3854–3864.

Simas, D.L., Chagas, F.O., and da Silva, A.J. (2025). Plant-pathogen interaction: chemical mediation in plant signalling and defence. *Phytochemistry Reviews* 1–17.

Singh, D. (2022). Juggling with reactive oxygen species and antioxidant defense system – a coping mechanism under salt stress. *Plant Stress* 5: 100093.

Slavik, K.M. and Kranzusch, P.J. (2023). CBASS to cGAS-STING: the origins and mechanisms of nucleotide second messenger immune signaling. *Annual Review of Virology* 10 (1): 423–453.

Snoeck, S., Guayazán-Palacios, N., and Steinbrenner, A.D. (2022). Molecular tug-of-war: plant immune recognition of herbivory. *Plant Cell* 34 (5): 1497–1513.

Snoeck, S., Johanndrees, O., Nürnberger, T., and Zipfel, C. (2025). Plant pattern recognition receptors: from evolutionary insight to engineering. *Nature Reviews Genetics* 26 (4): 268–278.

Soltabayeva, A., Dauletova, N., Serik, S. et al. (2022). Receptor-like kinases (LRR-RLKs) in response of plants to biotic and abiotic stresses. *Plants* 11 (19): 2660.

Song, H., Ji, X., Wang, M. et al. (2024). Genome-wide identification and expression analysis of the Dof gene family reveals their involvement in hormone response and abiotic stresses in sunflower (Helianthus annuus L.). *Gene* 910: 148336.

Sosnowski, J., Truba, M., and Vasileva, V. (2023). The impact of auxin and cytokinin on the growth and development of selected crops. *Agriculture* 13 (3): 724.

Spoel, S.H. and Dong, X. (2024). Salicylic acid in plant immunity and beyond. *Plant Cell* 36 (5): 1451–1464.

Stegmann, M., Zecua-Ramirez, P., Ludwig, C. et al. (2022). RGI-GOLVEN signaling promotes cell surface immune receptor abundance to regulate plant immunity. *EMBO Reports* 23 (5): e53281.

Stintzi, A., Weber, H., Reymond, P. et al. (2001). Plant defense in the absence of jasmonic acid: the role of cyclopentenones. *Proceedings of the National Academy of Sciences of the United States of America* 98 (22): 12837–12842.

Stout, M.J., Thaler, J.S., and Thomma, B.P.H.J. (2006). Plant-mediated interactions between pathogenic microorganisms and herbivorous arthropods. *Annual Review of Entomology* 51: 663–689.

Subeshan, B., Atayo, A., and Asmatulu, E. (2024). Machine learning applications for electrospun nanofibers: a review. *Journal of Materials Science* 59 (31): 14095–14140.

Sun, J.Q., Jiang, H.L., and Li, C.Y. (2011). Systemin/jasmonate-mediated systemic defense signaling in tomato. *Molecular Plant* 4 (4): 607–615.

Sun, W., Shahrajabian, M.H., Kuang, Y., and Wang, N. (2024). Amino acids biostimulants and protein hydrolysates in agricultural sciences. *Plants* 13 (2): 210.

DeFalco, T.A. and Zipfel, C. (2021). Molecular mechanisms of early plant pattern-triggered immune signaling. *Molecular Cell* 81 (17): 3449–3467.

Takken, F.L., Albrecht, M., and Tameling, W.I. (2006). Resistance proteins: molecular switches of plant defence. *Current Opinion in Plant Biology* 9 (4): 383–390.

Tan, S.T., Abas, M., Verstraeten, I. et al. (2020). Salicylic acid targets protein phosphatase 2A to attenuate growth in plants. *Current Biology* 30: 381–395.e8.

Tan, W., Nian, H., Tran, L.S.P. et al. (2024). Small peptides: novel targets for modulating plant-rhizosphere microbe interactions. *Trends in Microbiology* 32: 1072–1083.

Tanaka, K. and Heil, M. (2021). Damage-Associated Molecular Patterns (DAMPs) in plant innate immunity: applying the danger model and evolutionary perspectives. *Annual Review of Phytopathology* 59: 53–75.

Tang, D., Wang, G., and Zhou, J.M. (2017). Receptor kinases in plant-pathogen interactions: more than pattern recognition. *Plant Cell* 29 (4): 618–637.

Tena, G. (2021). PTI and ETI are one. *Nature Plants* 7 (12): 1527–1527.

Thompson, M.D., Percy, M.E., Cole, D.E. et al. (2024). G protein-coupled receptor (GPCR) gene variants and human genetic disease. *Critical Reviews in Clinical Laboratory Sciences* 61 (5): 317–346.

Tian, W., Hou, C., Ren, Z. et al. (2019). A calmodulin-gated calcium channel links pathogen patterns to plant immunity. *Nature* 572: 131–135.

Tian, J., Jiang, L., Li, H. et al. (2024). The dual role of the DREAM/G2M pathway in non-tumorigenic immortalization of senescent cells. *Federation of European Biochemical Societies* 14 (2): 331–343.

Todd, J.N.A., Carreón-Anguiano, K.G., Islas-Flores, I., and Canto-Canché, B. (2022). Microbial effectors: key determinants in plant health and disease. *Microorganisms* 10 (10): 1980.

Vaishnav, D. and Chowdhury, P. (2023). Types and function of phytohormone and their role in stress. In: *Plant Abiotic Stress Responses and Tolerance Mechanisms* (ed. S. Hussain, T.H. Awan, E.A. Waraich, and M. Iqbal), 1–18. IntechOpen.

Val-Torregrosa, B., Bundó, M., and San Segundo, B. (2021). Crosstalk between nutrient signalling pathways and immune responses in rice. *Agriculture* 11 (8): 747.

Van Doorn, W.G. (2011). Classes of programmed cell death in plants, compared to those in animals. *Journal of Experimental Botany* 62 (14): 4749–4761.

Van Wees, S.C., De Swart, E.A., Van Pelt, J.A. et al. (2000). Enhancement of induced disease resistance by simultaneous activation of salicylate- and jasmonate-dependent defense pathways in *Arabidopsis thaliana*. *Proceedings of the National Academy of Sciences of the United States of America* 97 (15): 8711–8716.

Vanella, R., Kovacevic, G., Doffini, V. et al. (2022). High-throughput screening, next generation sequencing and machine learning: advanced methods in enzyme engineering. *Chemical Communications* 58 (15): 2455–2467.

Verhage, A., van Wees, S.C., and Pieterse, C.M. (2010). Plant immunity: it's the hormones talking, but what do they say? *Plant Physiology* 154 (2): 536–540.

Vos, I.A., Pieterse, C.M.J., and Van Wees, S.C.M. (2013). Costs and benefits of hormone-regulated plant defences. *Plant Pathology* 62: 43–55.

Waadt, R., Seller, C.A., Hsu, P.K. et al. (2022). Plant hormone regulation of abiotic stress responses. *Nature Reviews. Molecular Cell Biology* 23 (10): 680–694.

Wang, W., Bale, S., Wei, J. et al. (2022). Fibroblast A20 governs fibrosis susceptibility and its repression by DREAM promotes fibrosis in multiple organs. *Nature Communications* 13 (1): 6358.

Wang, M., Jiang, M., Wang, Q. et al. (2025). An in vitro bioinspired approach to enhance the bioactivity of titanium implants via electrophoretic deposition and biomimetic mineralization of type I collagen. *Biomaterials Advances* 167: 214110.

Wang, Y., Mostafa, S., Zeng, W., and Jin, B. (2021). Function and mechanism of jasmonic acid in plant responses to abiotic and biotic stresses. *International Journal of Molecular Sciences* 22 (16): 8568.

Wang, J., Song, Q., Yang, T. et al. (2024a). The challenges and breakthroughs in the development of diagnostic monoclonal antibodies. *View* 5 (4): 20240017.

Wang, P., Wu, T., Cheng, Y. et al. (2024b). The phytocytokine systemin enhances postharvest tomato fruit resistance to *Botrytis cinerea*. *Postharvest Biology and Technology* 210: 112738.

Wang, G., Zhang, G., and Wu, M. (2016). CLE peptide signaling and crosstalk with phytohormones and environmental stimuli. *Frontiers in Plant Science* 6: 1211.

War, A.R., Paulraj, M.G., War, M.Y., and Ignacimuthu, S. (2011). Role of salicylic acid in induction of plant defense system in chickpea (*Cicer arietinum* L.). *Plant Signaling & Behavior* 6 (11): 1787–1792.

Wasternack, C. and Hause, B. (2019). The missing link in jasmonic acid biosynthesis. *Nature Plants* 5: 776–777.

Wasternack, C. and Song, S. (2017). Jasmonates: biosynthesis, metabolism, and signaling by proteins activating and repressing transcription. *Journal of Experimental Botany* 68 (6): 1303–1321.

Wei, Q., He, F., Rao, J. et al. (2024). Targeting non-classical autophagy-dependent ferroptosis and the subsequent HMGB1/TfR1 feedback loop accounts for alleviating solar dermatitis by senkyunolide I. *Free Radical Biology & Medicine* 223: 263–280.

Weiss, J., Gado-Benarroch, L., and Egea-Cortines, M. (2005). Genetic control of floral size and proportions. *The International Journal of Developmental Biology* 49: 513–525.

Wu, L., Chen, H., Curtis, C., and Fu, Z.Q. (2014). Go in for the kill: How plants deploy effector-triggered immunity to combat pathogens. *Virulence* 5 (7): 710–721.

Wu, B., Qi, F., and Liang, Y. (2023). Fuels for ROS signaling in plant immunity. *Trends in Plant Science* 28 (10): 1124–1131.

Xi, L., Wu, X., Wang, J. et al. (2024). Receptor kinase signaling of BRI1 and SIRK1 is tightly balanced by their interactomes as revealed from domain-swap chimaera in AE-MS approaches. *Molecular & Cellular Proteomics* 23 (11): 100857.

Xia, C., Miranda, J., Mendoza-Cozatl, D. et al. (2025). Decoding long-distance communication under mineral stress: Advances in vascular signalling and molecular tools for plant resilience. *Plant, Cell & Environment* 48 (7): 4781–4802.

Xiao, Y., Sun, G., Yu, Q. et al. (2024). A plant mechanism of hijacking pathogen virulence factors to trigger innate immunity. *Science* 383: 732–739.

Xiao, F., Zhou, H., and Lin, H. (2025). Decoding small peptides: regulators of plant growth and stress resilience. *Journal of Integrative Plant Biology* 67 (3): 596–631.

Xie, H., Zhao, W., Li, W. et al. (2022). Small signaling peptides mediate plant adaptions to abiotic environmental stress. *Planta* 255: 72.

Yamaguchi, K., Yamada, K., and Kawasaki, T. (2013). Receptor-like cytoplasmic kinases are pivotal components in pattern recognition receptor-mediated signaling in plant immunity. *Plant Signaling & Behavior* 8: e25662.

Yang, Y.X., Ahammed, G.J., Wu, C. et al. (2015). Crosstalk among Jasmonate, Salicylate and ethylene signaling pathways in plant disease and immune responses. *Current Protein & Peptide Science* 16: 450–461.

Yang, C., Dolatabadian, A., and Fernando, W.D. (2022). The wonderful world of intrinsic and intricate immunity responses in plants against pathogens. *Canadian Journal of Plant Pathology* 44: 1–20.

Yang, J., Duan, G., Li, C. et al. (2019). The crosstalks between jasmonic acid and other plant hormone signaling highlight the involvement of jasmonic acid as a core component in plant response to biotic and abiotic stresses. *Frontiers in Plant Science* 10: 1349.

Yang, R., Su, C., Xue, Z. et al. (2025). Combination of PAMP-induced peptide signaling and its regulator SpWRKY65 boosts tomato resistance to Phytophthora infestans. *The Plant Journal* 121: e70098.

Yaschenko, A.E., Alonso, J.M., and Stepanova, A.N. (2024). Arabidopsis as a model for translational research. *Plant Cell* 37 (5): koae065.

Yu, Y., Gui, Y., Li, Z. et al. (2022). Induced systemic resistance for improving plant immunity by beneficial microbes. *Plants* 11: 386.

Yu, X.Q., Niu, H.Q., Liu, C. et al. (2024). PTI-ETI synergistic signal mechanisms in plant immunity. *Plant Biotechnology Journal* 22: 2113–2128.

Yuan, R., Zhang, J., Zhou, J., and Cong, Q. (2025). Recent progress and future challenges in structure-based protein-protein interaction prediction. *Molecular Therapy* 33: 2252–2268.

Zada, A., Lv, M., and Li, J. (2024). Molecular lesions in BRI1 and its orthologs in the plant kingdom. *International Journal of Molecular Sciences* 25: 8111.

Zelman, A.K. and Berkowitz, G.A. (2023). Plant elicitor peptide (Pep) signaling and pathogen defense in tomato. *Plants* 12: 2856.

Zhang, Z., Han, H., Zhao, J. et al. (2025). Peptide hormones in plants. *Molecular Horticulture* 5: 7.

Zhang, Z.C., He, B., Sun, S. et al. (2021b). The phytotoxin COR induces transcriptional reprogramming of photosynthetic, hormonal and defence networks in tomato. *Plant Biology* 23: 69–79.

Zhang, L., Koller, J., Ip, C.K. et al. (2021a). Lack of neuropeptide FF signalling in mice leads to reduced repetitive behavior, altered drinking behavior, and fuel type selection. *The FASEB Journal* 35: e21980.

Zhang, C., Xie, Y., He, P., and Shan, L. (2024a). Unlocking nature's defense: Plant pattern recognition receptors as guardians against pathogenic threats. *Molecular Plant-Microbe Interactions* 37: 73–83.

Zhang, Y., Xu, S., Ding, P. et al. (2010). Control of salicylic acid synthesis and systemic acquired resistance by two members of a plant-specific family of transcription factors. *Proceedings of the National Academy of Sciences of the United States of America* 107: 18220–18225.

Zhang, H., Zhang, H., and Lin, J. (2020). Systemin-mediated long-distance systemic defense responses. *The New Phytologist* 226: 1573–1582.

Zhang, P., Zhao, J., Zhang, W. et al. (2024b). Sulfated peptides: key players in plant development, growth, and stress responses. *Frontiers in Plant Science* 15: 1474111.

Zhao, B., Liu, Q., Wang, B., and Yuan, F. (2021). Roles of phytohormones and their signaling pathways in leaf development and stress responses. *Journal of Agricultural and Food Chemistry* 69: 3566–3584.

Zhiqi, H., Tingyi, W., Dongdong, C. et al. (2024). Leucine-rich repeat protein family regulates stress tolerance and development in plants. *Rice Science* 32 (1): 32–43.

Zhou, D., Chen, X., Chen, X. et al. (2023a). Plant immune receptors interact with hemibiotrophic pathogens to activate plant immunity. *Frontiers in Microbiology* 14: 1252039.

Zhou, X., Gao, H., Zhang, X. et al. (2023b). Plant extracellular self-DNA inhibits growth and induces immunity via the jasmonate signaling pathway. *Plant Physiology* 192: 2475–2491.

Zhou, K., Wu, F., Deng, L. et al. (2024). Antagonistic systemin receptors integrate the activation and attenuation of systemic wound signaling in tomato. *Developmental Cell* 60 (4): 535–550.

19

Utilization of *Bacillus*-Derived Secondary Metabolites in Sustainable Agriculture: A Natural Defense Mechanism against Fungal Pathogens

Pooja Parmar[1,2], *Abhishek Prajapati*[1,2], *Bhavesh Parmar*[2], *and Chaitanya Kumar Jha*[1,2]

[1] *Department of Microbiology and Biotechnology, Gujarat University, Ahmedabad, Gujarat, India*
[2] *Microbiology Department, Gujarat Arts and Science College, Ahmedabad, Gujarat, India*

19.1 Introduction

Phytodiseases pose a significant threat to global food production, causing an annual loss of 16% of the world's major crop yields, with fungal phytopathogens responsible for 70–80% of these diseases, leading to a substantial decrease in crop yield and quality (Rodrigo et al. 2022). Plant disease management has leaned on developing disease-resistant crop varieties, applying chemical pesticides, and implementing strategic control measures for last many years (Gao et al. 2016). While these methods have played a pivotal role in increasing agricultural productivity and crop quality, the excessive reliance on chemical pesticides has led to environmental degradation, soil and water contamination, and the emergence of resistant pathogens (Massawe et al. 2018). This has stimulated the search for sustainable and eco-friendly alternatives to manage fungal diseases in crops effectively.

Currently, researchers are exploring plant growth–promoting rhizobacteria (PGPR) as an eco-friendly measure for controlling phytodiseases (Agbodjato and Babalola 2024). Among all the biocontrol agents studied, *Bacillus* species stand out as the most potent biocontrol agents because of their ability to produce numerous antifungal secondary metabolites and thrive in many different environments.

The *Bacillus* genus entails a diverse group of rod-shaped, endospore-forming, and catalase-positive bacteria that thrive in a wide range of ecological habitats, including soil, the rhizosphere, compost, and fermented foods (Zhang et al. 2023). Unlike conventional fungicides, *Bacillus*-derived secondary metabolites unveil complex and versatile mechanisms of action, making them highly effective in controlling fungal infections without nurturing resistance development. Approximately 53% of the FDA-approved natural product–based medicines originate from microorganisms (Abdel-Nasser et al. 2024).

Bacillus species exhibit both direct and indirect mechanisms to protect plants against fungal phytopathogens. Directly, they produce a wide range of antimicrobial secondary

Peptide Hormones in Plant Development and Stress Responses, First Edition. Edited by Vijay Pratap Singh and Samiksha Singh.

metabolites, including iturins, fengycins, surfactins, bacillomycin, difficidin, bacillaene, macrolactins, mycosubtilin, and zwittermicin A, which impede fungal growth by disrupting cell membranes, interfering with metabolism, and preventing spore germination (Kaspar et al. 2019). Furthermore, they produce hydrolytic enzymes and volatile organic compounds (VOCs) that disrupt fungal cell walls and suppress pathogen proliferation. Indirectly, *Bacillus* increases plant defense through induced systemic resistance (ISR), competes with pathogens for nutrients and space, improves soil structure, and promotes plant growth by producing phytohormones such as indole-3-acetic acid (IAA) and solubilizing key minerals (Choudhary and Johri 2009). These comprehensive interactions establish *Bacillus* as an effective and sustainable biocontrol agent in modern agriculture.

The progress of biotechnology, metagenomics, and microbial fermentation techniques has accelerated the discovery and industrial production of *Bacillus*-based biofungicides. Many commercial formulations based on the power of *Bacillus* secondary metabolites are now available and are considered realistic alternatives to chemical fungicides by growers (Wang et al. 2024). The integration of these biological control agents in existing agricultural practices complies with the principles of sustainable agriculture, reducing dependence on synthetic chemicals and ensuring long-term crop protection and environmental safety.

With the demand for environmentally friendly plant disease control options growing globally, uncovering the potential of *Bacillus*-derived secondary metabolites has become more crucial. This chapter explores the biosynthetic pathways, structural diversity, and agricultural applications so far discovered in plant specialized metabolites, and discusses their potential contribution to the future of sustainable crop protection (Figure 19.1).

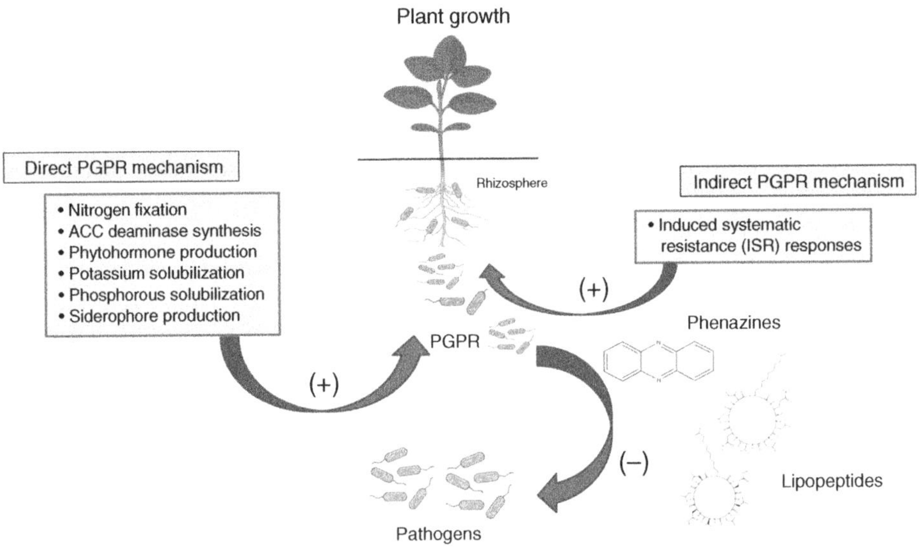

Figure 19.1 Mechanistic overview of PGPR-mediated biocontrol strategies against phytopathogens in the rhizosphere.

19.2 Unlocking the Biochemical Arsenal of *Bacillus*: Classification of Secondary Metabolites

Bacteria are trailblazers in the generation of chemical diversity, producing a diverse range of secondary metabolites. Surpassing basic survival, these bacteria contribute to the production of secondary metabolites—molecules that shield against antagonists, foster symbiosis, and impact their vicinal in many ways. Among the microbiota, *Bacillus* are renowned for their high productivity of diverse bioactive compounds that exhibit strong inhibitory bioactivities toward fungi, bacteria, and plants (Abdel-Nasser et al. 2024). Bacterial secondary metabolites are indispensable for their biological fitness in the natural environment, as well as a vast industrial tool for modern agriculture and medicinal advances (Mitra et al. 2023). *Bacillus* spp. produce a plethora of secondary metabolites as outlined in the table below.

19.2.1 Biosynthesis of Nonribosomal Peptides

Bacillus-derived nonribosomal peptide exhibits a broad range of structures (Sansinenea and Ortiz 2011) and biological potency, accounting for their significance in agriculture, biocontrol, and plant growth promotion by ISR (Patil et al. 2024). Unlike ribosomally synthesized peptides, which depend on mRNA templates, nonribosomal peptide synthetases (NRPS) assemble peptides de novo through a fixed sequence of modules that determine the synthesis of the final product (Roongsawang et al. 2011). Biosynthesis of NRPs is catalyzed by modular NRPSs as demonstrated in Figure 19.2. NRPSs are multidomain proteins that operate as assembly line-like machines on which each module represents one amino acid building block for the peptide chain (Koglin and Walsh 2009).

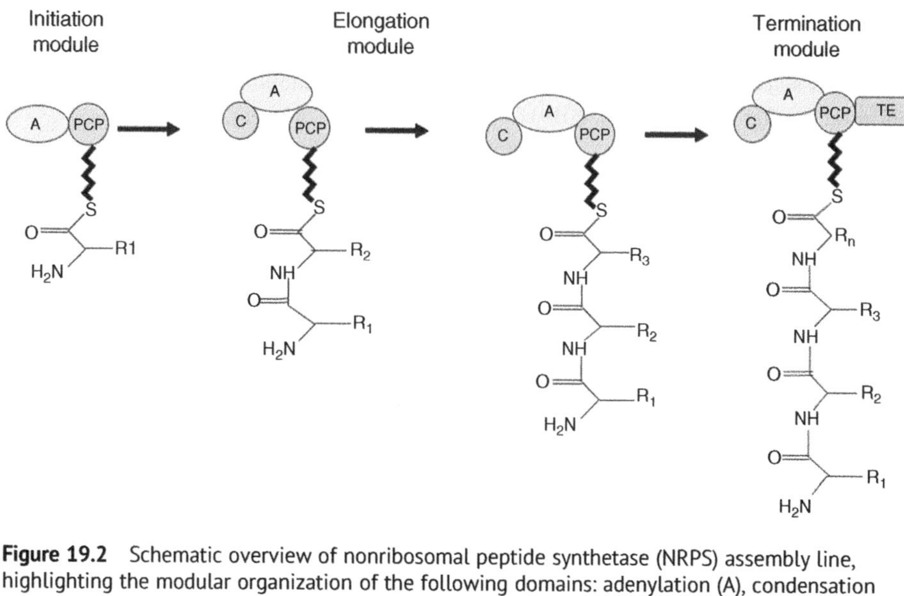

Figure 19.2 Schematic overview of nonribosomal peptide synthetase (NRPS) assembly line, highlighting the modular organization of the following domains: adenylation (A), condensation (C), peptidyl carrier protein (PCP), and termination (TE) domain with thioesterase (TE) activity.

Biosynthesis of nonribosomal peptides occurs via the function of the catalytic unit, referred to as a module. The order of modules in NRPSs is typically co-linear with the order of amino acids in the final peptide product. Each module adds a specific monomer, with the sequence and modification of these monomers determining the final structure and function of the peptide (Harwood et al. 2018). Each module of an NRPS comprises three core domains: a loading module, multiple elongation modules, and a termination module (Donadio et al. 2007).

NRPS assembles peptides autonomously using a predetermined sequence of modules that direct the synthesis of the end product, in contrast to ribosomally produced peptides, which depend on mRNA templates (Roongsawang et al. 2011).The adenylation domain detects the appropriate amino acid or hydroxy acid substrate and activates it at the expense of ATP, forming an acyl-adenylate intermediate. The activated amino acid is covalently attached to the phosphopantetheine (Ppant) arm of the adjacent thiolation (T) or peptidyl carrier protein (PCP) domain through a thioester bond, thus priming the module for peptide elongation. Thiolation (T) domain assures that the growing peptide remains tethered to the NRPS complex throughout the biosynthetic process (Sansinenea and Ortiz 2011). Then the Cdomain catalyzes amide or ester bond formation between the substrate on its T domain and the peptidyl intermediate bound to the T domain of the preceding module. This reaction assures a linear and successive elongation of the peptide chain, following the co-linearity rule, where the sequence of modules in the NRPS complex aligns to the order of amino acids in the final peptide product (Donadio et al. 2007). The last stage of nonribosomal peptide biosynthesis entails the termination and release of the mature peptide, which is catalyzed by the C-terminal thioesterase (Te) domain found in the termination module (Harwood et al. 2018). NRPs display vast structural diversity and can exist in multiple forms, such as in linear chains, branched structures, and circular frameworks (Iqbal et al. 2023). NRPs can be generated either through a multienzyme thio-template mechanism or autonomous of such a system (Sumi et al. 2015). When produced via the thio-template mechanism, they typically consist of 2–50 amino acids along with additional structural elements, such as fatty acid chains. In contradiction, NRPs generated without a multienzyme template tend to be smaller (Caulier et al. 2019).

19.2.1.1 Surfactin: Biosynthesis and Genetic Regulation

As one kind of cyclic lipopeptide (CLP) and a powerful biosurfactant, surfactin is mainly produced by *Bacillus subtilis* and has the same lipid ring as the analogous CLPs lichenysin (secreted by *Bacillus licheniformis*) and pumilacidin (secreted by *Bacillus pumilus*) (Chen et al. 2015). These CLPs exhibit similar structure except for small variations in type and chain length of the amino acids, as well as fatty acids, which are responsible for their different biological activities. The chemical structure of surfactin is composed of a seven-amino acid (L-asparagine/L-aspartate, L-leucine, L-glutamate, L-leucine, L-valine, D-leucine, and D-leucine) peptide loop that is linked to a hydrophobic β-hydroxy fatty acid chain of variable length depending on the isoform (Zhen et al. 2023). Surfactin is the best known among the natural biosurfactants because it is one of the most powerful biosurfactants since it can lower the surface tension of water from 72 to ~27 mN/m at a low concentration of 20 μM (Grangemard et al. 2001).

In *Bacillus*, the srfAA-AD operon is responsible for surfactin biosynthesis and consists of NRPS and a thioesterase during its synthesis. It is a 27-kb operon containing three genes, *srfAA*, *srfAB*, and *srfAC*, each with seven catalytic modules capable of recognizing and condensing different amino acids (Nakano et al. 1988) (Figure 19.3). Module 1: L-Glu; modules 2, 3, 6, and 7: L-Leu; module 4: L-Val; and module 5: L-Asp. The ultimate cyclization and release of surfactin is catalyzed by the type II thioesterase gene *srfAD* (Koglin et al. 2008). The *sfp* gene downstream of this operon encodes a phosphopantetheinyl transferase (PPTase) involved in the activation of surfactin biosynthesis, disruption of which abrogates surfactin production (Quadri et al. 1998; Mootz et al. 2001).

The operon is predominantly controlled via the ComA-ComP two-component system that activates transcription of the srfAA-AD operon (Okada et al. 2015) with ComX as the extracellular pheromone accumulating and activating the ComP, which, under the influence of phosphorylated ComA, phosphorylates ComA to ComA-P (Jacques 2011).The Rap-Phr system further modulates this control by repressing ComA through Rap proteins and counteracting Rap-inhibition with Phr peptides to increase surfactin expression (Auchtung et al. 2006).

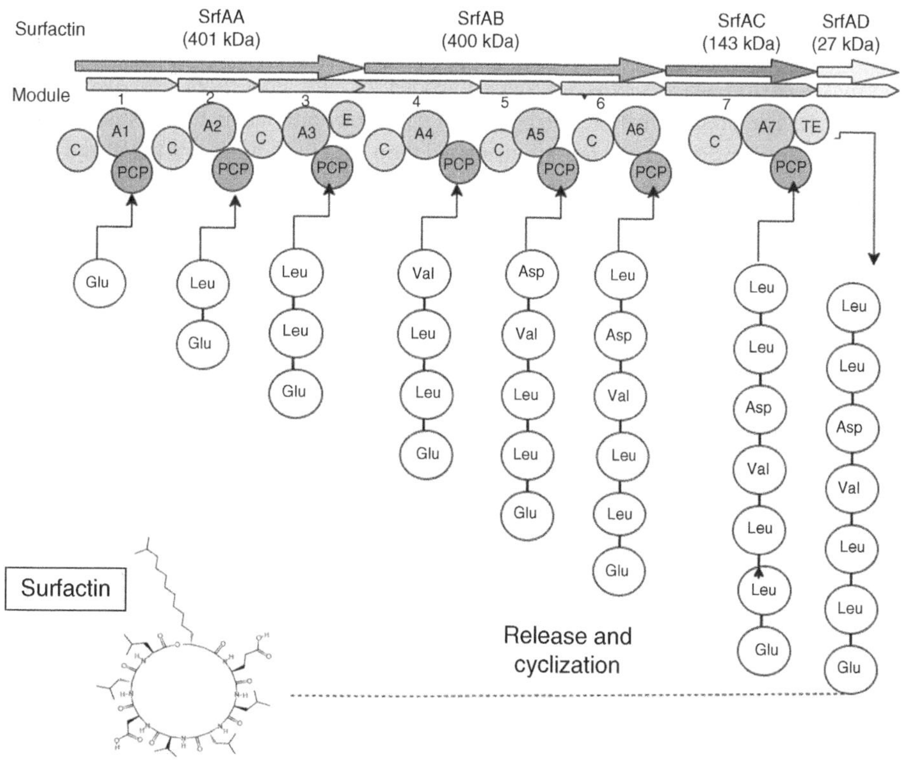

Figure 19.3 Genetic organization and domain architecture of the surfactin synthetase (*srfA*) operon in *Bacillus* spp. The operon comprises four genes—*srfAA*, *srfAB*, *srfAC*, and *srfAD*—encoding nonribosomal peptide synthetase (NRPS) modules with condensation (C), adenylation (A), and peptidyl carrier protein (PCP) domains responsible for the activation and incorporation of specific amino acids. The *srfAD* gene encodes a thioesterase (TE) domain that catalyzes the release and cyclization of the lipopeptide surfactin.

Furthermore, regulatory systems include the PhoR-PhoP system, which enhances surfactin production under phosphate limitation (Salzberg et al. 2015), the DegU-DegS system, which differentially regulates protease and lipopeptide biosynthesis (Ogura et al. 2001) and negatively affects surfactin production (Wu et al. 2019). Other key regulators, such as CodY, Spx, and PerR, suppress surfactin expression by competing for the srfAA promoter region, while Spo0A inhibits surfactin by enhancing competence and spore formation (Klausmann et al. 2021). In *Bacillus*, mutations or deletions in genes such as *comX*, *phrC*, *degU*, *codY*, and *spo0A* have demonstrated increased surfactin production, highlighting the complex network regulating surfactin biosynthesis (Coutte et al. 2015).

In the study by Armenova et al. (2024), *Bacillus velezensis* R22 was found to inhibit the growth of multiple fungal phytopathogens, notably *Botrytis cinerea* and *Phytophthora infestans,* through the production of surfactin. Similarly, Caretta et al. (2025) reported that surfactin produced by *Bacillus subtilis* also displayed strong antifungal activity against *Monilinia* and *Aspergillus* spp.in apple and strawberry.

19.2.1.2 Lichenysin Synthetase

Lichenysin, an effective biosurfactant produced by many strains of *Bacillus licheniformis*, is a CLP structurally identical to surfactin but with a modified heptapeptide sequence, typically L-Gln–L-Leu–D-Leu–L-Val/Ile–L-Asp–D-Leu–L-Leu/Ile (Madslien et al. 2013). Its biosynthesis occurs via an NRPS system encoded by the *licA*, *licB*, and *licC* genes (Figure 19.4). The process includes activation of β-hydroxy fatty acids by LicA, succeeded by polypeptide extension via condensation, adenylation, and PCP domains of LicB and LicC (Yakimov and Golyshin 1997). Cyclization and liberation of lichenysin are catalyzed by a thioesterase (TE) domain (Konz et al. 1999). Biosynthesis is controlled by environmental factors like temperature, nutrient availability, and oxygen levels (Konz et al. 1997).

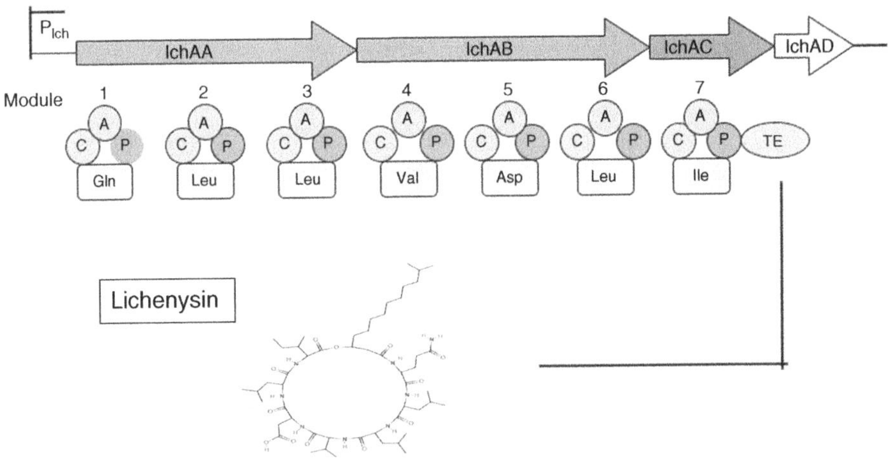

Figure 19.4 Domain organization of the lichenysin synthetase operon, consisting of *lchAA, lchAB, lchAC, and lchAD*. These genes encode NRPS modules with functional domains including condensation (C), adenylation (A), peptidyl carrier protein (PCP), and termination (TE), orchestrating the nonribosomal biosynthesis and cyclization of lichenysin.

19.2.1.3 Iturin: Biosynthesis and Genetic Regulation

The iturin group encompasses a varied family of cyclic heptapeptide rings defined by the presence of a C14-C17 aliphatic β-amino fatty acid. These lipopeptides exhibit a unique chiral arrangement of L- and D-amino acids in the sequence (LDDLLDL) and are circularized through the formation of an amide bond between the N-terminal β-amino fatty acid and the C-terminal peptide. This group consists of various members, such as iturins (A, C, D, and E), bacillomycins (D, F, L, and Lc), and mycosubtilin, along with other forms named after their bacterial origin, such as mojavensin. A defining characteristic shared by all members of the iturin group is the presence of a β-amino fatty acid linked to the conserved amino acid triad Asn-Tyr-Asn, contributing to their potent antifungal and membrane-disrupting properties (Duitman et al. 1999; Moyne et al. 2004). Iturins are synthesized by anNRPS operon, nearly 38 kbp in length, consisting of four key genes: *ituA*, *ituB*, *ituC*, and *ituD* (Figure 19.5). The first step is the activation of specific amino acids by the enzyme encoded by the *ituA* gene, which functions as a fatty acid ligase and aminoacyl-tRNA synthetase specific for iturin biosynthesis. This enzyme triggers amino acids by attaching them to adenosine monophosphate (AMP), forming aminoacyl-AMP intermediates that serve as substrates for subsequent reactions (Chen et al. 2006). The next step, cyclization, is catalyzed by the enzyme encoded by the *ituB* gene, which forms amide bonds between the activated amino acids, resulting in the cyclic peptide structure. This arrangement of hydrophobic and hydrophilic amino acids contributes to the amphiphilic properties of iturins (Gutiérrez-Chávez et al. 2021). The last step in iturin production is carried out by the enzyme encoded by the *ituD* gene. This enzyme attaches a hydrophobic (water-repelling) lipid tail to the cyclic peptide, helping the iturin molecule stick to cell membranes. This step increases the

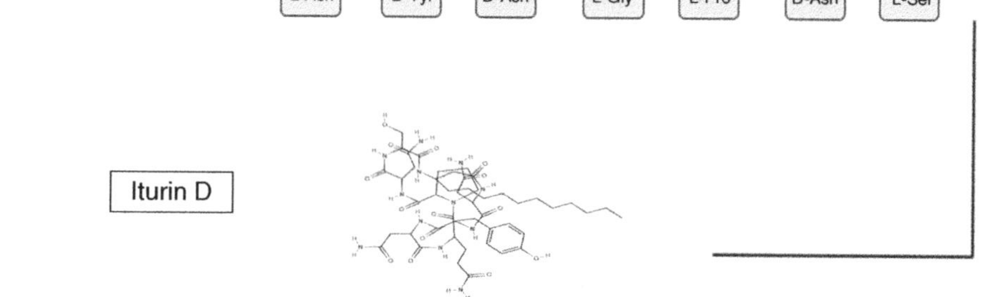

Figure 19.5 Genetic organization of the iturin operon in *Bacillus* spp. The operon comprises *ituA*, *ituB*, *ituC*, and *ituD*. The operon includes *ituA*, *ituB*, and *ituC*, encoding modular NRPS enzymes composed of functional domains: A (adenylation), C (condensation), peptidyl carrier protein (PCP), and TE (termination).

effectiveness of iturins by improving their ability to interact with and disrupt fungal membranes (Chen et al. 2006).

Iturin biosynthesis is firmly regulated by a complex network of regulatory proteins that respond to environmental conditions, cell density, and nutrient availability. The response regulator ComA plays a central role in quorum sensing by becoming phosphorylated in response to signaling peptides like ComX and activating the transcription of iturin biosynthetic genes (Dang et al. 2019). Similarly, DegU, part of a two-component regulatory system, enhances transcription when phosphorylated (Yu et al. 2023), while CodY, a transcriptional regulator sensing nutrient availability, represses iturin biosynthesis under nutrient-rich conditions (Vahidinasab et al. 2022).

19.2.1.3.1 Bacillomycin Synthetase

Bacillomycin D is synthesized by an NRPS system transcribed by the bmy operon, a 37.2 kb gene cluster comprising *bmyD*, *bmyA*, *bmyB*, and *bmyC* (Figure 19.6). The bam and bmy gene clusters, liable for bacillomycin biosynthesis, are highly similar, with bam identified in *Bacillus subtilis* AU195 and bmy in *Bacillus amyloliquefaciens* FZB42 (Koumoutsi et al. 2004). The first gene, *bmyD*, encodes a malonyl coenzyme A transacylase, similar to fabD, which is involved in fatty acid synthesis. The genes *bmyA* (3982 amino acids), *bmyB* (5633 amino acids), and *bmyC* (2619 amino acids) are arranged in a manner identical to the genes in the mycosubtilin and iturin A operons, sharing extensive sequence similarity. These genes transcribe enzymes with domains responsible for condensation, adenylation, and thiolation, essential for peptide assembly. The C-terminal thioesterase (TE) domain of

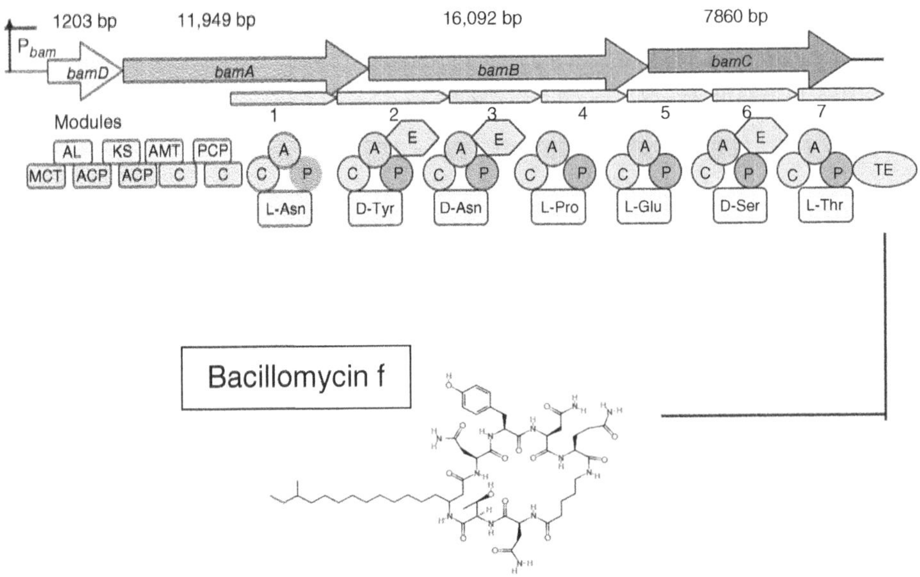

Figure 19.6 Domain organization of the bacillomycin synthetase operon, consisting of *bam A, bam B, bam C, and bam D.* These genes encode NRPS modules with functional domains including condensation (C), adenylation (A), peptidyl carrier protein (PCP), and termination (TE), orchestrating the nonribosomal biosynthesis and cyclization of bacillomycin.

bmyC plays a critical role in cyclizing and releasing the mature bacillomycin D molecule. The expression of the bmy operon is driven by a single sigma(A)-dependent promoter, P(bmy), and is enhanced by the small regulatory protein DegQ in its natural host. Entire transcriptional activation of bmy requires the involvement of global regulators DegU and ComA, with DegU directly binding to two upstream sites of the bacillomycin D promoter to regulate its expression effectively (Lv et al. 2024).

19.2.1.4 Fengycin: Biosynthesis and Genetic Regulation

Fengycins A and B are potent antifungal CLPs that vary by a single amino acid in their peptide ring structure. Fengycin A includes D-alanine (D-Ala), while fengycin B has a D-valine (D-Val) in the same position (Vanittanakom et al. 1986). Their unique structure comprises of a rare octapeptide ring formed through macrolactone synthesis between the hydroxyl group of tyrosine (Tyr-OH) and the acyl group of isoleucine (Ile). The lipid tail, which aids in their amphiphilic nature and bioactivity, is attached to the N-terminal amino acid through an amide bond with a β-hydroxy fatty acid. Fengycin synthetase (Fen) contains five NRPS subunits: FenC (287 kDa), FenD (290 kDa), FenE (286 kDa), FenA (406 kDa), and FenB (146 kDa) (Wu et al. 2007). These NRPS modules operate chronologically to add specific amino acids into the growing peptide chain, followed by cyclization and modification steps that introduce structural diversity in fengycins (Figure 19.7).

The regulation of fengycin biosynthesis is strictly controlled by more than one global regulator that responds to various environmental and physiological cues. CodY acts as a repressor, inhibiting fengycin production under nutrient-rich conditions (Zhang et al. 2020) while the ComA-ComP quorum-sensing system activates its biosynthesis in response to increased cell density (Deng et al. 2013). DegU, a key response regulator, fine-tunes fengycin synthesis by adapting to environmental changes (Wang et al. 2015), whereas AbrB indirectly influences production by modulating the regulatory effects of CodY and DegU (Tan et al.

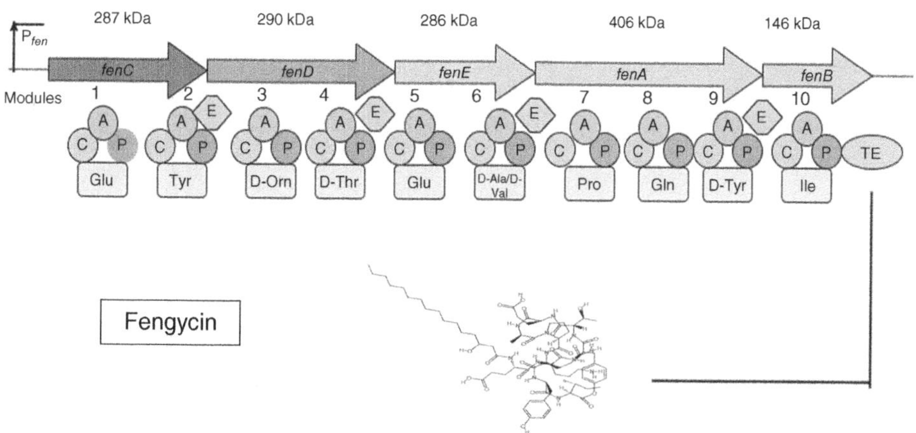

Figure 19.7 Domain organization of the fengycin synthetase operon, consisting of *fen A*, *fen B*, *fen C*, *fen D*, and *fen E*. These genes encode NRPS modules with functional domains including condensation (C), adenylation (A), peptidyl carrier protein (PCP), and termination (TE), orchestrating the nonribosomal biosynthesis and cyclization of fengycin.

2022). Moreover, the YycFG two-component system monitors cell envelope stress and dynamically adjusts fengycin biosynthesis to maintain cellular homeostasis (Ranjan et al. 2023).

19.2.1.5 Plipastatin Synthetase

Plipastatins, closely related to fengycins, were initially discovered in *Bacillus cereus* as potent inhibitors of phospholipase A2 before their presence was identified in *B. subtilis* (Müller et al. 2014). Structurally, plipastatins A1, A2, B1, and B2 are nearly identical to their corresponding fengycin analogs, with the primary distinction being the inversion of the stereocenters of the two tyrosine (ty3 and ty9) residues (Volpon et al. 2000). The *ppsABCDE* gene cluster, spanning 384 kb, encodes five peptide synthetases involved in plipastatin biosynthesis: PpsA (289 kDa), PpsB (290 kDa), PpsC (287 kDa), PpsD (407 kDa), and PpsE (145 kDa) (Harwood et al. 2018) (Figure 19.8).

19.2.1.6 Kurstakin: Biosynthesis and Genetic Regulation

Kurstakin is a CLP synthesized by *Bacillus thuringiensis* (Bt). It is composed of a β-hydroxy fatty acid linked to a heptapeptide ring structure and was first identified from *Bacillus thuringiensis* subsp. *kurstaki* strain HD-1 by Hathout et al. (2000). Structural elucidation confirmed that kurstakins possess a heptapeptide sequence (Thr-Gly-Ala-Ser-His-Gln-Gln), with an amide bond between the fatty acid chain and the first threonine residue, and a unique lactone linkage between the serine at position 4 and the C-terminal glutamine at position 7.

The biosynthesis of kurstakin in *Bacillus cereus* and *Bacillus thuringiensis* was foreseen through bioinformatics approaches, with Bumpus et al. (2009)using mass spectrometry–based proteomics to identify *NRPS* and *PKS* gene clusters. The kurstakin biosynthetic cluster consists of three key genes—*krsA*, *krsB*, and *krsC*—encoding large multifunctional

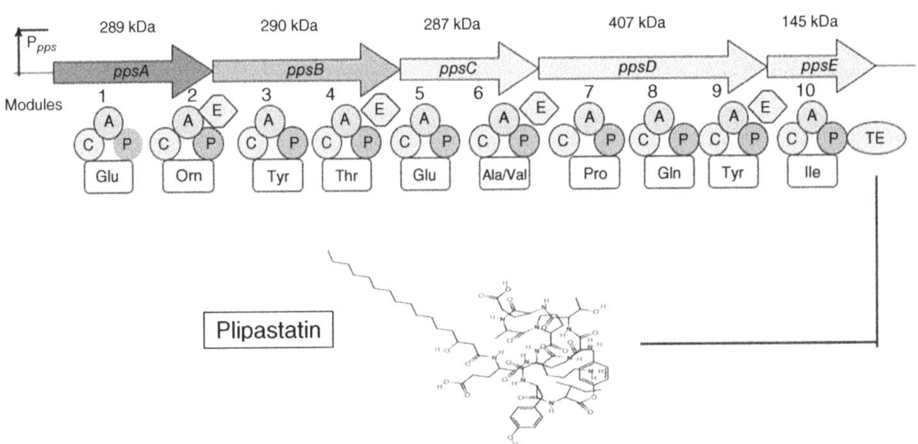

Figure 19.8 Domain organization of the plipastatin synthetase operon, consisting of *ppsA, ppsB, ppsC, ppsD*, and *ppsE*. These genes encode NRPS modules with functional domains including condensation (C), adenylation (A), peptidyl carrier protein (PCP), and termination (TE), orchestrating the nonribosomal biosynthesis and cyclization of plipastatin.

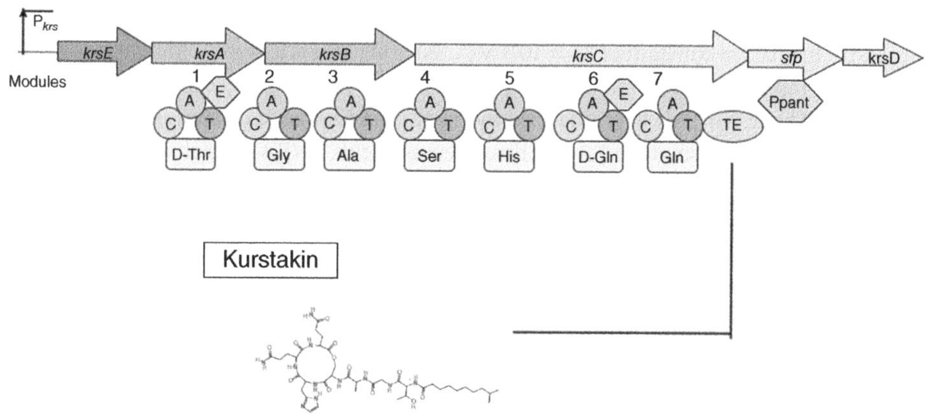

Figure 19.9 Domain organization of the kurstakin synthetase operon, consisting of *krs A*, *krs B*, *krs C*, *krs D, and krs E*. These genes encode NRPS modules with functional domains including condensation (C), adenylation (A), peptidyl carrier protein (PCP), and termination (TE), orchestrating the nonribosomal biosynthesis and cyclization of kurstakin.

proteins (KrsA, KrsB, and KrsC) that together form the complete synthetase (Abderrahmani et al. 2011) (Figure 19.9). KrsA retains a single module (m1) responsible for bridging a fatty acid to threonine using a sequence of condensation, adenylation, and thiolation domains, trailed by an epimerization domain that transforms L-Thr to D-Thr (Rausch et al. 2007). KrsB consists of two modules (m2 and m3), enabling the addition of glycine and alanine, while KrsC encompasses four modules (m4–m7) that sequentially incorporate serine, histidine, and two glutamine residues. The concluding module (m7) contains a thioesterase domain that catalyzes the release and likely the cyclization of the peptide, ultimately producing the mature lipopeptide (Kopp and Marahiel 2007). The predicted peptide sequence is D-Thr-Gly/Ala/Gln/Glu-Ala-Ser-X-D-Gln-Gln. Additional genes within the kurstakin biosynthetic cluster include krsD, which encodes a secondary thioesterase (TEII) involved in regenerating misacylated PCPs, and krsE, identified by (Dubois et al. 2012), which encodes a putative efflux protein potentially facilitating the secretion of the lipopeptide.

The regulation of kurstakin production in *Bacillus thuringiensis* Bt407 has been partially elucidated, with transcriptomic analyses revealing that the four genes—*krsEABC*—form a cluster whose transcription is activated by the NprR–NprX quorum-sensing system during the late stationary phase (Dubois et al. 2012). NprR is a quorum sensor that is activated upon binding with its cognate signaling peptide, NprX. The NprR–NprX system operates as a typical Gram-positive quorum-sensing mechanism, where the pro-signaling peptide NprX is exported from the bacterial cell, processed into its active form (presumably a heptapeptide), and subsequently reimported into the cell, where it binds to NprR, enabling the recognition of its DNA target (Perchat et al. 2011).

19.2.1.7 Bacillibactin Synthetase

Some *Bacillus* strains, including *B. subtilis, B. velezensis, B. licheniformis*, produce a high-affinity catechol-based hexadentate triscatecholamide siderophore called bacillibactin as a major siderophore which is essential for scavenging Fe^{3+} from the environment for the growth and metabolism of the bacterium (Fukushima et al. 2013). And it does so with an impressive formation constant of $Kf = 10^{45}\ M^{-1}$. Bacillibactin is a highly efficient iron scavenger that is required for respiration, DNA synthesis, and protection against oxidative stress (Cassat and Skaar 2013). It is synthesized by the *dhbACEBF* gene cluster, in which *DhbC* and *DhbA* cyclize chorismate to 2,3-dihydroxybenzoic acid (DHBA), which is further incorporated into the trimer by the action of anNRPS complex, which consists of the proteins DhbE, DhbB, and DhbF (Figure 19.9). Once synthesized, bacillibactin is exported by the YmfD and Mta efflux machinery, and once bound to Fe^{3+}, it is reimported into the cell by the FeuABC (Miethke et al. 2006; Miethke et al. 2008).

In the cells, BesA esterase hydrolyzes the Fe^{3+}-loaded bacillibactin, liberating iron for cellular metabolism. Bacillibactin not only enhances the survival of bacteria in iron-poor environments but also plays a role in the pathogenicity of pathogenic *Bacillus* species, suggesting that it is one of the key regulators that can be targeted for biocontrol purposes (Miethke et al. 2006; Fazle Rabbee and Baek 2020).

19.2.2 Biosynthesis of Polyketides

Polyketides (PKs) are bioactive secondary metabolites synthesized by microorganisms such as *Bacillus subtilis* and related species. These compounds are known to have a wide range of biological activities such as antibacterial, immunosuppressive, and antifungal activities, and functionally resemble fatty acid synthases (FASs) (Fazle Rabbee and Baek 2020). PK biosynthesis is catalyzed by polyketide synthases (PKSs), multienzyme systems that structurally and functionally resemble FASs (Figure 19.10). PKSs are enzymes that use acyl-CoA precursors, such as malonate and methylmalonate, to initiate condensation reactions for the biosynthesis of polyketide chains. The PKS system is characterized by a modular system, which is made up of a series of elongation modules that are sandwiched between an initiation and termination domain (Smith and Tsai 2007).

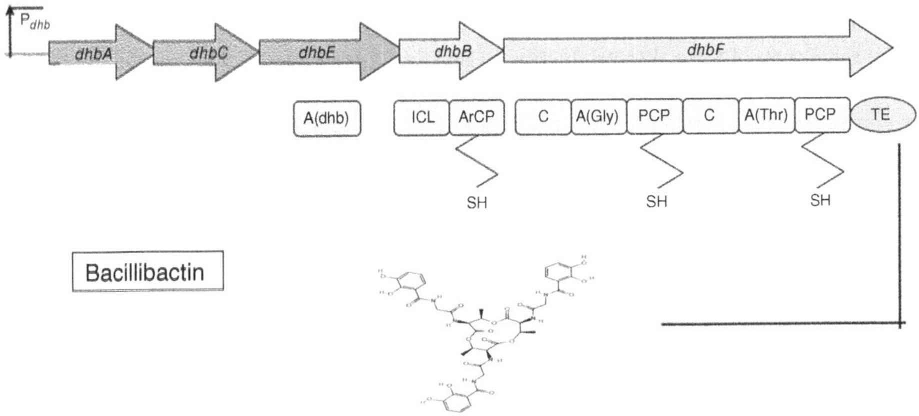

Figure 19.10 Domain organization of the bacillibactin synthetase operon, consisting of *dhb A*, *dhb B*, *dhb C*, *dhb E*, and *dhb F*. These genes encode NRPS modules with functional domains including condensation (C), adenylation (A), peptidyl carrier protein (PCP), and termination (TE), orchestrating the nonribosomal biosynthesis and cyclization of bacillibactin.

The biosynthetic pathway is initiated at the loading module that lacks a functional ketosynthase (KS) domain, but harbors an acyltransferase (AT) domain, which brings in and activates the starter unit that is then transferred to the acyl carrier protein (ACP) domain. This domain acts as a transit arm that delivers the substrate to the adjacent catalytic site. The number of elongation modules, at least three core domains—AT, KS, and ACP—support the stepwise elongation of the polyketide chain. The AT domain chooses and transports the right chain extender unit, malonyl-CoA or methylmalonyl-CoA being the most common, to the ACP domain. The KS domain mediates the decarboxylative Claisen condensation, thereby attaching the newly added extender unit to the growing polyketide intermediate that is tethered to the ACP domain of the upstream module (Hertweck 2009).

In addition to these main enzymatic activities, individual PKS modules can also host ancillary domains that introduce structural variety to the final product. At a hydroxyl group, the DH domain removes water molecules, resulting in the introduction of a double bond, and the ER domain then reduces that double bond to a single bond. Some modules also contain a methylase (M) domain that installs site-specific methylation, affecting the lipophilicity and bioactivity of the polyketide (Donadio et al. 2007; Hertweck 2009).

Biosynthesis ends at the termination module, where the thioesterase (TE) domain is critical in the release of the polyketide product. The TE domain catalyzes either macrolactonization, forming cyclic PKs or hydrolysis, yielding linear PKs (Cane and Walsh 1999). Structural complexity concerning PKs is defined in the order of arrangements in the biosynthetic modules, enzymatic alterations, and domain interactions. This modular biosynthesis supports the ecological availability of PKs exhibiting diverse bioactivities significant in pharmaceuticals and antifungal medicine (Donadio et al. 2007).

19.2.2.1 Bacillaene Synthetase

Bacillaene is a polyene antibiotic and monocarboxylic acid amide synthesized from *Bacillus subtilis* and *Bacillus amyloliquefaciens*. In bacillaene's structure, there is a linear molecule comprising a highly conjugated hexaene moiety, which is attached to α-hydroxy carboxylic acid and conjugated triene via two amide bonds (Aleti et al. 2015). Bacillaene biosynthetic pathway has been explained in *Bacillus amyloliquefaciens* FZB42, which is subclassed by bae, a hybrid type I PKS-NRPS gene cluster. The bae gene cluster is made of five long open ORFs: baeJ, baeL, baeM, baeN, baeR (Chen et al. 2006) (Figure 19.11). In baeJ, the first and second adenylation domains incorporate α-hydroxy-isocaproic acid and glycine, and the third adenylation domain of baeN incorporates alanine (Calderone et al. 2006). Furthermore, modules 4, 8, and 14 are regionally located between neighboring genes. Three short ORFs upstream of baeJ, named baeC, baeD, and baeE, encode acyltransferase (AT) domains that are distinct and specifically load malonyl-CoA (Calderone et al. 2006).

19.2.2.2 Macrolactin Synthetase

Macrolactins are a class of over 30 different 24-membered macrolactones featuring three diene moieties within the carbon skeleton (Figure 19.12). These compounds have been extracted from different bacterial havens, which include *Bacillus* species like *B. subtilis* and *B. amyloliquefaciens*. The cyclic macrolactins in *Bacillus* species are synthesized through the mln PKS gene cluster consisting of 9 operons (mlnA-mln). Their mln gene cluster has 11 KS domains containing a sufficient number of genes, which bifunctionally utilize malonate and acetate as the only building blocks (Schneider et al. 2007).

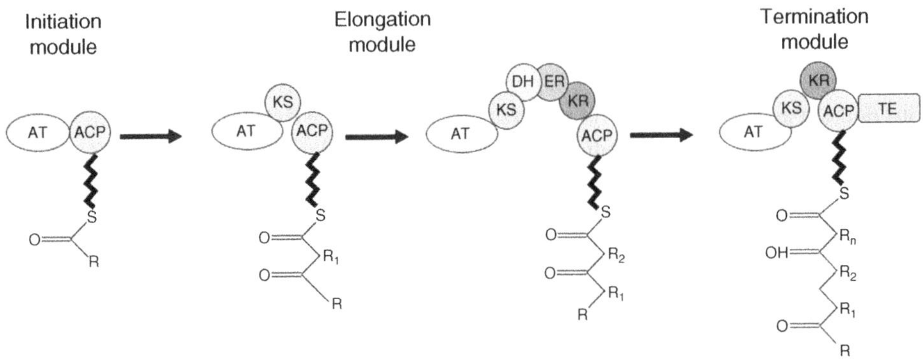

Figure 19.11 Schematic representation of a typical polyketide synthase (PKS) gene cluster involved in polyketide biosynthesis. The operon encodes modular enzymes composed of essential catalytic domains: KS (ketosynthase), AT (acyltransferase), ACP (acyl carrier protein), and optional KR (ketoreductase), DH (dehydratase), and ER (enoyl reductase) domains for structural modification. A terminal TE (thioesterase) domain facilitates the release and cyclization of the final product.

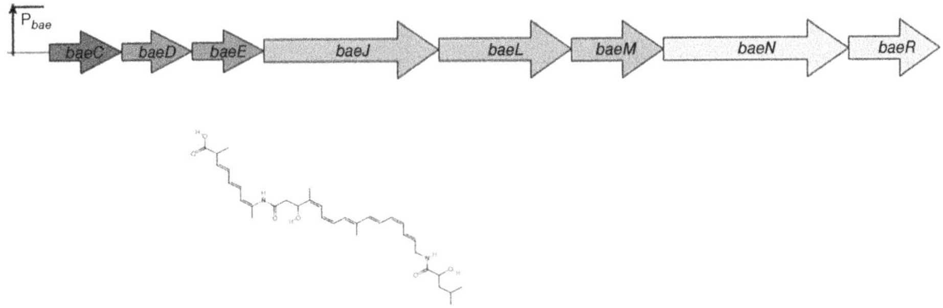

Figure 19.12 Schematic representation of the bacillaene biosynthetic operon and chemical structure.

19.2.2.3 Difficidin Synthetase

Dificidin has an unsaturated 22-carbon cyclic polyene backbone and a phosphate group, which is rare for secondary metabolites (Figure 19.13). Its structural variant, oxydifficidin, has an extra hydroxyl group on position 5 (Wilson et al. 1987). Difficidin biosynthesis is governed by the dif gene cluster, which contains 14 open reading frames (difA to N and difY) (Piel 2010). Plant-associated *Bacillus amyloliquefaciens* synthesizes difficidin, which is highly active against the fire blight pathogen *Erwinia amylovora*. It also effectively manages disease caused by *Pectobacterium carotovorum* and *Xanthomonas oryzae* as well (Olishevska et al. 2019).

19.2.3 Biosynthesis of Ribosomally Synthesized Peptides

Ribosomally synthesized peptides (RPs) are typically formed from short precursors (approximately 100 amino acids), which are covalently modified into functional forms through

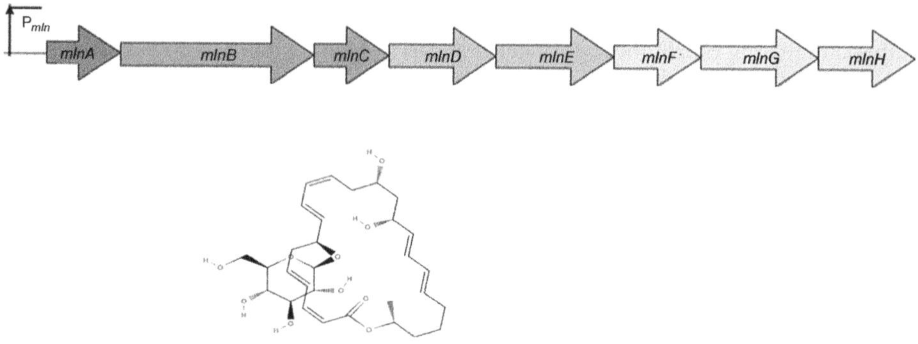

Figure 19.13 Schematic representation of the macrolactin biosynthetic operon and chemical structure.

additional enzymatic steps (Oman and Van Der Donk 2010). A multitude of enzymes is responsible for these modifications and the plethora of chemical structures that they create. These peptides were initially referred to as bacteriocins, which are described as low molecular weight compounds (Field et al. 2007). Alongside the bacteriocins are ribosomally synthesized antimicrobial peptides such as subtilin, amylolysin, and microcin-like peptides, which have antibacterial and antifungal properties (Caulier et al. 2019). Abriouel et al. (2011) divided ribosomally synthesized bacteriocins of *Bacillus* species into three classes. Lantipeptides of class I are enzymatically modified, that is, they are posttranslationally enhanced to be more stable and antimicrobial than their precursors. Nonderivative, antimicrobial peptides like pediocin- and thuricin-like peptides make up class II. These nonmodified, small molecules are heat-stable and have strong antimicrobial activity. Class III includes the larger, more linear peptides (Abriouel et al. 2011).

19.2.3.1 Class I: Posttranslationally Modified Peptides (Lantipeptides)

Lantipeptides (or lantibiotics of class I), peptides with posttranslational modifications, are referred to as heat-stable small-size antimicrobial peptides which are 19–38 amino acids long with lanthionine and 3-methyllanthionine residues. They are divided into two types: type A and type B (Sumi et al. 2015; Willey and Van Der Donk 2007). Type A lantibiotics are more globular and are uncharged or slightly anionic, marked by rigidity. Of note are subtilin

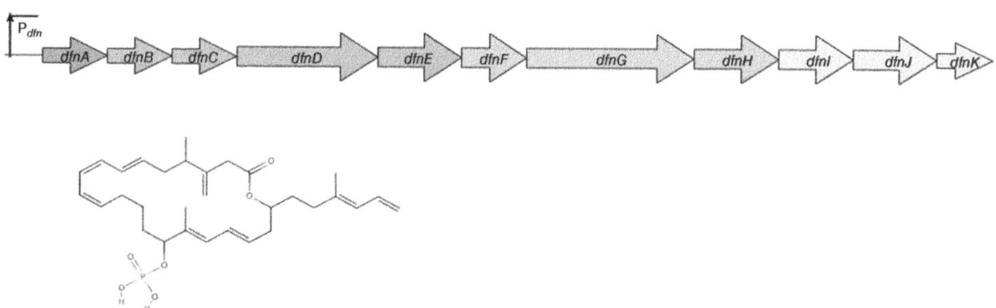

Figure 19.14 Schematic representation of the difficidin biosynthetic operon and chemical structure.

and ericin from *Bacillus subtilis* (type A) which have pore-forming and peptidoglycan biosynthesis-inhibiting capabilities (Abriouel et al. 2011). Mersacidin (type B) is an E-II form globular lantibiotic targeting lipid II to inhibit cell wall synthesis. Highly modified Paenibacillin from *Paenibacillus polymyxa* shows broad-spectrum activity through forming membrane pores. *B. subtilis* contains Subtilosin A, a macrocyclic lantibiotic which is noteworthy for its surpassing anionic quality and rupturing membranes by situating with lipid head groups. The genes accountable for their synthesis comprise structural, modification, immunity, transport, and regulation elements, which reveal the complexity and therapeutic promise of lantibiotics (Olishevska et al. 2019).

19.2.3.2 Class II: Nonmodified Peptides

In the framework suggested by Abriouel et al. (2011), class II bacteriocins from *Bacillus* include small peptides (0.77–10 kDa) which are nonmodified, linear, and thermally stable. They are divided into three subclasses: pediocin-like, thuricin-like, and other linear peptides. Coagulin, which is synthesized by *Bacillus coagulans* I4 and SRCAM peptides (SRCAM 37, SRCAM 602, and SRCAM 1580) produced by *Bacillus circulans* and *Paenibacillus polymyxa* strains, also fall into this subclass alongside other like SRCAM 37 and 602. Their structure contains YGNGVXC, which is an N-terminal motif necessary for antimicrobial function (Abriouel et al. 2011). The Coagulin gene, which codes for a peptide of 44 amino acids (~4.6 kDa), is associated with a cluster residing on a 14 kb plasmid consisting of four genes, where coaB provides immunity, and coaC and coaD serve as building blocks for an ABC transporter type secretion system (Le Marrec et al. 2000).

19.2.3.3 Class III: Large Proteins

Class III bacteriocins comprise large proteins with phospholipase activities, such as megacins A-216 and A-19213, produced by strains of *Bacillus megaterium* (Abriouel et al. 2011). These megacins are divided into type A based on their mechanism of action and their induction by UV light (Von Tersch and Carlton 1983). Megacin A-216, which corresponds to the megA gene, is a phospholipase A2 with a 66 kDa weight and an ORF of 293 amino acids. In addition to megA, there is an ORF of immunity protein with a sequence of 91 amino acids, with other genes, including one resembling RNA polymerase factors are responsible for the induction of megacin (Kiss et al. 2008).

19.3 Volatile Metabolites: A Biocontrol Strategy of *Bacillus*

Research recognizes the value of microbial volatile organic compounds (mVOCs) produced by bacteria as biological control agents against complex fungal pathogens of plants (Hassan et al. 2017). The antifungal phytotoxic compounds such as fusarin have unique properties, such as low molecular weight, high vapor pressure, lipophilicity, and volatility, enabling their evaporation and diffusion through air and soil at ambient temperatures, and therefore, do not need to come into contact with the fungi directly in order to exert effectiveness. Bacterial VOCs encompass an extensive array of compounds, which includes not only alcohols like 2-methylbutan-1-ol, ethanol, and 2-phenylethanol, but also esters, aldehydes, cyclic hydrocarbons, organic acids, ketones, benzenoids, pyrazines, terpenes, and other sulfur and nitrogen compounds (Buzzini et al. 2003; Fialho et al. 2011; Schmidt et al. 2015).

However, the most potent antifungal activity has been attributed to a few specific compounds: caryophyllene, dimethyl disulfide (DMDS), dimethyl trisulfide (DMTS), hydrogen cyanide, 2-methylpyrazine, S-methyl thioacetate, 2,3,6-trimethylphenol, various methyl ketones such as nonan-2-one, decan-2-one, dodecan-2-one, benzonitrile, and acetoin (Schmidt et al. 2015; Tyc et al. 2015; He et al. 2020).

Importantly, bacterial VOCs are created in minimal amounts and have complex compositions. Their action is generally not due to a single compound but rather due to the combined effects of several volatile molecules whose interaction broadens the antifungal activity. This synergy enhances the strength and scope of antifungal action, and mVOCs act as a multifaceted healer, making them a strong and hopeful contender for eco-friendly solutions in combating plant diseases (Poveda 2021).

Several strains have broad-spectrum antifungal activities that are potent against various plant fungal pathogens. Concerning this, Liu et al. (2008) described the antifungal activity of two strains of *Bacillus pumilus*, BSH-4 and ZB13, which were retrieved from the cucumber rhizosphere. These mVOC-producing strains had strong growth inhibition on a broad range of phytopathogenic fungi such as *Sclerotinia sclerotiorum, Botrytis cinerea, Alternaria brassicae, Alternaria solani, Ascochyta citrullina, Fusarium oxysporum, Fusarium graminearum, Cercospora kikuchii, Rhizoctonia solani, Phoma arachidicola,* and *Verticillium dahliae*. He et al. (2020) studied the fungi-inhibitory activity of VOCs produced by two *Bacillus* strains, *Bacillus methylotrophicus* BCN2, *Bacillus thuringiensis* BCN10, and reported that such VOCs inhibited the mycelial growth of all five target postharvest fungal pathogens, *Fusarium oxysporum, Botryosphaeria* sp., *Penicillium expansum, Trichoderma atroviride,* and *Colletotrichum gloeosporioides*.

As noted earlier, research about other species of *Bacillus* indicates that mVOCs possess antifungal properties and help trigger ISR within host plants. Tahir et al. (2017) reported that the VOCs of *Bacillus subtilis* SYST2 not only stimulated growth but also caused systemic resistance (ISR) in tobacco against *Ralstonia solanacearum*. These VOCs increased the biomass, leaf area, and photosynthesis while notably decreasing the disease's symptoms. Moreover, some ISR bioactive VOCs such as albuterol and 1,3-propanediol were recognized as active contributors to this specific resistance.

Dutta et al. (2025) investigated the effects of *Bacillus vallismortis* strain EXTN-1 VOCs on plant growth and discovered that priming tobacco seeds with these VOCs boosted seedling growth when grown in a greenhouse. In the GC–MS analysis, several key growth-promoting compounds were identified as VOCs, such as acetoin, 1-butanol, and 2,3-butanedione, as well as heneicosane and benzaldehyde.

19.4 Whispers in the Rhizosphere: ISR as a Silent Defense Strategy by *Bacillus*

Besides phytopathogen antagonistic activities, *Bacillus* spp. also promote enhanced and sustained systemic defense mechanisms in plants. ISR refers to a state of heightened readiness in a plant that is activated by beneficial microorganisms like PGPR and results in the plant sensing and being able to resist a broad array of pathogens and pests. Plants orchestrate ISR through the signaling pathways of jasmonic acid (JA) and ethylene (ET) (Yu et al. 2022).

Plants also have, in addition to ISR, another well-defined type of defense mechanism known as systemic acquired resistance (SAR). SAR differs from ISR with respect to the type of elicitors and the signaling pathways engaged. While SAR is often triggered by virulent, avirulent, or even nonpathogenic microbes, the establishment of SAR requires some time to accumulate salicylic acid (SA) and produce pathogenesis-related (PR) proteins such as chitinases and glucanases. On the other hand, ISR is not accompanied by the accumulation of SA or PR proteins but is predominantly mediated by JA and ET signaling pathways, thus making it a complementary defense strategy in plants (Yu et al. 2022).

Induced systemic resistance of *Bacillus* species entails a sophisticated and delicate interplay between the plant and the microbe whereby the plant's basal defense is heightened while the pathogen is neither directly attacked nor neutralized. It starts when beneficial *Bacillus* spp. such as *B. subtilis*, *B. amyloliquefaciens*, or *B. pumilus* are known to colonize plant roots (Gómez-Gómez 2004). These bacteria synthesize different bioactive compounds such as CLPs (surfactin, fengycin, and iturin), VOCs, and other signaling molecules classified as elicitors. Plant receptors recognize these elicitors and intracellular signaling cascades are activated. In contrast to the SA-dependent route characteristic of SAR, ISR brought about by *Bacillus* is predominantly reliant on JA and ET pathways of signaling (Ashraf et al. 2025). These hormones control the expression of downstream genes associated with defense responses in plants. This mechanism of immune priming fails to activate PR proteins under nonchallenging conditions. Instead, it puts the plant on standby for a swift and robust countermeasure when a real pathogen strikes (Yan et al. 2002). In addition, ISR from *Bacillus* is systemic in that the resistance is expressed not just at the roots, where recognition occurs, but also in other parts of the plant, such as the leaves and stems. Significantly, the host ISR response and effectiveness are dependent on the strain of *Bacillus* and plant genotype (Pieterse et al. 2014).

The application of PGPR as an activator of systemic resistance in different field crops has been documented under field conditions in recent years. Considerable efforts have been made to induce ISR with PGPR in plants. ISR through PGPR has been documented in a large number of crops, such as soybean (Kloepper et al. 2004), tomato (Akram et al. 2015), rice (Chandler et al. 2015), pepper (Yang et al. 2009), wheat (Burkhanova et al. 2017), chili (Bharathi et al. 2004),tomato and hot pepper (Ramamoorthy et al. 2002), and mango (Vivekananthan et al. 2004).

The induction of systemic resistance (ISR) by *Bacillus* species is known to occur in various plant-pathogen interactions. Enebak and Carey (2000) described that some strains of *Bacillus pumilus*, in particular INR7, SE52, SE34, and SE49, were able to induce systemic resistance in *Pinus taeda*, and thus suppress fusiform rust by *Cronartium quercuum*. In a similar vein, Wei et al. (1996) noted that the application of *B. pumilus* during the growing season significantly reduced the incidence of anthracnose disease of cucumber caused by *Colletotrichum orbiculare*, further underscoring the impact of ISR in managing disease. Wu et al. (2018) later showed that *Bacillus amyloliquefaciens* FZB42 induces systemic resistance in *Nicotiana benthamiana* and controls *Phytophthora nicotianae* leaf disease. Chowdhury et al. (2015) demonstrated that *Bacillus amyloliquefaciens* subsp. *plantarum* induces defense mechanisms in lettuce against bottom rot inflicted by *Rhizoctonia solani* through lipopeptide cyclic defenses. *Bacillus megaterium* de Bary TRS-4, which was isolated from the tea rhizosphere, induces resistance to brown root rot caused by *Fomes lamaoensis*, while significantly improving plant growth under in vivo conditions in tea plants (Chakraborty et al. 2006).

19.5 Multifunctional Roles of *Bacillus* Secondary Metabolites in Agriculture

Bacillus species are noted as PGPR because of their bioactive compounds that aid in plant protection and productivity enhancement. The antibiotic metabolites manufactured by these bacteria undermine phytopathogens through a multitude of biocontrol strategies and mechanisms, such as producing antibiotics and cell wall–degrading enzymes, along with eliciting plant immune responses (Jha and Saraf 2015). Besides the pathogen antagonism, *Bacillus* strains aid in plant health through numerous and diverse plant growth-promotion mechanisms, including the formation of phytohormones, solubilizing nutrients, increasing abiotic stress tolerance, and improving soil structure (Tariq and Ahmed 2024). These various functions lessen the reliance on synthetic agrochemicals and also help mitigate the adverse impacts of climate change, advancing strategic international initiatives for sustainable agricultural development. These applications will elaborate on the biocontrol strategies first and then on the other contributions to plant growth and soil health.

19.5.1 Suppression of Fungal Phytopathogens

The interruption of fungal phytopathogens is among the most documented and elaborative uses of *Bacillus* secondary metabolites in phytopathogenic bioagriculture. A variety of antifungal compounds, especially lipopeptides such as iturins, surfactins, and fengycins, along with PKs like bacillaene, macrolactin, and difficidin, are produced by *Bacillus* species and wield significant antifungal activity as summarized in Table 19.1. Figueiredo et al. (2025), supporting the biocontrol ability of *Bacillus* species, showed that *Bacillus velezensis* CNPMS-22 strongly suppresses the cultivation of *Fusarium oxysporum* and *Sclerotinia sclerotiorum* through the synthesis of antifungal lipopeptides and PKs and improves certain growth parameters of the plants. This *B. velezensis's* multifunctionality reinforces the notion that it is a biocontrol agent as well as a plant growth-promoting rhizobacterium or PGPR. In her latest study, Campos-Avelar et al. (2025) studied *Bacillus cabrialesii* subsp. *cabrialesii* Strain TE5 and analyzed its biocontrol capability against the spot blotch wheat pathogen, *Bipolaris sorokiniana*. Thisstudy showed that the strain suppressed the fungal growth by producing antifungal secondary metabolites and lytic enzymes while promoting wheat plant growth, enhancing disease resistance, and improving overall plant vigor.

19.5.2 *Bacillus*-Based Solutions for Disease, Pest, and Weed Control

Besides their strong antifungal activities, secondary metabolites of *Bacillus* (B-SMs) provide multifunctional protection against plant viruses, bacteria, algae, pests, and weeds. Such prowess is evident in a study by Wang et al. (2023), where *Bacillus subtilis* strain L2 exhibited strong antiviral activity for a swine pathogenic DNA virus (PRV) by blocking viral replication through mechanisms interfering with binding, entry, and intracellular dissemination.

The works of Markelova and Chumak (2025)showed that surfactin has strong antimicrobial properties, as some studies have shown that surfactin inhibits *Staphylococcus aureus* adhesion and biofilm formation by blockade of biofilm matrix polysaccharide production and downregulation of major biofilm-forming genes, *icaA* and *icaD*. Furthermore, it restrained the adhesion as well as biofilm formation of a number of foodborne pathogens,

Table 19.1 Comprehensive summary of *Bacillus* secondary metabolites used in plant disease management. It includes associated plants, key genes, targeted fungi, and antifungal activities contributing to biocontrol.

Bacillus spp.	Secondary metabolite	Crop	Targeted fungus	Antifungal mechanism	Gene(s) involved	Reference
Bacillus subtilis	Surfactin	Potato	*Fusarium foetens*	Disruption of cell membrane	*srfA*, *sfp*	Liu et al. (2023), Bóka et al. (2019)
Bacillus subtilis	Iturin	Tomato	*Botrytis cinerea*	Pore formation in fungal membrane	*ituA*, *ituD*	Ongena and Jacques (2008)
Bacillus subtilis	Fengycin	Rice	*Rhizoctonia solani*	Inhibition of fungal growth	*fenA*, *fenB*	Sansinenea and Ortiz (2011)
Bacillus subtilis	Bacillomycin D	Wheat	*Fusarium graminearum*	Membrane disruption	*bamA*, *bamB*	Caulier et al. (2019)
Bacillus subtilis	Bacilysin	Wheat, rice	*Fusarium oxysporum*	Inhibits cell wall synthesis	*bacA*, *bacB*	Caulier et al. (2019)
Bacillus subtilis	Subtilin	Cereals, legumes	*Alternaria* spp.	Inhibits spore germination *spaS*, *spar*	*spaS*, *spa*R	Harwood et al. (2018)
Bacillus subtilis	Mycosubtilin	Rice, soybean	*Magnaporthe oryzae*	Alters fungal membrane permeability *mycA*, *mycB*	*mycA*, *mycB*	Kaspar et al. (2019)
Bacillus subtilis	Subtilosin	Vegetables	*Phytophthora* spp.	Membrane damage and pore formation	*sboA*, *sboB*	Harwood et al. (2018)
Bacillus subtilis	Plipastatin	Multiple crops	*Fusarium* spp.	Membrane disruption	*pps*	Tsuge et al. (1999)
Bacillus subtilis	Kurstakin	Tomato	*Fusarium oxysporum*	Membrane disruption	*krs*	Hussain et al. (2025)
Bacillus subtilis	Bacillibactin	Pepper	*Fusarium oxysporum*	Iron sequestration	*dhbACEBF*	Kiesewalter et al. (2021), Yu et al. (2011)

Bacillus subtilis	Gageopeptide	Wheat	*Magnaporthe oryzae*	Inhibition of fungal growth		Chakraborty et al. (2020)
Bacillus subtilis BS-58	Macrolactin A	*Amaranthus hypochondriacus*	*Fusarium oxysporum, Rhizoctonia solani*	Biocontrol and inhibition	*mlnA*	Pandey et al. (2023)
Bacillus amyloliquefaciens	Bacillibactin	Maize	*Fusarium* spp.	Iron chelation to limit fungal growth *dhbA*, *dhbB*	*dhbA*, *dhbB*	Caulier et al. (2019)
Bacillus amyloliquefaciens GKT04	Surfactin, fengycin, iturin	Multiple crops	*Fusarium oxysporum*	Disruption of cell membrane and inhibition of sporulation	*srfAA*, *fenA*, *ituD*	Tian et al. (2021)
Bacillus amyloliquefaciens S76-3	Iturin A, plipastatin A	Wheat	*Fusarium graminearum*	Inhibition of fungal growth by disrupting membrane integrity, *ituA*, *ppsA*	*ituA*, *ppsA*	Gong et al. (2015)
Bacillus amyloliquefaciens 83	Bacillomycin D	Mango	*Colletotrichum gloeosporioides*	Inhibition by disrupting fungal growth and viability at different physiological stages	b*my*	Luna-Bulbarela et al. (2018) Koumoutsi et al. (2004)
B. cereus	Zwittermycin A	*R. solani*, *P. aphanidermatum*, and *F. Oxysporum*	*alfalfa*	Inhibits fungal growth by disrupting cellular processes, specifically DNA transcription and replication.	*zma*	Moshe et al. (2023) Silo-Suh et al. (1998)
B. thuringiensis	Fengycin	*Colletotrichum gloeosporioides*	*Pepper*	Disrupting the cell membrane	*fen*	Kim et al. (2004)
Bacillus licheniformis	Surfactin	*Magnaporthe grisea*	*Rice*	Disrupting the cell membrane	*srf*	Tendulkar et al. (2007) Shleeva et al. (2023)

(*Continued*)

Table 19.1 (Continued)

Bacillus spp.	Secondary metabolite	Crop	Targeted fungus	Antifungal mechanism	Gene(s) involved	Reference
B. cereus	Fengycin	*Sclerotinia sclerotiorum*	*Sunflower*	Disrupting the cell membrane	*fen*	Deleu et al. (2008)
B. subtilisB47	Iturin A_2	Tomato	*Fusarium oxysporium, Aspergillus flavus*	Interrupt activity of fungal lipopeptide	*itu*	YE et al. (2012)
B. amyloliquefaciens	Surfactin, iturin, fengycin, a bacillibactin, bacilysin, bacillaene, difficidin, and macrolactin	Wheat	*Fusarium*	Pore formation in cell wall and cell membrane by disrupting the lipid bilayer	*srf, itu*	Dunlap et al. (2013)
B. vallismortis	Surfactin and fengycin	Chilli pepper	*Phytophthora capsici and Colletotrichum acutatum*	Cell-wall disintegration	*srf, fen*	Park et al. (2013)
B. subtilis	Fengycin and bacillomycin	Cucumber	*Fusarium oxysporum* f. sp. *Cucumerinum*	Inhibit mycelial growth and conidial germination of the fungal pathogen	*fen, bmy*	Cao et al. (2012)

including *Listeria monocytogenes*, *Enterobacter sakazakii*, and *Salmonella enteritidis*, while carrying out significant antibiofilm and antimicrobial action against *Legionella pneumophila* (Loiseau et al. 2015; Nitschke et al. 2009). In addition, Surfactin-15 F was shown to obliterate preformed biofilms of *Staphylococcus epidermidis* biofilms, significantly reducing their structural integrity and viability (Jardak et al. 2024).

19.5.3 Soil Remediation and Restoration of Soil Health

Bacillus species are important for soil cleansing and restoring soil health because they can produce a variety of metabolites, including compounds useful in bioremediation. Wu et al. (2019) demonstrated that *Bacillus megaterium* isolated from *Pennisetum americanum* and *Pennisetum purpureum* had the ability to remediate metals such as cadmium (Cd) and significantly enhanced root and shoot length. Saran et al. (2020) studied *Bacillus proteolyticus* isolated from *Helianthus annuus* and their ability to bioremediate cadmium (Cd) and lead (Pb) contamination. The study demonstrated these strains of the genus *Bacillus* were able to considerably decrease the concentration of heavy metals in the soil, which resulted in significant biomass production.

19.6 Conclusion and Future Prospects

As discussed above, *Bacillus* species are well known for their fungi static activity against a wide variety of phytopathogenic fungi due to the numerous antifungal secondary metabolites (SMs) they produce. These metabolites are capable of inhibiting fungal growth and, at the same time, inducing systemic resistance in the plants. A critical factor in their potential for biocontrol antifungal activity is the biosynthetic gene clusters (BGCs) such as *itu*, *fen*, *srf*, and also *bam*, which control the biosynthesis, regulation, and efflux of these bioactive elements that build these antifungals. There are future prospects in using advanced tools and mining genomes, engineering metabolism, and synthetic biology to increase the aforementioned attributes and enhance the biocidal activity. Besides, there is some promise of discovering new SMs with antifungal activity using metagenomics and high-throughput screening.

References

Abdel-Nasser, A., Badr, A.N., Fathy, H.M. et al. (2024). Antifungal, antiaflatoxigenic, and cytotoxic properties of bioactive secondary metabolites derived from *Bacillus* species. *Scientific Reports* 14 (1): 16590.

Abderrahmani, A., Tapi, A., Nateche, F. et al. (2011). Bioinformatics and molecular approaches to detect NRPS genes involved in the biosynthesis of kurstakin from *Bacillus thuringiensis*. *Applied Microbiology and Biotechnology* 92: 571–581.

Abriouel, H., Franz, C.M., Omar, N.B., and Gálvez, A. (2011). Diversity and applications of *Bacillus* bacteriocins. *FEMS Microbiology Reviews* 35 (1): 201–232.

Agbodjato, N.A. and Babalola, O.O. (2024). Promoting sustainable agriculture by exploiting plant growth-promoting rhizobacteria (PGPR) to improve maize and cowpea crops. *PeerJ* 12: e16836.

Akram, W., Anjum, T., and Ali, B. (2015). Searching ISR determinant/s from *Bacillus* subtilis IAGS174 against Fusarium wilt of tomato. *BioControl* 60: 271–280.

Aleti, G., Sessitsch, A., and Brader, G. (2015). Genome mining: prediction of lipopeptides and polyketides from Bacillus and related Firmicutes. *Computational and Structural Biotechnology Journal* 13: 192–203.

Armenova, N., Petrova, P., Gerginova, M., Krumova, E., Kaynarov, D., Velkova, L., ... Petrov, K. (2024). Bacillus velezensis R22 inhibits the growth of multiple fungal phytopathogens by producing surfactin and four fengycin homologues. *Biotechnology & Biotechnological Equipment*, 38(1). https://doi.org/10.1080/13102818.2024.2313072

Ashraf, M.H., Hussain, N., Saleem, M.Z., and Haqqi, R. (2025). Biocontrol mechanisms by root-associated Bacillus species. In: *Microbial Biocontrol Techniques: Importance in Ensuring Food Security* (ed. A. Kumar and M.K. Solanki), 155–177. Springer Nature Singapore: Singapore.

Auchtung, J.M., Lee, C.A., and Grossman, A.D. (2006). Modulation of the ComA-dependent quorum response in *Bacillus subtilis* by multiple Rap proteins and Phr peptides. *Journal of Bacteriology* 188 (14): 5273–5285.

Bharathi, R., Vivekananthan, R., Harish, S. et al. (2004). Rhizobacteria-based bio-formulations for the management of fruit rot infection in chillies. *Crop Protection* 23 (9): 835–843.

Bóka, B., Manczinger, L., Kocsubé, S. et al. (2019). Genome analysis of a *Bacillus subtilis* strain reveals genetic mutations determining biocontrol properties. *World Journal of Microbiology and Biotechnology* 35: 1–14.

Bumpus, S.B., Evans, B.S., Thomas, P.M. et al. (2009). A proteomics approach to discovering natural products and their biosynthetic pathways. *Nature Biotechnology* 27 (10): 951–956.

Burkhanova, G.F., Veselova, S.V., Sorokan', A.V. et al. (2017). Strains of *Bacillus* ssp. regulate wheat resistance to Septoria nodorum Berk. *Applied Biochemistry and Microbiology* 53: 346–352.

Buzzini, P., Martini, A., Cappelli, F. et al. (2003). A study on volatile organic compounds (VOCs) produced by tropical ascomycetous yeasts. *Antonie Van Leeuwenhoek* 84: 301–311.

Calderone, C.T., Kowtoniuk, W.E., Kelleher, N.L. et al. (2006). Convergence of isoprene and polyketide biosynthetic machinery: isoprenyl-S-carrier proteins in the pksX pathway of *Bacillus subtilis*. *Proceedings of the National Academy of Sciences* 103 (24): 8977–8982.

Campos-Avelar, I., Jaime, M.F.G., Sandoval, P.H.M. et al. (2025). *Bacillus cabrialesii* subsp. *cabrialesii* Strain TE5: apromising biological control bacterium against the causal agent of spot blotch in wheat. *Plants* 14 (2): 209.

Cane, D.E. and Walsh, C.T. (1999). The parallel and convergent universes of polyketide synthases and nonribosomal peptide synthetases. *Chemistry & Biology* 6 (12): R319–R325.

Cao, Y., Xu, Z., Ling, N. et al. (2012). Isolation and identification of lipopeptides produced by *B. subtilis* SQR 9 for suppressing Fusarium wilt of cucumber. *Scientia Horticulturae* 135: 32–39.

Caretta, T., Unterholzner, M., Alzate Zuluaga, M.Y. et al. (2025). Antifungal activity of surfactin produced by ***Bacillus subtilis*** against phytopathogenic fungi in apple and strawberry: An in vitro and in vivo study. *International Journal of Food Science and Technology* 60.

Cassat, J.E. and Skaar, E.P. (2013). Iron in infection and immunity. *Cell Host & Microbe* 13 (5): 509–519.

Caulier, S., Nannan, C., Gillis, A. et al. (2019). Overview of the antimicrobial compounds produced by members of the *Bacillus subtilis* group. *Frontiers in Microbiology* 10: 302.

Chakraborty, U., Chakraborty, B., and Basnet, M. (2006). Plant growth promotion and induction of resistance in Camellia sinensis by *Bacillus megaterium*. *Journal of Basic Microbiology* 46 (3): 186–195.

Chakraborty, M., Mahmud, N.U., Gupta, D.R. et al. (2020). Inhibitory effects of linear lipopeptides from a marine *Bacillus subtilis* on the wheat blast fungus *Magnaporthe oryzae Triticum*. *Frontiers in Microbiology* 11: 665.

Chandler, S., Van Hese, N., Coutte, F. et al. (2015). Role of cyclic lipopeptides produced by *Bacillus subtilis* in mounting induced immunity in rice (*Oryza sativa* L.). *Physiological and Molecular Plant Pathology* 91: 20–30.

Chen, W.C., Juang, R.S., and Wei, Y.H. (2015). Applications of a lipopeptide biosurfactant, surfactin, produced by microorganisms. *Biochemical Engineering Journal* 103: 158–169.

Chen, X.H., Vater, J., Piel, J. et al. (2006). Structural and functional characterization of three polyketide synthase gene clusters in *Bacillus amyloliquefaciens* FZB 42. *Journal of Bacteriology* 188 (11): 4024–4036.

Choudhary, D.K. and Johri, B.N. (2009). Interactions of *Bacillus* spp. and plants–with special reference to induced systemic resistance (ISR). *Microbiological Research* 164 (5): 493–513.

Chowdhury, S.P., Uhl, J., Grosch, R. et al. (2015). Cyclic lipopeptides of *Bacillus amyloliquefaciens* subsp. plantarum colonizing the lettuce rhizosphere enhance plant defense responses toward the bottom rot pathogen Rhizoctonia solani. *Molecular Plant-Microbe Interactions* 28 (9): 984–995.

Coutte, F., Niehren, J., Dhali, D. et al. (2015). Modeling leucine's metabolic pathway and knockout prediction improving the production of surfactin, a biosurfactant from *Bacillus subtilis*. *Biotechnology Journal* 10 (8): 1216–1234.

Dang, Y., Zhao, F., Liu, X. et al. (2019). Enhanced production of antifungal lipopeptide iturin A by *Bacillus amyloliquefaciens* LL3 through metabolic engineering and culture conditions optimization. *Microbial Cell Factories* 18: 1–14.

Deleu, M., Paquot, M., and Nylander, T. (2008). Effect of fengycin, a lipopeptide produced by *Bacillus subtilis*, on model biomembranes. *Biophysical Journal* 94 (7): 2667–2679.

Deng, X., Tian, Y., Niu, Q. et al. (2013). The ComP-ComA quorum system is essential for "Trojan horse" like pathogenesis in *Bacillus nematocida*. *PLoS One* 8 (10): e76920.

Donadio, S., Monciardini, P., and Sosio, M. (2007). Polyketide synthases and nonribosomal peptide synthetases: the emerging view from bacterial genomics. *Natural Product Reports* 24 (5): 1073–1109.

Dubois, T., Faegri, K., Perchat, S. et al. (2012). Necrotrophism is a quorum-sensing-regulated lifestyle in *Bacillus thuringiensis*. *PLoS Pathogens* 8 (4): e1002629.

Duitman, E.H., Hamoen, L.W., Rembold, M. et al. (1999). The mycosubtilin synthetase of *Bacillus subtilis* ATCC6633: a multifunctional hybrid between a peptide synthetase, an amino transferase, and a fatty acid synthase. *Proceedings of the National Academy of Sciences* 96 (23): 13294–13299.

Dunlap, C.A., Bowman, M.J., and Schisler, D.A. (2013). Genomic analysis and secondary metabolite production in *Bacillus amyloliquefaciens* AS 43.3: a biocontrol antagonist of Fusarium head blight. *Biological Control* 64 (2): 166–175.

Dutta, S., Balaraju, K., Oh, S.Y. et al. (2025). Plant growth promotion via priming with volatile organic compounds emitted from *Bacillus vallismortis* strain EXTN-1. *Frontiers in Microbiology* 15: 1524888.

Enebak, S.A. and Carey, W.A. (2000). Evidence for induced systemic protection to fusiform rust in loblolly pine by plant growth-promoting rhizobacteria. *Plant Disease* 84 (3): 306–308.

Fazle Rabbee, M. and Baek, K.H. (2020). Antimicrobial activities of lipopeptides and polyketides of *Bacillus velezensis* for agricultural applications. *Molecules* 25 (21): 4973.

Fialho, M.B., Moraes, M.H.D.D., Tremocoldi, A.R., and Pascholati, S.F. (2011). Potential of antimicrobial volatile organic compounds to control *Sclerotinia sclerotiorum* in bean seeds. *Pesquisa Agropecuária Brasileira* 46: 137–142.

Field, D., Cotter, P., Hill, C., and Ross, R.P. (2007). Bacteriocin biosynthesis, structure, and function. *Research and Applications in Bacteriocins* 4: 5–41.

Figueiredo, J.E.F., Diniz, G.D.F.D., Marins, M.S. et al. (2025). Bacillus velezensis CNPMS-22 as biocontrol agent of pathogenic fungi and plant growth promoter. *Frontiers in Microbiology* 16: 1522136.

Fukushima, T., Allred, B.E., Sia, A.K. et al. (2013). Gram-positive siderophore-shuttle with iron-exchange from Fe-siderophore to apo-siderophore by *Bacillus cereus* YxeB. *Proceedings of the National Academy of Sciences* 110 (34): 13821–13826.

Gao, H., Qi, G., Yin, R. et al. (2016). *Bacillus cereus* strain S2 shows high nematicidal activity against *Meloidogyne incognita* by producing sphingosine. *Scientific Reports* 6 (1): 28756.

Gómez-Gómez, L. (2004). Plant perception systems for pathogen recognition and defence. *Molecular Immunology* 41 (11): 1055–1062.

Gong, A.D., Li, H.P., Yuan, Q.S. et al. (2015). Antagonistic mechanism of iturin A and plipastatin A from *Bacillus amyloliquefaciens* S76-3 from wheat spikes against *Fusarium graminearum*. *PLoS One* 10 (2): e0116871.

Grangemard, I., Wallach, J., Maget-Dana, R., and Peypoux, F. (2001). Lichenysin: a more efficient cation chelator than surfactin. *Applied Biochemistry and Biotechnology* 90: 199–210.

Gutiérrez-Chávez, C., Benaud, N., and Ferrari, B.C. (2021). The ecological roles of microbial lipopeptides: where are we going? *Computational and Structural Biotechnology Journal* 19: 1400–1413.

Harwood, C.R., Mouillon, J.M., Pohl, S., and Arnau, J. (2018). Secondary metabolite production and the safety of industrially important members of the *Bacillus subtilis* group. *FEMS Microbiology Reviews* 42 (6): 721–738.

Hassan, N., Nakasuji, S., Elsharkawy, M.M. et al. (2017). Biocontrol potential of an endophytic *Streptomyces* sp. strain MBCN152-1 against *Alternaria brassicicola* on cabbage plug seedlings. *Microbes and Environments* 32 (2): 133–141.

Hathout, Y., Ho, Y.P., Ryzhov, V. et al. (2000). Kurstakins: a new class of Lipopeptides isolated from *Bacillus thuringiensis*. *Journal of Natural Products* 63 (11): 1492–1496.

He, C.N., Ye, W.Q., Zhu, Y.Y., and Zhou, W.W. (2020). Antifungal activity of volatile organic compounds produced by *Bacillus methylotrophicus* and *Bacillus thuringiensis* against five common spoilage fungi on loquats. *Molecules* 25 (15): 3360.

Hertweck, C. (2009). The biosynthetic logic of polyketide diversity. *Angewandte Chemie International Edition* 48 (26): 4688–4716.

Hussain, S., Ali, M., Ghazy, A.H. et al. (2025). Identification of antifungal lipopeptides from *Bacillus subtilis* Sh-17 targeting *Fusarium oxysporum* f. sp. *lycopersici*. *Chemical and Biological Technologies in Agriculture* 12 (1): 29.

Iqbal, S., Begum, F., Rabaan, A.A. et al. (2023). Classification and multifaceted potential of secondary metabolites produced by *Bacillus subtilis* group: a comprehensive review. *Molecules* 28 (3): 927.

Jacques, P. (2011). Surfactinn and other lipopeptides from *Bacillus* spp. In: *Biosurfactants: From Genes to Applications (Microbiology Monographs*, vol. 20 (ed. G. Soberón-Chávez), 57–91. Springer: Berlin & Heidelberg https://doi.org/10.1007/978-3-642-14490-5_3.

Jardak, M., Lami, R., Saadaoui, O. et al. (2024). Control of Staphylococcus epidermidis biofilm by surfactins of an endophytic bacterium Bacillus sp. 15 F. *Enzyme and Microbial Technology* 180: 110477.

Jha, C.K. and Saraf, M. (2015). Plant growth promoting rhizobacteria (PGPR): a review. *Journal of Agricultural Research and Development* 5 (2): 108–119.

Kaspar, F., Neubauer, P., and Gimpel, M. (2019). Bioactive secondary metabolites from *Bacillus subtilis*: a comprehensive review. *Journal of Natural Products* 82 (7): 2038–2053.

Kiesewalter, H.T., Lozano-Andrade, C.N., Wibowo, M. et al. (2021). Genomic and chemical diversity of *Bacillus subtilis* secondary metabolites against plant pathogenic fungi. *Msystems* 6 (1): 10–1128.

Kim, P.I., Bai, H., Bai, D. et al. (2004). Purification and characterization of a lipopeptide produced by *Bacillus thuringiensis* CMB26. *Journal of Applied Microbiology* 97 (5): 942–949.

Kiss, A., Balikó, G., Csorba, A. et al. (2008). Cloning and characterization of the DNA region responsible for Megacin A-216 production in *Bacillus megaterium* 216. *Journal of Bacteriology* 190 (19): 6448–6457.

Klausmann, P., Hennemann, K., Hoffmann, M. et al. (2021). *Bacillus subtilis* high cell density fermentation using a sporulation-deficient strain for the production of surfactin. *Applied Microbiology and Biotechnology* 105 (10): 4141–4151.

Kloepper, J.W., Ryu, C.M., and Zhang, S. (2004). Induced systemic resistance and promotion of plant growth by *Bacillus* spp. *Phytopathology* 94 (11): 1259–1266.

Koglin, A., Löhr, F., Bernhard, F. et al. (2008). Structural basis for the selectivity of the external thioesterase of the surfactin synthetase. *Nature* 454 (7206): 907–911.

Koglin, A. and Walsh, C.T. (2009). Structural insights into nonribosomal peptide enzymatic assembly lines. *Natural Product Reports* 26 (8): 987–1000.

Konz, D., Doekel, S., and Marahiel, M.A. (1999). Molecular and biochemical characterization of the protein template controlling biosynthesis of the lipopeptide lichenysin. *Journal of Bacteriology* 181 (1): 133–140.

Konz, D., Klens, A., Schörgendorfer, K., and Marahiel, M.A. (1997). The bacitracin biosynthesis operon of Bacillus licheniformis ATCC 10716: molecular characterization of three multi-modular peptide synthetases. *Chemistry & Biology* 4 (12): 927–937.

Kopp, F. and Marahiel, M.A. (2007). Macrocyclization strategies in polyketide and nonribosomal peptide biosynthesis. *Natural Product Reports* 24 (4): 735–749.

Koumoutsi, A., Chen, X.H., Henne, A. et al. (2004). Structural and functional characterization of gene clusters directing nonribosomal synthesis of bioactive cyclic lipopeptides in *Bacillus amyloliquefaciens* strain FZB42. *Journal of bacteriology* 186 (4): 1084–1096. https://doi.org/10.1128/JB.186.4.1084-1096.2004.

Le Marrec, C., Hyronimus, B., Bressollier, P. et al. (2000). Biochemical and genetic characterization of coagulin, a new antilisterial bacteriocin in the pediocin family of bacteriocins, produced by *Bacillus coagulans* I4. *Applied and Environmental Microbiology* 66 (12): 5213–5220.

Liu, L., Jin, X., Lu, X. et al. (2023). Mechanisms of surfactin from *Bacillus subtilis* SF1 against *Fusarium foetens*: a novel pathogen inducing potato wilt. *Journal of Fungi* 9 (3): 367.

Liu, W.W., Wei, M.U., Zhu, B.Y. et al. (2008). Antagonistic activities of volatiles from four strains of *Bacillus spp.* and *Paenibacillus* spp. against soil-borne plant pathogens. *Agricultural Sciences in China* 7 (9): 1104–1114.

Loiseau, C., Schlusselhuber, M., Bigot, R. et al. (2015). Surfactin from *Bacillus subtilis* displays an unexpected anti-Legionella activity. *Applied Microbiology and Biotechnology* 99: 5083–5093.

Luna-Bulbarela, A., Tinoco-Valencia, R., Corzo, G. et al. (2018). Effects of bacillomycin D homologues produced by *Bacillus amyloliquefaciens* 83 on growth and viability of Colletotrichum gloeosporioides at different physiological stages. *Biological Control* 127: 145–154.

Lv, Z., Li, R., Shi, C. et al. (2024). The efficient synthesis strategy of bacillomycin D in *Bacillus amyloliquefaciens* fmbJ. *Process Biochemistry* 145: 131–138.

Madslien, E.H., Rønning, H.T., Lindbäck, T. et al. (2013). Lichenysin is produced by most Bacillus licheniformis strains. *Journal of Applied Microbiology* 115 (4): 1068–1080.

Markelova, N. and Chumak, A. (2025). Antimicrobial activity of *Bacillus* cyclic lipopeptides and their role in the host adaptive response to changes in environmental conditions. *International Journal of Molecular Sciences* 26 (1): 336.

Massawe, V.C., Hanif, A., Farzand, A. et al. (2018). Volatile compounds of endophytic *Bacillus* spp. have biocontrol activity against *Sclerotinia sclerotiorum*. *Phytopathology* 108 (12): 1373–1385.

Miethke, M., Klotz, O., Linne, U. et al. (2006). Ferri-bacillibactin uptake and hydrolysis in *Bacillus subtilis*. *Molecular Microbiology* 61 (6): 1413–1427.

Miethke, M., Schmidt, S., and Marahiel, M.A. (2008). The major facilitator superfamily-type transporter YmfE and the multidrug-efflux activator Mta mediate bacillibactin secretion in *Bacillus subtilis*. *Journal of Bacteriology* 190 (15): 5143–5152.

Mitra, R., Ghosh, S., Mukherjee, G., and Acharya Chowdhury, A. (2023). Secondary metabolites: treasure trove for future medicine. In: *Plant Specialized Metabolites: Phytochemistry, Ecology and Biotechnology* (ed. J.-M. Mérillon and K.G. Ramawat), 1–45. Cham: Springer Nature Switzerland.

Mootz, H.D., Finking, R., and Marahiel, M.A. (2001). 4′-Phosphopantetheine transfer in primary and secondary metabolism of *Bacillus subtilis*. *Journal of Biological Chemistry* 276 (40): 37289–37298.

Moshe, M., Gupta, C.L., Jain, R.M. et al. (2023). Comparative genomics of *Bacillus cereus sensu lato* spp. biocontrol strains in correlation to in-vitro phenotypes and plant pathogen antagonistic capacity. *Frontiers in Microbiology* 14: 996287.

Moyne, A.L., Cleveland, T.E., and Tuzun, S. (2004). Molecular characterization and analysis of the operon encoding the antifungal lipopeptide bacillomycin D. *FEMS Microbiology Letters* 234 (1): 43–49.

Müller, S., Strack, S.N., Hoefler, B.C. et al. (2014). Bacillaene and sporulation protect *Bacillus subtilis* from predation by Myxococcus xanthus. *Applied and Environmental Microbiology* 80 (18): 5603–5610.

Nakano, M.M., Marahiel, M.A., and Zuber, P. (1988). Identification of a genetic locus required for biosynthesis of the lipopeptide antibiotic surfactin in *Bacillus subtilis*. *Journal of Bacteriology* 170 (12): 5662–5668.

Nitschke, M., Araújo, L.V., Costa, S.G.V.A.O. et al. (2009). Surfactin reduces the adhesion of food-borne pathogenic bacteria to solid surfaces. *Letters in Applied Microbiology* 49 (2): 241–247.

Ogura, M., Yamaguchi, H., Yoshida, K.I. et al. (2001). DNA microarray analysis of *Bacillus subtilis* DegU, ComA and PhoP regulons: an approach to comprehensive analysis of *B. subtilis* two-component regulatory systems. *Nucleic Acids Research* 29 (18): 3804–3813.

Okada, M., Nakamura, Y., Hayashi, S. et al. (2015). Chemical structure and biological activity of a quorum sensing pheromone from *Bacillus subtilis* subsp. natto. *Bioorganic & Medicinal Chemistry Letters* 25 (19): 4293–4296.

Olishevska, S., Nickzad, A., and Déziel, E. (2019). *Bacillus* and *Paenibacillus* secreted polyketides and peptides involved in controlling human and plant pathogens. *Applied Microbiology and Biotechnology* 103: 1189–1215.

Oman, T.J. and Van Der Donk, W.A. (2010). Follow the leader: the use of leader peptides to guide natural product biosynthesis. *Nature Chemical Biology* 6 (1): 9–18.

Ongena, M. and Jacques, P. (2008). *Bacillus lipopeptides*: versatile weapons for plant disease biocontrol. *Trends in Microbiology* 16 (3): 115–125.

Pandey, C., Prabha, D., Negi, Y.K. et al. (2023). Macrolactin A mediated biocontrol of *Fusarium oxysporum* and *Rhizoctonia solani* infestation on *Amaranthus hypochondriacus* by *Bacillus subtilis* BS-58. *Frontiers in Microbiology* 14: 1105849.

Park, J.W., Balaraju, K., Kim, J.W. et al. (2013). Systemic resistance and growth promotion of chili pepper induced by an antibiotic producing *Bacillus vallismortis* strain BS07. *Biological Control* 65 (2): 246–257.

Patil, N., Raghu, S., Mohanty, L. et al. (2024). Rhizosphere bacteria isolated from medicinal plants improve rice growth and induce systemic resistance in host against pathogenic fungus. *Journal of Plant Growth Regulation* 43 (3): 770–786.

Perchat, S., Dubois, T., Zouhir, S. et al. (2011). A cell–cell communication system regulates protease production during sporulation in bacteria of the Bacillus cereus group. *Molecular Microbiology* 82 (3): 619–633.

Piel, J. (2010). Biosynthesis of polyketides by trans-AT polyketide synthases. *Natural Product Reports* 27 (7): 996–1047.

Pieterse, C.M., Zamioudis, C., Berendsen, R.L. et al. (2014). Induced systemic resistance by beneficial microbes. *Annual Review of Phytopathology* 52 (1): 347–375.

Poveda, J. (2021). Beneficial effects of microbial volatile organic compounds (MVOCs) in plants. *Applied Soil Ecology* 168: 104118.

Quadri, L.E., Weinreb, P.H., Lei, M. et al. (1998). Characterization of Sfp, a *Bacillus subtilis* phosphopantetheinyl transferase for peptidyl carrier protein domains in peptide synthetases. *Biochemistry* 37 (6): 1585–1595.

Ramamoorthy, V., Raguchander, T., and Samiyappan, R. (2002). Enhancing resistance of tomato and hot pepper to Pythium diseases by seed treatment with fluorescent pseudomonads. *European Journal of Plant Pathology* 108: 429–441.

Ranjan, A., Rajput, V.D., Prazdnova, E.V. et al. (2023). Nature's antimicrobial arsenal: non-ribosomal peptides from PGPB for plant pathogen biocontrol. *Fermentation* 9 (7): 597.

Rausch, C., Hoof, I., Weber, T. et al. (2007). Phylogenetic analysis of condensation domains in NRPS sheds light on their functional evolution. *BMC Evolutionary Biology* 7: 1–15.

Rodrigo, S., García-Latorre, C., & Santamaria, O. (2022). Metabolites Produced by Fungi against Fungal Phytopathogens: Review, *Implementation and Perspectives. Plants*, 11(1), 81. https://doi.org/10.3390/plants11010081

Roongsawang, N., Washio, K., and Morikawa, M. (2011). Diversity of nonribosomal peptide synthetases involved in the biosynthesis of lipopeptide biosurfactants. *International Journal of Molecular Sciences* 12 (1): 141–172.

Salzberg, L.I., Botella, E., Hokamp, K. et al. (2015). Genome-wide analysis of phosphorylated PhoP binding to chromosomal DNA reveals several novel features of the PhoPRmediated phosphate limitation response in *Bacillus subtilis*. *Journal of Bacteriology* 197: 1492–1506.

Sansinenea, E. and Ortiz, A. (2011). Secondary metabolites of soil *Bacillus* spp. *Biotechnology Letters* 33: 1523–1538.

Saran, A., Imperato, V., Fernandez, L. et al. (2020). Phytostabilization of polluted military soil supported by bioaugmentation with PGP-trace element tolerant bacteria isolated from Helianthus petiolaris. *Agronomy* 10 (2): 204.

Schmidt, R., Cordovez, V., De Boer, W. et al. (2015). Volatile affairs in microbial interactions. *The ISME Journal* 9 (11): 2329–2335.

Schneider, K., Chen, X.H., Vater, J. et al. (2007). Macrolactin is the polyketide biosynthesis product of the pks2 cluster of *Bacillus amyloliquefaciens* FZB42. *Journal of Natural Products* 70 (9): 1417–1423.

Shleeva, M.O., Kondratieva, D.A., and Kaprelyants, A.S. (2023). Bacillus licheniformis: a producer of antimicrobial substances, including antimycobacterials, which are feasible for medical applications. *Pharmaceutics* 15 (7): 1893.

Silo-Suh, L.A., Stabb, E.V., Raffel, S.J., and Handelsman, J. (1998). Target range of zwittermicin A, an aminopolyol antibiotic from Bacillus cereus. *Current Microbiology* 37: 6–11.

Smith, S. and Tsai, S.C. (2007). The type I fatty acid and polyketide synthases: a tale of two megasynthases. *Natural Product Reports* 24 (5): 1041–1072.

Sumi, C.D., Yang, B.W., Yeo, I.C., and Hahm, Y.T. (2015). Antimicrobial peptides of the genus Bacillus: a new era for antibiotics. *Canadian Journal of Microbiology* 61 (2): 93–103.

Tahir, H.A.S., Gu, Q., Wu, H. et al. (2017). Effect of volatile compounds produced by *Ralstonia solanacearum* on plant growth promoting and systemic resistance inducing potential of *Bacillus volatiles*. *BMC Plant Biology* 17: 1–16.

Tan, W., Yin, Y., and Wen, J. (2022). Increasing fengycin production by strengthening the fatty acid synthesis pathway and optimizing fermentation conditions. *Biochemical Engineering Journal* 177: 108235.

Tariq, A. and Ahmed, A. (2024). Phosphate solubilization potential of PSB: an advance approach to enhance phosphorous availability for phytostimulation. *Environmental Science and Pollution Research* 31 (44): 56174–56193.

Tendulkar, S.R., Saikumari, Y.K., Patel, V. et al. (2007). Isolation, purification and characterization of an antifungal molecule produced by Bacillus licheniformis BC98, and its effect on phytopathogen Magnaporthe grisea. *Journal of Applied Microbiology* 103 (6): 2331–2339.

Tian, D., Song, X., Li, C. et al. (2021). Antifungal mechanism of *Bacillus amyloliquefaciens* strain $GKT0_4$ against Fusarium wilt revealed using genomic and transcriptomic analyses. *Microbiology* 10 (3): e1192.

Tsuge, K., Ano, T., Hirai, M. et al. (1999). The genes *degQ*, *pps*, and *lpa-8* (*sfp*) are responsible for conversion of *Bacillus subtilis* 168 to plipastatin production. *Antimicrobial Agents and Chemotherapy* 43 (9): 2183–2192.

Tyc, O., Zweers, H., De Boer, W., and Garbeva, P. (2015). Volatiles in inter-specific bacterial interactions. *Frontiers in Microbiology* 6: 1412.

Vahidinasab, M., Adiek, I., Hosseini, B. et al. (2022). Characterization of *Bacillus velezensis* UTB96, demonstrating improved lipopeptide production compared to the strain *B. velezensis* FZB42. *Microorganisms* 10 (11): 2225.

Vanittanakom, N., Loeffler, W., Koch, U., and Jung, G. (1986). Fengycin-a novel antifungal lipopeptide antibiotic produced by *Bacillus subtilis* F-29-3. *The Journal of Antibiotics* 39 (7): 888–901.

Vivekananthan, R., Ravi, M., Saravanakumar, D. et al. (2004). Microbially induced defense related proteins against postharvest anthracnose infection in mango. *Crop Protection* 23 (11): 1061–1067.

Volpon, L., Besson, F., and Lancelin, J.M. (2000). NMR structure of antibiotics plipastatins A and B from *Bacillus subtilis* inhibitors of phospholipase A2. *FEBS Letters* 485 (1): 76–80.

Von Tersch, M.A. and Carlton, B.C. (1983). Bacteriocin from *Bacillus megaterium* ATCC 19213: comparative studies with megacin A-216. *Journal of Bacteriology* 155 (2): 866–871.

Wang, P., Guo, Q., Ma, Y. et al. (2015). DegQ regulates the production of fengycins and biofilm formation of the biocontrol agent *Bacillus subtilis* NCD-2. *Microbiological Research* 178: 42–50.

Wang, X., Hao, G., Zhou, M. et al. (2023). Secondary metabolites of *Bacillus subtilis* L2 show antiviral activity against pseudorabies virus. *Frontiers in Microbiology* 14: 1277782. https://doi.org/10.3389/fmicb.2023.1277782.

Wang, J., Zhang, X., Zhang, Y. et al. (2024). Formulation of the microbial agent *Bacillus amyloliquefaciens* NCPSJ7 and its application against *Botrytis cinerea* in tomato. *European Journal of Plant Pathology* 169 (3): 611–623.

Wei, G., Kloepper, J.W., and Tuzun, S. (1996). Induced systemic resistance to cucumber diseases and increased plant growth by plant growth-promoting rhizobacteria under field conditions. *Phytopathology* 86 (3): 221–224.

Willey, J.M. and Van Der Donk, W.A. (2007). Lantibiotics: peptides of diverse structure and function. *Annual Review of Microbiology* 61 (1): 477–501.

Wilson, K.E., Flor, J.E., Schwartz, R.E. et al. (1987). Difficidin and oxydifficidin: novel broad spectrum antibacterial antibiotics produced by *bacillus subtilis* II. Isolation and physico-chemical characterization. *The Journal of Antibiotics* 40 (12): 1682–1691.

Wu, C.Y., Chen, C.L., Lee, Y.H. et al. (2007). Nonribosomal synthesis of fengycin on an enzyme complex formed by fengycin synthetases. *Journal of Biological Chemistry* 282 (8): 5608–5616.

Wu, L., Huang, Z., Li, X. et al. (2018). Stomatal closure and SA-, JA/ET-signaling pathways are essential for *Bacillus amyloliquefaciens* FZB42 to restrict leaf disease caused by *Phytophthora nicotianae* in *Nicotiana benthamiana*. *Frontiers in Microbiology* 9: 847.

Wu, J., Kamal, N., Hao, H. et al. (2019). Endophytic *Bacillus megaterium* BM18-2 mutated for cadmium accumulation and improving plant growth in hybrid Pennisetum. *Biotechnology Reports* 24: e00374.

Wu, Q., Zhi, Y., and Xu, Y. (2019). Systematically engineering the biosynthesis of a green biosurfactant surfactin by *Bacillus subtilis* 168. *Metabolic Engineering* 52: 87–97.

Yakimov, M.M. and Golyshin, P.N. (1997). ComA-dependent transcriptional activation of lichenysin A synthetase promoter in *Bacillus subtilis* cells. *Biotechnology Progress* 13 (6): 757–761.

Yan, Z., Reddy, M.S., Ryu, C.M. et al. (2002). Induced systemic protection against tomato late blight elicited by plant growth-promoting rhizobacteria. *Phytopathology* 92 (12): 1329–1333.

Yang, J.W., Yu, S.H., and Ryu, C.M. (2009). Priming of defense-related genes confers root-colonizing bacilli-elicited induced systemic resistance in pepper. *The Plant Pathology Journal* 25 (4): 389–399.

Ye, Y.F., Li, Q.Q., Gang, F.U. et al. (2012). Identification of antifungal substance (Iturin A_2) produced by *Bacillus subtilis* B47 and its effect on southern corn leaf blight. *Journal of Integrative Agriculture* 11 (1): 90–99.

Ye, M., Liu, Z., Gao, C. et al. (2022). Iturin A rescued STb-R-induced pork skeletal muscle growth restriction through the hypothalamic-pituitary-mTORC1 growth axis. *Animals* 12 (12): 1568.

Yu, X., Ai, C., Xin, L., and Zhou, G. (2011). The siderophore-producing bacterium, *Bacillus subtilis* CAS15, has a biocontrol effect on Fusarium wilt and promotes the growth of pepper. *European Journal of Soil Biology* 47 (2): 138–145.

Yu, Y., Gui, Y., Li, Z. et al. (2022). Induced systemic resistance for improving plant immunity by beneficial microbes. *Plants* 11 (3): 386.

Yu, C., Qiao, J., Ali, Q. et al. (2023). degQ associated with the degS/degU two-component system regulates biofilm formation, antimicrobial metabolite production, and biocontrol activity in Bacillus velezensis DMW_1. *Molecular Plant Pathology* 24 (12): 1510–1521.

Zhang, F., Huo, K., Song, X. et al. (2020). Engineering of a genome-reduced strain *Bacillus amyloliquefaciens* for enhancing surfactin production. *Microbial Cell Factories* 19: 1–13.

Zhang, N., Wang, Z., Shao, J. et al. (2023). Biocontrol mechanisms of Bacillus: improving the efficiency of green agriculture. *Microbial Biotechnology* 16 (12): 2250–2263.

Zhen, C., Ge, X.F., Lu, Y.T., and Liu, W.Z. (2023). Chemical structure, properties and potential applications of surfactin, as well as advanced strategies for improving its microbial production. *AIMS Microbiology* 9 (2): 195–217.

20

Regulation of Seed Size and Yield in Plant by Peptides Hormones

Deva Andrews Arulsoosairaj[1], *Kanagaraj Muthu-Pandian Chanthini*[2], *and Sengottayan Senthil-Nathan*[1]

[1] *Division of Bio-pesticides and Environmental Toxicology, Sri Paramakalyani Centre for Excellence in Environmental Sciences, Manonmaniam Sundaranar University, Tirunelveli, Tamil Nadu, India*
[2] *Department of Pharmacology, Saveetha Institute of Medical and Technical Sciences (SIMATS), Chennai, Tamil Nadu, India*

20.1 Introduction

Seed size is a significant factor in determining plant quality of growth and farm yield in agriculture. It is essential to the health, survival, and utilization of seedlings at the early stages of development in intensive agriculture as well as in natural communities (Reed et al. 2022). Large seeds contain higher amounts of starch, protein, and lipids that contribute to the strengthening and aggressiveness of seedlings for light, water, and other nutrients (Aguirre et al. 2018). As previously stated, the size of the seed is not just a trait but the outcome of a number of developmental processes intricately organized at the genetic, cellular, and physiological levels. This trait is critical in determining the productivity of a number of crops, especially cereals, legumes, and oilseeds (Dwivedi et al. 2021). Because the seed participates in determining the grain weight, it directly influences the yield per hectare. For example, in wheat and rice, differences in seed length or width are accompanied by substantial increases in yield (Khush 2013). Seed size is also a major commercial characteristic influencing milling quality, market grade, and nutritional content, and is thus a crop target in breeding projects.

The correlation between yield and seed size is not uniform; negotiations are usually involved. For instance, seed size per plant can be raised through reduction in seed number per plant as a result of resource limitation, and this determines overall yield (Clark et al. 2007). Knowing how seed sizes are regulated is important for plant developmental biology as a basic science. It is also important for planning sustainable ways to achieve higher overall yields. Moreover, food demand is increasing globally as a result of population growth, dietary change, and shifting climate factors. Production of food needs to increase by an estimated 60–70% by 2050 to support more than 9 billion people (Hall et al. 2017). This critical requirement for increased crop production, without further environmental degradation, has put pressure on plant scientists to look for novel biological regulators.

Peptide Hormones in Plant Development and Stress Responses, First Edition. Edited by Vijay Pratap Singh and Samiksha Singh.

Over decades, studies on plant hormones provided the basis for how growth and development at the whole plant level are controlled by some hormones produced in the plant system. These hormones have assigned roles during embryogenesis, endosperm formation, seed filling, and maturation (Locascio et al. 2014). Regulation of plant growth regulators or phytohormones like auxins, gibberellins (GAs), cytokinins (CKs), abscisic acid (ABA), ethylene, jasmonates (JAs), and brassinosteroids (BRs) is crucial for the physiological well-being and reproductive success of plants (Samanta and Roychoudhury 2025). Auxins regulate embryo formation and its tissue orientation. They can initiate or inhibit germination based on their concentration and how they interact with other hormones like ABA (Matilla 2020). GAs are effective stimulators of germination. They induce embryo growth, soften the endosperm, and help in syntheses enzymes necessary to mobilize energy (Tuan et al. 2018). CKs induce cell division and differentiation. They also facilitate seed breaking of dormancy, particularly under light (Jameson and Song 2016). ABA is important during the later stages of seed development. It significantly prevents germination to ensure dormancy and limits growth under stressful conditions (Vishal and Kumar 2018).

Ethylene promotes germination by inhibiting ABA metabolites and aiding radicle protrusion, mainly under stress conditions. JAs delay the germination and interact with ABA and GA signaling to maintain stress dormancy (Linkies and Leubner-Metzger 2012). BRs improve germination by increasing GA signaling and reducing ABA activity, helping plants cope with environmental stress (Ahanger et al. 2018). These traditional hormones usually function via complex feedback and cross-talk mechanisms to produce complex regulatory networks. For example, the interaction between auxins and GAs regulates coordination between embryonic growth and nutrient mobilization. However, these hormones usually work at systemic levels and lack the spatial accuracy needed for some developmental adaptations (Mazzoni-Putman et al. 2021). Their broad action range can also make them hard to control without causing unintended effects. With the increasing abiotic stress from climate change, managing hormones has become essential for stable plant yields. The classical hormones are tiny signal molecules that regulate cell processes and transitions in development (Bhattacharya 2022). Due to this limitation, scientists have been fascinated with a new class of regulators called peptide hormones, which gain context-dependent and consistent control during seed development.

Over the last two decades, the understanding of plant hormonal signaling has been greatly revolutionized. We now recognize peptide hormones to be significant regulators of development (Santner et al. 2009). The peptides were first related to functions in root development, meristem control, and defense mechanisms. Currently, scientists are progressively focusing on peptide hormones for their activities in reproductive and seed developmental processes (Hsiao and Yamada 2020). These peptides are short which typically consisting of 5–20 amino acids in their active forms, coded from longer precursor genes (Klein et al. 2018). These are modified after translation and cleaved when activated. Unlike regular hormones that circulate through the bloodstream and produce effects diffusely, peptide hormones are secreted locally. The hormones are strongly spatially specific and help the accurate management of developmental processes within microscopic tissue domains (Al-Suhaimi 2022). They respond through membrane-bound receptor-like kinases (RLKs) or receptor-like proteins (RLPs) with a strong binding affinity to the peptides. This interaction triggers phosphorylation cascades and responses in gene expression (Liang and

Zhou 2018). The very first discovered peptide hormone was CLAVATA3 (CLV3), which is a member of the CLAVATA3/EMBRYO SURROUNDING REGION-related (CLE) family and acts to regulate shoot meristem stem cell maintenance. It was not long before other peptides like inflorescence deficient in abscission (IDA) and phytosulfokine (PSK) were found to regulate organ shedding and cell proliferation, respectively (Lu and Xiao 2024). Recently, C-terminally encoded peptide (CEP), rapid alkalinization factor (RALF), and Casparian strip integrity factor (CIF) members have been found to be involved in the perception of nutrients, pH regulation, and maintenance of tissue borders (Yadav 2024). They are being unraveled to be involved in seed activity as well. For example, CLE8 is involved in embryo and endosperm development in early stages, assisting in regulating their coordinated development. CEPs regulate shoot-root communication during times of fluctuating nitrogen supply, thereby governing resource allocation during reproductive growth (Taleski et al. 2024).

RALF peptides control cell wall acidification and expansion. Thus it is affecting seed filling to reach the final size (Araya et al. 2016). These findings advance our understanding of peptide hormones as key factors that collaborate with and in some ways modify standard hormonal systems. Their integration into the existing network of hormones in the plants has added to the multifunctionality and precision that governs plant developmental processes, particularly in the enhancement of seed development and yield improvement. This chapter aims to provide a consolidated and analytic insight into the role of peptide hormones in regulating seed size and yield in plants. Their recognition enriches the narrative of plant developmental biology on a micro scale as it presents another layer to how plant structures and processes are modulated in a given region.

20.2 Peptide Hormones in Plants

Plant peptide hormones are yet another distinctive set of signal molecules consisting of short amino acid sequences typically between 5 and 20 residues in their mature active state. Unlike the larger or enzyme-like regulators that are protein hormones, the peptide hormones are typically secreted from large precursor proteins referred to as prepropeptides (Frayn 2013). These precursors are cleaved, folded, and posttranslationally modified to yield bioactive peptides that can possibly function as exceedingly selective messengers between plant tissue cells. They can function locally at or near the point of synthesis, typically influencing proximal cells in a tissue-specific fashion. In opposition to the diffusible, systemic activity of traditional hormones like auxins or GAs, peptide hormones tend to act through specific ligand–receptor interactions (Novák et al. 2017). They are secreted out to the extracellular space (apoplast) and sensed by receptor kinases embedded in the plasma membrane of neighboring cells. This local and highly specialized action has the ability to provide tight regulation of cell events such as division, differentiation, and growth, which are central to seed germination and organ development (Sanchez-Corrionero et al. 2023). Peptide hormones are readily organized yet complicated in specificity; they have a tendency to regulate one process or set of developmental options. Though brief in size, they function in an extremely broad array of biological processes ranging from meristem homeostasis to reproductive organ abscission and stress signaling (Sawicki et al. 2015).

20.2.1 Peptide Hormone Classification and Its Role

Plant peptide hormones are not a monotonous class. They contain a number of structurally and functionally distinct families, each defined by sequence motifs, processing needs, and biological functions (Zhang et al. 2025). The families of peptides and their particular functions are presented in Table 20.1.

Table 20.1 The peptide families with specified receptor and related function in the plant system.

Families	Example peptides	Amino acid (approximate size)	Binding receptors(s)	Key functions	References
CLAVATA3/ ESR-related (CLE)	CLV3, CLE40, CLE41/ TDIF	~12–14	CLV1, CLV2-CRN, BAM1/2/3, RPK2, ACR4	Postgermination, stem cell regulation, meristem maintenance, vascular development	Fiume (2010)
C-terminally encoded peptide (CEP)	AtCEP1, MtCEP1, OsCEP5/6	~15	CEPR1, CEPR2	Nitrogen signaling, root development, nodulation, and aiding seedling establishment	Taleski et al. (2018)
Phytosulfokine (PSK)	PSK-α, PSK-γ/δ/ε	5	PSKR1, PSKR2	Cell proliferation during germination, growth, stress tolerance	Li et al. (2024)
Rapid alkalinization factor (RALF)	RALF1, RALF23, RALF4/19	~49–50	FERONIA, ANX1/2, BUPS1/2, LLG1–3	pH modulation, immunity, pollen tube integrity, and facilitating radicle emergence	Zhang et al. (2023)
Systemin (SYS)	Tomato systemin	18	SYR1, SYR2, SR160, SlSERKs	Wound response, jasmonic acid signaling	Zhou et al. (2025)
Plant elicitor peptide (PEP)	AtPEP1, ZmPEP1	~23	PEPR1, PEPR2	DAMP-triggered immunity, defense amplification during germination	Safaeizadeh and Boller (2019)
Inflorescence deficient in abscission (IDA/IDL)	AtIDA, IDL1-IDL8	~20	HAE, HSL2	Organ abscission, cell separation during radicle emergence	Shi et al. (2019)

Table 20.1 (Continued)

Families	Example peptides	Amino acid (approximate size)	Binding receptors(s)	Key functions	References
Casparian strip integrity factor (CIF)	CIF1, CIF2	~21	SGN3 (GSO1), SGN1	Casparian strip integrity in root endodermis	Cheng et al. (2025)
Root meristem growth factor (RGF/GLV/CLEL)	RGF1, GLV1	~13–14	RGI1-RGI5, SERKs	Root stem cell niche maintenance postgermination, PLT stabilization	Yamoune et al. (2021)
Epidermal patterning factor (EPF/EPFL)	EPF1, EPFL9	~45–50	ER, ERL1/2	Stomatal development, epidermal patterning	Hunt et al. (2010)
LURE peptides (LURE)	LURE1	~60–70	PRK6, MDIS1, MIK1	Pollen tube guidance	Higashiyama and Yang (2017)
S-Locus cysteine-rich protein (SCR/SP11)	SCR/SP11	~50–60	SRK	Self-incompatibility in pollen	Fobis-Loisy et al. (2012)
Thionin (THI)	Thionin	~45	Unknown	Antimicrobial activity	Höng et al. (2021)
Defensin (DEF)	PDF1.2	~45–55	Unknown	Pathogen defense	Zhao et al. (2025)
Plant sulfated peptide (PSY)	PSY1	~18–20	PSY1R	Early cell expansion and immune priming	Zhang et al. (2024a, b)
Pathogen-induced peptide (PIP)	PIP1	~23	PEPRs	Defense signaling	Najafi et al. (2020)
Plant natriuretic peptide (PNP)	AtPNP-A	~45–50	PNP-R1	Water and ion homeostasis	Lee et al. (2020)
***Zea mays* immune peptide (ZIP1)**	ZIP1	~30–40	Unknown	Maize immune signaling	Ziemann (2017)
Germin-like protein-related peptide (GRI)	GRI	~50–60	Unknown	Cell death regulation	Cruz (2019)
MicroRNA-encoded peptides (miPEPs)	miPEP171d, miPEP858a	~20–30	Unknown	Root development, flavonoid biosynthesis	Ormancey et al. (2021)

Together, these collections of peptides form a complex and precisely regulated signaling system that integrates the surroundings, the stage of development, and hormonal interplay. Certainly, the number of identified peptide hormones is increasing with the identification of newer families and novel peptides due to numerous transcriptomic, proteomic, and genetic analyses conducted on a wide variety of model and crop plant species (Zhang et al. 2025).

20.2.2 Biosynthesis, Posttranslational Modifications, and Transport of Peptides

Plant peptide hormone biosynthesis begins with the transcription of genes that produce peptides. This transcription results in a longer inactive precursor, a prepropeptide. The precursor peptides undergo N-terminal modifications where they are tagged with signal peptides, which lead them to the secretion pathway. While signal sequences are removed during the reception of the endoplasmic reticulum (ER), the more refined Golgi apparatus can perform further modifications (Gancheva et al. 2019). The propeptides are cleaved proteolytically at specific recognition sites by metacaspases or subtilisin-like proteases to generate mature peptides. Posttranslational modifications (PTMs) play a vital role in the activity, stability, and receptor-binding capacity of peptide hormones (Boon et al. 2020). Tyrosine sulfation, proline hydroxylation, and arabinosylation are some common modifications. For instance, sulfation of two tyrosine residues is necessary for the bioactivity of PSK. Without such a modification, the peptide is not effectively bound to its receptor. Hydroxylation of proline residues of CLE peptides also affects the structural conformation and biological activity of the peptides. After maturation, the peptides are released to the apoplastic space where they diffuse a short distance to bind onto their cognate receptors on adjacent cells (Wang and Fiers 2010). The localized character of this system of signaling imparts specificity and prevents the general systemic effects as opposed to the classical hormones that circulate via vascular tissues. Some examples of long-distance peptide transport have been noted, such as in the case of CEP peptides that are able to transport through the xylem from shoots to roots, thus taking part in systemic nutrient signaling (Okamoto et al. 2016).

20.2.3 Peptide–Receptor Interactions

Peptide hormones act by binding to plasma membrane-bound receptors, which are predominantly members of the leucine-rich repeat receptor-like kinase (LRR-RLK) family (Zhong and Qu 2019). The receptors have extracellular domains that bind specific peptide motifs, a transmembrane domain that keeps them bound on the membrane, and a cytoplasmic kinase domain that triggers intracellular signal transduction once they bind to the ligand. Peptide and its binding receptor are represented in Table 20.1. Spatial extent of peptide action in most cases is determined by receptor expression patterns and localization. This is especially the case in seed development, when different tissues like embryo, endosperm, and maternal integuments need to communicate with one another in a highly specific manner to initiate coordinated growth (Figueiredo and Sharma 2025). At early phases of seed germination, plants utilize a sophisticated network of peptide hormones in mediating transitions of development and defense preparedness. PSK and plant sulfated peptide (PSY) families are involved in the commencement of cell expansion and division. PSK activates its receptors PSKR1 and PSKR2 to enhance embryo tissue growth, whereas PSY peptides induce immune priming and early cell enlargement through the action of PSY1R, especially during imbibition and radicle emergence (Li et al. 2024). Peptides such as RALF help by regulating the pH of the cell wall. This makes it easier for the seed coat to loosen, allowing the radicle to grow. This is controlled by the FERONIA receptor, which also controls the stress response during germination (Cheung 2024). PEPs and miPEPs, on the other hand, prepare seedlings for stress. PEPs activate PEPR1 and PEPR2 to improve immune readiness. In contrast, miPEPs influence gene expression for root and shoot growth by triggering the transcription of specific miRNAs (Xiao et al. 2025).

Morphological developments in germinating seeds are also elicited by IDA. They control the separation of cells through HAE and HSL2 receptors that could be involved in rupture of the seed coat (Kumpf et al. 2013). Root meristem growth factor (RGF) peptides are also involved in stabilizing critical transcription factors such as PLT, which preserves stem cell integrity needed for root organization during postgermination. Even though not directly involved in stimulating germination, families like CLE and CEP regulate early seedling growth through establishing meristem identity and regulating root systems according to nutrient supply (Delay 2015). Overall, these peptide families make up a dynamic signaling network wherein growth, immunity, and environmental plasticity converge to create developmental outputs like seed size and rate of maturation. There have also been discovered novel noncanonical receptors and receptor complexes, which suggest that the perception of the peptide hormones is more malleable and modular than may have been expected. The increasing peptide-receptor pairs indicate a dynamic and intricate signaling network, including internal developmental signals and external environmental stimuli. Some other peptides like THI and DEF are antimicrobial peptides (AMPs) and no receptors have yet been found (Mookherjee et al. 2020). *Zea mays* ZIP1 triggers salicylic acid–mediated defense. GRI has a role in priming responses. miPEPs regulate pri-miRNA expression through intracellular action, not through membrane-bound receptors (Jie et al. 2021).

20.2.4 Interaction among Peptide Hormones and Classical Phytohormones

Even though both peptide hormones and classic phytohormones have roles to play in the development of plant seeds, their actions differ greatly in their scope and specificity. Phytohormones, which include auxin, cytokinin, and gibberellin, are small organic molecules that are capable of transporting themselves in the vascular plant tissues and, therefore, have the ability to influence growth from a distance. They frequently act through nuclear receptors and transcriptional reprogramming that alter multiple developmental pathways (Gasperini and Howe 2024). In contrast, peptide hormones act through membrane-bound receptors and are typically restricted to localizing short-range signaling. This allows for higher spatial and precise control. Their effects are often binary, turning specific methods on or off in well-defined contexts such as boundary formation, problem layout, or organ abscission. Besides these differences, peptide and standard hormones oftentimes interact (Adams 2019). Specifically, PSK and BRs jointly promote cell proliferation, while RALF peptides modulate auxin-induced proton efflux by altering apoplastic pH. Such interactions contribute to the fine-tuning of developmental processes, including those that order source development and render. The coexistence of both types of hormones allows plants to maintain a balance between systemic coordination and localized control (Bhat 2021). This dual-layered sign structure is specifically useful during reproductive and source growth, where the timing and position of pitted events need to work meticulously and orderly. As research advances, it is becoming increasingly clear that peptide hormones are not secondary messengers or auxiliary regulators. Alternatively, they are essential signaling molecules that full complement and frequently conduct the activity of standard phytohormones in stylish developmental methods such as equally source organization and development.

20.3 A Hormonal Role in Seed Development and Yield

Angiosperm seed development starts shortly after double fertilization and continues toward developmentally regulated embryogenesis, endosperm formation, and seed coat formation. These are all based on genetically regulated time frames and hormonally regulated processes with a basis for ultimate seed yield and size potential. Embryogenesis begins with gamete fusion and terminates with zygotic patterning and division. Polarization at the first molecular level, meristem identity, and layering of tissues are all procedures that occur in embryonic growth and ultimately control the structural architecture of the plant subsequently (Brevini and Georgia 2012). Endosperm growth is characterized by the formation of a distinct triploid tissue providing the developing embryo with sustenance. The endosperm forms as a syncytium, becomes cellularized, and forms a reservoir of nutrients. At last, the seed coat, being maternal integuments, becomes lignified as well as dehydrated to enclose the embryo and control dormancy (Olsen and Becraft 2013). For all these, each one of them is individually regulated by hormonal signals, i.e. auxins, CKs, and GAs, coordinating the timing of development with internal as well as external cues. Mutations in the signaling pathway interfering with these signals would most probably result in aborted seeds or smaller seeds, thereby evidencing the function of hormonal control of seed development. Finally, hormonal peptides are end regulators, controllers, gap fillers for hormonal cross talk, space communication, standby functions, and environmental responsiveness of classical hormones like auxins, CKs, GAs, and ABA (Yetgin et al. 2025). Their function is most important in intercellular and positional seed germination and yield determination. The hormonal control is stringent at this stage, and the process of developing the seeds becomes sturdy, efficient, and in sync with the reproduction success (Kurepa and Smalle 2023).

20.3.1 Role of Hormones in Each Proliferation Developmental Stage

Auxins establish concentration gradients during embryogenesis to induce apical-basal axis formation and cotyledon outgrowth. However, they also stimulate cell division and differentiation in the meristematic zones of the embryo (Su et al. 2011). GAs regulate elongation and mobilization of reserves, whereas ABA plays a crucial role in initiating dormancy and desiccation tolerance (Vishal and Kumar 2018). Meanwhile, the peptides CLE8 and RGF1 control cell division rates and meristem initiation. Tracheary element differentiation inhibitory factor (TDIF) ensures development and tissue patterning, positive regulation of the proper cell fate decisions (Leasure and He 2012). During endosperm development, where the embryo's nutritional supply is controlled, the cross talk between auxin and peptide hormones such as CLE and RALF peptides allows cellularization and deposition of storage proteins (Chevalier 2012). ABA helps control starch synthesis, thus stopping seeds from sprouting too early (Zhao et al. 2024). Also, regular hormones like jasmonic acid and ethylene affect the seed coat development, which changes secondary metabolism and makes the seed coat hard. The timing of when the seed coat gets woody is super important because if it happens too early or too late, the seed might not be as healthy. The formation of the seed coat is during which the spatial control of cell expansion and differentiation is necessary and such regulation in then likely to account for the action of the signaling peptides in the LRR-RKs. For example, the RGF family of hormones is known to affect epidermal patterning and could be involved in seed coat development (Matilla 2019). In addition, TWS1, an *Arabidopsis* peptide, controls communication between the embryo and seed coat during development (Fiume et al. 2016). In the postfertilization phase, peptide hormones play a role in vascular patterning and nutrient distribution by managing phloem unloading and source-sink dynamics. This directly affects seed filling and the final seed size.

20.3.2 Yield Determination in Plant Cycle

The yield of seed-bearing plants depends on the size of the seed and the effectiveness of four reproductive processes, such as flowering, pollination, seed set, and seed filling. Flowering, the transition from vegetative to reproductive growth, is regulated by two main factors: the photoperiodic signals and endogenous factors. The hormones include FLOWERING LOCUS T (FT), GAs, and several EPFL peptides (Azevedo et al. 2025). The identification and development of all floral organs is critical for successful pollination. FT and peptide signals are synthesized in specific tissues, which means FT in leaf phloem companion cells, and peptides, including CLE and PSK, in organs such as the root or shoot. They then travel throughout tissues to control development (Lu and Xiao 2024). FT is transported through the plant's phloem systemically, moving all the way to the shoot tip called the shoot apical meristem. There, it interacts with proteins like FD and 14-3-3 to help turn the meristem into a floral one, starting the process of flowering. In a similar way, small peptide hormones often bind to receptor kinases on cell surfaces outside the cell, start internal signals that control how cells divide, specialize, or form different organs (Mishra et al. 2024). This shows that plants use a common way of sending signals over long distances. The signals often come from outside the cell and act somewhere else, relying on specific protein interactions to make sure the message is accurate. Plus, environmental signals can be integrated into these pathways to influence plant development. Sometimes, FT and certain peptide hormones work together to regulate the activity of the meristem or respond to stress. For example, peptides like CLE25 can move through the plant during drought conditions, similar to how FT performs during changes in day length (Ren et al. 2024). Despite their differences in size and origin, these signaling systems have a similar overall aspect.

During pollination, pollen lands on the stigma and grows a pollen tube toward the ovule. This process is not determined only by hormones like auxins and GAs. Small peptides also have important functions. LURE peptides, which are released by ovule cells in order to direct the pollen tube to the ovule itself and allow proper fertilization. Following fertilization, hormonal signals permit the ovule to be transformed into a seed. This starts development and supports the growth of the embryo and endosperm (Qu et al. 2015). The RALF family of peptides helps in regulating pollen tube elongation and abounding that is needed for successful fertilization. The growth from fertilized ovules to mature seeds is controlled by hormones that compete with one another (Kim et al. 2021). Peptides such as IDA and their receptors regulate regions from which organs differentiate, indirectly influencing normal fertilization and seed formation (Shi et al. 2019). During seed filling, hormones like auxin, cytokinin, and peptides regulate the starch, proteins, and fat accumulation, which is responsible for the final size of the seeds and yield. Nutrient availability and hormone signals are key regulators during seed filling. Peptides like CEP help to allow plants to absorb and transport nitrogen and enhance seed filling. Besides, peptides like CLE and PSK can drive blood vessel development and nutrient delivery to the seeds to support their growth (Ren et al. 2024). The vital harvest in plants is largely determined by a series of highly coordinated developmental processes starting from flowering and ending with seed filling and maturation. All these processes are controlled by the elaborate interactions among hormone signals and genetic programs, such as peptide hormones. Taken together, each process from flowering to seed filling is strictly controlled by collective plant hormones and an increasing number of peptide signals. This suggests their function for maximum seed yield. Failure of any process due to environmental stress or genes can intensely lower yield. Therefore, determining how hormones control these processes is important to increasing crop production.

20.4 Specific Peptide Hormones Regulating Seed Size and Yield

20.4.1 CLE Peptides

CLE peptides play an important role in plant growth and development. They have been found to control the formation of large seeds, the growth of embryos, and the overall yield of seeds. They perform through binding to some receptors, often via LRR-RLKs like CLV1, CLV2, and CRN. In the model plant *Arabidopsis*, scientists have found that CLE8 is a key signal in the embryo area. It has the potential to control embryo size by delaying cell growth from an early point, during seed formation (Yu et al. 2025). The mode through which CLE8 signals is through coreceptor proteins CRN and CLV2, which combine to make a signaling complex. This configuration keeps embryo and adjacent tissue growth in check, preventing it from the effect of excessive water (Yamaguchi et al. 2016). In rice, various CLE genes like *OsCLE6* and *OsCLE16* are associated with alterations in the grain number and panicle structure, the area where the seeds are contained. The genes have a direct effect on the number of seeds a plant can yield (Hu et al. 2021). When researchers investigated what might happen if these genes are overactive or silenced, they discovered that disturbing CLE signaling may cause larger embryos and additional seeds overall. That means that CLE peptides should prevent seed development. Because of this, scientists consider CLE peptides as promising targets for crop improvement. By modifying either how much of these peptides is produced or how their receptors are activated, it might be possible to grow bigger seeds and boost yields. This could really open up new opportunities in agriculture to increase food production (Wang et al. 2025).

20.4.2 CEP Peptides

It plays a key role in many aspects of plant growth and functions like shaping roots, signaling nitrogen levels, and helping shoots and roots communicate. Recent studies have even shown that, although indirectly, it helps influence seed size and how much a plant can produce by managing how nutrients are spread out and how growth areas called meristems work. Typically, these signals called CEPs, occur when the plant lacks nutrients, especially nitrogen. They perform like messengers that move from the roots to the shoots, instructing the plant to modify its growth (Taleski et al. 2018). In plant *Arabidopsis*, these CEPs bind to specific receptors called CEPR1 and CEPR2 in the phloem. The phloem is part of the nutrient transport system of a plant. This signaling process allows roots and shoots to share information about their nitrogen needs. It enables the plant to develop in equilibrium and have efficient sharing of resources. Accessing nutrients during flowering and seed formation is important as it affects the sugars and amino acids that seeds use for growth (Taleski et al. 2024). Scientists have found that changing these CEP pathways in rice and maize can help plants use nitrogen more efficiently and distribute biomass well. Rice plants, for example, if genetically engineered to produce additional CEP-related proteins, can take up greater amounts of nitrogen. This leads to increased seed and a higher number of grain formation.

It also improves energy allocation from photosynthesis and accelerates growth in the reproductive areas. Also, CEP peptides bind to proteins called NIN-like proteins (NLPs) and TGA1 transcription factors. They help adjust hormone pathways such as auxin and cytokinin which are important for seed filling and the development of the seed's nutrient-rich part called the endosperm (Ren et al. 2024). While CEPs do not directly control embryo or endosperm development like CLE8 does, they enhance the movement of nutrients and help balance hormones inside the plant. This helps the plant to grow larger seeds and a higher yield (Fiume and Fletcher 2012). Although CEP peptides do not directly affect seed size by acting on the embryo, they play an important role in controlling resource flow and hormone levels of the plant. This makes them useful for indirectly increasing seed yield. As a result, scientists consider CEP signaling as a promising target for breeding and biotech efforts aimed at producing more productive crops, both in ideal conditions and under stress.

20.4.3 RALF Peptides

RALF peptides are small, cysteine-rich signaling molecules secreted by plants that influence many important processes, including cell growth, reproduction, and defense. The recent research has emphasized their key role in reproductive development. They help in the regulation of pollen tube development and the fertilization process, hence influencing seed development, size, and crop yield. RALF4 and RALF19 are essential to the plant processes of pollen tube for plants such as *Arabidopsis*. Mutations of these peptides decrease the process of fertilization and the production of seeds. The RALF-FERONIA (FER) receptor complex is likewise important in fertilization for male–female interaction. FER controls the restructuring of the cell wall and hormone signaling during early seed development and seed growth. RALF-FER signaling inhibits cells from elongating, keeping the growth intact while holding the structure together. The moment this process is terminated, fertilization does not work, and seed growth becomes smaller (Zhang et al. 2020). In addition, a recent study by Zhong et al. (2024) indicates that RALF peptides are associated after fertilization to regulate the embryo and endosperm development that is necessary in seed size determination at maturity. However, direct research on modifying RALF genes to enhance yield is still in its early stages; their crucial roles in the process of fertilization and cell expansion make them a possible thing for methods aiming to improve seed size and quantity.

20.4.4 IDA and IDL Peptides

In seed plant reproduction, IDA and IDL peptide signals are essential for the timely release of floral organs. These signals influence the efficiency of pollination, fertilization, and seed development (Stø et al. 2015). The peptides also play a key role in the timely release of mature seeds, which helps with active dispersal. Recent studies on tomato fruits and canola seeds suggest that IDA and IDL peptide signals can be changed to improve the retention of fruits and seeds. This change could reduce preharvest losses and stabilize overall yield (Wang et al. 2023; Lu and Xiao 2024). The IDA and IDL signals help determine when seeds are released. Disrupting these signals can cause seeds to detach too early or stay attached too long, both of which could affect the reproduction and crop yield. These peptides are crucial for maintaining the timing of seed production and overall crop yield.

20.4.5 PSK Peptides

Current research indicates that PSK signaling is crucial in seed germination and seed development. It has been shown to stimulate growth and increase cell numbers in the ovule and the developing embryo, which affects seed size (Shen et al. 2023). For example, the study by Stührwohldt et al. (2015) found that *Arabidopsis* plants with more PSKR1 could produce larger seeds and longer siliques. This indicates that PSK actually enhances the development of the reproductive organs. Otherwise, the PSK signal crisis causes the plants to have significantly lower and smaller seeds which indicates the function of PSK in successful reproductive processes. Qin et al. (2023) also state that PSK signaling influences ovule and embryo development through regulation of cell division in the embryo sac and adjacent tissue. This control provides better management of the seed's ability to receive the nutrients necessary to reach the desired weight at maturity. In rice, PSK control genes were activated early in the seed development stages. This suggests that PSK functions in an analogous way in other plants that have been bioengineered to improve grain size and yield (Shen et al. 2023). PSK peptides have also been reported to function in performance with plant hormones like auxin and BRs to promote growth in the meristem.

20.5 Molecular Mechanisms of Action of Peptide–Receptor–Induced Functioning

Understanding the molecular details of how peptide hormones affect seed size and crop yield is important for applying these findings to improve crops. These hormones function through stable signaling pathways. They involve specific receptor binding, message transmission, and gene regulation. This process is associated with traditional plant hormones (Li et al. 2025; Blázquez et al. 2020).

20.5.1 Precursor Processing and Secretion of Peptides

Peptide hormones start as large, inactive forms that have a signal part at the front. This part guides the peptide to the ER. Inside, helper molecules fold it correctly. In both the ER and Golgi, specific enzymes called endoproteases cut the precursor at specific spots to free the active bit (Benham 2012). Then, changes like adding sulfate, hydroxyl, and sugar groups make the hormone more stable and better at binding receptors. Once prepared, the peptides are stored in secretory vesicles and moved to the cell surface. When released, some stick to the cell wall, creating a local stash to help send signals around. Precursors that are a part of seed growth, like CLE8, take the same route. The CLE8 precursors are found in the ovules and endosperm. Once they mature in the ER and Golgi, they are released as active CLE8, which controls embryonic cell growth and patterning in the early stages of seed formation (Yu et al. 2025).

20.5.2 Perception by Cell-Surface Receptors

The interaction between receptor and ligand molecules is precisely controlled and is crucial for peptide signals functioning in seed tissues (Figure 20.1). Plants recognize peptide hormones using LRR-RKs found on their cell surfaces. These receptors have unique outer parts called LRR or malectin domains that capture the ligands. Inside the receptors is a kinase domain that initiates the next steps in signaling (Chakraborty et al. 2019). For example, CLE peptides are detected by the receptor CLV1 along with helper proteins like CLV2 and CORYNE. These create a complex that leads the plant's meristem and embryo development (Song et al. 2022). Likewise, PSK peptides bind to PSKR1, initiating pathways for

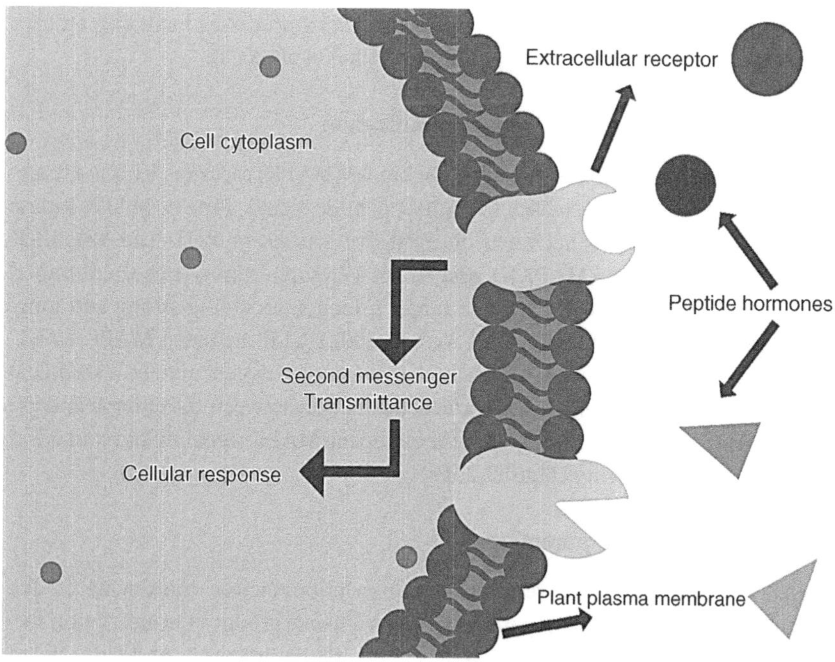

Figure 20.1 Peptide–receptor interaction and signal transmission on the cell surface.

cell growth and seed development (Li et al. 2024). There also exist seed-specific receptor kinases on the zygote's surface that recognize small peptides that regulate embryo development. When such peptides interact with their LRR-RKs, they initiate localized signals that are responsible for normal embryo development and seed growth (Chakraborty et al. 2019).

20.5.3 Activation of Receptor and Early Phosphorylation

When peptides bind to their receptors, the receptor kinases are activated by adding phosphate groups to each other's activation loops. This process is called transphosphorylation. This change creates specific sites where phospho-serine and phospho-threonine amino acids can attach, functioning as docking spots. Receptor-like cytoplasmic kinases (RLCKs) like BIK1 in PEP signaling and RIPK in FERONIA-RALF signaling are drawn to these locations. Once they dock, RLCKs also become phosphorylated, which assists them in sending the signal deeper into the cell. During seed development, signals from peptides like PSK bind to their receptors on embryonic cells. This causes rapid phosphorylation of PSKR1 and near RLCKs. This early activation step increases cell division within the embryo and supports the growth time of the developing seed coat and endosperm, which is crucial for determining the final seed size (Jose et al. 2020).

20.5.4 Generation of Second Messengers

When a peptide is bound to its receptor, RLKs initiate cascades of phosphorylation that transmit the message into the cytoplasm of the cell. These signals usually move through MAPK or CDPK-containing pathways. These kinases help transfer the message to target proteins. Secondary messengers such as reactive oxygen species (ROS), calcium ions (Ca^{2+}), and cyclic nucleotides highly contribute to controlling the activity of these proteins downstream. In seed growth, these pathways control crucial processes including embryo development, cell expansion, and nutrient transport (Ravi et al. 2023).

20.5.5 MAP Kinase Cascades and Signal Amplification

Second messengers start mitogen-activated protein kinase (MAPK) cascades. This is a three-member process that suddenly increases the initial peptide signal. First, a MAPK kinase kinase (MAPKKK) like YODA is activated by upstream kinases or ROS. The MAPKKK phosphorylates a MAPK kinase (MAPKK), and the MAPKK phosphorylates the terminal MAPK. And phosphorylated MAPKs activate target proteins, providing strong and long-lasting downstream responses (Kyriakis and Avruch 2001). CLE-mediated MAPK activation in ovule integuments and embryonal tissue at early stages triggers maternal seed-coat cell proliferation. Faster rates of cell division and coordination between the embryo and the maternal tissues and a coordinate regulation of these by the MAPK signaling have an effect on the final seed size and viability (Matilla 2019).

20.5.6 Transcriptional Reprogramming

Phosphorylated MAPKs and Ca^{2+} responsive transcription factors translocate to the nucleus to reprogram gene expression patterns. They influence groups of transcription factors, such as WRKYs in immune signaling, NACs in abscission processes, and basic Helix-Loop-Helix (bHLHs) in developmental pathways. This action leads to the upregulation of defense genes or cell wall–modifying enzymes (Thilakarathne et al. 2025). Chromatin remodelers and histone modifiers are also brought in, making sure that the transcriptional landscape continues to be dynamic and responsive to current environmental or developmental cues. In seeds, transcription factors activated by peptides like the WUSCHEL-RELATED HOMEOBOX (WOX) and LEAFY COTYLEDON (LEC) families are triggered. For example, CLE8-induced MAPK action induces LEC2 and ABI3, the master regulators of embryo identity and storage reserve deposition, thus coordinating the transition from cell proliferation to maturation in the seed developing (Lepiniec et al. 2018).

20.5.7 Regulation of Feedback and Termination of Signal

To prevent overstimulation, plants use numerous negative feedback loops. Receptors on the cell surface are internalized and degraded by lysosomes. Specific protein phosphatases like PP2Cs, which dephosphorylate receptors and MAPKs, abolish kinase activity quickly (Wu et al. 2025a, b). Endogenous peptide antagonists or shortened decoys also compete with bioactive peptides for binding to the receptor, suppressing signal intensity. Feedback controllers such as CLE decoy peptides are sequestered in the endosperm to arrest CLE-mediated proliferation when the embryo reaches a predetermined size during seed development. This ensures that resources are used for storage reserve deposition and desiccation tolerance rather than uncontrollable cell division (Wang et al. 2025).

20.5.8 Cross Talk with Classical Phytohormones

Peptide hormones never act alone but work synergistically or antagonistically with traditional phytohormones to regulate complicated physiological effects (Dodueva et al. 2021). CLE or RALF peptides usually shape auxin gradients in establishing polarity during embryogenesis. BRs combine with PSK signaling to induce cell enlargement and overall organ growth. GA and ABA, essential for seed dormancy and maturation, are influenced by peptide signals that regulate their sensitivity and production (Zolkiewicz and Gruszka 2022). For example, peptides of IDA involved in seed and abscission release interact with ethylene and ABA signals (Mesejo et al. 2021). The jasmonic acid and salicylic acid pathways also cross talk with PEP and RALF signals during pathogen infection to manage a response that is defense-growth balanced. These interactions make a complex regulatory network that maintains hormonal balance and ensures coordinated development of seeds and tissues (Roychowdhury et al. 2025).

20.6 Biotechnological Methods for Peptide Hormone Editing

Over the last few years, biotechnological changes have unlocked new possibilities for regulating plant growth and productivity. Peptide hormones are key regulatory molecules. They play an important role in seed development, organ size, and overall yield quality.

20.6.1 CRISPR/Cas9 Gene Editing of Peptide-Coding Genes

The development of genome editing tools like CRISPR/Cas9 has transformed functional genomics in plants. CRISPR/Cas9 allows for the introduction of accurate site-directed mutations in peptide hormone–encoding genes. It allows scientists to build knockouts or gain-of-function alleles to study and use their functions in seed development (Koyama et al. 2021). For example, disrupting the *CLE8* gene that controls embryogenesis through CRISPR editing in *Arabidopsis* has given insights into embryo patterning and the control of seed size (Verma et al. 2022). Targeting RALF peptide–coding genes or their receptors also influences cell expansion mechanisms, which directly contribute to final seed size. Notably, CRISPR/Cas9 editing provides a marker-free, clean way to produce genetically improved lines with good reproductive yield. This method does not introduce foreign DNA, making it a desirable option for regulatory approval and public acceptance.

20.6.2 Overexpression and Silencing of Peptides Hormones

Functional analysis of peptide hormones has, to a large extent, been facilitated by overexpression and RNAi-induced silencing technologies in model plants, including *Arabidopsis thaliana* and *Oryza sativa* (Naim et al. 2024). Gain-of-function and loss-of-function assays have given early insights into how specific peptide hormones function during seed morphogenesis and yield. CLE gene overexpression has shown them to be able to alter meristem behavior toward deformed organogenesis and seed development. CEP peptides RNAi suppression in rice, however, has shown their role in the regulation of nutrient signaling and thereby influencing reproductive success and grain production (Humphrey and Brooks 2008). These approaches allow researchers to investigate gene function in planta, under controlled expression and frequently under inducible or tissue-preferred promoters to address developmental response with low pleiotropic effects. Exogenous characterization and chemical synthesis of biologically active peptide hormones are new methods for studying and regulating plant development.

20.6.3 Using Synthetic Peptides to Modulate Development

Synthetic peptides that mimic endogenous signal molecules can be applied to seeds, flowers, or roots externally to temporarily activate specified signaling pathways (Ren et al. 2024). For instance, synthetic CLE peptides have been found to inhibit meristem growth, whereas exogenous RALF peptides control cell wall pH and specify seed size. This is most convenient for temporary experiments, which do not allow the employment of transgenics and support rapid screening of activities. It also represents the potential to construct peptide-based agrochemicals or seed coatings to trigger productivity without genetic transformation (Fedoreyeva 2023).

20.6.4 Gene Stacking for Yield Improvement

Single-gene alteration has a predisposition toward restricted or temporary expression of characteristics because of intricate compensatory mechanisms within plants (Birchler and Veitia 2021). Gene stacking, pyramiding multiple peptide hormone–related genes with synergistic activity, is, therefore, becoming an effective approach to enable stable and improved yield performance. As an example, co-overexpression of CLE peptides and PSK can stimulate cell division and embryogenesis, resulting in increased seed number and size. Gene stacking approaches are especially useful for multigenic traits such as yield, regulated by multifactorial interactions involving developmental and environmental signals (Khan et al. 2021). Breakthroughs in cloning parts and synthetic biology have made it easier to build multigene cassettes. These can work under specific regulatory control to express properly in time and space (Naseri and Koffas 2020).

20.6.5 Tissue-Specific Promoters for Targeted Expression

Changing peptide hormones at the whole-plant level can cause unwanted side effects, such as changes in root structure or reduced fertility. To avoid these issues, researchers are using tissue-specific or stage-specific promoters to control the expression of peptide-coding genes or receptors in specific seed tissues (Gonzalez 2015). For example, using embryo-specific promoters to express CLE or RALF genes allows for better control over embryo development without interfering with vegetative growth. Scientists also use ovule or endosperm-specific promoters to express yield-regulating peptides at specific stages of development. This method allows restricted action and minimizes off-target effects (Tian et al. 2020). Overall, this approach increases the precision of genetic manipulation and meets biosafety standards. It is suitable for commercial crop breeding programs.

20.7 Peptide Hormones and Environment Regulation of Yield

Peptide hormones play a crucial role in plant development. They help the plants respond to changing environmental conditions. In areas where factors like drought, salinity, and nutrient availability vary lead to crop yield drops significantly. Peptide hormones support plants to adapt their development, particularly in processes related to seed formation and yield (Khan 2025). These hormones work through signaling pathways that allow communication between roots, shoots, and reproductive tissues. This combination of environmental signals and the plant's hormonal systems increases flexibility and develops reproductive success during tough conditions.

20.7.1 Hormonal Response to Abiotic Stresses

Abiotic stresses like drought, salinity, and nutrient stresses are higher challenges for plant growth and reproduction. Peptide hormones are essential to sense and respond to these stresses. For example, CEPs increase when there is a lack of nitrogen in the roots. These peptides move to the shoot to support nitrogen uptake and distribution, which protects against reproductive development (Taleski et al. 2018). RALF peptides help maintain pH levels when salinity stress occurs. They keep membranes intact and cells alive during seed development. During drought, peptide-receptor complexes like CLE25 play a role in signaling water stress over long distances. They initiate stomatal closure by inducing ABA (Chowdhary and Songachan 2025). This process helps protect water during flowering and seed filling. These examples exhibit that peptide hormones are not only growth factors. They also act as signals that shift developmental pathways based on available resources. Nanocarrier-based delivery systems and microbial production pipelines.

20.7.2 Role in Adaptive Developmental Plasticity

Peptide hormones are beneficial in developmental plasticity or the capability of plant systems to alter developmental adaptations as a reaction to environmental stimuli. In cases of restriction to nutrients and exogenous salinity, peptide hormones play a main role in controlling the balance between vegetative and reproductive growth (Nadeem et al. 2023). CLE peptides play a significant role in meristem control and expression levels are regulated under environmental stress. Adjusting the level of expression will regulate when organs develop and when growth for reproduction is determined. By regulating the plant's vegetative and reproductive growth, one can prevent the resource-expensive reproductive process of flowering or seed setting when resources are not adequate anymore (Yu et al. 2025). Likewise, IDA-like peptides are involved in organ detachment and abscission, processes that may be induced by stress. In stressful conditions, IDA and IDL signaling allow resource reallocation by selectively aborting flowers or seeds in order to optimize seed quality over yield (Wang et al. 2023).

20.7.3 Root-Shoot Signaling Peptides under Stress Conditions

The role of peptide hormones during environmental stress in root-to-shoot communication is one of the most important. When there is a nutrient deficiency, such as nitrogen and phosphate, root CKs are transported through the xylem to the shoot, where they are detected through LRR-RKs such as CEPR1/2 (Xie et al. 2022). This communication leads to systemic changes in transcriptional response that lead to the alteration of nutrient acquisition and allocation of available energy toward seed development. Furthermore, RALF peptides affect apoplastic pH and freezing calcium ion flux in shoot tissues to regulate stomatal activity, maintenance of osmotic status, and integrity of reproductive structures (Chowdhary and Songachan 2025). Root-shoot signaling pathway mediated through peptides is an important process to mediate local perception of nutrients and global developmental response subject to seed maturation.

20.7.4 Combining Reproductive Output with Environmental Cues

Peptide signaling associates environmental factors with the reproductive stage of plant development, which directly impacts seed yield. When plants experience stress, changes in peptide signaling can cause delayed blooming, reduced pollen viability, inhibited fertilization, or altered seed filling rates (Qu et al. 2015). For example, IDL peptides affect the aging of floral organs and the timing of seed dispersal during stress, while CLE and RALF peptides influence ovule development and fertilization efficiency (Perkins et al. 2025). Additionally, peptide hormones often work together with plant hormones like ABA, GAs, and ethylene. During drought, for example, CLE25 not only works alone but also increases ABA production. This creates a survival feedback loop that controls reproductive output. This type of cooperation between hormones and peptides helps plants respond effectively to environmental conditions while balancing their growth and development needs with external factors.

20.8 Agricultural Applications and Translational Potential of Peptides

Peptide hormones play a key role in agricultural biotechnology. They are significant for plant growth, development, and stress response. Their small size and unique targeting, along with their ability to undergo PTMs, provide specific advantages for altering plant physiology at the molecular level.

20.8.1 Synthetic Peptide Models in the Field

Using modern technology, it is now possible to synthesize and use synthetic peptide hormones or their equivalent molecules on crops in the field. Such synthetic peptides are applicable by spraying onto plants, seed coating, or administered through minute-sized carriers known as nanocarriers (Khan et al. 2022). For example, peptides like CLE and RALF have been tested to influence root growth and help plants absorb nutrients better in real-world conditions (Xie et al. 2022). Besides, certain peptides that target roots, such as RGF and GOLVEN Peptides (GLV), are important for keeping stem cells active and encouraging roots to grow longer. Peptides like CEP respond to nutrient levels and help improve how roots branch out (Hsiao and Yamada 2020). All these actions help plants gather water and nutrients more effectively, especially in poorer soils. What's great about these methods is that they can deliver quick results without changing the plant's genes, making it a non-GMO way to boost crop yields.

20.8.2 Peptide Hormone Targets in Crop Improvement Programs

Peptide hormone genes are now increasingly included as markers or targets in traditional and molecular breeding strategies. For instance, manipulation of CLE8, CEP1, or PSK in crops such as rice, maize, and soybean has had a positive impact on seed size, seed number, and stress susceptibility in some cases (Yu et al. 2025). Traditional breeding and molecular breeding strategies allow for such manipulations to be facilitated through CRISPR/Cas9 and transgenics, which allow for precise editing or overexpression of peptide genes, respectively. This enables customization of developmental traits to drive yield and adaptable capacity (Saikia et al. 2020).

20.8.3 Integration with Smart Farming and Precision Agriculture

Peptide signaling networks can be integrated into digital farming systems using biosensors and predictive models. Understanding how peptides react to environmental signals like drought and nitrogen levels can help farmers make real-time decisions (Shrestha and Wei 2024). Smart farming tools could use biosensors to measure peptide activity levels in tissues. This would enable targeted management of water, fertilizer needs, and harvesting schedules. Synthetic biology methods might even allow for the expression of peptides on demand in response to remote triggers (Singla 2021).

20.8.4 Challenges in Delivery, Stability, and Cost

Although promising, some challenges limit the large-scale implementation of peptide-based approaches. Peptides are susceptible to degradation by environmental and plant proteases. Stabilizing formulations or encapsulation methods are thus a requirement (Rafferty et al. 2016). Moreover, scaling up the synthesis of bioactive peptides is expensive, and field delivery systems need optimization regarding efficacy, dosage, and even application uniformity.

20.8.5 Biosafety and Regulatory Aspects

Using peptide hormones, especially those made synthetically or through genetic engineering, requires following strict biosafety rules (Zane et al. 2021). We need to check for unexpected side effects, whether they cause allergies, and how long they stick around in the environment. The rules depend on how the peptides are used as a plant spray, seed treatment, or a genetic change. It's really important to carefully weigh the benefits against the risks and get everyone involved to get public support and approval (Uma et al. 2024).

20.9 Future Directions

Peptide hormones have great potential to improve crop productivity and quality. Future research may investigate finding yield-related peptides, improving signaling pathways, and producing peptide-based biostimulants. Understanding how they interact with hormones and stress signals could also increase crop resilience. This supports sustainable and precision agriculture to meet global food needs.

20.9.1 Unspecified Pairs of Peptides and Receptors

Even though research on peptide hormones has grown substantially, many peptide ligands and their corresponding receptors remain unidentified, especially in nonmodel plants. Discovering the novel peptide–receptor interactions by methods like yeast two-hybrid screening, co-immunoprecipitation, and transcriptome-based ligand-receptor prediction will reveal new pathways for seed trait modulation (Lu 2020). It will also assist in defining species-specific peptide signaling pathways important for targeted plant breeding.

20.9.2 The Function of Peptides in Seed–Microbiome Interactions

Current research indicates that peptide hormones can assist in communication between germinating seeds and the nearby microbiome. AMPs, signaling peptides, and defense-related small proteins influence microbial colonization. This influences seedling vigor, germination, and initial plant immunity (Ghanbarzadeh et al. 2024). Future research can analyze how altering peptide expression can selectively enhance the beneficial microbes and decrease pathogens. This could enhance overall seed health.

20.9.3 Using Nanoscale Tools to Deliver Peptides

Traditional use of peptides in the field is constrained by instability and low uptake. Peptide-loaded nanocarriers, hydrogels, or nanoparticle sprays based on nanotechnology delivery systems could be used for protecting peptides from degradation and site-specific release in reproductive tissues or in seed compartments. These technologies also allow precision delivery based on developmental stage or environmental cues, optimizing peptide hormone effectiveness under field conditions (Patil et al. 2015).

20.9.4 Peptide Library Screening at High Efficiency

Functional genomics and synthetic biology make possible the design and screening of large peptide libraries to identify useful traits like enhanced seed size, drought tolerance, or nutrient use efficiency (Nguyễn et al. 2024). High-throughput phenotyping platforms, CRISPR-mediated peptide-reporter fusions, and in planta screening methods might expedite the identification of bioactive peptides; this methodology might uncover novel yield regulators beyond established peptide families.

20.9.5 Moving Toward Customized Seed Traits with Designer Peptides

Peptide research aims to create "designer peptides" with specific functions for crops or environmental applications. These peptides can be modified to improve receptor binding, stability, or even carry two functions, such as promoting growth and conferring stress resistance (Wu 2023). Integrating omics data, machine learning, and molecular modeling, we are able to generate predictive models for the design of peptides. This facilitates accurate control over seed development, allocation of nutrients, and reproductive timing (Zhang et al. 2024a, b).

20.9.6 Toward Sustainable Agriculture

Peptide hormones are unique molecules that contribute toward achieving sustainability targets by minimizing the application of agrochemicals. They increase plant stress resistance and provide food security in a dynamic environment. Peptides might play a key role in future crop management systems with high improvements (Khan 2025).

20.10 Conclusion

Plant biology and agricultural biotechnology are developing because of peptide hormones. These small signaling peptides were once considered for more traditional phytohormones. Now, they are recognized as crucial growth regulators. They are important for seed development, immunity, stress resilience, and yield in many plant species. The practical uses of peptide hormones are growing increasingly as research progresses. This includes synthetic biostimulants, gene-edited receptor systems, nanocarrier-based delivery systems, and microbial production pipelines. These technologies can potentially reduce chemical inputs, enhance climate resilience, and promise food security in a changing world. In the future, the intersection of peptide signaling with synthetic biology, CRISPR-based trait engineering, and omics-driven discovery will likely give rise to a new era of intelligent agriculture where crops are not only bred for yield but also programmed for adaptability, efficiency, and ecological balance. Peptide hormones are thus becoming strategic seed improvement enhancers that offer specific control over size, viability, and yield. With our increasing knowledge, these molecules are being developed from matters of theoretical interest into practical reality, opening the way for smart and resource-saving cultivation practices.

References

Adams, M.E. (2019). Peptides as chemical signals: hormones to transmitters. In: *Cockroaches as Models for Neurobiology: Applications in Biomedical Research* (ed. I. Huber, E.P. Masler, and B.R. Rao), 3–33. CRC Press.

Aguirre, M., Kiegle, E., Leo, G., and Ezquer, I. (2018). Carbohydrate reserves and seed development: an overview. *Plant Reproduction* 31 (3): 263–290.

Ahanger, M.A., Ashraf, M., Bajguz, A., and Ahmad, P. (2018). Brassinosteroids regulate growth in plants under stressful environments and crosstalk with other potential phytohormones. *Journal of Plant Growth Regulation* 37 (4): 1007–1024.

Al-Suhaimi, E.A. (2022). Introduction to Endocrinology. In: *Emerging concepts in endocrine structure and functions* (ed. E.A. Al-Suhaimi), 1–24. Singapore: Springer Nature Singapore.

Araya, T., von Wirén, N., and Takahashi, H. (2016). CLE peptide signaling and nitrogen interactions in plant root development. *Plant Molecular Biology* 91 (6): 607–615.

Azevedo, L.M., de Oliveira, R.R., and Chalfun-Junior, A. (2025). The role of FT/TFL1 clades and their hormonal interactions to modulate plant architecture and flowering time in perennial crops. *Plants* 14 (6): 923.

Benham, A.M. (2012). Protein secretion and the endoplasmic reticulum. *Cold Spring Harbor Perspectives in Biology* 4 (8): a012872.

Bhat, A. (2021). Getting to the root of rapid growth inhibition and mechanical signal transduction in *Arabidopsis thaliana* roots. Plant Biology, Doctor of Philosophy, February 25, 2021. Getting to the root of rapid growth inhibition and mechanical signal transduction in Arabidopsis thaliana roots – Blacklight

Bhattacharya, A. (2022). Plant growth hormones in plants under low-temperature stress: a review. In: *Physiological Processes in Plants under Low Temperature Stress*, 517–627. Singapore: Springer Singapore.

Birchler, J.A. and Veitia, R.A. (2022). One hundred years of gene balance: how stoichiometric issues affect gene expression, genome evolution, and quantitative traits. *Cytogenetic and Genome Research* 161 (10–11): 529–550.

Blázquez, M.A., Nelson, D.C., and Weijers, D. (2020). Evolution of plant hormone response pathways. *Annual Review of Plant Biology* 71 (1): 327–353.

Boon, L., Ugarte-Berzal, E., Vandooren, J., and Opdenakker, G. (2020). Protease propeptide structures, mechanisms of activation, and functions. *Critical Reviews in Biochemistry and Molecular Biology* 55 (2): 111–165.

Brevini, T.A. and Georgia, P. (2012). *Gametogenesis, Early Embryo Development and Stem Cell Derivation*. Springer Science & Business Media.

Chakraborty, S., Nguyen, B., Wasti, S.D., and Xu, G. (2019). Plant leucine-rich repeat receptor kinase (LRR-RK): structure, ligand perception, and activation mechanism. *Molecules* 24 (17): 3081.

Cheng, M., Hu, H., Chen, R. et al. (2025). Identification of the Casparian strip integrity factor (CIF) gene family in Brassica napus and functional prediction of mature CIF small peptides. *BMC Plant Biology* 25 (1): 767.

Cheung, A.Y. (2024). FERONIA: a receptor kinase at the core of a global signaling network. *Annual Review of Plant Biology* 75: 345–375.

Chevalier, E. (2012). *Implication des peptides RALFs dans les communications cellulaires lors du développement du gamétophyte femelle chez Solanum chacoense et Arabidopsis thaliana*. Universite de Montreal (Canada).

Chowdhary, N.A. and Songachan, L.S. (2025). Plant peptides involved in ROS signalling and biotic and abiotic stress responses. *International Journal of Peptide Research and Therapeutics* 31 (3): 53.

Clark, C.J., Poulsen, J.R., Levey, D.J., and Osenberg, C.W. (2007). Are plant populations seed limited? A critique and meta-analysis of seed addition experiments. *The American Naturalist* 170 (1): 128–142.

Cruz, W.T.D. (2019). Enzymatic and structural studies of peruvianin-I: the first Germin-like protein with proteolytic activity. 115 f. Tese (Doutorado em Bioquímica) - Universidade Federal do Ceará, Fortaleza, 2019. http://repositorio.ufc.br/handle/riufc/79104

Delay, C. (2015). The roles of C-TERMINALLY ENCODED PEPTIDES in Arabidopsis root development. Openresearch-repository library. Australin National University. https://doi.org/10.25911/5d78d5a1604e7

Dodueva, I., Lebedeva, M., and Lutova, L. (2021). Dialog between kingdoms: enemies, allies and peptide phytohormones. *Plants* 10 (11): 2243.

Dwivedi, S.L., Spillane, C., Lopez, F. et al. (2021). First the seed: genomic advances in seed science for improved crop productivity and food security. *Crop Science* 61 (3): 1501–1526.

Fedoreyeva, L.I. (2023). Molecular mechanisms of regulation of root development by plant peptides. *Plants* 12 (6): 1320.

Figueiredo, D.D. and Sharma, R.A. (2025). The mysterious dialogue between the embryo, endosperm, and seed coat, and its implications on seed traits. *Journal of Experimental Botany* 00 (0): eraf320, 1–eraf315. doi: 10.1093/jxb/eraf320.

Fiume, E. (2010). Expression analysis of the CLE signaling gene family in Arabidopsis thaliana and functional characterization of CLE8 in seed development (Doctoral dissertation, UC Berkeley).

Fiume, E. and Fletcher, J.C. (2012). Regulation of Arabidopsis embryo and endosperm development by the polypeptide signaling molecule CLE8. *The Plant Cell* 24 (3): 1000–1012.

Fiume, E., Guyon, V., Remoué, C. et al. (2016). TWS1, a novel small protein, regulates various aspects of seed and plant development. *Plant Physiology* 172 (3): 1732–1745.

Fobis-Loisy, I., Ivanov, R., and Gaude, T. (2012). The S-locus cysteine-rich protein (SCR): a small peptide with a high impact on the evolution of flowering plants. In: *Plant Signaling Peptides* (ed. H.R. Irving and C. Gehring), 77–92. Berlin, Heidelberg: Springer.

Frayn, K.N. (2013). *Metabolic Regulation: A Human Perspective*. John Wiley & Sons.

Gancheva, M.S., Malovichko, Y.V., Poliushkevich, L.O. et al. (2019). Plant peptide hormones. *Russian Journal of Plant Physiology* 66 (2): 171–189.

Gasperini, D. and Howe, G.A. (2024). Phytohormones in a universe of regulatory metabolites: lessons from jasmonate. *Plant Physiology* 195 (1): 135–154.

Ghanbarzadeh, Z., Mohagheghzadeh, A., and Hemmati, S. (2024). The roadmap of plant antimicrobial peptides under environmental stress: from farm to bedside. *Probiotics and Antimicrobial Proteins* 16 (6): 2269–2304.

Gonzalez, T.L. (2015). Tight, inducible gene expression using alternatively spliced exons from plants (Doctoral dissertation, UC Berkeley).

Hall, C., Dawson, T.P., Macdiarmid, J.I. et al. (2017). The impact of population growth and climate change on food security in Africa: looking ahead to 2050. *International Journal of Agricultural Sustainability* 15 (2): 124–135.

Higashiyama, T. and Yang, W.C. (2017). Gametophytic pollen tube guidance: attractant peptides, gametic controls, and receptors. *Plant Physiology* 173 (1): 112–121.

Höng, K., Austerlitz, T., Bohlmann, T., and Bohlmann, H. (2021). The thionin family of antimicrobial peptides. *Plops One* 16 (7): e0254549.

Hsiao, Y.C. and Yamada, M. (2020). The roles of peptide hormones and their receptors during plant root development. *Genes* 12 (1): 22.

Hu, J., Huang, L., Chen, G. et al. (2021). The elite alleles of OsSPL4 regulate grain size and increase grain yield in rice. *Rice* 14 (1): 90.

Humphrey, T. and Brooks, G. (ed.) (2008). *Cell Cycle Control: Mechanisms and Protocols*, vol. 296. Springer Science & Business Media.

Hunt, L., Bailey, K.J., and Gray, J.E. (2010). The signalling peptide EPFL9 is a positive regulator of stomatal development. *New Phycologist* 186: 609–614.

Jameson, P.E. and Song, J. (2016). Cytokinin: a key driver of seed yield. *Journal of Experimental Botany* 67 (3): 593–606.

Jie, M., Feng, T., Huang, W. et al. (2021). Subcellular localization of miRNAs and implications in cellular homeostasis. *Genes* 12 (6): 856.

Jose, J., Ghantasala, S., and Roy Choudhury, S. (2020). *Arabidopsis* transmembrane receptor-like kinases (RLKs): a bridge between extracellular signal and intracellular regulatory machinery. *International Journal of Molecular Sciences* 21 (11): 4000.

Khan, N. (2025). Decoding the long-term impacts of genetic modifications in hormone pathways on plant physiology and ecosystem stability. *Environmental Reviews*, (ja) 33: 1–20.

Khan, S.U., Khan, M.H.U., Ahmar, S., and Fan, C. (2021). Comprehensive study and multipurpose role of the CLV3/ESR-related (CLE) genes family in plant growth and development. *Journal of Cellular Physiology* 236 (4): 2298–2317.

Khan, M.N., Khan, Z., Zhu, L. et al. (2022). Impact of chitosan and chitosan-based metallic nanoparticles on the regulation of plant hormones. In: *Role of Chitosan and Chitosan-Based Nanomaterials in Plant Sciences* (ed. S. Kumar and S.V. Madihally), 343–356. Elsevier.

Khush, G.S. (2013). Strategies for increasing the yield potential of cereals: case of rice as an example. *Plant Breeding* 132 (5): 433–436.

Kim, M.J., Jeon, B.W., Oh, E. et al. (2021). Peptide signaling during plant reproduction. *Trends in Plant Science* 26 (8): 822–835.

Klein, T., Eckhard, U., Dufour, A. et al. (2018). Proteolytic cleavage—mechanisms, function, and "Omic" approaches for a near-ubiquitous posttranslational modification. *Chemical Reviews* 118 (3): 1137–1168.

Koyama, T., Terhzaz, S., Naseem, M.T. et al. (2021). A nutrient-responsive hormonal circuit mediates an inter-tissue program regulating metabolic homeostasis in adult *Drosophila*. *Nature Communications* 12 (1): 5178.

Kumpf, R.P., Shi, C.L., Larrieu, A. et al. (2013). Floral organ abscission peptide IDA and its HAE/HSL2 receptors control cell separation during lateral root emergence. *Proceedings of the National Academy of Sciences* 110 (13): 5235–5240.

Kurepa, J. and Smalle, J.A. (2023). Plant hormone modularity and the survival-reproduction trade-off. *Biology* 12 (8): 1143.

Kyriakis, J.M. and Avruch, J. (2001). Mammalian mitogen-activated protein kinase signal transduction pathways activated by stress and inflammation. *Physiological Reviews* 81: 807–869.

Leasure, C.D. and He, Z.H. (2012). CLE and RGF family peptide hormone signaling in plant development. *Molecular Plant* 5 (6): 1173–1175.

Lee, K.P., Liu, K., Kim, E.Y. et al. (2020). Plant natriuretic peptide A and its putative receptor PNP-R2 antagonize salicylic acid–mediated signaling and cell death. *The Plant Cell* 32 (7): 2237–2250.

Lepiniec, L., Devic, M., Roscoe, T.J. et al. (2018). Molecular and epigenetic regulations and functions of the LAFL transcriptional regulators that control seed development. *Plant Reproduction* 31 (3): 291–307.

Li, Y., Di, Q., Luo, L., and Yu, L. (2024). Phytosulfokine peptides, their receptors, and functions. *Frontiers in Plant Science* 14: 1326964.

Li, X., Zhu, L., Wang, H. et al. (2025). Peptide hormone-mediated regulation of plant development and environmental adaptability. *Advanced Science* 12 (34): e06590.

Liang, X. and Zhou, J.M. (2018). Receptor-like cytoplasmic kinases: central players in plant receptor kinase–mediated signaling. *Annual Review of Plant Biology* 69 (1): 267–299.

Linkies, A. and Leubner-Metzger, G. (2012). Beyond gibberellins and abscisic acid: how ethylene and jasmonates control seed germination. *Plant Cell Reports* 31 (2): 253–270.

Locascio, A., Roig-Villanova, I., Bernardi, J., and Varotto, S. (2014). Current perspectives on the hormonal control of seed development in Arabidopsis and maize: a focus on auxin. *Frontiers in Plant Science* 5: 412.

Lu, L. (2020). Antimicrobial and immune-modulation action within the RNase A superfamily using macrophage infection models. https://ddd.uab.cat/record/241289

Lu, S. and Xiao, F. (2024). Small peptides: orchestrators of plant growth and developmental processes. *International Journal of Molecular Sciences* 25 (14): 7627.

Matilla, A.J. (2019). Seed coat formation: its evolution and regulation. *Seed Science Research* 29 (4): 215–226.

Matilla, A.J. (2020). Auxin: hormonal signal required for seed development and dormancy. *Plants* 9 (6): 705.

Mazzoni-Putman, S.M., Brumos, J., Zhao, C. et al. (2021). Auxin interactions with other hormones in plant development. *Cold Spring Harbor Perspectives in Biology* 13 (10): a039990.

Mesejo, C., Marzal, A., Martínez-Fuentes, A. et al. (2021). On how auxin, ethylene and IDA-peptide relate during mature Citrus fruit abscission. *Scientia Horticulturae* 278: 109855.

Mishra, S., Levengood, H., Fan, J., and Zhang, C. (2024). Plants under stress: exploring physiological and molecular responses to nitrogen and phosphorus deficiency. *Plants* 13 (22): 3144.

Mookherjee, N., Anderson, M.A., Haagsman, H.P., and Davidson, D.J. (2020). Antimicrobial host defence peptides: functions and clinical potential. *Nature Reviews Drug Discovery* 19 (5): 311–332.

Nadeem, M., Yahya, M., Tong, J. et al. (2023). Improving nitrogen acquisition and utilization through root architecture remodelling: insight from legumes. *Journal of Plant Growth Regulation* 42 (9): 5295–5310.

Naim, M.D., Alamin, M., Mosharof, M.P. et al. (2024). Major components of RNAi gene families in Oryza sativa cultivar Kitaake: in-silico discovery and characterization. *Heliyon* 10 (22): e40395.

Najafi, J., Brembu, T., Vie, A.K. et al. (2020). PAMP-induced secreted peptide 3 modulates immunity in Arabidopsis. *Journal of Experimental Botany* 71 (3): 850–864.

Naseri, G. and Koffas, M.A. (2020). Application of combinatorial optimization strategies in synthetic biology. *Nature Communications* 11 (1): 2446.

Nguyễn, C.T., Gammatantrawet, N., Susawaengsup, C. et al. (2024). Drought-ready plant resilience: harnessing nano-biotechnology techniques for swift screening and selection of organic crop varieties. *South African Journal of Botany* 169: 553–566.

Novák, O., Napier, R., and Ljung, K. (2017). Zooming in on plant hormone analysis: tissue-and cell-specific approaches. *Annual Review of Plant Biology* 68: 323–348.

Okamoto, S., Tabata, R., and Matsubayashi, Y. (2016). Long-distance peptide signaling essential for nutrient homeostasis in plants. *Current Opinion in Plant Biology* 34: 35–40.

Olsen, O.A. and Becraft, P.W. (2013). Endosperm development. In: *Seed Genomics* (ed. P.W. Becraft), 43–62. Wiley.

Ormancey, M., Guillotin, B., San Clemente, H. et al. (2021). Use of microRNA-encoded peptides to improve agronomic traits. *Plant Biotechnology Journal* 19 (9): 1687–1689.

Patil, S., Vhora, I., Amrutiya, J. et al. (2015). Role of nanotechnology in delivery of protein and peptide drugs. *Current Pharmaceutical Design* 21 (29): 4155–4173.

Perkins, C.J., Pryze, K. and Palanivelu, R., 2025. Peptide signaling in flowering plants: insights into reproductive thermotolerance. *Journal of Experimental Botany*, p.eraf192; 76 (19): 5666–5681. doi: https://doi.org/10.1093/jxb/eraf192. PMID: 40350712.

Qin, Z., Wu, Y.N., Li, S., and Zhang, Y. (2023). Signaling between sporophytic integuments and developing female gametophyte during ovule development. *Plant Science* 335: 111829.

Qu, L.J., Li, L., Lan, Z., and Dresselhaus, T. (2015). Peptide signalling during the pollen tube journey and double fertilization. *Journal of Experimental Botany* 66 (17): 5139–5150.

Rafferty, J., Nagaraj, H., McCloskey, A. et al. (2016). Peptide therapeutics and the pharmaceutical industry: barriers encountered translating from the laboratory to patients. *Current Medicinal Chemistry* 23 (37): 4231–4259.

Ravi, B., Foyer, C.H., and Pandey, G.K. (2023). The integration of reactive oxygen species (ROS) and calcium signalling in abiotic stress responses. *Plant, Cell & Environment* 46 (7): 1985–2006.

Reed, R.C., Bradford, K.J., and Khanday, I. (2022). Seed germination and vigor: ensuring crop sustainability in a changing climate. *Heredity* 128 (6): 450–459.

Ren, G., Zhang, Y., Chen, Z. et al. (2024). Research progress of small plant peptides on the regulation of plant growth, development, and abiotic stress. *International Journal of Molecular Sciences* 25 (7): 4114.

Roychowdhury, R., Hada, A., Biswas, S. et al. (2025). Jasmonic acid (JA) in plant immune response: unravelling complex molecular mechanisms and networking of defence signalling against pathogens. *Journal of Plant Growth Regulation* 44 (1): 89–114.

Safaeizadeh, M. and Boller, T. (2019). Differential and tissue-specific activation pattern of the AtPROPEP and AtPEPR genes in response to biotic and abiotic stress in Arabidopsis thaliana. *Plant Signaling & Behavior* 14 (5): e1590094.

Saikia, B., Singh, S., Debbarma, J. et al. (2020). Multigene CRISPR/Cas9 genome editing of hybrid proline rich proteins (HyPRPs) for sustainable multi-stress tolerance in crops: the review of a promising approach. *Physiology and Molecular Biology of Plants* 26 (5): 857–869.

Samanta, S. and Roychoudhury, A. (2025). Molecular crosstalk of jasmonate with major phytohormones and plant growth regulators during diverse stress responses. *Journal of Plant Growth Regulation* 44 (1): 62–88.

Sanchez-Corrionero, A., Sánchez-Vicente, I., Arteaga, N. et al. (2023). Fine-tuned nitric oxide and hormone interface in plant root development and regeneration. *Journal of Experimental Botany* 74 (19): 6104–6118.

Santner, A., Calderon-Villalobos, L.I.A., and Estelle, M. (2009). Plant hormones are versatile chemical regulators of plant growth. *Nature Chemical Biology* 5 (5): 301–307.

Sawicki, M., Aït Barka, E., Clément, C. et al. (2015). Cross-talk between environmental stresses and plant metabolism during reproductive organ abscission. *Journal of Experimental Botany* 66 (7): 1707–1719.

Shen, X., Stührwohldt, N., and Lin, C. (2023). The research process of PSK biosynthesis, signaling transduction, and potential applications in *Brassica napus*. *Plants* 12 (17): 3075.

Shi, C.L., Alling, R.M., Hammerstad, M., and Aalen, R.B. (2019). Control of organ abscission and other cell separation processes by evolutionary conserved peptide signaling. *Plants* 8 (7): 225.

Shrestha, M.M. and Wei, L. (2024). Perspectives on the roles of real time nitrogen sensing and IoT integration in smart agriculture. *Journal of the Electrochemical Society* 171 (2): 027526.

Singla, P. (2021). Plant synthetic biology: A paradigm shift targeting stress mitigation, reduction of ecological footprints and sustainable transformation in agriculture. In: *Plant Stress Biology: Strategies and Trends* (ed. B. Giri and M.P. Sharma), 435–489. Singapore: Springer Singapore.

Song, X.F., Hou, X.L., and Liu, C.M. (2022). CLE peptides: critical regulators for stem cell maintenance in plants. *Planta* 255 (1): 5.

Stø, I.M., Orr, R.J., Fooyontphanich, K. et al. (2015). Conservation of the abscission signaling peptide IDA during Angiosperm evolution: withstanding genome duplications and gain and loss of the receptors HAE/HSL2. *Frontiers in Plant Science* 6: 931.

Stührwohldt, N., Dahlke, R.I., Kutschmar, A. et al. (2015). Phytosulfokine peptide signaling controls pollen tube growth and funicular pollen tube guidance in Arabidopsis thaliana. *Physiologia Plantarum* 153 (4): 643–653.

Su, Y.H., Liu, Y.B., and Zhang, X.S. (2011). Auxin–cytokinin interaction regulates meristem development. *Molecular Plant* 4 (4): 616–625.

Taleski, M., Imin, N., and Djordjevic, M.A. (2018). CEP peptide hormones: key players in orchestrating nitrogen-demand signalling, root nodulation, and lateral root development. *Journal of Experimental Botany* 69 (8): 1829–1836.

Taleski, M., Jin, M., Chapman, K. et al. (2024). CEP hormones at the nexus of nutrient acquisition and allocation, root development, and plant–microbe interactions. *Journal of Experimental Botany* 75 (2): 538–552.

Thilakarathne, A.S., Liu, F., and Zou, Z. (2025). Plant signaling hormones and transcription factors: key regulators of plant responses to growth, development, and stress. *Plants (Basel)* 14 (7): 1070.

Tian, R., Paul, P., Joshi, S., and Perry, S.E. (2020). Genetic activity during early plant embryogenesis. *Biochemical Journal* 477 (19): 3743–3767.

Tuan, P.A., Kumar, R., Rehal, P.K. et al. (2018). Molecular mechanisms underlying abscisic acid/gibberellin balance in the control of seed dormancy and germination in cereals. *Frontiers in Plant Science* 9: 668.

Uma, G.S., Mahanta, D.K., and Sharma, L. (2024). Advancing forest insect pest management: a focus on biotechnological approaches. In: *Biotechnological Approaches for Sustaining Forest Trees and Their Products* (ed. T. Dennis Thomas, M.K. Razdan, and A. Kumar), 347–383. Singapore: Springer Nature Singapore.

Verma, S., Attuluri, V.P.S., and Robert, H.S. (2022). Transcriptional control of Arabidopsis seed development. *Planta* 255 (4): 90.

Vishal, B. and Kumar, P.P. (2018). Regulation of seed germination and abiotic stresses by gibberellins and abscisic acid. *Frontiers in Plant Science* 9: 838.

Wang, G. and Fiers, M. (2010). CLE peptide signaling during plant development. *Protoplasma* 240 (1): 33–43.

Wang, S., Liang, Y., and Jiang, Y. (2025). Plant CLE peptides: functions, challenges, and future prospects. *Plant Hormones* 1 (1): e006.

Wang, P., Wu, T., Jiang, C. et al. (2023). Brt9SIDA/IDALs as peptide signals mediate diverse biological pathways in plants. *Plant Science* 330: 111642.

Wu, K. (2023). *Developing novel biosensing elements for molecular diagnostics* (Doctoral dissertation).

Wu, B., Liang, H., Lv, J. et al. (2025a). Key regulators of seed germination: kinases and phosphatases. *Seeds* 4 (3): 30.

Wu, C., Munir, R., Li, F. et al. (2025b). From biosynthesis to signaling: unveiling the multifaceted roles of phytosulfokine peptide in plants. *Journal of Experimental Botany* 76 (19): eraf230, 5682–eraf5697.

Xiao, F., Zhou, H., and Lin, H. (2025). Decoding small peptides: Regulators of plant growth and stress resilience. *Journal of Integrative Plant Biology* 67 (3): 596–631.

Xie, H., Zhao, W., Li, W. et al. (2022). Small signaling peptides mediate plant adaptions to abiotic environmental stress. *Planta* 255 (4): 72.

Yadav, V. (2024). Endogenous Peptides Involved in Plant Growth and Development. In: *Plant Elicitor Peptides: New Tool for Sustainable Agriculture*, 113–130. Singapore: Springer Nature Singapore.

Yamaguchi, Y.L., Ishida, T., and Sawa, S. (2016). CLE peptides and their signaling pathways in plant development. *Journal of Experimental Botany* 67 (16): 4813–4826.

Yamoune, A., Cuyacot, A.R., Zdarska, M., and Hejatko, J. (2021). Hormonal orchestration of root apical meristem formation and maintenance in Arabidopsis. *Journal of Experimental Botany* 72 (19): 6768–6788.

Yetgin, A., Srivastava, R.K., and Mandal, N. (2025). Insights into plant hormone signaling networks for environmental responses. In: *Mitigation and Adaptation Strategies Against Climate Change in Natural Systems* (ed. R.K. Srivastava and A. Chakraborty), 505–523. Cham: Springer Nature Switzerland.

Yu, Y., Zhang, S., and Chang, F. (2025). CLE peptides: key regulators of plant development and stress adaption. *Seed Biology* 4 (1): 0.

Zane, D., Feldman, P.L., Sawyer, T., Sobol, Z., and Hawes, J., 2021. Development and regulatory challenges for peptide therapeutics. *International Journal of Toxicology*, vol. 40, no. 2, pp. 108–124, doi: https://doi.org/10.1177/1091581820977846.

Zhang, Z., Han, H., Zhao, J. et al. (2025). Peptide hormones in plants. *Molecular Horticulture* 5 (1): 7.

Zhang, X., Ibrahim, Z., Khaskheli, M.B. et al. (2024a). Integrative approaches to abiotic stress management in crops: combining bioinformatics educational tools and artificial intelligence applications. *Sustainability* 16 (17): 7651.

Zhang, R., Shi, P.T., Zhou, M. et al. (2023). Rapid alkalinization factor: function, regulation, and potential applications in agriculture. *Stress Biology* 3 (1): 16.

Zhang, X., Yang, Z., Wu, D., and Yu, F. (2020). RALF–FERONIA signaling: linking plant immune response with cell growth. *Plant Communications* 1 (4): 100084.

Zhang, P., Zhao, J., Zhang, W. et al. (2024b). Sulfated peptides: key players in plant development, growth, and stress responses. *Frontiers in Plant Science* 15: 1474111.

Zhao, J., He, Y., Zhang, H., and Wang, Z. (2024). Advances in the molecular regulation of seed germination in plants. *Seed Biology* 3 (1): 0.

Zhao, C., Liu, X., Zhou, A. et al. (2025). Transcriptome analysis of cabbage near-isogenic lines reveals the involvement of the plant defensin gene PDF1. 2 in *Fusarium wilt* resistance. *International Journal of Molecular Sciences* 26 (8): 3770.

Zhong, S. and Qu, L.J. (2019). Peptide/receptor-like kinase-mediated signaling involved in male–female interactions. *Current Opinion in Plant Biology* 51: 7–14.

Zhong, S., Zhao, P., Peng, X. et al. (2024). From gametes to zygote: mechanistic advances and emerging possibilities in plant reproduction. *Plant Physiology* 195 (1): 4–35.

Zhou, K., Wu, F., Deng, L. et al. (2025). Antagonistic systemin receptors integrate the activation and attenuation of systemic wound signaling in tomato. *Developmental Cell* 60 (4): 535–550.e8.

Ziemann, S. (2017). Identifizierung und Charakterisierung des Mais-spezifischen Signalpeptids Zip1 und dessen Aktivierung durch Cysteinproteasen. https://open.uni-marburg.de/handle/10.17192/z2017.0668

Zolkiewicz, K. and Gruszka, D. (2022). Glycogen synthase kinases in model and crop plants–from negative regulators of brassinosteroid signaling to multifaceted hubs of various signaling pathways and modulators of plant reproduction and yield. *Frontiers in Plant Science* 13: 939487.

21

Peptide Hormones as Potential Candidates for Food Security

Sanjoy Majumder[1], Gagan Kumar Panigrahi[2], Kunja Bihari Satapathy[3], and Annapurna Sahoo[4]

[1] *School of Biotechnology, Centurion University of Technology and Management, Bhubaneswar, Odisha, India*
[2] *Department of Zoology, School of Applied Sciences, Centurion University of Technology and Management, Bhubaneswar, Odisha, India*
[3] *Department of Botany, School of Applied Sciences, Centurion University of Technology and Management, Bhubaneswar, Odisha, India*
[4] *Department of Zoology, Jatni College, Utkal University, Khordha, Odisha, India*

21.1 Introduction

As the global population is expected to reach close to 9.8 billion by 2050, unprecedented demand for food arises (https://www.un.org/en/desa). Food security is one of the greatest challenges in the twenty-first century. Ensuring food for the global population is becoming an increasingly difficult task. As the population grows, ensuring equitable access to food resources also becomes critically important, as currently, 1 in every 11 people worldwide faced hunger in 2023. As per UN reports dated 24 July2024, more than 733 million people globally suffer from hunger. The rapid growth of the population creates a challenge for global food provision. The existing agricultural systems are already under strain, and many regions cannot produce enough food to meet local demand (Bai et al. 2022; Bock et al. 2022). Urbanization and changes in dietary preferences further exacerbate the problem, increasing the demand for resource-intensive products. Moreover, currently climate change is a serious threat and challenge for the human race. The combination of population surge and difficulties due to climate change and urbanization has left us with very limited arable land, so there is a need for innovative agriculture (Magazzino et al. 2023). There is a need to develop new ways to boost crop yield to tackle these critical situations. Plant peptide hormones are essential for maintaining the growth of plants, tissue differentiation, nutrient uptake responses, and also for the interactions between plants and beneficial microbes in the soil (Ogawa et al. 2008; Roy et al. 2018; Lu and Xiao 2024a; Zhang et al. 2025). It has been seen that some peptide hormones, specifically, can induce cell division or elongation, root development, along with stress tolerance (e.g. drought or salinity) (Yamada and Sawa 2013; Patel et al. 2018; Hsiao and Yamada 2020). In this chapter, we explored the possibility of plant peptide hormones as a tool to deal with food security problems via genetic manipulation and biotechnological applications. Adverse conditions such as climate change, soil deterioration, and also shortage of farmlandhave made it necessary to implement innovative

Peptide Hormones in Plant Development and Stress Responses, First Edition. Edited by Vijay Pratap Singh and Samiksha Singh.

solutions for increasing agricultural production (Furtak and Wolińska 2023). To deal with these issues, biotechnology has emerged as an indispensable resource to address these problems, such as boosting the resilience of crop growth, improving the productivity of livestock, and minimizing food waste (Garland and Curry 2022; Shahid et al. 2024). Use of peptide hormones could be very helpful in agricultural productivity, which in turn aiding in global food security.

21.2 Plant Peptide Hormones as Crucial Players in Agricultural Productivity

Phytohormones, significant regulators of plant growth and development, are increasingly recognized for their multifaceted roles in enhancing crop resilience against environmental stresses (Wang et al. 2016; Xie et al. 2022; Dao et al. 2022) (Figure 21.1). Peptide hormones have a vital role in promoting sustainable agricultural practices, which in turn can contribute tothe increasing global demand for food. These small peptide hormones can act on regulating a number of physiological processes in plants, ranging from growth and development to stress responses (Kim et al. 2021; Fedoreyeva 2023; Ren et al. 2024). It is essential to continue extensive research in this field to make the fullest use of their potential in modern agriculture in terms of improved crop yields as well as crop resilience. These molecules are crucial for maintaining plant health because these small signaling molecules responsible for regulating key physiological processes vital for its life (Chen et al. 2023; Hirakawa and Sawa 2019). This regulation enables plants to survive in adverse environmental conditions such as drought, salinity, and nutrient deficiencies (Oldroyd and Leyser 2020; Panigrahi et al. 2021a; Kang et al. 2022). For instance, phytosulfokine (PSK) peptide hormones are known to be crucial for cell proliferation and fruit development, leading to increased crop yields (Ding et al. 2023; van de Sande et al. 2024; Li et al. 2024). An important additional class of peptides is C-terminally encoded peptides (CEPs), which help in the handling limited resources and allow plants to optimize nutrient allocation to ensure that they achieve the maximum growth potential and also mediate stress responses (Aggarwal et al. 2020; Qiu et al. 2022; Pan et al. 2024). In addition, plant peptide hormones also have an influence on the enhancement of natural defense mechanisms against different pathogenic organisms, including fungi, bacteria, and pests (Albert 2013; Panigrahi and Satapathy 2020; Okamoto et al. 2022). Such physicochemical actions by peptide hormones help plants to limit their dependency on chemical pesticides, encouraging an eco-friendly farming that is friendlier to our environment and human health (Feng et al. 2023; Zhang et al. 2023). Sulfated peptide hormones regulate growth and a wide variety of developmental processes and intricately modulate immunity to pathogens (Kaeothip et al. 2013; Arsova et al. 2018; Sang et al. 2025). These peptides boost the innate immune responses of plants, making them part of sustainable agriculture. The integration of plant peptide hormone research with compatible and sustainable agricultural practices is essential for enhancing global food security. This research is significant in the war against hunger and malnutrition because, by supporting efforts to increase crop resilience and productivity, it helps in solving the problem of food scarcity, which is a serious global issue.

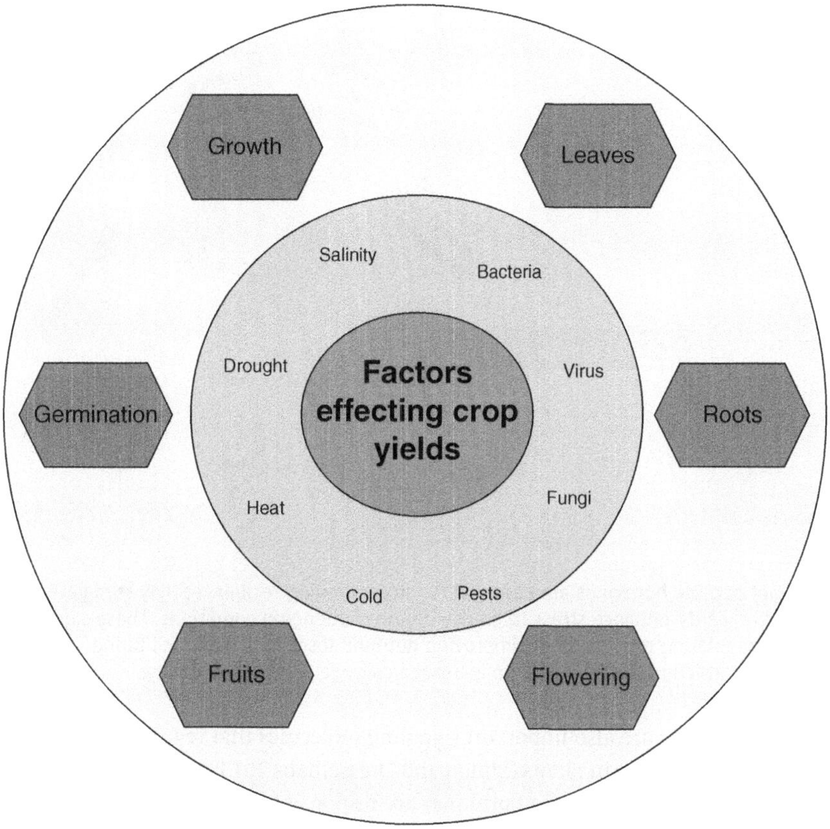

Figure 21.1 Different biotic and abiotic stressors, such as drought, salinity, extreme temperatures, pathogens, and pests, negatively affect the growth and development of plants. These stress factors can lead to noticeable damage in different plant parts, such as leaf, root, flowers, fruits, seeds, and also overall plant health.

These peptide molecules have a positive effect on different physiological processes and can support improved growth rate, tolerance to stress, and nutrient utilization (Furumizu and Aalen 2023; Ji et al. 2025) (Figure 21.2).The addition of peptide hormones into agricultural procedures would not only cut our dependence on chemical fertilizers and pesticides significantly, but could also vastly decrease the reliance on enormous orchards that survive through chemical use (Priya et al. 2023). This increasingly lessens the likelihood of environmental pollution while building healthier ecosystems. In addition, the peptide hormones can enhance the resistance of crops to high levels of environmental stresses like drought and salinity, and also to extreme temperatures (Panigrahi and Satapathy 2020a; Johns et al. 2021; Kuromori et al. 2022).Therefore, the use of these peptide hormones can contribute to a reduction of the footprint of agriculture and corresponding sustainable agricultural food production systems that provide a stable means of food supply. As the world is challenged by climate change and a greater global population, the strategic application of peptide hormones may well be a part of the way forward to develop resilient and sustainable agricultural practices that ensure food availability for all.

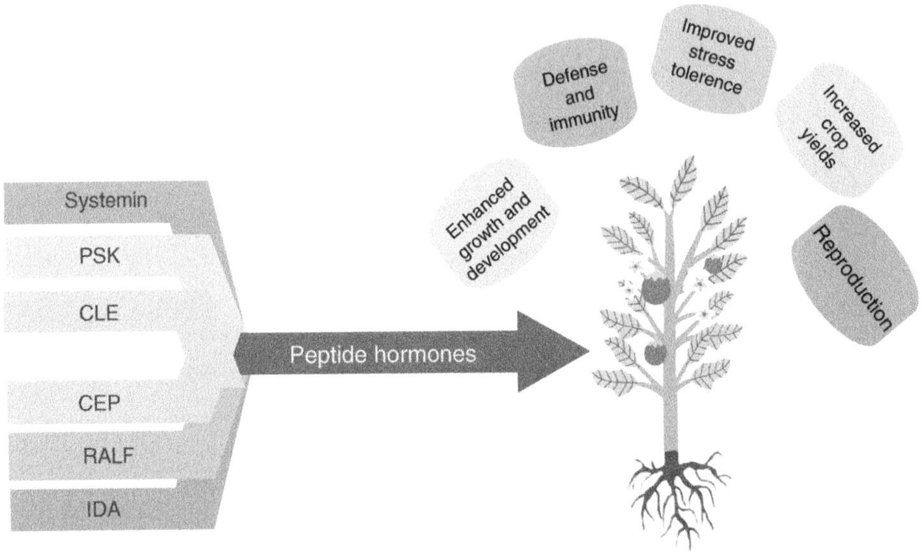

Figure 21.2 Plant peptide hormones are essential for strengthening resilience; they stimulate growth and improve yields, enhance stress tolerance during challenging conditions. These signaling compounds aid in regulating root structure, improving nutrient absorption, and facilitating reproductive growth, which ultimately leads to increased crop yield and productivity.

Plant peptide hormones are also important signaling molecules that regulate the flowering and reproductive processes in plants (Simon and Dresselhaus 2015; Zhong and Qu 2019; Stührwohldt and Schaller 2019). These hormones are responsible for orchestrating the formation of floral organs, which is crucial for successful pollination and fertilization (Haruta et al. 2014; Gao et al. 2023; Lu and Xiao 2024b). Plant peptide hormones also modulate pistil interactions, which are necessary for fertilization and for controlling the way pollen identifies and germinates on the pistil (Stintzi and Schaller 2022; Patro et al. 2024; Baillie et al. 2024). In addition, these hormones help in dispersing the seeds by influencing the maturation of fruits and seeds. Plant peptide hormones help by integrating multifaceted environmental signals with intricate hormonal pathways during the plant reproductive process, thus creating an overall effective process that promotes crop yield, improves fertility, and boosts seed production.

21.3 Functional Dynamics of Peptide Hormones in Plant Development and Resilience

In both shoot and root systems of plants, the CLAVATA3/embryo surrounding region (CLE) peptide hormones are critical regulators of stem cell maintenance (Xu et al. 2013; Kim et al. 2017; Selby and Jones 2023; Bashyal et al. 2024). CLE peptides allow the correct balance of stem cells and thus help build the architecture of roots and shoots, necessary for maximum nutrient and water absorption from the soil (Carbonnel et al. 2022; Zhou et al.

2024; Ávalos-Rangel et al. 2024). They also have an influence on legumes, where they help the plant form nodules and therefore increase the plant's capability to fix atmospheric nitrogen (critical for enriching the soil with nutrients and enhancing the crop yield) (Ohkubo et al. 2017; de Bang et al. 2017). Essentially, they are crucial in maintaining stem cell populations and in influencing the root structure, which significantly supports maximizing nutrient absorption and improving resilience to environmental challenges (Leasure and He 2012). The root growth and elongation are important for the efficient uptake of water and nutrients and their regulation by rapid alkalinization factor (RALF) peptide hormones is critical (Blackburn et al. 2020; Tost et al. 2021; Sui et al. 2023). Given the prevalence of stresses due to climate change, this capability is particularly important to agricultural productivity. PSK peptides are also known to stimulate cell proliferation and differentiation that eventually increase plant biomass production (Sauter 2015; Wu et al. 2019; Millar et al. 2019). These hormones also help in enhancing plant regeneration and tissue culture efficiency, which are critical to the crop-breeding program intent on raising yield. PSK may contribute to promoting healthy cell and tissue growth of plants that are more robust, better prepared to face environmental adversities (Matsubayashi et al. 2006; Friso and van Wijk 2015). CEP hormones are needed for nitrogen uptake and nitrogen use efficiency optimization and also homeostasis in plants (Qiu et al. 2022; Xu et al. 2023). The CEPs take part in root architecture remodeling so that plants can acclimatize better to nutrient-deficient soils and thus increase the yield even under suboptimal conditions (Ma et al. 2024; Taleski et al. 2024). In fact, among the CEP peptides, CEP1 especially has a significant influence on improving the absorption of nutrients, mainly nitrogen (Taleski et al. 2018). CEP peptides are valuable for boosting crop performance in nutrient-deficient soils, making them promising targets for crop improvement. Control of fruit and seed abscission is a key event in agricultural productivity, which is greatly influenced by INFLORESCENCE DEFICIENT IN ABSCISSION (IDA) peptides (Ventimilla et al. 2021). These peptides control the way fruits and seeds are dropped, through which fruits and seeds lead to reproductive success by influencing seed development. Additionally, IDA peptides aid in fruit ripening and harvesting synchronization, increasing yields, and enhancing marketability of crops (Estornell et al. 2015).

Hydroxyproline-rich systemin (HypSys) is involved in the plant's defense mechanisms against herbivores and pathogens, leading to the reduction of the yield losses caused by biotic stress (Schardon et al. 2016; Xu et al. 2015; Andersen et al. 2018). Systemin also plays an important role in orchestrating wound-induced defense responses alongside these peptides (Ryan and Pearce 2003; Zhang et al. 2016; Cirillo et al. 2022).First, the systemin produced by the wounding acts as a signaling molecule, which is transmitted to the adjacent site through the apoplast and phloem to activate the JA cascade reaction pathway (Lyapina et al. 2024; Combest et al. 2021). Second, JA and JA-Ile induced by systemin act as signals and are transported to adjacent sites for defensive responses (Ruan et al. 2019; Yu et al. 2024).Plant peptides hormones have been shown to be very powerful agents to increase yield in low-yield crop production environments (Lv and Yang 2023). These signaling molecules are important factors to regulate many physiological processes during plant growth and development and in response to environmental stress (Xiao and Zhou 2023). In response to climate change–induced unprecedented challenges such as drought, heat stress, and nutrient deficiency, peptide hormones have a significant ability that modulate stress responses (Reichardt et al. 2020). Improving the crops' stress resilience means these hormones will also provide farmers with the means to adapt to the changing environment and retain productivity (Vanyushin et al. 2017; Das et al. 2025; He et al. 2024). Eventually, such fluctuating environmental conditions necessitate harnessing the potential of plant peptide hormones to get climate-resilient crops.

21.3.1 Plant Growth and Development

Peptide hormones act as signal molecules that allow cell-to-cell communication. These hormones are mostly small posttranslationally modified peptides with different structures and functions that are involved in different physiological processes (Gancheva et al. 2019; Chaulagain and Frugoli 2021; Corcilius et al. 2017). There are a number of families of plant peptide hormones, of which CLE, CEP, and PSK have been identified, which play distinct roles (Pečenková et al. 2024). For instance, CLE peptides are critical for controlling diverse developmental functions, including in root development, shoot architecture, and meristem maintenance (Wang et al. 2016; Qian et al. 2018; Carbonnel et al. 2023; Chapman et al. 2024). Numerous studies have shown that by manipulating the signaling pathways of CLE peptides, agricultural benefits such as greater crop yields and better crop adaptation to changing environmental conditions can be achieved. In general, CEP peptides are associated with nutrient acquisition and stress responses. The special feature of these peptides is that they can pass through the plant's xylem, transporting long-distance signaling from the roots to the shoots (Hastwell et al. 2014; Nakayama et al. 2017; Takahashi et al. 2020). This gives them the ability to convey critical information about nutrient availability, water stress, and communication of roots with beneficial soil microorganisms, all of which impact plant growth and environmental resilience. However, PSK peptides possess pleiotropic effects on cell growth and differentiation (Liu et al. 2022a; Wulf et al. 2024). Therefore, they are critical in stimulating cell proliferation, in promoting tracheary element differentiation, which is an essential step in the transport of water and the stimulation of somatic embryogenesis, which is key to plant tissue culture and propagation (Zhang et al. 2019; Fukuda and Hardtke 2020; Narasimhan and Simon 2022). Overall, these peptide hormones are important to plant development and plant adaptation to different environments and therefore they are the key focus of plant biology and agricultural research. Peptide hormones are important mediators of plant–microbe interactions and have profound effects on symbiotic relationships with nitrogen-fixing bacteria, such as *Rhizobium*, as well as mycorrhizal fungi, etc. (Phour et al. 2020; Sahoo et al. 2024; Zhang et al. 2024;). These natural symbiotic relationships between plants and beneficial soil microbes makeagriculture less dependent on synthetic fertilizers,improve soil health, and achieve environmental sustainability (Lepetit and Brouquisse 2023). Use of bioengineered peptide hormones–based biostimulants derived from these hormones may further reduce agricultural inputs by enhancing the use of nutrients and increasing yields of crops.

In recent decades, small peptides that are secreted into extracellular spaces have been recognized as key signaling molecules involved in plant development and stress responses (Murphy et al. 2012; Rzemieniewski et al. 2024). These are peptide hormones that play an important role in a number of developmental processes, including their role in the specific regulation of cell differentiation of shoot and root meristems (Kang et al. 2022a; Song et al. 2022; Fedoreyeva2023). For instance, the signaling peptide CLAVATA3 (CLV3) in the model organism Arabidopsis is found to be essential for maintenance of the shoot apical meristem (SAM) (Shinohara and Matsubayashi 2010; Reid et al. 2011; Pečenková et al. 2024). In addition, a range of environmental conditions, which includes nutrient deficiencies, drought stress, salinity stress, and pathogen attacks, trigger systemic responses in plants more indirectly, through the action of these hormone-like peptides (Kondo et al. 2011). One such example is the *Arabidopsis* drought stress-induced synthesis of the CLV3/EMBRYO SURROUNDING REGION-RELATED 25 (CLE25) peptide, used to signal dehydration stress change from the roots to the shoots (Yamaguchi et al. 2016; Stührwohldt et al. 2021; Endo and Fukuda 2024). Activating of this signal stimulates the ABA biosynthetic genes expression through an ABA cellular signaling pathway with a minimal number of meristem receptors, permitting this plant to adapt to water scarcity (Takahashi et al. 2018; Boschiero et al. 2019; Lee et al. 2019). More research into these putative signaling peptides is needed to deepen our understanding of plant behavior and adaptation at the molecular level. Such insights can help to deliver better agricultural practices as well as crop development with higher levels of climate resilience.

During different stages of development, these signaling peptides are required to guide plant growth, accommodation to their environment, and to assemble more complex structures like roots, stems, leaves, and flowers. CLV3 and related CLE peptides are important group of these peptide hormones (Wang and Fiers 2010; Couto and Zipfel 2016; Greifenhagen et al. 2024). SAM, the region that produces new stems and leaves, depends on CLVs and CLE peptides to maintain its ability to conduct the process of stem cell maintenance (Je et al. 2018; Laffont et al. 2020; Scott et al. 2025). By relaying specific molecular signals, these peptides ensure that the stem cells in the meristem remain undifferentiated (Ryan et al. 2002; Matsuzaki et al. 2010). It is important because this undifferentiated state is a continuous source of stem cells and it differentiates into specialized tissues, which can be used to develop different plant organs as and when required (Mens et al. 2021; Ren et al. 2019; Fernández-Fernández et al. 2023). In addition to the role they play in stem cell regulation, peptide hormones are involved in other areas of development in plants. Coordination of timing and formation of different organs is in the hands of these in order to grow a well-rooted plant and the roots will also be anchored in the soil and will also absorb and conduct water or essential nutrients. Proper regulation of peptide hormones is necessary for maintaining developmental pathways and also for the appropriate functioning of plant organs (Tavormina et al. 2015; Hu et al. 2022). Peptide hormones help plants optimize growth patterns and developmental responses to prevent diminishing within their ecological niches, adapt to fluctuating conditions, and achieve reproductive success, thereby ensuring their survival and propagationacross generations (Liu et al. 2021).

Specifically, the molecular genetics of phytohormone-mediated events has made significant progress in advancing our knowledge of the molecular mechanisms behind those events in plants (Okuda 2021; Guillou et al. 2022). Several studies have unraveled the intricate cross talks among multiple phytohormones, demonstrated how they interact under different situations, such as growth regulation, developmental processes, and reaction with biotic and abiotic stressors. Recently, it was found that many peptide hormones, which act as signaling molecules, have a central role in systemic regulation of such plant responses to changing ambient conditions. They allow long-distance transport and signaling in different parts of the plant for maximizing the overall plant fitness and adaptability (Furumizu et al. 2021; Yoshida et al. 2021). Besides classical phytohormones as auxins, cytokinins, gibberellins, ethylene, and abscisic acid, there has been an increased focus on a number of small peptide hormones that act as signaling molecules. These peptide hormones (e.g. PSKs, CLAVATA) are involved in systemic regulation of stress adaptation and developmental pathways, allowing plants to deal with environmental challenges (van Wijk et al. 2023).

21.3.2 Stress Tolerance

While classical plant hormones such as auxins, gibberellins, and cytokinins mainly control plant growth and development, peptide hormones have a special role as vital intercellular messengers that precisely control plants' adaptation to numerous environmental stresses, such as drought, salinity, extreme temperatures, and pathogen attack (Lindsey 2001; Müller and Harrison 2019).

With climate change causing a rise in various biotic and abiotic factors such as extended drought, soil salinity, extreme temperature fluctuations, and increased pathogen attack, peptide hormones such as RALF and PSY (plant peptides containing sulfated tyrosine) are critical in the regulation of plant stress responses (Chen et al. 2020; Zhang et al. 2024a). It has been revealed by research that RALF can cause quick alkalinization of surrounding soil and enhance the availability of nutrients, while PSY promotes root growth and improves plant tolerance of saline conditions (Bedinger et al. 2010; Lebedeva et al. 2020; Das and Panigrahi 2025; Bhalla et al. 2025). With the advances in genetic engineering and the use of peptide-based treatment, scientists are improving crop tolerance to these adverse conditions. Such advancements in peptide hormone–based agricultural activity may reduce reliance on chemical fertilizers and pesticides. This approach has the added benefit of maintaining productivity while maintaining sustainability in difficult conditions, which adhere to those of sustainable agriculture and food security in an ecologically responsible way.

21.3.2.1 Abiotic Stress

Abiotic stresses such as drought, high salinity, and wide fluctuations in temperature levels deeply affect crop productivity. These stresses interfere with physiological processes, such as the balance of water, uptake of nutrients, photosynthesis, and overall metabolism of the plant (Oldroyd and Leyser 2020). These adverse effects are averted by peptide hormones that act as pivotal players in inducing protective mechanisms and adaptive responses that enable plants to cope with stressful situations (Mosher et al. 2013; Yuan and Wang 2021). Additionally, peptide hormones influence the expression of stress-related genes involved in environmental stress tolerance (Liu et al. 2022). Some peptide hormones activate specific transcription factors that control the biosynthesis of osmolytes and protective proteins, that is, they stabilize cellular structures and maintain cellular functions in stress conditions.For example, PSK has been associated with increased expression of genes required for osmotic stress tolerance. This enhances root growth under drought conditions in an attempt to boost plant resilience and survival rates in harsh environments.

21.3.2.2 Biotic Stress

On the other hand, crops are also often damaged by biotic stresses arising from interaction with pathogenic organisms (fungal, bacterial, viral) and also herbivore attack (Lin et al. 2023). In response to these threats, the plant defense mechanisms include peptide hormones such as peptide pheromones and chitinases (Zelman and Berkowitz 2023; Kim et al. 2021). They help plants to produce complex immune responses such as enhanced protection at cellular walls, formation of antimicrobial compounds, and induction of the systemic acquired resistance system in the plant (Panigrahi et al. 2021; Sahoo et al. 2023; Ghosh and Roychoudhury 2023). In addition, plant peptide hormones are major players in the adaptation of crops to resist biotic stresses, i.e. pathogen attacks (Jung et al. 2020; Zhang et al. 2022). Peptide hormones have been shown to interact with many receptor proteins to start the cascade of defensive reactions in plants, which renders them impregnable to diseases and cuts down on the use of chemicals for protection.

Plants under arid conditions often experience serious difficulty in securing sufficient water to ensure their survival and growth. An example of one effective adaptation mechanism is that longer roots may develop, which have access to deeper water sources than shorter root systems (Oh et al. 2018). Plant peptide hormones are important for the process of root elongation, so that the plant is better prepared when drought stress approaches. These hormones are recognized by cell membrane receptors. These receptors, upon binding with specific peptide hormone molecules, activate downstream signaling pathways that trigger physiological and developmental responses, mainly enhancing the plant's stress tolerance (Motomitsu et al. 2015; Nguyen and Kim 2023). To give an example, it has been shown that in the model organism *Arabidopsis thaliana*, CLE peptides act as signals for dehydration stress responses which result in physiological changes that allow the plant to become more resilient in water-scarce environments.

Largely, the application of peptide hormones with precision agriculture technology is a promising method of optimizing agricultural practices (Zhang et al. 2023). Together, these hormonal treatments can be combined with real-time collection of data on soil health, moisture conditions, and the real needsof the plants, which can be used to do targeted intervention by farmers. It is not only for maximizing crop productivity but also for optimizing the use of resources, and hence minimizing waste as well as reducing the environmental footprint of agricultural activities (Setty et al. 2024). Such advanced strategies give the capacity to make crops more climate change resilience and water scarcity tolerant (Hirayama and Mochida 2022). New evidence is emerging that plant peptide hormones are outstanding regulators of stress tolerance in agricultural food crops. These advancements in agricultural practices hold promise for global food security.

21.4 Disease Resistance

Plant peptide hormones have an essential and multifaceted role in strengthening disease resistance against pathogens (Panigrahi and Satapathy 2021; Pastor-Fernández et al. 2023). These little but potent signaling molecules start many of the important processes to fight against infections (Toruño et al. 2016). These peptides act to trigger the production of antimicrobial compounds, including the phytoalexins and pathogenesis-related proteins, that directly inhibit the growth and spread of pathogens upon pathogen recognition (Igarashi et al. 2012). In addition, they trigger the deposition of structural proteins and polysaccharides in cell walls to reinforce the cell walls, thus limiting pathogen entry. In some more advanced cases of infection, these hormones cause programmed cell death of infected tissue in localized infected regions to limit pathogen growth and prevent systemic infection. These responses are mediated by a number of specific peptide hormones. For example, systemins are signals for most of the resistance-enhancing pathways primarily against herbivorous pests, while the RALF peptides have been linked to both stress and growth regulation (Andersen et al. 2018; Panigrahi et al. 2024; Žárský et al. 2024). Phytosulfokines, peptides enhance cell growth, stimulate immune response, while the plant elicitor peptides (Peps) bind to specific pattern recognition receptors present on the plant cell surface (Yang et al. 2001; Hou et al. 2021). As a result, these signaling events lead to a cascade of gene activation resulting in the plant's systemic acquired resistance (Hirakawa and Sawa 2019; Zelman et al. 2024). Furthermore, the crosstalk between these peptide hormones and other signaling molecules, such as jasmonic acid and salicylic acid, underscores a complex network of plant defense strategies (Matsuzaki et al. 2010). Plant peptide hormones greatly contribute to innate immunity and provide plants with mechanisms to survive and adapt to a broad range of pathogenic challenges and maintain fitness in different adverse environments (Salvi et al. 2021).

21.5 Engineering Peptide Hormone Production for Agricultural Applications

Peptide hormones engineering for improved food crop yields is a modern biotechnology approach based on the principles of molecular biology and genetic engineering aimed at improving various aspects of growth of plants as well as stress tolerance and efficient nutrient use (Roy et al. 2022;Luo et al. 2023; Chen et al. 2023). These small signaling molecules are crucial in regulating important physiological processes of seed germination, root development, flowering, as well as responses to both biotic (e.g. pest attacks and pathogenic infections) and abiotic (e.g. extreme temperature, drought, and salinity) stresses (Mogg 2020; Fang et al. 2024; Zhong et al. 2022). Researchers are now modifying the structural and expression patterns of these peptides with enhanced resilience to unfavorable environmental conditions. Additionally, CLE peptides essential for root architecture regulation can be manipulated so that root growth occurs in nutrient-rich zones of the soil, increasing nutrients and water uptake efficiency (Xiong et al. 2022; Nakagami et al. 2024; Chang and Xiao 2025). However, recombinant production methods make the possibility of large-scale application of these hormones, with resultant improved crop performance on marginal lands. Advancements in synthetic biology and CRISPR-based genome editing technologies have increased the ability to design and introduce optimized peptide hormone pathways in various crop species (Gan and Ling 2022; Rao et al. 2022; Amoah et al. 2024). The ability to target specific traits very precisely gives the crops the ability to maintain high yields under fluctuating climatic conditions. Research utilizing CRISPR or other gene-editing tools to develop crops with improved tolerance to environmental stresses is needed. In addition to increasing the yield of potential of food crops, these peptides not only reduce reliance on synthetic chemical pesticides and fertilizers but also shorten the life cycle involved in the production of these products. Such a shift facilitates more sustainable, eco-friendly yet less expensive and socially as well as environmentally friendly agricultural practices that can contribute to global food security and address environmental issues in this field (Zhang et al. 2023; Kobayashi et al. 2024; Setty et al. 2024). Finally, the use of peptide hormone engineering into crop development is a promising path for future agri innovations that give farmers enough tools to face adverse situations like climate change and global population explosion.

21.6 Advances in Research of Peptide Hormones

Plant peptide hormones have recently gained much attention owing to their important role in the regulation of plant growth, development, and stress responses (Ren et al. 2024). Instead of their small sizes, which can be as short as a few amino acids, these peptides play important roles in crucial signaling to carry out many physiological and developmental processes (Hirakawa et al. 2017). For example, CLAVATA3, SUNN, and other peptide hormones provide input to regulate stem cell populationsand control organ formation (De Coninck and De Smet 2016). In addition, it has been demonstrated how these peptide hormones work with receptor-like kinases and other signaling pathways, further enabling us to understand plant biology on a more comprehensive level. Moreover, there isgrowing interest in research on plant peptide hormones in agricultural biotechnology (Badosa et al. 2022). Continued exploration of the peptide signaling pathways, development of synthetic peptide analogs that mimic natural peptides,and advances in genetic-engineering techniques are expected to advance plant biotechnology in the coming years and enable cropsto survive challenging environmental conditions (Pruitt et al. 2017).

21.7 Sustainable Agricultural Practices

Plant peptide hormones are important for accomplishing sustainable development goals (SDGs), most especially those on food security (SDG 2) and good health and well-being (SDG 3) (von Braun et al. 2023). These small powerful signaling molecules regulate a wide range of physiological processes such as growth promotion, stress response mechanisms, and defenses against biotic and abiotic stresses (Waadt et al. 2022). As peptide hormones act as natural growth enhancers, they are valuable in sustainable agricultural practices (Fang et al. 2021; Schaller 2024). Application of these hormones can lead to a decrease in the use of synthetic fertilizers and pesticides, decreasing chemical runoff and environmental degradation. For instance, application of recombinant PSK peptides has demonstrated increasing root biomass and nutrient uptake efficiency, increasing crop yields, and also maintaining ecological stability (Ntim 2021; Li et al. 2024) (Figure 21.3). Integration of plant peptide hormones with advanced agricultural and biotechnological practices

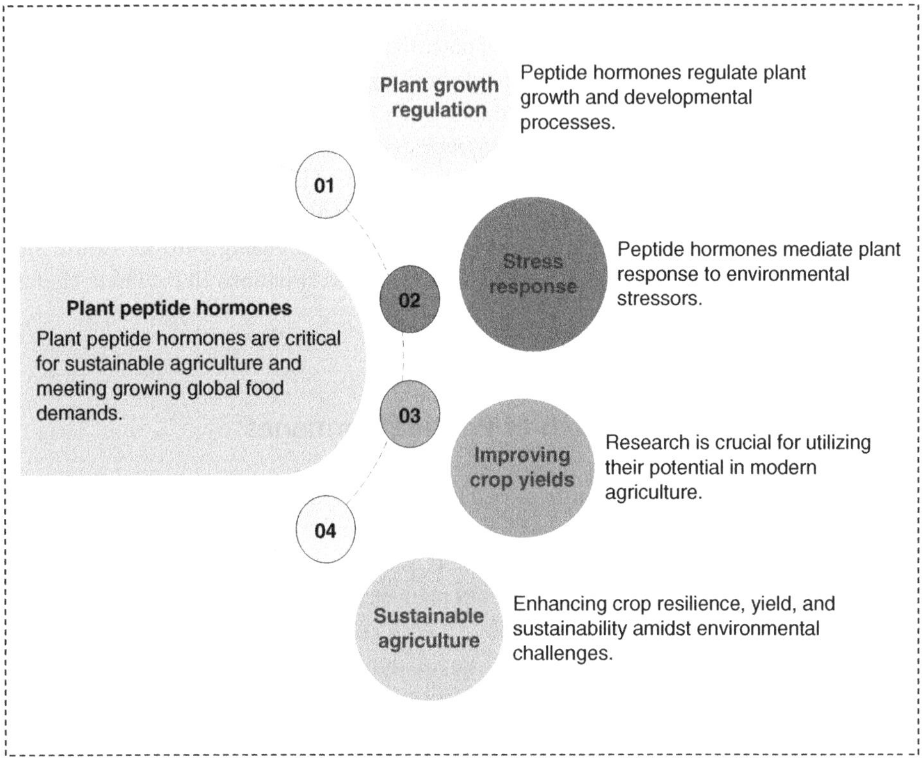

Figure 21.3 Plant peptide hormones play vital roles in regulating growth, development, and responses to stress, making them important contributors to sustainable agriculture. Progress in peptide engineering allows for the creation of synthetic analogs that can replicate natural peptides, providing novel options for enhancing crop performance and resilience.

can enhance crop resilience and reduce environmental footprint, and also align with several SDGs, which in turn lead to future agricultural practices that are more sustainable and more secure (Amoah et al. 2024; Setty et al. 2024).

21.8 Challenges in Agricultural Productivity and Defining the Role of Peptide Hormones

In the twenty-first century, the grand challenge facing scientists is to optimize climate change adaptation to maintain agricultural productivity, food security, and environmental protection. Research and application of peptide hormones are limited by several factors. For instance, signaling pathways of many plant peptide hormones are poorly understood. However, owing to highly interconnected pathways involving multiple receptors that link with classical phytohormones such as auxin and gibberellin, it is difficult to define their individual roles. This is a clear indication that there is a need for precise application strategies. Unlike conventional phytohormones, peptide hormones often require more sophisticated methods in order for them to be absorbed well and take action inside plant tissues. Unraveling complex signaling pathways on how these molecules interact with each other and plant cells can also open the way to such innovative, future agronomic solutions as a means to secure food for future generations. Yet there are several noteworthy challenges to the practical application of plant peptide hormones in agricultural settings. Solutions to these problems and the development of new research approaches may facilitate the full exploitation of peptide hormones by the agricultural community as tools to ensure food security and environmental sustainability.

21.9 Conclusions and Future Perspectives

By bringing research on peptide hormones together with advanced technologies, the use of precision farming and environmentally friendly methodologies, it is possible to bring about a balance between providing food for the world while preserving natural vital resources. However, collaborative research among the plant scientists, the agronomists, and biotechnologists will be key to uncovering the complete potential for plant peptide hormone applications in current conditions, with outcomes so that we may continue producing food in a sustainable, efficient, and equitable manner to feed the ever-increasing population globally. The role of plant peptide hormones toward the achievement of global food security has promising future prospects. Whole-genome sequencing technologies are revolutionizing our understanding of plant biology since they enabled the discovery of specific genes coding for plant peptide hormones and their respective receptors. Moreover, RNA sequencing (RNAseq) is a powerful tool that provides information regarding the stress-responsive expression of genes under different types of environmental stressors, like drought, salinity, and extreme temperature fluctuation. The ability to understand expression profiles of specific peptide hormones allows scientists to decide how the plant responses to stress conditions at the different stages. Plant peptide hormones coupled with new technologies offer great opportunities for addressing the problem of food resources for a growing population of people. These hormones greatly influence the process of growth, development, and stress responses in plants. Understanding these mechanisms would enable scientists to develop crops that not only are climate smart but also drought tolerant and flood tolerant as well as able to resist the range of diseases caused by microbial pathogens and pests. It is a groundbreaking approach for the creation of high-yielding crop varieties that will be able to satisfy the ever-increasing global food demand. If we cultivate crops that need less input of chemicals, i.e. much less inputs of fertilizers and pesticides, we also have a high chance of significantly reducing the harmfulness to the environment. Largely, research is the essence of future agricultural sustainability and global food security; hence, it is an indispensable part of our sustainable agricultural production goal. Currently, there are many advanced molecular biology techniques available that can be used for the comprehensive study of different signaling pathways of plants with more precision. For instance, with the use of advanced omics technologies such as genomics, transcriptomics, proteomics, and metabolomics, the molecular mechanism of the functioning of the plant peptide hormones can be finely decoded. These technologies can assist in identifying key candidates to improve growth, tolerance to stress, nutritional profiles, and thus, more sustainable agricultural practices. The plant peptide hormones offer a promising opportunity in agriculture science to provide a significant boost to global food security. Some of these molecules have important functions to regulate plant growth, to increase tolerance to abiotic and biotic stresses, and to promote the uptake of soil nutrients, making them useful tools to achieve sustainable agriculture. More research into the elaborate mechanisms of action of these peptide hormones is needed to take full advantage of them to tackle global food challenges. Therefore, more work at the molecular level is required to reveal the complexity of signaling networks that govern peptide hormone interactions and establish long-term effects on various plant communities. Moreover, the synthesis and application of plant peptide hormones will need to be developed at affordable costs with potential for scaling up. Such technologies' exploration could also result in inexpensive formulations that are easily included in farmers' practical uses.

References

Aggarwal, S., Kumar, A., Jain, M. et al. (2020). C-terminally encoded peptides (CEPs) are potential mediators of abiotic stress response in plants. *Physiology and Molecular Biology of Plants* 26: 2019–2033.

Albert, M. (2013). Peptides as triggers of plant defence. *Journal of Experimental Botany* 64: 5269–5279.

Amoah, P., Oumarou Mahamane, A.R., Byiringiro, M.H. et al. (2024). Genome editing in sub-Saharan Africa: a game-changing strategy for climate change mitigation and sustainable agriculture. *GM Crops & Food* 15 (1): 279–302. https://doi.org/10.1080/21645698.2024.2411767.

Andersen, E.J., Ali, S., Byamukama, E. et al. (2018). Disease resistance mechanisms in plants. *Genes* 9 (7): 339. https://doi.org/10.3390/genes9070339.

Arsova, B., Watt, M., and Usadel, B. (2018). Monitoring of plant protein post-translational modifications using targeted proteomics. *Frontiers in Plant Science* 9: 1168.

Ávalos-Rangel, A., Ruiz-Herrera, L.F., Gutiérrez-Alanis, D. et al. (2024). CLE14 peptide impairs root tip regeneration and callogenesis in Arabidopsis. *Journal of Plant Growth Regulation* 43 (5): 1458–1465.

Badosa, E., Planas, M., Feliu, L. et al. (2022). Synthetic peptides against plant pathogenic bacteria. *Microorganisms* 10 (9): 1784. https://doi.org/10.3390/microorganisms10091784.

Bai, Y., Herforth, A., and Masters, W.A. (2022). Global variation in the cost of a nutrient-adequate diet by population group: an observational study. *Lancet Planet. Health* 6 (1): e19–e28. https://doi.org/10.1016/S2542-5196(21)00285-0.

Baillie, A.L., Sloan, J., Qu, L.J., and Smith, L.M. (2024). Signalling between the sexes during pollen tube reception. *Trends in Plant Science* 29 (3): 343–354.

de Bang, T.C., Lundquist, P.K., Dai, X. et al. (2017). Genome-wide identification of Medicago peptides involved in macronutrient responses and nodulation. *Plant Physiology* 175 (4): 1669–1689.

Bashyal, S., Gautam, C.K., and Müller, L.M. (2024). CLAVATA signaling in plant-environment interactions. *Plant Physiology* 194 (3): 1336–1357. https://doi.org/10.1093/plphys/kiad591.

Bedinger, P.A., Pearce, G., and Covey, P.A. (2010). RALFs: peptide regulators of plant growth. *Plant Signaling & Behavior* 5 (11): 1342–1346. https://doi.org/10.4161/psb.5.11.12954.

Bhalla, H., Sudarsanam, K., Srivastava, A., and Sankaranarayanan, S. (2025). Structural insights into the recognition of RALF peptides by FERONIA receptor kinase during Brassicaceae pollination. *Plant Molecular Biology* 115 (1): 1–6.

Blackburn, M.R., Haruta, M., and Moura, D.S. (2020). Twenty years of progress in physiological and biochemical investigation of RALF peptides. *Plant Physiology* 182 (4): 1657–1666.

Bock, A.-K., Bontoux, L., and Rudkin, J. (2022). *Concepts for a Sustainable EU Food System: Reflections from a Participatory Process*. Publications Office of the European Union, LU https://data.europa.eu/doi/10.2760/381319.

Boschiero, C., Lundquist, P.K., Roy, S. et al. (2019). Identification and functional investigation of genome-encoded, small, secreted peptides in plants. *Current Protocols in Plant Biology* 4 (3): e20098.

von Braun, J., Afsana, K., Fresco, L.O., and Hassan, M.H.A. (2023). Science for Transformation of Food Systems: Opportunities for the UN Food Systems Summit. In: *Science and Innovations for Food Systems Transformation* (ed. J. von Braun, K. Afsana, L.O. Fresco, and M.H.A. Hassan). Cham: Springer https://doi.org/10.1007/978-3-031-15703-5_50.

Carbonnel, S., Cornelis, S., and Hazak, O. (2023). The CLE33 peptide represses phloem differentiation via autocrine and paracrine signaling in Arabidopsis. *Communications Biology* 6 (1): 588.

Carbonnel, S., Falquet, L., and Hazak, O. (2022). Deeper genomic insights into tomato CLE genes repertoire identify new active peptides. *BMC Genomics* 23 (1): 756.

Chang, S. and Xiao, F. (2025). Comprehensive review of plant small signaling peptides: from stress adaptation mechanisms to practical solutions for crop resilience. *International Journal of Biological Macromolecules* 299, 139971.

Chapman, K., Taleski, M., Frank, M., and Djordjevic, M.A. (2024). C-TERMINALLY ENCODED PEPTIDE (CEP) and cytokinin hormone signaling intersect to promote shallow lateral root angles. *Journal of Experimental Botany* 75 (2): 631–641.

Chaulagain, D. and Frugoli, J. (2021). The regulation of nodule number in legumes is a balance of three signal transduction pathways. *International Journal of Molecular Sciences* 22 (3): 1117.

Chen, Y.L., Fan, K.T., Hung, S.C., and Chen, Y.R. (2020). The role of peptides cleaved from protein precursors in eliciting plant stress reactions. *The New Phytologist* 225 (6): 2267–2282. https://doi.org/10.1111/nph.16241.

Chen, H., Song, Y., Li, H. et al. (2023). Enhancing the adaptability of tea plants (*Camellia sinensis* L.) to high-temperature stress with small peptides and biosurfactants. *Plants (Basel, Switzerland)* 12 (15): 2817. https://doi.org/10.3390/plants12152817.

Cirillo, V., Molisso, D., Aprile, A.M. et al. (2022). Systemin peptide application improves tomato salt stress tolerance and reveals common adaptation mechanisms to biotic and abiotic stress in plants. *Environmental and Experimental Botany* 199: 104865.

Combest, M.M., Moroz, N., Tanaka, K. et al. (2021). St PIP1, a PAMP-induced peptide in potato, elicits plant defenses and is associated with disease symptom severity in a compatible interaction with potato virus Y. *Journal of Experimental Botany* 72 (12): 4472–4488.

Corcilius, L., Hastwell, A.H., Zhang, M. et al. (2017). Arabinosylation modulates the growth-regulating activity of the peptide hormone CLE40a from soybean. *Cell Chemical Biology* 24 (11): 1347–1355.

Couto, D. and Zipfel, C. (2016). Regulation of pattern recognition receptor signalling in plants. *Nature Reviews Immunology* 16 (9): 537–552.

Dao, T.Q., Weksler, N., Liu, H.M.H. et al. (2022). Interactive CLV3, CLE16 and CLE17 signaling mediates stem cell homeostasis in the Arabidopsis shoot apical meristem. *Development* 149 (19): dev200787.

Das, R. and Panigrahi, G.K. (2025). Messenger RNA surveillance: current understanding, regulatory mechanisms, and future implications. *Molecular Biotechnology* 67 (2): 393–409. https://doi.org/10.1007/s12033-024-01062-4.

Das, R., Singh, S.P., Sahoo, A. et al. (2025). Synergistic stress of zinc oxide nanoparticles and pH triggers gill damage and mortality in zebrafish. *Nanotechnology for Environmental Engineering* 10 (1): 1–11.

De Coninck, B. and De Smet, I. (2016). Plant peptides – taking them to the next level. *Journal of Experimental Botany* 67 (16): 4791–4795. https://doi.org/10.1093/jxb/erw309.

Ding, S., Lv, J., Hu, Z. et al. (2023). Phytosulfokine peptide optimizes plant growth and defense via glutamine synthetase GS2 phosphorylation in tomato. *The EMBO Journal* 42 (6): e111858.

Endo, S. and Fukuda, H. (2024). A cell-wall-modifying gene-dependent CLE26 peptide signaling confers drought resistance in Arabidopsis. *PNAS Nexus* 3 (2): pgae049. https://doi.org/10.1093/pnasnexus/pgae049.

Estornell, L.H., Wildhagen, M., Pérez-Amador, M.A. et al. (2015). The IDA peptide controls abscission in *Arabidopsis* and *Citrus*. *Frontiers in Plant Science* 6: 1003.

Fang, Y., Chang, J., Shi, T. et al. (2021). Evolution of RGF/GLV/CLEL peptide hormones and their roles in land plant growth and regulation. *International Journal of Molecular Sciences* 22 (24): 13372. https://doi.org/10.3390/ijms222413372.

Fang, H., Zuo, J., Ma, Q. et al. (2024). Phytosulfokine promotes fruit ripening and quality via phosphorylation of transcription factor DREB2F in tomato. *Plant Physiology* 194 (4): 2739–2754.

Fedoreyeva, L.I. (2023). Molecular mechanisms of regulation of root development by plant peptides. *Plants* 12 (6): 1320. https://doi.org/10.3390/plants12061320.

Feng, Y.Z., Zhu, Q.F., Xue, J. et al. (2023). Shining in the dark: the big world of small peptides in plants. *aBIOTECH* 4 (3): 238–256. https://doi.org/10.1007/s42994-023-00100-0.

Fernández-Fernández, Á.D., Stael, S., and Van Breusegem, F. (2023). Mechanisms controlling plant proteases and their substrates. *Cell Death and Differentiation* 30 (4): 1047–1058. https://doi.org/10.1038/s41418-023-01120-5.

Friso, G. and van Wijk, K.J. (2015). Posttranslational protein modifications in plant metabolism. *Plant Physiology* 169 (3): 1469–1487. https://doi.org/10.1104/pp.15.01378.

Fukuda, H. and Hardtke, C.S. (2020). Peptide signaling pathways in vascular differentiation. *Plant Physiology* 182 (4): 1636–1644.

Furtak, K. and Wolińska, A. (2023). The impact of extreme weather events as a consequence of climate change on the soil moisture and on the quality of the soil environment and agriculture–a review. *Catena* 231: 107378.

Furumizu, C. and Aalen, R.B. (2023). Peptide signaling through leucine-rich repeat receptor kinases: insight into land plant evolution. *New Phytologist* 238 (3): 977–982.

Furumizu, C., Krabberød, A.K., Hammerstad, M. et al. (2021). The sequenced genomes of nonflowering land plants reveal the innovative evolutionary history of peptide signaling. *The Plant Cell* 33 (9): 2915–2934.

Gan, W.C. and Ling, A.P.K. (2022). CRISPR/Cas9 in plant biotechnology: applications and challenges. *Biotechnologia* 103 (1): 81–93. https://doi.org/10.5114/bta.2022.113919.

Gancheva, M.S., Malovichko, Y.V., Poliushkevich, L.O. et al. (2019). Plant peptide hormones. *Russian Journal of Plant Physiology* 66: 171–189.

Gao, Q., Wang, C., Xi, Y. et al. (2023). RALF signaling pathway activates MLO calcium channels to maintain pollen tube integrity. *Cell Research* 33 (1): 71–79.

Garland, S. and Curry, H.A. (2022). Turning promise into practice: crop biotechnology for increasing genetic diversity and climate resilience. *PLoS Biology* 20 (7): e3001716. https://doi.org/10.1371/journal.pbio.3001716.

Ghosh, P. and Roychoudhury, A. (2023). Plant peptides involved in abiotic and biotic stressresponses and Reactive Oxygen Species (ROS) signaling. *Journal of Plant Growth Regulation* 43 (5): 1410–1427.

Greifenhagen, A., Ruwe, H., Zimmer, V. et al. (2024). The peptide hormone *PjCLE1* stimulates haustorium formation in the parasitic plant *Phtheirospermum japonicum*. *Proceedings of the National Academy of Sciences* 121 (42): e2414582121.

Guillou, M.C., Balliau, T., Vergne, E. et al. (2022). The PROSCOOP10 gene encodes two extracellular hydroxylated peptides and impacts flowering time in Arabidopsis. *Plants* 11 (24): 3554.

Haruta, M., Sabat, G., Stecker, K. et al. (2014). A peptide hormone and its receptor protein kinase regulate plant cell expansion. *Science(New York, N.Y.)* 343 (6169): 408–411. https://doi.org/10.1126/science.1244454.

Hastwell, A.H., Gresshoff, P.M., and Ferguson, B.J. (2014). The structure and activity of nodulation-suppressing CLE peptide hormones of legumes. *Functional Plant Biology* 42 (3): 229–238.

He, L., Wu, L., and Li, J. (2024). Sulfated peptides and their receptors: key regulators of plant development and stress adaptation. *Plant Communications*.

Hirakawa, Y. and Sawa, S. (2019). Diverse function of plant peptide hormones in local signaling and development. *Current Opinion in Plant Biology* 51: 81–87. https://doi.org/10.1016/j.pbi.2019.04.005.

Hirakawa, Y., Torii, K.U., and Uchida, N. (2017). Mechanisms and strategies shaping plant peptide hormones. *Plant & Cell Physiology* 58 (8): 1313–1318. https://doi.org/10.1093/pcp/pcx069.

Hirayama, T. and Mochida, K. (2022). Plant hormonomics: a key tool for deep physiological phenotyping to improve crop productivity. *Plant and Cell Physiology* 63 (12): 1826–1839.

Hou, S., Liu, D., and He, P. (2021). Phytocytokines function as immunological modulators of plant immunity. *Stress Biology* 1: 1–14.

Hsiao, Y.C. and Yamada, M. (2020). The roles of peptide hormones and their receptors during plant root development. *Genes* 12 (1): 22. https://doi.org/10.3390/genes12010022.

Hu, C., Zhu, Y., Cui, Y. et al. (2022). A CLE–BAM–CIK signalling module controls root protophloem differentiation in Arabidopsis. *New Phytologist* 233 (1): 282–296.

Igarashi, D., Tsuda, K., and Katagiri, F. (2012). The peptide growth factor, phytosulfokine, attenuates pattern-triggered immunity. *The Plant Journal* 71 (2): 194–204.

Je, B.I., Xu, F., Wu, Q. et al. (2018). The CLAVATA receptor FASCIATED EAR2 responds to distinct CLE peptides by signaling through two downstream effectors. *eLife* 7: e35673.

Ji, C., Li, H., Zhang, Z. et al. (2025). The power of small signaling peptides in crop and horticultural plants. *The Crop Journal* 13: 656–667.

Johns, S., Hagihara, T., Toyota, M., and Gilroy, S. (2021). The fast and the furious: rapid long-range signaling in plants. *Plant Physiology* 185 (3): 694–706.

Jung, H.W., Panigrahi, G.K., Jung, G.-Y. et al. (2020). PAMP-triggered immunity involves proteolytic degradation of core nonsense-mediated mRNA decay factors during early defense response. *The Plant Cell* 32 (4): 1081–1101. https://doi.org/10.1105/tpc.19.00631.

Kaeothip, S., Ishiwata, A., and Ito, Y. (2013). Stereoselective synthesis of Arabidopsis CLAVATA3 (CLV3) glycopeptide, unique protein post-translational modifications of secreted peptide hormone in plant. *Organic & Biomolecular Chemistry* 11 (35): 5892–5907.

Kang, J., Peng, Y., and Xu, W. (2022). Crop root responses to drought stress: molecular mechanisms, nutrient regulations, and interactions with microorganisms in the rhizosphere. *International Journal of Molecular Sciences* 23 (16): 9310. https://doi.org/10.3390/ijms23169310.

Kang, J., Wang, X., Ishida, T. et al. (2022a). A group of CLE peptides regulates de novo shoot regeneration in *Arabidopsis thaliana*. *New Phytologist* 235 (6): 2300–2312.

Kim, J.S., Jeon, B.W., and Kim, J. (2021). Signaling peptides regulating abiotic stress responses in plants. *Frontiers in Plant Science* 12: 704490.

Kim, H.J., Wu, C.Y., Yu, H.M. et al. (2017). Dual CLAVATA3 Peptides in Arabidopsis Shoot Stem Cell Signaling. *Journal of Plant Biology = Singmul Hakhoe chi* 60 (5): 506–512. https://doi.org/10.1007/s12374-017-0083-2.

Kobayashi, R., Ohkubo, Y., Izumi, M. et al. (2024). Integration of shoot-derived polypeptide signals by root TGA transcription factors is essential for survival under fluctuating nitrogen environments. *Nature Communications* 15 (1): 6903.

Kondo, T., Yokomine, K., Nakagawa, A., and Sakagami, Y. (2011). Analogs of the CLV3 peptide: synthesis and structure-activity relationships focused on proline residues. *Plant & Cell Physiology* 52 (1): 30–36. https://doi.org/10.1093/pcp/pcq146.

Kuromori, T., Fujita, M., Takahashi, F. et al. (2022). Inter-tissue and inter-organ signaling in drought stress response and phenotyping of drought tolerance. *The Plant Journal* 109 (2): 342–358.

Laffont, C., Ivanovici, A., Gautrat, P. et al. (2020). The NIN transcription factor coordinates CEP and CLE signaling peptides that regulate nodulation antagonistically. *Nature Communications* 11 (1): 3167.

Leasure, C.D. and He, Z.H. (2012). CLE and RGF family peptide hormone signaling in plant development. *Molecular Plant* 5 (6): 1173–1175.

Lebedeva, M., Azarakhsh, M., Yashenkova, Y., and Lutova, L. (2020). Nitrate-induced CLE peptide systemically inhibits nodulation in *Medicago truncatula*. *Plants* 9 (11): 1456.

Lee, Z.H., Hirakawa, T., Yamaguchi, N., and Ito, T. (2019). The roles of plant hormones and their interactions with regulatory genes in determining meristem activity. *International Journal of Molecular Sciences* 20 (16): 4065.

Lepetit, M. and Brouquisse, R. (2023). Control of the rhizobium–legume symbiosis by the plant nitrogen demand is tightly integrated at the whole plant level and requires inter-organ systemic signaling. *Frontiers in Plant Science* 14: 1114840.

Li, J., Huang, Y., Yu, X. et al. (2024). Identification and application of CLE peptides for drought resistance in Solanaceae crops. *Journal of Agricultural and Food Chemistry.*.

Lin, Z.B., Zhao, Y.X., and Yang, W.C. (2023). A balanced mechanism between plant growth and stress response. *Advanced Agrochem* 2 (2): 105–106.

Lindsey, K. (2001). Plant peptide hormones: the long and the short of it. *Current Biology* 11 (18): R741–R743.

Liu, D., Shen, Z., Zhuang, K. et al. (2022). Systematic annotation reveals CEP function in tomato root development and abiotic stress response. *Cells* 11 (19): 2935.

Liu, Q., Wu, K., Song, W. et al. (2022a). Improving crop nitrogen use efficiency toward sustainable green revolution. *Annual Review of Plant Biology* 73 (1): 523–551.

Liu, Y., Zuo, T., Qiu, Z. et al. (2021). Genome-wide identification reveals the function of CEP peptide in cucumber root development. *Plant Physiology and Biochemistry* 169: 119–126.

Lu, S., andXiao, F. (2024a). Current advances in small peptides during plant growth and development. DOI:https://doi.org/10.20944/preprints202406.0156.v1

Lu, S. and Xiao, F. (2024b). Small peptides: orchestrators of plant growth and developmental processes. *International Journal of Molecular Sciences* 25 (14): 7627. https://doi.org/10.3390/ijms25147627.

Luo, Z., Wang, J., Li, F. et al. (2023). The small peptide CEP1 and the NIN-like protein NLP1 regulate NRT2. 1 to mediate root nodule formation across nitrate concentrations. *The Plant Cell* 35 (2): 776–794.

Lv, Q.W. and Yang, Y.F. (2023). The biological functions of peptide signaling in plant and the advances on its utilization for crop improvement. Yi chuan =. *Hereditas* 45 (9): 813–828. https://doi.org/10.16288/j.yczz.23-169.

Lyapina, I., Ganaeva, D., Rogozhin, E.A. et al. (2024). Comparative analysis of small secreted peptide signaling during defense response: insights from vascular and non-vascular plants. *bioRxiv, 2024-05.*

Ma, Y., Zhang, Y., Xu, J. et al. (2024). Research on the mechanisms of phytohormone signaling in regulating root development. *Plants* 13 (21): 3051.

Magazzino, C., Cerulli, G., Shahzad, U., and Khan, S. (2023). The nexus between agricultural land use, urbanization, and greenhouse gas emissions: novel implications from different stages of income levels. *Atmospheric Pollution Research* 14 (9): 101846.

Matsubayashi, Y., Ogawa, M., Kihara, H. et al. (2006). Disruption and overexpression of Arabidopsis phytosulfokine receptor gene affects cellular longevity and potential for growth. *Plant Physiology* 142 (1): 45–53.

Matsuzaki, Y., Ogawa-Ohnishi, M., Mori, A., and Matsubayashi, Y. (2010). Secreted peptide signals required for maintenance of root stem cell niche in Arabidopsis. *Science* 329: 1065–1067.

Mens, C., Hastwell, A.H., Su, H. et al. (2021). Characterisation of *Medicago truncatula* CLE34 and CLE35 in nitrate and rhizobia regulation of nodulation. *New Phytologist* 229 (5): 2525–2534.

Millar, A.H., Heazlewood, J.L., Giglione, C. et al. (2019). The scope, functions, and dynamics of posttranslational protein modifications. *Annual Review of Plant Biology* 70 (1): 119–151.

Mogg, S. (2020). Investigating the diversity of CLE proteins and their function in *Arabidopsis thaliana*. The University of Manchester (United Kingdom).

Mosher, S., Seybold, H., Rodriguez, P. et al. (2013). The tyrosine-sulfated peptide receptors PSKR1 and PSY1R modify the immunity of Arabidopsis to biotrophic and necrotrophic pathogens in an antagonistic manner. *The Plant Journal* 73 (3): 469–482.

Motomitsu, A., Sawa, S., and Ishida, T. (2015). Plant peptide hormone signalling. *Essays in Biochemistry* 58: 115–131. https://doi.org/10.1042/bse0580115.

Müller, L.M. and Harrison, M.J. (2019). Phytohormones, miRNAs, and peptide signals integrate plant phosphorus status with arbuscular mycorrhizal symbiosis. *Current Opinion in Plant Biology* 50: 132–139.

Murphy, E., Smith, S., and De Smet, I. (2012). Small signaling peptides in Arabidopsis development: how cells communicate over a short distance. *The Plant Cell* 24 (8): 3198–3217.

Nakagami, S., Kajiwara, T., Tsuda, K., and Sawa, S. (2024). CLE peptide signaling in plant-microbe interactions. *Frontiers in Plant Science* 15: 1481650.

Nakayama, T., Shinohara, H., Tanaka, M. et al. (2017). A peptide hormone required for Casparian strip diffusion barrier formation in Arabidopsis roots. *Science* 355 (6322): 284–286.

Narasimhan, M. and Simon, R. (2022). Spatial range, temporal span, and promiscuity of CLE-RLK signaling. *Frontiers in Plant Science* 13: 906087.

Nguyen, Q.T.C. and Kim, J. (2023). PSY–PSYR peptide–receptor pairs control the trade-off between plant growth and stress response. *Plant Signaling & Behavior* 18 (1): https://doi.org/10.1080/15592324.2023.2260638.

Ntim, T. (2021). Synthesis of Phytosulfokine analogs as probes for studying plant signaling and molecular trafficking (Master's thesis, East Tennessee State University).

Ogawa, M., Shinohara, H., Sakagami, Y., and Matsubayashi, Y. (2008). Arabidopsis CLV3 peptide directly binds CLV1 ectodomain. *Science* 319 (5861): 294–294.

Oh, E., Seo, P.J., and Kim, J. (2018). Signaling peptides and receptors coordinating plant root development. *Trends in Plant Science* 23 (4): 337–351.

Ohkubo, Y., Tanaka, M., Tabata, R. et al. (2017). Shoot-to-root mobile polypeptides involved in systemic regulation of nitrogen acquisition. *Nature Plants* 3 (4): 1–6.

Okamoto, S., Kawasaki, A., Makino, Y. et al. (2022). Long-distance translocation of CLAVATA3/ESR-related 2 peptide and its positive effect on roots sucrose status. *Plant Physiology* 189 (4): 2357–2367. https://doi.org/10.1093/plphys/kiac227.

Okuda, S. (2021). Molecular mechanisms of plant peptide binding to receptors. *Peptides* 144: 170614.

Oldroyd, G.E. and Leyser, O. (2020). A plant's diet, surviving in a variable nutrient environment. *Science* 368 (6486): eaba0196.

Pan, X., Deng, Z., Wu, R. et al. (2024). Identification of CEP peptides encoded by the tobacco (*Nicotiana tabacum*) genome and characterization of their roles in osmotic and salt stress responses. *Plant Physiology and Biochemistry* 209: 108525.

Panigrahi, G.K., Sahoo, A., and Satapathy, K.B. (2021). Insights to plant immunity: defense signaling to epigenetics. *Physiological and Molecular Plant Pathology* 113: 1–7. https://doi.org/10.1016/j.pmpp.2020.101568.

Panigrahi, G.K., Sahoo, A., and Satapathy, K.B. (2021a). Differential expression of selected Arabidopsis resistant genes under abiotic stress conditions. *Plant Science Today* 8 (4): 859–864. https://doi.org/10.14719/pst.2021.8.4.1213.

Panigrahi, G.K., Sahoo, A., and Satapathy, K.B. (2024). The processing body component varicose plays a multiplayer role towards stress management in Arabidopsis. *Plant Physiology Reports* 1–10. https://doi.org/10.1007/s40502-023-00778-w.

Panigrahi, G.K. and Satapathy, K.B. (2020). Sacrificed surveillance process favours plant defense: a review. *Plant Archives* 20 (1): 2551–2559.

Panigrahi, G.K. and Satapathy, K.B. (2020a). Formation of Arabidopsis poly(A)-specific ribonuclease associated processing bodies in response to pathogenic infection. *Plant Archives* 20 (2): 4907–4912.

Panigrahi, G.K. and Satapathy, K.B. (2021). Pseudomonas syringaepv. syringae Infection Orchestrates the Fate of the Arabidopsis J domain containing Cochaperone and decapping protein factor 5. *Physiological and Molecular Plant Pathology* 113 (101598): 1–9. https://doi.org/10.1016/j.pmpp.2020.101598.

Pastor-Fernández, J., Sánchez-Bel, P., Flors, V. et al. (2023). Small signals lead to big changes: the potential of peptide-induced resistance in plants. *Journal of Fungi* 9 (2): 265.

Patel, N., Mohd-Radzman, N.A., Corcilius, L. et al. (2018). Diverse peptide hormones affecting root growth identified in the *Medicago truncatula* secreted peptidome. *Molecular & Cellular Proteomics* 17 (1): 160–174.

Patro, A.K., Panigrahi, G.K., Majumder, S. et al. (2024). Nonsense-mediated mRNA decay: physiological significance, mechanistic insights and future implications. *Pathology – Research and Practice* 264: 155677. https://doi.org/10.1016/j.prp.2024.155677.

Pečenková, T., Potocký, M., and Stegmann, M. (2024). More than meets the eye: knowns and unknowns of the trafficking of small secreted proteins in Arabidopsis. *Journal of Experimental Botany* erae172.

Phour, M., Sehrawat, A., Sindhu, S.S., and Glick, B.R. (2020). Interkingdom signaling in plant-rhizomicrobiome interactions for sustainable agriculture. *Microbiological Research* 241: 126589.

Priya, A.K., Alagumalai, A., Balaji, D., and Song, H. (2023). Bio-based agricultural products: a sustainable alternative to agrochemicals for promoting a circular economy. *RSC Sustainability* 1 (4): 746–762.

Pruitt, R.N., Joe, A., Zhang, W. et al. (2017). A microbially derived tyrosine-sulfated peptide mimics a plant peptide hormone. *New Phytologist* 215 (2): 725–736.

Qian, P., Song, W., Yokoo, T. et al. (2018). The CLE9/10 secretory peptide regulates stomatal and vascular development through distinct receptors. *Nature Plants* 4 (12): 1071–1081.

Qiu, Z., Zhuang, K., Liu, Y. et al. (2022). Functional characterization of C-terminally encoded peptide (CEP) family in *Brassica rapa* L. *Plant Signaling & Behavior* 17 (1): 2021365.

Rao, Y., Yang, X., Pan, C. et al. (2022). Advance of clustered regularly interspaced short palindromic repeats-Cas9 system and its application in crop improvement. *Frontiers in Plant Science* 13: 839001. https://doi.org/10.3389/fpls.2022.839001.

Reichardt, S., Piepho, H.P., Stintzi, A., and Schaller, A. (2020). Peptide signaling for drought-induced tomato flower drop. *Science* 367 (6485): 1482–1485.

Reid, D.E., Ferguson, B.J., and Gresshoff, P.M. (2011). Inoculation-and nitrate-induced CLE peptides of soybean control NARK-dependent nodule formation. *Molecular Plant-Microbe Interactions* 24 (5): 606–618.

Ren, S.C., Song, X.F., Chen, W.Q. et al. (2019). CLE25 peptide regulates phloem initiation in Arabidopsis through a CLERK-CLV2 receptor complex. *Journal of Integrative Plant Biology* 61 (10): 1043–1061.

Ren, G., Zhang, Y., Chen, Z. et al. (2024). Research progress of small plant peptides on the regulation of plant growth, development, and abiotic stress. *International Journal of Molecular Sciences* 25 (7): 4114. https://doi.org/10.3390/ijms25074114.

Roy, S., Griffiths, M., Torres-Jerez, I. et al. (2022). Application of synthetic peptide CEP1 increases nutrient uptake rates along plant roots. *Frontiers in Plant Science* 12: 793145.

Roy, S., Lundquist, P., Udvardi, M., and Scheible, W.R. (2018). Small and mighty: peptide hormones in plant biology. *Plant Cell* 30 (tc118tt0718).

Ruan, J., Zhou, Y., Zhou, M. et al. (2019). Jasmonic acid signaling pathway in plants. *International Journal of Molecular Sciences* 20 (10): 2479. https://doi.org/10.3390/ijms20102479.

Ryan, C.A. and Pearce, G. (2003). Systemins: a functionally defined family of peptide signals that regulate defensive genes in Solanaceae species. *Proceedings of the National Academy of Sciences of the United States of America, 100 Suppl* 2 (Suppl 2): 14577–14580. https://doi.org/10.1073/pnas.1934788100.

Ryan, C.A., Pearce, G., Scheer, J., and Moura, D.S. (2002). Polypeptide hormones. *The Plant Cell* 14 (suppl_1): S251–S264.

Rzemieniewski, J., Leicher, H., Lee, H.K. et al. (2024). CEP signaling coordinates plant immunity with nitrogen status. *Nature Communications* 15 (1): 1–16.

Sahoo, A., Satapathy, K.B., and Panigrahi, G.K. (2023). Ectopic expression of disease resistance protein promotes resistance against pathogen infection and drought stress in Arabidopsis. *Physiological and Molecular Plant Pathology* 124 (101949): 1–7. https://doi.org/10.1016/j.pmpp.2023.101949.

Sahoo, A., Satapathy, K.B., and Panigrahi, G.K. (2024). Security check: plant immunity under temperature surveillance. *Journal of Plant Biochemistry and Biotechnology* 33: 1–4. https://doi.org/10.1007/s13562-023-00846-0.

Salvi, P., Manna, M., Kaur, H. et al. (2021). Phytohormone signaling and crosstalk in regulating drought stress response in plants. *Plant Cell Reports* 40: 1305–1329.

van de Sande, J.W., Streefkerk, D.E., Immink, R.G. et al. (2024). Phytosulfokine peptide library: chemical synthesis and biological evaluation on protoplast regeneration. *New Journal of Chemistry* 48 (18): 8055–8063.

Sang, T., Zhang, Z., Liu, G., and Wang, P. (2025). Navigating the landscape of plant proteomics. *Journal of Integrative Plant Biology*..

Sauter, M. (2015). Phytosulfokine peptide signalling. *Journal of Experimental Botany* 66 (17): 5161–5169.

Schaller, A. (ed.) (2024). *Plant Peptide Hormones and Growth Factors*. Humana Press.

Schardon, K., Hohl, M., Graff, L. et al. (2016). Precursor processing for plant peptide hormone maturation by subtilisin-like serine proteinases. *Science* 354 (6319): 1594–1597.

Scott, T., Correa-Lozano, A., and Foo, E. (2025). Upstairs, downstairs-conserved and divergent CLAVATA signalling in shoot meristem development and root symbioses. *Annals of Botany* 136 (5-6): 1397–1406.

Selby, R. and Jones, D.S. (2023). Complex peptide hormone signaling in plant stem cells. *Current Opinion in Plant Biology* 75: 102442.

Setty, J., Singh, P., Tantuway, G., and Vijai, P. (2024). Recentadvancement on peptide research and their application in eco-agriculture. In: *Plant Elicitor Peptides: New Tool for Sustainable Agriculture* (ed. S. Singh and R. Mehrotra), 269–281. Springer Nature Singapore: Singapore.

Shahid, M., Shafi, Z., Ilyas, T. et al. (2024). Crosstalk between phytohormones and pesticides: Insights into unravelling the crucial roles of plant growth regulators in improving crop resilience to pesticide stress. *Scientia Horticulturae* 338: 113663.

Shinohara, H. and Matsubayashi, Y. (2010). Arabinosylated glycopeptide hormones: new insights into CLAVATA3 structure. *Current Opinion in Plant Biology* 13: 515–519.

Simon, R. and Dresselhaus, T. (2015). Peptides take centre stage in plant signaling. *Preface. Journal of Experimental Botany* 66 (17): 5135–5138. https://doi.org/10.1093/jxb/erv376.

Song, X.F., Hou, X.L., and Liu, C.M. (2022). CLE peptides: critical regulators for stem cell maintenance in plants. *Planta* 255: 1–17.

Stintzi, A. and Schaller, A. (2022). Biogenesis of post-translationally modified peptide signals for plant reproductive development. *Current Opinion in Plant Biology* 69: 102274.

Stührwohldt, N., Bühler, E., Sauter, M., and Schaller, A. (2021). Phytosulfokine (PSK) precursor processing by subtilase SBT3. 8 and PSK signaling improve drought stress tolerance in Arabidopsis. *Journal of Experimental Botany* 72 (9): 3427–3440.

Stührwohldt, N. and Schaller, A. (2019). Regulation of plant peptide hormones and growth factors by post-translational modification. *Plant Biology (Stuttgart, Germany)* 21 (Suppl 1): 49–63. https://doi.org/10.1111/plb.12881.

Sui, J., Xiao, X., Yang, J. et al. (2023). The rubber tree RALF peptide hormone and its receptor protein kinase FER implicates in rubber production. *Plant Science* 326: 111510.

Takahashi, F., Kuromori, T., Urano, K. et al. (2020). Drought stress responses and resistance in plants: From cellular responses to long-distance intercellular communication. *Frontiers in Plant Science* 11: 556972.

Takahashi, F., Suzuki, T., Osakabe, Y. et al. (2018). A small peptide modulates stomatal control via abscisic acid in long-distance signalling. *Nature* 556 (7700): 235–238.

Taleski, M., Imin, N., and Djordjevic, M.A. (2018). CEP peptide hormones: key players in orchestrating nitrogen-demand signalling, root nodulation, and lateral root development. *Journal of Experimental Botany* 69 (8): 1829–1836.

Taleski, M., Jin, M., Chapman, K. et al. (2024). CEP hormones at the nexus of nutrient acquisition and allocation, root development, and plant–microbe interactions. *Journal of Experimental Botany* 75 (2): 538–552.

Tavormina, P., De Coninck, B., Nikonorova, N. et al. (2015). The plant peptidome: an expanding repertoire of structural features and biological functions. *The Plant Cell* 27 (8): 2095–2118. https://doi.org/10.1105/tpc.15.00440.

Toruño, T.Y., Stergiopoulos, I., and Coaker, G. (2016). Plant-pathogen effectors: cellular probes interfering with plant defenses in spatial and temporal manners. *Annual Review of Phytopathology* 54 (1): 419–441.

Tost, A.S., Kristensen, A., Olsen, L.I. et al. (2021). The PSY peptide family—expression. *Modification and Physiological Implications. Genes* 12 (2): 218. https://doi.org/10.3390/genes12020218.

Vanyushin, B.F., Ashapkin, V.V., and Aleksandrushkina, N.I. (2017). Regulatory peptides in plants. *s* 82: 89–94.

Ventimilla, D., Velázquez, K., Ruiz-Ruiz, S. et al. (2021). *IDA (INFLORESCENCE DEFICIENT IN ABSCISSION)-like peptides and HAE (HAESA)*-like receptors regulate corolla abscission in *Nicotiana benthamiana* flowers. *BMC Plant Biology* 21 (1): 226.

Waadt, R., Seller, C.A., Hsu, P.K. et al. (2022). Plant hormone regulation of abiotic stress responses. *Nature Reviews Molecular Cell Biology* 23 (10): 680–694.

Wang, G. and Fiers, M. (2010). CLE peptide signaling during plant development. *Protoplasma* 240 (1–4): 33–43. https://doi.org/10.1007/s00709-009-0095-y.

Wang, G., Zhang, G., and Wu, M. (2016). CLE peptide signaling and crosstalk with phytohormones and environmental stimuli. *Frontiers in Plant Science* 6: 1211.

van Wijk, K.J., Leppert, T., Sun, Z. et al. (2023). Detection of the *Arabidopsis* proteome and its post-translational modifications and the nature of the unobserved (dark) proteome in PeptideAtlas. *Journal of Proteome Research* 23 (1): 185–214.

Wu, H., Zheng, R., Hao, Z. et al. (2019). Cunninghamia lanceolata PSK peptide hormone genes promote primary root growth and adventitious root formation. *Plants* 8 (11): 520.

Wulf, K., Sun, J., Wang, C. et al. (2024). The role of CLE peptides in the suppression of Mycorrhizal colonization of tomato. *Plant and Cell Physiology* 65 (1): 107–119.

Xiao, F. and Zhou, H. (2023). Plant salt response: Perception, signaling, and tolerance. *Frontiers in Plant Science* 13: 1053699.

Xie, H., Zhao, W., Li, W. et al. (2022). Small signaling peptides mediate plant adaptions to abiotic environmental stress. *Planta* 255 (4): 72.

Xiong, L., Huang, Y., Liu, Z. et al. (2022). Small EPIDERMAL PATTERNING FACTOR-LIKE2 peptides regulate awn development in rice. *Plant Physiology* 190 (1): 516–531. https://doi.org/10.1093/plphys/kiac278.

Xu, C., Liberatore, K.L., MacAlister, C.A. et al. (2015). A cascade of arabinosyltransferases controls shoot meristem size in tomato. *Nature Genetics* 47 (7): 784–792.

Xu, T.T., Song, X.F., Ren, S.C., and Liu, C.M. (2013). The sequence flanking the N-terminus of the CLV3 peptide is critical for its cleavage and activity in stem cell regulation in Arabidopsis. *BMC Plant Biology* 13: 1–12.

Xu, K., Tian, D., Wang, T. et al. (2023). Small secreted peptides (SSPs) in tomato and their potential roles in drought stress response. *Molecular Horticulture* 3 (1): 17.

Yamada, M. and Sawa, S. (2013). The roles of peptide hormones during plant root development. *Current Opinion in Plant Biology* 16 (1): 56–61.

Yamaguchi, Y.L., Ishida, T., and Sawa, S. (2016). CLE peptides and their signaling pathways in plant development. *Journal of Experimental Botany* 67 (16): 4813–4826.

Yang, H., Matsubayashi, Y., Nakamura, K., and Sakagami, Y. (2001). Diversity of Arabidopsis genes encoding precursors for phytosulfokine, a peptide growth factor. *Plant Physiology* 127 (3): 842–851.

Yoshida, T., Fernie, A.R., Shinozaki, K., and Takahashi, F. (2021). Long-distance stress and developmental signals associated with abscisic acid signaling in environmental responses. *The Plant Journal* 105 (2): 477–488.

Yu, X., Huang, Z., Cheng, Y. et al. (2024). Comparative genomics screens identify a novel small secretory peptide, SlSolP12, which activates both local and systemic immune response in tomatoes and exhibits broad-spectrum activity. *Journal of Agricultural and Food Chemistry* 72 (33): 18507–18519.

Yuan, B. and Wang, H. (2021). Peptide signaling pathways regulate plant vascular development. *Frontiers in Plant Science* 12: 719606. https://doi.org/10.3389/fpls.2021.719606.

Žárský, V., Nielsen, M.E., and Blatt, M.R. (2024). Formins, cell wall integrity, ROP guanine exchange factors, secretion regulators, and small secreted peptides in plant cell exocytosis and defence. *Journal of Experimental Botany* 75 (12): 3665–3667.

Zelman, A.K. and Berkowitz, G.A. (2023). Plant Elicitor Peptide (Pep) signaling and pathogen defense in tomato. *Plants (Basel, Switzerland)* 12 (15): 2856. https://doi.org/10.3390/plants12152856.

Zelman, A.K., Ma, Y., and Berkowitz, G.A. (2024). Pathogen elicitor peptide (pep), systemin, and their receptors in tomato: sequence analysis sheds light on standing disagreements about biotic stress signaling components. *BMC Plant Biology* 24 (1): 728.

Zhang, P., Gao, W., Guo, L. et al. (2024a). Functional characterization of plant Peptide-Containing Sulfated Tyrosine (PSY) Family in Wheat (*Triticum aestivum* L.). *International Journal of Molecular Sciences* 25 (23): 12663.

Zhang, H., Han, Z., Song, W., and Chai, J. (2016). Structural insight into recognition of plant peptide hormones by receptors. *Molecular Plant* 9 (11): 1454–1463.

Zhang, Z., Han, H., Zhao, J. et al. (2025). Peptide hormones in plants. *Molecular Horticulture* 5 (1): 7. https://doi.org/10.1186/s43897-024-00134-y.

Zhang, S., Li, C., Si, J. et al. (2022). Action mechanisms of effectors in plant-pathogen interaction. *International Journal of Molecular Sciences* 23 (12): 6758.

Zhang, F., Rosental, L., Ji, B. et al. (2024). Metabolite-mediated adaptation of crops to drought and the acquisition of tolerance. *The Plant Journal* 118 (3): 626–644.

Zhang, L., Shi, X., Zhang, Y. et al. (2019). CLE9 peptide-induced stomatal closure is mediated by abscisic acid, hydrogen peroxide, and nitric oxide in *Arabidopsis thaliana*. *Plant, Cell & Environment* 42 (3): 1033–1044.

Zhang, Y.M., Ye, D.X., Liu, Y. et al. (2023). Peptides, new tools for plant protection in eco-agriculture. *Advanced Agrochem* 2 (1): 58–78.

Zhong, S., Li, L., Wang, Z. et al. (2022). RALF peptide signaling controls the polytubey block in Arabidopsis. *Science* 375 (6578): 290–296.

Zhong, S. and Qu, L.J. (2019). Peptide/receptor-like kinase-mediated signaling involved in male–female interactions. *Current Opinion in Plant Biology* 51: 7–14.

Zhou, Y., Zheng, J., Wu, H. et al. (2024). A novel toolbox to record CLE peptide signaling. *Frontiers in Plant Science* 15: 1468763.

Index

Peptide Hormones in Plant Development and Stress Responses, First Edition. Edited by Vijay Pratap Singh and Samiksha Singh.

Printed and bound by CPI Group (UK) Ltd, Croydon, CR0 4YY
17/09/2026
14964300-0006